20. $\displaystyle\int \frac{du}{u\sqrt{a+bu}} = \begin{cases} \dfrac{1}{\sqrt{a}}\ln\left|\dfrac{\sqrt{a+bu}-\sqrt{a}}{\sqrt{a+bu}+\sqrt{a}}\right| + C & \text{if } a > 0 \\[3mm] \dfrac{2}{\sqrt{-a}}\tan^{-1}\sqrt{\dfrac{a+bu}{-a}} + C & \text{if } a < 0 \end{cases}$

21. $\displaystyle\int \frac{du}{u^n\sqrt{a+bu}} = -\frac{\sqrt{a+bu}}{a(n-1)u^{n-1}} - \frac{b(2n-3)}{2a(n-1)}\int \frac{du}{u^{n-1}\sqrt{a+bu}}$

22. $\displaystyle\int \frac{\sqrt{a+\text{'}}}{}$

23. $\displaystyle\int \frac{\sqrt{a+}}{u^n}{}_{-(n-1)}\int \frac{\sqrt{a+bu}\,du}{u^{n-1}}$

Integrals Containing $a^2 \pm u^2$

24. $\displaystyle\int \frac{du}{a^2+u^2} = \frac{1}{a}\tan^{-1}\frac{u}{a} + C$

25. $\displaystyle\int \frac{du}{a^2-u^2} = \frac{1}{2a}\ln\left|\frac{u+a}{u-a}\right| + C = \begin{cases} \dfrac{1}{a}\tanh^{-1}\dfrac{u}{a} + C & \text{if } |u| < a \\[3mm] \dfrac{1}{a}\coth^{-1}\dfrac{u}{a} + C & \text{if } |u| > a \end{cases}$

26. $\displaystyle\int \frac{du}{u^2-a^2} = \frac{1}{2a}\ln\left|\frac{u-a}{u+a}\right| + C = \begin{cases} -\dfrac{1}{a}\tanh^{-1}\dfrac{u}{a} + C & \text{if } |u| < a \\[3mm] -\dfrac{1}{a}\coth^{-1}\dfrac{u}{a} + C & \text{if } |u| > a \end{cases}$

Integrals Containing $\sqrt{u^2 \pm a^2}$

In formulas 27 through 38, we may replace

$\ln(u + \sqrt{u^2+a^2})$ by $\sinh^{-1}\dfrac{u}{a}$

$\ln|u + \sqrt{u^2-a^2}|$ by $\cosh^{-1}\dfrac{u}{a}$

$\ln\left|\dfrac{a + \sqrt{u^2+a^2}}{u}\right|$ by $\sinh^{-1}\dfrac{a}{u}$

27. $\displaystyle\int \frac{du}{\sqrt{u^2\pm a^2}} = \ln|u + \sqrt{u^2\pm a^2}| + C$

28. $\displaystyle\int \sqrt{u^2\pm a^2}\,du = \frac{u}{2}\sqrt{u^2\pm a^2} \pm \frac{a^2}{2}\ln|u + \sqrt{u^2\pm a^2}| + C$

29. $\displaystyle\int u^2\sqrt{u^2\pm a^2}\,du = \frac{u}{8}(2u^2\pm a^2)\sqrt{u^2\pm a^2}$
$\displaystyle - \frac{a^4}{8}\ln|u + \sqrt{u^2\pm a^2}| + C$

30. $\displaystyle\int \frac{\sqrt{u^2+a^2}\,du}{u} = \sqrt{u^2+a^2} - a\ln\left|\frac{a+\sqrt{u^2+a^2}}{u}\right| + C$

31. $\displaystyle\int \frac{\sqrt{u^2-a^2}\,du}{u} = \sqrt{u^2-a^2} - a\sec^{-1}\left|\frac{u}{a}\right| + C$

32. $\displaystyle\int \frac{\sqrt{u^2\pm a^2}\,du}{u^2} = -\frac{\sqrt{u^2\pm a^2}}{u} + \ln|u + \sqrt{u^2\pm a^2}| + C$

33. $\displaystyle\int \frac{u^2\,du}{\sqrt{u^2\pm a^2}} = \frac{u}{2}\sqrt{u^2\pm a^2} - \frac{\pm a^2}{2}\ln|u + \sqrt{u^2\pm a^2}| + C$

34. $\displaystyle\int \frac{du}{u\sqrt{u^2+a^2}} = -\frac{1}{a}\ln\left|\frac{a+\sqrt{u^2+a^2}}{u}\right| + C$

35. $\displaystyle\int \frac{du}{u\sqrt{u^2-a^2}} = \frac{1}{a}\sec^{-1}\left|\frac{u}{a}\right| + C$

36. $\displaystyle\int \frac{du}{u^2\sqrt{u^2\pm a^2}} = -\frac{\sqrt{u^2\pm a^2}}{\pm a^2 u} + C$

37. $\displaystyle\int (u^2\pm a^2)^{3/2}\,du = \frac{u}{8}(2u^2\pm 5a^2)\sqrt{u^2\pm a^2}$
$\displaystyle + \frac{3a^4}{8}\ln|u + \sqrt{u^2\pm a^2}| + C$

38. $\displaystyle\int \frac{du}{(u^2\pm a^2)^{3/2}} = \frac{u}{\pm a^2\sqrt{u^2\pm a^2}} + C$

Integrals Containing $\sqrt{a^2 - u^2}$

39. $\displaystyle\int \frac{du}{\sqrt{a^2-u^2}} = \sin^{-1}\frac{u}{a} + C$

40. $\displaystyle\int \sqrt{a^2-u^2}\,du = \frac{u}{2}\sqrt{a^2-u^2} + \frac{a^2}{2}\sin^{-1}\frac{u}{a} + C$

41. $\displaystyle\int u^2\sqrt{a^2-u^2}\,du = \frac{u}{8}(2u^2-a^2)\sqrt{a^2-u^2} + \frac{a^4}{8}\sin^{-1}\frac{u}{a} + C$

42. $\displaystyle\int \frac{\sqrt{a^2-u^2}\,du}{u} = \sqrt{a^2-u^2} - a\ln\left|\frac{a+\sqrt{a^2-u^2}}{u}\right| + C$
$\displaystyle = \sqrt{a^2-u^2} - a\cosh^{-1}\frac{a}{u} + C$

43. $\displaystyle\int \frac{\sqrt{a^2-u^2}\,du}{u^2} = -\frac{\sqrt{a^2-u^2}}{u} - \sin^{-1}\frac{u}{a} + C$

(continued on the back)

Calculus

with analytic geometry

BRIEF EDITION

Calculus

with analytic geometry

BRIEF EDITION

HOWARD ANTON
Drexel University

JOHN WILEY & SONS
New York   Chichester   Brisbane   Toronto

Library of Congress Cataloging in Publication Data:

Anton, Howard
    Calculus, analytic geometry

    Includes index.
    1. Calculus. 2. Geometry, Analytic.   I. Title
QA303.A53 1981   515   81-1609
ISBN 0-471-09443-9   AACR2

Printed in the United States of America

10 9 8 7 6 5 4 3

To
My wife Pat
My children Brian, David, and Lauren
My mother Shirley

In memory of
Stephen Girard (1750-1831) benefactor

preface

This text is designed for a course in single-variable calculus. The material presented here consists of the first thirteen chapters of *Calculus with Analytic Geometry,* which covers both functions of one variable and functions of several variables. My intent is to present the concepts in the clearest possible way. The primary effort has been devoted to *teaching* the subject matter with a level of rigor suitable for the mainstream calculus audience.

Anyone who has taught the calculus sequence knows that the course contains so much material that it is impossible for an instructor to spend an adequate amount of classroom time on every topic. For this reason a calculus textbook bears a major portion of the burden in the teaching process. I am hopeful that this book will be one that can be relied on for sound, clear, and complete explanations, thereby freeing valuable classroom time for the instructor.

FEATURES

Precalculus Material Because of the vast amount of material to be covered, it is desirable to spend as little time as possible on precalculus topics. However, it is a fact that freshmen have a wide variety of educational backgrounds and different levels of preparedness. Therefore, I have included an optional first chapter devoted to precalculus material. However, it is written in enough detail to enable the instructor to feel confident in moving quickly through these preliminaries.

Trigonometry Deficiencies in trigonometry plague many students throughout the entire calculus sequence. Therefore, I have included a substantial trigonometry review in the appendix. It is more detailed than most such reviews because I feel that students will appreciate having this material readily available. The review is broken into two units: the first to be mastered before reading Section 1.4 and the second before Section 3.3.

Illustrations This calculus text is more heavily illustrated than most. It is my experience that beginners in mathematics frequently have difficulty extracting concepts from mathematical formulas, yet when the right picture is presented the concept becomes clear immediately. For this reason, I have chosen to take full advantage of modern two-color typography and the most up-to-date illustrative techniques. There are approximately 1000 illustrations designed to clarify and enhance the exposition. (Most calculus books have half this number.)

Pedagogy I have devoted special effort to the explanations of more difficult concepts. In places where students traditionally have trouble, I have tried to give the reader some foothold on the problem. At times I have simply used artwork designed to focus the student's thinking process in the right direction; at other times I have paraphrased ideas informally in a way that cuts through the technical roadblocks; and at still other times I have simply broken the discussion into smaller, more understandable pieces than generally found in most calculus texts.

Rigor Where possible, theory is presented precisely in a style tailored for freshmen. However, in those places where precision conflicts with clarity, I have presented informal intuitive discussions. Whenever this occurs, there is a clear indication that the arguments given are not intended as formal proof. I have tried to make a clear distinction between rigorous mathematics and informal developments. Theory involving $\delta\epsilon$ arguments has been placed in optional sections so it can be avoided if desired.

Flexibility This book is designed to ensure maximum flexibility. There will be no difficulty in permuting chapters in any reasonable way.

Order of Presentation The order in which topics are presented is fairly standard, with two exceptions: derivatives of trigonometric functions are introduced early (Chapter 3) and a discussion of first order separable and linear differential equations is given in the chapter on logarithmic and exponential functions (Chapter 7). This placement of differential equations material allows us to give some nice applications of logarithms and exponentials immediately, and also helps meet the needs of those engineering and science students who will encounter this material in other courses taken concurrently with calculus.

Applications An abundance of applications to physics, engineering, biology, population growth, chemistry, and economics appear throughout the text.

Exercises Each exercise set begins with routine drill problems and progresses gradually toward problems of greater difficulty. I have tried to construct well-balanced exercise sets with more variety than is available in most calculus texts. Answers, including art, are given for the odd-numbered problems, and each chapter contains a set of review problems to help the student consolidate his or her mastery of the chapter.

Supplements   A student solutions manual containing worked-out solutions to the odd-numbered problems is available.

Reviewing and Class Testing   All of the material in this text has been refined and polished through extensive class testing and reviewing. In addition to the author, each section has been read by *seven* reviewers to ensure smoothness of the exposition and mathematical accuracy.

HOWARD ANTON

acknowledgments

Reviewers It was my good fortune to have the advice and guidance of many fine reviewers. The knowledge and skills that they shared with me have greatly enhanced this book. For their contributions, I thank:

Dennis DeTurck, *University of Pennsylvania*
Garret J. Etgen, *University of Houston*
William R. Fuller, *Purdue University*
Douglas W. Hall, *Michigan State University*
Albert Herr, *Drexel University*
Robert Higgins, *Quantics Corporation*
Harvey B. Keynes, *University of Minnesota*
Leo Lampone, *Ketron Corporation*
Stanley M. Lukawecki, *Clemson University*
Melvin J. Maron, *University of Louisville*
Mark A. Pinsky, *Northwestern University*
Donald R. Sherbert, *University of Illinois*
William F. Trench, *Drexel University*
Richard C. Vile, *Eastern Michigan University*
Irving Drooyan, *Los Angeles Pierce College*
Charles Denlinger, *Millersville State College*
Edith Ainsworth, *University of Alabama*

Exercises I gratefully acknowledge the fine work of Melvin J. Maron who helped prepare the end-of-chapter supplementary exercises. Thanks also go to Albert Herr, Robert Higgins, Leo Lampone, and Hal Schwalm for working the exercises.

Computer Graphics The excellent computer graphics were prepared by Robert Conley, James Warner, and Richard Yuskaitis of Precision Visuals, P.O. Box 1185, Boulder, Colorado, 80306.

Typing The fine appearance of this text is due in part to the careful typing and formatting of the final manuscript by Technitype, 305 Provincetown Circle, Cherry Hill, New Jersey, 08034. I am especially grateful to Kathleen R. McCabe, who personally handled the project and magically transformed my rough work into polished copy. A word of thanks is also due to Susan Gershuni who typed preliminary versions of the manuscript.

Wiley Staff I give special thanks to the entire Wiley staff for the care and cooperation they have shown in the production and marketing of this text. In particular, I thank the following people with whom I had the good fortune to work personally:

Elaine Rauschal, *Production Manager*
Charlotte Shelby, *Production Manager*
Eugene T. Patti, *Manuscript Editor*
John Balbalis, *Consulting Artist*
Rafael H. Hernandez, *Assistant Design Director*
Arthur R. Beck, *Marketing Manager*
Rhoda Villanueva, *Advertising Copy Chief*
Peter Robinson, *Senior Copywriter*

Finally, I wish to acknowledge the major role of my editor, Gary W. Ostedt. He conceived the idea for this book and then guided its development with imagination, creativity, and infectious enthusiasm. During the many years it has taken to complete this project, he has played the role of editor and critic, but most of all—friend.

H. A.

table of contents

introduction

Calculus is the mathematical tool used to analyze changes in physical quantities. It was developed in the seventeenth century to study four major classes of scientific and mathematical problems of the time:

1. Find the tangent to a curve at a point.

2. Find the length of a curve, the area of a region, and the volume of a solid.

3. Find the maximum or minimum value of a quantity—for example, the maximum and minimum distances of a planet from the sun, or the maximum range attainable for a projectile by varying its angle of fire.

4. Given a formula for the distance traveled by a body in any specified amount of time, find the velocity and acceleration of the body at any instant. Conversely, given a formula that specifies the acceleration or velocity at any instant, find the distance traveled by the body in a specified period of time.

These problems were attacked by the greatest minds of the seventeenth century, culminating in the crowning achievements of Gottfried Wilhelm Leibniz and Isaac Newton—the creation of calculus.

Gottfried Wilhelm Leibniz (1646–1716)

This gifted genius was one of the last people to have mastered most major fields of knowledge—an impossible accomplishment in our own era of specialization. He was an expert in law, religion, philosophy, literature, politics, geology, metaphysics, alchemy, history, and mathematics.

Leibniz was born in Leipzig, Germany. His father, a professor of moral philosophy at the University of Leipzig, died when Leibniz was six years old.

The precocious boy then gained access to his father's library and began reading voraciously on a wide range of subjects, a habit that he maintained throughout his life. At age 15 he entered the University of Leipzig as a law student and by the age of 20 received a doctorate from the University of Altdorf. Subsequently, Leibniz followed a career in law and international politics, serving as counsel to kings and princes.

During his numerous foreign missions, Leibniz came in contact with outstanding mathematicians and scientists who stimulated his interest in mathematics—most notably, the physicist Christian Huygens. In mathematics Leibniz was self-taught; he learned the subject by reading papers and journals. As a result of this fragmented mathematical education, Leibniz often duplicated the results of others. This ultimately led to a raging conflict over who invented calculus, Leibniz or Newton? The argument over this question engulfed the scientific circles of England and Europe. Most scientists on the continent supported Leibniz, while those in England supported Newton. The conflict was unfortunate, and both sides lost in the end. The continent lost the benefit of Newton's discoveries in astronomy and physics for more than 50 years, and England became a second-rate country mathematically for a long period because its mathematicians were hampered by complexities resulting from Newton's inferior calculus notation. It is of interest to note that Newton and Leibniz never went to the lengths of vituperation of their advocates—both were sincere admirers of each other's work. The fact is that both men invented calculus independently. Leibniz invented it 10 years after Newton, in 1685, but he published his results 20 years before Newton published his own work on the subject.

Leibniz never married. He was moderate in his habits, quick tempered, but easily appeased, and charitable in his judgment of other people's work. In

Gottfried Leibniz
(Culver Pictures)

Isaac Newton
(Culver Pictures)

spite of his great achievements, Leibniz never received the honors showered on Newton. He spent his final years as a lonely embittered man. At his funeral there was one mourner, his secretary. An eyewitness stated, "He was buried more like a robber than what he really was—an ornament of his country."

Isaac Newton (1642–1727)

Newton was born in the village of Woolsthorpe, England. His father died before Newton was born and his mother raised the boy on the family farm. As a youth he showed little evidence of his later brilliance, except for an unusual talent with mechanical devices—he apparently built a working water clock and a toy flour mill powered by a mouse. In 1661 he entered Trinity College in Cambridge with a deficiency in geometry. Fortunately, Newton caught the eye of Isaac Barrow, a gifted mathematician and teacher. Under Barrow's guidance Newton immersed himself in mathematics and science. Newton graduated without any special distinction. Because the plague was spreading rapidly through London, Newton returned to his home in Woolsthorpe and stayed there during the years of 1665 and 1666. In those two momentous years the entire framework of modern science was miraculously created in Newton's mind—he discovered calculus, recognized the underlying principles of planetary motion and gravity, and determined that "white" sunlight was composed of all colors, red to violet. For some reason he kept his discoveries to himself. In 1667 he returned to Cambridge to obtain his Master's degree. After graduating, he became a teacher at Trinity. Then in 1669 Newton succeeded his teacher, Isaac Barrow, to the Lucasian chair of mathematics at Trinity, one of the most honored chairs of mathematics in the world. Thereafter, brilliant discoveries flowed from Newton steadily. He formulated the law of gravitation and used it to explain the motion of the moon, the planets, and the tides; he formulated basic theories of light, thermodynamics, and hydrodynamics; and he devised and constructed the first modern reflecting telescope.

Throughout his life Newton was hesitant to publish his major discoveries; he revealed them only to a select circle of friends, perhaps because of a fear of criticism or controversy. In 1687, only after intense coaxing by the astronomer, Edmond Halley (Halley's comet), did Newton publish his masterpiece, *Philosophae Naturalis Principia Mathematica* (The Mathematical Principles of Natural Philosophy). This work is generally considered to be the most important and influential scientific book ever written. In it Newton explained the workings of the solar system and formulated the basic laws of motion that to this day are fundamental in engineering and physics. However, not even the pleas of his friends could convince Newton to publish his discovery of calculus. Only after Leibniz published his results did Newton relent and publish his own work on calculus.

After 35 years as a professor, Newton suffered depression and a nervous breakdown. He gave up research in 1695 to accept a position as warden and later master of the London mint. During the 25 years that he worked at the mint, he did virtually no scientific or mathematical work. He was knighted in 1705 and on his death was buried in Westminster Abbey with all the honors his country could bestow. It is interesting to note that Newton was a learned theologian. He viewed the primary value of his work to be its support of the existence of God. Throughout his life he worked passionately to date biblical events by relating them to astronomical phenomena. He was so consumed with this passion that he wasted years searching the Book of Daniel for clues to the end of the world and the geography of hell.

Newton described his brilliant accomplishments as follows, "I seem to have been only like a boy playing on the seashore and diverting myself in now and then finding a smoother pebble or prettier shell than ordinary, whilst the great ocean of truth lay all undiscovered before me."

Calculus

with analytic geometry

BRIEF EDITION

1 coordinates, graphs, lines

1.1 REAL NUMBERS, SETS, AND INEQUALITIES (A REVIEW)

Since numbers and their properties play a fundamental role in calculus, we will begin by reviewing some terminology and facts about them.

The simplest numbers are the ***integers:***

$$\ldots, \; -4, \; -3, \; -2, \; -1, \; 0, \; 1, \; 2, \; 3, \; 4, \ldots$$

With the exception that division by zero is ruled out, ratios of integers are called ***rational numbers.*** Examples are

$$\frac{2}{3}, \frac{0}{9}, \frac{6}{2}, \frac{17}{1000}, \; -\frac{5}{2}\left(=\frac{-5}{2}=\frac{5}{-2}\right)$$

Observe that every integer is also a rational number since an integer p can be written as the ratio

$$p = \frac{p}{1}$$

Division by zero is ruled out since we would want to be able to express the relationship

$$y = \frac{p}{0}$$

in the alternative form

$$0 \cdot y = p$$

However, if p is different from zero, this equation is contradictory. Moreover, if p is equal to zero, this equation is satisfied by any number y, which means that the ratio $p/0$ does not have a unique value—a situation that is mathematically unsatisfactory. For these reasons such symbols as

$$\frac{p}{0} \quad \text{and} \quad \frac{0}{0}$$

are not assigned a value; they are said to be **undefined.**

The early Greeks believed that the size of every physical quantity could, in theory, be represented by a rational number. They reasoned that the size of a physical quantity must consist of a certain whole number of units plus some fraction m/n of an additional unit. This idea was shattered in the fifth century B.C. by Hippasus of Metapontum,* who demonstrated by geometric methods that the hypotenuse of the triangle in Figure 1.1.1 cannot be expressed as a ratio of integers. This dramatic discovery demonstrated the existence of **irrational numbers;** that is, numbers not expressible as ratios of integers. Other examples of irrational numbers are

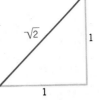

Figure 1.1.1

$$1 + \sqrt{2}, \ \sqrt{3}, \ \sqrt[3]{7}, \ \pi, \ \cos 19°$$

The proof that π is irrational is difficult and evaded mathematicians for centuries; it was finally proved in 1761 by J. H. Lambert[†] (see page 3).

Rational and irrational numbers can be distinguished by their decimal representations. Rational numbers have *repeating decimals;* that is, from some point the decimal consists entirely of zeros or else a fixed finite string of digits are repeated over and over. For example,

$$\tfrac{1}{2} = 0.5000 \ldots \quad \tfrac{8}{25} = 0.32000 \ldots \quad \tfrac{12}{4} = 3.000 \ldots$$
$$\tfrac{4}{3} = 1.3333 \ldots \quad \tfrac{3}{11} = 0.272727 \ldots \quad \tfrac{5}{7} = 0.714285714285714285 \ldots$$

Irrational numbers are represented by nonrepeating decimals.

▶ Example 1

Since π is irrational, the decimal

$$\pi = 3.14159265358979323845 \ldots$$

does not begin to repeat from some point. Since the decimal

---

*HIPPASUS OF METAPONTUM (circa 500 B.C.). A Greek Pythagorean philosopher. According to legend, Hippasus made his discovery at sea and was thrown overboard by fanatic Pythagoreans because his result contradicted their doctrine. The discovery of Hippasus is one of the most fundamental in the entire history of science.

$$0.10100100010000100000.1 \ldots$$

does not repeat (the number of zeros between the ones keeps growing), it represents an irrational number. ◄

In 1637 René Descartes†† published a philosophical work called *Discourse on the Method of Rightly Conducting the Reason.* In the back of that book were three appendices that purported to show how the "method" could be applied to concrete examples. The first two appendices were minor works that endeavored to explain the behavior of lenses and the movement of shooting stars. The third appendix, however, was an inspired stroke of genius; it was described by the nineteenth century British philosopher John Stuart Mill as, "The greatest single step ever made in the progress of the exact sciences." In that appendix René Descartes linked together two branches of mathematics, algebra and geometry. Descartes' work evolved into a new subject called ***analytic geometry;*** it gave a way of describing algebraic formulas by means of geometric curves and, conversely, a description of geometric curves using algebraic formulas.

In analytic geometry, the key step is to establish a correspondence between real numbers and points on a line. To accomplish this, we arbitrarily choose one direction along the line to be called ***positive*** and the other ***negative.*** It is usual to mark the positive direction with an arrowhead as shown in Figure 1.1.2. Next, we choose an arbitrary reference point on the line to be called the ***origin,*** and select a unit of length for measuring distances.

Figure 1.1.2

$$\xleftarrow{\hspace{2em}} - \underset{\text{Origin}}{\big|} \hspace{6em} + \xrightarrow{\hspace{2em}}$$

With each real number we can associate a point on the line as follows:

---

†JOHANN HEINRICH LAMBERT (1728–1777). A Swiss-German scientist, Lambert taught at the Berlin Academy of Sciences. In addition to his work on irrational numbers, Lambert wrote landmark books on geometry, the theory of cartography, and perspective in art. His influential book on geometry foreshadowed the discovery of modern non-Euclidean geometry.

†† RENÉ DESCARTES (1596–1650). Descartes, a French aristocrat, was the son of a government official. He graduated from the University of Poitiers with a law degree at age 20. After a brief probe into the pleasures of Paris he became a military engineer, first for the Dutch Prince of Nassau and then for the German Duke of Bavaria. It was during his service as a soldier that Descartes began to pursue mathematics seriously and develop his analytic geometry. After the wars, he returned to Paris where he stalked the city as an eccentric, wearing a sword in his belt and a plumed hat. He lived in leisure, never arose before 11 a.m., and dabbled in the study of human physiology, philosophy, glaciers, meteors, and rainbows. He eventually moved to Holland, where he published his *Discourse on the Method,* and finally to Sweden where he died while serving as tutor to Queen Christina. Descartes is regarded as a genius of the first magnitude. In addition to major contributions in mathematics and philosophy, he is considered, along with William Harvey, to be a founder of modern physiology.

(a) Associate with each positive number r the point that is a distance of r units in the positive direction from the origin.

(b) Associate with each negative number $-r$ the point that is a distance of r units in the negative direction from the origin.

(c) Associate the origin with the number 0.

The real number corresponding to a point on the line is called the **coordinate** of the point and the line is called a **coordinate line** or sometimes the **real line.**

▶ Example 2

In Figure 1.1.3 we have marked the approximate location of the points whose coordinates are -3, -1.75, $-\frac{1}{2}$, $\sqrt{2}$, π, and 4. ◀

Figure 1.1.3

It is evident from the way in which real numbers and points on a coordinate line are related that each real number corresponds to a single point and each point corresponds to a single real number. This fact is described by stating that the real numbers and the points on a coordinate line are in **one-to-one correspondence.**

Of particular importance to us is the fact that the real numbers are *ordered,* that is, given any two numbers a and b, exactly one of the following is true:

a is less than b
b is less than a
a is equal to b

Assuming we know what positive numbers are, we can define the order symbols $<$ (less than) and $\leq$ (less than or equal to) as follows.

1.1.1 DEFINITION   If a and b are real numbers, then

$a < b$ means $b - a$ is positive
$a \leq b$ means $a < b$ or $a = b$

The inequality $a < b$, which is read *a is less than b* can also be written as $b > a$, which is read *b is greater than a.* Geometrically the inequality $a < b$ states that b falls to the right of a on a coordinate line (Figure 1.1.4). Similarly $a \leq b$, which is read *a is less than or equal to b,* can also be written as $b \geq a$, which is read, *b is greater than or equal to a.*

Figure 1.1.4

► Example 3

The following inequalities are correct:

$$3 < 8, \quad -7 < 1.5, \quad -12 \le -\pi, \quad 5 \le 5$$
$$8 \ge 3, \quad 1.5 > -7, \quad -\pi > -12, \quad 5 \ge 5$$ ◄

A number b will be called **nonnegative** if $0 \le b$, and **nonpositive** if $b \le 0$.
The following properties of inequalities are frequently used in calculus.
We omit the proofs.

1.1.2 THEOREM (a) *If $a < b$ and $b < c$ then $a < c$.*
(b) *If $a < b$ then $a + c < b + c$.*
(c) *If $a < b$ and $c < d$ then $a + c < b + d$.*
(d) *If $a < b$ then $ac < bc$ when c is positive and $ac > bc$ when c is negative.*
(e) *If a and b are both positive or both negative and $a < b$ then $1/a > 1/b$.*

REMARK. These five properties remain true if $<$ and $>$ are replaced by $\le$
and $\ge$, respectively.

► Example 4

To paraphrase property (b), *an inequality remains true if the same number is
added to both sides.*
  For example, adding 7 to both sides of the inequality

$$-2 < 6 \tag{1}$$

yields the valid inequality

$$5 < 13$$

Similarly, adding -8 to both sides of (1) yields the valid inequality

$$-10 < -2$$ ◄

► Example 5

To paraphrase property (c), *inequalities in the same direction can be added.*
  For example, adding

$$2 < 6 \quad \text{and} \quad -8 < 5$$

yields the valid inequality

$$-6 < 11$$ ◄

► Example 6

To paraphrase property (*d*), *if both sides of an inequality are multiplied by a positive number, the direction of the inequality remains the same, and if both sides of an inequality are multiplied by a negative number, the direction of the inequality is reversed.*
   For example, multiplying the inequality

$$-4 < 7$$

by 2 yields

$$-8 < 14$$

and multiplying by -3 yields

$$12 > -21$$ ◄

► Example 7

To paraphrase property (*e*), *taking reciprocals reverses the direction of an inequality* involving two positive or two negative numbers.
   For example, taking reciprocals on both sides of the inequality

$$3 < 7$$

yields

$$\frac{1}{3} > \frac{1}{7}$$

and taking reciprocals on both sides of the inequality

$$-8 < -6$$

yields

$$-\frac{1}{8} > -\frac{1}{6}$$ ◄

1.1.3 DEFINITION   If *a*, *b*, and *c* are real numbers, we will write

$$a < b < c$$

when $a < b$ and $b < c$.

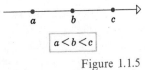

$$a < b < c$$

Figure 1.1.5

Geometrically, $a < b < c$ states that on a coordinate line, *b* falls to the right of *a* and *c* falls to the right of *b* (Figure 1.1.5).

The symbol $a < b \leq c$ means $a < b$ and $b \leq c$. We leave it to the reader to deduce the meanings of such symbols as $a \leq b < c$, $a \leq b \leq c$, $a < b < c < d$, and so on.

1.1.4 DEFINITION   A *set* is a collection of objects; the objects are called the ***elements*** or ***members*** of the set.

In this text we will be concerned primarily with sets whose members are real numbers. One way of describing a set is to list its elements between braces. Thus, the set of all positive integers less than 5 can be written

$$\{1, 2, 3, 4\}$$

and the set of all positive even integers can be written

$$\{2, 4, 6, \ldots\}$$

where the dots are used to indicate that only a portion of the members have been explicitly listed and the rest of the members are obtained by continuing the pattern.

When it is inconvenient or impossible to list the members, it is sufficient to define a set by stating a property common only to its members. For example,

the set of all rational numbers
the set of all real numbers x such that $2x^2 - 4x + 1 = 0$
the set of all real numbers between 2 and 3

As an alternative to such verbal descriptions of sets, we can use the notation

$$\{x: \underline{\hspace{3em}}\}$$

which is read, "the set of all x such that." Where the blue line is placed, we state the property that specifies the set.

▶ Example 8

(a) $\{x: x \text{ is a rational number}\}$ is read, "the set of all x such that x is a rational number."

(b) $\{x: x \text{ is a real number satisfying } 2x^2 - 4x + 1 = 0\}$ is read, "the set of all x such that x is a real number satisfying $2x^2 - 4x + 1 = 0$."

(c) $\{x: x \text{ is a real number between 2 and 3}\}$ is read, "the set of all x such that x is a real number between 2 and 3." ◀

REMARK. When it is clear that the members of a set are real numbers, we will omit the reference to this fact. Thus the sets in Example 8 might be written

$$\{x : x \text{ is rational}\}$$
$$\{x : 2x^2 - 4x + 1 = 0\}$$
$$\{x : 2 < x < 3\}$$

To indicate that an element a is a member of a set A, we write

$$a \in A$$

which is read "a is an element of A" or "a belongs to A." To indicate that the element a is *not* a member of the set A, we write

$$a \notin A$$

which is read, "a is not an element of A" or "a does not belong to A."

▶ Example 9

Let $A = \{x : x \text{ is rational}\}$ and $B = \{x : 2 < x < 4\}$, then

$$\frac{3}{4} \in A, \quad -2 \in A, \quad \pi \notin A, \quad -\sqrt{2} \notin A$$
$$2.5 \in B, \quad \pi \in B, \quad -1 \notin B, \quad 4 \notin B \qquad ◀$$

Sometimes sets arise that have no members. For example,

$$\{x : x^2 < 0\}$$

A set with no members is called an *empty set* or a *null set* and is denoted by the symbol $\emptyset$. Thus,

$$\emptyset = \{x : x^2 < 0\}$$

Of special interest are certain sets of real numbers called *intervals.* Geometrically, an interval is a line segment. For example, if $a < b$, then the *closed interval* from a to b is the set

$$\{x : a \leq x \leq b\}$$

and the *open interval* from a to b is the set

$$\{x : a < x < b\}$$

These sets are pictured in Figure 1.1.6.

Figure 1.1.6

REMARK. A closed interval includes both its endpoints (indicated by solid dots in Figure 1.1.6), while an open interval excludes both endpoints (indicated by open dots in Figure 1.1.6). Closed and open intervals are usually denoted by the symbols $[a, b]$ and (a, b), where

$$[a, b] = \{x : a \leq x \leq b\}$$
$$(a, b) = \{x : a < x < b\}$$

A square bracket [or] indicates that the endpoint is included while a rounded bracket (or) indicates that the endpoint is excluded.

An interval can include one endpoint and exclude the other. Such intervals are called **half-open** (or sometimes **half-closed**). For example,

$$[a, b) = \{x : a \leq x < b\}$$
$$(a, b] = \{x : a < x \leq b\}$$

An interval can extend indefinitely in either the positive or negative direction (see Figure 1.1.7). The intervals in Figure 1.1.7 are denoted by

$$(a, +\infty) = \{x : x > a\}$$

and

$$(-\infty, b] = \{x : x \leq b\}$$

where the symbol $+\infty$ ("plus infinity") indicates the interval extends indefinitely in the positive direction and the symbol $-\infty$ ("minus infinity") indicates the interval extends indefinitely in the negative direction.

Figure 1.1.7

Table 1.1.1 gives a complete list of the kinds of intervals possible.

Table 1.1.1

INTERVAL NOTATION	SET NOTATION	GEOMETRIC PICTURE
$[a, b]$	$\{x : a \le x \le b\}$	
(a, b)	$\{x : a < x < b\}$	
$[a, b)$	$\{x : a \le x < b\}$	
$(a, b]$	$\{x : a < x \le b\}$	
$(-\infty, b]$	$\{x : x \le b\}$	
$(-\infty, b)$	$\{x : x < b\}$	
$[a, +\infty)$	$\{x : x \ge a\}$	
$(a, +\infty)$	$\{x : x > a\}$	
$(-\infty, +\infty)$	$\{x : x \text{ is a real number}\}$	

1.1.5 DEFINITION Two sets A and B are said to be *equal* if they have the same elements, in which case we write

$$A = B$$

▶ Example 10

$$\{x : x^2 = 1\} = \{-1, 1\}, \qquad \{\pi, 0, 3\} = \{3, \pi, 0\}$$
$$\{x : x^2 < 9\} = \{x : -3 < x < 3\}$$ ◀

1.1.6 DEFINITION If every member of a set A is also a member of set B, then we say A is a *subset* of B and write

$$A \subset B$$

In addition, we will agree that the empty set $\varnothing$ is a subset of every set.

▶ Example 11

$$\{-2, 4\} \subset \{-2, 1, 0, 4\}$$
$$\{x : x \text{ is rational}\} \subset \{x : x \text{ is a real number}\}$$
$$\varnothing \subset A \text{ (for every set } A)$$ ◀

REMARK. If $A \subset B$ and $B \subset A$, then $A = B$. (Why?)

1.1.7 DEFINITION If *A* and *B* are two given sets, then the set of all elements belonging to both *A* and *B* is called the ***intersection*** of *A* and *B*; it is denoted by

$$A \cap B$$

1.1.8 DEFINITION If *A* and *B* are two given sets, then the set of all elements belonging to *A* or *B* or both is called the ***union*** of *A* and *B*; it is denoted by

$$A \cup B$$

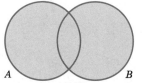

▶ Example 12

In Figure 1.1.8, *A* is the set of points inside the left circle and *B* the set of points inside the right circle. The sets $A \cap B$ and $A \cup B$ are shown as shaded regions. ◄

Figure 1.1.8

▶ Example 13

Sketch the sets:

(a) $[-4, 1) \cup (3, 6)$
(b) $(-3, 2] \cup (1, 7]$

Solution. See Figure 1.1.9. ◄

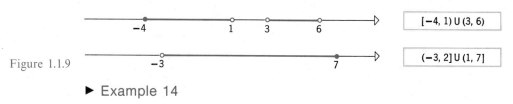

Figure 1.1.9

▶ Example 14

Sketch the set

$$(-4, 3] \cap (2, 6)$$

Solution. The set $(-4, 3] \cap (2, 6)$ consists of all points common to the intervals $(-4, 3]$ and $(2, 6)$. As shown in Figure 1.1.10, the intersection is the interval $(2, 3]$. ◄

Figure 1.1.10

In the following examples we use Theorem 1.1.2 to solve some inequalities; the solutions are expressed in terms of intervals. The first example points out a subtle problem in logic that we shall encounter throughout the text.

▶ Example 15

Solve

$$2x - 3 < 7$$

(That is, find all real numbers satisfying the inequality.)

Solution. If x is any solution, then

$$2x - 3 < 7 \quad \text{[Given]}$$
$$2x \phantom{{}- 3} < 10 \quad \text{[We added 3 to both sides.]}$$
$$x \phantom{{}- 3} < 5 \quad \text{[We multiplied both sides by } \tfrac{1}{2}.\text{]}$$

At this point we are tempted to conclude that the solutions of $2x - 3 < 7$ consist of all x less than 5, that is, all x in the interval $(-\infty, 5)$. While this conclusion is correct, it is premature. To see why, let us examine the logic of our argument. We have proved: *If x is a solution, then $x < 5$.* In other words, if S denotes the set of solutions to $2x - 3 < 7$, we have shown

$$S \subset (-\infty, 5) \tag{2}$$

We have *not* yet shown that

$$S = (-\infty, 5) \tag{3}$$

If we can prove that

$$(-\infty, 5) \subset S \tag{4}$$

then by the remark after Example 11, (3) will follow from (2) and (4).

To prove (4) we must show that every member of $(-\infty, 5)$ is also a member of S. That is, we must assume $x < 5$ and deduce that $2x - 3 < 7$. We do this as follows:

$$x \phantom{{}- 3} < 5 \quad \text{[Given.]}$$
$$2x \phantom{{}- 3} < 10 \quad \text{[We multiplied both sides by 2.]}$$
$$2x - 3 < 7 \quad \text{[We added } -3 \text{ to both sides.]}$$

Thus the solutions of $2x - 3 < 7$ form the interval $(-\infty, 5)$. ◀

REMARK. In the last example, compare the steps needed to prove (2) with the steps needed to prove (4):

Steps to Prove (2)	Steps to Prove (4)
$2x - 3 < 7$	$x < 5$
$2x < 10$	$2x < 10$
$x < 5$	$2x - 3 < 7$

In the two parts the steps are the same, but in opposite orders. In any "two-stage" mathematical argument where the steps in the second stage are the same as those in the first stage, but in the opposite order, it is common to give only the steps for the first stage and then conclude the proof by stating that *the steps are reversible*.

▶ Example 16

Solve

$$3 + 7x \leq 2x - 9$$

Solution.

$3 + 7x \leq 2x - 9$	[Given.]
$7x \leq 2x - 12$	[We added -3 to both sides.]
$5x \leq -12$	[We added $-2x$ to both sides.]
$x \leq -\frac{12}{5}$	[We multiplied both sides by $\frac{1}{5}$.]

$-\frac{12}{5}$

Figure 1.1.11

Since the steps are reversible, the set of solutions is $(-\infty, -\frac{12}{5}]$; this solution set is shown in Figure 1.1.11. ◀

▶ Example 17

Solve

$$7 \leq 2 - 5x < 9$$

Solution.

$7 \leq 2 - 5x < 9$	[Given.]
$5 \leq -5x < 7$	[We added -2 to each member.]
$-1 \geq x > -\frac{7}{5}$	[We multiplied each member by $-\frac{1}{5}$.]

$-\frac{7}{5}$ -1

Figure 1.1.12

Since the steps are reversible, the set of solutions is $(-\frac{7}{5}, -1]$; this solution set is shown in Figure 1.1.12. ◀

REMARK. Observe that in Example 17 the inequalities were reversed when we multiplied by the negative quantity $-\frac{1}{5}$.

▶ Example 18

Solve

$$\frac{2x - 5}{x - 2} < 1$$

Solution. As a first step it would be helpful to multiply both sides by $x - 2$. However, depending on whether $x - 2$ is positive or negative, the direction of the inequality will remain the same or get reversed. To account for both possibilities, we must consider two cases.

Case 1: $x - 2$ positive; that is, $x > 2$.

$$\frac{2x - 5}{x - 2} < 1 \qquad \text{[Given.]}$$

$$2x - 5 < x - 2 \qquad \text{[We multiplied both sides by the } positive \text{ quantity } x - 2.]$$

$$x - 5 < -2 \qquad \text{[We added } -x \text{ to both sides.]}$$

$$x < 3 \qquad \text{[We added 5 to both sides.]}$$

Since the steps are reversible the solutions for Case 1 consist of all x such that

$$x > 2 \qquad \text{and} \qquad x < 3$$

that is, all x in the interval $(2, 3)$.

Case 2: $x - 2$ negative; that is, $x < 2$.

$$\frac{2x - 5}{x - 2} < 1 \qquad \text{[Given.]}$$

$$2x - 5 > x - 2 \qquad \text{[We multiplied both sides by the } negative \text{ quantity } x - 2.]$$

$$x - 5 > -2 \qquad \text{[We added } -x \text{ to both sides.]}$$

$$x > 3 \qquad \text{[We added 5 to both sides.]}$$

Since the steps are reversible, the solutions for Case 2 consist of all x such that

$$x < 2 \qquad \text{and} \qquad x > 3$$

Figure 1.1.13

There are no such x, so that this case adds no additional solutions to those found in Case 1. Thus the entire set of solutions is $(2, 3)$, which is shown in Figure 1.1.13. ◄

REMARK. In the last example we ignored the case $x - 2 = 0$ because it is obvious that $x = 2$ is not a solution. (The left side of the given inequality would involve a division by zero.)

▶ Example 19

Solve

$$(x + 2)(x - 5) > 0$$

Solution. If x is a solution, then the factors $x + 2$ and $x - 5$ have the same sign; that is,

$$x + 2 > 0 \quad \text{and} \quad x - 5 > 0$$

or

$$x + 2 < 0 \quad \text{and} \quad x - 5 < 0$$

Let us consider the two cases.

Case 1: $x + 2 > 0$ and $x - 5 > 0$.
In this case x must satisfy $x > -2$ and $x > 5$; these inequalities both hold when $x > 5$, that is, when x lies in the interval $(5, +\infty)$.

Case 2: $x + 2 < 0$ and $x - 5 < 0$.
In this case x must satisfy $x < -2$ and $x < 5$; these inequalities both hold when $x < -2$, that is, when x is in the interval $(-\infty, -2)$.

Figure 1.1.14

Combining the solutions from the two cases, we obtain all x in the two intervals $(-\infty, -2)$ and $(5, +\infty)$ or, equivalently, all x in the set $(-\infty, -2) \cup (5, +\infty)$; this set is shown in Figure 1.1.14. ◄

▶ Exercise Set 1.1

1. In each part, sketch on a coordinate line all values of x that satisfy the stated condition.
   (a) $x \leq 4$ (b) $x \geq -3$ (c) $-1 \leq x \leq 7$
   (d) $x^2 = 9$ (e) $x^2 \leq 9$ (f) $x^2 \geq 9$.

2. In parts (a)–(d), sketch on a coordinate line all values of x, if any, that satisfy the stated conditions.
   (a) $x > 4$ and $x \leq 8$
   (b) $x \leq 2$ or $x \geq 5$

(c) $x > -2$ and $x \geq 3$
(d) $x \leq 5$ and $x > 7$.

3. Among the terms, *integer, rational,* and *irrational,* which ones apply to the given number?
   (a) $-\frac{3}{4}$ (b) 0   (c) $\frac{24}{8}$
   (d) 0.25   (e) $-\sqrt{16}$ (f) $2^{1/2}$
   (g) 0.020202 . . .   (h) 7.000 . . .

4. Which of the terms, *integer, rational,* and *irrational,* apply to the given number?
   (a) 0.31311311131111 . . .   (b) 0.729999 . . .
   (c) 0.376237623762 . . .   (d) $17\frac{4}{3}$.

5. Which of the following are always correct if $a \leq b$?
   (a) $a - 3 \leq b - 3$ (b) $-a \leq -b$
   (c) $3 - a \leq 3 - b$ (d) $6a \leq 6b$
   (e) $a^2 \leq ab$ (f) $a^3 \leq a^2 b$.

6. Which of the following are always correct if $a \leq b$ and $c \leq d$?
   (a) $a + 2c \leq b + 2d$ (b) $a - 2c \leq b - 2d$
   (c) $a - 2c \geq b - 2d$.

7. The repeating decimal 0.137137137 . . . can be expressed as a ratio of integers by writing

   $$x = 0.137137137 \ldots$$
   $$1000x = 137.137137137 \ldots$$

   and subtracting to obtain $999x = 137$ or $x = \frac{137}{999}$. Use this idea, where needed, to express the following decimals as ratios of integers.
   (a) 0.123123123 . . .   (b) 12.7777 . . .
   (c) 38.07818181 . . .   (d) 0.4296000 . . .

8. Show that the repeating decimal 0.99999 . . . represents the number 1. Since 1.000 . . . is also a decimal representation of 1, this problem shows that a real number can have two different decimal representations. [*Hint:* Use the technique of Exercise 7.]

9. For what values of a are the following inequalities valid?
   (a) $a \leq a$ (b) $a < a$.

10. If $a \leq b$ and $b \leq a$, what can you say about a and b?

11. (a) If $a < b$ is true, does it follow that $a \leq b$ must also be true?

(b) If $a \leq b$ is true, does it follow that $a < b$ must also be true?

12. In each part, list the elements in the set.
    (a) $\{x : x^2 - 5x = 0\}$
    (b) $\{x : x$ is an integer satisfying $-2 < x < 3\}$.

13. In each part, express the set in the notation $\{x : \underline{\hspace{1cm}}\}$.
    (a) $\{1, 3, 5, 7, 9, \ldots\}$
    (b) the set of even integers
    (c) the set of irrational numbers
    (d) $\{7, 8, 9, 10\}$.

14. Let $A = \{1, 2, 3\}$. Which of the following are equal to A?
    (a) $\{0, 1, 2, 3\}$ (b) $\{3, 2, 1\}$
    (c) $\{x : (x - 3)(x^2 - 3x + 2) = 0\}$.

15. Which of the following sets are empty?
    (a) $\{x : x^2 + 1 = 0\}$
    (b) $\{x : x^2 - 1 = 0\}$
    (c) $\{x : x > 3 $ and $ x < 3\}$
    (d) $\{x : x \geq 3 $ and $ x \leq 3\}$.

16. List all subsets of
    (a) $\{a_1, a_2, a_3\}$ (b) $\varnothing$.

17. In Figure 1.1.15, let
    $S = $ the set of points inside the square
    $T = $ the set of points inside the triangle
    $C = $ the set of points inside the circle
    and let a, b, and c be the points shown. Answer the following as true or false.
    (a) $T \subset C$ (b) $T \subset S$ (c) $a \notin T$
    (d) $a \notin S$
    (e) $b \in T$ and $b \in C$
    (f) $a \in C$ or $a \in T$
    (g) $c \in T$ and $c \notin C$.

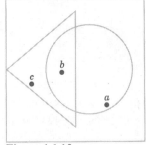

Figure 1.1.15

18. In each part, sketch the set on a coordinate line.
   (a) $[-3, 2] \cup [1, 4]$
   (b) $[4, 6] \cup [8, 11]$
   (c) $(-4, 0) \cup (-5, 1)$
   (d) $[2, 4) \cup (4, 7)$
   (e) $(-2, 4) \cap (0, 5]$
   (f) $[1, 2.3) \cup (1.4, \sqrt{2})$
   (g) $(-\infty, -1) \cup (-3, +\infty)$
   (h) $(-\infty, 5) \cap [0, +\infty)$.

In Exercises 19–40, solve the inequality and sketch the solution on a coordinate line.

19. $3x - 2 < 8$.
20. $\frac{1}{5}x + 6 \geq 14$.
21. $4 + 5x \leq 3x - 7$.
22. $2x - 1 > 11x + 9$.
23. $3 \leq 4 - 2x < 7$.
24. $-2 \geq 3 - 8x \geq -11$.
25. $\dfrac{x}{x - 3} < 4$.
26. $\dfrac{x}{8 - x} \geq -2$.
27. $\dfrac{3x + 1}{x - 2} < 1$.
28. $\dfrac{\frac{1}{2}x - 3}{4 + x} > 1$.
29. $\dfrac{4}{2 - x} \leq 1$.
30. $\dfrac{3}{x - 5} \leq 2$.
31. $x^2 > 9$.
32. $x^2 \leq 5$.
33. $(x - 4)(x + 2) > 0$.
34. $(x - 3)(x + 4) < 0$.
35. $x^2 - 9x + 20 \leq 0$.
36. $2 - 3x + x^2 \geq 0$.
37. $\dfrac{2}{x} < \dfrac{3}{x - 4}$.
38. $\dfrac{1}{x + 1} \geq \dfrac{3}{x - 2}$.
39. $x^3 - x^2 - x - 2 > 0$.
40. $x^3 - 3x + 2 \leq 0$.
41. Prove the following results about sums of rational and irrational numbers:

   (a)   rational + rational = rational
   (b)   rational + irrational = irrational.

42. Prove the following results about products of rational and irrational numbers:
   (a)   rational · rational = rational
   (b)   rational · irrational = irrational   (provided the rational factor is nonzero).

43. Show that the sum or product of two irrational numbers can be rational or irrational.

44. Classify the following as rational or irrational and justify your conclusion.
   (a) $3 + \pi$ (b) $\frac{3}{4}\sqrt{2}$
   (c) $\sqrt{8}\sqrt{2}$ (d) $\sqrt{\pi}$
   (See Exercises 41 and 42.)

45. Prove: The average of two rational numbers is a rational number, but the average of two irrational numbers can be rational or irrational.

46. Can a rational number satisfy $10^x = 3$?

47. Solve: $8x^3 - 4x^2 - 2x + 1 < 0$.

48. Solve: $12x^3 - 20x^2 \geq -11x + 2$.

49. Prove: If a, b, c, and d are positive numbers such that $a < b$ and $c < d$, then $ac < bd$. (This result gives conditions under which inequalities can be "multiplied together.")

50. Show that rational numbers are represented by repeating decimals. [*Hint:* Examine the long division process that converts a ratio of integers into a decimal.]

1.2 ABSOLUTE VALUE

1.2.1 DEFINITION   The *absolute value* or *magnitude* of a real number a is denoted by $|a|$ and is defined by

$$\begin{cases} |a| = a & \text{if} \quad a \geq 0 \\ |a| = -a & \text{if} \quad a < 0 \end{cases}$$

▶ Example 1

$\lvert 5 \rvert = 5$	[since $5 \geq 0$]
$\lvert -\frac{4}{7} \rvert = -(-\frac{4}{7}) = \frac{4}{7}$	[since $-\frac{4}{7} < 0$]
$\lvert 0 \rvert = 0$	[since $0 \geq 0$]

◀

REMARK. Symbols such as $+a$ and $-a$ are deceptive since it is tempting to conclude that $+a$ is positive and $-a$ is negative. However, these conclusions are unwarranted since a itself can represent either a positive or negative number. In fact, if a itself is negative, then $-a$ is positive and $+a$ is negative. With this comment in mind, it should be evident that $\lvert a \rvert \geq 0$ for any number a.

Recall that a number whose square is a is called a **square root** of a. In algebra it is learned that every nonnegative real number has exactly one *nonnegative* square root; we denote this square root by $\sqrt{a}$. For example, the number 9 has two square roots, -3 and 3. Since 3 is the nonnegative square root we have

$$\sqrt{9} = 3$$

REMARK. Readers who were previously taught to write $\sqrt{9} = \pm 3$ are advised to stop doing so, since it is incorrect.

It is another common error to write $\sqrt{a^2} = a$. Although this equality is correct when a is nonnegative, it is false for negative a. For example, if $a = -4$, then

$$\sqrt{a^2} = \sqrt{(-4)^2} = \sqrt{16} = 4 \neq a$$

A result that is correct for all a is given in the following theorem.

1.2.2 THEOREM *For any real number a,*

$$\sqrt{a^2} = \lvert a \rvert$$

Proof. Since $a^2 = (+a)^2 = (-a)^2$, the numbers $+a$ and $-a$ are square roots of a^2. If $a \geq 0$, then $+a$ is the nonnegative square root of a^2 and if $a < 0$, then $-a$ is the nonnegative square root of a^2. Since $\sqrt{a^2}$ denotes the nonnegative square root of a^2, we have

$$\begin{cases} \sqrt{a^2} = +a & \text{if} \quad a \geq 0 \\ \sqrt{a^2} = -a & \text{if} \quad a < 0 \end{cases}$$

That is, $\sqrt{a^2} = \lvert a \rvert$. ▮

Some basic properties of absolute value are listed in the following theorem.

1.2.3 THEOREM *If a and b are real numbers and n is an integer, then*

(a) $-|a| \leq a \leq |a|$

(b) $|ab| = |a|\,|b|$

(c) $\left|\dfrac{a}{b}\right| = \dfrac{|a|}{|b|}$

(d) $|a^n| = |a|^n$

We will prove parts (a) and (b) only.

Proof of (a). If $a \geq 0$, then we can write

$$-a \leq a \leq a \qquad \text{and} \qquad |a| = a$$

from which it follows that

$$-|a| \leq a \leq |a|$$

If $a < 0$, we can write

$$a \leq a \leq -a \qquad \text{and} \qquad -a = |a|$$

from which it again follows that

$$-|a| \leq a \leq |a|$$

Thus statement (a) holds in all cases.

Proof (b). Using Theorem 1.2.2

$$|ab| = \sqrt{(ab)^2} = \sqrt{a^2b^2} = \sqrt{a^2}\sqrt{b^2} = |a|\,|b|.$$

REMARK. In words, parts (b) and (c) of this theorem state that *the absolute value of a product is the product of the absolute values and the absolute value of a ratio is the ratio of the absolute values.*

The notion of absolute value arises naturally in distance problems. On a coordinate line, let A and B be points with coordinates a and b. Because distance is always nonnegative, the distance d between A and B is

$$d = b - a$$

(a)

(b)

Figure 1.2.1

when B is to the right of A (Figure 1.2.1a), and

$$d = a - b$$

when B is to the left of A (Figure 1.2.1b).

In the first case, $b - a$ is positive, so we can write

$$d = b - a = |b - a|$$

and in the second case, $b - a$ is negative, so we can write

$$d = a - b = -(b - a) = |b - a|$$

Thus, regardless of whether B is to the right or left of A, the distance d between A and B is

$$d = |b - a|$$

This formula is useful when the relative positions of A and B are unknown.

For any real number b we can write

$$|b| = |b - 0|$$

Therefore, the absolute value of a number b can be interpreted geometrically as its distance from the origin on a coordinate line. For example, if $|b| = 5$, then b lies 5 units from the origin, that is, $b = 5$ or $b = -5$.

▶ Example 2

Solve $|x - 3| = 4$.

Solution 1 (geometric). The solution consists of all x that are 4 units away from the point 3. There are two such x values, $x = 7$ and $x = -1$ (Figure 1.2.2).

Figure 1.2.2

Solution 2 (algebraic). Depending on whether $x - 3$ is positive or negative, the equation $|x - 3| = 4$ can be written as

$$x - 3 = 4 \quad \text{or} \quad -(x - 3) = 4$$

Solving these two equations gives

$$x = 7 \quad \text{and} \quad x = -1$$

which agrees with the solution obtained geometrically. ◀

▶ Example 3

Solve

$$|x - 3| < 4$$

Solution. The solution consists of all x whose distance from 3 is less than 4 units, that is, all x satisfying

$$-1 < x < 7$$

This is the interval $(-1, 7)$ shown in Figure 1.2.3. ◀

Figure 1.2.3

The following general result will be of importance in later sections.

1.2.4 THEOREM *For any real numbers x and a and any positive number k*

(a) $|x| < k$ *if and only if* $-k < x < k$

(b) $|x - a| < k$ *if and only if* $a - k < x < a + k$

Part (*a*) states that $|x| < k$ if and only if x is within k units of the origin (Figure 1.2.4*a*) and part (*b*) states that $|x - a| < k$ if and only if x is within k units of a (Figure 1.2.4*b*).

Figure 1.2.4 (*a*) (*b*)

REMARK. Theorem 1.2.4 is also true if we replace $<$ by $\leq$.

The following examples illustrate additional techniques for solving equations and inequalities involving absolute values.

▶ Example 4

Solve

$$|x + 4| > 2$$

Solution. The given inequality can be rewritten

$$|x - (-4)| > 2$$

Thus the solution consists of all x whose distance from -4 is greater than 2 units. This is the set

$$(-\infty, -6) \cup (-2, +\infty)$$

shown in Figure 1.2.5. ◀

Figure 1.2.5

▶ Example 5

Solve

$$|3x - 2| = |5x + 4|$$

Solution. The equation will be satisfied if either

$$3x - 2 = 5x + 4 \quad \text{or} \quad 3x - 2 = -(5x + 4)$$

Solving the first equation gives $x = -3$ and solving the second equation gives $x = -\frac{1}{4}$. Thus $x = -3$ and $x = -\frac{1}{4}$ are the solutions. ◀

It is *not* generally true that $|a + b| = |a| + |b|$. For example, if $a = 2$ and $b = -3$, then $a + b = -1$, so that

$$|a + b| = |-1| = 1$$

whereas

$$|a| + |b| = |2| + |-3| = 2 + 3 = 5$$

The following theorem shows that *the absolute value of a sum is always less than or equal to the sum of the absolute values.*

1.2.5 THEOREM   *For any real numbers, a and b,*

The Triangle Inequality

$$|a + b| \leq |a| + |b|$$

Proof. From Theorem 1.2.3*a*.

$$-|a| \leq a \leq |a| \qquad \text{and} \qquad -|b| \leq b \leq |b|$$

Adding these inequalities yields

$$-(|a| + |b|) \leq a + b \leq (|a| + |b|) \tag{1}$$

Using Theorem 1.2.4*a* with $\leq$ rather than $<$, with $x = a + b$, and with $k = |a| + |b|$, it follows from (1) that

$$|a + b| \leq |a| + |b|$$

▶ Exercise Set 1.2

1. Compute $|x|$ if
   (a) $x = 7$ (b) $x = -\sqrt{2}$
   (c) $x = k^2$ (d) $x = -k^2$.

2. Rewrite $\sqrt{(x - 6)^2}$ without using a square root or absolute value sign.

3. Verify $\sqrt{a^2} = |a|$ for $a = 7$ and $a = -7$.

4. Verify the inequalities $-|a| \leq a \leq |a|$ for $a = 2$ and $a = -5$.

5. Let A and B be points with coordinates a and b. In each part find the distance between A and B.
   (a) $a = 9, b = 7$ (b) $a = 2, b = 3$
   (c) $a = -8, b = 6$ (d) $a = \sqrt{2}, b = -3$
   (e) $a = -11, b = -4$ (f) $a = 0, b = -5$.

6. Is the equality $\sqrt{a^4} = a^2$ valid for all values of a? Explain.

7. Let A and B be points with coordinates a and b. In each part, use the given information to find b.
   (a) $a = -3$, B is to the left of A, and $|b - a| = 6$.
   (b) $a = -2$, B is to the right of A, and $|b - a| = 9$.
   (c) $a = 5$, $|b - a| = 7$, and $b > 0$.

8. Let E and F be points with coordinates e and f. In each part, determine whether E is to the left or to the right of F on a coordinate line.
   (a) $f - e = 4$ (b) $e - f = 4$
   (c) $f - e = -6$ (d) $e - f = -7$.

In Exercises 9–16 solve for x.

9. $|6x - 2| = 7$.   10. $|3 + 2x| = 11$.

11. $|6x - 7| = |3 + 2x|$.   12. $|4x + 5| = |8x - 3|$.

13. $|9x| - 11 = x$.   14. $2x - 7 = |x + 1|$.

15. $\left|\dfrac{x + 5}{2 - x}\right| = 6$.   16. $\left|\dfrac{x - 3}{x + 4}\right| = 5$.

In Exercises 17–26, solve for x and express the solution in terms of intervals.

17. $|x + 6| < 3$.   18. $|7 - x| \leq 5$.

19. $|5 - 2x| \geq 4$.   20. $|7x + 1| > 3$.

21. $|x + 3| < |x - 8|$.   22. $|3x| \leq |2x - 5|$.

23. $|4x| \geq |7 - 6x|$.   24. $|2x + 1| > |x - 5|$.

25. $\left|\dfrac{x - \frac{1}{2}}{x + \frac{1}{2}}\right| < 1$.   26. $\left|\dfrac{3 - 2x}{1 + x}\right| \leq 4$.

27. For which values of α is $\sqrt{(\alpha^2 - 5\alpha + 6)^2} = \alpha^2 - 5\alpha + 6$?

28. Verify the triangle inequality $|a + b| \leq |a| + |b|$ (Theorem 1.2.5) for

(a) $a = 3$, $b = 4$ (b) $a = -2$, $b = 6$

(c) $a = -7$, $b = -8$ (d) $a = -4$, $b = 4$.

29. Under what conditions will equality hold in the triangle inequality, $|a + b| \leq |a| + |b|$ (Theorem 1.2.5)?

30. Solve the problem in Example 4 algebraically.

31. Prove: $|a - b| \leq |a| + |b|$.

32. Prove: $|a| - |b| \leq |a - b|$.

33. Prove: $||a| - |b|| \leq |a - b|$. [*Hint:* Use Exercise 32.]

34. Find the smallest value of M such that

$$\left|\frac{1}{x}\right| \leq M \text{ for all } x \text{ in the interval } [2, 7].$$

35. Find the smallest value of M such that

$$\left|\frac{1}{x + 7}\right| \leq M$$

for all x in the interval $(-4, 2)$.

36. Use the triangle inequality to find a value of M such that

$$|x^3 - 2x + 1| \leq M$$

for all x in the interval $(-2, 3)$.

37. Find a value of M such that

$$\left|\frac{x + 3}{x - 3}\right| \leq M$$

for all x in the interval $[-\frac{3}{4}, \frac{1}{4}]$.

1.3 COORDINATE PLANES; DISTANCE; CIRCLES

Just as points on a line can be placed in one-to-one correspondence with the real numbers, so points in a plane can be placed in one-to-one correspondence with pairs of real numbers by using two coordinate lines. To do this, we construct two perpendicular coordinate lines that intersect at their origins; for convenience, we make one of the lines horizontal with its positive direction to the right and the other line vertical with its positive direction upward (Figure 1.3.1). The two lines are called *coordinate axes;* the horizon-

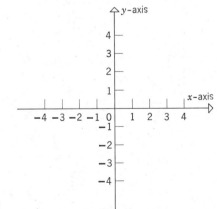

Figure 1.3.1

tal line is called the **x-axis**, the vertical line is called the **y-axis**, and the coordinate axes together form what is called a ***Cartesian coordinate system*** or sometimes a ***rectangular coordinate system.*** The point of intersection of the coordinate axes is denoted by O and is called the ***origin*** of the coordinate system.

REMARK. Throughout this text we will assume that the same unit of measurement is used on both coordinate axes (see Exercise 40).

A plane in which a rectangular coordinate system has been introduced is called a ***coordinate plane*** or an ***xy-plane.*** We will now show how to establish a one-to-one correspondence between points in a coordinate plane and pairs of real numbers. If P is a point in a coordinate plane, then we draw two lines through P, one perpendicular to the x-axis and one perpendicular to the y-axis. If the first line intersects the x-axis at the point with coordinate a and the second line intersects the y-axis at the point with coordinate b, then we associate the pair

$$(a, b)$$

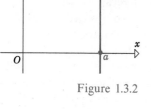

Figure 1.3.2

with the point P (Figure 1.3.2). The number a is called the **x-coordinate** or **abscissa** of P and the number b is called the **y-coordinate** or ***ordinate*** of P; we say that P is the point with ***coordinates*** (a, b) and denote the point by $P(a, b)$.

▶ Example 1

In Figure 1.3.3 we have located the points

$$P(2, 5), \qquad Q(-4, 3), \qquad R(-5, -2), \qquad S(4, -3) \qquad ◄$$

Figure 1.3.3

Figure 1.3.4

Figure 1.3.5

From the above construction, each point in a coordinate plane determines a unique pair of numbers. Conversely, starting with a pair of numbers (a, b) we can construct lines perpendicular to the x-axis and y-axis at the points with coordinates a and b, respectively; the intersection of these lines determines a unique point P in the plane whose coordinates are (a, b) (Figure 1.3.2). Thus we have a one-to-one correspondence between pairs of real numbers and points in a coordinate plane.

REMARK. Since the order in which the members of a set are listed does not matter, the set $\{a, b\}$ and the set $\{b, a\}$ are the same. However, the pair of real numbers (a, b) and the pair of real numbers (b, a) represent different points (unless $a = b$), so that order is important. For this reason a pair (a, b) of real numbers is often called an **ordered pair.**

The coordinate axes divide the plane into four parts, called **quadrants.** These quadrants are numbered from one to four as shown in Figure 1.3.4. As shown in Figure 1.3.5 it is easy to determine the quadrant in which a point lies from the signs of its coordinates. A point with two positive coordinates $(+, +)$ lies in Quadrant I, a point with a negative x-coordinate and a positive y-coordinate $(-, +)$ lies in Quadrant II, and so on.

Occasionally, we will use letters other than x and y to label coordinate axes. Figure 1.3.6 shows a uv-plane and a tQ-plane. The first letter in the name of the plane refers to the horizontal axis and the second to the vertical axis.

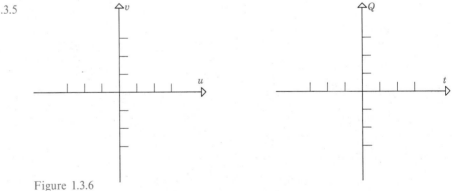

Figure 1.3.6

In the last section we showed that the distance between points a and b on a coordinate line is $|b - a|$. It follows that the distance d between two points $A(x_1, y)$ and $B(x_2, y)$ on a horizontal line in the xy-plane is

$$d = |x_2 - x_1|$$

Figure 1.3.7   Figure 1.3.8

(Figure 1.3.7) and the distance d between two points $A(x, y_1)$ and $B(x, y_2)$ on a vertical line in the xy-plane is

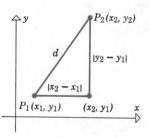

Figure 1.3.9

$$d = |y_2 - y_1|$$

(See Figure 1.3.8.)

To find the distance d between any two points $P_1(x_1, y_1)$ and $P_2(x_2, y_2)$ in the xy-plane we apply the Theorem of Pythagoras to the triangle shown in Figure 1.3.9. This yields

$$d = \sqrt{|x_2 - x_1|^2 + |y_2 - y_1|^2}$$

Since $|x_2 - x_1|^2 = (x_2 - x_1)^2$ and $|y_2 - y_1|^2 = (y_2 - y_1)^2$, we are led to the following result.

1.3.1 THEOREM   *The distance d between two points (x_1, y_1) and (x_2, y_2) in a coordinate plane is given by*

$$d = \sqrt{(x_2 - x_1)^2 + (y_2 - y_1)^2} \tag{1}$$

▶ Example 2

The distance between $(-1, 2)$ and $(3, 4)$ is

$$d = \sqrt{(3 - (-1))^2 + (4 - 2)^2} = \sqrt{4^2 + 2^2} = \sqrt{20} \qquad ◀$$

On a coordinate line, the average $\frac{1}{2}(a + b)$ is the midpoint of the line segment joining a and b (Figure 1.3.10a). In the xy-plane the midpoint of the

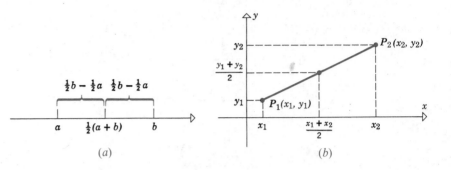

Figure 1.3.10   (a)   (b)

line segment joining two points (x_1, y_1) and (x_2, y_2) is obtained by averaging the coordinates of the endpoints. More precisely:

$$\begin{bmatrix} \textit{The midpoint of the line segment} \\ \textit{joining } P_1(x_1, y_1) \textit{ and } P_2(x_2, y_2) \end{bmatrix} = \left(\frac{x_1 + x_2}{2}, \frac{y_1 + y_2}{2} \right) \qquad (2)$$

(See Figure 1.3.10b.) The proof is left as an exercise.

▶ Example 3

From (2), the midpoint of the line segment joining the points $(3, -4)$ and $(7, 2)$ is

$$\left(\frac{3 + 7}{2}, \frac{-4 + 2}{2} \right) = (5, -1) \qquad ◀$$

Rectangular coordinate systems enable us to describe equations geometrically. To explain how, we need some preliminary ideas. Assume we have constructed a rectangular coordinate system and we are given an equation involving only two variables, x and y; for example,

$$5xy = 2, \qquad x^2 + 2y^2 = 7, \qquad \text{or} \qquad y = \frac{1}{1 - x}$$

By a **solution** of such an equation we mean an ordered pair of real numbers (a, b) such that the equation is satisfied when we substitute $x = a$ and $y = b$.

▶ Example 4

The pair $(3, 2)$ is a solution of

$$6x - 4y = 10$$

since the equation is satisfied when we substitute $x = 3$ and $y = 2$. However, the pair $(2, 0)$ is not a solution since the equation is not satisfied when we substitute $x = 2$ and $y = 0$.   .   ◀

The set of all solutions of an equation is called its **solution set.** If we locate or **plot** all members of the solution set in a rectangular coordinate system, we obtain a set of points called the **graph** of the equation.

▶ Example 5

Sketch the graph of

$$y = x^2$$

Solution. The solution set of $y = x^2$ has infinitely many members, so that it is impossible to plot them all. However, some sample members of the solution set can be obtained by substituting some arbitrary x values into the right side of $y = x^2$ and solving for the associated values of y. Some typical computations are given in Table 1.3.1.

Table 1.3.1

x	0	1	2	3	-1	-2	-3
$y = x^2$	0	1	4	9	1	4	9
(x, y)	$(0, 0)$	$(1, 1)$	$(2, 4)$	$(3, 9)$	$(-1, 1)$	$(-2, 4)$	$(-3, 9)$

In Figure 1.3.11 the points listed in Table 1.3.1 are connected by a smooth curve. This curve approximates the graph of $y = x^2$. ◀

Figure 1.3.11

Figure 1.3.12

REMARK. It is important to keep in mind that the sketch in Figure 1.3.11 is only an *approximation* to the graph of $y = x^2$ based on plotting a few points. It is conceivable that the smooth curve connecting the plotted points does not accurately describe the true graph. For example, isn't it logically possible that the graph of $y = x^2$ oscillates between the points we have plotted and looks like the dark blue curve in Figure 1.3.12? Although we might look for such oscillations by plotting additional points between those already obtained, we can never resolve the problem with certainty through point plotting since we will never be sure how the true graph behaves *between* our plotted points. In later sections we will show how calculus can be used to determine the true shape of a graph.

CIRCLES   If (x_0, y_0) is a fixed point in the plane, then the circle of radius r centered at (x_0, y_0) is the set of all points in the plane whose distance from (x_0, y_0) is r (Figure 1.3.13). Thus a point (x, y) will lie on this circle if and only if

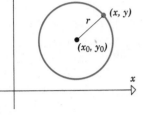

$$\sqrt{(x - x_0)^2 + (y - y_0)^2} = r$$

or equivalently

$$(x - x_0)^2 + (y - y_0)^2 = r^2 \tag{3}$$

Figure 1.3.13   This is called the **standard equation of a circle.**

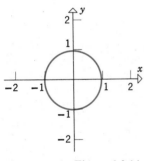

▶ Example 6

Sketch the graph of

$$x^2 + y^2 = 1$$

Solution. This equation results when $x_0 = y_0 = 0$ and $r = 1$ in (3). Thus the graph is a circle of radius 1 centered at the origin (Figure 1.3.14). This is often called the **unit circle.**   ◀

Figure 1.3.14

▶ Example 7

Find an equation for the circle of radius 5 centered at $(-3, 2)$.

Solution. From (3) with $x_0 = -3$, $y_0 = 2$, and $r = 5$ we obtain

$$(x + 3)^2 + (y - 2)^2 = 25$$

or, if preferred, we can square out the terms to obtain

$$x^2 + y^2 + 6x - 4y - 12 = 0$$ ◀

By squaring the terms in (3) and simplifying, we obtain an equation of the form

$$x^2 + y^2 + dx + ey + f = 0 \tag{4}$$

where d, e, and f are constants. Conversely, if the equation of a circle is given in form (4), we can obtain its standard equation by completing the squares.

▶ Example 8

Find the center and radius of the circle with equation

$$x^2 + y^2 - 8x + 2y + 8 = 0$$

Solution. First, group the x-terms, group the y-terms, and take the constant to the right side:

$$(x^2 - 8x) + (y^2 + 2y) = -8$$

Next add the appropriate constant within each set of parentheses to complete the square, and add the same constants to the right side to maintain equality:

$$(x^2 - 8x + 16) + (y^2 + 2y + 1) = -8 + 16 + 1$$

or

$$(x - 4)^2 + (y + 1)^2 = 9$$

Thus from (3), the circle has center $(4, -1)$ and radius 3.   ◀

An equation of form (4) does not always have a circle as its graph. On completing the squares we obtain

$$(x - x_0)^2 + (y - y_0)^2 = k$$

If $k > 0$ then the graph is a circle with center (x_0, y_0) and radius $\sqrt{k}$. If $k = 0$ the only solution of the equation is $x = x_0, y = y_0$; thus the graph is the single point (x_0, y_0).

If $k < 0$ then the equation has no solutions and there is no graph. In summary, we have the following result.

1.3.2 THEOREM   *An equation of the form*

$$x^2 + y^2 + dx + ey + f = 0$$

represents a circle, or a point, or else has no graph.

The last two cases in Theorem 1.3.2 are called ***degenerate cases.***

▶ Exercise Set 1.3

1. Draw a rectangular coordinate system and locate the points
   (a) $(3, 4)$ (b) $(-2, 5)$
   (c) $(-2.5, -3)$ (d) $(1.7, -2)$

2. In each part, draw a rectangular coordinate system and sketch the set of points whose coordinates (x, y) satisfy the given conditions.
   (a) $x = 0$ (b) $y = 0$
   (c) $y < 0$ (d) $x \geq 1$ and $y \leq 2$
   (e) $x = y$ (f) $x \geq y$
   (g) $x = 3$ (h) $y = -1$
   (i) $|x| = 5$.

3. Find the fourth vertex of the rectangle, three of whose vertices are $(-1, 4)$, $(6, 4)$, and $(-1, 9)$.

4. In each part, the points lie on a horizontal or vertical line. State whether the line is horizontal or vertical, and find the distance between the points.
   (a) $A(9, 2)$, $B(7, 2)$
   (b) $A(2, -6)$, $B(3, -6)$
   (c) $A(-8, 1)$, $B(6, 1)$
   (d) $A(-4, \sqrt{2})$, $B(-4, -3)$
   (e) $A(3, -11)$, $B(3, -4)$
   (f) $A(0, 0)$, $B(0, -5)$.

In Exercises 5–8, find
(a) the distance between A and B
(b) the midpoint of the line segment joining A and B.

5. $A(7, 1)$, $B(1, 9)$.

6. $A(2, 5)$, $B(-1, 1)$.

7. $A(-2, -6)$, $B(-7, -4)$.

8. $A(2, 0)$, $B(-3, 6)$.

9. Show that the triangle with vertices $(5, -2)$, $(6, 5)$, $(2, 2)$ is isosceles.

10. Show that the points $(-1, 7)$, $(5, 4)$, $(2, -2)$, and $(-4, 1)$ are vertices of a square.

11. Show that $(0, -2)$, $(-4, 8)$, and $(3, 1)$ lie on a circle with center $(-2, 3)$.

12. Show that $(1, 3)$, $(4, 2)$, and $(-2, -6)$ are vertices of a right triangle and specify the vertex at which the right angle occurs.

13. In each part determine if the given ordered pair (x, y) is a solution of $x^2 - 2x + y = 4$.
    (a) $(0, 4)$ (b) $(-3, 7)$
    (c) $(\frac{1}{2}, \frac{19}{4})$ (d) $(1 + \sqrt{5 - t}, t)$.

In Exercises 14–21, sketch the graph of the equation. (A hand calculator will be helpful in some of these problems.)

14. $y = 2x - 3$.   15. $y = \sqrt{x}$.

16. $y = |x|$.   17. $y - \dfrac{1}{x} = 0$.

18. $x^3 = y$.   19. $xy = 1$.

20. $y = -\sqrt{x + 1}$.   21. $x^3 - 9x - y = 0$.

22. Sketch the graph of $y^2 = 3x$ and explain how this graph is related to the graphs of $y = \sqrt{3x}$ and $y = -\sqrt{3x}$.

23. Sketch the graph of $(x - y)(x + y) = 0$ and explain how it is related to the graphs of $x - y = 0$ and $x + y = 0$.

24. Graph $F = \frac{9}{5}C + 32$ in a CF-coordinate system.

25. Graph $u = 3v^2$ in a uv-coordinate system.

26. Graph $Y = 4X + 5$ in a YX-coordinate system.

In Exercises 27–30, find the standard equation for the circle satisfying the given conditions.

27. Center $(3, -2)$; radius $= 4$.

28. Center $(5, 8)$; circle is tangent to the y-axis.

29. Center $(-3, -4)$; circle passes through the origin.

30. A diameter has endpoints $(6, 1)$ and $(-2, 3)$.

In Exercises 31–36, determine if the equation represents a circle, a point, or no graph. If the equation represents a circle, find the center and radius.

31. $x^2 + y^2 - 2x - 4y - 11 = 0$.

32. $x^2 + y^2 + 8x + 8 = 0$.

33. $x^2 + y^2 + 2x + 2y + 2 = 0$.

34. $4x^2 + 4y^2 - 16x - 24y = 9$.

35. $9x^2 + 9y^2 = 1$.

36. $x^2 + y^2 - 10x - 2y + 29 = 0$.

37. Find an equation whose graph is the perpendicular bisector of the line segment connecting $(-2, 1)$ and $(4, -3)$.

38. Use the distance formula in Theorem 1.3.1 to prove that $(1, 1)$, $(-2, -8)$, and $(4, 10)$ lie on a straight line.

39. Find k, given that $(2, k)$ is equidistant from $(3, 7)$ and $(9, 1)$.

40. Where in this section did we use the fact that the same unit of measure was used on both coordinate axes?

41. Prove: The midpoint of the hypotenuse of a right triangle is equidistant from the three vertices. [*Hint:* Let a and b denote the lengths of the sides and introduce coordinate axes so that the vertices are $(0, 0)$, $(a, 0)$, and $(0, b)$.]

1.4 SLOPE OF A LINE

Before starting this section, readers who need to review trigonometry are advised to read Unit 1 of the trigonometry review in Appendix 1.

In this section we will discuss ways to measure the "steepness" or "slope" of a line in the plane. The ideas we develop here will be important when we discuss equations and graphs of straight lines.

The steepness of the line segment connecting two distinct points $P_1(x_1, y_1)$ and $P_2(x_2, y_2)$ can be measured by traveling from one end of the line segment to the other and relating the vertical change in height to the horizontal change in distance. To illustrate, compare the line segments in Figures 1.4.1*a* and 1.4.1*b* as we travel from P_1 to P_2. In the two cases, the horizontal change $x_2 - x_1$ is the same. However, the corresponding vertical change $y_2 - y_1$ is larger for the steeper line segment, which means that the quotient,

$$\frac{y_2 - y_1}{x_2 - x_1}$$

is larger for the steeper line segment. This suggests that we can use this quotient as a numerical measure of steepness.

Figure 1.4.1   (*a*)   (*b*)

1.4.1 DEFINITION If $P_1(x_1, y_1)$ and $P_2(x_2, y_2)$ are distinct points such that $x_1 \neq x_2$, then the number m given by the formula

$$m = \frac{y_2 - y_1}{x_2 - x_1} \tag{1}$$

is called the ***slope of the line segment connecting P_1 and P_2.***

Formula (1) does not apply to line segments parallel to the y-axis. For such segments $x_2 = x_1$ and (1) would involve a division by zero; therefore, for line segments parallel to the y-axis we say the slope is **undefined,** or the line segment has **infinite slope.**

REMARK. In light of the equality

$$\frac{y_2 - y_1}{x_2 - x_1} = \frac{y_1 - y_2}{x_1 - x_2}$$

it does not matter which endpoint is labeled P_1 and which is labeled P_2 when we apply (1) to compute the slope of a line segment.

▶ Example 1

Let L be the line segment connecting $(6, 2)$ and $(9, 8)$. If we label $(6, 2)$ as P_1 and $(9, 8)$ as P_2, then from (1) the slope is

$$m = \frac{8 - 2}{9 - 6} = 2$$

Similarly, the slope of the line segment connecting the points $(2, 9)$ and $(4, 3)$ is

$$m = \frac{3 - 9}{4 - 2} = -3 \qquad ◀$$

It is clear geometrically that two line segments on the same straight line will be equally "steep" and consequently have the same slope. This is the content of the following theorem.

1.4.2 THEOREM *On a nonvertical line, all line segments have the same slope.*

Proof. Let $P_1(x_1, y_1)$ and $P_2(x_2, y_2)$ be distinct points on a nonvertical line L and let $P_1'(x_1', y_1')$ and $P_2'(x_2', y_2')$ be another pair of distinct points on L. We will show that the slope

$$m = \frac{y_2 - y_1}{x_2 - x_1}$$

of the line segment joining P_1 and P_2 is equal to the slope

$$m' = \frac{y_2' - y_1'}{x_2' - x_1'}$$

of the line segment joining P_1' and P_2'. (We will assume the points are ordered as in Figure 1.4.2. The proofs of the remaining cases are similar and will be omitted for brevity.) The triangles $P_1 Q P_2$ and $P_1' Q' P_2'$ in Figure 1.4.2 are similar, so that the lengths of corresponding sides are proportional. Therefore

$$\frac{y_2 - y_1}{x_2 - x_1} = \frac{y_2' - y_1'}{x_2' - x_1'}$$

or, equivalently,

$$m = m'$$

We define the **slope** of a nonvertical line L to be the common slope of all line segments on L. The slope of a vertical line is **undefined.**

Figure 1.4.2

The slope of a line has a useful interpretation. Imagine a particle moving left to right along a line L from a point $P_1(x_1, y_1)$ to a point $P_2(x_2, y_2)$. As shown in Figure 1.4.3, the particle moves $y_2 - y_1$ units in the y-direction as it travels $x_2 - x_1$ units in the x-direction. From (1) these quantities are related by

$$y_2 - y_1 = m(x_2 - x_1) \tag{2}$$

Figure 1.4.3

Figure 1.4.4

Figure 1.4.5

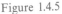

where m is the slope of the line. Thus (2) states that the movement in the y-direction is proportional to the movement in the x-direction and the slope m is the constant of proportionality. For this reason, the slope m is said to measure the *rate* at which y changes with respect to x along L.

▶ Example 2

In Example 1, we showed that the line determined by the points $P_1(6, 2)$ and $P_2(9, 8)$ has slope $m = 2$. This means that a particle traveling left to right along this line *rises* two units for every unit it moves in the positive x-direction (Figure 1.4.4). ◀

▶ Example 3

In Example 1, we showed that the line determined by the points $P_1(2, 9)$ and $P_2(4, 3)$ has slope $m = -3$. This means that a particle traveling left to right along this line *falls* three units for every unit it moves in the positive x-direction (Figure 1.4.5). ◀

▶ Example 4

Construct a line through $(3, 4)$ with slope -2.

Solution. A particle traveling left to right along the line *falls* 2 units for every unit it travels in the positive x-direction. Thus, if it begins at $P_1(3, 4)$ and travels so that x increases by one unit it will end up at $P_2(4, 2)$. The line through P_1 and P_2 is the one we want (Figure 1.4.6). ◀

In Figure 1.4.7 we have sketched several lines with varying slopes. Observe that lines with positive slope are inclined upward to the right, while lines with negative slope are inclined downward to the right.

Figure 1.4.6

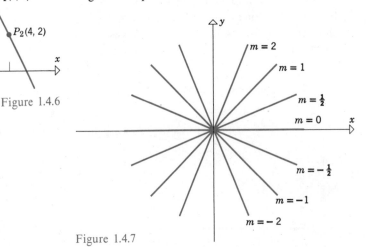

Figure 1.4.7

The slope of a line is related to the angle the line makes with the positive *x*-axis. To establish this relationship we need the following definition.

1.4.3 DEFINITION For a line *L* not parallel to the *x*-axis, the ***angle of inclination*** is the smallest angle ϕ measured counterclockwise from the direction of the positive *x*-axis to *L* (Figure 1.4.8). For a line parallel the *x*-axis, we take $\phi = 0$.

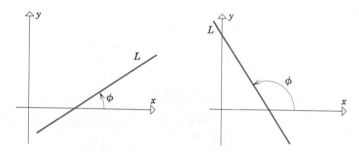

Figure 1.4.8

REMARK. In degree measure the angle of inclination satisfies $0° \leq \phi < 180°$ and in radian measure it satisfies $0 \leq \phi < \pi$.

The following theorem relates the slope of a line to its angle of inclination.

1.4.4 THEOREM *For a line not parallel to the y-axis, the slope and angle of inclination are related by*

$$m = \tan \phi \tag{3}$$

Proof. If the line is horizontal, then $m = 0$ and $\phi = 0°$. Thus

$$\tan \phi = \tan 0 = 0 = m$$

so (3) holds in this case. If the line is not horizontal, let $(x_0, 0)$ be its point of intersection with the *x*-axis, and construct a circle of radius 1 centered at this point. Because the angle of inclination of the line is ϕ, the line will intersect this circle at the point $(x_0 + \cos \phi, \sin \phi)$ (Figure 1.4.9). (This follows from Theorem 3 in Unit 1 of the trigonometry review in Appendix 1.)

Figure 1.4.9

Since the points $(x_0, 0)$ and $(x_0 + \cos\phi, \sin\phi)$ lie on L, we can use them to compute the slope m. This gives

$$m = \frac{\sin\phi - 0}{(x_0 + \cos\phi) - x_0} = \frac{\sin\phi}{\cos\phi} = \tan\phi$$

which proves (3). ▨

REMARK. If the line L is parallel to the y-axis, then $\phi = \frac{1}{2}\pi$ and $\tan\phi$ in (3) is undefined. This agrees with the fact that the slope m is also undefined in this case.

▶ Example 5

Find the angle of inclination for a line of slope $m = 1$ and also for a line of slope $m = -1$.

Solution. If $m = 1$, then from (3), $\tan\phi = 1$ so that $\phi = \frac{1}{4}\pi$ (or in degree measure $\phi = 45°$). If $m = -1$, then from (3), $\tan\phi = -1$; from this equality and the fact that $0 \leq \phi < \pi$ we obtain $\phi = \frac{3}{4}\pi$ or, in degree measure, $\phi = 135°$ (Figure 1.4.10). ◀

Figure 1.4.10

As a consequence of Theorem 1.4.4, we obtain the following basic result.

1.4.5 THEOREM *Two nonvertical lines are parallel if and only if they have the same slope.*

Figure 1.4.11

Proof. If L_1 and L_2 are nonvertical parallel lines, then their angles of inclination ϕ_1 and ϕ_2 are equal since two parallel lines cut by a transversal have equal corresponding angles (Figure 1.4.11). Thus

$$\text{slope } L_1 = \tan\phi_1 = \tan\phi_2 = \text{slope } L_2$$

Conversely, if L_1 and L_2 have the same slope m, then

$$m = \tan\phi_1 = \tan\phi_2$$

This relationship and the fact that

$$0 \leq \phi_1 < \pi, \qquad 0 \leq \phi_2 < \pi$$

imply that ϕ_1 and ϕ_2 are equal. Thus, L_1 and L_2 are parallel (Figure 1.4.11).

The next theorem shows how slopes can be used to determine whether two lines are perpendicular.

1.4.6 THEOREM *Two nonvertical lines are perpendicular if and only if the product of their slopes is -1; equivalently, lines with slopes m_1 and m_2 are perpendicular if and only if*

$$m_2 = -\frac{1}{m_1} \tag{4}$$

Proof. Suppose two nonvertical perpendicular lines, L_1 and L_2, have angles of inclination ϕ_1 and ϕ_2 and slopes m_1 and m_2, respectively. Assume L_1 has the smaller angle of inclination (Figure 1.4.12), so that

$$\phi_2 = \phi_1 + \frac{\pi}{2}$$

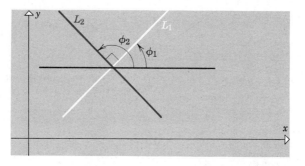

Figure 1.4.12

Thus

$$m_2 = \tan \phi_2 = \tan \left(\phi_1 + \frac{\pi}{2} \right) = \frac{\sin \left(\phi_1 + \frac{\pi}{2} \right)}{\cos \left(\phi_1 + \frac{\pi}{2} \right)}$$

$$= \frac{\sin \phi_1 \cos \frac{\pi}{2} + \cos \phi_1 \sin \frac{\pi}{2}}{\cos \phi_1 \cos \frac{\pi}{2} - \sin \phi_1 \sin \frac{\pi}{2}}$$

$$= -\frac{\cos \phi_1}{\sin \phi_1} = -\frac{1}{\sin \phi_1 / \cos \phi_1} = -\frac{1}{\tan \phi_1} = -\frac{1}{m_1}$$

which establishes (4). The proof of the converse, that is, that L_1 and L_2 are perpendicular if (4) holds, is discussed in Exercise 25. ▮

▶ Example 6

Use slopes to show that the points $A(1, 3)$, $B(3, 7)$, and $C(7, 5)$ are vertices of a right triangle.

Solution. The line through A and B has slope

$$m_1 = \frac{7 - 3}{3 - 1} = 2$$

and the line through B and C has slope

$$m_2 = \frac{5 - 7}{7 - 3} = -\frac{1}{2}$$

Since $m_1 m_2 = -1$, the line through A and B is perpendicular to the line through B and C; thus ABC is a right triangle (Figure 1.4.13). ◀

Figure 1.4.13

▶ Exercise Set 1.4

1. Find the slope of the line through
   (a) $(-1, 2)$ and $(3, 4)$
   (b) $(5, 3)$ and $(7, 1)$
   (c) $(4, \sqrt{2})$ and $(-3, \sqrt{2})$
   (d) $(-2, -6)$ and $(-2, 12)$.

2. Use a calculator or the trigonometric tables in the appendix, where necessary, to find the slope of the line whose angle of inclination is
   (a) $7°$ (b) $\dfrac{5\pi}{6}$ (c) $117°$
   (d) $47°$ (e) 0.5760.

3. Use a calculator or the trigonometric tables in the appendix, where necessary to find (to the nearest degree) the angle of inclination of a line having slope:
   (a) $m = \frac{1}{2}$ (b) $m = 1$ (c) $m = 2$
   (d) $m = 57$ (e) $m = -\frac{1}{2}$ (f) $m = -1$
   (g) $m = -\sqrt{3}$ (h) $m = -57$.

4. Draw the line through $(4, 2)$ with slope
   (a) $m = 3$ (b) $m = -2$ (c) $m = -\frac{3}{4}$.

5. Draw the line through $(-1, -2)$ with slope
   (a) $m = \frac{2}{5}$ (b) $m = -1$ (c) $m = \sqrt{2}$.

6. Let L be a line of slope $m = 2$. Determine if the given line L' is parallel to L, perpendicular to L, or neither.
   (a) L' is the line through $(2, 4)$ and $(4, 8)$
   (b) L' is the line through $(2, 4)$ and $(6, 2)$
   (c) L' is the line through $(1, 5)$ and $(2, -3)$.

7. A particle, initially at $(1, 2)$, moves along a line of slope $m = 3$ to a new position (x, y).
   (a) Find y if $x = 5$.   (b) Find x if $y = -2$.

8. A particle, initially at $(7, 5)$, moves along a line of slope $m = -2$ to a new position (x, y).
   (a) Find y if $x = 9$.   (b) Find x if $y = 12$.

9. Let the point $(3, k)$ lie on the line of slope $m = 5$ through $(-2, 4)$; find k.

10. Find the slopes of the sides of the triangle with vertices $(-1, 2)$, $(6, 5)$ and $(2, 7)$.

11. Use slopes to determine if the given points lie on a line.
    (a) $(1, 1)$, $(-2, -5)$, and $(0, -1)$
    (b) $(-2, 4)$, $(0, 2)$, and $(1, 5)$.

12. An equilateral triangle has one vertex at the origin, another on the x-axis, and the third in the first quadrant. Find the slopes of its sides.

13. Show that if (x, y) lies on the line of slope $m = 2$ passing through $(0, 3)$, then $y = 2x + 3$.

14. Use slopes to show that $(3, 1)$, $(6, 3)$, and $(2, 9)$ are vertices of a right triangle.

15. Use slopes to show that $(3, -1)$, $(6, 4)$, $(-3, 2)$, and $(-6, -3)$ are vertices of a parallelogram.

16. Find x and y if the line through $(0, 0)$ and (x, y) has slope $\frac{1}{2}$, and the line through (x, y) and $(7, 5)$ has slope 2.

17. Given two intersecting lines, let L_2 be the line with the larger angle of inclination ϕ_2, and let L_1 be the line with the smaller angle of inclination ϕ_1. We define the **angle θ between L_1 and L_2** by

    $$\theta = \phi_2 - \phi_1$$

    (a) Prove geometrically that θ is the angle pictured in Figure 1.4.14. [*Remark:* The angle θ

Figure 1.4.14

is the smallest positive angle through which L_1 can be rotated until it coincides with L_2.]

    (b) Prove: If L_1 and L_2 are not perpendicular, then

    $$\tan \theta = \frac{m_2 - m_1}{1 + m_1 m_2}$$

18. Use the result of Exercise 17 to find the tangent of the angle between the lines whose slopes are
    (a) 1 and 3   (b) -2 and 4
    (c) $-\frac{1}{2}$ and $-\frac{2}{3}$.

19. Use the result of Exercise 17 to find the tangent of the angle between the lines whose slopes are
    (a) $\frac{4}{5}$ and $\frac{1}{3}$ (b) -0.7 and 5
    (c) -6 and -2.

20. Use the result of Exercise 17 and a hand calculator or the trigonometric tables in Appendix 3 to find, to the nearest degree, the angle between the lines whose slopes are given in Exercise 18.

21. Repeat the directions of Exercise 20 for the lines whose slopes are given in Exercise 19.

22. Use the result of Exercise 17 to find, to the nearest degree, the interior angles of the triangle whose vertices are $(4, -3)$, $(-3, -1)$, and $(6, 6)$.

23. Use the result of Exercise 17 to find the slope of the line that bisects the angle A of the triangle whose vertices are $A(0, 2)$, $B(-8, 8)$, and $C(8, 6)$.

24. Let L be a line of slope $m = -2$. Use the result of Exercise 17 to find the slope of a line K such that the angle between K and L is $45°$. (Two solutions.)

25. Complete the proof of Theorem 1.4.6 by showing: If L_1 and L_2 are lines whose slopes satisfy $m_1 m_2 = -1$, then L_1 and L_2 are perpendicular. [*Hint:* One slope is positive and the other negative; if $m_1 > 0$ and $m_2 < 0$, then $0 < \phi_1 < \frac{1}{2}\pi$ and $\frac{1}{2}\pi < \phi_2 < \pi$. Use this and the identity $\tan(\phi_1 + \frac{1}{2}\pi) = -1/\tan \phi_1$.]

1.5 EQUATIONS OF STRAIGHT LINES

In this section we are concerned with recognizing those equations whose graphs are straight lines and finding equations for lines specified geometrically.

VERTICAL LINES   A line parallel to the y-axis meets the x-axis at some point $(a, 0)$. Every point on this line has an x-coordinate of a and conversely every point with an x-coordinate of a lies on the line (Figure 1.5.1). Thus,

1.5.1 THEOREM   *The vertical line passing through $(a, 0)$ is represented by the equation*

$$x = a \tag{1}$$

▶ Example 1

The graph of $x = -2$ is the vertical line passing through $(-2, 0)$.   ◀

LINES DETERMINED   A nonvertical line can be determined by specifying a point on the line and
BY POINT AND SLOPE   the slope; the slope determines how the line is tilted and the point pins down the location. Let us try to find an equation for the line L passing through $P_1(x_1, y_1)$ and having slope m.

$\qquad$ If $P(x, y)$ is any point on L, other than P_1, then the slope of L can be obtained from the points $P(x, y)$ and $P_1(x_1, y_1)$; this gives

$$m = \frac{y - y_1}{x - x_1}$$

Every point on the line L has an x-coordinate of a

Figure 1.5.1

which can be rewritten as

$$y - y_1 = m(x - x_1) \tag{2}$$

With the possible exception of (x_1, y_1) we have shown that every point on L satisfies (2). But $x = x_1, y = y_1$ satisfies (2), so that all points on L satisfy (2). We leave it as an exercise to show that every point satisfying (2) lies on L. To summarize:

1.5.2 THEOREM   *The line passing through $P_1(x_1, y_1)$ and having slope m is given by the equation*

$$y - y_1 = m(x - x_1) \tag{3}$$

*This is called the **point-slope form** of the line.*

▶ Example 2

Find the point-slope form of the line through $(4, -3)$ with slope 5.

Solution. Substituting the values $x_1 = 4$, $y_1 = -3$, and $m = 5$ in (3) yields the point-slope form

$$y + 3 = 5(x - 4)$$ ◀

LINES DETERMINED
BY SLOPE AND
y-INTERCEPT

A nonvertical line crosses the *y*-axis at some point $(0, b)$. If we use this point in the point-slope form of its equation, we obtain

$$y - b = m(x - 0)$$

which we can rewrite as

$$y = mx + b$$

The number b in this equation is called the **y-intercept** of the line; it is the *y*-coordinate of the point where the line crosses the *y*-axis.
      To summarize:

1.5.3 THEOREM   *The line with y-intercept b and slope m is given by the equation*

$$y = mx + b \tag{4}$$

*This is called the **slope-intercept form** of the line.*

REMARK. The slope-intercept form is of particular importance because it expresses y in terms of x.

▶ Example 3

Comparing

$$y = 4x + 7$$

to (4), we have $m = 4$ and $b = 7$, so that the equation represents a line crossing the *y*-axis at $(0, 7)$ with slope 4. ◀

HORIZONTAL LINES   By definition, the angle of inclination of a horizontal line is $\phi = 0$. Thus, the slope of a horizontal line is

$$m = \tan \phi = \tan 0 = 0$$

Substituting $m = 0$ in (4) yields the following result.

1.5.4 THEOREM *The horizontal line passing through $(0, b)$ is represented by the equation*

$$y = b \tag{5}$$

▶ Example 4

The equation $y = -7$ represents the horizontal line passing through $(0, -7)$.
◀

LINES DETERMINED
BY TWO POINTS

If $P_1(x_1, y_1)$ and $P_2(x_2, y_2)$ are distinct points on a nonvertical line, then the slope of the line is

$$m = \frac{y_2 - y_1}{x_2 - x_1}$$

Substituting this expression in (3) we obtain the following result.

1.5.5 THEOREM *The nonvertical line determined by the points $P_1(x_1, y_1)$ and $P_2(x_2, y_2)$ can be represented by the equation*

$$y - y_1 = \frac{y_2 - y_1}{x_2 - x_1}(x - x_1) \tag{6}$$

*This is called the **two-point form** of the line.*

▶ Example 5

Find the slope-intercept form of the line passing through $(3, 4)$ and $(2, -5)$.

Solution. Letting $(x_1, y_1) = (3, 4)$ and $(x_2, y_2) = (2, -5)$ and substituting in (6), we obtain the two-point form

$$y - 4 = \frac{-5 - 4}{2 - 3}(x - 3)$$

which can be written

$$y - 4 = 9(x - 3)$$

Solving for y yields the slope-intercept form

$$y = 9x - 23$$ ◄

THE GENERAL
EQUATION OF A LINE

An equation expressible in the form

$$Ax + By + C = 0 \tag{7}$$

where A, B, and C are constants and A and B are not both zero, is called a *first degree equation* in x and y.

► Example 6

Comparing $7x + 2y - 3 = 0$ to (7), we see that this is a first degree equation in x and y with $A = 7$, $B = 2$, and $C = -3$. ◄

The following theorem shows that the first degree equations in x and y are precisely the equations whose graphs in the xy-plane are straight lines.

1.5.6 THEOREM *Every first degree equation in x and y has a straight line as its graph and, conversely, every straight line can be represented by a first degree equation in x and y.*

Proof. Let

$$Ax + By + C = 0 \tag{8}$$

be any first degree equation in x and y. To prove that the graph of this equation is a straight line, we distinguish between two cases, $B = 0$ and $B \neq 0$.

If $B \neq 0$, we can solve (8) for y in terms of x to obtain

$$y = -\frac{A}{B}x - \frac{C}{B}$$

But this is the slope-intercept form (4) of the line with

$$m = -\frac{A}{B} \quad \text{and} \quad b = -\frac{C}{B}$$

If $B = 0$, then $A \neq 0$ since A and B are not both zero in a first degree equation in x and y. Thus (8) reduces to

$$Ax + C = 0$$

which can be rewritten as

$$x = -\frac{C}{A}$$

But this is the equation of a line parallel to the y-axis. Thus, in either case, the graph of (8) is a straight line.

Conversely, consider any straight line in the xy-plane. If the line is vertical, then it has an equation of the form $x = a$. This can be rewritten as $x - a = 0$, which is of form (7) with

$$A = 1, \qquad B = 0, \qquad \text{and } C = -a$$

If the line is not vertical, it can be expressed in slope-intercept form

$$y = mx + b$$

This can be rewritten as

$$mx - y + b = 0$$

which is of form (7) with

$$A = m, \qquad B = -1, \qquad \text{and } C = b$$

Because every straight line has a first degree equation in x and y and every first degree equation in x and y has a straight line for its graph, (7) is sometimes called the **general equation** of a line or a **linear equation** in x and y.

▶ Example 7

Graph the equation

$$3x - 4y + 12 = 0 \qquad (9)$$

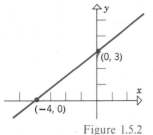

Figure 1.5.2

Solution. Since this is a linear equation in x and y, its graph is a straight line. Thus to sketch the graph we need only plot two points on the graph and draw the line through them. It is particularly convenient to plot the points where the line crosses the coordinate axes. The line crosses the y-axis when $x = 0$ and crosses the x-axis when $y = 0$. Substituting $x = 0$ in (9) and solving for y, we obtain the intersection $(0, 3)$ with the y-axis. Substituting $y = 0$ in (9) and solving for x, we obtain the intersection $(-4, 0)$ with the x-axis. The graph of (9) is given in Figure 1.5.2. ◀

▶ Exercise Set 1.5

1. Graph the equations
   (a) $2x + 5y = 15$ (b) $x = 3$
   (c) $y = -2$ (d) $y = 2x - 7$.

2. Graph the equations
   (a) $\dfrac{x}{3} - \dfrac{y}{4} = 1$ (b) $x = -8$
   (c) $y = 0$ (d) $x = 3y + 2$.

3. Graph the equations
   (a) $y = 2x - 1$ (b) $y = 3$
   (c) $y = -2x$.

4. Graph the equations
   (a) $y = 2 - 3x$ (b) $y = \frac{1}{4}x$
   (c) $y = -\sqrt{3}$.

5. Find the slope and y-intercept of
   (a) $y = 3x + 2$ (b) $y = 3 - \frac{1}{4}x$
   (c) $3x + 5y = 8$ (d) $y = 1$
   (e) $\dfrac{x}{a} + \dfrac{y}{b} = 1$.

6. Find the slope and y-intercept of
   (a) $y = -4x + 2$ (b) $x = 3y + 2$
   (c) $\dfrac{x}{2} + \dfrac{y}{3} = 1$ (d) $y - 3 = 0$
   (e) $a_0 x + a_1 y = 0$.

7. To the nearest degree, find the angle of inclination of
   (a) $y = \sqrt{3}x + 2$ (b) $y + 2x + 5 = 0$.

8. To the nearest degree, find the angle of inclination of
   (a) $3y = 2 - \sqrt{3}x$ (b) $y - 4x + 7 = 0$.

In Exercises 9–20, find the slope-intercept form of the line satisfying the given conditions.

9. Slope $= -2$, y-intercept $= 4$.

10. $m = 5$, $b = -3$.

11. The line is parallel to $y = 4x - 2$ and has y-intercept 7.

12. The line is perpendicular to $y = 5x + 9$ and has y-intercept 6.

13. The line passes through $(2, 4)$ and $(1, -7)$.

14. The line passes through $(-3, 6)$ and $(-2, 1)$.

15. The line has angle of inclination $\phi = \frac{1}{6}\pi$ and y-intercept -3.

16. The line has angle of inclination $\phi = \frac{2}{3}\pi$ and passes through the point $(1, 2)$.

17. The y-intercept is 2 and the x-intercept is -4.

18. The y-intercept is b and the x-intercept is a.

19. The line is perpendicular to the y-axis and passes through $(-4, 1)$.

20. The line is parallel to $y = -5$ and passes through $(-1, -8)$.

21. Find an equation for the line that passes through $(5, -2)$ and has angle of inclination $\phi = \frac{1}{2}\pi$.

22. Find an equation for the line along the y-axis.

23. In each part, classify the lines as parallel, perpendicular, or neither.
    (a) $y = 4x - 7$ and $y = 4x + 9$
    (b) $y = 2x - 3$ and $y = 7 - \frac{1}{2}x$
    (c) $5x - 3y + 6 = 0$ and $10x - 6y + 7 = 0$
    (d) $Ax + By + C = 0$ and $Bx - Ay + D = 0$
    (e) $y - 2 = 4(x - 3)$ and $y - 7 = \frac{1}{4}(x - 3)$.

24. In each part, classify the lines as parallel, perpendicular, or neither.
    (a) $y = -5x + 1$ and $y = 3 - 5x$
    (b) $y - 1 = 2(x - 3)$ and $y - 4 = -\frac{1}{2}(x + 7)$
    (c) $4x + 5y + 7 = 0$ and $5x - 4y + 9 = 0$
    (d) $Ax + By + C = 0$ and $Ax + By + D = 0$
    (e) $y = \frac{1}{2}x$ and $x = \frac{1}{2}y$.

25. In each part, find the point of intersection of the lines.
    (a) $2x + 3y = 5$ and $y = -1$
    (b) $4x + 3y = -2$ and $5x - 2y = 9$.

26. In each part, find the point of intersection of the lines.
    (a) $6x - 9y = 7$ and $x = -\frac{2}{3}$
    (b) $6x - 2y = -3$ and $-8x + 3y = 5$.

27. Find the distance from the point $(2, 1)$ to the line $4x - 3y + 10 = 0$. [*Hint:* Find the foot of the perpendicular dropped from the point to the line.]

28. Find the distance from the point $(8, 4)$ to the line $5x + 12y - 36 = 0$. [*Hint:* See the hint in Exercise 27.]

29. Use the method described in Exercise 27 to prove that the distance d from (x_0, y_0) to the line $Ax + By + C = 0$ is

$$d = \frac{|Ax_0 + By_0 + C|}{\sqrt{A^2 + B^2}}$$

30. Use the formula in Exercise 29 to solve Exercise 27.

31. Use the formula in Exercise 29 to solve Exercise 28.

32. Is the graph of

$$\frac{y}{x} = 1$$

a line? Explain.

33. Prove: The midpoint of the line segment joining (a_1, b_1) and (a_2, b_2) is the point

$$\left(\frac{a_1 + a_2}{2}, \frac{b_1 + b_2}{2} \right)$$

34. Use the result in Exercise 33 to find
    (a) the midpoint of the line segment joining $(2, 8)$ and $(-4, 6)$;
    (b) the equation of the perpendicular bisector of the line segment in part (a).

35. Repeat Exercise 34 for the line segment joining $(5, -1)$ and $(4, 8)$.

36. Prove: For any triangle, the perpendicular bisectors of the sides meet at a point. [*Hint:* See Exercises 33 and 34; also position the triangle with one vertex on the y-axis and the opposite side on the x-axis, so that the vertices are $(0, a)$, $(b, 0)$ and $(c, 0)$.

37. Find the point on the line $4x - 2y + 3 = 0$ that is equidistant from $(3, 3)$ and $(7, -3)$.

38. In physical problems linear equations involving variables other than x and y often arise. In each part, determine if the equation is linear.
    (a) $3\alpha - 2\beta = 5$
    (b) $A = 2000(1 + 0.06t)$
    (c) $A = \pi r^2$
    (d) $E = mc^2$ (c constant)
    (e) $V = C(1 - rt)$ (r and C constant)
    (f) $V = \frac{1}{3}\pi r^2 h$ (r constant)
    (g) $V = \frac{1}{3}\pi r^2 h$ (h constant).

39. There are two common systems for measuring temperature, Celsius and Fahrenheit. Water freezes at $0°$ Celsius and $32°$ Fahrenheit; it boils at $100°$ Celsius and $212°$ Fahrenheit.
    (a) Assuming that the Celsius temperature C and the Fahrenheit temperature F are related by a linear equation, find the equation.
    (b) What is the slope of the line relating F and C if F is plotted on the horizontal axis?

40. Prove: If (x, y) satisfies Equation (2) then the point $P(x, y)$ lies on the line with slope m passing through $P_1(x_1, y_1)$.

▶ REVIEW EXERCISES

In Exercises 1–5, use interval notation to describe the set of all values of x (if any) that satisfy the given inequalities.

1. (a) $-3 < x \leq 5$
   (b) $-1 < x^2 \leq 9$
   (c) $x^2 \geq \frac{1}{4}$.

2. (a) $|2x + 1| > 5$
   (b) $|x^2 - 9| \geq 7$
   (c) $1 \leq |x| \leq 3$.

3. (a) $2x^2 - 5x > 3$
   (b) $x^2 - 5x + 4 \leq 0$.

4. (a) $\frac{x}{1 - x} \geq 3$
   (b) $\frac{2x + 3}{x} \geq x$.

5. (a) $\frac{|x| - 1}{|x| - 2} \leq 0$
   (b) $|x - 1| \leq 2|x + 2|$.

6. Among the terms: integer, rational, irrational, which ones apply to the given number?
   (a) $\sqrt{4/9}$ (b) 2^{-2}
   (c) $-4^{1/3}$ (d) 0.87
   (e) $-4^{1/2}$ (f) $0.1010010001 \cdots$
   (g) 3.222 (h) $3/(-1)$.

7. (a) Find values of a and b such that $a < b$, but $a^2 > b^2$.
   (b) If $a < b$, what additional assumptions on a and b are required to ensure that $a^2 < b^2$?

8. Which of the following are true for all sets A and B?
   (a) $A \subset (A \cap B)$ (b) $(A \cap B) \subset A$
   (c) $\phi \subset A$ (d) $A \subset (A \cup B)$
   (e) $(A \cap B) \subset (A \cup B)$
   (f) either $A \subset B$ or $B \subset A$
   (g) $A \in B$.

9. Prove: $|x| \le \sqrt{x^2 + y^2}$ and $|y| \le \sqrt{x^2 + y^2}$. Interpret this geometrically.

In Exercises 10–14, draw a rectangular coordinate system and sketch the set of points whose coordinates (x, y) satisfy the given conditions.

10. (a) $y = 0$ and $x > 0$
    (b) $2x - y \le 3$.

11. (a) $xy = x^2$
    (b) $y(x - 1) = x^2 - 1$.

12. (a) $y = (x^3 - 1)/(x - 1)$
    (b) $y > x^2 - 9$.

13. (a) $y^2 - 6y + x^2 - 2x - 6 \ge 0$
    (b) $x + |y - 2| = 1$.

14. (a) $|x| + |y| = 4$
    (b) $|x| - |y| = 4$.

15. Where does the parabola $y = x^2$ intersect the line $y - 2 = x$?

In Exercises 16–19, sketch the graph of the given equation.

16. $xy + 4 = 0$.

17. $y = |x - 2|$.

18. $y = \sqrt{4 - x^2}$.

19. $y = x(x - 2)$.

In Exercises 20–24, find the standard equation for the circle satisfying the given conditions.

20. The circle centered at $(3, -2)$ and tangent to the line $y = 1$.

21. The circle centered at $(1, 2)$ and passing through the point $(4, -2)$.

22. The circle centered on the line $x = 2$ and passing through the points $(1, 3)$ and $(3, -11)$.

23. The circle of radius 5 tangent to the lines $y = 7$ and $x = 6$.

24. The circle of radius 13 that passes through the origin and the point $(0, -24)$.

In Exercises 25–28, determine if the equation represents a circle, a point, or has no graph. If it represents a circle, find the center and radius.

25. $x^2 + y^2 + 4x + 2y + 5 = 0$.

26. $4x^2 + 4y^2 - 4x + 8y + 1 = 0$.

27. $x^2 + y^2 - 3x + 2y + 4 = 0$.

28. $3x^2 + 3y^2 - 5x + 7y + 3 = 0$.

29. In each part, find an equation for the line through A and B, the distance between A and B, and the coordinates of the midpoint of the line segment joining A and B.
    (a) $A(3, 4)$, $B(-3, -4)$
    (b) $A(3, 4)$, $B(3, -4)$
    (c) $A(3, 4)$, $B(-3, 4)$
    (d) $A(3, 4)$, $B(4, 3)$.

30. Show that the point $(8, 1)$ is *not* on the line through the points $(-3, -2)$ and $(1, -1)$.

31. Where does the circle of radius 5 centered at the origin intersect the line of slope $-3/4$ through the origin?

32. Fahrenheit and Celsius temperatures are related by $F - 32 = 9C/5$. What temperature is the same in both Fahrenheit and Celsius?

33. Find the inclination angle of the line whose equation is
    (a) $x = 3$ (b) $y = -2$
    (c) $2x + 2y = 1$ (d) $\sqrt{3}x - y = 4$.

34. Find the slope of the line whose angle of inclination is
   (a) $30°$ (b) $120°$ (c) $90°$.

In Exercises 35–37, find the slope-intercept form of the line satisfying the stated conditions.

35. The line through $(2, -3)$ and $(4, -3)$.

36. The line with x-intercept -2 and angle of inclination $\phi = 45°$

37. The line parallel to $x + 2y = 3$ that passes through the origin.

38. Find the equation of the perpendicular bisector of the line segment joining $A(-2, -3)$ and $B(1, 1)$.

In Exercises 39–41, find equations of the lines L and L' and determine their point of intersection.

39. L passes through $(1, 0)$ and $(-1, 4)$.
   L' is perpendicular to L and has y-intercept -3.

40. L passes through $(-2, 0)$ and $(-2, 3)$.
   L' passes through $(-1, 4)$ and is perpendicular to L.

41. L has slope $2/5$ and passes through $(3, 1)$.
   L' has x-intercept $-8/3$ and y-intercept -4.

42. Consider the triangle with vertices $A(5, 2)$, $B(1, -3)$, and $C(-3, 4)$. Find the point-slope form of the line containing:
   (a) the median from C to AB
   (b) the altitude from C to AB.

43. Use slopes to show that the points $(5, 6)$, $(-4, 3)$, $(-3, -2)$, and $(6, 1)$ are vertices of a parallelogram. Is it a rectangle?

44. For what value of k (if any) will the line $2x + ky = 3k$ satisfy the stated condition?
   (a) have slope 3
   (b) have y-intercept 3
   (c) be parallel to the x-axis
   (d) pass through $(1, 2)$.

2 functions and limits

2.1 FUNCTIONS

In this section we consider one of the most fundamental concepts in all of mathematics, the notion of a *function*. Historically, the term "function" was first used by Leibniz in 1673 to denote the dependence of one quantity on another. To illustrate:

1. The area A of a circle depends on its radius r by the equation $A = \pi r^2$; we say that "A is a function of r."

2. The velocity of a ball dropped from a height increases with time until it hits the ground. Thus the velocity v depends on the time t and we say that "v is a function of t."

3. At a fixed point on earth, the wind speed w varies with the time t. Thus, "w is a function of t."

4. In a bacteria culture, the number of bacteria present after one hour of growth depends on the number present initially; we say that "the size of the bacteria population after one hour is a function of the initial population size."

In order to discuss functions or relationships between quantities without stating specific formulas, the Swiss mathematician, Leonhard Euler* (see page 52), developed the ingenious idea of using a letter of the alphabet such as f to denote a function or relationship. Thus, by writing

$$y = f(x)$$

(read "y equals f of x"), we convey the idea that y is a function of x, that is, the value of y depends on the value of x. To indicate that the wind speed w

at a fixed point on earth depends on the time t at which it is measured, we can write

$$w = f(t) \tag{1}$$

There is nothing special about the letter f; any symbol can be used to denote a function; thus (1) could have been written

$$w = F(t), \qquad w = f_1(t), \qquad w = g(t), \qquad \text{or} \qquad w = \phi(t)$$

By having a variety of symbols available to denote functions we are able to distinguish between different possible relationships. For example, at a point in Alaska the wind speed w will have a certain dependence on the time t, say,

$$w = g_1(t)$$

while at a point in Florida the wind speed w will have a different dependence on the time t, say

$$w = g_2(t)$$

---

*LEONHARD EULER (1707–1783). Euler was probably the most prolific mathematician who ever lived. It has been said that, "Euler wrote mathematics as effortlessly as most men breathe." He was born in Basel, Switzerland, and was the son of a Protestant minister who had himself studied mathematics. Euler's genius developed early. He attended the University of Basel, where by age 16 he obtained both a Bachelor of Arts degree and a Master's degree in philosophy. While at Basel, Euler had the good fortune to be tutored one day a week in mathematics by a distinguished mathematician, Johann Bernoulli. At the urging of his father, Euler then began to study theology. The lure of mathematics was too great, however, and by age 18 Euler had begun to do mathematical research. Nevertheless, the influence of his father and his theological studies remained, and throughout his life Euler was a deeply religious, unaffected person. At various times Euler taught at St. Petersburg Academy of Sciences (in Russia), the University of Basel, and the Berlin Academy of Sciences. Euler's energy and capacity for work were virtually boundless. His collected works form about 60 to 80 quarto sized volumes and it is believed that much of his work has been lost. What is particularly astonishing is that Euler was blind for the last 17 years of his life, and this was one of his most productive periods! Euler's flawless memory was phenomenal. Early in his life he memorized the entire Aeneid by Virgil and at age 70 could not only recite the entire work, but could also state the first and last sentence on each page of the book from which he memorized the work. His ability to solve problems in his head was beyond belief. He worked out in his head major problems of lunar motion that baffled Isaac Newton and once did a complicated calculation in his head to settle an argument between two students whose computations differed in the fiftieth decimal place.

Euler's main contribution was the systematization of mathematics. Following the development of calculus by Leibniz and Newton, results in mathematics developed helter-skelter at a rapid pace. Euler's genius gave coherence to the mathematical landscape. He was the first mathematician to bring the full power of calculus to bear on problems from physics. He made major contributions to virtually every branch of mathematics as well as to the theory of optics, planetary motion, electricity, magnetism, and general mechanics.

The idea of using a symbol to denote a function is a relatively sophisticated concept. In (1), for example, the letters w and t denote measurable *physical* quantities, wind speed and time. The letter f, however, does not denote a physical quantity; it stands for a *dependence* of one physical quantity on another; this is a completely *abstract* idea.

Since the time of Euler and Leibniz the original idea of a function has evolved into the following more precise and more general mathematical concept:

2.1.1 DEFINITION A *function* is a rule that assigns to each element in a set A one and only one element in a set B.

In general the sets A and B need not be sets of real numbers; however, for the time being we will only be concerned with functions for which A and B are both subsets of the real numbers.

To see how Definition 2.1.1 relates to the historical concept of a function, consider the relationship

$$w = f(t) \tag{2}$$

where w denotes the wind speed at a point and t denotes the time of measurement. Let A be the set of all possible t values and let B be the set of all possible w values. With each instant t in time there is associated by (2) a unique wind speed w; that is, the function f in (2) assigns to each element t in A a unique element w in B. This is precisely the definition of a function as stated in Definition 2.1.1.

The set A in Definition 2.1.1 is called the ***domain*** of the function. If x is an element in the domain of a function f, then the element that f associates with x is denoted by the symbol $f(x)$ (read "f of x") and is called the ***image of x under f*** or the ***value of f at x*** (Figure 2.1.1). The set of all possible values of $f(x)$ as x varies over the domain is called the ***range*** of f.

Most often the rule for obtaining the value of a function is given by means of a formula.

The element that f associates with x is denoted by $f(x)$

Figure 2.1.1

▶ Example 1

The formula

$$f(x) = x^3$$

tells us that the value of f at x is x^3. Thus

$$f(2) = 2^3 = 8, \quad f(-4) = (-4)^3 = -64, \quad \text{and} \quad f(0) = 0^3 = 0$$

◀

► Example 2

If

$$\phi(x) = 2x^2 - 1$$

then

$$\phi(4) = 2(4)^2 - 1 = 31, \qquad \phi(t) = 2t^2 - 1$$
$$\phi(k + 1) = 2(k + 1)^2 - 1 = 2k^2 + 4k + 1 \qquad ◄$$

REMARK. If x is an element in the domain of a function f, the definition of a function (Definition 2.1.1) requires that f assign one and only one value to x. This means that a function cannot be "multiple-valued." For example, the expression

$$\pm \sqrt{x}$$

does not define a function of x since it assigns two values to each positive x.

If the rule for evaluating a function is given by a formula, and if there is no mention of the domain, it is understood that the domain consists of all real numbers for which the formula makes sense and yields a real value.

► Example 3

Let

$$g(x) = \sqrt{x - 1}$$

When $x < 1$, $g(x)$ is imaginary; thus the domain of g is $[1, +\infty)$. As x varies over this interval, $g(x)$ varies from 0 to $+\infty$; therefore the range of g is $[0, +\infty)$. ◄

► Example 4

Let

$$h(x) = \frac{1}{(x - 1)(x - 3)}$$

When $x = 1$ or $x = 3$, $h(x)$ is undefined since division by zero is not allowed; otherwise $h(x)$ yields real values. Thus the domain of h consists of all real numbers other than $x = 1$ and $x = 3$. In interval notation the domain is $(-\infty, 1) \cup (1, 3) \cup (3, +\infty)$. (See Figure 2.1.2.) ◄

Figure 2.1.2

Sometimes it is necessary to specify restrictions on the domain even though the formula for the function makes sense elsewhere.

▶ Example 5

Assume it costs 12 cents to manufacture a certain kind of resistor and let $f(x)$ be the cost in dollars for manufacturing x such resistors. The formula for $f(x)$ is

$$f(x) = 0.12x \tag{3}$$

If we were to specify no restrictions on the domain of this function, then the domain would be $(-\infty, +\infty)$ since the formula in (3) makes sense for all real x. Physically, however, x must be a nonnegative integer; thus we must restrict the domain to the set $\{0, 1, 2, \cdots\}$. We do this by writing

$$f(x) = 0.12x, \qquad x = 0, 1, 2, \ldots \tag{4}$$

From (4) it is to be understood that the formula $f(x) = 0.12x$ only applies when $x = 0, 1, 2, \ldots$; for other values of x, $f(x)$ is undefined. Thus

$$f(3) = 0.12(3) = 0.36 \qquad \text{and} \qquad f(\tfrac{3}{2}) \text{ is undefined} \qquad ◀$$

As the next example shows, functions are sometimes specified by formulas that have been "pieced together."

▶ Example 6

The cost of a taxi ride in a certain metropolitan area is 75 cents for any ride up to and including one mile. After one mile the rider pays an additional amount at the rate of 50 cents per mile. If $f(x)$ is the total cost in dollars for a ride x miles long, then the value of $f(x)$ is

$$0.75 \quad \text{if} \quad 0 < x \le 1 \qquad \left(\begin{array}{l}\$0.75 \text{ for a ride up to}\\ \text{and including one mile}\end{array}\right)$$

and

$$0.75 + 0.50(x - 1) \quad \text{if} \quad x > 1 \qquad \left(\begin{array}{l}\$0.75 \text{ for the first}\\ \text{mile plus \$0.50 a mile}\\ \text{for each mile after the}\\ \text{first.}\end{array}\right)$$

We express this by writing

$$f(x) = \begin{cases} 0.75, & 0 < x \le 1 \\ 0.75 + 0.50(x - 1), & 1 < x \end{cases} \qquad\qquad ◀$$

Although most functions are specified by formulas, this is not the only possibility. Any description that tells us what values f assigns to the points in its domain will suffice. One possibility is to use a table relating the values of x and $f(x)$.

▶ Example 7

At 10 a.m. on a weekday, a two-minute call is made between two stations in New Jersey. Table 2.1.1, obtained from the New Jersey telephone directory, shows how the cost of such a call varies with the distance between the stations. If $f(x)$ is the cost in dollars for a call between stations x miles apart, then the function f is completely specified by Table 2.1.1. For example,

$$f(22) = 0.25, \qquad f(114) = 0.60, \qquad \text{and} \qquad f(64) = 0.40 \qquad ◀$$

Table 2.1.1

DISTANCE IN MILES	more than	0	10	15	20	25	32	48	64	80	96	112
	up to and including	10	15	20	25	32	48	64	80	96	112	more
COST IN DOLLARS		0.10	0.15	0.20	0.25	0.30	0.35	0.40	0.45	0.50	0.55	0.60

If x is an element in the domain of a function f, then by forming the ordered pair of numbers

$$(x, f(x))$$

we match each x with the value of f at x. For the function $f(x) = x^3$, a few such ordered pairs would be

$$(2, 8), \qquad (-3, -27), \qquad (0, 0), \qquad \text{and} \qquad (\tfrac{1}{2}, \tfrac{1}{8})$$

As illustrated in the next example, a function can be specified by giving the set of all such possible ordered pairs.

▶ Example 8

Let f be a function with domain $\{1, 2, 3, 4, 5\}$ and let the set C of all ordered pairs $(x, f(x))$ be

$$C = \{(1, 0), (2, -1), (3, 0), (4, 7), (5, \tfrac{1}{7})\}$$

The set C completely specifies the function, since it tells us that

$$f(1) = 0, \quad f(2) = -1, \quad f(3) = 0, \quad f(4) = 7, \quad \text{and} \quad f(5) = \tfrac{1}{7} \quad \blacktriangleleft$$

In the symbol $f(x)$, we can think of x as a quantity that can be varied arbitrarily over the domain of f. As x varies, so does the number $f(x)$, but with one difference; where the value of x can be varied arbitrarily within the domain of f, the value of $f(x)$ is determined once the value of x is specified. For this reason, x is sometimes called the **independent variable** and $f(x)$ the **dependent variable.** In many problems it is convenient to introduce a single letter such as y to stand for the dependent variable. Thus, if we write

$$y = f(x) \tag{5}$$

then x is the independent variable and y is the dependent variable.

▶ Example 9

If $f(x) = 9x^3 - 2x + 7$, then the independent variable is x and

$$y = 9x^3 - 2x + 7$$

is the dependent variable. ◀

The following definition will enable us to study functions geometrically.

2.1.2 DEFINITION We define the **graph of a function f** to be the graph of the equation

$$y = f(x)$$

▶ Example 10

Sketch the graph of

$$f(x) = x + 2$$

Solution. By definition, the graph of the function $x + 2$ is the graph of the equation

$$y = x + 2$$

Figure 2.1.3

which is a line of slope 1 and y-intercept 2 (Figure 2.1.3). ◀

▶ Example 11

Sketch the graph of

$$f(x) = |x|$$

Solution. By definition, the graph of $f(x) = |x|$ is the graph of the equation

$$y = |x|$$

or equivalently

$$y = \begin{cases} x, & x \geq 0 \\ -x, & x < 0 \end{cases}$$

The graph coincides with the line $y = x$ for $x \geq 0$ and with the line $y = -x$ for $x < 0$ (Figure 2.1.4). ◀

Graph of |x|

Figure 2.1.4

▶ Example 12

Sketch the graph of

$$h(x) = \frac{x^2 - 4}{x - 2} \tag{6}$$

Solution. The function h can be rewritten as

$$h(x) = \frac{(x - 2)(x + 2)}{(x - 2)}$$

Algebraically, it is desirable to cancel the $x - 2$ factors, but we must exercise some care. If we simply cancel the factors and write

$$h(x) = x + 2 \tag{7}$$

then we would erroneously alter the domain of h. To see this, observe that the domain of the function h in (6) consists of all x other than $x = 2$, whereas the domain of the function h in (7) consists of all x. In fact, when $x = 2$, the function h in (7) has the value

$$h(2) = 2 + 2 = 4$$

whereas for the original function h in (6),

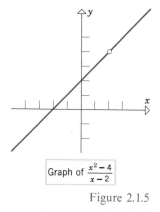

Graph of $\frac{x^2-4}{x-2}$

Figure 2.1.5

$$h(2) = \frac{2^2 - 4}{2 - 2} = \frac{0}{0}$$

which is undefined.

To cancel the $x - 2$ factors in (6) and not alter the domain of h, we must restrict the domain in (7) and write

$$h(x) = x + 2, \qquad x \neq 2 \tag{8}$$

The function h in (8) is identical to the function f in Example 10, except that h is undefined at $x = 2$. Thus the graph of h is identical to the graph of f (Figure 2.1.3), except that the graph of h has a hole in it above $x = 2$ (Figure 2.1.5). ◄

► Example 13

Sketch the graph of

$$\phi(x) = \begin{cases} x + 2, & x \neq 2 \\ 6, & x = 2 \end{cases}$$

Figure 2.1.6

Solution. The function ϕ is identical to the function f in Example 10, except at $x = 2$, where we have

$$\phi(2) = 6 \qquad \text{and} \qquad f(2) = 4$$

Thus the graph of f (Figure 2.1.3) is identical to the graph of ϕ (Figure 2.1.6) except that the graph of ϕ has a point separated from the line at $x = 2$. ◄

► Example 14

Sketch the graph of

$$g(x) = \begin{cases} 1, & x \leq 2 \\ x + 2, & x > 2 \end{cases}$$

Solution. By definition, the graph of g is the graph of

$$y = \begin{cases} 1, & x \leq 2 \\ x + 2, & x > 2 \end{cases}$$

Thus for $x \leq 2$, the value of y is always 1 and for $x > 2$, y is given by

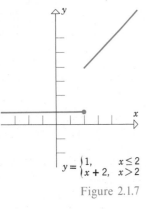

$$y = x + 2$$

which is the straight line graphed in Example 10. The graph of g is shown in Figure 2.1.7. ◄

$y = \begin{cases} 1, & x \le 2 \\ x+2, & x>2 \end{cases}$

Figure 2.1.7

REMARK. In Figure 2.1.7, we used the heavy dot above $x = 2$ to emphasize that the value $g(2) = 1$ lies on the horizontal line and not on the inclined line.

In each of the last five examples we found the graph of a *given* function. We shall now consider the converse problem.

2.1.3 PROBLEM *Given a curve in the xy-plane, does there exist a function f whose graph is the given curve?*

This problem is investigated in the following examples.

▶ Example 15

Show that the graph of

$$3x^2 - 2y = 1 \tag{9}$$

is also the graph of $f(x)$ for some function f.

Solution. Equation (9) can be rewritten in the equivalent form

$$y = \tfrac{1}{2}(3x^2 - 1)$$

so that the graph of (9) is also the graph of the function

$$f(x) = \tfrac{1}{2}(3x^2 - 1) \qquad ◄$$

▶ Example 16

The curve in Figure 2.1.8*a* cannot be the graph of any function of x. To see why, consider the vertical line in Figure 2.1.8*b*; this line intersects the curve at the two points, (a, b) and (a, c). If the curve were the graph of

$$y = f(x) \tag{10}$$

for some function f, then, since (a, b) and (a, c) lie on the curve, (10) would imply that

$$b = f(a) \qquad \text{and} \qquad c = f(a)$$

But this is impossible since f cannot assign two different values to a. Thus, there is no function f whose graph is the curve in Figure 2.1.8a. ◄

Figure 2.1.8   (a)   (b)

This example illustrates the following general result, which we will call the **vertical line test.**

2.1.4 THE VERTICAL LINE TEST   *A curve is the graph of $f(x)$ for some function f if and only if no vertical line intersects the curve more than once.*

▶ Example 17

Is the graph of the circle

$$x^2 + y^2 = 25 \tag{11}$$

also the graph of $f(x)$ for some function f?

Solution. Since some vertical lines intersect the circle more than once (see Figure 2.1.9), the circle is not the graph of any function.

We can also deduce this result algebraically by observing that (11) can be written as

$$y = \pm \sqrt{25 - x^2} \tag{12}$$

But the right side of (12) is not a function of x since it is "multiple-valued." Thus (11) is not equivalent to an equation of the form

$$y = f(x) \qquad ◄$$

We conclude this section with some useful terminology. If the graph of an equation in x and y is also the graph of some function f, then we say, the equation defines **y as a function of x;** if in addition, the equation is written in the form $y = f(x)$ we say that equation defines y **explicitly** as a function of x.

Figure 2.1.9

If an equation defines y as a function of x, but is written in some form other than $y = f(x)$, then we say that the equation defines y *implicitly* as a function of x. These ideas will be studied in more detail later.

▶ Example 18

From Example 15,

$$3x^2 - 2y = 1 \tag{13}$$

defines y implicitly as a function of x. When (13) is rewritten as

$$y = \tfrac{1}{2}(3x^2 - 1)$$

we have an equation that defines y explicitly as a function of x. ◀

▶ Exercise Set 2.1

1. Given that $f(x) = 3x^2 + 2$, find
   (a) $f(-2)$ (b) $f(4)$ (c) $f(0)$
   (d) $f(-\sqrt{3})$ (e) $f(t)$.

2. Given that $g(x) = \dfrac{x+1}{x-1}$, find

   (a) $g(1.1)$ (b) $g(\tfrac{1}{4})$ (c) $g(\sqrt[3]{5}+1)$
   (d) $g(\pi)$ (e) $g(a-1)$.

3. Given that

   $$\phi(x) = \begin{cases} \dfrac{1}{x}, & x \geq 3 \\[2mm] 2x, & x < 3 \end{cases}$$

   find
   (a) $\phi(2)$ (b) $\phi(-4)$ (c) $\phi(3)$
   (d) $\phi(3.1)$ (e) $\phi(2.9)$.

4. Given that $f(x) = \sqrt{4x-3}$, find
   (a) $f(\tfrac{3}{4})$ (b) $f(-\tfrac{3}{4})$ (c) $f(0)$
   (d) $f(3)$ (e) $f(5.1)$.

5. Given that $H(x) = x/|x|$, find
   (a) $H(3)$ (b) $H(7.1)$ (c) $H(-6)$
   (d) $H(-\pi)$ (e) $H(0)$.

In Exercises 6–17, find the domain of the given function.

6. $f(x) = \dfrac{1}{x-3}$.

7. $f(x) = \dfrac{1}{5x+7}$.

8. $g(x) = \sqrt{x^2 - 3}$.

9. $g(x) = \sqrt{(x-1)(x+2)}$.

10. $\phi(x) = \sqrt{x^2 + 3}$.

11. $\phi(x) = \dfrac{x}{\sqrt{|x| + 1}}$.

12. $H(x) = x^2, \; x < -3$.

13. $H(x) = \sqrt{x}, \; x \geq 5$.

14. $F(x) = \begin{cases} \sqrt{x}, & x \geq 2 \\ \dfrac{1}{x-2}, & x < 2 \end{cases}$

15. $F(x) = \begin{cases} \sqrt{-x}, & x \leq -3 \\ 0, & x \geq 2 \end{cases}$

16. $f(x) = \dfrac{(x-2)(x-1)}{x-1}$.

17. $f(x) = \dfrac{(x-1)(x-3)}{x-3}$.

In Exercises 18–33, sketch the graph of the function.

18. $f(x) = 2x + 1$.

19. $f(x) = 3x - 2$.

20. $G(x) = x, \qquad 1 \leq x \leq 2$.

21. $G(x) = x - 2, \qquad -1 \leq x \leq 1$.

22. $h(x) = x^2 - 3$.

23. $h(x) = (x - 2)^2$.

24. $\phi(x) = \dfrac{x}{|x|}$.

25. $\phi(x) = \dfrac{|x - 2|}{x - 2}$.

26. $g(x) = \begin{cases} x^2, & x \neq 4 \\ 0, & x = 4 \end{cases}$

27. $g(x) = \begin{cases} x - 1, & x \neq 1 \\ 3, & x = 1 \end{cases}$

28. $F(x) = \sqrt{x(x-1)}$.

29. $F(x) = \sqrt{x^2 + 3x}$.

30. $f(x) = \begin{cases} x + 2, & x \leq 3 \\ x + 4, & x > 3 \end{cases}$

31. $f(x) = \begin{cases} x^2, & x > 1 \\ 2, & x \leq 1 \end{cases}$

32. $h(x) = \begin{cases} 1, & 0 < x \leq 1 \\ 3, & 1 < x \leq 2 \\ -1, & 2 < x \leq 3 \\ 0, & \text{elsewhere} \end{cases}$

33. $h(x) = \begin{cases} -2, & -2 \leq x < -1 \\ 1, & -1 \leq x < 0 \\ 2, & 0 \leq x < 1 \\ 0, & \text{elsewhere} \end{cases}$

34. Let $g(x) = x - 3$ and let

$$f(x) = \begin{cases} \dfrac{x^2 - 9}{x + 3}, & x \neq -3 \\ k, & x = -3 \end{cases}$$

Find k so that $f(x) = g(x)$ for all x.

35. Criticize the following statement: The function

$$\frac{1 - 1/x}{1 + 1/x}$$

can be simplified by multiplying numerator and denominator by x to obtain

$$\frac{1 - 1/x}{1 + 1/x} = \frac{x - 1}{x + 1}$$

How would you write the statement to make it accurate?

36. Simplify the following functions by canceling factors, but be sure not to alter the domain.

(a) $f(x) = \dfrac{x^2 - 4}{x + 2}$

(b) $f(x) = \dfrac{(x + 2)(x^2 - 1)}{(x + 2)(x + 1)}$

(c) $f(x) = \dfrac{x^2 + x}{x}$

(d) $f(x) = \dfrac{x + 1 + \sqrt{x + 1}}{\sqrt{x + 1}}$.

37. In each part express y explicitly as a function of x.

    (a) $x^2y - 1 = 0$ (b) $x = \dfrac{1-y}{1+y}$

    (c) $y^2 + 2xy + x^2 = 0.$

38. Show that $y^2 + 3xy + x^2 = 0$ does not define y implicitly as a function of x.

39. In Figure 2.1.10, which curves are graphs of functions of x?

40. Consider the function f graphed in Figure 2.1.11. In each part, find all values of x satisfying the given condition.

    (a) $f(x) = 0$ (b) $f(x) = 3$ (c) $f(x) \geq 0.$

Figure 2.1.11

    (a)   (b)

    (c)   (d)

    (e)   (f)

Figure 2.1.10

2.2   OPERATIONS ON FUNCTIONS; CLASSIFYING FUNCTIONS

    Just as numbers can be added, subtracted, multiplied, and divided to produce other numbers, there is a useful way of adding, subtracting, multiplying, and dividing functions to produce other functions. These operations are defined as follows:

2.2.1 DEFINITION Given functions f and g, their **sum $f + g$**, **difference $f - g$**, **product $f \cdot g$**, and **quotient f/g**, are defined by

$$(f + g)(x) = f(x) + g(x)$$
$$(f - g)(x) = f(x) - g(x)$$
$$(f \cdot g)(x) \quad = f(x) \cdot g(x)$$
$$(f/g)(x) \quad = f(x)/g(x)$$

For the functions $f + g$, $f - g$, and $f \cdot g$ the domain is defined to be the intersection of the domains of f and g, and for f/g the domain is this intersection with the points where $g(x) = 0$ excluded.

▶ Example 1

Let f and g be the functions

$$f(x) = \sqrt{5 - x} \quad \text{and} \quad g(x) = \sqrt{x - 3}$$

Then the formulas for $f + g$, $f - g$, $f \cdot g$, and f/g are

$$(f + g)(x) = f(x) + g(x) = \sqrt{5 - x} + \sqrt{x - 3} \qquad (1)$$
$$(f - g)(x) = f(x) - g(x) = \sqrt{5 - x} - \sqrt{x - 3} \qquad (2)$$
$$(f \cdot g)(x) \quad = f(x) \cdot g(x) \quad = \sqrt{5 - x}\sqrt{x - 3} \qquad (3)$$
$$(f/g)(x) \quad = f(x)/g(x) \quad = \sqrt{5 - x}/\sqrt{x - 3} \qquad (4)$$

Since the domain of f is $(-\infty, 5]$ and the domain of g is $[3, +\infty)$, the domain of $f + g$, $f - g$ and $f \cdot g$ is

$$[3, 5] \qquad (5)$$

because this is the intersection of $(-\infty, 5]$ and $[3, +\infty)$. Since $g(x) = 0$ when $x = 3$, we must exclude this point to obtain the domain of f/g; thus the domain of f/g is

$$(3, 5] \qquad (6)$$

◀

REMARK. In the last section we agreed that when a function is given by a formula and there is no mention of the domain, it is understood that the domain consists of all real numbers where the formula makes sense and yields real values. Since the set in (5) is precisely where formulas (1), (2), and

(3) make sense and yield real values, the formulas themselves specify the domain for us. Similarly the formula in (4) specifies that (6) is the domain of f/g since this is precisely where the formula makes sense and yields real values. However, as the next example shows, the formula for $f + g, f - g,$ $f \cdot g,$ or f/g does not always suffice to specify the domain.

▶ Example 2

Let

$$f(x) = 3\sqrt{x} \qquad \text{and} \qquad g(x) = \sqrt{x}$$

Find $f \cdot g$.

Solution. Since the domain of f is $[0, +\infty)$ and the domain of g is $[0, +\infty)$, the domain of $f \cdot g$ is also

$$[0, +\infty)$$

since this is the intersection of the domains of f and g. The formula for $f \cdot g$ is

$$(f \cdot g)(x) = f(x) \cdot g(x) = (3\sqrt{x}) \cdot (\sqrt{x}) = 3x \tag{7}$$

Since the formula in (7) makes sense and yields real values for all x, the formula alone will not correctly describe the domain of $f \cdot g$ for us; we must write

$$(f \cdot g)(x) = 3x, \qquad x \geq 0 \qquad\qquad ◀$$

Sometimes we will write f^2 to denote the product $f \cdot f$. For example, if $f(x) = 8x$, then

$$f^2(x) = (f \cdot f)(x) = f(x) \cdot f(x) = (8x) \cdot (8x) = 64x^2$$

Similarly, we will denote $f^2 \cdot f$ by f^3, $f^3 \cdot f$ by f^4, and so on.

2.2.2 DEFINITION   If f is a function and k is a real number, the function kf is defined by

$$(kf)(x) = k \cdot f(x)$$

and the domain of kf is the same as the domain of f.

▶ Example 3

Let f be the function given by

$$f(x) = \sqrt{x^2 - 3}$$

Then the function πf is given by

$$(\pi f)(x) = \pi f(x) = \pi \sqrt{x^2 - 3}$$ ◀

Sometimes a given function has the same effect as two simpler functions evaluated in succession. For example, consider the function h given by

$$h(x) = (x + 1)^2$$

To evaluate $h(x)$ we first compute $x + 1$ and then square the result. In other words, if we consider the functions g and f given by

$$g(x) = x + 1$$

and

$$f(x) = x^2$$

then

$$h(x) = (x + 1)^2 = [g(x)]^2 = f(g(x))$$

Thus, h has the same effect as g and f evaluated successively. Loosely speaking, h is "composed" of the two functions f and g. This idea is formalized in the following definition.

2.2.3 DEFINITION Given two functions f and g, the *composition of f with g,* denoted by $f \circ g$, is the function defined by

$$(f \circ g)(x) = f(g(x))$$

where the domain of $f \circ g$ consists of all x in the domain of g for which $g(x)$ is in the domain of f.

▶ Example 4

Find $f \circ g$ if

$$f(x) = x^2 + 3 \quad \text{and} \quad g(x) = \sqrt{x}$$

Solution. Since the domain of g is $[0, +\infty)$ and the domain of f is $(-\infty, +\infty)$, the domain of $f \circ g$ consists of all x in $[0, +\infty)$ such that $g(x) = \sqrt{x}$ lies in $(-\infty, +\infty)$; thus the domain of $f \circ g$ is $[0, +\infty)$. Since

$$f(g(x)) = [g(x)]^2 + 3 = (\sqrt{x})^2 + 3 = x + 3$$

the composition of f with g is given by

$$(f \circ g)(x) = x + 3, \qquad x \geq 0 \qquad \blacktriangleleft$$

▶ Example 5

Let $f(x) = x - 1$ and $g(x) = \sqrt{x}$. Find

(a) $f \circ g$ (b) $g \circ f$

Solution (a). Since the domain of g is $[0, +\infty)$ and the domain of f is $(-\infty, +\infty)$, the domain of $f \circ g$ consists of all x in $[0, +\infty)$ such that $g(x) = \sqrt{x}$ lies in $(-\infty, +\infty)$, that is, all x in $[0, +\infty)$. Since

$$f(g(x)) = g(x) - 1 = \sqrt{x} - 1$$

we obtain

$$(f \circ g)(x) = \sqrt{x} - 1 \tag{8}$$

In (8) there is no need to indicate that the domain is $[0, +\infty)$ since this is precisely the set where $\sqrt{x} - 1$ is defined and yields real values.

Solution (b). The domain of $g \circ f$ consists of all x in the domain of f such that $f(x)$ lies in the domain of g. Since the domain of f is $(-\infty, +\infty)$ and the domain of g is $[0, +\infty)$, the domain of $g \circ f$ consists of all x in $(-\infty, +\infty)$ such that $f(x) = x - 1$ lies in $[0, +\infty)$; thus the domain is $[1, +\infty)$. Since

$$g(f(x)) = \sqrt{f(x)} = \sqrt{x - 1}$$

we have

$$(g \circ f)(x) = \sqrt{x - 1} \tag{9}$$

[As in part (a), there is no need to indicate in (9) that the domain is $[1, +\infty)$, since this is precisely the set where $\sqrt{x - 1}$ is defined and yields real values.]

$\blacktriangleleft$

CLASSIFICATION OF
FUNCTIONS

A function that assigns the same value to every member of its domain is called a **constant function.**

▶ Example 6

The function f given by

$$f(x) = 3$$

is constant since it assigns the value 3 to every real number x. By definition, the graph of f is the graph of the equation $y = f(x)$; that is,

$$y = 3$$

(See Figure 2.2.1.)   ◀

The graph of
$f(x) = 3$

Figure 2.2.1

REMARK. The constant function that assigns the value c to each real number is sometimes called **the constant function c.** Thus, the function f in Example 6 is *the constant function* 3.

A function of the form cx^n, where c is a constant and n is a nonnegative integer, is called a **monomial in x.** Examples are

$$2x^3, \qquad \pi x^7, \qquad 4x^0 \ (= 4), \qquad -6x, \qquad \text{and} \qquad x^{17}$$

The functions

$$4x^{1/2} \qquad \text{and} \qquad x^{-3}$$

are *not* monomials because the powers of x are not nonnegative integers.

A function that is expressible as the sum of finitely many monomials in x is called a **polynomial in x.** Examples are

$$x^3 + 4x + 7, \qquad 17 - \tfrac{2}{3}x, \qquad 9, \qquad \text{and} \qquad x^5$$

Also

$$(x^2 - 4)^3$$

is a polynomial in x because it is expressible as a sum of monomials (by carrying out the cubing operation). In general, f is a polynomial in x if it is expressible in the form

$$f(x) = a_0 + a_1 x + a_2 x^2 + \cdots + a_n x^n$$

where n is a nonnegative integer and $a_0, a_1, a_2, \ldots, a_n$ are real constants. The domain of a polynomial is $(-\infty, +\infty)$.

A polynomial is called **linear** if it has the form

$$a_0 + a_1 x \qquad \text{(where } a_1 \neq 0),$$

quadratic if it has the form

$$a_0 + a_1 x + a_2 x^2 \qquad \text{(where } a_2 \neq 0)$$

and **cubic** if it has the form

$$a_0 + a_1 x + a_2 x^2 + a_3 x^3 \qquad \text{(where } a_3 \neq 0)$$

A function that is expressible as a ratio of two polynomials is called a **rational function.** For example,

$$f(x) = \frac{x^5 - 2x^2 + 1}{x^2 - 4}$$

defines a rational function. In general, f is a rational function if it is expressible in the form

$$f(x) = \frac{a_0 + a_1 x + \cdots + a_n x^n}{b_0 + b_1 x + \cdots + b_m x^m}$$

The domain of f consists of all x where the denominator differs from zero.

An **explicit algebraic function** is a function that can be evaluated using finitely many additions, subtractions, multiplications, divisions, and root extractions. For example,

$$f(x) = x^{2/3} = (\sqrt[3]{x})^2 \qquad \text{and} \qquad g(x) = \frac{(x-3)\sqrt[4]{x}}{x^5 + \sqrt{x^2 + 1}}$$

define explicit algebraic functions of x. Also, every polynomial and every rational function is an explicit algebraic function.

All remaining functions fall into two categories, **implicit algebraic functions** and **transcendental functions.** We shall not define these terms, but instead refer the interested reader to a classic book in calculus, G. H. Hardy* (see page 71), *A Course of Pure Mathematics,* Cambridge Press, 1958 (10th edition). Among the transcendental functions already familiar to the reader are the trigonometric functions.

For reference, we have sketched in Figure 2.2.2 the graphs of various functions we will encounter frequently. In later sections we will develop calculus methods for obtaining these curves. However, for now, try to convince yourself that these graphs are correct by plotting points in the case $a = 1$.

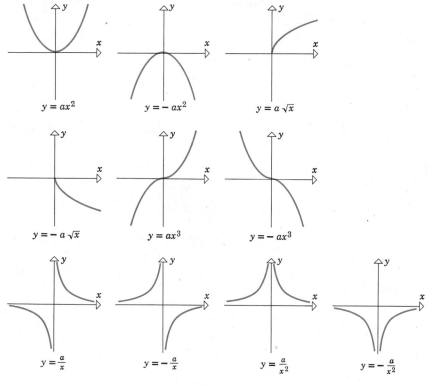

Figure 2.2.2

(*a* is any positive constant)

---

*G. H. HARDY (1877–1947). Hardy was a world-renowned British mathematician. He taught at Cambridge and Oxford universities and was a prolific researcher who produced over 300 research papers. He received numerous medals and honorary degrees for his accomplishments. Hardy's book, *A Course of Pure Mathematics,* gave the first rigorous English exposition of functions and limits for the college undergraduate; it had a great impact on university teaching. Hardy had a rebellious spirit; he once listed among his most ardent wishes: (1) to prove the Riemann hypothesis (a famous unsolved mathematical problem), (2) to make a brilliant play in a crucial cricket match, (3) to prove the nonexistence of God, and (4) to murder Mussolini.

In Table 2.2.1 we have listed four operations whose effect is to translate the graph of a function f.

Table 2.2.1

OPERATION	EFFECT	EXAMPLE
Add a positive constant c to $f(x)$. $(f(x) + c)$	Translates the graph of $f(x)$ up c units.	$y = x^2 + 2$ $y = x^2$
Subtract a positive constant c from $f(x)$. $(f(x) - c)$	Translates the graph of $f(x)$ down c units.	$y = x^2$ $y = x^2 - 2$
Replace x by $x - c$. (c positive)	Translates the graph of $f(x)$ right c units.	$y = (x - 2)^2$ $y = x^2$
Replace x by $x + c$. (c positive)	Translates the graph of $f(x)$ left c units.	$y = (x + 2)^2$ $y = x^2$

Figure 2.2.3

▶ Example 7

Sketch the graph of $f(x) = \sqrt{x + 3}$.

Solution. Translate the graph of $\sqrt{x}$ left three units (Figure 2.2.3). ◀

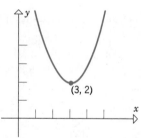

Figure 2.2.4

▶ Example 8

Sketch the graph of $f(x) = x^2 - 6x + 11$.

Solution. Complete the square on the first two terms:

$$f(x) = (x^2 - 6x + 9) + 2 = (x - 3)^2 + 2$$

In this form we see that the graph may be obtained by translating the graph of x^2 right three units because of the $x - 3$, and up two units because of the $+2$ (Figure 2.2.4). ◀

The method in the last example can be used to show that the graph of every quadratic polynomial $f(x) = ax^2 + bx + c$ is a translation of the graph of ax^2. These curves, called **parabolas,** will be studied later in more detail.

▶ Exercise Set 2.2

1. Let $f(x) = x^2 + 1$. Find
   (a) $f(t)$
   (b) $f(t + 2)$
   (c) $f(x + 2)$
   (d) $f\left(\dfrac{1}{x}\right)$
   (e) $f(x + h)$
   (f) $f(-x)$
   (g) $f(\sqrt{x})$
   (h) $f(3x)$.

2. Let $g(x) = \sqrt{x}$. Find
   (a) $g(5s + 2)$
   (b) $g(\sqrt{x} + 2)$
   (c) $3g(5x)$
   (d) $\dfrac{1}{g(x)}$
   (e) $g(g(x))$
   (f) $g^2(x)$
   (g) $g\left(\dfrac{1}{\sqrt{x}}\right)$
   (h) $g((x - 1)^2)$.

In Exercises 3–8, find
   (a) $f + g$,
   (b) $f - g$,
   (c) $f \cdot g$,
   (d) f/g,
   (e) $f \circ g$,
   (f) $g \circ f$,
   (g) kf
where k is a constant.

3. $f(x) = 2x, g(x) = x^2 + 1$.
4. $f(x) = 3x - 2, g(x) = |x|$.

5. $f(x) = \sqrt{x + 1}, g(x) = x - 2$.
6. $f(x) = \dfrac{x}{1 + x^2}, g(x) = \dfrac{1}{x}$.
7. $f(x) = \sqrt{x - 2}, g(x) = \sqrt{x - 3}$.
8. $f(x) = x^3, g(x) = \dfrac{1}{\sqrt[3]{x}}$.

9. Let h be defined by $h(x) = 2x - 5$. Find
   (a) $h \circ h$
   (b) h^2.

In Exercises 10–15, express f as a composition of two functions, that is, find functions g and h such that $f = g \circ h$. (Each exercise has more than one correct solution.)

10. $f(x) = x^2 + 1$.
11. $f(x) = \sqrt{x + 2}$.
12. $f(x) = \dfrac{1}{x - 3}$.
13. $f(x) = (x - 5)^7$.
14. $f(x) = a + bx$.
15. $f(x) = |x^2 - 3x + 5|$.
16. Is it ever true that $f \circ g = g \circ f$? Is it always true that $f \circ g = g \circ f$?

17. Let $f(x) = \dfrac{1}{x}$.

    (a) If $g(x) = x^2 + 1$, show that $f \circ g$ is defined for all x even though f is not defined when $x = 0$.

    (b) Can you find a different function g such that $f \circ g$ is defined for all x?

    (c) What property must a function g have in order for $f \circ g$ to be defined for all x?

For the functions in Exercises 18–21, state which of the following terms apply: monomial, polynomial, rational function, explicit algebraic function.

18. (a) $3x^7$ (b) $4x^{1/7}$

    (c) $\dfrac{1}{5x^6}$ (d) $2x^3 - 1$.

19. (a) $2x^{1/3} + 1$ (b) x^{-2}

    (c) $x^{-1/2}$ (d) $(x - 3)^{12}$.

20. (a) $x^2 \sqrt{x^3 - 3}$ (b) $\dfrac{x+1}{x+2}$

    (c) $\pi^{-2} + 1$ (d) $|x|$.

21. (a) $\sqrt{x^2 + \sqrt{x}}$ (b) $\dfrac{x^3 - 2x + 1}{x^2 - 9}$

    (c) $\sqrt{\pi} - 3$ (d) $|x - 2|$.

22. A function f is called **even** if $f(-x) = f(x)$ for each x in the domain of f, or **odd** if $f(-x) = -f(x)$ for each such x. In each part, classify the function as even, odd, or neither.

    (a) $f(x) = x^2$ (b) $f(x) = x^3$

    (c) $f(x) = |x|$ (d) $f(x) = x + 1$

    (e) $f(x) = \dfrac{x^5 - x}{1 + x^2}$ (f) $f(x) = 2$.

23. In Figure 2.2.5 we have sketched part of the graph of a function f (the part to the right of the y-axis.) Complete the graph assuming

    (a) f is an even function

    (b) f is an odd function

Figure 2.2.5

(See Exercise 22 for the definitions of even and odd functions.)

24. Classify the functions graphed in Figure 2.2.6 as even, odd, or neither. (See Exercise 22 for the definitions of even and odd.)

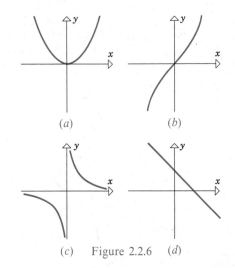

    (c)   Figure 2.2.6   (d)

25. Can a function be both even and odd? (See Exercise 22 for the definitions of even and odd functions.)

26. Let a be a constant and suppose $f(a - x) = f(a + x)$ for all x. What geometric property must the graph of f have?

27. Let

$$f(x) = \sqrt{x - 4}$$

and

$$g(x) = \frac{1}{2}x + 1, \qquad x \geq 6$$

Find $f \circ g$.

28. Let

$$f(x) = \begin{cases} 5x, & x \leq 0 \\ -x, & 0 < x \leq 8 \\ \sqrt{x}, & x > 8 \end{cases} \quad \text{and} \quad g(x) = x^3$$

Find $f \circ g$.

29. Prove or disprove: For any three functions f, g, and h, $f \circ (g \circ h) = (f \circ g) \circ h$.

2.3 INTRODUCTION TO CALCULUS: TANGENTS AND VELOCITY

Calculus centers around the following two fundamental problems.

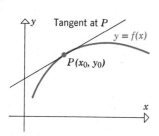

Figure 2.3.1

THE TANGENT PROBLEM. Given a function f and a point $P(x_0, y_0)$ on its graph, find the equation of the tangent to the graph at P (Figure 2.3.1).

THE AREA PROBLEM. Given a function f, find the area between the graph of f and an interval $[a, b]$ on the x-axis (Figure 2.3.2).

Traditionally, that portion of calculus arising from the tangent problem is called **differential calculus** and that portion arising from the area problem is called **integral calculus.** As we shall see, however, the tangent and area problems are closely related, so that the distinction between differential calculus and integral calculus is often hard to discern. We will consider the tangent problem first.

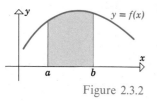

Figure 2.3.2

In plane geometry, a line is called **tangent** to a circle if it meets the circle at precisely one point (Figure 2.3.3a). However, this definition is not satisfactory for other kinds of curves. In Figure 2.3.3b the line meets the curve exactly once, yet is not a tangent, and in Figure 2.3.3c the line is tangent yet meets the curve more than once. It is important to define mathematically the concept of a tangent so that the definition applies not only to circles, but to other curves as well. To motivate the appropriate definition, we must approach the tangent concept in a different way. Consider the tangent to a circle at a point P and consider a secant line from P to another point Q on the circle. If, as shown in Figure 2.3.4, we move the point Q along the circle toward P, then the position of the secant line moves closer and closer to the position of the tangent at P. In other words, the secant lines "approach" the tangent line as Q approaches P along the circle. Using this idea we will be able to define the notion of a tangent so that it applies to curves other than circles.

Figure 2.3.3

Figure 2.3.4

Figure 2.3.5

Figure 2.3.6

Consider a point P on a curve in the xy-plane. To define the tangent to the curve at P it suffices to define the *slope* of the tangent at P since the slope and the point P together can be used to obtain the equation of the tangent (from the point-slope form of a line.) If Q is any point on the curve different from P, the line through P and Q is called a **secant line** for the curve (Figure 2.3.5). As with the circle, intuition suggests that if we move the point Q along the curve toward P, the secant line will rotate toward a "limiting" position (Figure 2.3.6). The line T occupying this limiting position we define to be the **tangent line** at P.

With this *intuitive* understanding of the tangent we can now discuss how to determine the slope of the tangent line. Let us suppose that $P(x_0, y_0)$ is a point on the graph of a function f, and assume we want to obtain the slope m_{tan} of the tangent to the graph at P. If we let $Q(x_1, y_1)$ be any point on the graph of f, different from P, then as shown in Figure 2.3.7a the slope m_{sec} of the secant line joining P and Q is

$$m_{\text{sec}} = \text{slope of } PQ = \frac{y_1 - y_0}{x_1 - x_0} \tag{1a}$$

But (x_0, y_0) and (x_1, y_1) lie on the graph of f so that $y_0 = f(x_0)$ and $y_1 = f(x_1)$. Thus, (1a) can be rewritten as

$$m_{\text{sec}} = \frac{f(x_1) - f(x_0)}{x_1 - x_0} \tag{1b}$$

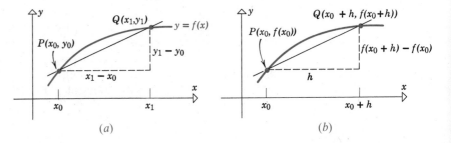

Figure 2.3.7 (a) (b)

For computational purposes, it is desirable to rewrite (1b) in another form by letting

$$h = x_1 - x_0$$

so that

$$x_1 = x_0 + h$$

In terms of h, (1b) becomes

$$m_{sec} = \frac{f(x_0 + h) - f(x_0)}{h} \qquad (2)$$

(Figure 2.3.7b). As Q approaches P along the graph of f, or equivalently as $h = x_1 - x_0$ gets closer and closer to zero, the secant line through P and Q approaches the tangent line at P. Thus the slope of the secant line m_{sec} approaches the slope of the tangent line m_{tan}. Therefore, (2) suggests the definition

$$m_{tan} = \begin{array}{c} \text{limiting value} \\ \text{as } h \text{ approaches} \\ \text{zero of} \end{array} \frac{f(x_0 + h) - f(x_0)}{h} \qquad (3)$$

This discussion would be complete if it were not for the fact that the notion of a "limiting value" is an intuitive idea, not yet mathematically defined. One of our primary objectives is to give a precise definition of the term "limit" and thereby complete the definition of a tangent. However, an intuitive under-standing of these notions will suffice in the following example.

▶ Example 1

Find the slope and the equation of the tangent line to the graph of $f(x) = x^2$ at the point $P(3, 9)$. (See Figure 2.3.8.)

Solution. From (2), we have

$$m_{sec} = \frac{f(x_0 + h) - f(x_0)}{h}$$

In this problem $x_0 = 3$ so that

$$m_{sec} = \frac{f(3 + h) - f(3)}{h}$$

or, since $f(x) = x^2$,

$$m_{sec} = \frac{(3 + h)^2 - (3)^2}{h}$$

$$m_{sec} = \frac{9 + 6h + h^2 - 9}{h}$$

or

$$m_{sec} = 6 + h, \qquad h \neq 0 \qquad (4)$$

y

$f(x) = x^2$

$P(3, 9)$

x

Figure 2.3.8

If we now let h approach 0, then the value of m_{sec} in (4) will approach 6. Thus the slope of the tangent line at $P(3, 9)$ is

$$m_{tan} = 6$$

Using this slope and the point $P(3, 9)$, the point-slope form of the tangent line is

$$(y - 9) = 6(x - 3)$$

and the slope-intercept form is

$$y = 6x - 9 \qquad\qquad \blacktriangleleft$$

While the tangent problem is of interest as a matter of pure geometry, much of the impetus for studying this problem arose in the seventeenth century when scientists recognized that many problems involving objects moving with variable velocity could be reduced to problems involving tangents. To see why this is so, we need to examine critically the meaning of the word "velocity."

If a car travels 75 miles over a straight road in a 3 hour period, then we say that the average velocity of the car is 25 miles per hour (25 mi/hr). More generally, the *average velocity* of an object moving in *one direction* along a line is

$$\text{average velocity} = \frac{\text{distance traveled}}{\text{time elapsed}}$$

Obviously, if an object travels with an average velocity of 25 mi/hr during a 3 hour trip, it need not travel at a fixed velocity of 25 mi/hr; sometimes it may speed up and sometimes it may slow down.

While average velocity is useful for some purposes, it is not always significant in physical problems. For example, if a moving car strikes a tree, the damage sustained is not determined by the average velocity up to the time of impact, but rather the instantaneous velocity at the precise moment of impact.

A clear understanding of instantaneous velocity evaded scientists until the advent of calculus in the seventeenth century. The subtlety of this concept was nicely described by Morris Kline* (see page 79), who wrote,

> In contrasting average velocity with instantaneous velocity we implicitly utilize a distinction between interval and instant. . . . An average velocity is one that concerns what happens over an interval of time—3 hours, 5 seconds, one-half second, and so forth. The interval may be small or large, but it does represent the passage of a definite amount of time. We use the word instant, however, to state the fact that something happens so fast that no time elapses. The event is

momentary. When we say, for example, that it is 3 o'clock, we refer to an instant, a precise moment. If the lapse of time is pictured by length along a line, then an interval (of time) is represented by a line segment, whereas an instant corresponds to a point. The notion of an instant, although it is used in everyday life, is strictly a mathematical idealization.

Our ways of thinking about real events cause us to speak in terms of instants and velocity at an instant, but closer examination shows that the concept of velocity at an instant presents difficulties. Average velocity, which is simply the distance traveled during some interval of time divided by that amount of time, is easily calculated. Suppose, however, that we try to carry over this process to instantaneous velocity. The distance an automobile travels in one instant is 0 and the time that elapses during one instant is also 0. Hence the distance divided by the time is 0/0, which is meaningless. Thus, although instantaneous velocity is a physical reality, there seems to be a difficulty in calculating it, and unless we can calculate it, we cannot work with it mathematically.

In order to really understand the concept of instantaneous velocity and work with it mathematically we must think small! Suppose we are interested in the instantaneous velocity of a car at a certain instant of time, say exactly 5 seconds after it starts to move along a straight road. Although the car may be moving with varying velocity it is evident that over a short interval of time, say 0.1 sec, the velocity does not vary much. Thus we make only a small error if we approximate the instantaneous velocity after 5 seconds by the *average velocity* over the *interval* from 5 seconds to 5.1 seconds. This average velocity can be calculated by measuring the distance traveled during the time interval between 5 seconds and 5.1 seconds and then dividing by the time elapsed, which is 0.1 second. Thus, even though the instantaneous velocity after 5 sec cannot be calculated directly, it can be approximated by an average velocity that can be determined from physical measurements.

The error that results from approximating instantaneous velocity by average velocity can be reduced by shrinking the time interval over which the average velocity is computed. For example, we would expect that the average velocity between 5 sec and 5.01 sec better approximates the instantaneous velocity after 5 sec than does the average velocity between 5 sec and 5.1 sec, because the time interval is smaller. Thus, by computing average velocity over smaller and smaller time intervals, we expect to get closer and closer to the instantaneous velocity.

---

* MORRIS KLINE (1908–   ) American mathematician, scholar, and educator. Kline has made numerous contributions to mathematical thought, written extensively on education, especially mathematics education, and has taught, lectured, and served as consultant throughout his very active career. He is the author of many popular books including *Mathematical Thought from Ancient to Modern Times* and *Why Johnny Can't Add: The Failure of the New Mathematics.* I wish to thank him for permission to use the above quotation, which is taken from *Calculus: An Intuitive and Physical Approach,* Wiley, New York, 1977, p. 17.

We stated earlier that there is a relationship between the tangent problem and physical problems involving objects moving with varying velocity. To illustrate this relationship we will consider the simplest kind of motion, a particle moving in one direction along a line. For example, the particle might idealize a rock dropped toward earth from a height, a billiard ball moving across a table, or a car traveling on a straight track.

For convenience, introduce a coordinate line along the path of the particle. Let the positive direction be the direction of motion, and imagine that a clock is keeping track of the elapsed time t, starting with $t = 0$ initially (Figure 2.3.9). After t units of time, the particle will be at a certain distance s from the origin. As t changes, so does s; thus s is a function of t, say,

$$s = f(t) \tag{5}$$

Figure 2.3.9   Elapsed time

For example, if a rock, initially at rest, is dropped from a height not too far above the surface of the earth, we will see later that the relationship between s and t is approximately

$$s = 16t^2 \tag{6}$$

where t is the number of seconds elapsed from the time the rock is released and s is the distance in feet from the starting point.

If we graph (5) with the t-axis horizontal and the s-axis vertical, we obtain a *position versus time curve* (Figure 2.3.10). Using this curve, we can interpret average velocity and instantaneous velocity geometrically. For example, suppose we are interested in the average velocity of the particle over the time interval from t_0 to t_1. At time t_0 the particle is at some distance s_0 from the origin and at time t_1 it is at some distance s_1. Over the time interval from t_0 to t_1 the distance traveled is

$$s_1 - s_0$$

and the time elapsed is

$$t_1 - t_0$$

Figure 2.3.10

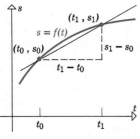

Figure 2.3.11

so that the average velocity during the interval is given by

$$\text{average velocity} = \frac{s_1 - s_0}{t_1 - t_0} \tag{7}$$

However, the points (t_0, s_0) and (t_1, s_1) lie on the position versus time curve, so that expression (7) is also the slope of the secant line connecting these points (Figure 2.3.11). In other words:

Geometric Interpretation of Average Velocity
For a particle moving in one direction on a straight line, the average velocity between time t_0 and t_1 is represented geometrically by the slope of the secant line connecting (t_0, s_0) and (t_1, s_1) on the position versus time curve.

If we choose t_1 close to t_0 then the average velocity between times t_0 and t_1 closely approximates the instantaneous velocity at time t_0; moreover, as we move t_1 closer and closer to t_0, intuition suggests that these approximations will approach the exact value of the instantaneous velocity at t_0. However, as t_1 moves toward t_0, the point (t_1, s_1) moves toward (t_0, s_0) on the position versus time curve (Figure 2.3.12) so that the slope of the secant line between (t_0, s_0) and (t_1, s_1) approaches the slope of the tangent line at (t_0, s_0). Since the slope of the secant line represents the average velocity between times t_0 and t_1, we conclude that:

Geometric Interpretation of Instantaneous Velocity
For a particle moving in one direction on a straight line, the instantaneous velocity at time t_0 is represented geometrically by the slope of the tangent line at (t_0, s_0) on the position versus time curve.

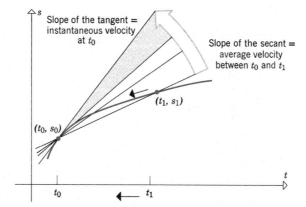

Figure 2.3.12

▶ Example 2

If a rock is dropped from a height not too far above the surface of the earth, then it is shown in physics that the relationship between the distance traveled the time elapsed is approximately

$$s = 16t^2 \tag{8}$$

where t is the number of seconds elapsed from the time the rock is released and s is its distance in feet from the starting point.

(a) Find the instantaneous velocity of the rock after t_0 seconds.
(b) Find the instantaneous velocity of the rock after 3 seconds.

Solution (*a*). Let s_0 be the distance traveled after t_0 seconds and s_1 the distance traveled after t_1 seconds. From (8),

$$s_0 = 16t_0^2 \quad \text{and} \quad s_1 = 16t_1^2$$

so that the average velocity in the time interval from t_0 to t_1 seconds is

$$\text{average velocity} = \frac{s_1 - s_0}{t_1 - t_0} = \frac{16t_1^2 - 16t_0^2}{t_1 - t_0} \tag{9}$$

For convenience, let

$$h = t_1 - t_0$$

so that (9) can be rewritten

$$\text{average velocity} = \frac{16(t_0 + h)^2 - 16t_0^2}{h}$$

or after squaring and simplifying

$$\text{average velocity} = \frac{32ht_0 + 16h^2}{h}$$

or

$$\text{average velocity} = 32t_0 + 16h, \quad h \neq 0 \tag{10}$$

As t_1 approaches t_0, the average velocity between t_0 and t_1 approaches the instantaneous velocity at t_0. But when t_1 approaches t_0, $h = t_1 - t_0$ approaches 0, so that from (10) we obtain

$$\text{instantaneous velocity} = 32t_0 \tag{11}$$

Since (8) presumes that distance is measured in feet and time in seconds, (11) tells us that the instantaneous velocity after t_0 seconds is $32t_0$ feet per second (ft/sec).

Solution (b). Letting $t_0 = 3$ in (11), we find that the instantaneous velocity of the rock after 3 seconds is $32(3) = 96$ ft/sec. ◄

► Exercise Set 2.3

1. Let $f(x) = \frac{1}{2}x^2$.
   (a) Find the slope of the secant line between those points on the graph of f for which $x = 3$ and $x = 4$.
   (b) Use the method of Example 1 to find the slope and equation for the tangent to the graph of f at the point where $x = 3$.
   (c) Sketch the graph of f together with the secant and tangent lines from (a) and (b).

2. Let $f(x) = x^3$.
   (a) Find the slope of the secant line between those points on the graph of f for which $x = 1$ and $x = 2$.
   (b) Use the method of Example 1 to find the slope and equation for the tangent to the graph of f at the point where $x = 1$.
   (c) Sketch the graph of f together with the secant and tangent lines from (a) and (b).

3. Let $f(x) = \frac{1}{x}$.
   (a) Find the slope of the secant line between those points on the graph of f for which $x = 2$ and $x = 3$.
   (b) Use the method of Example 1 to find the slope and equation for the tangent to the graph of f at the point where $x = 2$.
   (c) Sketch the graph of f together with the secant and tangent lines from (a) and (b).

4. Let $f(x) = \frac{1}{x^2}$.
   (a) Find the slope of the secant line between those points on the graph of f at which $x = 1$ and $x = 2$.
   (b) Use the method of Example 1 to find the slope and equation for the tangent to the graph of f at the point where $x = 1$.
   (c) Sketch the graph of f together with the secant and tangent lines from (a) and (b).

5. Let $f(x) = x^3$.
   (a) Use the method of Example 1 to show that the slope of the tangent to the graph of f at the point where $x = x_0$ is $3x_0^2$.
   (b) Use the result in (a) to find the equation of the tangent to the graph of f at the point where $x = 5$.
   (c) Use the result in (a) to find the equation of the tangent to the graph of f at the point where $x = x_0$.

6. Let $f(x) = \frac{1}{x}$.
   (a) Use the method of Example 1 to show that the slope of the tangent to the graph of f at the point where $x = x_0$ is $-1/x_0^2$.
   (b) Use the result in (a) to find the equation of the tangent to the graph of f at the point where $x = -7$.
   (c) Use the result in (a) to find the equation of the tangent to the graph of f at the point where $x = x_0$.

7. Figure 2.3.13 shows the position versus time curve for a certain particle moving on a straight line.
   (a) Is the particle moving faster at time t_0 or time t_2? Explain.
   (b) At the origin, the tangent is horizontal. What does this tell us about the initial velocity of the particle?
   (c) Is the particle speeding up or slowing down in the interval $[t_0, t_1]$? Explain.
   (d) Is the particle speeding up or slowing down in the interval $[t_1, t_2]$? Explain.

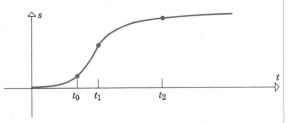

Figure 2.3.13

8. An automobile, initially at rest, begins to move along a straight track. The velocity increases continually until suddenly the driver sees a concrete barrier in the road and applies the brakes sharply at time t_0. The car decelerates rapidly, but it is too late—the car crashes into the barrier at time t_1 and instantaneously comes to rest. Sketch a position versus time curve that might represent the motion of the car.

9. If a particle moves at constant velocity, what can you say about its position versus time curve?

10. A rock is dropped from a height of 576 ft and falls toward earth in a straight line. In t sec the rock drops $16t^2$ ft.
    (a) How long does it take for the rock to hit the ground?
    (b) What is the average velocity of the rock during the time it is falling?

(c) What is the average velocity of the rock for the first 3 sec?
(d) What is the velocity of the rock when it hits the ground?

11. During the first 40 sec of a rocket flight, the rocket is propelled straight up so that in t sec it reaches a height of $5t^3$ ft.
    (a) How high does the rocket travel in 40 sec?
    (b) What is the average velocity of the rocket during the first 40 sec?
    (c) What is the average velocity of the rocket during the first 135 ft of its flight?
    (d) What is the velocity of the rocket at the end of 40 sec?

12. A particle moves on a line away from its initial position so that after t hr its distance is $s = 3t^2 + t$ mi from its initial position.
    (a) Find the average velocity of the particle over the interval $[1, 3]$.
    (b) Find the instantaneous velocity at $t = 1$.

13. A particle moves in one direction along a straight line so that after t min its distance is $s = 6t^4$ ft from the origin.
    (a) Find the average velocity of the particle over the interval $[2, 4]$.
    (b) Find the instantaneous velocity at $t = 2$.

14. A car is traveling on a straight road that is 120 mi long. For the first 100 mi the car travels at an average velocity of 50 mph. Show that no matter how fast the car travels for the final 20 mi it cannot bring the average velocity up to 60 mph for the entire trip.

15. Let $f(x) = x^2$. If we approximate the slope of the tangent at the point $(x_0, f(x_0))$ by the slope of the secant line between $(x_0, f(x_0))$ and $(x_1, f(x_1))$, show that the error is $|x_1 - x_0|$. (By error we mean $|m_{tan} - m_{sec}|$, where m_{tan} is the slope of the tangent line and m_{sec} the slope of the secant line.)

2.4 LIMITS (AN INTUITIVE INTRODUCTION)

In the last section we saw that the concepts of tangent and instantaneous velocity ultimately rest on the notion of a "limit" or "value approached by" a function. In this section as well as the next few we will investigate the notion of limit in more detail. Our development of limits in this text proceeds in three stages:

1. First we discuss limits intuitively.

2. Then we discuss methods for computing limits.

3. Finally, we give a precise mathematical discussion of limits.

Limits are used to describe how a function behaves as the independent variable moves toward a certain value. To illustrate, consider the function

$$f(x) = \frac{\sin x}{x}$$

Although this function is not defined at $x = 0$, it still makes sense to ask what happens to the values of $f(x)$ as x moves along the x-axis toward $x = 0$. To answer this question, we used a calculator to obtain values of $f(x)$ at a succession of points moving along the *positive* x-axis toward $x = 0$. The results, which appear in Table 2.4.1, suggest that the values of $f(x)$ approach 1. We call the number 1 the *limit* of

$$\frac{\sin x}{x}$$

as x approaches 0 from the right side, and we write

$$\lim_{x \to 0^+} \frac{\sin x}{x} = 1 \tag{1}$$

In this expression, "lim" tells us that we are computing a limit; the symbol $x \to 0$ tells us that we are letting x approach 0; and the label "+" on $x \to 0$ tells us that x is approaching zero from the *right* side. We can also ask what happens to the values of

$$f(x) = \frac{\sin x}{x}$$

as x approaches 0 from the left side. From Table 2.4.2 [or from the fact that $f(-x) = f(x)$] it is evident that the values of $f(x)$ again approach 1. We denote this by writing

$$\lim_{x \to 0^-} \frac{\sin x}{x} = 1 \tag{2}$$

Table 2.4.1

x	$\dfrac{\sin x}{x}$
1.0	0.84147
0.9	0.87036
0.8	0.89670
0.7	0.92031
0.6	0.94107
0.5	0.95885
0.4	0.97355
0.3	0.98507
0.2	0.99335
0.1	0.99833
0.01	0.99998

Table 2.4.2

x	$\dfrac{\sin x}{x}$
−1.0	0.84147
−0.9	0.87036
−0.8	0.89670
−0.7	0.92031
−0.6	0.94107
−0.5	0.95885
−0.4	0.97355
−0.3	0.98507
−0.2	0.99335
−0.1	0.99833
−0.01	0.99998

In (2), the label "−" on $x \to 0$ tells us that we are computing the limit as x approaches zero from the *left* side.

REMARK. It is important to keep in mind that the limits given in (1) and (2) are really just guesses based on numerical evidence. It is conceivable that if we continue the calculator computations in Tables 2.4.1 and 2.4.2 for values of x still closer to 0, the pattern of values for $(\sin x)/x$ might change and approach some value different from 1 or perhaps approach no number at all. Moreover, no matter how much we enlarge our tables, this problem will persist since we can never be certain what happens for values of x not included in the table. To be certain that the limits in (1) and (2) are really correct, we need mathematical proof, and to give a mathematical proof, we need a precise mathematical definition of a limit. These questions will be taken up later. However, in this section we will rely completely on our intuition to obtain limits. Our main objective at this time is to introduce the mathematical notation associated with limits and to develop our intuitive understanding of limits geometrically.

If the value of $f(x)$ approaches the number L_1 as x approaches x_0 from the right side, we write

$$\lim_{x \to x_0^+} f(x) = L_1 \tag{3}$$

If the value of $f(x)$ approaches the number L_2 as x approaches x_0 from the left side, we write

$$\lim_{x \to x_0^-} f(x) = L_2 \tag{4}$$

Expression (3) is read, "the limit of $f(x)$ as x approaches x_0 from the right equals L_1" and expression (4) is read, "the limit of $f(x)$ as x approaches x_0 from the left equals L_2."

If the limit from the left side is the same as the limit from the right side, say

$$\lim_{x \to x_0^-} f(x) = \lim_{x \to x_0^+} f(x) = L$$

then we write

$$\lim_{x \to x_0} f(x) = L \tag{5}$$

Equation (5) is read, "the limit of $f(x)$ as x approaches x_0 equals L."

▶ Example 1

From (1) and (2)

$$\lim_{x \to 0^+} \frac{\sin x}{x} = \lim_{x \to 0^-} \frac{\sin x}{x} = 1$$

Therefore,

$$\lim_{x \to 0} \frac{\sin x}{x} = 1 \qquad \blacktriangleleft$$

The following examples illustrate limits geometrically.

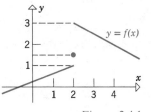

Figure 2.4.1

▶ Example 2

Let f be the function whose graph is shown in Figure 2.4.1.
As x approaches 2 from the left, $f(x)$ approaches 1, so

$$\lim_{x \to 2^-} f(x) = 1$$

As x approaches 2 from the right, $f(x)$ approaches 3, so that

$$\lim_{x \to 2^+} f(x) = 3 \qquad \blacktriangleleft$$

Since $f(2) = 1.5$, this example shows that the value of a function at a point, and the left- and right-hand limits at the point can all be different.

▶ Example 3

Let f_1 and f_2 be the functions with graphs as shown in Figure 2.4.2. The functions f_1 and f_2 are identical to the function f in Example 2 except at the point $x = 2$; there we have $f(2) = 1.5$ and $f_1(2) = 1$, while $f_2(2)$ is undefined. However, even though the functions f, f_1, and f_2 behave differently *at* $x = 2$, their limits as x approaches 2 are the same, that is,

$$\lim_{x \to 2^-} f(x) = \lim_{x \to 2^-} f_1(x) = \lim_{x \to 2^-} f_2(x) = 1$$

and

$$\lim_{x \to 2^+} f(x) = \lim_{x \to 2^+} f_1(x) = \lim_{x \to 2^+} f_2(x) = 3 \qquad \blacktriangleleft$$

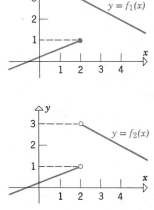

Figure 2.4.2

This example illustrates a general principle about limits that can be stated loosely as follows:

The limit of a function as the independent variable approaches a point does not depend on the value of the function *at* the point.

Thus, if we alter the value of a function f only at $x = x_0$, then we do not affect either

$$\lim_{x \to x_0^-} f(x) \qquad \text{or} \qquad \lim_{x \to x_0^+} f(x)$$

Function f_2 in Example 3 illustrates another idea: namely, that the left- and right-hand limits can have values at a point where the function itself is undefined.

Figure 2.4.3

▶ Example 4

Let f be the function whose graph is shown in Figure 2.4.3. From the graph we see that

$$\lim_{x \to 3^-} f(x) = 2 \qquad \text{and} \qquad \lim_{x \to 3^+} f(x) = 2$$

so that

$$\lim_{x \to 3} f(x) = 2$$

The graph also shows that $f(3) = 2$. Thus the value of a function at a point and the limits as x approaches that point may be equal. ◀

▶ Example 5

Let f be the function whose graph is shown in Figure 2.4.4. This graph is intended to convey the idea that as x approaches 6 from the right side, $f(x)$ oscillates between -3 and 3 infinitely often and with increasing frequency. Since $f(x)$ approaches no fixed value as x approaches 6 from the right, we say that

$$\lim_{x \to 6^+} f(x) \ does \ not \ exist$$

On the other hand, as x approaches 6 from the left side, $f(x)$ approaches 1, so that

$$\lim_{x \to 6^-} f(x) = 1 \qquad \blacktriangleleft$$

Figure 2.4.4

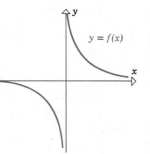

Figure 2.4.5

► Example 6

Let f be the function with graph as shown in Figure 2.4.5. As x approaches 0 from the right side, $f(x)$ gets larger and larger without bound and consequently approaches no fixed value. Thus

$$\lim_{x \to 0^+} f(x)$$

does not exist. In this case we would write

$$\lim_{x \to 0^+} f(x) = +\infty \qquad (6)$$

to indicate that the limit fails to exist because $f(x)$ is increasing without bound.

As x approaches 0 from the left side, $f(x)$ becomes more and more negative without bound and consequently approaches no fixed value. Thus

$$\lim_{x \to 0^-} f(x)$$

does not exist. In this case we would write

$$\lim_{x \to 0^-} f(x) = -\infty \qquad (7)$$

to indicate that the limit fails to exist because $f(x)$ is decreasing without bound. ◄

► Example 7

Let f and g be the functions with graphs as shown in Figure 2.4.6. From the

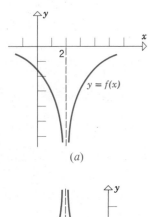

(a)

(b)

Figure 2.4.6

graph of f we see that

$$\lim_{x \to 2^-} f(x) = -\infty \quad \text{and} \quad \lim_{x \to 2^+} f(x) = -\infty$$

We can indicate these two results more succinctly by writing

$$\lim_{x \to 2} f(x) = -\infty$$

From the graph of g we see that

$$\lim_{x \to -3^-} g(x) = +\infty \quad \text{and} \quad \lim_{x \to -3^+} g(x) = +\infty$$

More briefly, we will write

$$\lim_{x \to -3} g(x) = +\infty \qquad \blacktriangleleft$$

So far we have used limits to describe how a function behaves as the independent variable approaches a fixed point on the x-axis. However, limits can also be used to describe how a function behaves as the independent variable moves "indefinitely far" from the origin along the x-axis. If x is allowed to increase without bound, then we write $x \to +\infty$ (read, "x approaches plus infinity"), and if x is allowed to decrease without bound then we write $x \to -\infty$ (read, "x approaches minus infinity").

▶ Example 8

Let f be the function with graph as shown in Figure 2.4.7. As $x \to +\infty$ the

Figure 2.4.7

graph of f approaches the line $y = 4$ so that the value of $f(x)$ approaches 4. We denote this by writing

$$\lim_{x \to +\infty} f(x) = 4$$

As $x \to -\infty$ the graph of f tends toward the line $y = -1$ and so the value of $f(x)$ approaches -1. We denote this by writing

$$\lim_{x \to -\infty} f(x) = -1 \qquad \blacktriangleleft$$

► Example 9

Let f be the function with graph as shown in Figure 2.4.8. As x approaches $-\infty$, the value of $f(x)$ increases without bound and consequently approaches no fixed value. Thus

$$\lim_{x \to -\infty} f(x)$$

does not exist. In this case we would write

$$\lim_{x \to -\infty} f(x) = +\infty$$

to indicate that the limit fails to exist because $f(x)$ is increasing without bound.

Figure 2.4.8

As x approaches $+\infty$, the graph of f oscillates but tends toward the line $y = -2$. Thus the value of $f(x)$ approaches -2 as x approaches $+\infty$ and we write

$$\lim_{x \to +\infty} f(x) = -2 \qquad \blacktriangleleft$$

► Example 10

Let f be the function with graph as shown in Figure 2.4.9.

As x approaches $-\infty$, the value of $f(x)$ decreases without bound and consequently approaches no fixed value. Thus

$$\lim_{x \to -\infty} f(x)$$

does not exist. In this case we would write

$$\lim_{x \to -\infty} f(x) = -\infty$$

to indicate that the limit fails to exist because $f(x)$ is decreasing without bound.

As x approaches $+\infty$, the value of $f(x)$ oscillates between -3 and $+3$ and consequently approaches no fixed value. Thus

$$\lim_{x \to +\infty} f(x)$$

does not exist. There is no special notation used to describe limits that fail to exist because of oscillation.　　　◀

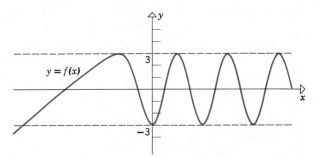

Figure 2.4.9

We conclude this section with some comments on terminology. The expressions

$$\lim_{x \to x_0^-} f(x) \qquad \text{and} \qquad \lim_{x \to x_0^+} f(x)$$

are called the **one-sided limits of $f(x)$ at x_0** and the expression

$$\lim_{x \to x_0} f(x)$$

is called the **two-sided limit of $f(x)$ at x_0**. As noted previously, the two-sided limit exists only when the one-sided limits exist and are equal. The two-sided limit does not exist if either of the one-sided limits does not exist or if the one-sided limits exist but are unequal.

▶ Exercise Set 2.4

1. For the function f graphed below, find
    (a) $\lim\limits_{x \to 3^-} f(x)$
    (b) $\lim\limits_{x \to 3^+} f(x)$
    (c) $\lim\limits_{x \to 3} f(x)$
    (d) $f(3)$
    (e) $\lim\limits_{x \to -\infty} f(x)$
    (f) $\lim\limits_{x \to +\infty} f(x)$.

2. For the function f graphed below, find
    (a) $\lim\limits_{x \to 2^-} f(x)$
    (b) $\lim\limits_{x \to 2^+} f(x)$
    (c) $\lim\limits_{x \to 2} f(x)$
    (d) $f(2)$
    (e) $\lim\limits_{x \to -\infty} f(x)$
    (f) $\lim\limits_{x \to +\infty} f(x)$.

3. For the function g graphed below, find
    (a) $\lim\limits_{x \to 4^-} g(x)$
    (b) $\lim\limits_{x \to 4^+} g(x)$
    (c) $\lim\limits_{x \to 4} g(x)$
    (d) $g(4)$
    (e) $\lim\limits_{x \to -\infty} g(x)$
    (f) $\lim\limits_{x \to +\infty} g(x)$.

4. For the function g graphed below, find
    (a) $\lim\limits_{x \to 0^-} g(x)$
    (b) $\lim\limits_{x \to 0^+} g(x)$
    (c) $\lim\limits_{x \to 0} g(x)$
    (d) $g(0)$
    (e) $\lim\limits_{x \to -\infty} g(x)$
    (f) $\lim\limits_{x \to +\infty} g(x)$.

5. For the function F graphed below, find
    (a) $\lim\limits_{x \to -2^-} F(x)$
    (b) $\lim\limits_{x \to -2^+} F(x)$
    (c) $\lim\limits_{x \to -2} F(x)$
    (d) $F(-2)$
    (e) $\lim\limits_{x \to -\infty} F(x)$
    (f) $\lim\limits_{x \to +\infty} F(x)$.

6. For the function F graphed below, find
    (a) $\lim\limits_{x \to 3^-} F(x)$
    (b) $\lim\limits_{x \to 3^+} F(x)$
    (c) $\lim\limits_{x \to 3} F(x)$
    (d) $F(3)$
    (e) $\lim\limits_{x \to -\infty} F(x)$
    (f) $\lim\limits_{x \to +\infty} F(x)$.

7. For the function ϕ graphed below, find

(a) $\lim\limits_{x \to -2^-} \phi(x)$ (b) $\lim\limits_{x \to -2^+} \phi(x)$

(c) $\lim\limits_{x \to -2} \phi(x)$ (d) $\phi(-2)$

(e) $\lim\limits_{x \to -\infty} \phi(x)$ (f) $\lim\limits_{x \to +\infty} \phi(x)$.

$y = \phi(x)$

8. For the function ϕ graphed below, find

(a) $\lim\limits_{x \to 4^+} \phi(x)$ (b) $\lim\limits_{x \to 4^-} \phi(x)$

(c) $\lim\limits_{x \to 4} \phi(x)$ (d) $\phi(4)$

(e) $\lim\limits_{x \to -\infty} \phi(x)$ (f) $\lim\limits_{x \to +\infty} \phi(x)$.

$y = \phi(x)$

9. For the function f graphed below find

(a) $\lim\limits_{x \to 3^-} f(x)$ (b) $\lim\limits_{x \to 3^+} f(x)$

(c) $\lim\limits_{x \to 3} f(x)$ (d) $f(3)$

(e) $\lim\limits_{x \to -\infty} f(x)$ (f) $\lim\limits_{x \to +\infty} f(x)$.

$y = f(x)$

10. For the function f graphed below, find

(a) $\lim\limits_{x \to 0^-} f(x)$ (b) $\lim\limits_{x \to 0^+} f(x)$

(c) $\lim\limits_{x \to 0} f(x)$ (d) $f(0)$

(e) $\lim\limits_{x \to -\infty} f(x)$ (f) $\lim\limits_{x \to +\infty} f(x)$.

$y = f(x)$

11. For the function G graphed below, find

(a) $\lim\limits_{x \to 0^-} G(x)$ (b) $\lim\limits_{x \to 0^+} G(x)$

(c) $\lim\limits_{x \to 0} G(x)$ (d) $G(0)$

(e) $\lim\limits_{x \to -\infty} G(x)$ (f) $\lim\limits_{x \to +\infty} G(x)$.

$y = G(x)$

12. For the function G graphed below, find

    (a) $\lim\limits_{x \to 0^-} G(x)$ (b) $\lim\limits_{x \to 0^+} G(x)$

    (c) $\lim\limits_{x \to 0} G(x)$ (d) $G(0)$

    (e) $\lim\limits_{x \to -\infty} G(x)$ (f) $\lim\limits_{x \to +\infty} G(x)$.

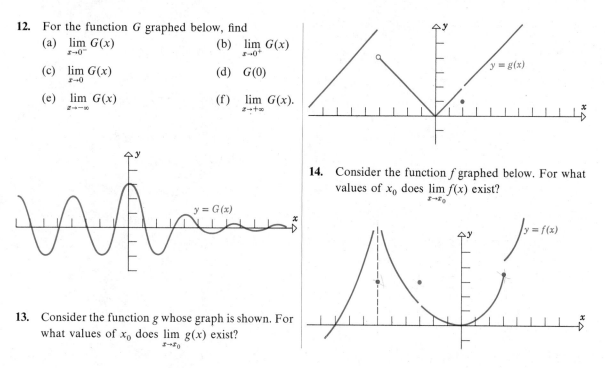

14. Consider the function f graphed below. For what values of x_0 does $\lim\limits_{x \to x_0} f(x)$ exist?

13. Consider the function g whose graph is shown. For what values of x_0 does $\lim\limits_{x \to x_0} g(x)$ exist?

2.5 LIMITS (COMPUTATIONAL TECHNIQUES)

In the last section we concentrated on the graphical interpretation of limits. In this section we will discuss techniques for finding limits directly from the formula for a function. Again our results will be based on intuition. Mathematical proofs of the results here will have to wait until we have a precise mathematical definition of a limit.

The limits discussed in our first four examples are the basic tools used to solve more complicated limit problems.

▶ Example 1

Since a constant function

$$f(x) = k$$

has the same value k everywhere, it follows that at each point a

$$\lim_{x \to a} k = k$$

and also

$$\lim_{x \to +\infty} k = \lim_{x \to -\infty} k = k$$

(Figure 2.5.1). These results can be summarized by the statement:

The limit of a constant function k at any point is k. ◀

Figure 2.5.1

► Example 2

The limit

$$\lim_{x \to a} x = a$$

is self-evident. For example,

$$\lim_{x \to 5} x = 5, \qquad \lim_{x \to 0} x = 0, \qquad \lim_{x \to -2} x = -2 \qquad ◀$$

► Example 3

Find the limits

$$\lim_{x \to 0^+} \frac{1}{x}, \qquad \lim_{x \to 0^-} \frac{1}{x}, \qquad \lim_{x \to 0} \frac{1}{x}$$

Solution. It will be evident from some simple calculations that none of these limits exist. As x approaches 0 from the right, the value of $1/x$ gets larger and larger without bound. For example, if x successively assumes values

$$x = 1, \; \frac{1}{10}, \; \frac{1}{100}, \; \frac{1}{1000}, \; \frac{1}{10,000}, \; \ldots$$

then $1/x$ successively assumes values

$$\frac{1}{x} = 1, \; 10, \; 100, \; 1000, \; 10,000, \; \ldots$$

Thus

$$\lim_{x \to 0^+} \frac{1}{x} = +\infty$$

As x approaches 0 from the left, the value of $1/x$ decreases (that is, becomes more and more negative) without bound. For example, if x successively assumes values

$$x = -1, \; -\frac{1}{10}, \; -\frac{1}{100}, \; -\frac{1}{1000}, \; -\frac{1}{10,000}, \; \ldots$$

then $1/x$ successively assumes values

$$\frac{1}{x} = -1, \; -10, \; -100, \; -1000, \; -10,000, \; \ldots$$

Thus

$$\lim_{x \to 0^-} \frac{1}{x} = -\infty$$

It follows that

$$\lim_{x \to 0} \frac{1}{x} \text{ does not exist}$$

since the one-sided limits as x approaches zero do not exist. Moreover, we cannot even write

$$\lim_{x \to 0} \frac{1}{x} = +\infty \qquad \text{or} \qquad \lim_{x \to 0} \frac{1}{x} = -\infty$$

(Why?)

The reader should check that the results in this example are consistent with the graph of $1/x$ (see Figure 2.2.2). ◀

▶ Example 4

Find the limits

$$\lim_{x \to +\infty} \frac{1}{x}, \qquad \lim_{x \to -\infty} \frac{1}{x}$$

Solution. These limits are also evident from some sample calculations. To investigate the case $x \to +\infty$, assume x takes on successive values

$$x = 1, 10, 100, 1000, 10{,}000, \ldots$$

Then $1/x$ successively assumes values

$$\frac{1}{x} = 1, \frac{1}{10}, \frac{1}{100}, \frac{1}{1000}, \frac{1}{10{,}000}, \ldots$$

It is clear from these values of $1/x$ that

$$\lim_{x \to +\infty} \frac{1}{x} = 0$$

To investigate the case $x \to -\infty$, assume x takes on successive values

$$x = -1, -10, -100, -1000, -10{,}000, \ldots$$

Then $1/x$ successively assumes values

$$\frac{1}{x} = -1, -\frac{1}{10}, -\frac{1}{100}, -\frac{1}{1000}, -\frac{1}{10{,}000}, \ldots$$

It is clear from these values of $1/x$ that

$$\lim_{x \to -\infty} \frac{1}{x} = 0$$

The reader should check that the results in this example are consistent with the graph of $1/x$ (see Figure 2.2.2). ◀

The results in our first four examples together with the fundamental properties of limits given in the following theorem can be used to help solve more complicated limit problems. Parts of this theorem are proved in Appendix 2.

2.5.1 THEOREM *Let* lim *stand for one of the limits* $\lim_{x \to a}$, $\lim_{x \to a^-}$, $\lim_{x \to a^+}$, $\lim_{x \to +\infty}$, *or* $\lim_{x \to -\infty}$. *If*

$L_1 = \lim f(x)$ *and* $L_2 = \lim g(x)$ *both exist, then*

(a) $\lim [f(x) + g(x)] = \lim f(x) + \lim g(x) = L_1 + L_2$.
(b) $\lim [f(x) - g(x)] = \lim f(x) - \lim g(x) = L_1 - L_2$.
(c) $\lim [f(x)g(x)] = \lim f(x) \lim g(x) = L_1 L_2$.
(d) $\lim \dfrac{f(x)}{g(x)} = \dfrac{\lim f(x)}{\lim g(x)} = \dfrac{L_1}{L_2}$ *if* $L_2 \neq 0$.
(e) $\lim \sqrt[n]{f(x)} = \sqrt[n]{\lim f(x)} = \sqrt[n]{L_1}$ *provided* $L_1 > 0$ *if n is even.*

In other words, this theorem tells us:

(a) *The limit of a sum is the sum of the limits.*
(b) *The limit of a difference is the difference of the limits.*
(c) *The limit of a product is the product of the limits.*
(d) *The limit of a quotient is the quotient of the limits provided the limit of the denominator is nonzero.*
(e) *The limit of an nth root is the nth root of the limit.*

REMARK. Although results (a) and (c) are stated for two functions f and g, these results hold as well for any finite number of functions; that is, if

$$\lim f_1(x), \lim f_2(x), \ldots, \lim f_n(x)$$

all exist, then

$$\lim [f_1(x) + f_2(x) + \cdots + f_n(x)] = \lim f_1(x) + \lim f_2(x) + \cdots + \lim f_n(x)$$

and

$$\lim [f_1(x) f_2(x) \cdots f_n(x)] = \lim f_1(x) \lim f_2(x) \cdots \lim f_n(x) \tag{1}$$

In particular, if $f_1, f_2, \ldots, f_n$ are all the same function f, then (1) reduces to

$$\lim [f(x)]^n = [\lim f(x)]^n \tag{2}$$

From (2) we obtain the useful result

$$\lim_{x \to a} x^n = [\lim_{x \to a} x]^n = a^n \tag{3}$$

For example,

$$\lim_{x \to 3} x^4 = 3^4 = 81$$

Another useful result follows from Example 1 and part (c) of Theorem 2.5.1. If k is a constant, then

$$\lim kf(x) = \lim k \lim f(x) = k \lim f(x) \tag{4}$$

The two end expressions in (4) tell us:

A constant factor can be moved through a limit sign.

▶ Example 5

Find

$$\lim_{x \to 5} (x^2 - 4x + 3)$$

Justify each step.

Solution.

$$\lim_{x \to 5} (x^2 - 4x + 3) = \lim_{x \to 5} x^2 - \lim_{x \to 5} 4x + \lim_{x \to 5} 3 \qquad \text{[Theorem 2.5.1}(a), (b)\text{]}$$

$$= \lim_{x \to 5} x^2 - 4 \lim_{x \to 5} x + \lim_{x \to 5} 3 \qquad \text{[Equation (4)]}$$

$$= 5^2 - 4(5) + 3 \qquad \text{[Equation (3); Example 1]}$$

$$= 8 \qquad \qquad \blacktriangleleft$$

▶ Example 6

Show that for any polynomial

$$p(x) = c_0 + c_1 x + \cdots + c_n x^n$$

and any real number a,

$$\lim_{x \to a} p(x) = c_0 + c_1 a + \cdots + c_n a^n = p(a)$$

Solution.

$$\lim_{x \to a} p(x) = \lim_{x \to a} (c_0 + c_1 x + \cdots + c_n x^n)$$

$$= \lim_{x \to a} c_0 + \lim_{x \to a} c_1 x + \cdots + \lim_{x \to a} c_n x^n$$

$$= \lim_{x \to a} c_0 + c_1 \lim_{x \to a} x + \cdots + c_n \lim_{x \to a} x^n$$

$$= c_0 + c_1 a + \cdots + c_n a^n$$

$$= p(a) \qquad \qquad \blacktriangleleft$$

REMARK. Example 6 states that for a polynomial $p(x)$, the limit as x approaches a is equal to the value of the polynomial at a. If we apply this result to the limit problem in Example 5, we can bypass the intermediate steps and write immediately

$$\lim_{x \to 5} (x^2 - 4x + 3) = 5^2 - 4(5) + 3 = 8$$

▶ Example 7

Find

$$\lim_{x \to 2} \frac{5x^3 + 4}{x - 3}$$

Solution.   Using the result of Example 6 and part (*d*) of Theorem 2.5.1 we obtain

$$\lim_{x \to 2} \frac{5x^3 + 4}{x - 3} = \frac{\lim_{x \to 2} (5x^3 + 4)}{\lim_{x \to 2} (x - 3)} = \frac{5 \cdot 2^3 + 4}{2 - 3} = -44 \qquad \blacktriangleleft$$

▶ Example 8

Find

$$\lim_{x \to 2} \frac{x^2 - 4}{x - 2}$$

Solution.   The method used in the previous example cannot be used here since part (*d*) of Theorem 2.5.1 does not apply (the limit of the denominator is 0). However, by factoring the numerator and canceling, we obtain

$$\lim_{x \to 2} \frac{x^2 - 4}{x - 2} = \lim_{x \to 2} \frac{(x - 2)(x + 2)}{x - 2}$$
$$= \lim_{x \to 2} (x + 2)$$
$$= 4 \qquad \blacktriangleleft$$

Though correct, the second equality in this calculation needs some justification. We pointed out in Example 12 of Section 2.1 that to cancel the $x - 2$ factors from

$$h(x) = \frac{x^2 - 4}{x - 2} = \frac{(x - 2)(x + 2)}{(x - 2)}$$

we must write

$$h(x) = x + 2, \qquad x \neq 2$$

in order not to alter the domain of h. However, the functions

$$f(x) = x + 2$$

and

$$h(x) = x + 2, \qquad x \neq 2$$

differ only at $x = 2$, so that their limits as x approaches 2 are the same. Thus

$$\lim_{x \to 2} \frac{(x - 2)(x + 2)}{x - 2} = \lim_{x \to 2} (x + 2)$$

even though

$$\frac{(x - 2)(x + 2)}{x - 2} \qquad \text{and} \qquad x + 2$$

are different functions.

▶ Example 9

Find

$$\lim_{x \to 4} \frac{2 - x}{(x - 4)(x + 2)}$$

Solution. Again, part (*d*) of Theorem 2.5.1 cannot be applied since the limit of the denominator is zero. There are no common factors in the numerator and denominator as there were in Example 8; we must use a different approach. This kind of problem is best attacked by looking at the one-sided limits.

First consider

$$\lim_{x \to 4^+} \frac{2 - x}{(x - 4)(x + 2)}$$

As x approaches 4 from the right, the numerator is a negative quantity approaching -2 and the denominator a positive quantity approaching 0. Consequently the ratio is a negative quantity that decreases without bound. That is,

$$\lim_{x \to 4^+} \frac{2 - x}{(x - 4)(x + 2)} = -\infty \tag{5}$$

Now consider

$$\lim_{x \to 4^-} \frac{2 - x}{(x - 4)(x + 2)}$$

As x approaches 4 from the left, the numerator is eventually a negative quantity approaching -2 and the denominator is a negative quantity approaching 0. Consequently the ratio is a positive quantity that increases without bound; that is,

$$\lim_{x\to 4^-} \frac{2-x}{(x-4)(x+2)} = +\infty \tag{6}$$

From (5) and (6)

$$\lim_{x\to 4} \frac{2-x}{(x-4)(x+2)}$$

does not exist.   ◀

Next we will consider methods for finding limits of rational functions as $x \to +\infty$ or $x \to -\infty$. But first we need some preliminary results.

In Example 4 we obtained the limits

$$\lim_{x\to +\infty} \frac{1}{x} = \lim_{x\to -\infty} \frac{1}{x} = 0$$

Consequently, for any positive integer n,

$$\lim_{x\to +\infty} \frac{1}{x^n} = \left(\lim_{x\to +\infty} \frac{1}{x}\right)^n = 0 \tag{7}$$

and

$$\lim_{x\to -\infty} \frac{1}{x^n} = \left(\lim_{x\to -\infty} \frac{1}{x}\right)^n = 0 \tag{8}$$

Results (7) and (8) are also evident from the graph of $1/x^n$, which has one of the two general shapes shown in Fig. 2.5.2, depending on whether n is even or odd.

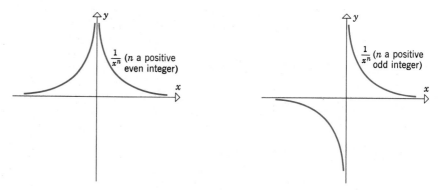

Figure 2.5.2

If we divide numerator and denominator of a rational function by the highest power of x that occurs in the function, then all the powers of x become constants or powers of $1/x$. As the next two examples show, this idea together with (7) and (8) can be used to find $\lim_{x \to +\infty}$ or $\lim_{x \to -\infty}$ for rational functions.

▶ Example 10

Find

$$\lim_{x \to +\infty} \frac{3x + 5}{6x - 8}$$

Solution. Divide the numerator and denominator by the highest power of x that occurs; this is $x^1 = x$. We obtain

$$\lim_{x \to +\infty} \frac{3x + 5}{6x - 8} = \lim_{x \to +\infty} \frac{3 + 5/x}{6 - 8/x} = \frac{\lim_{x \to +\infty} (3 + 5/x)}{\lim_{x \to +\infty} (6 - 8/x)}$$

$$= \frac{\lim_{x \to +\infty} 3 + \lim_{x \to +\infty} 5/x}{\lim_{x \to +\infty} 6 - \lim_{x \to +\infty} 8/x} = \frac{3 + 5 \lim_{x \to +\infty} 1/x}{6 - 8 \lim_{x \to +\infty} 1/x}$$

$$= \frac{3 + 5 \cdot 0}{6 - 8 \cdot 0} = \frac{1}{2} \qquad ◀$$

▶ Example 11

Find

$$\lim_{x \to -\infty} \frac{4x^2 - x}{2x^3 - 5}$$

Solution. Divide numerator and denominator by the highest power of x that occurs, namely x^3. We obtain

$$\lim_{x \to -\infty} \frac{4x^2 - x}{2x^3 - 5} = \lim_{x \to -\infty} \frac{4/x - 1/x^2}{2 - 5/x^3} = \frac{\lim_{x \to -\infty} (4/x - 1/x^2)}{\lim_{x \to -\infty} (2 - 5/x^3)}$$

$$= \frac{\lim_{x \to -\infty} 4/x - \lim_{x \to -\infty} 1/x^2}{\lim_{x \to -\infty} 2 - \lim_{x \to -\infty} 5/x^3} = \frac{4 \lim_{x \to -\infty} 1/x - \lim_{x \to -\infty} 1/x^2}{2 - 5 \lim_{x \to -\infty} 1/x^3}$$

$$= \frac{4 \cdot 0 - 0}{2 - 5 \cdot 0} = \frac{0}{2} = 0 \qquad ◀$$

► Example 12

Find

$$\lim_{x \to +\infty} \sqrt[3]{\frac{3x + 5}{6x - 8}}$$

Solution.

$$\lim_{x \to +\infty} \sqrt[3]{\frac{3x + 5}{6x - 8}} = \sqrt[3]{\lim_{x \to +\infty} \frac{3x + 5}{6x - 8}} \qquad \text{[Theorem 2.5.1, part (e)]}$$

$$= \sqrt[3]{\frac{1}{2}} \qquad \text{[Example 10]} \qquad \blacktriangleleft$$

► Example 13

Find

$$\lim_{x \to +\infty} \frac{3 - 2x^4}{x + 1}$$

Solution. Dividing numerator and denominator by x^4, we obtain

$$\lim_{x \to +\infty} \frac{3 - 2x^4}{x + 1} = \lim_{x \to +\infty} \frac{3/x^4 - 2}{1/x^3 + 1/x^4} \qquad (9)$$

Since the limit of the denominator in the right side of (9) is zero, we cannot apply Theorem 2.5.1. However, we can argue as follows. In the numerator, $3/x^4$ approaches 0 as $x \to +\infty$ so that

$$\frac{3}{x^4} - 2$$

is eventually negative and approaches -2. As $x \to +\infty$, the denominator is a positive quantity approaching 0. Consequently, as $x \to +\infty$, the ratio is eventually a negative quantity that decreases without bound. That is,

$$\lim_{x \to +\infty} \frac{3 - 2x^4}{x + 1} = \lim_{x \to +\infty} \frac{3/x^4 - 2}{1/x^3 + 1/x^4} = -\infty \qquad \blacktriangleleft$$

▶ Example 14

Find

$$\lim_{x \to +\infty} \frac{\sqrt{x^2 + 2}}{3x - 6}$$

Solution. Since we are letting $x \to +\infty$, it would be helpful to manipulate the function so that the powers of x become powers of $1/x$. To achieve this, we divide numerator and denominator by x. In the numerator, we write x as

$$x = \sqrt{x^2}$$

This is justified because x eventually assumes positive values as $x \to +\infty$, so that

$$\sqrt{x^2} = |x| = x$$

Thus

$$
\begin{aligned}
\lim_{x \to +\infty} \frac{\sqrt{x^2 + 2}}{3x - 6} &= \lim_{x \to +\infty} \frac{\sqrt{x^2 + 2}/\sqrt{x^2}}{(3x - 6)/x} \\
&= \lim_{x \to +\infty} \frac{\sqrt{1 + 2/x^2}}{3 - 6/x} \\
&= \frac{\lim_{x \to +\infty} \sqrt{1 + 2/x^2}}{\lim_{x \to +\infty} (3 - 6/x)} \\
&= \frac{\sqrt{\lim_{x \to +\infty} (1 + 2/x^2)}}{\lim_{x \to +\infty} (3 - 6/x)} \\
&= \frac{\sqrt{\lim_{x \to +\infty} 1 + 2 \lim_{x \to +\infty} 1/x^2}}{\lim_{x \to +\infty} 3 - 6 \lim_{x \to +\infty} 1/x} \\
&= \frac{\sqrt{1 + 2 \cdot 0}}{3 - 6 \cdot 0} \\
&= \frac{1}{3}
\end{aligned}
$$

◀

▶ Example 15

Find

$$\lim_{x \to -\infty} \frac{\sqrt{x^2 + 2}}{3x - 6}$$

Solution. The function here is the same as in Example 14. However, in this problem we want $\lim_{x \to -\infty}$ rather than $\lim_{x \to +\infty}$. As in Example 14, we divide numerator and denominator by x, but in the numerator we write x as

$$x = -\sqrt{x^2}$$

This is justified because x eventually assumes negative values as $x \to -\infty$, so that

$$\sqrt{x^2} = |x| = -x$$

We obtain

$$
\begin{aligned}
\lim_{x \to -\infty} \frac{\sqrt{x^2 + 2}}{3x - 6} &= \lim_{x \to -\infty} \frac{\sqrt{x^2 + 2}/(-\sqrt{x^2})}{3 - (6/x)} \\
&= \lim_{x \to -\infty} \frac{-\sqrt{1 + 2/x^2}}{3 - (6/x)} \\
&= -\frac{1}{3}
\end{aligned}
$$
◀

▶ Example 16

Find $\lim_{x \to 3} f(x)$ for

$$
f(x) = \begin{cases} x^2 - 5, & x \le 3 \\ \sqrt{x + 13}, & x > 3 \end{cases}
$$

Solution. As x approaches 3 from the left, the formula for f is

$$f(x) = x^2 - 5$$

so that

$$\lim_{x \to 3^-} f(x) = \lim_{x \to 3^-} (x^2 - 5) = 3^2 - 5 = 4$$

As x approaches 3 from the right, the formula for f is

$$f(x) = \sqrt{x + 13}$$

so that

$$\lim_{x \to 3^+} f(x) = \lim_{x \to 3^+} \sqrt{x + 13} = \sqrt{\lim_{x \to 3^+} (x + 13)} = \sqrt{16} = 4$$

Therefore

$$\lim_{x \to 3} f(x) = 4$$

since the one-sided limits are equal.   ◀

▶ Exercise Set 2.5

Find the limits in Exercises 1–48.

1. $\lim_{x \to 8} 7$.

2. $\lim_{x \to -\infty} (-3)$.

3. $\lim_{x \to 0^+} \pi$.

4. $\lim_{x \to -2} 3x$.

5. $\lim_{y \to 3^+} 12y$.

6. $\lim_{h \to +\infty} (-2h)$.

7. $\lim_{x \to 5} \sqrt{x^3 - 3x - 1}$.

8. $\lim_{x \to 0^-} (x^4 + 12x^3 - 17x + 2)$.

9. $\lim_{y \to -1} (y^6 - 12y + 1)$.

10. $\lim_{x \to 3} \dfrac{x^2 - 2x}{x + 1}$.

11. $\lim_{y \to 2^-} \dfrac{(y - 1)(y - 2)}{y + 1}$.

12. $\lim_{x \to 0} \dfrac{6x - 9}{x^3 - 12x + 3}$.

13. $\lim_{x \to 4} \dfrac{x^2 - 16}{x - 4}$.

14. $\lim_{t \to -2} \dfrac{t^3 + 8}{t + 2}$.

15. $\lim_{x \to 1^+} \dfrac{x^4 - 1}{x - 1}$.

16. $\lim_{x \to 2} \dfrac{x^2 - 4x + 4}{x^2 + x - 6}$.

17. $\lim_{x \to -1} \dfrac{x^2 + 6x + 5}{x^2 - 3x - 4}$.

18. $\lim_{t \to 1} \dfrac{t^3 + t^2 - 5t + 3}{t^3 - 3t + 2}$.

19. $\lim_{x \to +\infty} \dfrac{3x + 1}{2x - 5}$.

20. $\lim_{x \to +\infty} \dfrac{1}{x - 12}$.

21. $\lim_{y \to -\infty} \dfrac{3}{y + 4}$.

22. $\lim_{x \to +\infty} \dfrac{5x^2 + 7}{3x^2 - x}$.

23. $\lim_{x \to -\infty} \dfrac{x - 2}{x^2 + 2x + 1}$.

24. $\lim_{s \to +\infty} \sqrt[3]{\dfrac{3s^7 - 4s^5}{2s^7 + 1}}$.

25. $\lim_{x \to -\infty} \dfrac{\sqrt{5x^2 - 2}}{x + 3}$.

26. $\lim_{x \to +\infty} \dfrac{\sqrt{5x^2 - 2}}{x + 3}$.

27. $\lim_{y \to -\infty} \dfrac{2 - y}{\sqrt{7 + 6y^2}}$.

28. $\lim_{y \to +\infty} \dfrac{2 - y}{\sqrt{7 + 6y^2}}$.

29. $\lim_{x \to -\infty} \dfrac{\sqrt{3x^4 + x}}{x^2 - 8}$.

30. $\lim_{x \to +\infty} \dfrac{\sqrt{3x^4 + x}}{x^2 - 8}$.

31. $\lim_{x \to 3^+} \dfrac{x}{x - 3}$.

32. $\lim_{x \to 3^-} \dfrac{x}{x - 3}$.

33. $\lim_{x \to 3} \dfrac{x}{x - 3}$.

34. $\lim_{x \to 2^+} \dfrac{x}{x^2 - 4}$.

35. $\lim_{x \to 2^-} \dfrac{x}{x^2 - 4}$.

36. $\lim_{x \to 2} \dfrac{x}{x^2 - 4}$.

37. $\lim_{y \to 6^+} \dfrac{y + 6}{y^2 - 36}$.

38. $\lim_{y \to 6^-} \dfrac{y + 6}{y^2 - 36}$.

39. $\lim_{y \to 6} \dfrac{y + 6}{y^2 - 36}$.

40. $\lim_{x \to 4^+} \dfrac{3 - x}{x^2 - 2x - 8}$.

41. $\lim_{x \to 4^-} \dfrac{3 - x}{x^2 - 2x - 8}$.

42. $\lim_{x \to 4} \dfrac{3 - x}{x^2 - 2x - 8}$.

43. $\lim_{x \to +\infty} \dfrac{7 - 6x^5}{x + 3}$.

44. $\lim_{t \to -\infty} \dfrac{5 - 2t^3}{t^2 + 1}$.

45. $\lim_{t \to +\infty} \dfrac{6 - t^3}{7t^3 + 3}$.

46. $\lim_{x \to 0^+} \dfrac{x}{|x|}$.

47. $\lim_{x \to 0^-} \dfrac{x}{|x|}$.

48. $\lim_{x \to 3^-} \dfrac{1}{|x - 3|}$.

49. Let
$$f(x) = \begin{cases} x - 1, & x \le 3 \\ 3x - 7, & x > 3 \end{cases}$$

Find:

(a) $\lim\limits_{x \to 3^-} f(x)$ (b) $\lim\limits_{x \to 3^+} f(x)$ (c) $\lim\limits_{x \to 3} f(x)$.

50. Let
$$g(t) = \begin{cases} t^2, & t \ge 0 \\ t - 2, & t < 0 \end{cases}$$

Find:

(a) $\lim\limits_{t \to 0^-} g(t)$ (b) $\lim\limits_{t \to 0^+} g(t)$ (c) $\lim\limits_{t \to 0} g(t)$.

51. Let
$$h(x) = \begin{cases} x^2 - 2x + 1, & x \ne 3 \\ 7, & x = 3 \end{cases}$$

Find $\lim\limits_{x \to 3} h(x)$.

52. Let
$$F(x) = \begin{cases} \dfrac{x^2 - 9}{x + 3}, & x \ne -3 \\ k, & x = -3 \end{cases}$$

(a) Find k so that $F(-3) = \lim\limits_{x \to -3} F(x)$.

(b) With k assigned the value $\lim\limits_{x \to -3} F(x)$, show that $F(x)$ can be expressed as a polynomial.

53. (a) Explain why the following calculation is incorrect.
$$\lim_{x \to 0^+} \left(\frac{1}{x} - \frac{1}{x^2} \right) = \lim_{x \to 0^+} \frac{1}{x} - \lim_{x \to 0^+} \frac{1}{x^2}$$
$$= +\infty - (+\infty) = 0$$

(b) Show that $\lim\limits_{x \to 0^+} \left(\dfrac{1}{x} - \dfrac{1}{x^2} \right) = -\infty$.

54. Find

(a) $\lim\limits_{x \to 0} \dfrac{\sqrt{x + 3} - \sqrt{3}}{x}$

(b) $\lim\limits_{x \to +\infty} \dfrac{\sin x}{x}$

(c) $\lim\limits_{x \to 0} x^2 \sin \dfrac{1}{x}$

55. Let $r(x)$ be a rational function. Under what conditions is it true that
$$\lim_{x \to a} r(x) = r(a)?$$

56. Find
$$\lim_{x \to +\infty} \frac{c_0 + c_1 x + \cdots + c_n x^n}{d_0 + d_1 x + \cdots + d_m x^m}$$

where $c_n \ne 0$ and $d_m \ne 0$.
[*Hint:* Your answer will depend on whether $m < n$, $m = n$, or $m > n$.]

2.6 LIMITS: A RIGOROUS APPROACH (OPTIONAL)

This section gives a rigorous treatment of two-sided limits. Readers interested in pursuing the theory of one-sided and infinite limits can turn to Appendix 2, where we give a rigorous treatment of these topics and also prove some of the basic limit theorems.

Let a and L be real numbers. When we write
$$\lim_{x \to a} f(x) = L \tag{1}$$

we mean intuitively that the value of $f(x)$ approaches L as x approaches a from either side. To capture this concept mathematically, we must clarify the meanings of the phrases "x approaches a from either side" and "$f(x)$ approaches L."

The limit in (1) is intended to describe the behavior of f when x is near but *different from a*. The value of f at a is irrelevant to the limit. In fact, we have seen examples like

$$\lim_{x \to 0} \frac{\sin x}{x} = 1$$

where the function f is not even defined at the point a. Thus, when we say "x approaches a from either side," we mean that x assumes values arbitrarily close to a, on one side or the other, but does not actually take on the value a. On the other hand, when we say that "$f(x)$ approaches L," we want to allow the possibility that $f(x)$ may actually take on the value L.

Let us see if we can phrase these ideas more precisely. To say that $f(x)$ becomes arbitrarily close to L as x approaches a means that if we pick any positive number ϵ (no matter how small), eventually the difference between $f(x)$ and L will be less than ϵ as x approaches a. That is, eventually $f(x)$ will satisfy the condition

$$|f(x) - L| < \epsilon$$

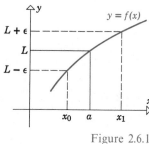

Figure 2.6.1

Figure 2.6.1 illustrates this idea. For the curve in that figure we have

$$\lim_{x \to a} f(x) = L$$

where a and L are the points shown. We picked an arbitrary positive number ϵ and indicated the interval

$$(L - \epsilon, L + \epsilon)$$

on the y-axis. As x approaches a, x will eventually lie between the points x_0 and x_1; when this occurs, $f(x)$ will lie in the interval $(L - \epsilon, L + \epsilon)$, that is, $f(x)$ will satisfy

$$L - \epsilon < f(x) < L + \epsilon$$

or equivalently

$$|f(x) - L| < \epsilon \tag{2}$$

Figure 2.6.2

(Figure 2.6.2). It is evident geometrically that no matter how small we make ϵ, eventually $f(x)$ will satisfy (2) as x approaches a.

For the curve in Figure 2.6.1, we have

$$f(a) = \lim_{x \to a} f(x) = L$$

Figure 2.6.3

But the equality of $f(a)$ and L is irrelevant to the limit discussion. As x approaches a, x is not allowed to take on the value a. Thus, if $f(a)$ is undefined as in Figure 2.6.3 or if $f(a)$ is defined, but unequal to L, as in Figure 2.6.4, it is still true that for an arbitrary positive value of ϵ, $f(x)$ will eventually satisfy (2) as x approaches a. In brief, the condition $|f(x) - L| < \epsilon$ need only be satisfied for x in the set $(x_0, a) \cup (a, x_1)$.

Before we can give a precise definition of limit, we need a preliminary observation. If the condition

Figure 2.6.4

$$|f(x) - L| < \epsilon \tag{3}$$

holds for all x in the set

$$(x_0, a) \cup (a, x_1) \tag{4}$$

then this condition holds as well for all x in any *subset* of (4). In particular, if we let δ be any positive number that is smaller than both

$$a - x_0 \qquad \text{and} \qquad x_1 - a$$

then

$$(a - \delta, a) \cup (a, a + \delta) \tag{5}$$

Figure 2.6.5

will be a subset of (4) (Figure 2.6.5) and condition (3) will hold for all x in this subset.

Since the set in (5) consists of all x satisfying

$$0 < |x - a| < \delta$$

we are led to the following definition.

2.6.1 DEFINITION   Let $f(x)$ be defined for all x in some open interval containing the number a, with the possible exception that $f(x)$ may or may not be defined at a. We will write

$$\lim_{x \to a} f(x) = L$$

if given any number $\epsilon > 0$, we can find a number $\delta > 0$ such that $f(x)$ satisfies

$$|f(x) - L| < \epsilon \qquad \text{whenever } x \text{ satisfies} \qquad 0 < |x - a| < \delta$$

▶ Example 1

Use Definition 2.6.1 to prove

$$\lim_{x \to 2} (3x - 5) = 1$$

Solution. We must show that given any positive number ϵ, we can find a positive number δ such that $f(x) = 3x - 5$ satisfies

$$|(3x - 5) - 1| < \epsilon \tag{6}$$

whenever x satisfies

$$0 < |x - 2| < \delta \tag{7}$$

To find δ, we can rewrite (6) as

$$|3x - 6| < \epsilon$$

or equivalently

$$3|x - 2| < \epsilon$$

or

$$|x - 2| < \frac{\epsilon}{3} \tag{8}$$

We must choose δ so that (8) is satisfied whenever (7) is satisfied. We can do this by taking

$$\delta = \frac{\epsilon}{3}$$

To prove that this choice for δ works, assume x satisfies (7). Since we are letting $\delta = \epsilon/3$, it follows from (7) that x satisfies

$$0 < |x - 2| < \frac{\epsilon}{3} \tag{9}$$

Thus (8) is satisfied since it is just the right-hand inequality in (9). This proves that $\lim_{x \to 2} (3x - 5) = 1$. ◄

As illustrated in Example 1, a limit proof proceeds as follows: We *assume* that an arbitrary positive number ϵ is given to us. Then we try to *find* a positive number δ such that the conditions in the definition are fulfilled. The idea is to find a formula for δ in terms of ϵ so that no matter what ϵ is chosen, we automatically obtain from the formula the required δ. (In Example 1, the formula was $\delta = \epsilon/3$.)

Example 1 is about as easy as a limit proof can get; most limit proofs require a little more algebraic and logical ingenuity. The reader who finds "δ-ϵ" discussions hard going should not become discouraged. The concepts and techniques are intrinsically difficult. In fact, a precise understanding of limits evaded the finest mathematical minds for centuries.

Note that the value of δ in Definition 2.6.1 is not unique. Once one value of δ is found that fulfills the requirements of the definition, any *smaller* positive value for δ will also fulfill these requirements. To see why, assume we have found a value of δ such that

$$|f(x) - L| < \epsilon \tag{10}$$

whenever x satisfies

$$0 < |x - a| < \delta \tag{11}$$

Let δ_1 be any positive number smaller than δ. If x satisfies

$$0 < |x - a| < \delta_1 \tag{12}$$

then we have

$$0 < |x - a| < \delta_1 < \delta$$

Thus x satisfies (11) and consequently $f(x)$ satisfies (10). Therefore δ_1 meets the requirements of Definition 2.6.1 also. For example, we found in Example 1 that $\delta = \epsilon/3$ fulfills the requirements of Definition 2.6.1. Consequently any smaller value for δ, such as $\delta = \epsilon/4$, $\delta = \epsilon/5$ or $\delta = \epsilon/6$ does also.

► Example 2

Prove that

$$\lim_{x \to 3} x^2 = 9$$

Solution. We must show that given any positive number ϵ we can find a positive number δ such that

$$|x^2 - 9| < \epsilon \tag{13}$$

whenever x satisfies

$$0 < |x - 3| < \delta \tag{14}$$

Because $|x - 3|$ occurs in (14), it will be helpful to rewrite (13) so that $|x - 3|$ appears as a factor on the left side. Therefore, we will rewrite (13) as

$$|x + 3|\,|x - 3| < \epsilon \tag{15}$$

If we can somehow assure that when x satisfies (14) the factor $|x + 3|$ remains less than some positive constant, say

$$|x + 3| < k \tag{16}$$

then on choosing

$$\delta = \frac{\epsilon}{k} \tag{17}$$

it will follow from (14) that

$$0 < |x - 3| < \frac{\epsilon}{k}$$

or

$$0 < k|x - 3| < \epsilon \tag{18}$$

From (16) and the right-hand inequality in (18) we will then have

$$|x + 3|\,|x - 3| < k|x - 3| < \epsilon$$

so that (15) will be satisfied, and the proof will be complete. As we shall now explain, condition (16) can be obtained by restricting the size of δ. We remarked in the discussion preceding this example that δ is not unique; once a value of δ is found, any *smaller* positive value for δ can also be used. At this point we do not have a formula relating δ and ϵ. However, let us agree in advance that if, for a given ϵ, δ turns out to exceed 1 we will choose the smaller value $\delta = 1$ instead. Thus we can assume in our discussion that δ satisfies

$$0 < \delta \leq 1 \qquad\qquad\qquad (19)$$

Assume that x satisfies (14). From (19) and the right side of (14) we obtain

$$|x - 3| < \delta \leq 1$$

so that

$$|x - 3| < 1$$

or equivalently

$$2 < x < 4$$

so

$$5 < x + 3 < 7$$

Therefore

$$|x + 3| < 7$$

Comparing the last inequality to (16) suggests $k = 7$; and from (17)

$$\delta = \frac{\epsilon}{k} = \frac{\epsilon}{7}$$

In summary, given $\epsilon > 0$ we choose

$$\delta = \frac{\epsilon}{7}$$

provided $\epsilon/7$ does not exceed 1. If $\epsilon/7$ exceeds 1 we choose

$$\delta = 1$$

In other words, we take δ to be the minimum of the numbers $\epsilon/7$ and 1. This is sometimes written

$$\delta = \min\left(\frac{\epsilon}{7}, 1\right) \qquad\qquad \blacktriangleleft$$

In this last example the reader may have wondered how we knew to make the restriction $\delta \leq 1$ as opposed to some other restriction like $\delta \leq \frac{1}{2}$ or $\delta \leq 5$. Actually our selection was completely arbitrary; any other restriction of the form $\delta \leq c$ would have worked equally well (Exercise 27).

▶ Example 3

Prove that

$$\lim_{x \to 1/2} \frac{1}{x} = 2$$

Solution. We must show that given $\epsilon > 0$ there exists a $\delta > 0$ such that

$$\left| \frac{1}{x} - 2 \right| < \epsilon \tag{20}$$

whenever x satisfies

$$0 < \left| x - \frac{1}{2} \right| < \delta \tag{21}$$

Because $|x - \frac{1}{2}|$ occurs in (21), it will be helpful to rewrite (20) so that $|x - \frac{1}{2}|$ appears as a factor on the left side. We do this as follows:

$$\left| \frac{1}{x} - 2 \right| = \left| \frac{2}{x} \left(\frac{1}{2} - x \right) \right|$$

$$= \left| \frac{2}{x} \right| \left| \frac{1}{2} - x \right| = \left| \frac{2}{x} \right| \left| x - \frac{1}{2} \right|$$

Thus (20) is equivalent to

$$\left| \frac{2}{x} \right| \left| x - \frac{1}{2} \right| < \epsilon \tag{22}$$

From here on, the procedure is similar to that used in the last example. By restricting δ we will try to make the factor $|2/x|$ satisfy

$$\left| \frac{2}{x} \right| < k \tag{23}$$

for some constant k when x satisfies (21). Then on choosing

$$\delta = \frac{\epsilon}{k} \tag{24}$$

we will have, from (21),

$$0 < \left| x - \frac{1}{2} \right| < \frac{\epsilon}{k}$$

or

$$0 < k\left|x - \frac{1}{2}\right| < \epsilon \qquad (25)$$

From (23) and the right-hand inequality in (25) it will then follow that

$$\left|\frac{2}{x}\right|\left|x - \frac{1}{2}\right| < k\left|x - \frac{1}{2}\right| < \epsilon$$

so that (22) will hold and the proof will be complete. Therefore, to finish the proof, we must show that δ can be restricted so that (23) holds for some k when x satisfies (21).

In Figure 2.6.6 we have sketched the graph of $|2/x|$ and marked the set of x values satisfying (21). In part (a) of the figure we took $\delta = \frac{1}{2}$ and in part (b) of the figure we took $\delta < \frac{1}{2}$.

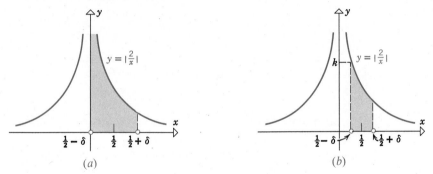

Figure 2.6.6

(a)   (b)

Part (a) of this figure makes it clear that if $\delta = \frac{1}{2}$ (or $\delta > \frac{1}{2}$) then the values of $|2/x|$ will have no upper bound, making it impossible to satisfy (23). However, part (b) shows that when $\delta < \frac{1}{2}$, there is an upper bound k to the values of $|2/x|$. Therefore, the arbitrary restriction

$$\delta \leq \frac{1}{4} \qquad (26)$$

will be satisfactory to assure that $|2/x|$ has an upper bound.

Assume that x satisfies (21). From (26) and the right side of (21), we obtain

$$\left|x - \frac{1}{2}\right| < \delta \leq \frac{1}{4}$$

so that

$$\left|x - \frac{1}{2}\right| < \frac{1}{4}$$

or equivalently

$$\frac{1}{4} < x < \frac{3}{4} \tag{27}$$

To estimate the size of $|2/x|$ we take reciprocals in (27) and multiply through by 2, which gives

$$8 > \frac{2}{x} > \frac{8}{3}$$

Thus

$$\left|\frac{2}{x}\right| < 8$$

Comparing the last inequality to (23) suggests $k = 8$; and from (24)

$$\delta = \frac{\epsilon}{k} = \frac{\epsilon}{8}$$

In summary, given $\epsilon > 0$ we can choose

$$\delta = \frac{\epsilon}{8}$$

provided this choice does not violate $\delta \le \frac{1}{4}$. If it does, we choose the value $\delta = \frac{1}{4}$ instead. In other words,

$$\delta = \min\left(\frac{\epsilon}{8}, \frac{1}{4}\right) \qquad \blacktriangleleft$$

▶ Example 4

Let

$$f(x) = \begin{cases} 1, & x > 0 \\ -1, & x < 0 \end{cases}$$

Prove that $\lim_{x \to 0} f(x)$ does not exist.

Solution. We will assume that there is a limit and obtain a contradiction. Assume there is a number L such that

$$\lim_{x \to 0} f(x) = L$$

Then given any $\epsilon > 0$ there exists a $\delta > 0$ such that

$$|f(x) - L| < \epsilon \qquad \text{whenever} \qquad 0 < |x - 0| < \delta$$

In particular, if we take $\epsilon = 1$, there is a $\delta > 0$ such that

$$|f(x) - L| < 1$$

whenever

$$0 < |x - 0| < \delta \tag{28}$$

But $x = \delta/2$ and $x = -\delta/2$ both satisfy (28) so that

$$\left| f\left(\frac{\delta}{2}\right) - L \right| < 1 \qquad \text{and} \qquad \left| f\left(-\frac{\delta}{2}\right) - L \right| < 1 \tag{29}$$

However $\delta/2$ is positive and $-\delta/2$ is negative, so

$$f\left(\frac{\delta}{2}\right) = 1 \qquad \text{and} \qquad f\left(-\frac{\delta}{2}\right) = -1$$

Thus (29) states that

$$|1 - L| < 1 \qquad \text{and} \qquad |-1 - L| < 1$$

or equivalently

$$0 < L < 2 \qquad \text{and} \qquad -2 < L < 0$$

But this is a contradiction, since no number L can satisfy these two conditions.   ◀

▶ Exercise Set 2.6

In Exercises 1–10, we are told that $\lim\limits_{x \to a} f(x) = L$ and we are given a value of ϵ. In each exercise, find a number δ such that $|f(x) - L| < \epsilon$ whenever $0 < |x - a| < \delta$.

1. $\lim\limits_{x \to 4} 2x = 8$; $\epsilon = 0.1$.

2. $\lim\limits_{x \to -2} \frac{1}{2}x = -1$; $\epsilon = 0.1$.

3. $\lim\limits_{x \to -1} (7x + 5) = -2$; $\epsilon = 0.01$.

4. $\lim\limits_{x \to 3} (5x - 2) = 13$; $\epsilon = 0.01$.

5. $\lim\limits_{x \to 2} \frac{x^2 - 4}{x - 2} = 4$; $\epsilon = 0.05$.

6. $\lim\limits_{x \to -1} \frac{x^2 - 1}{x + 1} = -2$; $\epsilon = 0.05$.

7. $\lim\limits_{x \to 4} x^2 = 16$; $\epsilon = 0.001$.

8. $\lim\limits_{x \to 9} \sqrt{x} = 3$; $\epsilon = 0.001$.

9. $\lim\limits_{x\to 5}\dfrac{1}{x}=\dfrac{1}{5}$; $\epsilon = 0.05$.

10. $\lim\limits_{x\to 0}|x| = 0$; $\epsilon = 0.05$.

In Exercises 11–22 use Definition 2.6.1 to prove that the given limit statement is correct.

11. $\lim\limits_{x\to 5} 3x = 15$.

12. $\lim\limits_{x\to 3}(4x - 5) = 7$.

13. $\lim\limits_{x\to 0}\dfrac{x^2 + x}{x} = 1$.

14. $\lim\limits_{x\to -3}\dfrac{x^2 - 9}{x + 3} = -6$.

15. $\lim\limits_{x\to 1} 2x^2 = 2$.

16. $\lim\limits_{x\to 3}(x^2 - 5) = 4$.

17. $\lim\limits_{x\to 1/3}\dfrac{1}{x} = 3$.

18. $\lim\limits_{x\to -2}\dfrac{1}{x + 1} = -1$.

19. $\lim\limits_{x\to 4}\sqrt{x} = 2$.

20. $\lim\limits_{x\to 6}\sqrt{x + 3} = 3$.

21. $\lim\limits_{x\to 1} f(x) = 3$, where $f(x) = \begin{cases} x + 2, & x \neq 1 \\ 10, & x = 1 \end{cases}$

22. $\lim\limits_{x\to 2}(x^2 + 3x - 1) = 9$.

23. Let $f(x) = \begin{cases} \dfrac{1}{8}, & x > 0 \\ -\dfrac{1}{8}, & x < 0 \end{cases}$

Use the method of Example 4 to prove that $\lim\limits_{x\to 0} f(x)$ does not exist.

24. Let $g(x) = \begin{cases} 1 + x, & x > 0 \\ x - 1, & x < 0 \end{cases}$. Prove that $\lim\limits_{x\to 0} g(x)$ does not exist.

25. Prove that $\lim\limits_{x\to 1}\dfrac{1}{x - 1}$ does not exist.

26. (a) In Definition 2.6.1 there is a condition requiring that $f(x)$ be defined for every x in an open interval containing a, except possibly at a itself. What is the purpose of this requirement?
    (b) Why is $\lim\limits_{x\to 0}\sqrt{x} = 0$ an incorrect statement?
    (c) Is $\lim\limits_{x\to 0.01}\sqrt{x} = 0.1$ a correct statement?

27. Prove the result in Example 2 under the assumption that $\delta \leq 2$ rather than $\delta \leq 1$.

▶ REVIEW EXERCISES

In Exercises 1–5, find the domain of f and then evaluate f (if defined) at the given values of x.

1. $f(x) = \sqrt{4 - x^2}$; $x = -\sqrt{2}, 0, \sqrt{3}$.

2. $f(x) = 1/\sqrt{(x - 1)^3}$; $x = 0, 1, 2$.

3. $f(x) = (x - 1)/(x^2 + x - 2)$; $x = 0, 1, 2$.

4. $f(x) = \sqrt{|x| - 2}$; $x = -3, 0, 2$.

5. $f(x) = \begin{cases} x^2 - 1, & x \leq 2 \\ \sqrt{x - 1}, & x > 2 \end{cases}$; $x = 0, 2, 4$.

In Exercises 6 and 7, find:

    (a) $f(x^2) - (f(x))^2$
    (b) $f(x + 3) - [f(x) + f(3)]$
    (c) $f(1/x) - 1/f(x)$
    (d) $(f \circ f)(x)$.

6. $f(x) = \sqrt{3 - x}$.

7. $f(x) = \dfrac{3 - x}{x}$.

In Exercises 8–15, sketch the graph of f and find its domain and range.

8. $f(x) = (x - 2)^2$.

9. $f(x) = -\pi$.

10. $f(x) = |2 - 4x|$.

11. $f(x) = \dfrac{x^2 - 4}{2x + 4}$.

12. $f(x) = \sqrt{-2x}$.

13. $f(x) = -\sqrt{3x + 1}$.

14. $f(x) = 2 - |x|$.

15. $f(x) = \dfrac{2x - 4}{x^2 - 4}$.

16. In each part, complete the square, and then find the range of f.
    (a) $f(x) = x^2 - 5x + 6$
    (b) $f(x) = -3x^2 + 12x - 7$.

17. Express $f(x)$ as a composite function $(g \circ h)(x)$ in two different ways.
    (a) $f(x) = x^6 + 3$
    (b) $f(x) = \sqrt{x^2 + 1}$
    (c) $f(x) = \sin(3x + 2)$.

In Exercises 18–20, find the equation of the tangent line at x_0.

18. $f(x) = x^2 - 6x;\ x_0 = -1$.

19. $f(x) = 3x^3 - 1;\ x_0 = 0$.

20. $f(x) = \sqrt{x};\ x_0 = \frac{1}{4}$.
    [*Hint:* $\sqrt{a} - \sqrt{b} = (a - b)/(\sqrt{a} + \sqrt{b})$]

In Exercises 21–22, $s = f(t)$ is the distance from the origin at time t of a particle moving along a coordinate line. Find the instantaneous velocity of the particle at time t_0.

21. $s(t) = 5t + 8;\ t_0 = 2$.

22. $s(t) = \dfrac{3}{(t + 1)^2};\ t_0 = 0$.

In Exercises 23–24, sketch the graph of f and find the indicated limits (if they exist).

23. $$f(x) = \begin{cases} 1/x, & x < 0 \\ x^2, & 0 \le x < 1 \\ 2, & x = 1 \\ 2 - x, & x > 1 \end{cases}$$
    (a) as $x \to -1$ (b) as $x \to 0$
    (c) as $x \to 1$ (d) as $x \to 0^+$
    (e) as $x \to 0^-$ (f) as $x \to 2^+$
    (g) as $x \to -\infty$ (h) as $x \to +\infty$.

24. $$f(x) = \begin{cases} 2, & x \le -1 \\ -x, & -1 < x < 0 \\ x/(2 - x), & 0 < x < 2 \\ 1, & x \ge 2 \end{cases}$$
    (a) as $x \to -1^+$ (b) as $x \to -1^-$
    (c) as $x \to -1$ (d) as $x \to 0$
    (e) as $x \to 2^+$ (f) as $x \to 2^-$
    (g) as $x \to 2$ (h) as $x \to -\infty$.

In Exercises 25–28, find $\lim\limits_{x \to a} f(x)$ (if it exists).

25. $f(x) = \sqrt{2 - x}$;
    $a = -2,\ 1,\ 2^-,\ 2^+,\ -\infty,\ +\infty$.

26. $f(x) = \begin{cases} (x - 2)/|x - 2|, & x \ne 2 \\ 0 & , \quad x = 2 \end{cases}$
    $a = 0,\ 2^-,\ 2^+,\ 2,\ -\infty,\ +\infty$.

27. $f(x) = (x^2 - 25)/(x - 5)$;
    $a = 0,\ 5^+,\ -5^-,\ 5,\ -5,\ -\infty,\ +\infty$.

28. $f(x) = (x + 5)/(x^2 - 25)$;
    $a = 0,\ 5^+,\ -5^-,\ -5,\ 5,\ -\infty,\ +\infty$.

3 differentiation

3.1 THE DERIVATIVE

Many physical phenomena involve changing quantities—the speed of a rocket, the inflation of currency, the number of bacteria in a culture, the shock intensity of an earthquake, the voltage of an electrical signal, and so forth. In this chapter we will develop the mathematical tools for expressing rates of change precisely. In the next chapter we will apply our results to some important physical problems.

Let us review some earlier ideas. Suppose $P(x_0, y_0)$ and $Q(x_1, y_1)$ are points on the graph of a function f. Then the secant line connecting P and Q has slope

$$m_{\text{sec}} = \frac{y_1 - y_0}{x_1 - x_0}$$

or since $y_0 = f(x_0)$ and $y_1 = f(x_1)$

$$m_{\text{sec}} = \frac{f(x_1) - f(x_0)}{x_1 - x_0} \tag{1}$$

If we let

$$h = x_1 - x_0$$

then

$$x_1 = x_0 + h$$

so (1) can be written

$$m_{\text{sec}} = \frac{f(x_0 + h) - f(x_0)}{h} \qquad \text{(See Figure 3.1.1.)}$$

Figure 3.1.1

As Q approaches P along the graph of f, x_1 approaches x_0 and consequently

$$h = x_1 - x_0$$

approaches 0. However, as Q approaches P, the secant line connecting P and Q approaches the tangent line at P. This leads us to the following definition.

3.1.1   DEFINITION   If $P(x_0, y_0)$ is a point on the graph of a function f, then the **tangent line** to the graph of f at P is defined to be the line through P having slope

$$m_{\text{tan}} = \lim_{h \to 0} \frac{f(x_0 + h) - f(x_0)}{h} \tag{2}$$

provided the limit exists.

Sometimes we will abbreviate our terminology and refer to the tangent line to $y = f(x)$ at $P(x_0, y_0)$ as the **tangent line at $x = x_0$.**
   The limit in Definition 3.1.1 is of fundamental importance.

3.1.2   DEFINITION   The **derivative** of a function f is the function f' defined by

$$f'(x) = \lim_{h \to 0} \frac{f(x + h) - f(x)}{h} \tag{3}$$

The domain of f' consists of all x where this limit exists.

REMARK.   The symbol $f'(x)$ is read "f prime of x."

   If x_0 lies in the domain of f', then we say that f is **differentiable at x_0.** It follows from (2) and (3) that if f is differentiable at x_0, the value of the derivative at x_0 is

$$f'(x_0) = \lim_{h \to 0} \frac{f(x_0 + h) - f(x_0)}{h} = m_{\text{tan}}$$

In other words, the derivative of f is a function whose value at $x = x_0$ is the slope of the tangent line to $y = f(x)$ at $x = x_0$. The domain of the derivative is the set of x values at which a tangent line to $y = f(x)$ exists.

▶ Example 1

Find the derivative of $f(x) = x^2 + 1$.

Solution. From Definition 3.1.2,

$$f'(x) = \lim_{h \to 0} \frac{f(x + h) - f(x)}{h}$$

$$= \lim_{h \to 0} \frac{[(x + h)^2 + 1] - [x^2 + 1]}{h}$$

$$= \lim_{h \to 0} \frac{x^2 + 2xh + h^2 + 1 - x^2 - 1}{h}$$

$$= \lim_{h \to 0} \frac{2xh + h^2}{h}$$

$$= \lim_{h \to 0} (2x + h)$$

$$= 2x$$
◀

▶ Example 2

In Example 1 we found the derivative of $f(x) = x^2 + 1$ to be $f'(x) = 2x$. Thus the slope of the tangent line to the graph f at the point $x = x_0$ is

$$f'(x_0) = 2x_0$$

For example, the slope of the tangent line to $y = x^2 + 1$ at $x = 2$ is

$$f'(2) = 2(2) = 4$$

The slope at $x = -2$ is

$$f'(-2) = 2(-2) = -4$$

and the slope at $x = 0$ is

$$f'(0) = 2(0) = 0 \qquad \text{(See Figure 3.1.2.)}$$
◀

Figure 3.1.2

▶ Example 3

Find $f'(x)$ if

$$f(x) = \frac{1}{x}$$

Solution. From Definition 3.1.2,

$$f'(x) = \lim_{h \to 0} \frac{f(x+h) - f(x)}{h}$$

$$= \lim_{h \to 0} \frac{\dfrac{1}{x+h} - \dfrac{1}{x}}{h}$$

$$= \lim_{h \to 0} \frac{x - (x+h)}{hx(x+h)}$$

$$= \lim_{h \to 0} \frac{-h}{hx(x+h)}$$

$$= \lim_{h \to 0} \frac{-1}{x(x+h)} = -\frac{1}{x^2} \qquad \blacktriangleleft$$

There are three common ways in which a function f can *fail* to be differentiable at a point. Loosely phrased, these may be classified as:

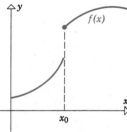

Figure 3.1.3

1. Breaks.
2. Corners.
3. Vertical tangents.

It is intuitively clear that if the graph of a function f has a "break" at $x = x_0$ (Figure 3.1.3) then the function cannot have a tangent at x_0. We shall prove this to be true after we make the term "break" more precise.

The graph of a function f has a "corner" at a point $P(x_0, f(x_0))$ if the graph of f is unbroken at P and the limiting position for the secant line joining P and Q depends on whether Q approaches P from the left or the right (Figure 3.1.4). At corners a tangent line does not exist since the slopes of the secant lines do not have a (two-sided) limit.

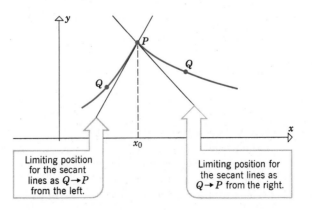

Limiting position for the secant lines as $Q \to P$ from the left.

Limiting position for the secant lines as $Q \to P$ from the right.

Figure 3.1.4

▶ Example 4

Show that

$$f(x) = |x|$$

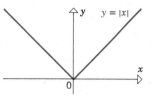

Figure 3.1.5

is not differentiable at $x = 0$.

Solution. This result is geometrically evident since the graph of $|x|$ has a corner at $x = 0$ (Figure 3.1.5). However, an analytic argument can be given as follows. From Definition 3.1.2

$$f'(0) = \lim_{h \to 0} \frac{f(0 + h) - f(0)}{h}$$

$$= \lim_{h \to 0} \frac{f(h) - f(0)}{h}$$

$$= \lim_{h \to 0} \frac{|h| - |0|}{h}$$

$$= \lim_{h \to 0} \frac{|h|}{h}$$

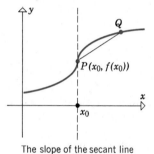

The slope of the secant line
tends toward $+\infty$ as
$Q \to P$

(a)

But

$$\frac{|h|}{h} = \begin{cases} 1, & h > 0 \\ -1, & h < 0 \end{cases}$$

so that

$$\lim_{h \to 0^-} \frac{|h|}{h} = -1 \quad \text{and} \quad \lim_{h \to 0^+} \frac{|h|}{h} = 1$$

Thus

$$f'(0) = \lim_{h \to 0} \frac{|h|}{h}$$

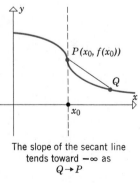

The slope of the secant line
tends toward $-\infty$ as
$Q \to P$

(b)

Figure 3.1.6

does not exist since the one-sided limits are not equal. Consequently $f(x) = |x|$ is not differentiable at $x = 0$. ◀

If the slope of the secant line joining P and Q tends toward $+\infty$ or $-\infty$ as Q approaches P along the graph of f, then f is not differentiable at x_0. Geometrically, such points occur where the secant lines tend toward a vertical limiting position (Figure 3.1.6).

The process of finding a derivative is called *differentiation.* It is often useful to think of differentiation as an operation which, when applied to a function

f, produces a new function f'. In the case where the independent variable is x the differentiation operation is often denoted by the symbol

$$\frac{d}{dx}[\quad]$$

which is read, ***the derivative with respect to x of***
For example,

$$\frac{d}{dx}[x^2 + 1]$$

stands for the derivative with respect to x of $x^2 + 1$
From Example 1,

$$\frac{d}{dx}[x^2 + 1] = 2x \qquad (4)$$

The d/dx notation allows us to write derivatives without using any function symbols like f or f'. This notation also makes it easy to write derivatives of functions involving variables other than x. For example, if we use u as the variable in (4) instead of x, we can write

$$\frac{d}{du}[u^2 + 1] = 2u$$

(The symbol d/du [ ] is read, "*the derivative with respect to u of*".)
With the d/dx notation it is awkward to denote the value of the derivative at a particular point; the prime notation is better for this purpose. As an illustration, we found in Example 2 that if

$$f(x) = x^2 + 1$$

then $f'(x_0) = 2x_0$. To express this in d/dx notation, we must write

$$\frac{d}{dx}[x^2 + 1]\bigg|_{x=x_0} = 2x_0$$

which is rather clumsy. In general

$$\frac{d}{dx}[\quad]\bigg|_{x=x_0}$$

denotes the value of the derivative with respect to x at $x = x_0$.

When we use a dependent variable

$$y = f(x)$$

we will usually denote the derivative as

$$\frac{dy}{dx}$$

and the value of the derivative at a point $x = x_0$ as

$$\frac{dy}{dx}\bigg|_{x=x_0}$$

For example, if $y = x^2 + 1$ then $\dfrac{dy}{dx} = 2x$ and $\dfrac{dy}{dx}\bigg|_{x=x_0} = 2x_0$

Later the symbols dy and dx will be defined separately. However, for the time being, dy/dx should not be regarded as a ratio; rather, it should be considered as a single symbol denoting the derivative.

REMARK.   Instead of writing

$$\frac{d}{dx}[\ \ ]$$

to denote the differentiation operation, some writers use

$$D_x[\ \ ]$$

We will not use this notation, however.

▶ Exercise Set 3.1

In Exercises 1–10, use Definition 3.1.2 to find $f'(x)$.

1. $f(x) = 3x^2$.
2. $f(x) = x^3$.
3. $f(x) = 5x - 2$.
4. $f(x) = mx + b$ (m, b constants).
5. $f(x) = \sqrt{x}$.
6. $f(x) = 4x^2 + x - 1$.
7. $f(x) = \dfrac{1}{x^2}$.
8. $f(x) = x^4$.

9. $f(x) = ax^2 + b$ (a, b constants).
10. $f(x) = \dfrac{1}{x + 1}$.

In Exercises 11–16, find $f'(a)$ and the equation of the tangent to the graph of f at the point where $x = a$.

11. f is the function in Exercise 1; $a = 3$.
12. f is the function in Exercise 2; $a = 2$.
13. f is the function in Exercise 3; $a = 0$.
14. f is the function in Exercise 4; $a = -1$.
15. f is the function in Exercise 5; $a = 9$.

16. f is the function in Exercise 6; $a = -2$.

17. Let $y = 4x^2 + 2$. Find

    (a) $\dfrac{dy}{dx}$ (b) $\dfrac{dy}{dx}\Big|_{x=1}$.

18. Let $y = \dfrac{5}{x} + 1$. Find

    (a) $\dfrac{dy}{dx}$ (b) $\dfrac{dy}{dx}\Big|_{x=-2}$.

In Exercises 19–26, sketch the graph of the derivative of the function whose graph is shown.

19.

20.

21.

22.

23.

24.

25.

26.

27. Show that $f(x) = \sqrt[3]{x - 2}$ is not differentiable at $x = 2$. Sketch the graph of f.

28. Show that

$$f(x) = \begin{cases} x^2 + 1, & x \le 1 \\ 2x & , & x > 1 \end{cases}$$

is differentiable at 1, but that

$$g(x) = \begin{cases} x^2 + 2, & x \le 1 \\ 3x & , & x > 1 \end{cases}$$

is not differentiable at 1. Sketch the graphs of f and g. [*Remark:* Geometrically, the results in this problem tell us that the graph of g has a "corner" at 1, but the graph of f does not.]

29. Use Definition 3.1.2 to find $f'(x)$ if

$$f(x) = \frac{1}{\sqrt{x}}$$

30. Use Definition 3.1.2 to find $f'(x)$ if

$$f(x) = x^{1/3}$$

3.2 TECHNIQUES OF DIFFERENTIATION

Up to now we have obtained derivatives directly from the definition. In this section we will develop some theorems and formulas that provide more efficient methods.

3.2.1 THEOREM *If f is a constant function, say $f(x) = c$ for all x, then $f'(x) = 0$.*

Proof.

$$f'(x) = \lim_{h \to 0} \frac{f(x + h) - f(x)}{h} = \lim_{h \to 0} \frac{c - c}{h}$$
$$= \lim_{h \to 0} 0 = 0 \quad \blacksquare$$

REMARK. The result in Theorem 3.2.1 is obvious geometrically. If $f(x) = c$ is a constant function, then the graph of f is a line parallel to the x-axis; consequently, the tangent at each point is a horizontal line. Therefore $f'(x) = m_{\tan} = 0$, since a horizontal line has slope 0.

In another notation, Theorem 3.2.1 states

$$\frac{d}{dx}[c] = 0$$

► Example 1

If $f(x) = 5$ for all x, then $f'(x) = 0$ for all x; that is,

$$\frac{d}{dx}[5] = 0$$

◄

3.2.2 THEOREM
The Power Rule

If n is a positive integer, then

$$\frac{d}{dx}[x^n] = nx^{n-1}$$

Proof. Let $f(x) = x^n$. Then

$$f'(x) = \lim_{h \to 0} \frac{f(x + h) - f(x)}{h}$$

$$= \lim_{h \to 0} \frac{(x + h)^n - x^n}{h}$$

Expanding $(x + h)^n$ by the binomial theorem, we obtain

$$f'(x) = \lim_{h \to 0} \frac{\left[x^n + nx^{n-1}h + \dfrac{n(n - 1)}{2!}x^{n-2}h^2 + \cdots + nxh^{n-1} + h^n \right] - x^n}{h}$$

$$= \lim_{h \to 0} \frac{nx^{n-1}h + \dfrac{n(n - 1)}{2!}x^{n-2}h^2 + \cdots + nxh^{n-1} + h^n}{h}$$

Canceling a factor of h we obtain

$$f'(x) = \lim_{h \to 0} \left[nx^{n-1} + \frac{n(n - 1)}{2!}x^{n-2}h + \cdots + nxh^{n-2} + h^{n-1} \right]$$

Every term but the first has a factor of h, and so approaches zero as $h \to 0$. Therefore

$$f'(x) = nx^{n-1}$$

REMARK. In words, *to differentiate x to a positive integer power, take the power and multiply it by x to the next lower integer power.*

► Example 2

$$\frac{d}{dx}[x^5] = 5x^4, \qquad \frac{d}{dx}[x] = 1 \cdot x^0 = 1, \qquad \frac{d}{dx}[x^{12}] = 12x^{11}$$

◄

3.2.3 THEOREM *Let c be a constant. If f is differentiable at x, then so is cf, and*

$$(cf)'(x) = cf'(x)$$

Proof.

$$(cf)'(x) = \lim_{h \to 0} \frac{(cf)(x + h) - (cf)(x)}{h}$$

$$= \lim_{h \to 0} \frac{cf(x + h) - cf(x)}{h}$$

$$= \lim_{h \to 0} c \left[\frac{f(x + h) - f(x)}{h} \right]$$

$$= c \lim_{h \to 0} \frac{f(x + h) - f(x)}{h} \qquad \begin{bmatrix} \text{A constant factor can be} \\ \text{moved through a} \\ \text{limit sign} \end{bmatrix}$$

$$= cf'(x) \quad \blacksquare$$

In another notation, Theorem 3.2.3 states

$$\frac{d}{dx}[cf(x)] = c\frac{d}{dx}[f(x)]$$

REMARK. In words, *a constant factor c can be moved through a derivative sign.*

▶ Example 3

$$\frac{d}{dx}[4x^8] = 4\frac{d}{dx}[x^8] = 4[8x^7] = 32x^7$$

$$\frac{d}{dx}[-x^{12}] = (-1)\frac{d}{dx}[x^{12}] = -12x^{11} \qquad\qquad ◀$$

3.2.4 THEOREM *If f and g are differentiable at x, then so is f + g, and*

$$(f + g)'(x) = f'(x) + g'(x)$$

Proof.

$$(f + g)'(x) = \lim_{h \to 0} \frac{(f + g)(x + h) - (f + g)(x)}{h}$$

$$= \lim_{h \to 0} \frac{[f(x + h) + g(x + h)] - [f(x) + g(x)]}{h}$$

$$= \lim_{h \to 0} \frac{[f(x + h) - f(x)] + [g(x + h) - g(x)]}{h}$$

$$= \lim_{h \to 0} \frac{f(x + h) - f(x)}{h} + \lim_{h \to 0} \frac{g(x + h) - g(x)}{h} \quad \left[\begin{array}{l} \text{The limit of a sum} \\ \text{is the sum of} \\ \text{the limits.} \end{array} \right]$$

$$= f'(x) + g'(x) \quad \blacksquare$$

Theorem 3.2.4 can be written in the alternate notation,

$$\frac{d}{dx}[f(x) + g(x)] = \frac{d}{dx}[f(x)] + \frac{d}{dx}[g(x)]$$

Moreover, by writing $f - g = f + (-1)g$ and then applying Theorems 3.2.3 and 3.2.4 it follows that

$$\frac{d}{dx}[f(x) - g(x)] = \frac{d}{dx}[f(x)] - \frac{d}{dx}[g(x)]$$

REMARK. In words, *the derivative of a sum equals the sum of the derivatives,* and *the derivative of a difference equals the difference of the derivatives.*

▶ Example 4

$$\frac{d}{dx}[x^4 + x^2] = \frac{d}{dx}[x^4] + \frac{d}{dx}[x^2] = 4x^3 + 2x$$

$$\frac{d}{dx}[x^4 - x^2] = \frac{d}{dx}[x^4] - \frac{d}{dx}[x^2] = 4x^3 - 2x$$

$$\frac{d}{dx}[6x^{11} + 9] = \frac{d}{dx}[6x^{11}] + \frac{d}{dx}[9] = 66x^{10} + 0 = 66x^{10} \quad \blacktriangleleft$$

The result in Theorem 3.2.4 can be extended to any finite number of functions. More precisely, if the functions $f_1, f_2, \ldots, f_n$ are all differentiable at x, then their sum is differentiable at x and

$$\frac{d}{dx}[f_1(x) + f_2(x) + \cdots + f_n(x)] = \frac{d}{dx}[f_1(x)] + \frac{d}{dx}[f_2(x)] + \cdots + \frac{d}{dx}[f_n(x)]$$

▶ Example 5

$$\frac{d}{dx}[3x^8 - 2x^5 + 6x + 1] = \frac{d}{dx}[3x^8] + \frac{d}{dx}[-2x^5] + \frac{d}{dx}[6x] + \frac{d}{dx}[1]$$

$$= 24x^7 - 10x^4 + 6 \quad \blacktriangleleft$$

3.2.5 THEOREM *If f and g are differentiable at x, then so is the product $f \cdot g$, and*

The Product Rule

$$(f \cdot g)'(x) = f(x)g'(x) + g(x)f'(x)$$

Proof.

$$(f \cdot g)'(x) = \lim_{h \to 0} \frac{(f \cdot g)(x + h) - (f \cdot g)(x)}{h}$$

$$= \lim_{h \to 0} \frac{f(x + h) \cdot g(x + h) - f(x) \cdot g(x)}{h}$$

If we add and subtract $f(x + h) \cdot g(x)$ in the numerator, we obtain

$$(f \cdot g)'(x) = \lim_{h \to 0} \frac{f(x + h)g(x + h) - f(x + h)g(x) + f(x + h)g(x) - f(x)g(x)}{h}$$

$$= \lim_{h \to 0} \left[f(x + h) \cdot \frac{g(x + h) - g(x)}{h} + g(x) \cdot \frac{f(x + h) - f(x)}{h} \right]$$

$$= \lim_{h \to 0} f(x + h) \cdot \lim_{h \to 0} \frac{g(x + h) - g(x)}{h} + \lim_{h \to 0} g(x) \cdot \lim_{h \to 0} \frac{f(x + h) - f(x)}{h}$$

$$= [\lim_{h \to 0} f(x + h)]g'(x) + [\lim_{h \to 0} g(x)]f'(x) \tag{1}$$

Consider the limit

$$\lim_{h \to 0} g(x)$$

occurring in the second term of (1). The expression $g(x)$ does not involve h so that $g(x)$ remains *constant* as $h \to 0$. Therefore

$$\lim_{h \to 0} g(x) = g(x) \tag{2}$$

Next consider the limit

$$\lim_{h \to 0} f(x + h)$$

which appears in the first term of (1). We can rewrite this limit as

$$\lim_{h \to 0} f(x + h) = \lim_{h \to 0} \left[\frac{f(x + h) - f(x)}{h} \cdot h + f(x) \right]$$

$$= \lim_{h \to 0} \frac{f(x + h) - f(x)}{h} \cdot \lim_{h \to 0} h + \lim_{h \to 0} f(x)$$

$$= f'(x) \cdot 0 + \lim_{h \to 0} f(x)$$

Since $f(x)$ does not involve h

$$\lim_{h \to 0} f(x) = f(x)$$

so that

$$\lim_{h \to 0} f(x + h) = f(x) \tag{3}$$

Substituting (2) and (3) into (1) yields

$$(f \cdot g)'(x) = f(x)g'(x) + g(x)f'(x)$$

The Product Rule can be written in the alternate notation,

$$\frac{d}{dx}[f(x)g(x)] = f(x)\frac{d}{dx}[g(x)] + g(x)\frac{d}{dx}[f(x)]$$

REMARK. In words, *the derivative of a product of two functions equals the first function times the derivative of the second plus the second function times the derivative of the first.*

▶ Example 6

Find $h'(x)$ if

$$h(x) = (4x^2 - 1)(7x^3 + x)$$

Solution. There are two methods. We can either use the Product Rule or we can multiply out the factors in $h(x)$ and then differentiate. We will give both methods.

Method I. (*Using the Product Rule.*)

$$h'(x) = \frac{d}{dx}[(4x^2 - 1)(7x^3 + x)]$$

$$= (4x^2 - 1)\frac{d}{dx}[7x^3 + x] + (7x^3 + x)\frac{d}{dx}[4x^2 - 1]$$

$$= (4x^2 - 1)(21x^2 + 1) + (7x^3 + x)(8x)$$

If desired, we can simplify this expression to obtain

$$h'(x) = 140x^4 - 9x^2 - 1$$

Method II. (*Multiplying First.*)

$$h(x) = (4x^2 - 1)(7x^3 + x) = 28x^5 - 3x^3 - x$$

Thus

$$h'(x) = \frac{d}{dx}[28x^5 - 3x^3 - x] = 140x^4 - 9x^2 - 1$$

which agrees with the result obtained using the product rule. ◀

3.2.6 THEOREM

The Quotient Rule

If f and g are differentiable at x and $g(x) \neq 0$, then f/g is differentiable at x and

$$\left(\frac{f}{g}\right)'(x) = \frac{g(x)f'(x) - f(x)g'(x)}{[g(x)]^2}$$

Proof.

$$\left(\frac{f}{g}\right)'(x) = \lim_{h \to 0} \frac{\left(\frac{f}{g}\right)(x + h) - \left(\frac{f}{g}\right)(x)}{h}$$

$$= \lim_{h \to 0} \frac{\dfrac{f(x + h)}{g(x + h)} - \dfrac{f(x)}{g(x)}}{h}$$

$$= \lim_{h \to 0} \frac{f(x + h) \cdot g(x) - f(x) \cdot g(x + h)}{h \cdot g(x) \cdot g(x + h)}$$

Adding and subtracting $f(x) \cdot g(x)$ in the numerator yields

$$\left(\frac{f}{g}\right)'(x) = \lim_{h \to 0} \frac{f(x + h) \cdot g(x) - f(x) \cdot g(x) - f(x) \cdot g(x + h) + f(x) \cdot g(x)}{h \cdot g(x) \cdot g(x + h)}$$

$$= \lim_{h \to 0} \frac{\left[g(x) \cdot \dfrac{f(x + h) - f(x)}{h}\right] - \left[f(x) \cdot \dfrac{g(x + h) - g(x)}{h}\right]}{g(x) \cdot g(x + h)}$$

$$= \frac{\displaystyle\lim_{h \to 0} g(x) \cdot \lim_{h \to 0} \frac{f(x + h) - f(x)}{h} - \lim_{h \to 0} f(x) \cdot \lim_{h \to 0} \frac{g(x + h) - g(x)}{h}}{\displaystyle\lim_{h \to 0} g(x) \cdot \lim_{h \to 0} g(x + h)}$$

$$= \frac{[\lim_{h \to 0} g(x)] \cdot f'(x) - [\lim_{h \to 0} f(x)] \cdot g'(x)}{\displaystyle\lim_{h \to 0} g(x) \cdot \lim_{h \to 0} g(x + h)} \tag{4}$$

Since the expressions $f(x)$ and $g(x)$ do not involve h,

$$\lim_{h \to 0} g(x) = g(x) \quad \text{and} \quad \lim_{h \to 0} f(x) = f(x) \tag{5}$$

By an argument like that used to derive (3) in the previous theorem, the reader will be able to show that

$$\lim_{h \to 0} g(x + h) = g(x) \tag{6}$$

Substituting (5) and (6) into (4) yields

$$\left(\frac{f}{g}\right)'(x) = \frac{g(x)f'(x) - f(x)g'(x)}{[g(x)]^2}$$

The quotient rule can be written in the alternate notation,

$$\frac{d}{dx}\left[\frac{f(x)}{g(x)}\right] = \frac{g(x)\dfrac{d}{dx}[f(x)] - f(x)\dfrac{d}{dx}[g(x)]}{[g(x)]^2}$$

REMARK. In words, *the derivative of a quotient of two functions is the denominator times the derivative of the numerator minus the numerator times the derivative of the denominator, all divided by the denominator squared.*

▶ Example 7

Find $h'(x)$ if

$$h(x) = \frac{5x^4 + x^2}{x^3 + 2}$$

Solution.

$$h'(x) = \frac{d}{dx}\left[\frac{5x^4 + x^2}{x^3 + 2}\right] = \frac{(x^3 + 2)\dfrac{d}{dx}[5x^4 + x^2] - (5x^4 + x^2)\dfrac{d}{dx}[x^3 + 2]}{(x^3 + 2)^2}$$

$$= \frac{(x^3 + 2)(20x^3 + 2x) - (5x^4 + x^2)(3x^2)}{(x^3 + 2)^2}$$

If necessary, the numerator can be simplified. ◀

In Theorem 3.2.2 we established the formula

$$\frac{d}{dx}[x^n] = nx^{n-1}$$

for *positive* integer values of n. The following theorem extends this formula so that it applies to *all* integer values of n. Eventually, we will show that this formula is valid for all real exponents.

3.2.7 THEOREM *If n is any integer, then*

$$\frac{d}{dx}[x^n] = nx^{n-1} \tag{7}$$

Proof. The result has already been established in the case where $n > 0$. If $n < 0$, then let

$$m = -n$$

and let

$$f(x) = x^{-m} = \frac{1}{x^m}$$

From Theorem 3.2.6,

$$f'(x) = \frac{d}{dx}\left[\frac{1}{x^m}\right] = \frac{x^m \cdot \frac{d}{dx}[1] - 1 \cdot \frac{d}{dx}[x^m]}{(x^m)^2}$$

Since $n < 0$, it follows that $m > 0$, so x^m can be differentiated using Theorem 3.2.2. Thus

$$f'(x) = \frac{x^m \cdot 0 - 1 \cdot mx^{m-1}}{x^{2m}} = -mx^{m-1-2m}$$

$$= -mx^{-m-1}$$

$$= nx^{n-1}$$

Formula (7) also holds in the case $n = 0$ since the formula reduces to

$$\frac{d}{dx}[1] = 0 \cdot x^{-1} = 0$$

which is correct by Theorem 3.2.1.

▶ Example 8

$$\frac{d}{dx}[x^{-9}] = -9x^{-9-1} = -9x^{-10}$$

$$\frac{d}{dx}\left[\frac{1}{x}\right] = \frac{d}{dx}[x^{-1}] = (-1)x^{-1-1} = -x^{-2} = -\frac{1}{x^2} \quad ◀$$

▶ Exercise Set 3.2

In Exercises 1–28, use the results of this section to find $f'(x)$.

1. $f(x) = 4x^7$.

2. $f(x) = -3x^{12}$.

3. $f(x) = 3x^8 + 2x + 1$.

4. $f(x) = \frac{1}{2}(x^4 + 7)$.

5. $f(x) = \pi^3$.

6. $f(x) = \sqrt{2}x + \frac{1}{\sqrt{2}}$.

7. $f(x) = -\frac{1}{3}(x^7 + 2x - 9)$.

8. $f(x) = \frac{x^2 + 1}{5}$.

9. $f(x) = ax^3 + bx^2 + cx + d$
    (a, b, c, d constants).

10. $f(x) = \frac{1}{a}\left(x^2 + \frac{1}{b}x + c\right)$
    (a, b, c constants).

11. $f(x) = -3x^{-8}$.

12. $f(x) = 7x^{-6}$.

13. $f(x) = x^{-3} + \frac{1}{x^7}$.

14. $f(x) = x + \frac{1}{x}$.

15. $f(x) = (3x^2 + 6)\left(2x - \frac{1}{4}\right)$.

16. $f(x) = (2 - x - 3x^3)(7 + x^5)$.

17. $f(x) = (x^3 + 7x^2 - 8)(2x^{-3} + x^{-4})$.

18. $f(x) = \left(\frac{1}{x} + \frac{1}{x^2}\right)(3x^3 + 27)$.

19. $f(x) = (x^5 + 2x)^2$.

20. $f(x) = (x^2 + x)^{-2}$.

21. $f(x) = \frac{3x}{2x + 1}$.

22. $f(x) = \frac{x^2 + 1}{3x}$.

23. $f(x) = \frac{3x^8 - 6x^2 + 1}{2x^7 + 4x + 2}$.

24. $f(x) = \frac{x^3 + 7x + 1}{x^3 - x}$.

25. $f(x) = \left(\frac{3x + 2}{x}\right)(x^{-5} + 1)$.

26. $f(x) = (2x^7 - x^2)\left(\frac{x - 1}{x + 1}\right)$.

27. $f(x) = 3 + \frac{1}{4}\left(\frac{x}{x + 1}\right)(x^{-2} + 5)$.

28. $f(x) = \frac{1}{5} + \frac{1}{3}\left(\frac{x - 1}{x + 1}\right)(1 + x^{-2})$.

In Exercises 29–34, the functions involve independent variables other than x. Use the results in this section to find the indicated derivative.

29. Find $\frac{d}{dt}[16t^2]$.

30. $c = 2\pi r$; find $\frac{dc}{dr}$.

31. $V(r) = \pi r^3$; find $V'(r)$.

32. Find $\frac{d}{d\alpha}[2\alpha^{-1} + \alpha]$.

33. $s = \frac{t}{t^3 + 7}$; find $\frac{ds}{dt}$.

34. Find $\frac{d}{d\lambda}\left[\frac{\lambda\lambda_0 + \lambda^6}{2 - \lambda_0}\right]$ (λ_0 is constant).

35. At which point(s) does the graph of $y = \frac{1}{3}x^3 - \frac{3}{2}x^2 + 2x$ have a horizontal tangent?

36. Find an equation for the line tangent to $y = (1 - x)/(1 + x)$ at the point where $x = 2$.

37. Find the values of a and b if the tangent to $y = ax^2 + bx$ at $(1, 5)$ has slope $m_{\tan} = 8$.

38. Find k if the curve $y = x^2 + k$ is tangent to the line $y = 2x$.

39. Use Theorems 3.2.3 and 3.2.4 to prove: If the functions f and g are differentiable at x, then $f - g$ is differentiable at x and $(f - g)' = f' - g'$.

40. (a) Let the functions f, g, and h be differentiable at x. By applying Theorem 3.2.5 twice, show that the product $f \cdot g \cdot h$ is differentiable at x and

$$(f \cdot g \cdot h)'(x) = f(x)g(x)h'(x) + f(x)g'(x)h(x) + f'(x)g(x)h(x)$$

(b) Conjecture the formula for differentiating a product of n functions.

41. Use the results of Exercise 40 to find:

(a) $\dfrac{d}{dx}\left[(2x + 1)\left(1 + \dfrac{1}{x}\right)(x^{-3} + 7)\right]$

(b) $\dfrac{d}{dx}[x^{-5}(x^2 + 2x)(4 - 3x)(2x^9 + 1)]$

(c) $\dfrac{d}{dx}[(x^7 + 2x - 3)^3]$

(d) $\dfrac{d}{dx}[(x^2 + 1)^{50}]$.

42. Prove: If f is differentiable at x and $f(x) \neq 0$, then $1/f(x)$ is differentiable at x and

$$\frac{d}{dx}\left[\frac{1}{f(x)}\right] = -\frac{f'(x)}{[f(x)]^2}$$

43. (a) In our proof of Theorem 3.2.5, we showed that

$$\lim_{h \to 0} f(x + h) = f(x)$$

[see Equation (3)]. The argument given used the hypothesis that f is differentiable at x. Find the fallacy in the following proof that makes no assumptions about f. As h approaches 0, $x + h$ approaches x; consequently, $f(x + h)$ approaches $f(x)$; that is

$$\lim_{h \to 0} f(x + h) = f(x)$$

(b) Let

$$f(x) = \begin{cases} x, & x \neq 1 \\ 3, & x = 1 \end{cases}$$

Show that

$$\lim_{h \to 0} f(x + h) \neq f(x)$$

when $x = 1$.

3.3 DERIVATIVES OF TRIGONOMETRIC FUNCTIONS

Readers who need to review trigonometry should read Unit II of the trigonometry review in Appendix 1 before beginning this section.

The main objective of this section is to find the derivatives of the trigonometric functions: $\sin x$, $\cos x$, $\tan x$, $\cot x$, $\sec x$, and $\csc x$. We assume throughout that *all angles are in radian measure*.

Let us first consider the problem of differentiating $\sin x$. From the definition of a derivative,

$$\frac{d}{dx}[\sin x] = \lim_{h \to 0} \frac{\sin(x+h) - \sin x}{h}$$

$$= \lim_{h \to 0} \frac{\sin x \cos h + \cos x \sin h - \sin x}{h}$$

$$= \lim_{h \to 0} \left[\sin x \left(\frac{\cos h - 1}{h} \right) + \cos x \left(\frac{\sin h}{h} \right) \right]$$

Since $\sin x$ and $\cos x$ do not involve h, they remain constant as $h \to 0$; thus

$$\lim_{h \to 0} (\sin x) = \sin x \qquad \text{and} \qquad \lim_{h \to 0} (\cos x) = \cos x$$

Consequently

$$\frac{d}{dx}[\sin x] = \sin x \cdot \lim_{h \to 0} \left(\frac{\cos h - 1}{h} \right) + \cos x \cdot \lim_{h \to 0} \left(\frac{\sin h}{h} \right) \qquad (1)$$

In Theorems 3.3.2 and 3.3.4 below we will show that

$$\lim_{h \to 0} \frac{\cos h - 1}{h} = 0 \qquad \text{and} \qquad \lim_{h \to 0} \frac{\sin h}{h} = 1 \qquad (2)$$

so that (1) reduces to

$$\frac{d}{dx}[\sin x] = \cos x \qquad (3)$$

The derivative of $\cos x$ is obtained similarly:

$$\frac{d}{dx}[\cos x] = \lim_{h \to 0} \frac{\cos(x+h) - \cos x}{h}$$

$$= \lim_{h \to 0} \frac{\cos x \cos h - \sin x \sin h - \cos x}{h}$$

$$= \lim_{h \to 0} \left[\cos x \cdot \left(\frac{\cos h - 1}{h} \right) - \sin x \cdot \left(\frac{\sin h}{h} \right) \right]$$

$$= \cos x \cdot \lim_{h \to 0} \left(\frac{\cos h - 1}{h} \right) - \sin x \cdot \lim_{h \to 0} \left(\frac{\sin h}{h} \right)$$

Thus from (2)

$$\frac{d}{dx}[\cos x] = -\sin x \qquad (4)$$

The derivatives of the remaining trigonometric functions can be obtained using the relationships

$$\tan x = \frac{\sin x}{\cos x} \qquad \cot x = \frac{\cos x}{\sin x}$$

$$\sec x = \frac{1}{\cos x} \qquad \csc x = \frac{1}{\sin x}$$

For example,

$$\frac{d}{dx}[\tan x] = \frac{d}{dx}\left[\frac{\sin x}{\cos x}\right]$$

$$= \frac{\cos x \cdot \frac{d}{dx}[\sin x] - \sin x \cdot \frac{d}{dx}[\cos x]}{\cos^2 x}$$

$$= \frac{\cos x \cdot \cos x - \sin x \cdot (-\sin x)}{\cos^2 x}$$

$$= \frac{\cos^2 x + \sin^2 x}{\cos^2 x} = \frac{1}{\cos^2 x} = \sec^2 x$$

Thus

$$\frac{d}{dx}[\tan x] = \sec^2 x \tag{5}$$

We leave the remaining formulas for the exercises:

$$\frac{d}{dx}[\cot x] = -\csc^2 x \tag{6}$$

$$\frac{d}{dx}[\sec x] = \sec x \tan x \tag{7}$$

$$\frac{d}{dx}[\csc x] = -\csc x \cot x \tag{8}$$

REMARK. The derivative formulas for the trigonometric functions should be memorized. An easy way of doing this is discussed in Exercise 37.

As we have seen, the derivative formulas for the trigonometric functions ultimately rest on the two limits stated in (2). We will conclude this section by proving these two fundamental results. Unfortunately the values of these limits cannot be obtained using the methods available so far. In fact, it is not even clear that these limits exist. To see the difficulty, consider the limit

$$\lim_{h \to 0} \frac{\cos h - 1}{h}$$

Intuition tells us that as $h \to 0$ the numerator and denominator both approach 0. As a result, there are two conflicting influences on the ratio. The numerator approaching 0 drives the magnitude of the ratio toward zero, while the denominator approaching 0 drives the magnitude of the ratio toward $+\infty$.

Whether or not the limit exists, and if so its value, depend on the precise way in which these influences offset one another. In a limit problem where the numerator and denominator both approach 0, it is often possible to obtain the limit by some algebraic manipulations. Unfortunately no such algebraic manipulations will work for the limits in (2); we must develop other techniques.

When it is difficult to find the limit of a function directly, it is sometimes possible to obtain the limit indirectly by "squeezing" the function between simpler functions whose limits are known. For example, suppose we are unable to calculate directly

$$\lim_{x \to 0} f(x)$$

but we are able to "squeeze" $f(x)$ between $g(x) = 1 - 2x^2$ and $h(x) = 1 + x^2$ by means of the inequalities

$$1 - 2x^2 \leq f(x) \leq 1 + x^2 \tag{9}$$

Since $1 + x^2$ and $1 - 2x^2$ both approach 1 as $x \to 0$, and since the value of $f(x)$ is always between $1 + x^2$ and $1 - 2x^2$, it is intuitively evident that $f(x)$ must also approach 1 as $x \to 0$. Thus,

$$\lim_{x \to 0} f(x) = 1$$

This result is also clear geometrically since the inequalities in (9) imply that the graph of f is "squeezed" or "pinched" between the graphs of $g(x) = 1 - 2x^2$ and $h(x) = 1 + x^2$ (Figure 3.3.1).

Figure 3.3.1

This idea is formalized in the following theorem, which is called the **pinching theorem** or sometimes the **squeezing theorem.** We omit the proof.

3.3.1 THEOREM

The Pinching Theorem

Let f, g, and h be functions satisfying

$$g(x) \leq f(x) \leq h(x)$$

for all x in some open interval containing the point a, with the possible exception that the inequalities need not hold at a. If g and h have the same limit as x approaches a, say

$$\lim_{x \to a} g(x) = \lim_{x \to a} h(x) = L$$

then f also has this limit as x approaches a, that is,

$$\lim_{x \to a} f(x) = L$$

REMARK. The pinching theorem remains true if $\lim_{x \to a}$ is replaced by $\lim_{x \to a^+}$ or $\lim_{x \to a^-}$. Moreover, for $\lim_{x \to a^+}$ the condition

$$g(x) \leq f(x) \leq h(x)$$

need only hold on an open interval extending to the right from a, and for $\lim_{x \to a^-}$ the condition need only hold on an open interval extending to the left from a. The pinching theorem can also be extended to limits of the form $\lim_{x \to -\infty}$ and $\lim_{x \to +\infty}$.

▶ Example 1

Use the pinching theorem to evaluate the limit

$$\lim_{x \to 0} x^2 \sin^2 \frac{1}{x}$$

Solution. If $x \neq 0$ we can write

$$0 \leq \sin^2 \frac{1}{x} \leq 1$$

Multiplying through by x^2 yields

$$0 \leq x^2 \sin^2 \frac{1}{x} \leq x^2$$

if $x \neq 0$. But

$$\lim_{x \to 0} 0 = \lim_{x \to 0} x^2 = 0$$

so by the pinching theorem

$$\lim_{x \to 0} x^2 \sin^2 \frac{1}{x} = 0 \qquad \blacktriangleleft$$

3.3.2 THEOREM
$$\lim_{h \to 0} \frac{\cos h - 1}{h} = 0$$

*Proof.** Construct the angle h in standard position (Figure 3.3.2). Since the terminal side of h intersects the unit circle at the point $P(\cos h, \sin h)$ and since h is measured in radians, the *signed* arc length along the unit circle from $B(1, 0)$ to $P(\cos h, \sin h)$ is h. (See Definition 4 of the trigonometry review in Appendix 1.)

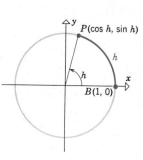

Figure 3.3.2

Thus the arc length along the circle from B to P is $|h|$. Since the straight-line distance between $B(1, 0)$ and $P(\cos h, \sin h)$ does not exceed the arc length along the circle between these points, we have

$$0 \leq \sqrt{(1 - \cos h)^2 + (0 - \sin h)^2} \leq |h|$$

On squaring and simplifying we get

$$0 \leq 2 - 2 \cos h \leq h^2 \qquad (10)$$

If h is *positive* then on division by $2h$, (10) becomes

$$0 \leq \frac{1 - \cos h}{h} \leq \frac{1}{2} h$$

But

$$\lim_{h \to 0^+} \frac{1}{2} h = 0 \qquad \text{and} \qquad \lim_{h \to 0^+} 0 = 0$$

so that

$$\lim_{h \to 0^+} \frac{1 - \cos h}{h} = 0 \qquad (11)$$

---

*This proof, due to Stephen Hoffman, was published in the *American Mathematical Monthly*, Vol. 67, (1960), pp. 671–672.

by the pinching theorem. If h is *negative* then on dividing (10) by $2h$ we obtain

$$\frac{1}{2}h \le \frac{1 - \cos h}{h} \le 0$$

But

$$\lim_{h \to 0^-} \frac{1}{2}h = 0 \quad \text{and} \quad \lim_{h \to 0^-} 0 = 0$$

so that

$$\lim_{h \to 0^-} \frac{1 - \cos h}{h} = 0 \tag{12}$$

by the pinching theorem. From (11) and (12) we can conclude

$$\lim_{h \to 0} \frac{1 - \cos h}{h} = 0$$

from which it follows that

$$\lim_{h \to 0} \frac{\cos h - 1}{h} = \lim_{h \to 0} \left[-\left(\frac{1 - \cos h}{h} \right) \right] = 0 \quad ▮$$

3.3.3 COROLLARY

$$\lim_{h \to 0} \cos h = 1$$

Proof.

$$\lim_{h \to 0} \cos h = \lim_{h \to 0} \left(1 + \frac{\cos h - 1}{h} \cdot h \right)$$

$$= \lim_{h \to 0} 1 + \lim_{h \to 0} \frac{\cos h - 1}{h} \cdot \lim_{h \to 0} h = 1 + 0 \cdot 0 = 1 \quad ▮$$

Before proving the last theorem we must develop the formula for the area of a circular sector. Consider a sector with radius r and a central angle of θ radians (Figure 3.3.3). If $\theta = 2\pi$, the sector is the full circle that has area πr^2. Thus, if we assume that the area A of an arbitrary sector is proportional to its central angle θ, we obtain

$$\frac{\text{area of sector}}{\text{area of circle}} = \frac{\text{central angle of sector}}{\text{central angle of circle}}$$

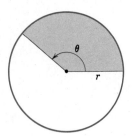

Figure 3.3.3

or

$$\frac{A}{\pi r^2} = \frac{\theta}{2\pi}$$

Solving for A we obtain the following formula.

Area of a sector with angle θ and radius r

$$A = \frac{1}{2}r^2\theta \qquad\qquad\qquad\qquad (13)$$

3.3.4 THEOREM $\displaystyle\lim_{h \to 0} \frac{\sin h}{h} = 1$

Proof. Assume that h satisfies

$$0 < h < \frac{\pi}{2}$$

and construct the angle h in standard position. The terminal side of h intersects the unit circle at $P(\cos h, \sin h)$ and intersects the vertical line through $B(1, 0)$ at the point $Q(1, \tan h)$ (Figure 3.3.4). From the figure

$$0 < \text{area of } \triangle OBP < \text{area of sector } OBP < \text{area of } \triangle OBQ$$

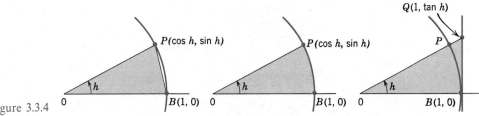

Figure 3.3.4

But

$$\text{area } \triangle OBP = \frac{1}{2} \text{ base} \cdot \text{altitude} = \frac{1}{2} \cdot 1 \cdot \sin h = \frac{1}{2}\sin h$$

$$\text{area sector } OBP = \frac{1}{2}(1)^2 \cdot h = \frac{1}{2}h \qquad\qquad \text{[Equation (13)]}$$

$$\text{area } \triangle OBQ = \frac{1}{2} \text{ base} \cdot \text{altitude} = \frac{1}{2} \cdot 1 \cdot \tan h = \frac{1}{2}\tan h$$

Therefore

$$0 < \frac{1}{2}\sin h < \frac{1}{2}h < \frac{1}{2}\tan h$$

Multiplying through by $2/\sin h$ yields

$$1 < \frac{h}{\sin h} < \frac{1}{\cos h}$$

and taking reciprocals yields

$$\cos h < \frac{\sin h}{h} < 1 \tag{14}$$

We have derived (14) under the assumption that $0 < h < \pi/2$. However (14) is also valid if $-\pi/2 < h < 0$ (Exercise 36) so that (14) holds for all h in the interval $(-\pi/2, \pi/2)$ except $h = 0$.

From Corollary 3.3.3 we have

$$\lim_{h \to 0} \cos h = 1$$

and since

$$\lim_{h \to 0} 1 = 1$$

we can apply the pinching theorem to (14) and conclude

$$\lim_{h \to 0} \frac{\sin h}{h} = 1$$

▶ Example 2

Evaluate $\displaystyle \lim_{\theta \to 0} \frac{\sin 2\theta}{\theta}$.

Solution.

$$\lim_{\theta \to 0} \frac{\sin 2\theta}{\theta} = \lim_{\theta \to 0} \frac{2 \sin \theta \cos \theta}{\theta} = 2 \lim_{\theta \to 0} \frac{\sin \theta}{\theta} \lim_{\theta \to 0} \cos \theta$$
$$= 2(1)(1) = 2$$

Alternate Solution. Make the substitution $\phi = 2\theta$. Since $\phi \to 0$ as $\theta \to 0$, we can write

$$\lim_{\theta \to 0} \frac{\sin 2\theta}{\theta} = \lim_{\phi \to 0} \frac{\sin \phi}{\tfrac{1}{2}\phi} = 2 \lim_{\phi \to 0} \frac{\sin \phi}{\phi}$$
$$= 2(1) = 2 \qquad \blacktriangleleft$$

▶ Exercise Set 3.3

In Exercises 1–10, find $f'(x)$.

1. $f(x) = 2 \cos x - 3 \sin x$.

2. $f(x) = \sin x \cos x$.

3. $f(x) = \dfrac{\sin x}{x}$.

4. $f(x) = x^2 \cos x$.

5. $f(x) = \sec x \tan x$.

6. $f(x) = \dfrac{1}{2} \sec x + 7 \tan x$.

7. $f(x) = \dfrac{\cot x}{1 + \csc x}$.

8. $f(x) = \dfrac{\csc x}{\tan x}$.

9. $f(x) = \sin^2 x + \cos^2 x$.

10. $f(x) = \dfrac{1}{\cot x}$.

11. Use Theorem 3.3.4 to prove that $\lim\limits_{h \to 0} \sin h = 0$.

In Exercises 12–23, find the limit if it exists.

12. $\lim\limits_{h \to 0} \dfrac{\sin h}{2h}$.

13. $\lim\limits_{\theta \to 0} \dfrac{\sin 3\theta}{\theta}$.

14. $\lim\limits_{x \to 0} \dfrac{\sin 6x}{\sin 8x}$.

15. $\lim\limits_{x \to 0} \dfrac{\tan 7x}{\sin 3x}$.

16. $\lim\limits_{\theta \to 0} \dfrac{\sin^2 \theta}{\theta}$.

17. $\lim\limits_{h \to 0} \dfrac{h}{\tan h}$.

18. $\lim\limits_{h \to 0} \dfrac{\sin h}{1 - \cos h}$.

19. $\lim\limits_{\theta \to 0} \dfrac{\theta^2}{1 - \cos \theta}$.

20. $\lim\limits_{x \to 0} \dfrac{x}{\cos(\frac{1}{2}\pi - x)}$.

21. $\lim\limits_{\theta \to 0} \dfrac{\theta}{\cos \theta}$.

22. $\lim\limits_{t \to 0} \dfrac{t^2}{1 - \cos^2 t}$.

23. $\lim\limits_{h \to 0} \dfrac{1 - \cos 5h}{\cos 7h - 1}$.

24. Find all points where the graph of f has a horizontal tangent.
   (a) $f(x) = \sin x$ (b) $f(x) = \tan x$
   (c) $f(x) = \sec x$ (d) $f(x) = \cos x$
   (e) $f(x) = \cot x$ (f) $f(x) = \csc x$.

25. Find the equation of the tangent to the graph of $\sin x$ at the point where:
   (a) $x = 0$ (b) $x = \pi$
   (c) $x = \dfrac{\pi}{4}$.

26. Find the equation of the tangent to the graph of $\tan x$ at the point where:
   (a) $x = 0$ (b) $x = \dfrac{\pi}{4}$
   (c) $x = -\dfrac{\pi}{4}$.

27. Derive the formulas
   (a) $\dfrac{d}{dx}[\cot x] = -\csc^2 x$
   (b) $\dfrac{d}{dx}[\sec x] = \sec x \tan x$
   (c) $\dfrac{d}{dx}[\csc x] = -\csc x \cot x$.

28. In each part, determine where f is differentiable.
   (a) $f(x) = \sin x$ (b) $f(x) = \cos x$
   (c) $f(x) = \tan x$ (d) $f(x) = \cot x$
   (e) $f(x) = \sec x$ (f) $f(x) = \csc x$
   (g) $f(x) = \dfrac{1}{1 + \cos x}$ (h) $f(x) = \dfrac{1}{\sin x \cos x}$
   (i) $f(x) = \dfrac{\cos x}{2 - \sin x}$.

29. (a) Show that $\lim\limits_{h \to 0} \dfrac{\tan h}{h} = 1$.
   (b) Use the result in (a) to help derive the formula $(d/dx)[\tan x] = \sec^2 x$ directly from the definition of a derivative.

30. In each part, find the limit by making the indicated substitution.

(a) $\lim\limits_{x \to +\infty} x \sin \dfrac{1}{x}$. $\left[\text{Let } t = \dfrac{1}{x}. \right]$

(b) $\lim\limits_{x \to -\infty} x \left(1 - \cos \dfrac{1}{x} \right)$. $\left[\text{Let } t = \dfrac{1}{x}. \right]$

(c) $\lim\limits_{x \to \pi} \dfrac{\pi - x}{\sin x}$. [Let $t = \pi - x$.]

31. Find $\lim\limits_{x \to 2} \dfrac{\cos (\pi/x)}{x - 2}$ by making appropriate substitutions.

32. Let f be a function that satisfies

$$1 - x^2 \le f(x) \le \cos x$$

for all x in $(-\pi/2, \pi/2)$. Does $\lim\limits_{x \to 0} f(x)$ exist? If so, find the limit. If not, explain why.

33. Let

$$f(x) = \begin{cases} 1 & \text{if } x \text{ is a rational number} \\ 0 & \text{if } x \text{ is an irrational number} \end{cases}$$

Use the pinching theorem to prove: $\lim\limits_{x \to 0} xf(x) = 0$.

34. Prove: If there are constants L and M such that

$$L \le f(x) \le M$$

for all x in some open interval containing 0, with the possible exception that the inequalities may not hold at 0, then

$$\lim\limits_{x \to 0} xf(x) = 0$$

35. State versions of the pinching theorem that apply to the limits $\lim\limits_{x \to +\infty} f(x)$ and $\lim\limits_{x \to -\infty} f(x)$. Draw some pictures to illustrate these results.

36. Prove that (14) holds if $-\pi/2 < h < 0$ by substituting $-h$ for h in (14).

37. Let us agree to call the functions $\cos x$, $\cot x$, and $\csc x$ the *cofunctions* of $\sin x$, $\tan x$, and $\sec x$, respectively. Convince yourself that the derivative of any cofunction can be obtained from the derivative of the corresponding function by introducing a minus sign and replacing each function in the derivative by its cofunction. Memorize the derivatives of $\sin x$, $\tan x$, and $\sec x$ and then use the above observation to deduce the derivatives of the cofunctions.

38. The derivative formulas for $\sin x$, $\cos x$, $\tan x$, $\cot x$, $\sec x$, and $\csc x$ were obtained under the assumption that x is measured in radians. This exercise shows that different (more complicated) formulas result if x is measured in degrees. Prove that if h and x are degree measures, then

(a) $\lim\limits_{h \to 0} \dfrac{\cos h - 1}{h} = 0$

(b) $\lim\limits_{h \to 0} \dfrac{\sin h}{h} = \dfrac{\pi}{180}$

(c) $\dfrac{d}{dx} [\sin x] = \dfrac{\pi}{180} \cos x$.

39. Evaluate $\lim\limits_{x \to \pi/4} \dfrac{\tan x - 1}{x - \pi/4}$.

3.4 Δ-NOTATION; DIFFERENTIALS

Often the key to solving or understanding a problem is to introduce the right notation. In this section, we pause in our development of differentiation methods to discuss some powerful notational tools.

If a variable changes from one value to another, then its final value minus its initial value is called the *increment* in the variable. It is traditional in calculus to denote an increment in a variable x by the symbol Δx ("delta x"). In this notation, "Δx" is not a product of "Δ" and "x." Rather, Δx is a single symbol representing the *change* in the value of x. Similarly, Δy, Δt, $\Delta \theta$, and so forth, denote increments in the variables y, t, and θ.

If $y = f(x)$, and x changes from an initial value x_0 to a final value x_1, then there is corresponding change in the value of y from $y_0 = f(x_0)$ to $y_1 = f(x_1)$. Stated another way, the increment

$$\Delta x = x_1 - x_0 \tag{1}$$

produces a corresponding increment

$$\Delta y = y_1 - y_0 = f(x_1) - f(x_0) \tag{2}$$

in y (Figure 3.4.1). Increments can be either positive or negative depending on the relative positions of the initial and final points. In Figure 3.4.1, Δx is positive since the final point x_1 is to the right of the initial point x_0. If x_1 were to the left of x_0, then Δx would be negative. Relation (1) can be rewritten

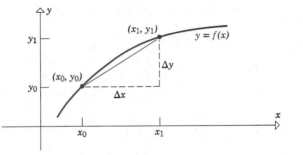

Figure 3.4.1

$$x_1 = x_0 + \Delta x$$

which states that the final value of x is the initial value of x plus the increment in x. Sometimes it is convenient to drop the zero subscript on x_0 and use x to denote the initial value as well as the name of the variable. When this is done, $x + \Delta x$ represents the final value of the variable. Similarly, the initial and final values of y may be denoted by y and $y + \Delta y$ rather than y_0 and y_1 (Figure 3.4.2).

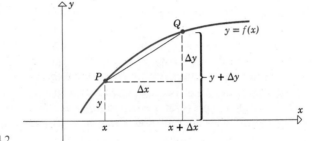

Figure 3.4.2

The Δ-notation provides a compact way of rewriting the derivative definition,

$$\frac{dy}{dx} = \lim_{h \to 0} \frac{f(x + h) - f(x)}{h} \tag{3}$$

Referring to Figure 3.4.2, the slope of the secant line joining the points P and Q is

$$\frac{\Delta y}{\Delta x}$$

As $\Delta x \to 0$, Q tends toward P along the curve $y = f(x)$, so that the slopes of the secant lines approach the slope of the tangent line at P; that is,

$$\frac{dy}{dx} = \lim_{\Delta x \to 0} \frac{\Delta y}{\Delta x} \tag{4}$$

This result is not unexpected since $\Delta y = f(x + \Delta x) - f(x)$, so (4) can be written

$$\frac{dy}{dx} = \lim_{\Delta x \to 0} \frac{f(x + \Delta x) - f(x)}{\Delta x} \tag{5}$$

This is simply a restatement of (3) with Δx used in place of h.

If it is undesirable to introduce a dependent variable y, then we write

$$\Delta f = f(x + \Delta x) - f(x)$$

and (4) as

$$f'(x) = \lim_{\Delta x \to 0} \frac{\Delta f}{\Delta x} \tag{6}$$

Expressions (3), (4), (5), and (6) provide four ways of writing the definition of a derivative. All are commonly used.

Up to now we have been viewing the expression dy/dx as a single symbol denoting the derivative with respect to x of $y = f(x)$. The symbols "dy" and "dx" in this expression are called *differentials;* so far they have no meaning by themselves. We will now show how to interpret these symbols so that dy/dx can be regarded as the ratio of dy and dx.

Figure 3.4.3

Figure 3.4.4

Let $P(x, y)$ be a fixed point on the graph of $y = f(x)$ in an xy-coordinate system, and let us introduce a new rectangular coordinate system with origin at P and coordinate axes parallel to the x- and y-axes. As indicated in Figure 3.4.3, we call the new vertical axis the dy-axis and the new horizontal axis the dx-axis. We might equally well use single letters like X and Y to denote these axes, but that would not serve our purpose.

Because the tangent line at P to the curve $y = f(x)$ passes through the origin of the $dx\ dy$-coordinate system, the equation of the tangent line in this coordinate system is

$$dy = m\ dx \tag{7}$$

where m is the slope of the tangent line. However, the xy-coordinate axes and the $dx\ dy$-coordinate axes are parallel, so that the tangent line has the same angle of inclination and slope in either system (Figure 3.4.4). Since we know the slope of the tangent line to be $f'(x)$ in the xy-system, this is also the slope in the $dx\ dy$-system. Thus (7) can be written as

$$dy = f'(x)\ dx \tag{8}$$

When $dx \neq 0$, we can rewrite (8) as

$$\frac{dy}{dx} = f'(x)$$

Geometrically, the differential dy in (8) represents the change in y that occurs when we start at (x, y) and travel *along the tangent line* at (x, y) until x changes an amount dx (Figure 3.4.5).

In contrast Δy represents the change in y that occurs when we start at (x, y) and travel *along the curve* $y = f(x)$ until x changes an amount Δx. A comparison of dy and Δy is shown in Figure 3.4.6 in the case where $dx = \Delta x$.

Figure 3.4.5

Figure 3.4.6

▶ Example 1

If $y = x^2$, then the relation $dy/dx = 2x$ can be written in the *differential form*

$$dy = 2x \, dx$$

When $x = 3$, this becomes

$$dy = 6 \, dx$$

This tells us that if we travel along the tangent to the graph at $x = 3$, then a change of dx units in x produces a change of $6 \, dx$ units in y. For example, if the change in x is

$$dx = 4$$

then the change in y along the tangent is

$$dy = 6(4) = 24$$

By comparison, if we start at $x = 3$ and travel along the curve $y = f(x) = x^2$ so that the change in x is

$$\Delta x = 4$$

then the change in y is

$$\Delta y = f(3 + \Delta x) - f(3) = f(7) - f(3) = 7^2 - 3^2 = 40 \qquad \blacktriangleleft$$

Just as

$$\frac{d}{dx}[\ \ ]$$

denotes the derivative of the expression inside the brackets, so

$$d[\ \ ]$$

denotes the differential of the expression in the brackets. For example

$$d[x^3] = 3x^2 \, dx$$
$$d[f(x)] = f'(x) \, dx$$

The basic rules of differentiation can be expressed in terms of differentials. In the following table, the differential formulas on the right result when the derivative formulas on the left are multiplied through by dx.

DERIVATIVE FORMULA	DIFFERENTIAL FORMULA
$\dfrac{dc}{dx} = 0$	$dc = 0$
$\dfrac{d}{dx}[cf] = c\dfrac{df}{dx}$	$d[cf] = c\,df$
$\dfrac{d}{dx}[f + g] = \dfrac{df}{dx} + \dfrac{dg}{dx}$	$d[f + g] = df + dg$
$\dfrac{d}{dx}[fg] = f\dfrac{dg}{dx} + g\dfrac{df}{dx}$	$d[fg] = f\,dg + g\,df$
$\dfrac{d}{dx}\left[\dfrac{f}{g}\right] = \dfrac{g\dfrac{df}{dx} - f\dfrac{dg}{dx}}{g^2}$	$d\left[\dfrac{f}{g}\right] = \dfrac{g\,df - f\,dg}{g^2}$

▶ Example 2

Find dy if $y = x \sin x$.

Solution.

$$
\begin{aligned}
dy = d[x \sin x] &= x\,d[\sin x] + (\sin x)\,d[x] \\
&= x(\cos x)\,dx + (\sin x)\,dx \qquad [\text{since } d[x] = 1 \cdot dx = dx] \\
&= (x \cos x + \sin x)\,dx \hspace{3cm} ◀
\end{aligned}
$$

It is evident geometrically from Figure 3.4.6 that $\Delta y - dy \to 0$ as $\Delta x \to 0$ (provided $dx = \Delta x$). This yields the useful approximation

$$\Delta y \approx dy \qquad [\text{when } \Delta x = dx \approx 0] \tag{9}$$

Approximation (9) has applications in the study of **error propagation.** Suppose a researcher measures a physical quantity. Because of inaccuracies in the instrumentation and other factors, the researcher will not usually obtain the exact value x of the quantity, but rather will obtain $x + \Delta x$, where Δx is a measurement error. This recorded value may then be used to calculate some other quantity y. In this way the measurement error Δx propagates to produce an error Δy in the calculated value of y.

▶ Example 3

The radius of a sphere is measured to be 50 inches with a maximum measurement error of ± 0.02 inches. Estimate the error in the computed volume of the sphere.

Solution. The volume of the sphere is

$$V = \tfrac{4}{3}\pi r^3 \tag{10}$$

so that an error of Δr in r yields an error ΔV in V given by

$$\Delta V = \tfrac{4}{3}\pi(r + \Delta r)^3 - \tfrac{4}{3}\pi r^3$$

If Δr is small and we let $dr = \Delta r$, then we can approximate ΔV by dV. Thus from (10)

$$\Delta V \approx dV = 4\pi r^2 \, dr = 4\pi r^2 \, \Delta r \tag{11}$$

Substituting $r = 50$ and $\Delta r = \pm 0.02$ in (11), we obtain

$$\Delta V \approx 4\pi(2500)(\pm 0.02) \approx \pm 628.32$$

Therefore, the maximum error in the volume is approximately ± 628.32 cubic inches. ◀

REMARK. In (11), r represents the exact value of the radius. Since the exact value of r was unknown, we substituted the measured value $r = 50$ to obtain ΔV. This is reasonable since the error Δr was assumed to be small.

If the exact value of a quantity is q and a measurement or calculation results in an error Δq, then $\Delta q/q$ is called the **relative error** in the measurement or calculation; when expressed as a percentage, $\Delta q/q$ is called the **percentage error.** As a practical matter, the exact value q is usually unknown, so that the measured or calculated value of q is used instead; and the relative error is approximated by dq/q. To illustrate, the measured radius of the sphere in Example 3 was 50 inches with a maximum error of ± 0.02 inches; thus the relative error in the radius is approximately $\pm 0.02/50 = \pm 0.0004$ or a percentage error of $\pm 0.04\%$.

RATES OF CHANGE   Up to now we have interpreted the derivative dy/dx as the slope of a tangent line to the curve $y = f(x)$. Derivatives also have interpretations as rates of change. To illustrate, consider the curve $y = f(x)$ in Figure 3.4.7. As we travel along that curve from P to Q, y increases slowly at first, then more rapidly and then more slowly again. However, at the end of the trip from P to Q, y has changed a total amount $\Delta y = 6$, while x has changed a total amount $\Delta x = 8$. Since

$$\frac{\Delta y}{\Delta x} = \frac{6}{8} = \frac{3}{4}$$

it follows that "on the average" y increases $\frac{3}{4}$ of a unit for each one unit increase in x. To say this another way, y is increasing at an average rate equal to $\frac{3}{4}$ the rate of increase of x between P and Q.

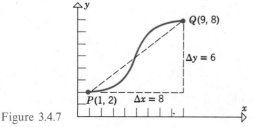

Figure 3.4.7

This discussion suggests the following definition:

3.4.1 DEFINITION If $P(x_0, y_0)$ and $Q(x_1, y_1)$ are points on the graph of $y = f(x)$, then we define

$$\frac{\Delta y}{\Delta x} = \frac{y_1 - y_0}{x_1 - x_0}$$

to be the ***average rate at which y changes with x*** between P and Q.

The ratio $\Delta y / \Delta x$ is, of course, nothing more than the slope of the secant line between P and Q.

Motivated by our earlier discussion of velocity, we may reasonably define the instantaneous rate at which y changes with x at a point $P(x_0, y_0)$ on the curve $y = f(x)$ as the limit of the average rate of change between $P(x_0, y_0)$ and $Q(x_1, y_1)$ as Q approaches P along the curve or, equivalently, as $\Delta x \to 0$. Since

$$\lim_{\Delta x \to 0} \frac{\Delta y}{\Delta x} = \frac{dy}{dx}$$

we are led to the following definition.

3.4.2 DEFINITION If $P(x_0, y_0)$ is a point on the graph of $y = f(x)$ and if f is differentiable at x_0, we define

$$\frac{dy}{dx}\bigg|_{x=x_0}$$

to be the ***instantaneous rate at which y changes with x*** at P.

▶ Example 4

On the curve

$$y = x^2$$

the instantaneous rate at which y changes with x at $P(3, 9)$ is

$$\frac{dy}{dx}\bigg|_{x=3} = 2x\bigg|_{x=3} = 6$$

Thus y is increasing at 6 times the rate of x at P.   ◀

▶ Example 5

For the line

$$y = mx + b$$

the derivative is constant,

$$\frac{dy}{dx} = m$$

Thus the instantaneous rate at which y varies with x along a straight line is the same at each point of the line. This constant rate of change is the slope of the line.   ◀

▶ Exercise Set 3.4

1. Let $y = x^2$.
    (a) Find Δy if $\Delta x = 1$ and the initial value of x is $x = 2$.
    (b) Find dy if $dx = 1$ and the initial value of x is $x = 2$.
    (c) Make a sketch of $y = x^2$ and show Δy and dy in the picture.

2. Repeat Exercise 1 with $x = 2$ as the initial value again, but $\Delta x = -1$ and $dx = -1$.

3. Let $y = \dfrac{1}{x}$.

    (a) Find Δy if $\Delta x = 0.5$ and the initial value of x is $x = 1$.

    (b) Find dy if $dx = 0.5$ and the initial value of x is $x = 1$.
    (c) Make a sketch of $y = 1/x$ and show Δy and dy in the picture.

4. Repeat Exercise 3 with $x = 1$ as the initial value again, but $\Delta x = -0.5$ and $dx = -0.5$.

In Exercises 5–8, find general formulas for dy and Δy.

5. $y = x^3$.

6. $y = 8x - 4$.

7. $y = x^2 - 2x + 1$.

8. $y = \sin x$.

In Exercises 9–12, find dy.

9. $y = 4x^3 - 7x^2 + 2x - 1$.

10. $y = \dfrac{1}{x^3 - 1}$.

11. $y = x \cos x$.

12. $y = \dfrac{1 - x^3}{2 - x}$.

13. Let $y = 2x^2 - 1$
    (a) Find the average rate at which y changes with x over the interval $[1, 4]$.
    (b) Find the instantaneous rate at which y changes with x at the point $x = 1$.

14. Let $y = \dfrac{1}{x^2 + 1}$.
    (a) Find the average rate at which y changes with x over the interval $[-1, 2]$.
    (b) Find the instantaneous rate at which y changes with x at the point $x = -1$.

15. Use the formula $A = \pi r^2$ for the area of a circle to find:
    (a) The average rate at which the area of a circle changes with r as the radius increases from $r = 1$ to $r = 2$.
    (b) The instantaneous rate at which the area changes with r when $r = 2$.

16. Use the formula $V = l^3$ for the volume of a cube of side l to find:

    (a) the average rate at which the volume of a cube changes with l as l increases from $l = 2$ to $l = 4$;
    (b) the instantaneous rate at which the volume of a cube changes with l when $l = 5$.

17. The side of a square is measured to be 10 feet with a maximum error of ± 0.1 feet.
    (a) Use differentials to estimate the error in the calculated area.
    (b) Estimate the percentage errors in the side and the area.

18. The side of a cube is measured to be 25 inches with a maximum percentage error of $\pm 2\%$. Use differentials to help estimate the percentage error in the volume of the cube.

19. The electrical resistance R of a certain wire is given by $R = k/r^2$ where k is a constant and r is the radius of the wire. Assuming that the radius r has a maximum error of $\pm 5\%$, use differentials to estimate the percentage error in R. (Assume k is exact.)

20. The volume of a sphere is to be computed from a measured value of its radius. Estimate the maximum permissible percentage error in the measurement if the percentage error in the volume must be kept within $\pm 3\%$. $[V = \frac{4}{3}\pi r^3$ is the volume of a sphere of radius r.]

3.5 THE CHAIN RULE

In this section we consider the following problem.

3.5.1 PROBLEM   If we know the derivatives of f and g, how can we use this information to find the derivative of the composition $f \circ g$?

For example, we know how to differentiate

$$f(x) = \sin x \quad \text{and} \quad g(x) = x^2 - 1$$

How can we use this knowledge to obtain the derivative of

$$(f \circ g)(x) = f(g(x)) = \sin(x^2 - 1)?$$

The idea is as follows. Introduce dependent variables

$$y = \sin(x^2 - 1) \quad \text{and} \quad u = x^2 - 1 \tag{1}$$

so that

$$y = \sin u$$

We know that

$$\frac{dy}{du} = \frac{d}{du}[\sin u] = \cos u \tag{2}$$

and

$$\frac{du}{dx} = \frac{d}{dx}[x^2 - 1] = 2x \tag{3}$$

We want to find

$$\frac{dy}{dx} = \frac{d}{dx}[\sin(x^2 - 1)]$$

If we view a derivative as a rate of change, then intuition suggests

$$\frac{dy}{du} \cdot \frac{du}{dx} = \frac{dy}{dx} \tag{4}$$

(For example, if y changes at 4 times the rate of u and u changes at 2 times the rate of x, then y changes at $4 \times 2 = 8$ times the rate of x.) Substituting (2) and (3) in (4) yields

$$\frac{dy}{dx} = (\cos u)(2x)$$

which may be expressed in terms of x alone using (1)

$$\frac{dy}{dx} = [\cos(x^2 - 1)]\, 2x = 2x \cos(x^2 - 1)$$

The ideas in this example are formalized in the following theorem.

3.5.2 THEOREM

The Chain Rule

If g is differentiable at the point x and f is differentiable at the point g(x), then the composition f ∘ g is differentiable at the point x. Moreover, if

$$y = f(g(x)) \qquad \text{and} \qquad u = g(x)$$

then

$$\frac{dy}{dx} = \frac{dy}{du} \cdot \frac{du}{dx} \tag{5}$$

Before proving this result we will consider some examples.

▶ **Example 1**

Find dy/dx if

$$y = 4 \cos (x^3)$$

Solution. Let $u = x^3$ so that

$$y = 4 \cos u$$

By the chain rule,

$$\frac{dy}{dx} = \frac{dy}{du} \cdot \frac{du}{dx} = \frac{d}{du}[4 \cos u] \cdot \frac{d}{dx}[x^3]$$
$$= (-4 \sin u) \cdot (3x^2)$$
$$= -12x^2 \sin (x^3) \qquad ◀$$

▶ **Example 2**

Find

$$\frac{d}{dx}[(x^3 + 7x + 1)^{35}]$$

Solution. Let

$$y = (x^3 + 7x + 1)^{35} \qquad \text{and} \qquad u = x^3 + 7x + 1$$

so that

$$y = u^{35}$$

By the chain rule,

$$\frac{d}{dx}[(x^3 + 7x + 1)^{35}] = \frac{dy}{dx} = \frac{dy}{du} \cdot \frac{du}{dx}$$

$$= \frac{d}{du}[u^{35}] \cdot \frac{d}{dx}[x^3 + 7x + 1]$$

$$= (35u^{34})(3x^2 + 7)$$

$$= 35(x^3 + 7x + 1)^{34}(3x^2 + 7) \qquad \blacktriangleleft$$

▶ Example 3

Find $f'(x)$ if

$$f(x) = \frac{1}{2x^4 - x^2 + 8}$$

Solution. First rewrite f as

$$f(x) = (2x^4 - x^2 + 8)^{-1}$$

Let

$$y = (2x^4 - x^2 + 8)^{-1} \qquad \text{and} \qquad u = 2x^4 - x^2 + 8$$

so that

$$y = u^{-1}$$

By the chain rule,

$$f'(x) = \frac{dy}{dx} = \frac{dy}{du} \cdot \frac{du}{dx} = \frac{d}{du}[u^{-1}] \cdot \frac{d}{dx}[2x^4 - x^2 + 8]$$

$$= (-u^{-2})(8x^3 - 2x)$$

$$= -(2x^4 - x^2 + 8)^{-2}(8x^3 - 2x)$$

$$= -\frac{8x^3 - 2x}{(2x^4 - x^2 + 8)^2}$$

[Note that $f(x)$ can also be differentiated without using the chain rule by treating f as a quotient and applying Theorem 3.2.6.] $\qquad \blacktriangleleft$

The chain rule,

$$\frac{dy}{dx} = \frac{dy}{du} \cdot \frac{du}{dx}$$

is easy to remember because the left side is exactly what results if we "cancel" the *du*'s on the right side. This "canceling" device provides a good way to remember the chain rule when variables other than *x*, *y*, and *u* are used.

▶ Example 4

Suppose

$$w = \tan x \qquad \text{and} \qquad x = 4t^3 + t$$

Find *dw/dt*.

Solution. In this case the chain rule takes the form

$$\frac{dw}{dt} = \frac{dw}{dx} \cdot \frac{dx}{dt}$$

$$= \frac{d}{dx}[\tan x] \cdot \frac{d}{dt}[4t^3 + t]$$

$$= (\sec^2 x)(12t^2 + 1)$$

$$= (12t^2 + 1)\sec^2(4t^3 + t) \qquad\qquad ◄$$

▶ Example 5

Find

$$\frac{d}{dx}[\sin^3 (9x + 1)]$$

Solution. Let

$$y = \sin^3 (9x + 1) \qquad \text{and} \qquad u = \sin (9x + 1)$$

so that

$$y = u^3$$

By the chain rule,

$$\frac{d}{dx}[\sin^3 (9x + 1)] = \frac{dy}{dx} = \frac{dy}{du} \cdot \frac{du}{dx} \qquad\qquad (6)$$

The calculation of

$$\frac{du}{dx} = \frac{d}{dx}[\sin (9x + 1)]$$

will itself require an application of the chain rule. To this end, let

$$v = 9x + 1$$

so that

$$u = \sin(9x + 1) = \sin v$$

Now du/dx can be obtained from the following form of the chain rule

$$\frac{du}{dx} = \frac{du}{dv} \cdot \frac{dv}{dx} \tag{7}$$

Substituting (7) into (6), we obtain

$$\frac{d}{dx}[\sin^3(9x + 1)] = \frac{dy}{dx} = \frac{dy}{du} \cdot \frac{du}{dv} \cdot \frac{dv}{dx}$$

$$= \frac{d}{du}[u^3] \cdot \frac{d}{dv}[\sin v] \cdot \frac{d}{dx}[9x + 1]$$

$$= 3u^2 \cdot \cos v \cdot 9$$

$$= 3\sin^2(9x + 1) \cdot \cos(9x + 1) \cdot 9$$

$$= 27\sin^2(9x + 1)\cos(9x + 1) \qquad \blacktriangleleft$$

REMARK. Once having mastered the chain rule, the reader should be able to dispense with some of the intermediate steps. For example, to differentiate

$$y = (3x^2 + 5)^{17}$$

we can mentally visualize this function as

$$y = u^{17}$$

where u stands for $3x^2 + 5$. Since we can differentiate both u^{17} and $3x^2 + 5$ in our heads, we can simply write

$$\frac{dy}{dx} = 17u^{16} \cdot 6x = 17(3x^2 + 5)^{16} \cdot 6x = 102x(3x^2 + 5)^{16}$$

With more practice the reader should even be able to dispense with the first step. Eventually the differentiation of a function like $\cos(x^3 + x)$ should be carried out by writing only

$$\frac{d}{dx}[\cos(x^3 + x)] = -\sin(x^3 + x) \cdot (3x^2 + 1)$$

We conclude this section with some comments about the proof of the chain rule.

Comments on the Proof of the Chain Rule (*Theorem 3.5.2*)

Let Δx be an increment in x and let Δu and Δy be the corresponding increments in g and $f \circ g$; that is,

$$\Delta u = g(x + \Delta x) - g(x)$$
$$\Delta y = f(g(x + \Delta x)) - f(g(x))$$

It follows that

$$\Delta u \to 0 \qquad \text{as} \qquad \Delta x \to 0$$

since

$$\lim_{\Delta x \to 0} \Delta u = \lim_{\Delta x \to 0} \frac{\Delta u}{\Delta x} \cdot \Delta x = \frac{du}{dx} \cdot 0 = 0$$

If we now make the assumption that $\Delta u \neq 0$, then we can complete the proof as follows:

$$\frac{dy}{dx} = \lim_{\Delta x \to 0} \frac{\Delta y}{\Delta x}$$

$$= \lim_{\Delta x \to 0} \left(\frac{\Delta y}{\Delta u} \cdot \frac{\Delta u}{\Delta x} \right)$$

$$= \lim_{\Delta x \to 0} \frac{\Delta y}{\Delta u} \cdot \lim_{\Delta x \to 0} \frac{\Delta u}{\Delta x}$$

$$= \lim_{\Delta u \to 0} \frac{\Delta y}{\Delta u} \cdot \lim_{\Delta x \to 0} \frac{\Delta u}{\Delta x} \qquad [\text{Since } \Delta u \to 0 \text{ as } \Delta x \to 0.]$$

$$= \frac{dy}{du} \cdot \frac{du}{dx}$$

Of course, this argument depends on the assumption that $\Delta u \neq 0$, which is not one of the hypotheses in the chain rule. In fact, it is quite possible for

$$\Delta u = g(x + \Delta x) - g(x)$$

to be zero even if $\Delta x \neq 0$. However, for many functions, the assumption $\Delta u \neq 0$ is valid. Although it is possible to prove the chain rule without the assumption $\Delta u \neq 0$, the proof is somewhat delicate and will be omitted.

▶ Exercise Set 3.5

In Exercises 1–12, find $f'(x)$.

1. $f(x) = (x^3 + 2x)^{37}$.

2. $f(x) = \sin^3 x$.

3. $f(x) = \tan(4x^2)$.

4. $f(x) = \dfrac{1}{(x^5 - x + 1)^9}$.

5. $f(x) = x^5 \sec\left(\dfrac{1}{x}\right)$.

6. $f(x) = \dfrac{\sin x}{\sec(3x + 1)}$.

7. $f(x) = \cos(\cos x)$.

8. $f(x) = \dfrac{1 + \csc x^2}{1 - \cot x^2}$.

9. $f(x) = (5x + 8)^{13}(x^3 + 7x)^{12}$.

10. $f(x) = \left(\dfrac{1 + x^2}{1 - x^2}\right)^{17}$.

11. $f(x) = \dfrac{(4x^2 - 1)^{-8}}{(2x + 1)^{-3}}$.

12. $f(x) = [1 + \sin^3(x^5)]^{12}$.

In Exercises 13–16, find an equation for the tangent to the graph at the specified point.

13. $y = x \cos 3x$, $x = \pi$.

14. $y = \sin(1 + x^3)$, $x = -3$.

15. $y = \sec^3\left(\dfrac{\pi}{2} - x\right)$, $x = -\dfrac{\pi}{2}$.

16. $y = \left(x - \dfrac{1}{x}\right)^3$, $x = 2$.

In Exercises 17–20, find dy/dx in two ways: first by using the chain rule and then by expressing y in terms of x and differentiating directly.

17. $x = 5t + 2$, $y = t^2$.

18. $x = \dfrac{1}{2}\pi - \theta$, $y = \sec \theta$.

19. $x = u^{-1}$, $y = 3 \sin^3 u$.

20. $x = \dfrac{\lambda}{1 - \lambda}$, $y = (1 + \lambda)^{12}$.

In Exercises 21–24, find the indicated derivative.

21. $y = \cot^3(\pi - \theta)$; find $\dfrac{dy}{d\theta}$.

22. $\lambda = \left(\dfrac{au + b}{cu + d}\right)^6$; find $\dfrac{d\lambda}{du}$ (a, b, c, d constants).

23. $\dfrac{d}{d\omega}[a \cos^2 \pi\omega + b \sin^2 \pi\omega]$ (a, b constants).

24. $x = \csc^2\left(\dfrac{\pi}{3} - y\right)$; find $\dfrac{dx}{dy}$.

25. (a) Show that
$$\dfrac{d}{dx}(|x|) = \begin{cases} 1, & x > 0 \\ -1, & x < 0 \end{cases}$$
   (b) Use the result in (a) and the chain rule to find
$$\dfrac{d}{dx}(|\sin x|) \quad \text{and} \quad \dfrac{d}{dx}(\sin |x|)$$
   for x in the interval $(-\pi, \pi)$.

26. Use the identity
$$\cos x = \sin\left(\dfrac{\pi}{2} - x\right)$$
   and the derivative formula for $\sin x$ to obtain the derivative formula for $\cos x$.

27. Find a formula for
$$\dfrac{d}{dx}[f(g(x))]$$
   which does not involve any dependent variables.

28. Find $f'(x^2)$ if $\dfrac{d}{dx}[f(x^2)] = x^2$.

29. Given that $f(0) = 1$, $f'(0) = 2$, $g(0) = 0$, and $g'(0) = 3$, find $(f \circ g)'(0)$.

30. Let $y = f_1(u)$, $u = f_2(v)$, $v = f_3(w)$, and $w = f_4(x)$. Express dy/dx in terms of dy/du, dw/dx, du/dv, and dv/dw.

3.6   IMPLICIT DIFFERENTIATION

In previous sections we found tangent lines to curves expressed in the form

$$y = f(x)$$

In this section we will show how to find tangent lines to curves whose equations do not express y explicitly as a function of x.

Consider the equation

$$xy = 1 \tag{1}$$

One way to obtain the derivative dy/dx is to rewrite this equation as

$$y = \frac{1}{x} \tag{2}$$

which implies that

$$\frac{dy}{dx} = \frac{d}{dx}\left[\frac{1}{x}\right] = -\frac{1}{x^2}$$

However, there is another possibility. We can differentiate both sides of (1) *before* solving for y in terms of x, treating y as a (temporarily unspecified) function of x. With this approach we obtain

$$\frac{d}{dx}[xy] = \frac{d}{dx}[1]$$

$$x\frac{d}{dx}[y] + y\frac{d}{dx}[x] = 0$$

$$x\frac{dy}{dx} + y = 0$$

$$\frac{dy}{dx} = -\frac{y}{x}$$

If we now substitute (2) into the last expression, we obtain

$$\frac{dy}{dx} = -\frac{1}{x^2}$$

which agrees with the previous result. This method of obtaining derivatives is called ***implicit differentiation.*** It is especially useful when it is inconvenient or impossible to solve explicitly for y in terms of x.

DERIVATIVES OF
RATIONAL POWERS
OF x

Earlier, we obtained the formula

$$\frac{d}{dx}[x^n] = nx^{n-1} \tag{3}$$

for integer values of n. By using implicit differentiation we will show that this formula holds for any rational power of x. More precisely, we will show that if r is a rational number then

$$\frac{d}{dx}[x^r] = rx^{r-1} \tag{4}$$

whenever x^r is defined. In our computations we will assume that x^r is differentiable; the justification for this assumption will be considered later.

Let $y = x^r$. Since r is a rational number, it can be expressed as a ratio of integers

$$r = \frac{m}{n}$$

Thus

$$y = x^r = x^{m/n}$$

can be written as

$$y^n = x^m$$

so that

$$\frac{d}{dx}[y^n] = \frac{d}{dx}[x^m]$$

On differentiating implicitly with respect to x and using (3) we obtain

$$ny^{n-1}\frac{dy}{dx} = mx^{m-1} \tag{5}$$

But

$$y^{n-1} = [x^{m/n}]^{n-1} = x^{m-(m/n)}$$

Thus (5) can be written as

$$nx^{m-(m/n)}\frac{dy}{dx} = mx^{m-1}$$

so that

$$\frac{dy}{dx} = \frac{m}{n}x^{(m/n)-1} = rx^{r-1}$$

which establishes (4). ▨

▶ Example 1

From (4)

$$\frac{d}{dx}[x^{4/5}] = \frac{4}{5}x^{(4/5)-1} = \frac{4}{5}x^{-1/5}$$

and

$$\frac{d}{dx}[(x^2+1)^{-3/4}] = -\frac{3}{4}(x^2+1)^{-(3/4)-1}(2x) = -\frac{3}{2}x(x^2+1)^{-7/4} \quad ◀$$

If $r = \frac{1}{2}$, then (4) states that

$$\frac{d}{dx}[x^{1/2}] = \frac{1}{2}x^{-1/2}$$

or in square root notation

$$\frac{d}{dx}[\sqrt{x}] = \frac{1}{2\sqrt{x}}$$

▶ Example 2

$$\frac{d}{dx}\left[\frac{x}{1+\sqrt{x}}\right] = \frac{(1+\sqrt{x})\frac{d}{dx}[x] - x\frac{d}{dx}[1+\sqrt{x}]}{(1+\sqrt{x})^2}$$

$$= \frac{(1+\sqrt{x}) - x\left(\frac{1}{2\sqrt{x}}\right)}{(1+\sqrt{x})^2} = \frac{1+\frac{1}{2}\sqrt{x}}{(1+\sqrt{x})^2} \quad ◀$$

▶ Example 3

$$\frac{d}{dx}[x^2 - x + 2]^{3/4} = \frac{3}{4}(x^2 - x + 2)^{-1/4} \cdot \frac{d}{dx}[x^2 - x + 2]$$

$$= \frac{3}{4}(x^2 - x + 2)^{-1/4}(2x - 1)$$

$$\frac{d}{dx}[\sin \sqrt{x^3 + 2}] = \cos \sqrt{x^3 + 2} \cdot \frac{d}{dx}(\sqrt{x^3 + 2})$$

$$= (\cos \sqrt{x^3 + 2}) \cdot \frac{1}{2\sqrt{x^3 + 2}} \cdot \frac{d}{dx}(x^3 + 2)$$

$$= \frac{3x^2}{2\sqrt{x^3 + 2}} \cos \sqrt{x^3 + 2} \qquad ◀$$

IMPLICIT FUNCTIONS   There are some theoretical considerations that must be taken into account when differentiating implicitly. To apply this technique, it must be assumed that y represents a differentiable function of x. Unless this assumption is warranted, the resulting calculations might be nonsense. For example, if we implicitly differentiate the equation

$$x^2 + y^2 + 1 = 0 \qquad (6)$$

we obtain

$$2x + 2y\frac{dy}{dx} = 0$$

or

$$\frac{dy}{dx} = -\frac{x}{y} \qquad (7)$$

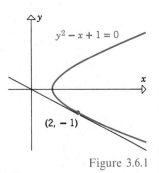

$y^2 - x + 1 = 0$

$(2, -1)$

Figure 3.6.1

However, the derivative in (7) is senseless because no function y satisfies (6). (The left side of the equation is always greater than zero.)

Sometimes implicit differentiation can be justified when the given equation is not the graph of a function of x. For example, suppose we are interested in finding the equation of the "tangent line" at $(2, -1)$ to the curve

$$y^2 - x + 1 = 0 \qquad \text{(See Figure 3.6.1.)} \qquad (8)$$

Since there exist vertical lines that cut this curve twice, the curve is not the graph of a function of x. Thus, strictly speaking, we are not even entitled to talk about a tangent line to the curve since tangent lines were defined only

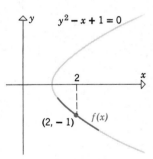

Figure 3.6.2

for graphs of functions. Suppose, however, that we can find some function f that is differentiable at $x = 2$ and whose graph matches our given curve over some "segment" extending on both sides of the point $(2, -1)$ (Figure 3.6.2). We can then *define* the tangent line at $(2, -1)$ to the given curve to be the tangent line to the graph of f at $(2, -1)$. Example 4 illustrates this idea.

▶ Example 4

If we "solve" (8) for y in terms of x we obtain

$$y = \sqrt{x-1} \quad \text{and} \quad y = -\sqrt{x-1}$$

Figure 3.6.3

As shown in Figure 3.6.3, the graph of the first equation is the upper half of the curve

$$y^2 - x + 1 = 0$$

(since $y \geq 0$), and the graph of the second is the lower half (since $y \leq 0$). Since $(2, -1)$ lies on the lower half of the curve, the slope $m_{\tan}$ of the tangent line at this point can be obtained by evaluating the derivative of

$$y = -\sqrt{x-1}$$

when $x = 2$; thus

$$\frac{dy}{dx} = \frac{d}{dx}[-\sqrt{x-1}] = -\frac{1}{2\sqrt{x-1}}$$

and

$$m_{\tan} = \frac{dy}{dx}\bigg|_{x=2} = -\frac{1}{2} \tag{9}$$

This same result can be obtained by implicit differentiation. Since we know that there is a function f, differentiable at $x = 2$, such that $y = f(x)$ matches the curve

$$y^2 - x + 1 = 0 \tag{10}$$

over a segment extending on both sides of the point $(2, -1)$, we can regard y in (10) as the dependent variable for $y = f(x)$, and we can differentiate (10) implicitly, treating y as a function of x.

We obtain

$$\frac{d}{dx}[y^2 - x + 1] = \frac{d}{dx}[0]$$

$$\frac{d}{dx}[y^2] - \frac{d}{dx}[x] + \frac{d}{dx}[1] = \frac{d}{dx}[0]$$

$$2y\frac{dy}{dx} - 1 = 0$$

Solving for dy/dx yields

$$\frac{dy}{dx} = \frac{1}{2y}$$

At $(2, -1)$ we have $y = -1$, so the slope of the tangent line there is

$$m_{\tan} = \frac{dy}{dx}\bigg|_{y=-1} = -\frac{1}{2}$$

which agrees with (9).　　　　　　　　　　　　　　　　　　◀

Before looking at more examples, let us introduce some terminology. Suppose that a curve C is the graph of an equation in x and y and $P(x_0, y_0)$ is any point on C. We will say that the equation defines y as an **implicit function of x in a neighborhood of (x_0, y_0)** if we can find some rectangle centered at (x_0, y_0), with sides parallel to the coordinate axes, and such that the portion of C contained within the rectangle is cut at most once by any vertical line (Figure 3.6.4). This portion of the curve C is the graph of a function f. If this function is differentiable at x_0, then we can calculate dy/dx at x_0 by implicit differentiation.

Unfortunately, it can be difficult to determine whether an equation defines y as a function of x in a neighborhood of a point and if so, whether the function is differentiable. We leave such questions for a course in advanced calculus. However, the following example should clarify the basic idea.

This portion of C is cut at most once by any vertical line.

Figure 3.6.4

▶ Example 5

The equation for the unit circle $x^2 + y^2 = 1$ does not define y as a function of x in a neighborhood of the point $P(1, 0)$, because the portion of the circle within any rectangle centered at P is cut twice by some vertical line (Figure 3.6.5a). The equation does, however, define y as a function of x in a neighborhood of the point $(1/\sqrt{2}, 1/\sqrt{2})$ (Figure 3.6.5b).

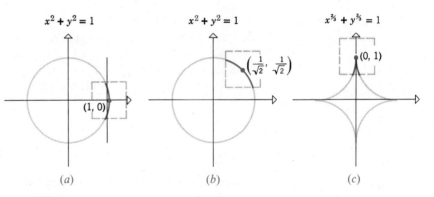

$x^2 + y^2 = 1$ $x^2 + y^2 = 1$ $x^{2/3} + y^{2/3} = 1$

$\left(\frac{1}{\sqrt{2}}, \frac{1}{\sqrt{2}}\right)$

(0, 1)

(1, 0)

Figure 3.6.5   (a)   (b)   (c)

The graph of the equation

$$x^{2/3} + y^{2/3} = 1$$

is called a **four-cusped hypocycloid** (Figure 3.6.5c). This equation defines y as a function of x in a neighborhood of the point (0, 1); however the function is not differentiable at $x = 0$, because of the corner or "cusp" at (0, 1). ◄

▶ Example 6

Use implicit differentiation to find the slope of the tangent line at (4, 0) to the graph of $7y^4 + x^3y + x = 4$ (11)

[Note that $x = 4$, $y = 0$ satisfies the equation so that (4, 0) is actually a point on the graph.]

Solution. Differentiating both sides of (11) with respect to x and treating y as a differentiable function of x we obtain

$$\frac{d}{dx}[7y^4 + x^3y + x] = \frac{d}{dx}[4]$$

$$\frac{d}{dx}[7y^4] + \frac{d}{dx}[x^3y] + \frac{d}{dx}[x] = 0$$

$$\frac{d}{dx}[7y^4] + \left(x^3\frac{dy}{dx} + y\frac{d}{dx}[x^3]\right) + \frac{d}{dx}[x] = 0$$

$$28y^3\frac{dy}{dx} + x^3\frac{dy}{dx} + 3yx^2 + 1 = 0$$

Solving for dy/dx we obtain

$$\frac{dy}{dx} = -\frac{3yx^2 + 1}{28y^3 + x^3}$$ (12)

Since we have no explicit expression for y in terms of x, we are forced to leave this derivative in terms of x and y. However, for purposes of finding the slope of the tangent line at $(x, y) = (4, 0)$ this causes no problem since we can substitute $x = 4$ and $y = 0$ in (12) to obtain

$$m_{\text{tan}} = \frac{dy}{dx}\bigg|_{\substack{x=4 \\ y=0}} = -\frac{1}{64}$$ ◀

▶ Example 7

Use implicit differentiation to find dy/dx if $5xy^2 + \pi \sin y = x$

Solution. Differentiating both sides with respect to x and treating y as a differentiable function of x, we obtain

$$\frac{d}{dx}[5xy^2] + \frac{d}{dx}[\pi \sin y] = \frac{d}{dx}[x]$$

$$\left(5x \frac{d}{dx}[y^2] + y^2 \frac{d}{dx}[5x]\right) + \pi \frac{d}{dx}[\sin y] = 1$$

$$10xy \frac{dy}{dx} + 5y^2 + (\pi \cos y)\frac{dy}{dx} = 1$$

Solving for dy/dx we obtain $\dfrac{dy}{dx} = \dfrac{1 - 5y^2}{10xy + \pi \cos y}$ (13)

Since we have no explicit expression for y in terms of x, we cannot express formula (13) in terms of x alone. ◀

▶ Exercise Set 3.6

In Exercises 1–10, find dy/dx.

1. $y = 2x^{3/4}$.

2. $y = (4x^2 - 1)^{-2/3}$.

3. $y = \sqrt{1 + x^3}$.

4. $y = (\cos x^2)^{-5/8}$.

5. $y = \tan \sqrt{x}$.

6. $y = \sqrt{1 + \cos^2 x}$.

7. $y = (x + 1/x)^{-1/7}$.

8. $y = \sqrt{1 + \sqrt{x}}$.

9. $y = \sqrt{x} + \sqrt[3]{3x} + \sqrt[4]{5x}$.

10. $y = [\sec(x^{-2/3} + 1)]^{4/5}$

In Exercises 11–24, find dy/dx by implicit differentiation.

11. $x^2 + y^2 = 100$.

12. $x^3 - y^3 = 6xy$.

13. $\dfrac{1}{y} + \dfrac{1}{x} = 1$.

14. $x^2 = \dfrac{x + y}{x - y}$.

15. $\sqrt{x} + \sqrt{y} = 8$.

16. $\sqrt{xy} + 1 = y$.

17. $(x^2 + 3y^2)^{35} = x$.

18. $xy^{2/3} + yx^{2/3} = x^2$.

19. $\sin y = x$.

20. $\cos xy = y$.

21. $\tan^3 (xy^2 + y) = x$.

22. $x^2 = \dfrac{\cot y}{1 + \csc y}$.

23. $\sqrt{1 + \sin^3 (xy^2)} = y$.

24. $\dfrac{xy^3}{1 + \sec y} = 1 + y^4$.

In Exercises 25–28, use implicit differentiation to find the slope of the tangent to the given curve at the specified point.

25. $x^3y + y^3x = 10; \ (1, 2)$.

26. $\sin xy = y$; $\left(\frac{\pi}{2}, 1\right)$.

27. $x^{2/3} - y^{2/3} - y = 1$; $(1, -1)$.

28. $\sqrt{3 + \tan xy} - 2 = 0$; $(\pi/12, 3)$.

In Exercises 29–33, find the value of dy/dx at the specified point in two ways: first by solving for y in terms of x (as in Example 4), and then by implicit differentiation.

29. $xy = 8$; $(2, 4)$.

30. $y^2 - x + 1 = 0$; $(10, 3)$.

31. $x^2 + y^2 = 1$; $\left(\frac{1}{\sqrt{2}}, -\frac{1}{\sqrt{2}}\right)$.

32. $\dfrac{1 - y}{1 + y} = x$; $(0, 1)$.

33. $y^2 - 3xy + 2x^2 = 4$; $(3, 2)$.

In Exercises 34–37, use implicit differentiation to find the specified derivative.

34. $\sqrt{u} + \sqrt{v} = 5$; du/dv.

35. $a^4 - t^4 = 6a^2t$; da/dt.

36. $y = \sin x$; dx/dy.

37. $a^2\omega^2 + b^2\lambda^2 = 1$ (a, b constants); $d\omega/d\lambda$.

38. Use implicit differentiation to show that the equation of the tangent to the curve $y^2 = kx$ at (x_0, y_0) is

$$y_0 y = \frac{k}{2}(x + x_0)$$

39. Find dy/dx if

$$2y^3 t + t^3 y = 1 \quad \text{and} \quad x = \sin t$$

40. At what point(s) is the tangent to the curve

$$y^2 = 2x^3$$

perpendicular to the line $4x - 3y + 1 = 0$?

41. Find equations for two lines through the origin that are tangent to the curve

$$x^2 - 4x + y^2 + 3 = 0$$

42. (a) Use implicit differentiation to find the slope of the tangent to the four-cusped hypocycloid $x^{2/3} + y^{2/3} = 1$ at $P(-\frac{1}{4}\sqrt{2}, \frac{1}{4}\sqrt{2})$. (See Figure 3.6.5c.)

   (b) Do part (a) by solving explicitly for y as a function of x and then differentiating.

   (c) Trace the curve in Figure 3.6.5c and sketch the tangent line at P.

   (d) Where does $x^{2/3} + y^{2/3} = 1$ not define y as an implicit function of x?

43. The graph of $8(x^2 + y^2)^2 = 100(x^2 - y^2)$, shown in Figure 3.6.6, is called a **lemniscate**.

   (a) Where does this equation not define y as an implicit function of x?

   (b) Use implicit differentiation to find the equation of the tangent to the curve at the point $(3, 1)$.

   (c) Trace the curve in Figure 3.6.6 and sketch the tangent at $(3, 1)$.

Figure 3.6.6

3.7 CONTINUITY

A moving physical object cannot vanish at some point and reappear someplace else to continue its motion. Thus we perceive the path of a moving object as a single, unbroken curve without gaps, jumps, or holes. Such curves can be described as "continuous." In this section we will express this intuitive idea mathematically and develop some properties of continuous curves.

Before we give any formal definitions, let us consider some of the ways in which curves can be "discontinuous." In Figure 3.7.1 we have graphed some curves which, because of their behavior at the point c, fail to be continuous. The curve in Figure 3.7.1a has a hole at the point c because the function f

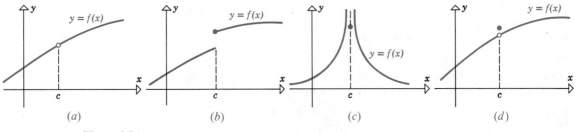

Figure 3.7.1

is undefined there. For the curves in Figures 3.7.1*b* and 3.7.1*c*, the function f is defined at c, but

$$\lim_{x \to c} f(x)$$

does not exist, thereby causing a break in the graph. For the curve in Figure 3.7.1*d*, the function f is defined at c, and the limit

$$\lim_{x \to c} f(x)$$

exists, yet the graph still has a break at the point c because

$$\lim_{x \to c} f(x) \neq f(c)$$

Based on this discussion, we see that there is a break or discontinuity in the graph of $y = f(x)$ at a point $x = c$ if any of the following conditions occur:

(i) The function f is undefined at c.

(ii) The limit $\lim_{x \to c} f(x)$ does not exist.

(iii) The function f is defined at c and the limit $\lim_{x \to c} f(x)$ exists, but the value of the function at c and the value of the limit at c are different.

This suggests the following definition:

3.7.1 DEFINITION   A function f is said to be ***continuous at a point c*** if the following conditions are satisfied:

1. $f(c)$ is defined.

2. $\lim_{x \to c} f(x)$ exists.

3. $\lim_{x \to c} f(x) = f(c)$.

If one or more of the conditions in this definition fails to hold, then f is called *discontinuous at c* and c is called a *point of discontinuity* of f. If f is continuous at all points of an open interval (a, b), then f is said to be *continuous on (a, b)*. A function that is continuous on $(-\infty, +\infty)$ is said to be *continuous everywhere* or simply *continuous*.

▶ Example 1

Let

$$f(x) = \frac{x^2 - 4}{x - 2} \quad \text{and} \quad g(x) = \begin{cases} \dfrac{x^2 - 4}{x - 2}, & x \neq 2 \\ 3, & x = 2 \end{cases}$$

Both f and g are discontinuous at 2; the function f because $f(2)$ is undefined, and the function g because $g(2) = 3$, while

$$\lim_{x \to 2} g(x) = \lim_{x \to 2} \frac{x^2 - 4}{x - 2} = \lim_{x \to 2} (x + 2) = 4$$

so that

$$\lim_{x \to 2} g(x) \neq g(2) \qquad \blacktriangleleft$$

▶ Example 2

It follows from Corollary 3.3.3 that

$$\lim_{x \to 0} \cos x = 1$$

Since $\cos (0) = 1$, it follows that $\cos x$ is continuous at 0. ◀

3.7.2 THEOREM *Polynomials are continuous functions.*

Proof. If p is a polynomial and c is any real number, then by Example 6 in Section 2.5

$$\lim_{x \to c} p(x) = p(c)$$

which proves the continuity of p at c. Since c is an arbitrary real number, p is continuous everywhere. ▌

The following basic result is an immediate consequence of Theorem 2.5.1.

3.7.3 THEOREM *If the functions f and g are continuous at c, then*

(a) $f + g$ *is continuous at c*
(b) $f - g$ *is continuous at c*
(c) $f \cdot g$ *is continuous at c*
(d) f/g *is continuous at c if $g(c) \neq 0$, and is discontinuous at c if $g(c) = 0$.*

We will prove part (*d*) and leave the remaining proofs as exercises.

Proof of (d). If $g(c) = 0$, then f/g is discontinuous at c because

$$\frac{f(c)}{g(c)}$$

is undefined.

Assume $g(c) \neq 0$. We must show that

$$\lim_{x \to c} \frac{f(x)}{g(x)} = \frac{f(c)}{g(c)} \tag{1}$$

Since f and g are continuous at c,

$$\lim_{x \to c} f(x) = f(c) \qquad \text{and} \qquad \lim_{x \to c} g(x) = g(c)$$

Thus, by Theorem 2.5.1(*d*)

$$\lim_{x \to c} \frac{f(x)}{g(x)} = \frac{\lim\limits_{x \to c} f(x)}{\lim\limits_{x \to c} g(x)} = \frac{f(c)}{g(c)}$$

which proves (1). ▪

▶ **Example 3**

Where is

$$h(x) = \frac{x^2 - 9}{x^2 - 5x + 6}$$

continuous?

Solution. The numerator and denominator of h are polynomials and, therefore, continuous everywhere by Theorem 3.7.2. Thus, by Theorem

3.7.3(d), the ratio is continuous everywhere except at the points where the denominator is zero. Since the solutions of

$$x^2 - 5x + 6 = 0$$

are $x = 2$ and $x = 3$, $h(x)$ is continuous everywhere except at these points.

◄

The result in Example 3 is a special case of the following general theorem. We leave the proof as an exercise.

3.7.4 THEOREM *A rational function is continuous everywhere except at the points where the denominator is zero.*

The following major theorem relates the notions of continuity and differentiability.

3.7.5 THEOREM *If f is differentiable at a point c, then f is also continuous at c.*

Proof. If we can show that

$$\lim_{x \to c} [f(x) - f(c)] = 0 \tag{2}$$

then the continuity of f at c will follow by writing

$$\lim_{x \to c} f(x) = \lim_{x \to c} [f(x) - f(c) + f(c)]$$
$$= \lim_{x \to c} [f(x) - f(c)] + \lim_{x \to c} f(c)$$
$$= 0 + f(c)$$
$$= f(c)$$

Let us prove (2). Since f is assumed to be differentiable at c, we know that the limit

$$\lim_{h \to 0} \frac{f(c + h) - f(c)}{h} = f'(c) \tag{3}$$

exists. Let us change the notation in (3) by defining

$$x = c + h \tag{4}$$

Thus

$$h = x - c$$

From (4), x approaches c as h approaches zero, so (3) can be written in the equivalent form

$$\lim_{x \to c} \frac{f(x) - f(c)}{x - c} = f'(c) \tag{5}$$

We can now use (5) to prove (2) by writing

$$
\begin{aligned}
\lim_{x \to c} [f(x) - f(c)] &= \lim_{x \to c} \left[\frac{f(x) - f(c)}{x - c} \cdot (x - c) \right] \\
&= \lim_{x \to c} \frac{f(x) - f(c)}{x - c} \cdot \lim_{x \to c} (x - c) \\
&= f'(c) \cdot 0 \\
&= 0
\end{aligned}
$$

▶ Example 4

Since the derivatives

$$\frac{d}{dx}[\sin x] = \cos x$$

$$\frac{d}{dx}[\cos x] = -\sin x$$

exist for all x, the functions $\sin x$ and $\cos x$ are differentiable everywhere; thus by Theorem 3.7.5, they are also continuous everywhere. ◀

▶ Example 5

Where is $\tan x$ continuous?

Solution. We can write

$$\tan x = \frac{\sin x}{\cos x}$$

Since the numerator and denominator are continuous everywhere by Example 4, the ratio will be continuous everywhere except at the points where the denominator is zero. But the solutions of

$$\cos x = 0$$

are

$$x = \pm \frac{\pi}{2},\ \pm \frac{3\pi}{2},\ \pm \frac{5\pi}{2},\ \ldots$$

so that tan x is continuous everywhere except at these points. ◀

Theorem 3.7.5 shows that differentiability at a point implies continuity at that point. The converse, however, is false—*a function may be continuous at a point but not differentiable there*. Whenever the graph of a function has a corner at a point, but no break or gap there, we have a point where the function is continuous, but not differentiable. For example, the function

$$f(x) = |x|$$

is continuous, but not differentiable at $x = 0$. (Figure 3.7.2.)

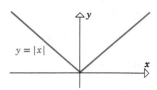

$y = |x|$

Figure 3.7.2

The relationship between continuity and differentiability was of great historical significance in the development of calculus. In the early nineteenth century mathematicians believed that the graph of a continuous function could not have too many points of nondifferentiability bunched up. They felt that if a continuous function had many points of nondifferentiability, these points, like the tips of a sawblade, would have to be separated from each other and joined by smooth curve segments (Figure 3.7.3).

Figure 3.7.3

This misconception was shattered by a series of discoveries beginning in 1834. In that year a Bohemian priest, philosopher, and mathematician named Bernhard Bolzano* discovered a procedure for constructing a continuous function that is not differentiable at any point. Later, in 1860, the

---

*BERNHARD BOLZANO (1781–1848). Bolzano, the son of an art dealer, was born in Prague, Bohemia (Czechoslovakia). He was educated at the University of Prague, and eventually won enough mathematical fame to be recommended for a mathematics chair there. However, Bolzano became an ordained Roman Catholic priest, and in 1805 he was appointed to a chair of Philosophy at the University of Prague. Bolzano was a man of great human compassion; he spoke out for educational reform, he voiced the right of individual conscience over government demands, and he lectured on the absurdity of war and militarism. His views so disenchanted Emperor Franz I of Austria that the emperor pressed the Archbishop of Prague to have Bolzano recant his statements. Bolzano refused and was then forced to retire in 1824 on a small pension. Bolzano's main contribution to mathematics was philosophical. His work helped convince mathematicians that sound mathematics must ultimately rest on rigorous proof rather than intuition. In addition to his work in mathematics, Bolzano investigated problems concerning space, force, and wave propogation.

great German mathematician, Karl Weierstrass,* produced the first formula for such a function. The graphs of such functions are impossible to draw; it is as if they oscillate so wildly that a corner occurs everywhere. Although these pathological functions rarely occur in practical problems, their discovery was of major importance because it made mathematicians distrustful of their geometric intuition and more reliant on precise mathematical proof.

3.7.6 THEOREM *Let* $\lim$ *stand for one of the limits* $\lim\limits_{x \to c}, \lim\limits_{x \to c^-}, \lim\limits_{x \to c^+}, \lim\limits_{x \to +\infty}, or \lim\limits_{x \to -\infty}$. *If* $\lim g(x) = L$ *and if the function* f *is continuous at* L, *then* $\lim f(g(x)) = f(L)$. *That is,* $\lim f(g(x)) = f(\lim g(x))$.

For those who have read Section 2.6, the optional proof of this result appears at the end of this section.

REMARK. In words, this theorem states that the limit symbol can be moved through a function sign provided the limit of the expression inside the function sign exists and the function is continuous at this limit.

▶ Example 6

Since the functions $\sin x$ and $\cos x$ are continuous everywhere, it follows from Theorem 3.7.6 that we can write

$$\lim \left[\sin \left(g(x) \right) \right] = \sin \left[\lim g(x) \right]$$

---

*KARL WEIERSTRASS (1815–1897). Weierstrass, the son of a customs officer, was born in Ostenfelde, Germany. As a youth Weierstrass showed outstanding skills in languages and mathematics. However, at the urging of his dominant father, Weierstrass entered the law and commerce program at the University of Bonn. To the chagrin of his family, the rugged and congenial young man concentrated instead on fencing and beer drinking. Four years later he returned home without a degree. In 1839 Weierstrass entered the Academy of Münster to study for a career in secondary education. There he met and studied under an excellent mathematician named Christof Gudermann. Gudermann's ideas greatly influenced the work of Weierstrass. After receiving his teaching certificate, Weierstrass spent the next 15 years in secondary education teaching German, geography, and mathematics. In addition he taught handwriting to small children. During this period much of Weierstrass's mathematical work was ignored because he was a secondary school teacher and not a college professor. Then in 1854 he published a paper of major importance, which created a sensation in the mathematics world and catapulted him to international fame overnight. He was immediately given an honorary Doctorate at the University of Königsberg and began a new career in college teaching at the University of Berlin in 1856. In 1859 the strain of his mathematical research caused him a temporary nervous breakdown and led to spells of dizziness that plagued him throughout his life. Weierstrass was a brilliant teacher and his classes overflowed with multitudes of auditors. In spite of his fame, he never lost his early beer-drinking congeniality and was always in the company of students, both ordinary and brilliant. Weierstrass was acknowledged as the leading mathematical analyst in the world. He and his students opened the door to the modern school of mathematical analysis.

and

$$\lim \left[\cos \left(g(x) \right) \right] = \cos \left[\lim g(x) \right]$$

if $\lim g(x)$ exists. For example,

$$\lim_{x \to \pi} \left[\sin \left(\frac{x^2}{\pi + x} \right) \right] = \sin \left[\lim_{x \to \pi} \left(\frac{x^2}{\pi + x} \right) \right] = \sin \frac{\pi}{2} = 1$$

$$\lim_{x \to +\infty} \left[\cos \left(\frac{\pi x^2 + 1}{x^2 + 3} \right) \right] = \cos \left[\lim_{x \to +\infty} \left(\frac{\pi x^2 + 1}{x^2 + 3} \right) \right] = \cos \left[\lim_{x \to +\infty} \left(\frac{\pi + (1/x^2)}{1 + (3/x^2)} \right) \right]$$

$$= \cos \pi = -1 \qquad \blacktriangleleft$$

▶ Example 7

Since $|x|$ is continuous everywhere (Exercise 24), Theorem 3.7.6 implies that

$$\lim |g(x)| = |\lim g(x)|$$

whenever $\lim g(x)$ exists. For example,

$$\lim_{x \to 3} |5 - x^2| = |\lim_{x \to 3} (5 - x^2)| = |-4| = 4 \qquad \blacktriangleleft$$

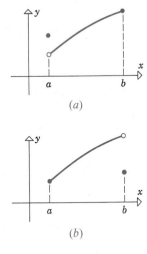

(a)

(b)

(c)

Figure 3.7.4

Figure 3.7.4 shows the graphs of three functions defined only on a closed interval $[a, b]$. Obviously the function shown in Figure 3.7.4a should be regarded as discontinuous at the left-hand endpoint a, the function in Figure 3.7.4b should be regarded as discontinuous at the right-hand endpoint b, and the function in Figure 3.7.4c should be regarded as continuous at both endpoints. However, the definition of continuity (Definition 3.7.1) does not apply at the endpoints, since the two-sided limits appearing in parts 2 and 3 of the definition makes no sense. At the left-hand endpoint the only sensible limit is the one-sided limit

$$\lim_{x \to a^+} f(x)$$

and at the right-hand endpoint the only sensible limit is the one-sided limit

$$\lim_{x \to b^-} f(x)$$

This suggests the following definitions.

3.7.7 DEFINITION   A function f is called *continuous from the left at the point c* if the conditions in the first column below are satisfied, and is called *continuous from the right at the point c* if the conditions in the second column are satisfied.

1. $f(c)$ is defined.	1'. $f(c)$ is defined.
2. $\lim\limits_{x \to c^-} f(x)$ exists.	2'. $\lim\limits_{x \to c^+} f(x)$ exists.
3. $\lim\limits_{x \to c^-} f(x) = f(c)$.	3'. $\lim\limits_{x \to c^+} f(x) = f(c)$.

▶ Example 8

If f denotes the function graphed in Figure 3.7.4a, then

$$f(a) \neq \lim_{x \to a^+} f(x) \quad \text{and} \quad f(b) = \lim_{x \to b^-} f(x)$$

Thus f is continuous from the left at b, but is not continuous from the right at a. ◀

When we state that a function f is *continuous on a closed interval* $[a, b]$ we will mean:

(i) f is continuous on (a, b),
(ii) f is continuous from the right at a,
(iii) f is continuous from the left at b.

The reader should have no trouble deducing the appropriate definitions of continuity on intervals of the form $[a, +\infty)$, $(-\infty, b]$, $[a, b)$, and $(a, b]$.

▶ Example 9

Show that

$$f(x) = \sqrt{9 - x^2}$$

is continuous on the closed interval $[-3, 3]$.

Solution.   Observe that the domain of f is the interval $[-3, 3]$. For c in the interval $(-3, 3)$ we have by Theorem 3.7.6

$$\lim_{x \to c} f(x) = \lim_{x \to c} \sqrt{9 - x^2} = \sqrt{9 - c^2} = f(c)$$

so that f is continuous on $(-3, 3)$. Also,

$$\lim_{x \to 3^-} f(x) = \lim_{x \to 3^-} \sqrt{9 - x^2} = 0 = f(3)$$

Figure 3.7.5

Figure 3.7.6

and

$$\lim_{x \to -3^+} f(x) = \lim_{x \to -3^+} \sqrt{9 - x^2} = 0 = f(-3)$$

so that f is continuous on $[-3, 3]$. ◄

▶ Example 10

Example 5 implies that

$$f(x) = \tan x$$

is continuous on the open interval $(-\pi/2, \pi/2)$. However,

$$f\left(-\frac{\pi}{2}\right) = \tan\left(-\frac{\pi}{2}\right) \quad \text{and} \quad f\left(\frac{\pi}{2}\right) = \tan\frac{\pi}{2}$$

are undefined, so $\tan x$ is not continuous on the closed interval $[-\pi/2, \pi/2]$. (See Figure 3.7.5.) ◄

If f is continuous on a closed interval $[a, b]$, and we draw a horizontal line crossing the y-axis between the numbers $f(a)$ and $f(b)$ (Figure 3.7.6), then it is geometrically obvious that this line will cross the curve $y = f(x)$ at least once over the interval $[a, b]$. Stated another way, a continuous function must assume every possible value between $f(a)$ and $f(b)$ as x varies from a to b. This idea is formalized in the following theorem.

3.7.8 THEOREM
Intermediate Value Theorem   *If f is continuous on a closed interval $[a, b]$ and C is any number between $f(a)$ and $f(b)$, inclusive, then there is at least one number x in the interval $[a, b]$ such that $f(x) = C$.*

(See Figure 3.7.6.) This theorem, though intuitively obvious, is not easy to prove. The proof can be found in most advanced calculus texts.

OPTIONAL

We conclude this section with a proof of Theorem 3.7.6 for those who have read Section 2.6. We will consider only the case where lim is the two-sided limit $\lim_{x \to c}$.

Proof. To prove that

$$\lim_{x \to c} f(g(x)) = f(L)$$

we must show that for every $\epsilon > 0$ there is a $\delta > 0$ such that $f(g(x))$ satisfies

$$|f(g(x)) - f(L)| < \epsilon \tag{6}$$

whenever x satisfies

$$0 < |x - c| < \delta \tag{7}$$

Since f is continuous at L,

$$\lim_{x \to L} f(x) = f(L)$$

If we use u as the variable rather than x, then this limit can be written

$$\lim_{u \to L} f(u) = f(L)$$

Thus, given $\epsilon > 0$ we can find a $\delta_1 > 0$ such that $f(u)$ satisfies

$$|f(u) - f(L)| < \epsilon \tag{8}$$

whenever u satisfies

$$|u - L| < \delta_1 \tag{9}$$

Since

$$\lim_{x \to c} g(x) = L$$

there exists a $\delta > 0$ such that

$$|g(x) - L| < \delta_1 \tag{10}$$

whenever x satisfies (7). To complete the proof let

$$u = g(x)$$

and assume that x satisfies (7). This implies that $g(x)$ satisfies (10), which implies that u satisfies (9). Therefore $f(u)$ satisfies (8), which implies that $f(g(x))$ satisfies (6). Thus, (6) is satisfied whenever (7) is satisfied.

► Exercise Set 3.7

In Exercises 1–4, let f be the function whose graph is shown. On which of the following intervals, if any, is f continuous?

(a) $[1, 3]$ (b) $(1, 3)$
(c) $[1, 2]$ (d) $(1, 2)$
(e) $[2, 3]$ (f) $(2, 3)$

On those intervals where f is discontinuous, state where the discontinuities occur.

1.

2.

3.
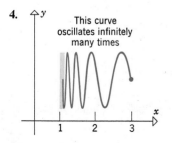

In Exercises 5–20, find the points of discontinuity, if any.

5. $f(x) = x^3 - 2x + 3$.

6. $f(x) = (x - 5)^{17}$.

7. $f(x) = \dfrac{x}{x^2 + 1}$.

8. $f(x) = \dfrac{x}{x^2 - 1}$.

9. $f(x) = \dfrac{x - 4}{x^2 - 16}$.

10. $f(x) = \dfrac{3x + 1}{x^2 + 7x - 2}$.

11. $f(x) = \sin(x^2 - 2)$.

12. $f(x) = \cos\left(\dfrac{x}{x - \pi}\right)$.

13. $f(x) = \cot x$.

14. $f(x) = \sec x$.

15. $f(x) = \csc x$.

16. $f(x) = \dfrac{1}{1 + \sin^2 x}$.

17. $f(x) = |\cos x|$.

18. $f(x) = \dfrac{x}{|x|}$.

19. $f(x) = \sqrt[3]{\dfrac{1}{x} - \dfrac{1}{x - 1}}$.

20. $f(x) = \sqrt{2 + \tan^2 x}$.

21. Find a value for the constant k that will make the function continuous.

(a) $f(x) = \begin{cases} 7x - 2, & x \le 1 \\ kx^2, & x > 1 \end{cases}$

(b) $f(x) = \begin{cases} \dfrac{\sin x}{x}, & x \ne 0 \\ k, & x = 0 \end{cases}$

22. On which of the following intervals is
$$f(x) = \dfrac{1}{\sqrt{x - 2}}$$
continuous?

(a) $[2, +\infty)$ (b) $(-\infty, +\infty)$
(c) $(2, +\infty)$ (d) $[1, 2)$.

4.

This curve oscillates infinitely many times

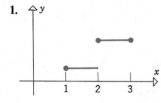

23. Prove that $f(x) = \sqrt{x}$ is continuous on $[0, +\infty)$.

24. Prove that $f(x) = |x|$ is continuous everywhere.

25. Prove that $\sin(g(x))$ is continuous at every point where $g(x)$ is continuous.

26. Prove: If g is continuous at c and f is continuous at $g(c)$, then $f \circ g$ is continuous at the point c.

27. Use Exercise 26 to prove that the composition of continuous functions is continuous.

28. Use Exercise 27 to prove that the following functions are continuous
    (a) $\sin(x^3 + 7x + 1)$ (b) $|\cos x|$
    (c) $\cos^3(x + 1)$.

29. Prove:
    (a) part (a) of Theorem 3.7.3
    (b) part (b) of Theorem 3.7.3
    (c) part (c) of Theorem 3.7.3.

30. Prove Theorem 3.7.4.

31. Let f and g be discontinuous at c. Give examples to show:
    (a) $f + g$ can be continuous or discontinuous at c.
    (b) $f \cdot g$ can be continuous or discontinuous at c.

32. **(For students who have read Section 2.6.)** Let f be defined at c. Prove that f is continuous at c if and only if, given $\epsilon > 0$ there exists a $\delta > 0$ such that $|f(x) - f(c)| < \epsilon$ whenever $|x - c| < \delta$.

In Exercises 33–35, you will have to determine whether a function f is differentiable at a point x_0 where the formula for f changes. Use the following result:

Theorem. *Let f be continuous at x_0 and suppose that*

$$\lim_{x \to x_0^+} f'(x) \qquad \text{and} \qquad \lim_{x \to x_0^-} f'(x)$$

exist. Then f is differentiable at x_0 if and only if these limits are equal. Moreover, in the case of equality

$$f'(x_0) = \lim_{x \to x_0^+} f'(x) = \lim_{x \to x_0^-} f'(x)$$

33. Let
$$f(x) = \begin{cases} x^2, & x \leq 1 \\ \sqrt{x}, & x > 1 \end{cases}$$

Determine if f is differentiable at $x = 1$. If so, find the value of the derivative there.

34. Let
$$f(x) = \begin{cases} x^3 + \dfrac{1}{16}, & x < \dfrac{1}{2} \\ \dfrac{3}{4}x^2, & x \geq \dfrac{1}{2} \end{cases}$$

Determine if f is differentiable at $x = \frac{1}{2}$. If so, find the value of the derivative there.

35. Find all points where f fails to be differentiable. Justify your answer.
    (a) $f(x) = |3x - 2|$
    (b) $f(x) = |x^2 - 4|$.

▶ REVIEW EXERCISES

In Exercises 1–4, use Definition 3.1.2 to find $f'(x)$.

1. $f(x) = kx$ (k constant).

2. $f(x) = (x - a)^2$ (a constant).

3. $f(x) = \sqrt{9 - 4x}$.

4. $f(x) = \dfrac{x}{x + 1}$.

5. Use Definition 3.1.2 to find $\dfrac{d}{dx}[|x|^3]\Big|_{x=0}$.

6. Suppose $f(x) = \begin{cases} x^2 - 1, & x \leq 1 \\ k(x - 1), & x > 1 \end{cases}$.
   For what values of k is f
   (a) continuous
   (b) differentiable?

7. Suppose $f(3) = -1$ and $f'(3) = 5$. Find an equation for the tangent line to the graph of f at $x = 3$.

8. Let $f(x) = x^2$. Show that for any distinct values of a and b, the slope of the tangent line to $y = f(x)$ at

$x = \frac{1}{2}(a + b)$ is equal to the slope of the secant line through the points (a, a^2) and (b, b^2).

9. Given the following table of values at $x = 1$ and $x = -2$, find the indicated derivatives in (a)–(l).

x	$f(x)$	$f'(x)$	$g(x)$	$g'(x)$
1	1	3	-2	-1
-2	-2	-5	1	7

(a) $\frac{d}{dx}[f^2(x) - 3g(x^2)]\Big|_{x=1}$

(b) $\frac{d}{dx}[f(x)g(x)]\Big|_{x=1}$

(c) $\frac{d}{dx}\left[\frac{f(x)}{g(x)}\right]\Big|_{x=-2}$

(d) $\frac{d}{dx}\left[\frac{g(x)}{f(x)}\right]\Big|_{x=-2}$

(e) $\frac{d}{dx}[f(g(x))]\Big|_{x=1}$

(f) $\frac{d}{dx}[f(g(x))]\Big|_{x=-2}$

(g) $\frac{d}{dx}[g(f(x))]\Big|_{x=-2}$

(h) $\frac{d}{dx}[g(g(x))]\Big|_{x=-2}$

(i) $\frac{d}{dx}[f(g(4 - 6x))]\Big|_{x=1}$

(j) $\frac{d}{dx}[g^3(x)]\Big|_{x=1}$

(k) $\frac{d}{dx}[\sqrt{f(x)}]\Big|_{x=1}$

(l) $\frac{d}{dx}[f(-\frac{1}{2}x)]\Big|_{x=-2}$

In Exercises 10–15, find $f'(x)$ and determine those values of x for which $f'(x) = 0$.

10. $f(x) = (2x + 7)^6(x - 2)^5$.

11. $f(x) = \dfrac{(x - 3)^4}{x^2 + 2x}$.

12. $f(x) = \sqrt{3x + 1}\,(x - 1)^2$.

13. $f(x) = \left(\dfrac{3x + 1}{x^2}\right)^3$.

14. $f(x) = \dfrac{3(5x - 1)^{1/3}}{3x - 5}$.

15. $f(x) = \sqrt{x}\,\sqrt[3]{x^2 + x + 1}$.

16. Suppose that $f'(x) = 1/x$ for all $x \neq 0$.
   (a) Use the chain rule to show that for any non-zero constant a, $d(f(ax))/dx = d(f(x))/dx$.
   (b) If $y = f(\sin x)$ and $v = f(1/x)$, find dy/dx and dv/dx.

In Exercises 17–26, find the indicated derivatives.

17. $\dfrac{d}{dx}\left(\dfrac{\sqrt{2}}{x^2} - \dfrac{2}{5x}\right)$.

18. $\dfrac{dy}{dx}$ if $y = \dfrac{3x^2 + 7}{x^2 - 1}$.

19. $\dfrac{dz}{dr}\Big|_{r=\frac{\pi}{6}}$ if $z = 4\sin^2 r \cos^2 r$.

20. $g'(2)$ if $g(x) = 1/\sqrt{2x}$.

21. $\dfrac{du}{dx}$ if $u = \left(\dfrac{x}{x - 1}\right)^{-2}$.

22. $\dfrac{dw}{dv}$ if $w = \sqrt[5]{v^3} - \sqrt[4]{v}$.

23. $d(\sec^2 x - \tan^2 x)/dx$.

24. $\dfrac{dy}{dx}\Big|_{x=\frac{\pi}{4}}$ if $y = \tan t$ and $t = \cos(2x)$.

25. $F'(x)$ if $F(x) = \dfrac{(1/x) + 2x}{\frac{1}{2}(1/x^2) + 1}$.

26. $\Phi'(x)$ if $\Phi(x) = \dfrac{x^2 - 4x}{5\sqrt{x}}$.

27. Find all values of x for which the tangent to $f(x) = x - (1/x)$ is parallel to the line $2x - y = 5$.

28. Find all values of x for which the tangent to $f(x) = 2x^3 - x^2$ is perpendicular to the line $x + 4y = 10$.

29. Find all values of x for which the tangent to $f(x) = (x + 2)^2$ passes through the origin.

30. Find all values of x for which the tangent to $f(x) = x - \sin 2x$ is horizontal.

31. Find all values of x for which the tangent to $f(x) = 3x - \tan x$ is parallel to the line $y - x = 2$.

In Exercises 32–34, find Δx, Δy, and dy.

32. $y = 1/(x - 1)$; x decreases from 2 to 1.5.

33. $y = \tan x$; x increases from $-\pi/4$ to 0.

34. $y = \sqrt{25 - x^2}$; x increases from 0 to 3.

35. Use a differential to approximate (a) $\sqrt[3]{-8.25}$ (b) $\cot 46°$.

36. The area of a right triangle whose hypotenuse is H centimeters is calculated using the formula $A = \frac{1}{4}H^2 \sin 2\theta$, where θ is one of the acute angles. Use differentials to approximate the error in calculating A if $H = 4$ centimeters (exactly) and $\theta = 30° \pm 15'$.

37. A 12-ft ladder leaning against a wall makes an angle θ with the floor. If the top of the ladder is h feet up the wall, express h in terms of θ and then use dh to estimate the change in h if θ changes from 60° to 59°.

38. Let V and S denote the volume and surface area of a cube. Find the rate of change of V with respect to S.

39. The amount of water in a tank t minutes after it has started to drain is given by $W = 100(t - 15)^2$ (gallons).
  (a) How fast is the water running out at the end of 5 minutes?
  (b) What is the average rate at which the water flows out during the first 5 minutes?

In Exercises 40–42, find dy/dx by implicit differentiation and use it to find the equation of the tangent line at the indicated points.

40. $(x + y)^3 + 3xy = -7$; $(-2, 1)$.

41. $xy^2 = \sin (x + 2y)$; $(0, 0)$.

42. $(x + y)^3 - 5x + y = 1$; at the point where the curve intersects the line $x + y = 1$.

43. Show that the curves whose equations are $2x^2 + 3y^2 = 5$ and $y^2 = x^3$ intersect at the point $(1, 1)$ and that their tangent lines are perpendicular there.

44. Show that for any point $P_0(x_0, y_0)$ on the circle $x^2 + y^2 = r^2$, the tangent line at P_0 is perpendicular to the radial line from the origin to P_0.

In Exercises 45–48, find the indicated limit if it exists.

45. $\lim\limits_{x \to 0} \dfrac{\tan ax}{\sin bx}$ $(a \neq 0, b \neq 0)$.

46. $\lim\limits_{x \to 0} \dfrac{\sin 3x}{\tan 3x}$.

47. $\lim\limits_{\theta \to 0} \dfrac{\sin 2\theta}{\theta^2}$.

48. $\lim\limits_{x \to 0} \dfrac{x \sin x}{1 - \cos x}$.

49. Given that the sine and cosine functions are continuous at $x = 0$, use a suitable trigonometric identity to deduce that for any x, $\lim\limits_{h \to 0} \sin (x + h) = \sin x$. What have you proved?

4 applications of differentiation

4.1 RELATED RATES

In this section we will study *related rates* problems. In such problems one tries to find the rate at which some quantity is changing by relating it to other quantities whose rates of change are known.

▶ Example 1

Assume that oil spilled from a ruptured tanker spreads in a circular pattern whose radius increases at a constant rate of 2 ft/sec. How fast is the area of the spill increasing when the radius of the spill is 60 ft?

Solution.

Let

t = number of seconds elapsed from the time of the spill
r = radius of the spill in feet after t seconds
A = area of the spill in square feet after t seconds

Figure 4.1.1

(See Figure 4.1.1.) At each instant the rate at which the radius is increasing with time is dr/dt, and the rate at which the area is increasing with time is dA/dt. We want to find

$$\left. \frac{dA}{dt} \right|_{r=60}$$

which is the rate at which the area of the spill is increasing at the instant when $r = 60$.

Since the radius increases at the constant rate of 2 ft/sec, we know

$$\frac{dr}{dt} = 2 \qquad (1)$$

for all t.

From the formula for the area of a circle we obtain

$$A = \pi r^2 \qquad (2)$$

Because A and r are functions of t, we can differentiate both sides of (2) with respect to t to obtain

$$\frac{dA}{dt} = 2\pi r \frac{dr}{dt}$$

Substituting (1) yields

$$\frac{dA}{dt} = 2\pi r(2) = 4\pi r$$

Thus, when $r = 60$ the area of the spill is increasing at the rate

$$\frac{dA}{dt}\bigg|_{r=60} = 4\pi(60) = 240\pi \ \text{ft}^2/\text{sec}$$

or approximately 754 ft²/sec. ◄

With only minor variations, the method used in Example 1 can be used to solve a variety of related rate problems. The method consists of four steps:

Step 1. Label the quantities that vary.
Step 2. Find an equation relating the quantity with the unknown rate of change to quantities whose rates of change are known.
Step 3. Differentiate both sides of this equation with respect to time and solve for the derivative that will give the unknown rate of change.
Step 4. Evaluate this derivative at the appropriate point.

Rocket

Camera to rocket distance

Elevation angle

3000 ft

Camera

Launching pad

Figure 4.1.2

In Figure 4.1.2 we have shown a camera mounted at a point 3000 ft from the base of a rocket launching pad. Let us assume that rocket rises vertically and the camera is to take a series of photographs. Because the rocket will be rising, the elevation angle of the camera will have to vary at just the right rate to keep the rocket in sight. Moreover, because the camera-to-rocket distance will be changing constantly, the camera focusing mechanism will also have to vary at just the right rate to keep the picture sharp. Our next two examples are concerned with these problems.

▶ Example 2

If the rocket shown in Figure 4.1.2 is rising vertically at 880 ft/sec when it is 4000 ft up, how fast is the camera-to-rocket distance changing at that instant?

Solution. Let

t = number of seconds elapsed from the time of launch
y = camera to rocket distance in feet after t seconds
x = height of rocket in feet after t seconds

Figure 4.1.3

At each instant the rate at which the camera-to-rocket distance is changing is dy/dt, and the rate at which the rocket is rising is dx/dt. We want to find

$$\frac{dy}{dt}\bigg|_{x=4000}$$

which is the rate the camera-to-rocket distance is changing at the instant when $x = 4000$.

From the Theorem of Pythagoras and Figure 4.1.3 we have

$$y^2 = x^2 + (3000)^2 \tag{3}$$

Because x and y are functions of t, we can differentiate both sides of (3) with respect to t to obtain

$$2y\frac{dy}{dt} = 2x\frac{dx}{dt}$$

or

$$\frac{dy}{dt} = \frac{x}{y}\frac{dx}{dt} \tag{4}$$

When $x = 4000$, it follows from (3) that $y = 5000$; moreover, we are given that $dx/dt = 880$ when $x = 4000$ so that from (4)

$$\frac{dy}{dt}\bigg|_{x=4000} = \frac{4000}{5000}(880) = 704 \text{ ft/sec} \qquad ◀$$

► Example 3

If the rocket shown in Figure 4.1.4 is rising vertically at 880 ft/sec when it is 4000 ft up, how fast must the camera elevation angle change at that instant to keep the rocket in sight?

Solution. Let

t = number of seconds elapsed from the time of launch
ϕ = camera elevation angle in radians after t seconds
x = height of rocket in feet after t seconds

Rocket

x

ϕ

3000 ft

Camera

Figure 4.1.4

(See Figure 4.1.4.) At each instant the rate at which the camera elevation angle must change is $d\phi/dt$, and the rate at which the rocket is rising is dx/dt. We want to find

$$\left. \frac{d\phi}{dt} \right|_{x=4000}$$

which is the rate the camera elevation angle must change at the instant when $x = 4000$.

From Figure 4.1.4 we see that

$$\tan \phi = \frac{x}{3000} \tag{5}$$

Because ϕ and x are functions of t, we can differentiate both sides of (5) with respect to t to obtain

$$(\sec^2 \phi) \frac{d\phi}{dt} = \frac{1}{3000} \frac{dx}{dt}$$

or

$$\frac{d\phi}{dt} = \frac{1}{3000 \sec^2 \phi} \frac{dx}{dt} \tag{6}$$

When $x = 4000$, it follows that

$$\sec \phi = \frac{5000}{3000} = \frac{5}{3}$$

5000

4000

ϕ

3000

Figure 4.1.5

(See Figure 4.1.5.) Moreover, we are given that $dx/dt = 880$ when $x = 4000$ so that from (6)

$$\left. \frac{d\phi}{dt} \right|_{x=4000} = \frac{1}{3000(\frac{5}{3})^2} \cdot 880 = \frac{66}{625} \approx 0.11 \text{ radian/sec} \qquad ◄$$

▶ Example 4

Suppose a liquid is to be cleared of sediment by pouring it through a cone-shaped filter as shown in Figure 4.1.6a. Assume the height of the cone is 16 in. and the radius at the base of the cone is 4 in. (Figure 4.1.6b.) If the liquid is flowing out of the cone at a constant rate of 2 in³/min, how fast is the depth of the liquid decreasing when the level is 8 in. deep?

Solution. Let

t = time elapsed from the initial observation (min)
V = volume of liquid in the cone at time t (in.³)
y = depth of the liquid in the cone at time t (in.)
x = radius of the liquid surface at time t (in.)

(See Figure 4.1.6b.) At each instant the rate at which the volume of liquid is changing is dV/dt and the rate at which the depth is changing is dy/dt. We want to find

$$\left. \frac{dy}{dt} \right|_{y=8}$$

which is the rate at which the depth is changing at the instant the depth is 8 in.

From the formula for the volume of a cone, the volume V and the depth y are related by

$$V = \tfrac{1}{3}\pi x^2 y \qquad (7)$$

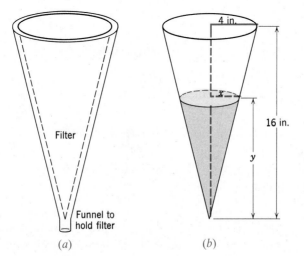

Figure 4.1.6   (a)   (b)

Since we are given that the volume of liquid decreases at a constant rate of 2 in.^{3}/min, we have

$$\frac{dV}{dt} = -2 \qquad (8)$$

(We must use a minus sign here because *V decreases* as *t* increases.) If we differentiate both sides of (7) with respect to *t*, the right side will involve the quantity dx/dt. Since we have no direct information about dx/dt, it is desirable to eliminate *x* from (7) before differentiating. This can be done using similar triangles. From Figure 4.1.6*b*, we see that

$$\frac{x}{y} = \frac{4}{16}$$

or

$$x = \frac{1}{4}y$$

Substituting this expression in (7) gives

$$V = \frac{\pi}{48}y^3 \qquad (9)$$

Differentiating both sides of (9) with respect to *t* we obtain

$$\frac{dV}{dt} = \frac{\pi}{48}\left(3y^2 \frac{dy}{dt}\right)$$

or

$$\frac{dy}{dt} = \frac{16}{\pi y^2}\frac{dV}{dt} \qquad (10)$$

Substituting (8) into (10), and letting $y = 8$ yields

$$\left.\frac{dy}{dt}\right|_{y=8} = \frac{16}{\pi(8)^2}(-2) = -\frac{1}{2\pi} \approx -0.16 \text{ in./min}$$

Thus, when $y = 8$ in., the depth of the liquid is decreasing at an approximate rate of 0.16 in./min. (The minus sign simply means that the depth *y* is decreasing as *t* is increasing.) ◀

Arteriosclerosis is a disease in which the openings in the arteries become narrow because of fatty deposits on the artery wall. In Figure 4.1.7 we have shown a cross section of an idealized circular artery with a uniformly thick fatty deposit on the wall. As the disease of arteriosclerosis progresses, the

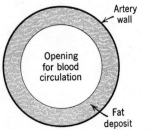

Artery wall

Opening for blood circulation

Fat deposit

Figure 4.1.7

thickness of the fat deposit increases, and the opening for blood circulation decreases. Ideally, a medical researcher would like to know the rate at which the area of the artery opening is decreasing with time. However, this depends on the rate at which the thickness of the fat deposit increases with time, and this is generally unknown because it depends on the individual's diet, exercise, smoking habits, and so forth. However, instead of considering the rate at which the area of the opening varies with time, the researcher can avoid this problem by studying the rate at which the area of the opening varies relative to the thickness of the fat deposit. This is illustrated in the next example.

▶ Example 5

Fat deposits are building up uniformly on the wall of an artery whose cross section is a circle of radius 1.5 cm. At what rate is the cross-sectional area of the artery opening changing relative to the thickness of the fat deposit when the deposit is 0.5 cm thick?

Solution. Let

$$A = \text{cross-sectional area (cm}^2\text{) of the artery opening}$$
$$h = \text{thickness (cm) of the fat deposit}$$

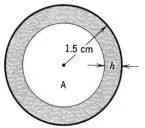

Figure 4.1.8

(See Figure 4.1.8.) The rate at which the area of the artery opening varies relative to the thickness of the fat deposit is dA/dh. Since we want to determine this rate when $h = 0.5$ cm, we must find

$$\frac{dA}{dh}\bigg|_{h=0.5}$$

From Figure 4.1.8 the radius of the artery opening is $1.5 - h$, so that the area of the artery opening is

$$A = \pi(1.5 - h)^2$$

Differentiating both sides with respect to h, we obtain

$$\frac{dA}{dh} = -2\pi(1.5 - h)$$

Thus

$$\frac{dA}{dh}\bigg|_{h=0.5} = -2\pi(1.5 - 0.5) = -2\pi \approx -6.28 \text{ cm}^2/\text{cm}$$

Therefore, when $h = 0.5$ the area of the artery opening is decreasing at a rate that is approximately 6.28 times as fast as the rate at which the fat deposit is thickening. ◀

▶ Exercise Set 4.1

1. Let A be the area of a square whose sides have length x, and assume that x varies with time.
   (a) How are dA/dt and dx/dt related?
   (b) At a certain instant the sides are 3 ft long and growing at a rate of 2 ft/min. How fast is the area growing at that instant?

2. Let A be the area of a circle whose radius is r, and assume that r varies with time.
   (a) How are dA/dt and dr/dt related?
   (b) At a certain instant the radius is 5 in. and is increasing at a rate of 2 in./sec. How fast is the area of the circle increasing at that instant?

3. Let V be the volume of a cylinder having height h and radius r, and assume that h and r vary with time.
   (a) How are dV/dt, dh/dt, and dr/dt related?
   (b) At a certain instant, the height is 6 in. and increasing at 1 in./sec, while the radius is 10 in. and decreasing at 1 in./sec. How fast is the volume changing at that instant? Is the volume increasing or decreasing at that instant?

4. Let l be the length of a diagonal of a rectangle whose sides have lengths x and y, and assume that x and y vary with time.
   (a) How are dl/dt, dx/dt, and dy/dt related?
   (b) If x increases at a constant rate of $\frac{1}{2}$ ft/sec and y decreases at a constant rate of $\frac{1}{4}$ ft/sec, how fast is the size of the diagonal changing when $x = 3$ ft and $y = 4$ ft? Is the diagonal increasing or decreasing at that instant?

5. Let θ (in radians) be an acute angle in a right triangle, and let x and y, respectively, be the lengths of the sides adjacent and opposite θ. Suppose also that x and y vary with time.
   (a) How are $d\theta/dt$, dx/dt, and dy/dt related?
   (b) At a certain instant, $x = 2$ units and is increasing at 1 unit/sec, while $y = 2$ units and is decreasing at $\frac{1}{4}$ unit/sec. How fast is θ changing at that instant? Is θ increasing or decreasing at that instant?

6. A 17 ft ladder is leaning against a wall. If the bottom of the ladder is pulled along the ground away from the wall at a constant rate of 5 ft/sec, how fast will the top of the ladder be moving down the wall when it is 8 ft above the ground?

7. Coffee is poured at a uniform rate of 2 cc/sec into a cup whose bowl is shaped like a truncated cone. If the upper and lower radii of the cup are 4 cm and 2 cm and the height of the cup is 6 cm, how fast will the coffee level be rising when the coffee is halfway up?

8. Wheat is poured through a chute at the rate of 10 ft³/min, and falls in a cone-shaped pile whose bottom radius is always half the altitude. How fast will the circumference of the base be increasing when the pile is 8 ft high?

9. A spherical balloon is inflated so that its volume is increasing at the rate of 3 ft³/min. How fast is the diameter of the balloon increasing when the radius is 1 ft?

10. A man 6 ft tall is walking at the rate of 3 ft/sec toward a street light 18 ft tall.

(a) At what rate is the size of his shadow changing?

(b) How fast is the tip of his shadow moving?

11. An aircraft is climbing at a 30° angle to the horizontal. How fast is the aircraft gaining altitude if its speed is 500 mph?

12. An aircraft is flying at a constant altitude with a constant speed of 600 mph. An antiaircraft missile is fired on a straight line perpendicular to the flight path of the aircraft so it will hit the aircraft at a point P. At the instant the aircraft is 2 miles from the impact point P the missile is 4 miles from P and flying at 1200 mph. At that instant, how rapidly is the distance between missile and aircraft decreasing?

13. Solve Exercise 12 under the assumption that the angle between the flight paths is 120° instead of the assumption that the paths are perpendicular. [*Hint:* Use the law of cosines.]

14. A stone dropped into a still pond sends out a circular ripple whose radius increases at a constant rate of 3 ft/sec. How rapidly is the area of the ripple increasing at the end of 10 sec?

15. For the camera and rocket shown in Figure 4.1.2, at what rate is the elevation angle changing when the rocket is 3000 ft up and rising vertically at 500 ft/sec?

16. For the artery shown in Figure 4.1.8, at what rate is the radius of the artery opening changing relative to the cross-sectional area of the fat deposit when the deposit is 0.2 cm thick?

17. A particle is moving along the curve whose equation is

$$\frac{xy^3}{1+y^2} = \frac{8}{5}$$

Assume that the x-coordinate is increasing at the rate of 6 units/sec when the particle is at the point (1, 2).

(a) At what rate is the y-coordinate of the point changing at that instant?

(b) Is the particle rising or falling at that instant?

18. A police helicopter is flying due north at 100 mph, and at a constant altitude of $\frac{1}{2}$ mile. Below, a car is traveling west on a highway at 75 mph. At the moment the helicopter crosses over the highway the car is 2 miles east of the helicopter.

(a) How fast is the distance between the car and helicopter changing at the moment the helicopter crosses the highway?

(b) Is the distance between car and helicopter increasing or decreasing at that moment?

19. On a certain clock the minute hand is 4 in. long and the hour hand is 3 in. long. How fast is the distance between the tips of the hands changing at 9 o'clock?

20. Water is stored in an inverted cone-shaped reservoir (vertex down). Assuming that the water evaporates at a rate proportional to the surface area exposed to the air, show that the depth of the water will decrease at a constant rate that does not depend on the dimensions of the reservoir.

21. A meteorite enters the earth's atmosphere and burns up at a rate that, at each instant, is proportional to its surface area. Assuming that the meteorite is always spherical, show that the radius decreases at a constant rate.

4.2 MAXIMUM AND MINIMUM VALUES OF A FUNCTION

Problems concerned with finding the "best" way to perform a task are called *optimization* problems. A large class of optimization problems can be reduced to finding the maximum or minimum value of a function and determining where the maximum or minimum value occurs. In this section we will develop some mathematical tools for solving such problems, and in the next section we will use these tools to solve some applied problems.

Given a function f, there are various questions we can ask about its maximum and minimum values; for example:

1. Does $f(x)$ have a maximum value?

2. If $f(x)$ has a maximum value, what is it?

3. If $f(x)$ has a maximum value, where does it occur?

These same questions could, of course, be asked about the minimum value of f. Sometimes we will be concerned with questions about the maximum or minimum values of a function f on a specified interval. Some typical examples are:

1. Does $f(x)$ have a maximum value on the interval $[a, b]$? On the interval (a, b)? On the interval $(0, +\infty)$?

2. If $f(x)$ has a minimum value on $[a, b]$, what is it?

3. If $f(x)$ has a minimum value on $[3, +\infty)$, where does it occur?

▶ Example 1

Figure 4.2.1

The function f graphed in Figure 4.2.1 has no maximum value. Its minimum value is 2, and the minimum occurs when $x = 3$. ◀

▶ Example 2

The function f graphed in Figure 4.2.2 has both maximum and minimum values on $[4, 10]$. The maximum value $f(x) = 5$ occurs when $x = 4$ and $x = 9$, while the minimum value $f(x) = 1$ occurs when $x = 7$. ◀

Figure 4.2.2

▶ Example 3

The function $f(x) = 2x + 1$ graphed in Figure 4.2.3 has a minimum, but no maximum value on $[0, 3)$. The minimum value is 1, and this occurs when $x = 0$. The reason why $f(x)$ has no maximum value is subtle, but important to understand. If we had been considering the interval $[0, 3]$ rather than $[0, 3)$, then $f(x)$ would have had a maximum value of 7 occurring at $x = 3$. However, the point $x = 3$ does not lie in the interval $[0, 3)$, so that 7 is *not* the maximum value on $[0, 3)$. But no number *less* than 7 can be the maximum either, for if M is any number less than 7, there are values of x in $[0, 3)$ where $f(x) > M$ (see Figure 4.2.4). Thus $f(x) = 2x + 1$ has no maximum value on $[0, 3)$. ◀

Figure 4.2.3   Figure 4.2.4

The following definitions formalize the notions of maximum and minimum values.

4.2.1 DEFINITION   A number M is called the ***maximum value*** or ***absolute maximum value*** of a function f if

1. $f(x) \leq M$ for all x in the domain of f.
2. $f(x) = M$ for at least one x in the domain of f.

4.2.2 DEFINITION   A number m is called the ***minimum value*** or ***absolute minimum value*** of a function f if

1. $f(x) \geq m$ for all x in the domain of f.
2. $f(x) = m$ for at least one x in the domain of f.

4.2.3 DEFINITION   A number k that is either a maximum or a minimum value of a function f is called an ***extreme value*** or ***absolute extreme value*** of f. Sometimes the terms ***extremum*** or ***absolute extremum*** are also used.

Often we will be concerned with the extreme values of f on some specified interval, rather than on the entire domain of f. The meaning of such terms as: ***maximum value of f on $[a, b]$*** or ***minimum value of f on (a, b)*** should be clear.

► Example 4

For the function f graphed in Figure 4.2.2, the extreme values (or extrema) on the interval $[4, 10]$ are $m = 1$ and $M = 5$. ◄

The following result gives conditions under which the existence of maximum and minimum values of a function is assured.

4.2.4 THEOREM *If a function f is continuous on a closed interval $[a, b]$, then f has both a*
Extreme-Value Theorem *maximum value and a minimum value on $[a, b]$.*

The proof of this theorem is surprisingly difficult, and will be omitted. However, the result is intuitively obvious if we imagine a particle moving along the graph of a continuous function over a closed interval $[a, b]$; during the trip the particle will have to pass through a highest point and a lowest point (Figure 4.2.5).

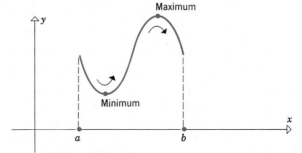

Figure 4.2.5

In the extreme-value theorem the hypotheses that f is continuous and the interval is closed are essential. If either hypothesis is violated the existence of maximum or minimum values cannot be guaranteed; this is shown in the next two examples.

► Example 5

Since $f(x) = 2x + 1$ is a polynomial, it is continuous everywhere. In particular, it is continuous on the interval $[0, 3)$. However, as shown in Example 3, the function f does not have a maximum value on $[0, 3)$. Thus the closed-interval hypothesis in Theorem 4.2.4 is essential. ◄

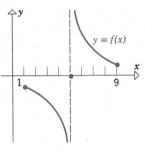

Figure 4.2.6

▶ **Example 6**

The function f graphed in Figure 4.2.6 is defined everywhere on the closed interval $[1, 9]$, yet it has neither a maximum nor minimum value on the interval. Since f has a point of discontinuity on the interval $[1, 9]$, this example shows that the continuity hypothesis in Theorem 4.2.4 is essential.
◀

▶ **Example 7**

The polynomial

$$f(x) = 2x^3 - 15x^2 + 36x$$

is continuous everywhere. In particular, it is continuous on the closed interval $[1, 5]$. Thus, the extreme value theorem tells us that f has both maximum and minimum values on $[1, 5]$. However, this theorem does not tell us what the maximum and minimum values are or where they occur.
◀

The extreme-value theorem is an example of what mathematicians call an *existence theorem;* it states conditions under which something exists, in this case maximum and minimum values for a function f. However, finding these maximum and minimum values is a separate problem. Before considering this problem we will introduce some useful terminology.

4.2.5 DEFINITION   A *critical point* for a function f is any value of x in the domain of f such that $f'(x) = 0$ or $f'(x)$ does not exist; the critical points where $f'(x) = 0$ are called *stationary points* of f.

▶ **Example 8**

For each of the functions graphed in Figure 4.2.7 x_0 is a critical point. In parts (a), (b), (c), and (d), x_0 is a stationary point because there is a horizontal tangent at x_0, while in the remaining parts x_0 is a critical point but not a stationary point because the derivative does not exist there.
◀

The following theorem is the key tool for finding the extreme values of a function.

4.2.6 THEOREM   *If a function f has an extreme value (either a maximum or a minimum) on an open interval (a, b), then the extreme value occurs at a critical point for f.*

Proof. We will prove the theorem in the case where the value is a maximum. The proof in the case where the value is a minimum is similar and is left as an exercise. If the maximum value occurs at x_0, then either f is differentiable

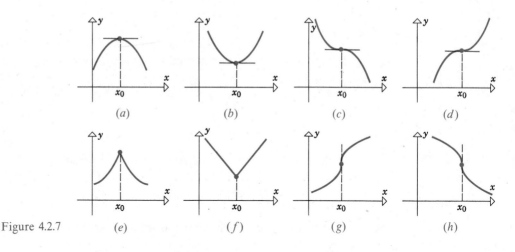

Figure 4.2.7

(a) (b) (c) (d)

(e) (f) (g) (h)

at x_0 or it is not. If it is not, then x_0 is a critical point for f and we are done. If f is differentiable at x_0, then we must show that $f'(x_0) = 0$. We will do this by showing that $f'(x_0) \geq 0$ and $f'(x_0) \leq 0$, from which it follows that $f'(x_0) = 0$. From the definition of a derivative we have

$$f'(x_0) = \lim_{h \to 0} \frac{f(x_0 + h) - f(x_0)}{h}$$

so that

$$f'(x_0) = \lim_{h \to 0^+} \frac{f(x_0 + h) - f(x_0)}{h} \tag{1}$$

and

$$f'(x_0) = \lim_{h \to 0^-} \frac{f(x_0 + h) - f(x_0)}{h} \tag{2}$$

Assume that h is sufficiently small so that $x_0 + h$ lies in the interval (a, b). Because $f(x_0)$ is the maximum value of f on (a, b), it follows that

$$f(x_0 + h) \leq f(x_0)$$

or equivalently

$$f(x_0 + h) - f(x_0) \leq 0$$

Thus, if h is negative

$$\frac{f(x_0 + h) - f(x_0)}{h} \geq 0 \qquad (3)$$

and if h is positive

$$\frac{f(x_0 + h) - f(x_0)}{h} \leq 0 \qquad (4)$$

But an expression that never assumes negative values cannot approach a negative limit and an expression that never assumes positive values cannot approach a positive limit, so that from (1), (2), (3), and (4) it follows that

$$f'(x_0) = \lim_{h \to 0^-} \frac{f(x_0 + h) - f(x_0)}{h} \geq 0$$

and

$$f'(x_0) = \lim_{h \to 0^+} \frac{f(x_0 + h) - f(x_0)}{h} \leq 0$$

Since $f'(x_0) \geq 0$ and $f'(x_0) \leq 0$, it must be that $f'(x_0) = 0$. ▨

This theorem can be used to locate the extreme values of a continuous function f on a *closed* interval $[a, b]$. For example, consider the possible locations of the maximum. Either the maximum occurs at an endpoint (Figure 4.2.8a) or it occurs in the open interval (a, b), in which case it occurs at a critical point (Figures 4.2.8b and 4.2.8c.) This suggests the following procedure:

(a)

Maximum occurs at an endpoint

(b)

Maximum occurs in the open interval (a, b) at a point x_0 where $f'(x_0) = 0$

(c)

Maximum occurs in the open interval (a, b) at a point x_0 where f is not differentiable

Figure 4.2.8

> *How to Find the Extreme Values of a Continuous Function f on a Closed Interval [a, b].*
>
> **Step 1.** Find the critical points of f in (a, b).
> **Step 2.** Evaluate f at the critical points and the endpoints a and b.
> **Step 3.** The largest of the values in Step 2 is the maximum and the smallest is the minimum.

► Example 9

Find the maximum and minimum values of

$$f(x) = 2x^3 - 15x^2 + 36x$$

on the interval [1, 5], and determine where they occur.

Solution. Since polynomials are differentiable everywhere, f is differentiable everywhere on the interval (1, 5). Thus, if an extreme value occurs on the open interval (1, 5), it must occur at a point where the derivative is zero. Since

$$f'(x) = 6x^2 - 30x + 36$$

the equation $f'(x) = 0$ becomes

$$6x^2 - 30x + 36 = 0$$

or

$$x^2 - 5x + 6 = 0$$

or

$$(x - 3)(x - 2) = 0$$

Thus there are two points in the interval (1, 5) where $f'(x) = 0$, namely $x = 2$ and $x = 3$. Evaluating f at these points and the endpoints, we have

$$f(1) = 2(1)^3 - 15(1)^2 + 36(1) = 23$$
$$f(2) = 2(2)^3 - 15(2)^2 + 36(2) = 28$$
$$f(3) = 2(3)^3 - 15(3)^2 + 36(3) = 27$$
$$f(5) = 2(5)^3 - 15(5)^2 + 36(5) = 55$$

Thus the minimum value is 23 and the maximum value is 55. The minimum occurs at $x = 1$ and the maximum occurs at $x = 5$. ◄

▶ Example 10

Find the extreme values of

$$f(x) = 6x^{4/3} - 3x^{1/3}$$

on the interval $[-1, 1]$ and determine where these values occur.

Solution. Differentiating, we obtain

$$f'(x) = 8x^{1/3} - x^{-2/3} = x^{-2/3}(8x - 1) = \frac{8x - 1}{x^{2/3}}$$

Thus, $f'(x) = 0$ when $x = \frac{1}{8}$ and $f'(x)$ does not exist when $x = 0$. It follows that the critical points of f are $x = 0$, $x = \frac{1}{8}$, both of which lie in the interval $[-1, 1]$. Evaluating f at these critical points and the endpoints we obtain Table 4.2.1.

Table 4.2.1

x	-1	0	$\dfrac{1}{8}$	1
$f(x)$	9	0	$-\dfrac{9}{8}$	3

Thus the minimum value of f on $[-1, 1]$ is $-\frac{9}{8}$, which occurs when $x = \frac{1}{8}$ and the maximum value of f on $[-1, 1]$ is 9, which occurs when $x = -1$. ◀

▶ Exercise Set 4.2

In Exercises 1–5 locate the critical points and classify them as stationary points or points of nondifferentiability.

1. (a) $f(x) = x^2 - 5x + 6$.
    (b) $f(x) = x^{2/3}$.

2. (a) $f(x) = (x + 2)^3$
    (b) $f(x) = \sqrt[3]{x + 2}$.

3. (a) $f(x) = \dfrac{x}{x^2 + 2}$
    (b) $f(x) = \cos 3x$.

4. (a) $f(x) = 3x^4 - 4x^3$
    (b) $f(x) = |\sin x|$.

5. (a) $f(x) = \sin^2 2x$, $0 < x < 2\pi$
    (b) $f(x) = x \tan x$, $-\pi/2 < x < \pi/2$.

For the functions in Exercises 6–19 find the maximum and minimum values on the given interval and state where the maximum and minimum values occur.

6. $8x - x^2$ on $[0, 6]$.

7. $4x^2 - 4x + 1$ on $[0, 1]$.

8. $2x^3 - 3x^2 - 12x$ on $[-2, 3]$.

9. $(x - 1)^3$ on $[0, 4]$.

10. $\dfrac{x}{x^2 + 2}$ on $[-1, 4]$.

11. $\dfrac{3x}{\sqrt{4x^2 + 1}}$ on $[-1, 1]$.

12. $(x^2 + x)^{2/3}$ on $[-2, 3]$.

13. $x^{2/3}(20 - x)$ on $[-1, 20]$.

14. $\sin x - \cos x$ on $[0, \pi]$.

15. $x - \tan x$ on $[-\pi/4, \pi/4]$.

16. $\sin^2 x + \cos x$ on $[-\pi, \pi]$.

17. $2 \sec x - \tan x$ on $[0, \pi/4]$.

18. $|6 - 4x|$ on $[-3, 3]$.

19. $1 + |9 - x^2|$ on $[-5, 1]$.

20. Let $f(x) = x^2 + px + q$. Find values of p and q such that $f(1) = 3$ is an extreme value of f on $[0, 2]$. Is this value a maximum or minimum?

21. Find the maximum and minimum values of

$$f(x) = 2 \sin 2x + \sin 4x$$

and state where the maximum and minimum values occur. [*Hint:* Since f is periodic, this problem can be solved by first finding the maximum and minimum values on an appropriate closed interval.]

22. Find the maximum and minimum values of

$$f(x) = 3 \cos \frac{x}{3} + 2 \cos \frac{x}{2}$$

[See the hint in the previous problem.]

23. Find the maximum and minimum values of

$$f(x) = \begin{cases} 4x - 2 & , x < 1 \\ (x - 2)(x - 3), & x \geq 1 \end{cases}$$

on $[\frac{1}{2}, \frac{7}{2}]$.

24. Prove that $\sin x \leq x$ for all x in the interval $[0, 2\pi]$. [*Hint:* Find the minimum value of $x - \sin x$ on the interval.]

25. Prove that

$$1 - \frac{x^2}{2} \leq \cos x$$

for all x in the interval $[0, 2\pi]$. [See the hint in the previous problem.]

26. (a) Sketch the graph of a function f that is continuous on the open interval $(-1, 1)$ and has both maximum and minimum values on the interval.

    (b) Sketch the graph of a function f that is defined everywhere on the open interval $(-1, 1)$ and is not continuous everywhere on $(-1, 1)$, yet has both maximum and minimum values on $(-1, 1)$.

    (c) Do the functions in parts (a) and (b) violate the extreme value theorem? Explain.

27. Prove: If $ax^2 + bx + c = 0$ has two distinct real roots, then the midpoint between these roots is a stationary point for $f(x) = ax^2 + bx + c$.

28. Prove Theorem 4.2.6 in the case where the extreme value is a minimum.

4.3 APPLIED MAXIMUM AND MINIMUM PROBLEMS

In this section we will show how the methods developed in the previous section can be used to solve some applied optimization problems.

▶ Example 1

An open box is to be made from a 16 in. by 30 in. piece of cardboard by cutting out squares of equal size from the four corners and bending up the

Figure 4.3.1   (*a*)   (*b*)

sides (Figure 4.3.1*a*). What size should the squares be to obtain a box with largest possible volume?

Solution. Let

x = length (in inches) of the sides of the squares to be cut out
V = volume (in cubic inches) of the resulting box

Because we are removing a square of side x from each corner, the resulting box will have dimensions $16 - 2x$ by $30 - 2x$ by x (Figure 4.3.1*b*). Since the volume of a box is the product of its dimensions, we have

$$V = (16 - 2x)(30 - 2x)x = 480x - 92x^2 + 4x^3 \qquad (1)$$

The variable x in this expression is subject to certain physical restrictions. Because x represents a length it cannot be negative and because the width of the cardboard is 16 in., we cannot cut out squares whose sides are more than 8 in. long. Thus the variable x in (1) must satisfy

$$0 \le x \le 8$$

We have thus reduced our problem to finding the value (or values) of x in the interval $[0, 8]$ for which (1) is maximum. Since the right side of (1) is a polynomial in x, it is continuous on the closed interval $[0, 8]$ and consequently we can use the methods developed in the previous section to find the maximum.

From (1) we obtain

$$\frac{dV}{dx} = 480 - 184x + 12x^2 = 4(120 - 46x + 3x^2)$$

Setting $dV/dx = 0$ yields

$$120 - 46x + 3x^2 = 0$$

which can be solved by the quadratic formula to obtain the critical points

$$x = \tfrac{10}{3} \quad \text{and} \quad x = 12$$

Since $x = 12$ falls outside the interval $[0, 8]$, the maximum value of V must occur either at the critical point $x = \tfrac{10}{3}$ or at one of the endpoints $x = 0$, $x = 8$. Substituting these values in (1) yields Table 4.3.1, which tells us that the greatest possible volume $V = \tfrac{19600}{27}$ in.$^3 \approx 726$ in.^{3} occurs when we cut out squares whose sides have length $\tfrac{10}{3}$ in. ◄

Table 4.3.1

x	0	$\dfrac{10}{3}$	8
V	0	$\dfrac{19600}{27}$	0

► Example 2

An offshore oil well is located in the ocean at a point W, which is 5 miles from the closest shorepoint A on a straight shoreline (Figure 4.3.2). The oil is to be piped to a shorepoint B that is 8 miles from A by piping it on a straight line underwater from W to some shorepoint P between A and B and then on to B via a pipe along the shoreline. If the cost of laying pipe is $100,000 per mile underwater and $75,000 per mile over land, where should the point P be located to minimize the cost of laying the pipe?

Figure 4.3.2

REMARK. Since the shortest distance between two points is a straight line, a pipeline directly from W to B uses the least amount of pipe. However, the pipe, being completely underwater, would be expensive to lay. At the other extreme, a pipeline from W to A to B uses the least amount of expensive underwater pipe, but uses the greatest total amount of pipe. Thus it seems plausible that by piping to some point P between A and B, one might incur less total cost than by piping to either of the extreme locations.

Solution. Let

$$x = \text{distance (in miles) between } A \text{ and } P$$
$$c = \text{cost (in thousands of dollars) for the entire pipeline}$$

From Figure 4.3.2 the length of pipe underwater is the distance between W and P. By the Theorem of Pythagoras, that length is

$$\sqrt{x^2 + 25} \text{ miles} \tag{2}$$

Also from Figure 4.3.2, the length of pipe over land is the distance between P and B, which is

$$8 - x \text{ miles} \tag{3}$$

From (2) and (3) it follows that the total cost c (in thousands of dollars) for the pipeline is

$$c = 100\sqrt{x^2 + 25} + 75(8 - x) \tag{4}$$

Because the distance between A and B is 8 miles, the distance x between A and P must satisfy

$$0 \le x \le 8$$

We have thus reduced our problem to finding the value (or values) of x in the interval $[0, 8]$ for which (4) is a minimum. Since c is a continuous function of x on the closed interval $[0, 8]$, we can use the methods developed in the previous section to find the minimum.

From (4) we obtain

$$\frac{dc}{dx} = \frac{100x}{\sqrt{x^2 + 25}} - 75 = 25\left(\frac{4x}{\sqrt{x^2 + 25}} - 3\right)$$

Setting $dc/dx = 0$ yields

$$\frac{4x}{\sqrt{x^2 + 25}} - 3 = 0 \tag{5}$$

or

$$4x = 3\sqrt{x^2 + 25}$$
$$16x^2 = 9(x^2 + 25)$$
$$7x^2 = 225$$
$$x = \pm\frac{15}{\sqrt{7}}$$

The number $-15/\sqrt{7}$ is not a root of (5) and must be discarded, leaving $x = 15/\sqrt{7}$ as the only critical point. Since this point lies in the interval $[0, 8]$, the minimum must occur at one of the points:

$$x = 0, \qquad x = 15/\sqrt{7}, \qquad x = 8$$

Substituting these values in (4) yields Table 4.3.2, which tells us that the least possible cost for the pipeline, approximately $c = 930.719$ (thousand dollars) = \$930,719 occurs when the point P is located at a distance of

$$\frac{15}{\sqrt{7}} \approx 5.67 \text{ miles}$$

from A. ◄

Table 4.3.2

x	0	$\dfrac{15}{\sqrt{7}}$	8
c	1100	$600 + 125\sqrt{7} \approx 930.719$	$100\sqrt{89} \approx 943.398$

► Example 3

Construct a rectangle of perimeter p whose area is as large as possible.

Solution. Let

$x = $ width of the rectangle
$y = $ length of the rectangle
$A = $ area of the rectangle

Then

$$A = xy \tag{6}$$

Since the perimeter of the rectangle is p, the variables x and y are related by the equation

$$p = 2x + 2y$$

Perimeter
$p = 2x + 2y$

Figure 4.3.3

or

$$y = \tfrac{1}{2}p - x \tag{7}$$

(See Figure 4.3.3.) Substituting (7) in (6) yields

$$A = x(\tfrac{1}{2}p - x) = \tfrac{1}{2}px - x^2 \tag{8}$$

Because x represents a length it cannot be negative, and because the two sides of length x cannot have a combined length exceeding the total perimeter, the variable x must satisfy

$$0 \le x \le \tfrac{1}{2}p$$

Thus we have reduced the problem to that of finding the value (or values) of x in $[0, \tfrac{1}{2}p]$ for which (8) is a maximum. Since A is a polynomial in x, it is continuous on the closed interval $[0, \tfrac{1}{2}p]$, and we can use the methods developed in the previous section to find the maximum.

From (8) we obtain

$$\frac{dA}{dx} = \tfrac{1}{2}p - 2x$$

Setting $dA/dx = 0$ we obtain

$$\tfrac{1}{2}p - 2x = 0$$

or

$$x = \tfrac{1}{4}p$$

Since this critical point lies in the interval $[0, \tfrac{1}{2}p]$, the maximum occurs at one of the points

$$x = 0, \qquad x = \tfrac{1}{4}p, \qquad x = \tfrac{1}{2}p$$

Substituting these values in (8) yields Table 4.3.3, which tells us that the maximum area, $A = \tfrac{1}{16}p^2$, occurs when $x = \tfrac{1}{4}p$. From (7) the corresponding value of y is $y = \tfrac{1}{4}p$ so that the rectangle of perimeter p with greatest area is a square with sides of length $\tfrac{1}{4}p$. ◀

Table 4.3.3

x	0	$\frac{1}{4}p$	$\frac{1}{2}p$
A	0	$\frac{1}{16}p^2$	0

The examples above illustrate the following 5-step procedure for solving many applied maximum and minimum problems.

Step 1. Label the quantities relevant to the problem.
Step 2. Find a formula for the quantity to be maximized or minimized.
Step 3. Using the conditions stated in the problem to eliminate variables, express the quantity to be maximized or minimized as a function of one variable.

Step 4. Find the interval of possible values for this variable from the physical restrictions in the problem.

Step 5. If applicable, use the techniques of the previous section to obtain the maximum or minimum.

(a)

(b)

Figure 4.3.4

▶ Example 4

Find the radius and height of the right-circular cylinder of largest volume that can be inscribed in a right-circular cone with radius 6 in. and height 10 in. (Figure 4.3.4*a*).

Solution. Let

r = radius (in inches) of the cylinder
h = height (in inches) of the cylinder
V = volume (in cubic inches) of the cylinder

The formula for the volume of the inscribed cylinder is

$$V = \pi r^2 h \qquad (9)$$

To eliminate one of the variables in (9) we need a relationship between r and h. Using similar triangles (Figure 4.3.4*b*) we obtain

$$\frac{10 - h}{r} = \frac{10}{6}$$

or

$$h = 10 - \tfrac{5}{3}r \qquad (10)$$

Substituting (10) into (9) we obtain

$$V = \pi r^2 (10 - \tfrac{5}{3}r) = 10\pi r^2 - \tfrac{5}{3}\pi r^3 \qquad (11)$$

which expresses V in terms of r alone. Because r represents a radius it cannot be negative, and because the radius of the inscribed cylinder cannot exceed the radius of the cone, the variable r must satisfy

$$0 \leq r \leq 6$$

Thus we have reduced the problem to that of finding the value (or values) of r in [0, 6] for which (11) is a maximum. Since V is a continuous function of r on [0, 6], the methods developed in the previous section apply.

From (11) we obtain

$$\frac{dV}{dr} = 20\pi r - 5\pi r^2 = 5\pi r(4 - r)$$

Setting $dV/dr = 0$ gives

$$5\pi r(4 - r) = 0$$

so that

$$r = 0 \qquad \text{and} \qquad r = 4$$

are critical points. Since these lie in the interval $[0, 6]$, the maximum must occur at one of the points

$$r = 0, \qquad r = 4, \qquad r = 6$$

Substituting these values in (11) yields Table 4.3.4, which tells us that the maximum volume $V = \frac{160}{3}\pi \approx 168$ in.³ occurs when the inscribed cylinder has radius 4 in. When $r = 4$ it follows from (10) that $h = \frac{10}{3}$. Therefore, the inscribed cone of largest volume has radius $r = 4$ in. and height $h = \frac{10}{3}$ in. ◀

Table 4.3.4

r	0	4	6
V	0	$\dfrac{160}{3}\pi$	0

MARGINAL ANALYSIS (AN APPLICATION TO ECONOMICS)

Economists and business people are interested in how changes in such variables as inventory, production, supply, advertising, and price affect other variables like profit, revenue, demand, inflation, and employment. Such problems are studied using *marginal analysis.* The word, *marginal,* is the economist's term for a rate of change or derivative.

Three functions of importance to an economist or a manufacturer are:

$C(x)$ = total cost of producing x units of a product during some time period.

$R(x)$ = total revenue received from selling x units of the product during the time period.

$P(x)$ = total profit obtained by selling x units of the product during the time period.

These are called respectively the *cost function, revenue function,* and *profit function.* If all units produced are sold, then these are related by,

$$P(x) = R(x) - C(x) \tag{12}$$
$$[\text{profit}] = [\text{revenue}] - [\text{cost}]$$

The derivatives $C'(x)$, $R'(x)$, and $P'(x)$ are called the **marginal cost, marginal revenue,** and **marginal profit.** If all units produced are sold, then a relationship between them is obtained by differentiating (12),

$$P'(x) = R'(x) - C'(x)$$

$$\begin{bmatrix} \text{marginal} \\ \text{profit} \end{bmatrix} = \begin{bmatrix} \text{marginal} \\ \text{revenue} \end{bmatrix} - \begin{bmatrix} \text{marginal} \\ \text{cost} \end{bmatrix}$$

The quantities $P'(x)$, $R'(x)$, and $C'(x)$ represent the instantaneous rates of change of profit, revenue, and cost with respect to x, where x is the amount of the product produced and sold.

In practice, $C'(x)$ is frequently interpreted as the cost of manufacturing the $(x + 1)$-th unit. Although this is not exact, it is usually a good approximation since

$$C'(x) = \lim_{\Delta x \to 0} \frac{C(x + \Delta x) - C(x)}{\Delta x} \approx \frac{C(x + 1) - C(x)}{1}$$
$$= C(x + 1) - C(x)$$

This approximation is reasonable since x is usually large and $\Delta x = 1$ can be considered close to zero by comparison. Since $C(x + 1)$ is the cost of producing $x + 1$ units and $C(x)$ is the cost of producing x units, it follows that $C'(x) \approx C(x + 1) - C(x)$ is the approximate cost of producing the $(x + 1)$-th unit. Similarly, $R'(x)$ is the approximate revenue received from selling the $(x + 1)$-th unit and $P'(x)$ is the approximate profit from selling the $(x + 1)$-th unit.

The total cost $C(x)$ of producing x units can be expressed as a sum

$$C(x) = a + M(x) \tag{13}$$

where a is a constant, called **overhead,** and $M(x)$ is a function representing **manufacturing cost.** The overhead, which includes such fixed costs as rent and insurance, does not depend on x; it must be paid even if nothing is produced. On the other hand the manufacturing cost $M(x)$, which includes such items as cost of materials and labor, depends on the number of items manufactured. It is shown in economics that with suitable simplifying assumptions, $M(x)$ can be expressed in the form

$$M(x) = bx + cx^2$$

Substituting this in (13) yields

$$C(x) = a + bx + cx^2 \tag{14}$$

▶ Example 5

A paint manufacturer determines that the total cost in dollars of producing x gallons of paint per day is

$$C(x) = 5000 + x + 0.001x^2$$

(a) Find the marginal cost when the production level is 500 gallons per day.
(b) Use the marginal cost to approximate the cost of producing the 501st gallon.
(c) Find the exact cost of producing the 501st gallon.

Solution.

(a) The marginal cost is $C'(x) = 1 + 0.002x$, so $C'(500) = 1 + (0.002)(500) = 2$.
(b) Since $C'(500) = 2$, the cost of producing the 501st gallon is approximately \$2.00.
(c) The total cost of producing 501 gallons is

$$C(501) = 5000 + 501 + 0.001(501)^2 = \$5752.001$$

and the total cost of producing 500 gallons is

$$C(500) = 5000 + 500 + 0.001(500)^2 = \$5750$$

so the exact cost of producing the 501st gallon is

$$C(501) - C(500) = \$2.001 \qquad \blacktriangleleft$$

If a manufacturing firm can sell all the items it produces for p dollars apiece, then its total revenue $R(x)$ (in dollars) will be

$$R(x) = px \tag{15}$$

and its total profit $P(x)$ (in dollars) will be

$$P(x) = [\text{total revenue}] - [\text{total cost}] = R(x) - C(x) = px - C(x)$$

Thus, if the cost function is given by (14),

$$P(x) = px - (a + bx + cx^2) \tag{16}$$

Depending on such factors as number of employees, amount of machinery available, economic conditions, and competition, there will be some upper limit l on the number of items a manufacturer is capable of producing and selling. Thus, during a fixed time period the variable x in (16) will satisfy

$$0 \le x \le l$$

By determining the value or values of x in $[0, l]$ that maximize (16), the firm can determine how many units of its product must be manufactured and sold to yield the greatest profit. This is illustrated in the following numerical example.

▶ Example 6

A liquid form of penicillin manufactured by a pharmaceutical firm is sold in bulk at a price of $200 per unit. If the total production cost (in dollars) for x units is

$$C(x) = 500,000 + 80x + 0.003x^2$$

and if the production capacity of the firm is at most 30,000 units in a specified time, how many units of penicillin must be sold in that time to maximize the profit?

Solution. Since the total revenue for selling x units is $R(x) = 200x$, the profit $P(x)$ on x units will be

$$P(x) = R(x) - C(x) = 200x - (500,000 + 80x + 0.003x^2) \qquad (17)$$

Since the production capacity is at most 30,000 units, x must lie in the interval $[0, 30,000]$. From (17)

$$\frac{dP}{dx} = 200 - (80 + 0.006x) = 120 - 0.006x$$

Setting $dP/dx = 0$ gives

$$120 - 0.006x = 0$$

or

$$x = 20,000$$

Since this critical point lies in the interval $[0, 30,000]$ the maximum profit must occur at one of the points

$$x = 0, \qquad x = 20,000 \qquad \text{or} \qquad x = 30,000$$

Substituting these values in (17) yields Table 4.3.5, which tells us that the maximum profit $P = \$700,000$ occurs when $x = 20,000$ units are manufactured and sold in the specified time. ◀

Table 4.3.5

x	0	20,000	30,000
$P(x)$	$-500,000$	700,000	400,000

▶ Exercise Set 4.3

1. Express the number 10 as a sum of two nonnegative terms whose product is as large as possible.

2. A square sheet of cardboard of side k is used to make an open box by the method described in Example 1. What size squares should be cut from the corners to obtain a box with largest possible volume?

3. A rectangular plot of land is to be fenced in using two kinds of fencing. Two opposite sides will use heavy-duty fencing selling for $3.00 a foot, while the remaining two sides will use standard fencing selling for $2.00 a foot. What are the dimensions of the rectangular plot of greatest area that can be fenced in at a cost of $6000?

4. A rectangle is to be inscribed in a right triangle having sides of length 6 in., 8 in., and 10 in. Find the dimensions of the rectangle with greatest area assuming the rectangle is positioned as in Figure 4.3.5a.

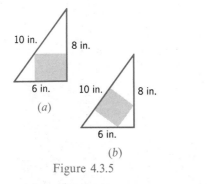

(a)

(b)

Figure 4.3.5

5. Solve the problem in Exercise 4 assuming the rectangle is positioned as in Figure 4.3.5b.

6. Assume that the operating cost of a certain truck (excluding driver's wages) is $12 + x/6$ cents per mile when the truck travels at x mph. If the driver earns $6.00 per hour, what is the most economical speed to operate the truck on a 400 mile turnpike where the minimum speed is 40 mph and the maximum speed is 70 mph?

7. A wire of length 12 in. is cut into two pieces, one of which is bent to form a circle and the other to form a square. How should the wire be cut if the sum of the enclosed areas is to be
(a) a maximum?
(b) a minimum?

8. A container with square base, vertical sides, and open top is to be made from 1000 ft² of material. Find the dimensions of the container with greatest volume.

9. A box-shaped wire frame consists of two identical wire squares whose vertices are connected by four straight wires of equal length (Figure 4.3.6). If the frame is to be made from a wire of length L, what should the dimensions be to obtain a box of greatest volume?

Figure 4.3.6

10. How should two nonnegative numbers be chosen so that their sum is 1 and the sum of their squares is:
(a) as large as possible?
(b) as small as possible?

11. (a) A chemical manufacturer sells sulphuric acid in bulk at a price of $100 per unit. If the daily total production cost in dollars for x units is

$$C(x) = 100,000 + 50x + 0.0025x^2$$

and if the daily production capacity is at most 7000 units, how many units of sulphuric acid must be manufactured and sold daily to maximize the profit?

(b) Would it benefit the manufacturer to expand the daily production capacity?

12. Find the dimensions of the rectangle of greatest area that can be inscribed in a semicircle of radius R.

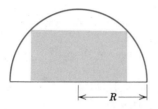

13. Find the height and radius of the cone of slant height L whose volume is as large as possible.

14. Find the dimensions of the right circular cylinder of largest volume that can be inscribed in a sphere of radius R.

15. Find the dimensions of the right circular cylinder of greatest surface area that can be inscribed in a sphere of radius R.

16. Show that the right circular cylinder of greatest volume that can be inscribed in a right circular cone has volume that is $\frac{4}{9}$ the volume of the cone (Figure 4.3.7).

Figure 4.3.7

17. A drainage channel is to be made so that its cross section is a trapezoid with equally sloping sides (Figure 4.3.8). If the sides and bottom all have a length of 5 ft, how should the angle θ be chosen to yield the greatest cross-sectional area? Assume $0 \leq \theta \leq \pi/2$.

Figure 4.3.8

18. A commercial cattle ranch currently allows 20 steers per acre of grazing land; on the average its steers weigh 2000 pounds at market. Estimates by the Agriculture Department indicate that the average market weight per steer will be reduced by 50 pounds for each additional steer added per acre of grazing land. How many steers per acre should be allowed in order for the ranch to get the largest possible total market weight for its cattle?

19. A cone is made from a circular sheet of radius R by cutting out a sector and gluing the cut edges of the remaining piece together (Figure 4.3.9). What is the maximum volume attainable for the cone? [*Hint:* Let ϕ be the central angle of the sector removed.]

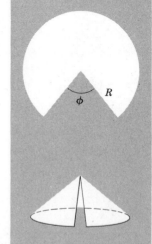

Figure 4.3.9

20. A man is on the bank of a river which is 1 mile wide. He wants to travel to a town on the opposite bank, but 1 mile upstream. He intends to row on a straight line to some point P on the opposite bank and then walk the remaining distance along the bank (Figure 4.3.10). To what point should he row in order to reach his destination in the least time if
(a) he can walk 5 miles per hour and row 3 miles per hour?
(b) he can walk 5 miles per hour and row 4 miles per hour?

Figure 4.3.10

21. The total cost of producing x units of a commodity per week is

$$C(x) = 200 + 4x + 0.1x^2$$

(a) Find the marginal cost when the production level is 100 units.
(b) Use the marginal cost to approximate the cost of producing the 101st unit.
(c) Find the exact cost of producing the 101st unit.
(d) Assuming that the commodity is sold for ten dollars per unit, find the marginal revenue and marginal profit functions.

22. A firm determines that x units of its product can be sold daily at p dollars per unit, where

$$x = 1000 - p$$

The cost of producing x units per day is

$$C(x) = 3000 + 20x$$

(a) Find the revenue function $R(x)$.

(b) Find the profit function $P(x)$.
(c) Assuming that the production capacity is at most 500 units per day, determine how many units the company must produce and sell each day to maximize the profit.
(d) Find the maximum profit.
(e) What price per unit must be charged to obtain the maximum profit?

23. Prove that $(1, 0)$ is the closest point on the curve $x^2 + y^2 = 1$ to $(2, 0)$.

24. We are given a straight fence 100 ft long and we wish to use 200 ft of additional fence to form a rectangular enclosure whose boundary contains all of the original fence. How should we lay out the additional fence to obtain an enclosure of maximum area? (This problem, due to V. L. Klee, Jr., appeared in *Selected Papers on Calculus*, Mathematical Association of America, 1969, p. 246.)

25. (a) Find the smallest value of M such that

$$|x^2 - 3x + 2| \leq M$$

for all x in the interval $[1, \frac{5}{2}]$.
(b) Find the largest value of m such that

$$|x^2 - 3x + 2| \geq m$$

for all x in the interval $[\frac{3}{2}, \frac{7}{4}]$.

26. At what point(s) in the interval $[0, \pi]$ are the graphs of

$$y = \tfrac{1}{2}x \qquad \text{and} \qquad y = \sin x$$

farthest apart?

27. Fermat's* principle in optics states that light traveling from one point to another follows that path for which the total travel time is minimum. In a uniform medium, the paths of "minimum time" and "shortest distance" turn out to be the same, so that light, if unobstructed, travels along a straight line. Assume we have a light source, a flat mirror, and an observer in a uniform medium. If a light ray leaves the source, bounces off the mirror, and travels on to the observer, then its path will consist of two line segments as shown in Figure 4.3.11. According to Fermat's principle, the path will be such that the total travel time t is minimum or, since the medium is uniform, the path will be such that the total distance traveled from A to P to B is as small as possible. Assuming that the minimum occurs when $dt/dx = 0$, show that the light ray will strike the mirror at the point P where the "angle of incidence" θ_1 equals the "angle of reflection" θ_2.

Figure 4.3.11

28. Fermat's* principle (Exercise 27) also explains why light rays traveling between air and water undergo bending or refraction. Imagine we have two uniform media (such as air and water) and a light ray traveling from a source A in one medium to an observer B in the other medium (Figure 4.3.12). It is known that light travels at a constant speed in a uniform medium, but more slowly in a dense medium (such as water) than in a thin medium (such as air.) Consequently, the path of shortest time from A to B is not necessarily a straight line,

but rather some broken line path A to P to B allowing the light to take greatest advantage of its higher speed through the thin medium. Snell's† law of refraction states that the path of the light ray will be such that

$$\frac{\sin \theta_1}{v_1} = \frac{\sin \theta_2}{v_2}$$

where v_1 is the speed of light in the first medium, v_2 is the speed of light in the second medium, and θ_1 and θ_2 are the angles shown in Figure 4.3.12. Show that this follows from the assumption that the path of minimum time occurs when $dt/dx = 0$.

Figure 4.3.12

---

*PIERRE DE FERMAT (1601–1665). Fermat, the son of a successful French leather merchant, was a lawyer who practiced mathematics as a hobby. He received a Bachelor of Civil Laws degree from the University of Orleans in 1631 and subsequently held various government positions, including a post as councillor to the Toulouse parliament. Although he was apparently financially successful, confidential documents of that time suggest that his performance in office and as a lawyer was poor, perhaps because he devoted so much time to mathematics. Throughout his life, Fermat fought all efforts to have his mathematical results published. He had the unfortunate habit of scribbling his work in the margins of books and often sent his results to friends without keeping copies for himself. As a result, he never received credit for many major achievements until his name was raised from obscurity in the mid-nineteenth century. It is now known that Fermat, simultaneously and independently of Descartes, developed analytic geometry. Unfortunately, Descartes and Fermat argued bitterly over various problems so that there was never any real cooperation between these two great geniuses.

Fermat solved many fundamental calculus problems. He obtained the first procedure for differentiating polynomials, and solved many important maximization, minimization, area, and

29. A farmer wants to walk at a constant rate from her barn to a straight river, fill her pail, and carry it to her house in the least time.

(a) Explain how this problem relates to Fermat's principle and the light-reflection problem in Exercise 27.

(b) Use the result of Exercise 27 to describe geometrically the best path for the farmer to take.

(c) Use part (b) to determine where the farmer should fill her pail if her house and barn are located as in Figure 4.3.13.

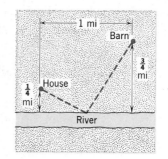

Figure 4.3.13

tangent problems. His work served to inspire Isaac Newton. Fermat is best known for his work in number theory, the study of properties and relationships between whole numbers. Fermat was the first mathematician to make substantial contributions to this field after the ancient Greek mathematician Diophantus. Unfortunately, none of Fermat's contemporaries appreciated his work in this area, a fact that eventually pushed Fermat into isolation and obscurity in later life.

One of the great mysteries of mathematics is shrouded in Fermat's work in number theory. In the margin of a book by Diophantus, Fermat scribbled that for values of n greater than 2, the equation $x^n + y^n = z^n$ has no nonzero integer solutions for x, y, and z. He stated, "I have discovered a truly marvelous proof of this, which however the margin is not large enough to contain." For the past 300 years the greatest mathematical geniuses have been unable to prove this result, even though it seems to be true. This result is now known as "Fermat's last theorem." In 1908 a prize of 100,000 German marks was offered for its solution. Although the prize is worthless because of post World War I inflation, it has never been won. Whether or not Fermat really proved this theorem is a mystery to this day.

In addition to his work in calculus and number theory Fermat was one of the founders of probability theory and made major contributions to the theory of optics. Outside mathematics, Fermat was a classical scholar of some note, was fluent in French, Italian, Spanish, Latin, and Greek, and he composed considerable Latin poetry.

† WILLEBRORD VAN ROIJEN SNELL (1591–1626) Dutch mathematician. Snell succeeded his father to the post of Professor of Mathematics at the University of Leiden in 1613. He is most famous for the result on light refraction that bears his name. Although this phenomenon was studied as far back as the ancient Greek astronomer Ptolemy, until Snell's work the relationship was incorrectly thought to be $\theta_1/v_1 = \theta_2/v_2$. Snell's law was published by Descartes in 1638 without giving proper credit to Snell. Snell also discovered a method for determining distances by triangulation and thereby founded the modern technique of mapmaking.

4.4 INTERVALS OF INCREASE AND DECREASE; CONCAVITY; HIGHER DERIVATIVES

Although point plotting is useful for determining the general shape of a graph, it has an inherent deficiency that limits its usefulness in problems where the shape of the graph must be known with absolute precision. By necessity, we can only plot finitely many points. Thus, no matter how many points are plotted, we can only guess at the shape of the graph *between* these points. For example, between two successive plotted points we are often forced to guess whether the graph exhibits a downward curvature (Figure 4.4.1*a*), an upward curvature (Figure 4.4.1*b*), or both (Figure 4.4.1*c*). In this section we will show how the derivative can be used to resolve such ambiguities.

Figure 4.4.1   (*a*)   (*b*)   (*c*)

The terms *increasing* and *decreasing* are used to describe the behavior of a function as we travel *left to right* along its graph. For example, the function graphed in Figure 4.4.2 can be described as increasing on the interval $(-\infty, 1)$, decreasing on the interval $(1, 2)$, and increasing again on the interval $(2, +\infty)$.

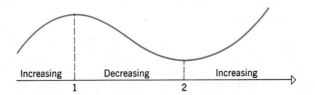

Figure 4.4.2   The following definitions express these intuitive ideas precisely.

4.4.1   DEFINITION   A function f is ***increasing*** on a given interval if

$$f(x_2) > f(x_1) \text{ whenever } x_2 > x_1$$

where x_1 and x_2 are points in the interval (Figure 4.4.3*a*).

4.4.2   DEFINITION   A function f is ***decreasing*** on a given interval if

$$f(x_2) < f(x_1) \text{ whenever } x_2 > x_1$$

where x_1 and x_2 are points in the interval (Figure 4.4.3*b*).

Figure 4.4.3   (*a*)   (*b*)

A function is said to be ***strictly monotone*** on a given interval if it is either increasing or decreasing on the interval.

Figure 4.4.4 suggests that if the graph of a function has tangent lines with positive slopes over an interval, then the function is increasing on that interval; and similarly, if the graph has tangent lines with negative slopes, the function is decreasing.

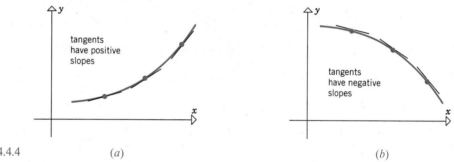

Figure 4.4.4   (*a*)   (*b*)

This intuitive observation suggests the following fundamental theorem whose proof will be deferred until Section 4.10.

4.4.3   THEOREM   (*a*)   *If $f'(x) > 0$ on an open interval* (*a, b*), *then f is increasing on* (*a, b*).
                        (*b*)   *If $f'(x) < 0$ on an open interval* (*a, b*), *then f is decreasing on* (*a, b*).

▶ Example 1

On which open intervals is the function

$$f(x) = \tfrac{1}{3}x^3 - 3x^2 + 8x$$

increasing? Decreasing?

Solution. Differentiating f we obtain

$$f'(x) = x^2 - 6x + 8 = (x - 2)(x - 4)$$

Thus f will be increasing when

$$(x - 2)(x - 4) > 0 \tag{1}$$

and decreasing when

$$(x - 2)(x - 4) < 0 \tag{2}$$

The signs of $(x - 2)$, $(x - 4)$, and $(x - 2)(x - 4)$ for various x are shown in Figure 4.4.5. From the bottom part of that figure we see that f is

increasing   if $x < 2$
decreasing   if $2 < x < 4$
increasing   if $x > 4$

The graph of f is shown in Figure 4.4.6. ◄

- - - 0 + + + + + + + + + + x

2
Sign of $(x - 2)$

- - - - - - 0 + + + + + + x

4
Sign of $(x - 4)$

+ + + 0 - - - 0 + + + + + + x

2   4
Sign of $(x - 2)(x - 4)$

Figure 4.4.5

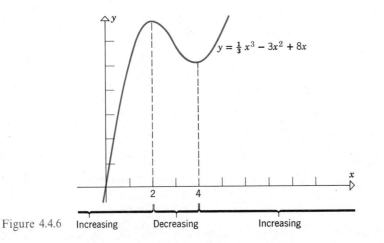

Figure 4.4.6   Increasing   Decreasing   Increasing

$y = \frac{1}{3}x^3 - 3x^2 + 8x$

Although the derivative of a function f can tell us where the graph of f is increasing or decreasing, f' may or may not reveal oscillations in the graph of f. If the oscillation is large enough that graph has an up and down movement (Figure 4.4.7a), then the derivative will exhibit sign changes. On the other hand, if the oscillation is slight, then the derivative may not change sign. For example, the tangent lines to the curve in Figure 4.4.7b all have positive slope. Thus, the derivative of f does not change sign even though an oscillation is present.

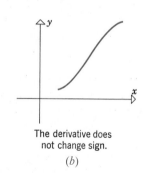

The derivative
changes sign.

The derivative does
not change sign.

Figure 4.4.7

(a)

(b)

How then, can we determine with certainty whether there are oscillations in the graph of a function? The key is in the intuitive observation that a single oscillation, no matter how slight, consists of two parts, a portion that is concave up, that is, "holds water" and a portion that is concave down, that is, "spills water" (Figure 4.4.8).

Figure 4.4.8

(a)

(b)

Figure 4.4.9

To investigate the problem of determining where a curve is concave up and where it is concave down consider the curves in Figure 4.4.9. The curve in part (a) is concave down. As we travel left to right along this curve, the tangent lines rotate clockwise so that the slopes of these tangent lines are constantly *decreasing*. By contrast, the curve in part (b) is concave up, and as we travel left to right along the curve the tangent lines rotate counterclockwise; thus the slopes of the tangent lines are constantly *increasing*.

These observations suggest the following definition.

4.4.4 DEFINITION   A function f is called ***concave up*** on a given interval if f' is increasing on the interval; it is called ***concave down*** on the interval if f' is decreasing on the interval.

If we let f'' denote the derivative of f', it follows from Theorem 4.4.3 that f' will be increasing on an open interval (a, b) if $f''(x) > 0$ for all x in (a, b), and f' will be decreasing on (a, b) if $f''(x) < 0$ for all x in (a, b). Thus, we have the following result.

4.4.5 THEOREM   (a)   If $f''(x) > 0$ on an open interval (a, b) then f is concave up on (a, b).
(b)   If $f''(x) < 0$ on an open interval (a, b) then f is concave down on (a, b).

▶ Example 2

Where is the function

$$f(x) = \tfrac{1}{3}x^3 - 3x^2 + 8x$$

concave up and where is it concave down?

Solution.   The derivative of f is

$$f'(x) = x^2 - 6x + 8$$

and the derivative of f' is

$$f''(x) = \frac{d}{dx}[f'(x)] = \frac{d}{dx}[x^2 - 6x + 8] = 2x - 6$$

Thus, at points where x satisfies

$$2x - 6 > 0 \tag{3}$$

f is concave up, while at points where x satisfies

$$2x - 6 < 0 \tag{4}$$

f is concave down. Since (3) holds when $x > 3$ and (4) holds when $x < 3$, we conclude:

f is concave up when $x > 3$
f is concave down when $x < 3$

The reader can check that these results agree with the graph of f shown in Figure 4.4.6.   ◀

The point $x = 3$ in Example 2 is of particular interest because it is the spot where the graph changes the direction of its concavity. There is a name for such points.

4.4.6 DEFINITION If a function f changes the direction of its concavity at a point x_0, then f is said to have an ***inflection point*** at x_0. More precisely, f has an inflection point at x_0 if on some open interval containing x_0

(i) f' is decreasing when $x < x_0$ and f' is increasing when $x > x_0$,

or

(ii) f' is increasing when $x < x_0$ and f' is decreasing when $x > x_0$.

(See Figure 4.4.10.)

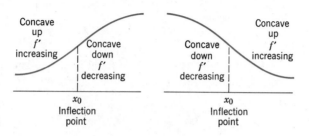

Figure 4.4.10

HIGHER
DERIVATIVES

The derivative f' of a function f is sometimes called the ***first derivative*** of f, and the derivative f'' of f' is called the ***second derivative*** of f. As long as we have differentiability, we can continue the process of differentiating derivatives to obtain third, fourth, fifth, and even higher derivatives of f. The successive derivatives of f are denoted by

$$f' \qquad \text{(the first derivative of } f)$$
$$f'' = (f')' \qquad \text{(the second derivative of } f)$$
$$f''' = (f'')' \qquad \text{(the third derivative of } f)$$
$$f^{(4)} = (f''')' \qquad \text{(the fourth derivative of } f)$$
$$f^{(5)} = (f^{(4)})' \qquad \text{(the fifth derivative of } f)$$
$$\vdots \qquad \vdots \qquad\qquad \vdots$$

Beyond the third derivative, it is too clumsy to continue using primes so that we switch from primes to integers in parentheses to denote the ***order*** of the derivative. In this notation it is easy to denote a derivative of arbitrary order by writing

$$f^{(n)} \qquad \text{(the } n\text{th derivative of } f)$$

Unlike the first and second derivatives, the derivatives of order three and higher have no obvious geometric significance. However, these derivatives will be of importance later in the text.

▶ Example 3

If

$$f(x) = 3x^4 - 2x^3 + x^2 - 4x + 2$$

then

$$f'(x) = 12x^3 - 6x^2 + 2x - 4$$
$$f''(x) = 36x^2 - 12x + 2$$
$$f'''(x) = 72x - 12$$
$$f^{(4)}(x) = 72$$
$$f^{(5)}(x) = 0$$
$$\vdots$$
$$f^{(n)}(x) = 0 \qquad (n \geq 5)$$

◀

In the case where f is a function of x, successive applications of the derivative operation are denoted as follows:

$$f'(x) = \frac{d}{dx}[f(x)]$$

$$f''(x) = \frac{d}{dx}\left[\frac{d}{dx}[f(x)]\right] = \frac{d^2}{dx^2}[f(x)]$$

$$f'''(x) = \frac{d}{dx}\left[\frac{d}{dx}\left[\frac{d}{dx}[f(x)]\right]\right] = \frac{d^3}{dx^3}[f(x)]$$

$$\vdots \qquad\qquad\qquad\qquad \vdots$$

etc.   etc.

In general we write

$$f^{(n)}(x) = \frac{d^n}{dx^n}[f(x)]$$

which is read, *the nth derivative of f with respect to x.*

When a dependent variable is involved, say

$$y = f(x)$$

then successive derivatives can be denoted by writing

$$y', \quad y'', \quad y''', \quad y^{(4)}, \ldots, y^{(n)}, \ldots$$

or

$$\frac{dy}{dx}, \frac{d^2y}{dx^2}, \frac{d^3y}{dx^3}, \frac{d^4y}{dx^4}, \ldots, \frac{d^ny}{dx^n}, \ldots$$

The following symbols denote values of derivatives at a particular point x_0; their meanings should be self-evident.

$$y''(x_0), \quad f^{(4)}(x_0), \quad \frac{d^3y}{dx^3}\bigg|_{x=x_0}, \quad \frac{d^2}{dx^2}[x^7 - x]\bigg|_{x=x_0}$$

▶ Example 4

$$\frac{d^2}{dx^2}[\sin x] = \frac{d}{dx}\left[\frac{d}{dx}(\sin x)\right] = \frac{d}{dx}[\cos x] = -\sin x$$

Thus

$$\frac{d^2}{dx^2}[\sin x]\bigg|_{x=\frac{1}{2}\pi} = -\sin\left(\frac{1}{2}\pi\right) = -1 \qquad ◀$$

Just as the first derivative of a function f can fail to exist at a point x_0, so one of the higher order derivatives can fail to exist at point x_0. In general, if $f'(x_0), f''(x_0), \ldots, f^{(n)}(x_0)$ all exist, we say that f is **n-times differentiable** at x_0.

▶ Example 5

Let

$$f(x) = x^{12/5}$$

then

$$f'(x) = \frac{12}{5}x^{7/5}, \qquad f''(x) = \frac{84}{25}x^{2/5}, \qquad f''' = \frac{168}{125}x^{-3/5}$$

Since $f'(0)$ and $f''(0)$ both exist, but $f'''(0)$ does not, f is twice differentiable at 0 but not three times differentiable at 0. ◀

▶ Example 6

Use implicit differentiation to find d^2y/dx^2 if $4x^2 - 2y^2 = 9$.

Solution. Differentiating both sides of $4x^2 - 2y^2 = 9$ implicitly yields

$$8x - 4y\frac{dy}{dx} = 0$$

from which we obtain

$$\frac{dy}{dx} = \frac{2x}{y} \tag{5}$$

Differentiating both sides of this equation implicitly yields

$$\frac{d^2y}{dx^2} = \frac{(y)(2) - (2x)(dy/dx)}{y^2}$$

Substituting (5) and simplifying yields

$$\frac{d^2y}{dx^2} = \frac{2y - 2x(2x/y)}{y^2} = \frac{2y^2 - 4x^2}{y^3} \qquad \blacktriangleleft$$

▶ Exercise Set 4.4

1. Find d^2y/dx^2.
   (a) $y = 7x^3 - 5x^2 + x$
   (b) $y = 12x^2 - 2x + 3$
   (c) $y = \sqrt{x + 3}$
   (d) $y = \tan x$.

2. Find y''.
   (a) $y = \sqrt[3]{2x^3 - 1}$
   (b) $y = \sin 4x$
   (c) $y = 1/x$
   (d) $y = \sec x$.

3. Find
   (a) $\dfrac{d^3}{dx^3}[\sin x \cos x]$
   (b) $\dfrac{d^3y}{dx^3}$, where $y = x^{5/3} - 4x^{2/3}$.

4. Find
   (a) $\dfrac{d^4}{dx^4}[\cos 3x]\Big|_{x=\frac{1}{3}\pi}$
   (b) $y^{(3)}(1)$, where $y = \sqrt{x}$.

5. Find
   (a) $f'''(2)$, where $f(x) = 3x^2 - 2$
   (b) $\dfrac{d^2y}{dx^2}\Big|_{x=1}$, where $y = \dfrac{x-1}{x+1}$.

6. Find d^2y/dx^2 by implicit differentiation.
   (a) $2xy - y^2 = 3$
   (b) $y + \sin y = x$.

7. Find y'' by implicit differentiation.
   (a) $x^{1/2} + y^{1/2} = 1$
   (b) $x \cos y = y$.

In Exercises 8–17, determine open intervals on which f is (a) increasing, (b) decreasing, (c) concave up, (d) concave down, and (e) find the location of all inflection points.

8. $f(x) = x^2 - 5x + 6$.
9. $f(x) = (x + 2)^3$.
10. $f(x) = 5 + 12x - x^3$.
11. $f(x) = 3x^4 - 4x^3$.
12. $f(x) = \dfrac{x}{x^2 + 2}$.
13. $f(x) = \cos x, \ 0 < x < 2\pi$.
14. $f(x) = \sin^2 2x, \ 0 < x < \pi$.
15. $f(x) = \tan x, \ -\pi/2 < x < \pi/2$.
16. $f(x) = x^{2/3}$.
17. $f(x) = \sqrt[3]{x} + 2$.

18. In each part sketch a continuous curve $y = f(x)$ with the stated properties.

(a) $f(2) = 4$, $f'(2) = 0$, $f''(x) < 0$ for all x

(b) $f(2) = 4$, $f'(2) = 0$, $f''(x) > 0$ for $x < 2$, $f''(x) < 0$ for $x > 2$

(c) $f(2) = 4$, $f''(x) > 0$ for $x > 2$, $f''(x) > 0$ for $x < 2$ and

$$\lim_{x \to 2^+} f'(x) = -\infty, \quad \lim_{x \to 2^-} f'(x) = +\infty$$

19. In each part sketch a continuous curve $y = f(x)$ with the stated properties.

(a) $f(2) = 4$, $f'(2) = 0$, $f''(x) > 0$ for all x

(b) $f(2) = 4$, $f'(2) = 0$, $f''(x) < 0$ for $x < 2$, $f''(x) > 0$ for $x > 2$

(c) $f(2) = 4$, $f''(x) < 0$ for $x > 2$, $f''(x) < 0$ for $x < 2$ and

$$\lim_{x \to 2^+} f'(x) = +\infty, \quad \lim_{x \to 2^-} f'(x) = -\infty$$

20. In each part sketch a continuous curve $y = f(x)$ with the stated properties.

(a) $f(2) = 4$, $f'(2) = 1$, $f''(x) < 0$ if $x < 2$, $f''(x) > 0$ if $x > 2$

(b) $f(2) = 4$, $f''(x) > 0$ if $x < 2$, $f''(x) < 0$ if $x > 2$ and

$$\lim_{x \to 2^-} f'(x) = +\infty, \quad \lim_{x \to 2^+} f'(x) = +\infty$$

(c) $f(2) = 4$, $f''(x) < 0$ if $x < 2$, $f''(x) < 0$ if $x > 2$ and

$$\lim_{x \to 2^-} f'(x) = 1, \quad \lim_{x \to 2^+} f'(x) = -1$$

21. In each part sketch a continuous curve $y = f(x)$ with the stated properties.

(a) $f(2) = 4$, $f''(x) < 0$ if $x < 2$, $f''(x) < 0$ if $x > 2$ and

$$\lim_{x \to 2^-} f'(x) = 0, \quad \lim_{x \to 2^+} f'(x) = +\infty$$

(b) $f(0) = 0$, $f(2) = 4$, $f''(x) = 0$ for all x.

22. Let $f(x) = (x - a)^3$.

(a) Find the stationary points, if any.

(b) Find the inflection points, if any.

(c) Sketch the graph of f.

23. Let $f(x) = (x - a)^4$.

(a) Find the stationary points, if any.

(b) Find the inflection points, if any.

(c) Sketch the graph of f.

24. Let $f(x) = ax^2 + bx + c$ $(a \neq 0)$.

(a) Find the stationary points, if any.

(b) Find the inflection points, if any.

(c) Sketch the possible shapes for the graph of f.

25. (a) Prove: If $f''(x)$ exists for each x in (a, b), then both f and f' are continuous on (a, b).

(b) What can be said about the continuity of f and its derivatives if $f^{(n)}(x)$ exists for each x in (a, b)?

26. Let $f(x) = x^8 - 2x + 3$ and $x_0 = 2$; find

$$\lim_{h \to 0} \frac{f'(x_0 + h) - f'(x_0)}{h}$$

27. (a) Show that $f(x) = x^{4/3}$ is differentiable at 0, but not twice differentiable at 0.

(b) Show that $f(x) = x^{7/3}$ is twice differentiable at 0, but not three times differentiable at 0.

(c) Find an exponent k such that $f(x) = x^k$ is $(n-1)$-times differentiable at 0, but not n-times differentiable at 0.

28. Prove: If f and g are twice differentiable at x then $f \cdot g$ is twice differentiable at x and

$$(f \cdot g)''(x) = f''(x) \cdot g(x) + 2f'(x) \cdot g'(x) + f(x) \cdot g''(x)$$

29. In each part compute f', f'', f''' and then conjecture the formula for $f^{(n)}$.

(a) $f(x) = \dfrac{1}{x}$

(b) $f(x) = \dfrac{1}{1 - 2x}$.

30. Find the kth derivative of x^n, where n is a positive integer.

31. Prove: A third degree polynomial $f(x) = ax^3 + bx^2 + cx + d$ $(a \neq 0)$ has exactly one inflection point.

32. Prove: An nth degree polynomial $f(x) = a_0 x^n + a_1 x^{n-1} + \cdots + a_n$ $(a_0 \neq 0)$ has at most $n - 1$ stationary points and at most $n - 2$ inflection points.

33. (a) Prove: If f and g are twice differentiable at x and c is any constant then cf and $f + g$ are twice differentiable at x and

$$\frac{d^2}{dx^2}[cf(x)] = c\,\frac{d^2}{dx^2}[f(x)]$$

$$\frac{d^2}{dx^2}[f(x) + g(x)] = \frac{d^2}{dx^2}[f(x)] + \frac{d^2}{dx^2}[g(x)]$$

(b) Do the results in part (a) generalize to nth derivatives? Justify your answer.

4.5 SKETCHING GRAPHS OF POLYNOMIALS AND RATIONAL FUNCTIONS

In this section we will show how to use the tools developed in the previous section to obtain graphs of polynomials and rational functions.

Polynomials are among the simplest functions to graph; the information obtained in the following steps is usually sufficient to obtain an accurate graph.

> ### How to Graph a Polynomial $P(x)$
> **Step 1.** Calculate $P'(x)$ and $P''(x)$.
> **Step 2.** From $P'(x)$ determine the stationary points and the intervals where P is increasing and decreasing.
> **Step 3.** From $P''(x)$ determine the inflection points and the intervals where P is concave up and concave down.
> **Step 4.** Plot a few well-chosen points to provide any additional information needed to complete the shape of the graph.

▶ Example 1

Sketch the graph of

$$y = x^3 - 3x + 2$$

Solution.

$$\frac{dy}{dx} = 3x^2 - 3 = 3(x - 1)(x + 1)$$

$$\frac{d^2y}{dx^2} = 6x$$

Sign of $(x - 1)$

Sign of $(x + 1)$

Sign of $\dfrac{dy}{dx}$

Figure 4.5.1

From the sign of dy/dx (Figure 4.5.1) the graph of f is increasing if $x < -1$, decreasing if $-1 < x < 1$, and increasing again when $x > 1$; there are stationary points at $x = -1$ and $x = 1$. This information is recorded in Figure 4.5.2. Since d^2y/dx^2 is positive when $x > 0$ and negative when $x < 0$, the graph of f is concave up when $x > 0$ and concave down when $x < 0$; there is an inflection point at $x = 0$, since f changes concavity there.

(a)

(b)

Figure 4.5.2

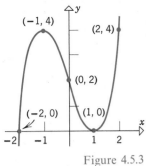

Figure 4.5.3

The concavity information is also recorded in Figure 4.5.2a. From the information obtained thus far we can deduce the basic shape of the graph (Figure 4.5.2b). All that remains is to refine this sketch by plotting some key points. In this case, the stationary points, the inflection point, and two additional points suffice to give an accurate sketch of the graph. From Table 4.5.1 and Figure 4.5.2 we obtain the final graph shown in Figure 4.5.3.

◀

Table 4.5.1

x	-2	-1	0	1	2
$y = x^3 - 3x + 2$	0	4	2	0	4

Recall that if $P(x)$ and $Q(x)$ are polynomials, then their ratio

$$f(x) = \frac{P(x)}{Q(x)}$$

is called a rational function of x. Graphing rational functions is complicated by the fact that discontinuities occur at points where $Q(x) = 0$. In Figure 4.5.4 we have sketched the graph of

$$f(x) = \frac{x}{x - 2}$$

This figure illustrates most of the characteristics that are typical of rational functions.

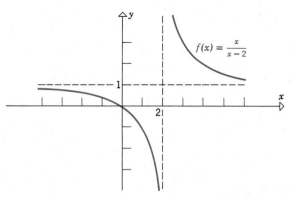

Figure 4.5.4

At $x = 2$ the denominator of $f(x)$ is zero, so that a discontinuity occurs in the graph at this point. Approaching the point $x = 2$ we have

$$\lim_{x \to 2^+} \frac{x}{x - 2} = +\infty \quad \text{and} \quad \lim_{x \to 2^-} \frac{x}{x - 2} = -\infty$$

The line $x = 2$ is called a *vertical asymptote* for the graph. Also

$$\lim_{x \to +\infty} \frac{x}{x - 2} = 1 \quad \text{and} \quad \lim_{x \to -\infty} \frac{x}{x - 2} = 1$$

so that the graph tends toward the line $y = 1$ as $x \to +\infty$ and as $x \to -\infty$. The line $y = 1$ is called a *horizontal asymptote* for the graph. For reference, let us formalize these ideas.

4.5.1 DEFINITION   A line $x = x_0$ is called a **vertical asymptote** for the graph of a function f if $f(x) \to +\infty$ or $f(x) \to -\infty$ as x approaches x_0 from the right or from the left. A line $y = L$ is called a **horizontal asymptote** for the graph of a function f if $f(x) \to L$ as $x \to +\infty$ or $x \to -\infty$.

In our discussion of graphing rational functions, we will assume for simplicity that the numerator and denominator have no common factors. For such functions, the information obtained in the following steps is usually sufficient to obtain an accurate sketch of the graph.

How to Graph a Rational Function $f(x) = \dfrac{P(x)}{Q(x)}$

Step 1. Find the values of x where $P(x) = 0$. At these values we have $f(x) = 0$, so that the graph intersects the x-axis at these points.

Step 2. Find the values of x where $Q(x) = 0$. At these values, $f(x)$ approaches $+\infty$ or $-\infty$, and the graph has a vertical asymptote.

Step 3. Compute $\lim\limits_{x \to +\infty} f(x)$ and $\lim\limits_{x \to -\infty} f(x)$. If either limit has a finite value L, then the line $y = L$ is a horizontal asymptote.

Step 4. The only places where $f(x)$ can change sign are at the points where the graph intersects the x-axis or has a vertical asymptote. Calculate a sample value of $f(x)$ in each of the open intervals determined by these points to see whether the graph is above or below the x-axis over the interval.

Step 5. From $f'(x)$ and $f''(x)$ determine the stationary points, inflection points, intervals of increase, decrease, upward concavity, and downward concavity.

Step 6. If needed, plot a few well-chosen points and determine if the graph crosses any of the horizontal asymptotes.

▶ Example 2

Sketch the graph of

$$f(x) = \frac{x^2 - 1}{x^3}$$

Solution. Setting $P(x) = x^2 - 1$ equal to zero yields $x = 1$ and $x = -1$ as the points where the curve intersects the x-axis.

Setting $Q(x) = x^3$ equal to zero yields the line $x = 0$ as the only vertical asymptote.

From the limits

$$\lim_{x \to +\infty} \frac{x^2 - 1}{x^3} = \lim_{x \to +\infty} \left(\frac{1}{x} - \frac{1}{x^3} \right) = 0 \tag{1}$$

and

$$\lim_{x \to -\infty} \frac{x^2 - 1}{x^3} = \lim_{x \to -\infty} \left(\frac{1}{x} - \frac{1}{x^3} \right) = 0 \tag{2}$$

it follows that the line $y = 0$ is the only horizontal asymptote.

If we mark on the x-axis, the location of the vertical asymptote and the places where the curve intersects the axis, then these points separate the x-axis into the open intervals

$$(-\infty, -1) \qquad (-1, 0) \qquad (0, 1) \qquad (1, +\infty)$$

Over each interval the graph remains above or below the x-axis. For each interval, we can determine which is the case by finding the sign of $f(x)$ at an arbitrary sample point in the interval.

At this stage we can make a rough sketch of the graph (Figure 4.5.5). Curve segments A and B in the figure were constructed by using Table 4.5.2 together with limits (1) and (2). Since the curve intersects the x-axis at -1

Table 4.5.2

INTERVAL	SAMPLE POINT	$f(x)$	CONCLUSION
$(-\infty, -1)$	$x = -2$	$f(-2) = -\frac{3}{8} < 0$	The graph is below the x-axis.
$(-1, 0)$	$x = -\frac{1}{2}$	$f\left(-\frac{1}{2}\right) = 6 > 0$	The graph is above the x-axis.
$(0, 1)$	$x = \frac{1}{2}$	$f\left(\frac{1}{2}\right) = -6 < 0$	The graph is below the x-axis.
$(1, +\infty)$	$x = 2$	$f(2) = \frac{3}{8} > 0$	The graph is above the x-axis.

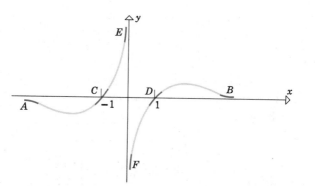

Figure 4.5.5

and 1, curve segments C and D follow from Table 4.5.2. Finally, since the y-axis is a vertical asymptote, curve segments E and F follow from Table 4.5.2.

It is conceivable that our rough sketch may have missed some inflection points or stationary points, but we can investigate this possibility using the first and second derivatives:

$$f'(x) = \frac{x^3(2x) - (x^2 - 1)(3x^2)}{(x^3)^2} = \frac{3 - x^2}{x^4}$$

$$f''(x) = \frac{x^4(-2x) - (3 - x^2)(4x^3)}{(x^4)^2} = \frac{2(x^2 - 6)}{x^5}$$

Since the denominator of $f'(x)$ cannot be negative, the sign of $f'(x)$ is determined by the sign of the numerator. A sign analysis of $f'(x)$ is shown in Figure 4.5.6. A sign analysis of $f''(x)$ is shown in Figure 4.5.7. From Figures 4.5.6 and 4.5.7 we deduce the results recorded in Figure 4.5.8.

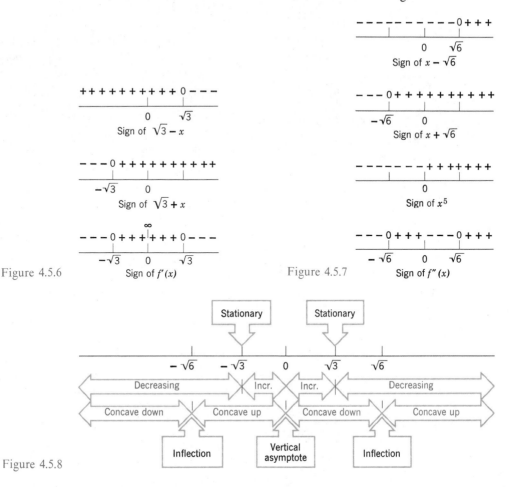

Figure 4.5.6

Figure 4.5.7

Figure 4.5.8

For our final sketch (Figure 4.5.9) we used our earlier rough sketch, and the data in Figure 4.5.8. We also plotted the stationary points and the inflection points. ◀

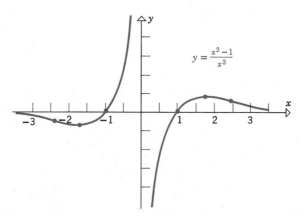

Figure 4.5.9

The work required to graph an equation can be reduced if the graph enjoys certain symmetry properties. A curve is called **symmetric about the x-axis** if for each point (x, y) on the curve, the point $(x, -y)$ is also on the curve (Figure 4.5.10a); geometrically, this makes the portion of curve below the x-axis the mirror image of the portion above. Similarly, a curve is **symmetric about the y-axis** if for each point (x, y) on the curve, the point $(-x, y)$ is also on the curve (Figure 4.5.10b). Finally, a curve is **symmetric about the origin** if for each point (x, y) on the curve, the point $(-x, -y)$ is also on the curve (Figure 4.5.10c); geometrically, this means that for each point P on the curve there is a companion point Q on the curve with origin bisecting the line segment joining P and Q.

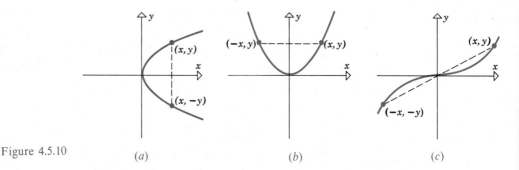

Figure 4.5.10   (a)   (b)   (c)

Symmetries can often be detected from the equation of a curve. For example, the graph of

$$y = x^2 + 1 \tag{3}$$

must be symmetric about the y-axis because for any point (x, y) whose coordinates satisfy (3), the coordinates of the point $(-x, y)$ also satisfy (3) since substituting these values in (3) yields

$$y = (-x)^2 + 1$$

or $y = x^2 + 1$ (Figure 4.5.11). This example suggests the following symmetry tests.

4.5.2 THEOREM

Symmetry Tests

(a) *A plane curve is symmetric about the y-axis if replacing x by −x in its equation produces an equivalent equation.*

(b) *A plane curve is symmetric about the x-axis if replacing y by −y in its equation produces an equivalent equation.*

(c) *A plane curve is symmetric about the origin if replacing x by −x and y by −y in the equation produces an equivalent equation.*

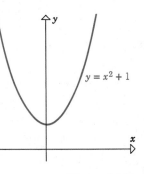

$y = x^2 + 1$

Figure 4.5.11

▶ Example 3

The curve

$$y = \frac{x^2 - 1}{x^3} \tag{4}$$

is symmetric about the origin since replacing y by $-y$ and x by $-x$ yields

$$-y = \frac{(-x)^2 - 1}{(-x)^3}$$

or

$$-y = -\frac{x^2 - 1}{x^3}$$

which reduces to the original equation on multiplying by -1. The given curve is not symmetric about the y-axis, however, since replacing x by $-x$ in (4) yields

$$y = \frac{(-x)^2 - 1}{(-x)^3}$$

or

$$y = -\frac{x^2 - 1}{x^3}$$

which is not equivalent to the original equation. Similarly, the graph is not symmetric about the x-axis. The graph is shown in Figure 4.5.9.   ◀

If it is determined that the graph of an equation is symmetric about the x-axis, then the entire graph can be obtained by sketching the portion of the graph for which $y \geq 0$, and then reflecting this portion about the x-axis. Similarly, if there is symmetry about the y-axis, the entire graph can be obtained by reflecting the portion of the graph for which $x \geq 0$ about the y-axis. Finally, if the graph is symmetric about the origin, the portion of the graph in any quadrant can be obtained from the portion of the graph in the quadrant diagonally opposite.

▶ Exercise Set 4.5

In Exercises 1–8, use the techniques illustrated in this section to sketch the graph of the given polynomial. Plot the stationary points and the inflection points.

1. $x^2 - 2x - 3$.

2. $1 + x - x^2$.

3. $x^3 + 3x^2 + 5$.

4. $x^2 - x^3$.

5. $(x - 1)^4$.

6. $(x - 1)^5$.

7. $x^4 + 2x^3 - 1$.

8. $x^5 - 4x^4 + 4x^3$.

In Exercises 9–17, use the techniques illustrated in this section to sketch the graph of the given rational function. Show any horizontal and vertical asymptotes, and plot the stationary points and inflection points. Take advantage of symmetry, when possible.

9. $\dfrac{2x}{x - 3}$.

10. $\dfrac{x}{x^2 - 1}$.

11. $\dfrac{x^2}{x^2 - 1}$.

12. $\dfrac{1}{(x - 1)^2}$.

13. $\dfrac{x}{1 + x^2}$.

14. $1 - \dfrac{1}{x}$.

15. $\dfrac{x - 1}{x - 2}$.

16. $\dfrac{1}{x^2 + 1}$.

17. $x^2 - \dfrac{1}{x}$.

18. (*Oblique Asymptotes*) If a rational function $P(x)/Q(x)$ is such that the degree of the numerator exceeds the degree of the denominator by *one*, then the graph of $P(x)/Q(x)$ will have an *oblique asymptote*, that is, an asymptote that is neither vertical nor horizontal. To see why, we perform the division of $P(x)$ by $Q(x)$ to obtain

$$\frac{P(x)}{Q(x)} = (ax + b) + \frac{R(x)}{Q(x)}$$

where $ax + b$ is the quotient and $R(x)$ is the remainder. Use the fact that the degree of the remainder $R(x)$ is less than the degree of the divisor $Q(x)$ to help prove:

$$\lim_{x \to +\infty} \left| \frac{P(x)}{Q(x)} - (ax + b) \right| = 0$$

$$\lim_{x \to -\infty} \left| \frac{P(x)}{Q(x)} - (ax + b) \right| = 0$$

These results tell us that the graph of the equation $y = P(x)/Q(x)$ "tends" toward the line (oblique asymptote) $y = ax + b$ as $x \to +\infty$ or $x \to -\infty$ (Figure 4.5.12).

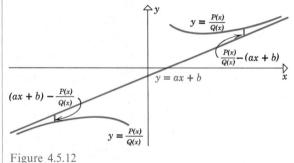

Figure 4.5.12

In Exercises 19–21, sketch the graph of the rational function. Show all vertical, horizontal, and oblique asymptotes (see Exercise 18).

19. $\dfrac{x^2 - 2}{x}$.

20. $\dfrac{x^2 - 2x - 3}{x + 2}$.

21. $\dfrac{(x - 2)^3}{x^2}$.

4.6 OTHER GRAPHING PROBLEMS

In this section we consider functions whose graphs have characteristics not found in the graphs of polynomials and rational functions.

4.6.1 DEFINITION   The graph of a function f is said to have a *vertical tangent line* at x_0 if f is continuous at x_0 and $|f'(x)|$ approaches $+\infty$ as $x \to x_0$.

▶ Example 1

Sketch the graph of $f(x) = \sqrt[3]{x}$.

Solution. The function $f(x) = \sqrt[3]{x} = x^{1/3}$ is continuous and

$$f'(x) = \frac{1}{3}x^{-2/3} = \frac{1}{3x^{2/3}}$$

$$f''(x) = \frac{1}{3}\left(-\frac{2}{3}\right)x^{-5/3} = -\frac{2}{9x^{5/3}}$$

The first derivative is positive for all x except $x = 0$, so that $f(x)$ is increasing on the intervals $(-\infty, 0)$ and $(0, +\infty)$. There are no stationary points since $f'(x) \neq 0$ for any x. However, $x = 0$ is a critical point since $f'(0)$ is undefined. At this point

$$\lim_{x \to 0} |f'(x)| = \lim_{x \to 0} \left| \frac{1}{3x^{2/3}} \right| = +\infty$$

so the graph of f has a vertical tangent line at $x = 0$.

The second derivative of f is negative if $x > 0$ and positive if $x < 0$, so that the graph of f is concave down on the interval $(0, +\infty)$ and concave up on the interval $(-\infty, 0)$ (Figure 4.6.1a).

Since

$$\lim_{x \to +\infty} f(x) = \lim_{x \to +\infty} \sqrt[3]{x} = +\infty$$

$$\lim_{x \to -\infty} f(x) = \lim_{x \to -\infty} \sqrt[3]{x} = -\infty$$

the graph of f has no horizontal asymptotes. Thus from Figure 4.6.1a and the points plotted from the following table, we obtain the sketch in Figure 4.6.1b.

x	-8	-1	0	1	8
$\sqrt[3]{x}$	-2	-1	0	1	2

◄

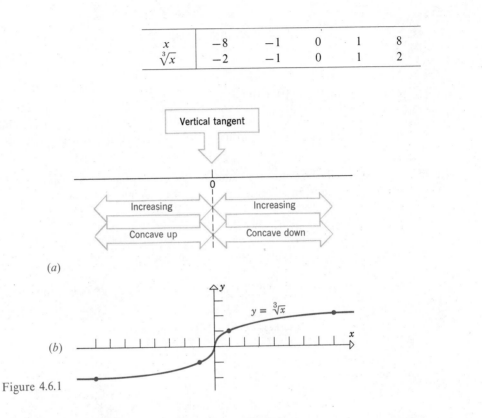

(a)

(b)

Figure 4.6.1

4.6.2 DEFINITION   The graph of a function f is said to have a *cusp* at x_0 if f is continuous at x_0 and $f'(x) \to +\infty$ as x approaches x_0 from one side, while $f'(x) \to -\infty$ as x approaches x_0 from the other side (Figure 4.6.2).

Figure 4.6.2

At a cusp there is a vertical tangent since $|f'(x)| \to +\infty$ as $x \to x_0$ from either side.

▶ Example 2

Sketch the graph of

$$f(x) = (x - 4)^{2/3}$$

Solution. The function f is continuous and

$$f'(x) = \frac{2}{3}(x - 4)^{-1/3} = \frac{2}{3(x - 4)^{1/3}}$$

$$f''(x) = -\frac{2}{9}(x - 4)^{-4/3} = -\frac{2}{9(x - 4)^{4/3}}$$

Since $f'(x) > 0$ when $x > 4$ and $f'(x) < 0$ when $x < 4$, the graph of f is increasing when $x > 4$ and decreasing when $x < 4$. There are no stationary points since $f'(x) \neq 0$ for any value of x. However, $x = 4$ is a critical point since $f'(4)$ does not exist. At this point

$$\lim_{x \to 4^-} f'(x) = \lim_{x \to 4^-} \frac{2}{3(x - 4)^{1/3}} = -\infty$$

$$\lim_{x \to 4^+} f'(x) = \lim_{x \to 4^+} \frac{2}{3(x - 4)^{1/3}} = +\infty$$

so that the graph of f has a cusp at $x = 4$ (Figure 4.6.3).

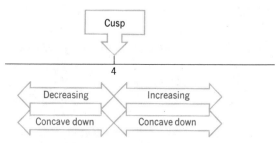

Figure 4.6.3

With the exception of the point $x = 4$, where $f''(x)$ is undefined, $f''(x)$ is negative. Thus, the graph of f is concave down if $x < 4$ and also concave down if $x > 4$ (Figure 4.6.3).

Since

$$\lim_{x \to -\infty} f(x) = \lim_{x \to -\infty} (x - 4)^{2/3} = +\infty$$

$$\lim_{x \to +\infty} f(x) = \lim_{x \to +\infty} (x - 4)^{2/3} = +\infty$$

the graph of f has no horizontal asymptotes. Thus from Figure 4.6.3 and the points plotted from the following table we obtain the graph in Figure 4.6.4.

◀

x	0	3	4	5	8
$y = (x - 4)^{2/3}$	$\sqrt[3]{16} \approx 2.52$	1	0	1	$\sqrt[3]{16} \approx 2.52$

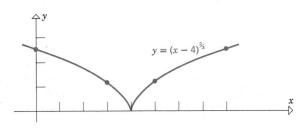

Figure 4.6.4

▶ Example 3

Sketch the graph of

$$f(x) = \sqrt{1 - x}$$

Solution. Since $f(x)$ is undefined when $x > 1$, the graph of f extends only over the interval $-\infty < x \leq 1$. Over this interval $f(x) = \sqrt{1 - x} = (1 - x)^{1/2}$ is continuous and

$$f'(x) = -\frac{1}{2}(1 - x)^{-1/2} = -\frac{1}{2\sqrt{1 - x}}$$

$$f''(x) = -\frac{1}{4}(1 - x)^{-3/2} = -\frac{1}{4(\sqrt{1 - x})^3}$$

For $x < 1$ both $f'(x)$ and $f''(x)$ are negative so that the graph of f is decreasing and concave down over the interval $(-\infty, 1)$. Also,

$$\lim_{x \to 1^-} f'(x) = \lim_{x \to 1^-} \left(-\frac{1}{2\sqrt{1 - x}} \right) = -\infty$$

so that the tangent lines tend toward a vertical position as x approaches 1 from the left.

From the above information and by plotting the points in the following table we are led to the graph in Figure 4.6.5.

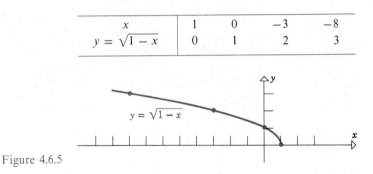

x	1	0	-3	-8
$y = \sqrt{1-x}$	0	1	2	3

Figure 4.6.5   ◀

▶ Exercise Set 4.6

In Exercises 1–18 use the techniques illustrated in this section to sketch the graph of the given function.

1. $(x-2)^{1/3}$.
2. $x^{1/4}$.
3. $x^{1/5}$.
4. $x^{2/5}$.
5. $x^{4/3}$.
6. $x^{-1/3}$.
7. $1 - x^{2/3}$.
8. $\sqrt{x+2}$.
9. $\sqrt{x^2 - 1}$.
10. $\sqrt[3]{x^2 - 4}$.
11. $2x + 3x^{2/3}$.
12. $4x - 3x^{4/3}$.
13. $x - \cos x$.
14. $x + \tan x$.
15. $\sin x + \cos x$.
16. $\sqrt{3}\cos x + \sin x$.
17. $\sin^2 x$, $0 \le x \le 2\pi$.
18. $x \tan x$, $-\frac{\pi}{2} < x < \frac{\pi}{2}$.

19. (a) How can one obtain the graph of $y = |f(x)|$ from the graph of $y = f(x)$?
    (b) Use the graph of $\sin x$ to obtain the graph of $|\sin x|$.

20. (a) How can one obtain the graph of $y = f(|x|)$ from the graph of $y = f(x)$?
    (b) Use the graph of $\sin x$ to obtain the graph of $\sin |x|$.

4.7 RELATIVE EXTREMA

In this section we will develop techniques for solving some optimization problems to which the extreme value theorem does not apply. In the next section we will use these tools to solve some applied problems.

The graphs of many functions form hills and valleys. The tops of the hills are called *relative maxima* and the bottoms of the valleys are called *relative minima* (Figure 4.7.1).

Figure 4.7.1

Just as the top of a hill on the earth's terrain need not be the highest point on earth, so a relative maximum need not be the highest point on the entire graph. However, relative maxima, like tops of hills, are the high points in their *immediate vicinity*. Similarly relative minima, like valley bottoms, are the low points in their *immediate vicinity*. These ideas are captured in the following definitions.

4.7.1 DEFINITION   A function f is said to have a ***relative maximum*** at x_0 if

$$f(x_0) \geq f(x)$$

for all x in some open interval containing x_0.

4.7.2 DEFINITION   A function f is said to have a ***relative minimum*** at x_0 if

$$f(x_0) \leq f(x)$$

for all x in some open interval containing x_0.

4.7.3 DEFINITION   A function f is said to have a ***relative extremum*** at x_0 if it has either a relative maximum or a relative minimum at x_0.

If the inequality $f(x_0) \geq f(x)$ holds for *all* x in the domain of f, then $f(x_0)$ is the largest value of f, and we say that f has a ***maximum*** or ***absolute maximum*** at x_0. In contrast, for f to have a relative maximum at x_0, this inequality need only hold on *some* open interval containing x_0. Similarly, f has a ***minimum*** or ***absolute minimum*** at x_0 if $f(x_0) \leq f(x)$ holds for all x in the domain of f.

▶ Example 1

The function f graphed in Figure 4.7.2 has a relative maximum at x_0 because there is an open interval containing x_0 on which $f(x_0) \geq f(x)$ holds. (See interval (a, b) in the figure.) However, f does not have an absolute maximum at x_0 because $f(x_0)$ is not the largest value for f. At the point x_1 the function f has both an absolute minimum and a relative minimum.   ◀

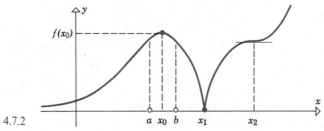

Figure 4.7.2

The following theorem, suggested by Figure 4.7.2, states that the relative extrema of a function always occur at critical points.

4.7.4 THEOREM *If f has a relative extremum at x_0, then x_0 is a critical point for f.*

Proof. If f has a relative maximum at x_0, then there is an open interval (a, b) containing x_0 on which $f(x_0) \geq f(x)$. Thus $f(x_0)$ is the absolute maximum value of f on (a, b). It now follows from Theorem 4.2.6 that x_0 is a critical point. The proof in the case of a relative minimum is similar. ▮

Theorem 4.7.4 does *not* state that a relative extremum occurs at every critical point. For example, the point x_2 in Figure 4.7.2 is a critical point at which there is no relative extremum. However, Theorem 4.7.4 does tell us that we need only look among the critical points to locate the relative extrema.

If a continuous curve is increasing on the left side of a point x_0 and decreasing on the right side of x_0, then obviously a relative maximum occurs at x_0. Similarly, if the curve is decreasing on the left side of x_0 and increasing on the right side, then a relative minimum occurs at x_0. These observations suggest the following test for locating relative extrema.

4.7.5 THEOREM
First Derivative Test

Suppose f is continuous at a critical point x_0.

(a) *If $f'(x) > 0$ on an open interval extending left from x_0 and $f'(x) < 0$ on an open interval extending right from x_0, then f has a relative maximum at x_0.*

(b) *If $f'(x) < 0$ on an open interval extending left from x_0 and $f'(x) > 0$ on an open interval extending right from x_0, then f has a relative minimum at x_0.*

(c) *If $f'(x)$ has the same sign [either $f'(x) > 0$ or $f'(x) < 0$] on an open interval extending left from x_0 and on an open interval extending right from x_0, then f does not have a relative extremum at x_0.*

REMARK. To paraphrase this theorem, *the relative extrema of a continuous function f occur at those critical points where f' changes sign.*

A formal proof of the theorem is given in Section 4.10.

▶ Example 2

Locate the relative extrema of

$$f(x) = 3x^{5/3} - 15x^{2/3}$$

Solution.

$$f'(x) = 5x^{2/3} - 10x^{-1/3} = 5x^{-1/3}(x - 2)$$

Since $f'(x)$ does not exist when $x = 0$, and $f'(x) = 0$ when $x = 2$, the critical points of f are 0 and 2.

As shown in Figure 4.7.3a, the sign of f' changes from positive to negative at 0 and from negative to positive at 2. Thus there is a relative maximum at 0 and a relative minimum at 2 (Figure 4.7.3b). ◀

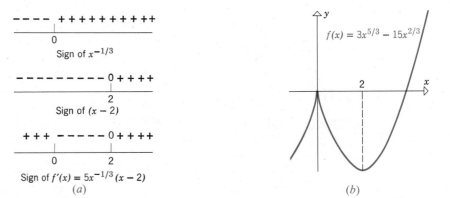

Figure 4.7.3 (a)

Sign of $x^{-1/3}$

Sign of $(x - 2)$

Sign of $f'(x) = 5x^{-1/3}(x - 2)$

(b)

$f(x) = 3x^{5/3} - 15x^{2/3}$

▶ Example 3

Locate the relative extrema of

$$f(x) = x^3$$

Figure 4.7.4

$f(x) = x^3$

Solution.

$$f'(x) = 3x^2$$

Solving $f'(x) = 0$ yields $x = 0$ as the only critical point. Since f' does not change sign at $x = 0$, $f(x) = x^3$ does not have a relative extremum at $x = 0$. (See Figure 4.7.4.) ◀

There is another test for relative extrema that is often easier to apply than the first derivative test. It is based on the geometric observation that a function has a relative maximum where its graph has a horizontal tangent line and is concave down (Figure 4.7.5), and a function has a relative minimum where its graph has a horizontal tangent line and is concave up. (See Figure 4.7.5).

$f'' < 0$
Concave down

$f'' > 0$
Concave up

Relative maximum

Relative minimum

Figure 4.7.5

4.7.6 THEOREM

Second Derivative Test

Suppose f is twice differentiable at a stationary point x_0.
(a) If $f''(x_0) > 0$, then f has a relative minimum at x_0.
(b) If $f''(x_0) < 0$, then f has a relative maximum at x_0.

A formal proof of this result is given in Section 4.10.

▶ Example 4

Locate and describe the relative extrema of

$$f(x) = x^4 - 2x^2$$

Solution.

$$f'(x) = 4x^3 - 4x = 4x(x - 1)(x + 1)$$
$$f''(x) = 12x^2 - 4$$

Solving $f'(x) = 0$ yields the stationary points $x = 0$, $x = 1$, and $x = -1$. Since

$$f''(0) = -4 < 0$$
$$f''(1) = 8 > 0$$
$$f''(-1) = 8 > 0$$

there is a relative maximum at $x = 0$ and there are relative minima at $x = 1$ and $x = -1$. ◀

REMARK. If f is not twice differentiable at the critical point x_0, or if $f''(x_0) = 0$ then we must either rely on the first derivative test or devise more imaginative techniques appropriate to the problem.

The following theorem is sometimes helpful for determining when a relative extremum is an absolute extremum.

4.7.7 THEOREM

Let f be continuous on an interval I and assume that f has exactly one relative extremum on I, say at x_0.
(a) If f has a relative minimum at x_0, then $f(x_0)$ is the absolute minimum value of f on the interval I.
(b) If f has a relative maximum at x_0, then $f(x_0)$ is the absolute maximum value of f on the interval I.

Although we will omit the proof, the result is easy to visualize. For example, if f has a relative maximum at x_0 and if $f(x_0)$ is *not* the absolute

Figure 4.7.6

maximum value for f on I, then the graph of f must make an upward turn somewhere on the interval I, thereby introducing a second relative extremum (Figure 4.7.6). Thus, if there is only one relative extremum on I, an absolute maximum value for f must also occur at x_0.

▶ Example 5

Find the absolute extreme values, if any, of

$$f(x) = x + \frac{1}{x}$$

on $(0, +\infty)$.

Solution. The function f is continuous on $(0, +\infty)$ because the only discontinuity is at 0, which is outside this interval. We have

$$f'(x) = 1 - \frac{1}{x^2}$$

so that $f'(x) = 0$ yields the critical points $x = 1$ and $x = -1$. Thus, 1 is the only critical point in the interval $(0, +\infty)$. Since

$$f''(x) = \frac{2}{x^3}$$

it follows that

$$f''(1) = 2 > 0$$

Thus, a relative minimum occurs at $x = 1$ by the second derivative test. Since this is the only relative extremum on $(0, +\infty)$, the absolute minimum value for f on $(0, +\infty)$ is

$$f(1) = 2$$

Since

$$\lim_{x \to +\infty} f(x) = \lim_{x \to +\infty} \left(x + \frac{1}{x} \right) = +\infty$$

there is no absolute maximum value for f on $(0, +\infty)$. ◀

► Exercise Set 4.7

In Exercises 1–4, find the relative extrema using: (a) the first derivative test, (b) the second derivative test.

1. $f(x) = 1 - 4x - x^2$.

2. $f(x) = 2x^3 - 9x^2 + 12x$.

3. $f(x) = \sin^2 x$, $0 < x < 2\pi$.

4. $f(x) = \frac{1}{2}x - \sin x$, $0 < x < 2\pi$.

In Exercises 5–22, use any method to find the relative extrema.

5. $f(x) = x^3 + 5x - 2$. 6. $f(x) = x^4 - 2x^2 + 7$.

7. $f(x) = x(x - 1)^2$. 8. $f(x) = x^4 + 2x^3$.

9. $f(x) = 2x^2 - x^4$. 10. $f(x) = (2x - 1)^5$.

11. $f(x) = x^{4/5}$. 12. $f(x) = 2x + x^{2/3}$.

13. $f(x) = \dfrac{x^2}{x^2 + 1}$. 14. $f(x) = \dfrac{x}{x + 2}$.

15. $f(x) = |x^2 - 4|$.

16. $f(x) = \begin{cases} 9 - x, & x \le 3 \\ x^2 - 3, & x > 3 \end{cases}$

17. $f(x) = \cos^2 x$.

18. $f(x) = \sqrt{3}x + 2\sin x$, $0 < x < 2\pi$.

19. $f(x) = \tan(x^2 + 1)$.

20. $f(x) = \dfrac{\sin x}{2 + \cos x}$, $0 < x < 2\pi$.

21. $f(x) = |\sin 2x|$, $0 < x < 2\pi$.

22. $f(x) = \cos 4x + 2\sin 2x$, $0 < x < \pi$.

In Exercises 23–26, find the maximum and minimum values of f on the given interval if they exist.

23. $f(x) = x^2 - 3x - 1$; $(-\infty, +\infty)$.

24. $f(x) = \dfrac{x}{x^2 + 1}$; $[0, +\infty)$.

25. $f(x) = \dfrac{x^2}{x + 1}$; $(-5, -1)$.

26. $f(x) = \dfrac{x + 3}{x - 3}$; $[-5, 5]$.

27. Let $f(x) = (x - a)^p$, where p is an integer greater than 1 and a is any real number. Find the relative extrema of f if

(a) p is even.
(b) p is odd.

28. What is the smallest possible slope for a tangent to $y = x^3 - 3x^2 + 5x$?

29. (a) Show that
$$f(x) = \frac{64}{\sin x} + \frac{27}{\cos x}$$
has a minimum value, but no maximum value on the interval $(0, \pi/2)$.

(b) Find the minimum value.

30. Prove: For every positive value of t the function $f(x) = x + t/x$ has a minimum value but no maximum value on $(0, +\infty)$.

31. Find a_0, a_1, and a_2 such that the graph of $f(x) = a_0 + a_1x + a_2x^2$ passes through the point $(0, 9)$ and f has a minimum value of 1 that occurs when $x = 2$.

32. Recall that the second derivative test (Theorem 4.7.6) does not apply if $f''(x_0) = 0$. Give examples that show that a function f can have a relative maximum at x_0, a relative minimum at x_0, or neither if $f''(x_0) = 0$.

33. Let h and g have relative maxima at x_0. Prove or disprove:

(a) $h + g$ has a relative maximum at x_0.
(b) $h - g$ has a relative maximum at x_0.

34. Sketch some curves that show that the three parts of Theorem 4.7.5 can be false without the assumption that f is continuous at x_0.

35. Let $f(x) = ax^2 + bx + c$, where $a > 0$. Prove that $f(x) \ge 0$ for all x if and only if $b^2 - 4ac \le 0$. [Hint: Find the minimum of $f(x)$.]

36. Prove that the minimum value of
$$f(x) = x^2 + \frac{16x^2}{(8 - x)^2}, \quad x > 8$$
occurs at $x = 4(2 + \sqrt[3]{2})$.

4.8 MORE APPLIED MAXIMUM AND MINIMUM PROBLEMS

In Section 4.3 we solved some applied optimization problems that could be reduced to maximizing or minimizing a continuous function on a *closed* interval [*a*, *b*]. In this section we will consider problems that require maximizing or minimizing functions over other kinds of intervals.

► Example 1

A closed cylindrical can is to hold 1 liter (1000 cm³) of liquid. How should we choose the height and radius to minimize the amount of material needed to manufacture the can?

Solution.   Let

$$h = \text{height (in cm) of the can}$$
$$r = \text{radius (in cm) of the can}$$
$$S = \text{surface area (in cm}^2\text{) of the can}$$

Assuming there is no waste or overlap, the amount of material needed for manufacture will be the same as the surface area of the can. Since the can can be made from two circular disks of radius *r* and a rectangular sheet with dimensions *h* by $2\pi r$ (Figure 4.8.1), the surface area will be

$$S = 2\pi r^2 + 2\pi rh \tag{1}$$

Figure 4.8.1

Area $2\pi r^2$

Area $2\pi rh$

We will now eliminate one of the variables in (1) so that *S* will be expressed as a function of one variable. Since the volume of the can is 1000 cm³, it follows from the formula

$$V = \pi r^2 h$$

for the volume of a cylinder that

$$1000 = \pi r^2 h \tag{2}$$

or

$$h = \frac{1000}{\pi r^2} \tag{3}$$

Substituting (3) in (1) yields

$$S = 2\pi r^2 + \frac{2000}{r} \tag{4}$$

Because r represents a radius it must be positive*; thus the variable r in (4) must satisfy

$$0 < r < +\infty$$

We have now reduced the problem to that of finding a value of r in $(0, +\infty)$ for which (4) is minimum (provided such a value exists). From (4) we obtain

$$\frac{dS}{dr} = 4\pi r - \frac{2000}{r^2} \tag{5}$$

Setting $dS/dr = 0$ gives

$$4\pi r - \frac{2000}{r^2} = 0$$

or

$$4\pi r^3 = 2000$$

or

$$r = \frac{10}{\sqrt[3]{2\pi}} \tag{6}$$

To see that an absolute minimum occurs for this value of r, we rewrite (5) as

$$\frac{dS}{dr} = \frac{4\pi}{r^2}\left(r^3 - \frac{1000}{2\pi}\right)$$

---

*Physically, we cannot manufacture a can of radius $r = 0$, so that this possibility must be excluded. There are two ways to do this: either require $r > 0$ throughout the problem, as we are doing here, or require only $r \geq 0$ throughout the problem and reject the solution $r = 0$ at the very end, if it happens to turn up. The latter method was used in Section 4.3. The choice of methods is a matter of taste.

Thus on $(0, +\infty)$, S is decreasing $(dS/dr < 0)$ for $r < 10/\sqrt[3]{2\pi}$ and S is increasing $(dS/dr > 0)$ for $r > 10/\sqrt[3]{2\pi}$, which tells us that an absolute minimum on $(0, +\infty)$ occurs at the value of r in (6). Thus the radius of the can of least surface area is

$$r = \frac{10}{\sqrt[3]{2\pi}}$$

and from (3) the corresponding height is

$$h = \frac{1000}{\pi(10/\sqrt[3]{2\pi})^2} = \frac{20}{\sqrt[3]{2\pi}}$$

Alternate Solution. The conclusion that an absolute minimum occurs at the value of r in (6) can be deduced from the second derivative test by noting that

$$\frac{d^2S}{dr^2} = 4\pi + \frac{4000}{r^3}$$

so that

$$\left.\frac{d^2S}{dr^2}\right|_{r=10/\sqrt[3]{2\pi}} = 4\pi + \frac{4000}{(10/\sqrt[3]{2\pi})^3} = 12\pi$$

Since the second derivative is positive, a relative minimum occurs at the critical point $r = 10/\sqrt[3]{2\pi}$. Because this is the only relative extremum of S in the interval $(0, +\infty)$, it follows from Theorem 4.7.7 that an absolute minimum value of S on $(0, +\infty)$ also occurs at this point. ◀

▶ Example 2

Find a point on the curve $y = x^2$ that is closest to the point $(18, 0)$.

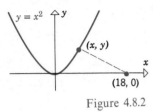

$y = x^2$

(x, y)

$(18, 0)$

Figure 4.8.2

Solution. The distance L between $(18, 0)$ and an arbitrary point (x, y) on the curve $y = x^2$ (Figure 4.8.2) is given by

$$L = \sqrt{(x - 18)^2 + (y - 0)^2}$$

Since (x, y) lies on the curve, x and y satisfy $y = x^2$; thus

$$L = \sqrt{(x - 18)^2 + x^4} \tag{7}$$

Because there are no restrictions on x, the problem reduces to finding a value of x in $(-\infty, +\infty)$ for which (7) is minimum, provided such a value exists.

In problems of minimizing or maximizing a distance, there is a trick that is helpful for simplifying the computations. It is based on the observation that the distance and the square of the distance have their maximum or minimum at the same point (see Exercise 20). Thus the minimum value of L in (7) and the minimum value of

$$S = L^2 = (x - 18)^2 + x^4 \tag{8}$$

occur at the same x value.

From (8),

$$\frac{dS}{dx} = 2(x - 18) + 4x^3 = 4x^3 + 2x - 36 \tag{9}$$

so that the critical points satisfy $4x^3 + 2x - 36 = 0$ or equivalently

$$2x^3 + x - 18 = 0 \tag{10}$$

To solve for x we will begin by searching for integer solutions. This task can be simplified by using the fact that all integer solutions, if there are any, to a polynomial equation with integer coefficients

$$a_0 + a_1x + a_2x^2 + \cdots + a_nx^n = 0$$

must be divisors of the constant term a_0. This result is usually proved in algebra courses. Thus, the only possible integer solutions of (10) are the divisors of -18: $\pm 1, \pm 2, \pm 3, \pm 6, \pm 9, \pm 18$. Successively substituting these values in (10) we find that $x = 2$ is a solution; therefore $x - 2$ is a factor of the left side of (10). After dividing by the factor $x - 2$ we can rewrite (10) as

$$(x - 2)(2x^2 + 4x + 9) = 0$$

Thus the remaining solutions of (10) satisfy the quadratic equation

$$2x^2 + 4x + 9 = 0$$

But these solutions are complex numbers (use the quadratic formula) so that $x = 2$ is the only real solution of (10) and consequently the only critical point. To determine the nature of this critical point we will use the second derivative test. From (9),

$$\frac{d^2S}{dx^2} = 12x^2 + 2$$

so

$$\frac{d^2S}{dx^2}\bigg|_{x=2} = 50 > 0$$

which shows that a relative minimum occurs at $x = 2$. Since $x = 2$ is the only relative extremum for L, it follows from Theorem 4.7.7 that an absolute minimum value of L also occurs at $x = 2$. Thus, the point on the curve $y = x^2$ closest to $(0, 18)$ is

$$(x, y) = (x, x^2) = (2, 4) \qquad \blacktriangleleft$$

▶ Exercise Set 4.8

1. Find two numbers whose sum is 20 and whose product is
   (a) maximum
   (b) minimum.

2. Find the dimensions of the rectangle of area A whose perimeter is
   (a) minimum
   (b) maximum.

3. A rectangular area of 1600 ft^2 is to be fenced off. Two opposite sides will use fencing costing $1 per foot and the remaining sides will use fencing costing $2 per foot. Find the dimensions of the rectangle of least cost.

4. A closed rectangular container with a square base is to have a volume of 3000 in.^{3} The material for the top and bottom of the container will cost $2 per in.^{2}, and the material for the sides will cost $3 per in.^{2} Find the dimensions of the container of least cost.

5. A closed rectangular container with a square base is to have a volume of 1000 cubic centimeters. It costs twice as much per square centimeter for the top and bottom as it does for the sides. Find the dimensions of the container of least cost.

6. A rectangular sheet of paper is to contain 72 in.^{2} of printed matter, with 2-in. margins at top and bottom and 1-in. margins on each side. What dimensions for the sheet will use the least paper?

7. A cone-shaped paper drinking cup is to hold 10 cm^3 of water. Find the height and radius of the cup that will require the least amount of paper.

8. A church window consisting of a rectangle topped by a semicircle is to have a perimeter p. Find the radius of the semicircle if the area of the window is to be maximum.

9. Find all points on the curve $x^2 - y^2 = 1$ closest to $(0, 2)$.

10. Find a point on the curve $x = 2y^2$ closest to $(0, 9)$.

11. Suppose a line L of variable slope passes through $(1, 3)$ and intersects the coordinate axes at the points $(a, 0)$ and $(0, b)$ where a and b are positive. Find the slope of L for which the area of the triangle with vertices $(a, 0)$, $(0, b)$, and $(0, 0)$ is
    (a) maximum
    (b) minimum.

12. In a certain chemical manufacturing process, the daily weight y of defective chemical output depends on the total weight x of all output according to the empirical formula

    $$y = 0.01x + 0.00003x^2$$

    where x and y are in pounds. If the profit is $100 per pound of nondefective chemical produced and the loss is $20 per pound of defective chemical produced, how many pounds of chemical should be produced daily to maximize the profit?

13. Find the points on the circle $x^2 + y^2 = 80$ that are closest to and farthest from $(1, 2)$.

14. A lamp is suspended above the center of a round table of radius r. How high above the table should the lamp be placed to achieve maximum illumination at the edge of the table? [Assume that the illumination I is directly proportional to the cosine of the angle of incidence ϕ of the light rays and inversely proportional to the square of the distance l from the light source (Figure 4.8.3).]

Figure 4.8.3

15. Where on the curve

$$y = \frac{1}{1 + x^2}$$

does the tangent line have the greatest slope?

16. Find the dimensions of the isosceles triangle of least area that can be circumscribed about a circle of radius R.

17. Find the height and radius of the right circular cone with least volume that can be circumscribed about a sphere of radius R.

18. If an unknown physical quantity x is measured n times, the measurements $x_1, x_2, \ldots, x_n$ often vary because of uncontrollable factors such as temperature, atmospheric pressure, and so forth. Thus a scientist is often faced with the problem of using n different observed measurements to obtain an estimate $\bar{x}$ of an unknown quantity x. One method for making such an estimate is based on the *least*

squares principle, which states that the estimate $\bar{x}$ should be chosen to minimize

$$s = (x_1 - \bar{x})^2 + (x_2 - \bar{x})^2 + \cdots + (x_n - \bar{x})^2$$

which is the sum of the squares of the deviations between the estimate $\bar{x}$ and the measured values. Show that the estimate resulting from the least squares principle is

$$\bar{x} = \frac{1}{n}(x_1 + x_2 + \cdots + x_n)$$

that is, $\bar{x}$ is the arithmetic average of the observed values.

19. A pipe of negligible thickness is to be carried horizontally around a corner from a hallway 8 ft wide into a hallway 4 ft wide. What is the maximum length that the pipe can have? [An interesting discussion of this problem in the case where the thickness of the pipe is not neglected is given by Norman Miller in the *American Mathematical Monthly,* vol. 56 (1949), pp. 177–179.]

20. Prove: If $f(x) \geq 0$ on an interval I and if $f(x)$ has a maximum value on I at x_0, then $\sqrt{f(x)}$ also has a maximum value at x_0. Similarly for minimum values. [*Hint:* Use the fact that $\sqrt{x}$ is an increasing function on the interval $[0, +\infty)$.]

21. Prove: If $P(x_1, y_1)$ is a fixed point, then the minimum distance between P and a point on the line

$$ax + by + c = 0$$

is

$$\frac{|ax_1 + by_1 + c|}{\sqrt{a^2 + b^2}}.$$

4.9 ROLLE'S THEOREM; MEAN VALUE THEOREM

In this section we will discuss a result called *the mean value theorem*. There are so many major consequences of this theorem that it is regarded as one of the most fundamental results in calculus.

We will begin with a special case of the mean value theorem, called ***Rolle's* theorem.*** Geometrically, Rolle's theorem states that between any two points where a "well-behaved" curve $y = f(x)$ crosses the x-axis, there must be at least one place where the tangent line to the curve is horizontal (Figure 4.9.1).

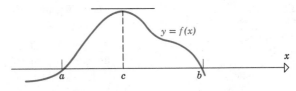

Figure 4.9.1

The precise statement of this result is as follows.

4.9.1 THEOREM

Rolle's Theorem

Let f be differentiable on (a, b) and continuous on $[a, b]$. If $f(a) = f(b) = 0$, then there is at least one point c in (a, b) where

$$f'(c) = 0$$

---

*MICHEL ROLLE (1652–1719) French mathematician. Rolle, the son of a shopkeeper, received only an elementary education. He married early and as a young man struggled hard to support his family on the meager wages of a transcriber for notaries and attorneys. In spite of his financial problems and minimal education, Rolle studied algebra and Diophantine analysis (a branch of number theory) on his own. Rolle's fortune changed drastically in 1682 when he published an elegant solution of a difficult, unsolved problem in Diophantine analysis. The public recognition of his achievement led to a patronage under minister Louvois, a job as an elementary mathematics teacher, and eventually to a short-termed administrative post in the Ministry of War. In 1685 he joined the Académie des Sciences, holding a very low-level position; it was not until 1699 that the Académie began paying him a regular salary. He remained there until he died of apoplexy in 1719.

While Rolle's forté was always Diophantine analysis, his most important work was a book on the algebra of equations, called *Traité d'algèbre,* published in 1690. In that book Rolle firmly established the notation $\sqrt[n]{a}$ [earlier written as $\sqrt{(n)\,a}$] for the nth root of a, and proved a polynomial version of the theorem that today bears his name. (Rolle's theorem was named by Giusto Bellavitis in 1846.) Ironically, Rolle was one of the most vocal early antagonists of calculus. He strove intently to demonstrate that it gave erroneous results and was based on unsound reasoning. He quarreled so vigorously on the subject that the Académie des Sciences was forced to intervene on several occasions. Among his several achievements, Rolle helped advance the currently accepted size order for negative numbers. Descartes, for example, viewed -2 as smaller than -5. Rolle preceded most of his contemporaries by adopting the current convention in 1691.

Proof. Either $f(x)$ is equal to zero for all x in $[a, b]$ or it is not. If it is, then $f'(x) = 0$ for all x in (a, b) since f is constant on (a, b). Thus, for any c in (a, b)

$$f'(c) = 0$$

If $f(x)$ is not equal to zero for all x in $[a, b]$, then there must be a point x in (a, b) where $f(x) > 0$ or $f(x) < 0$. We will consider the first case and leave the second as an exercise.

Since f is continuous on $[a, b]$, it follows from the extreme value theorem that f has a maximum value at some point c in $[a, b]$. Since $f(a) = f(b) = 0$ and $f(x) > 0$ at some point in (a, b), the point c cannot be an endpoint; it must lie in (a, b). By hypothesis, f is differentiable everywhere on (a, b). In particular it is differentiable at c so that

$$f'(c) = 0$$

by Theorem 4.2.6.

► **Example 1**

Let

$$f(x) = \sin x$$

The function $\sin x$ is both continuous and differentiable everywhere, so that $f(x) = \sin x$ is continuous on $[0, 2\pi]$ and differentiable on $(0, 2\pi)$. Moreover,

$$f(0) = \sin 0 = 0 \quad \text{and} \quad f(2\pi) = \sin 2\pi = 0$$

so that f satisfies the hypotheses of Rolle's theorem on the interval $[0, 2\pi]$. Thus, we are guaranteed that there is at least one point c in $(0, 2\pi)$ satisfying

$$f'(c) = \cos c = 0 \tag{1}$$

Because (1) is so simple, we can actually find the values of c; they are

$$c_1 = \frac{\pi}{2} \quad \text{and} \quad c_2 = \frac{3\pi}{2}$$

Sometimes, however, the function f is sufficiently complicated that it is difficult or impossible to find values for c. ◄

Rolle's theorem is a special case of a more general result, which states that between any two points A and B on a "well-behaved" curve $y = f(x)$, there must be at least one place where the tangent line to the curve is parallel to the secant line joining A and B (Figure 4.9.2). This result is called the **mean value theorem.**

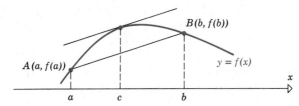

Figure 4.9.2

Noting that the slope of the secant line joining $A(a, f(a))$ and $B(b, f(b))$ is

$$\frac{f(b) - f(a)}{b - a}$$

and the slope of the tangent at c is $f'(c)$, the mean-value theorem can be stated precisely as follows.

4.9.2 THEOREM

Mean Value Theorem

Let f be differentiable on (a, b) and continuous on $[a, b]$. Then there is at least one point c in (a, b) where

$$f'(c) = \frac{f(b) - f(a)}{b - a} \tag{2}$$

The key to proving this theorem is the intuitive observation that the tangent line will be parallel to the secant line at the points where the vertical distance between the curve $y = f(x)$ and the secant line is greatest (Figure 4.9.2). Since the two-point form of the secant line joining $(a, f(a))$ and $(b, f(b))$ is

$$y - f(a) = \frac{f(b) - f(a)}{b - a}(x - a)$$

or equivalently

$$y = \frac{f(b) - f(a)}{b - a}(x - a) + f(a)$$

the difference $v(x)$ between the height of the graph of f and the height of the secant line is

$$v(x) = f(x) - \left[\frac{f(b) - f(a)}{b - a}(x - a) + f(a)\right] \tag{3}$$

Since the maximum vertical distance between $y = f(x)$ and the secant line is an extreme value for $v(x)$, we are led naturally to investigate the stationary points of v. Let c be any point where $v'(c) = 0$. From (3)

$$v'(x) = f'(x) - \frac{f(b) - f(a)}{b - a}$$

so that

$$v'(c) = f'(c) - \frac{f(b) - f(a)}{b - a}$$

or since $v'(c) = 0$,

$$f'(c) = \frac{f(b) - f(a)}{b - a}$$

Thus we have shown that (2) holds at any point c where $v'(c) = 0$. Therefore, to prove the mean value theorem we need only show that there is a point c in the open interval (a, b) where $v'(c) = 0$. This can be done as follows.

Proof of Theorem 4.9.2.   Since $f(x)$ is continuous on $[a, b]$ and differentiable on (a, b), so is the function $v(x)$ in (3). It is easy to check that

$$v(a) = 0 \qquad \text{and} \qquad v(b) = 0$$

so that $v(x)$ satisfies the hypotheses of Rolle's theorem on the interval $[a, b]$. Thus, there is a point c in (a, b) such that $v'(c) = 0$.

▶ Example 2

Let $f(x) = x^3 + 1$. Show that f satisfies the hypotheses of the mean-value theorem on the interval $[1, 2]$ and find all values of c in this interval whose existence is guaranteed by the theorem.

Solution.   Because f is a polynomial, f is continuous and differentiable everywhere. In particular, f is continuous on $[1, 2]$ and differentiable on $(1, 2)$, so that the hypotheses of the mean-value theorem are satisfied.
    If we take $a = 1$ and $b = 2$, then

$$f(a) = f(1) = 2, \qquad f(b) = f(2) = 9$$
$$f'(x) = 3x^2, \qquad f'(c) = 3c^2$$

so that the equation

$$f'(c) = \frac{f(b) - f(a)}{b - a}$$

becomes

$$3c^2 = 7$$

Of the two solutions,

$$c = \sqrt{7/3} \quad \text{and} \quad c = -\sqrt{7/3}$$

only the first is in the interval $(1, 2)$ so that $c = \sqrt{7/3}$ is the only number whose existence is guaranteed by the mean-value theorem. ◄

We stated earlier that the mean-value theorem is the starting point for many important results in calculus. We will conclude this section with two such results.

We know that the derivative of a constant function is zero. Using the mean-value theorem we will now prove the converse is also true.

4.9.3 THEOREM *If $f'(x) = 0$ for all x in an interval, then f is constant on the interval.*

Proof. Assume $f'(x) = 0$ for all x in an interval I. To prove that f is constant on I it suffices to show that f has the same value at any two points in I. Let a and b be arbitrary points in I with $a < b$. Since f is differentiable on I it is continuous on I and consequently the hypotheses of the mean-value theorem hold on the interval $[a, b]$. Thus, there is a number c in (a, b) such that

$$f'(c) = \frac{f(b) - f(a)}{b - a}$$

But $f'(c) = 0$ by hypothesis so that

$$f(b) = f(a)$$

which proves that f has the same value at any two points in I. ▮

4.9.4 COROLLARY *If $f'(x) = g'(x)$ for all x in an interval I then f and g differ by a constant on I, that is, there is a constant k such that*

$$f(x) - g(x) = k$$

for all x in I.

Proof. Let

$$h(x) = f(x) - g(x)$$

Then for every x in I

$$h'(x) = f'(x) - g'(x) = 0$$

Thus

$$h(x) = f(x) - g(x)$$

is constant on I by Theorem 4.9.3. ∎

This corollary has a useful geometric interpretation. It states that two functions with the same derivative at each point of an interval have "parallel" graphs over the interval (Figure 4.9.3).

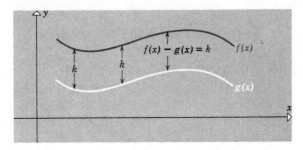

Figure 4.9.3

▶ Exercise Set 4.9

In Exercises 1–4, verify that the hypotheses of Rolle's theorem are satisfied on the given interval and find all values of c that satisfy the conclusion of the theorem.

1. $f(x) = x^2 - 6x + 8$; [2, 4].

2. $f(x) = x^3 - 3x^2 + 2x$; [0, 2].

3. $f(x) = \cos x$; $\left[\dfrac{\pi}{2}, \dfrac{3\pi}{2}\right]$.

4. $f(x) = \dfrac{x^2 - 1}{x - 2}$; [-1, 1].

In Exercises 5–8, verify that the hypotheses of the mean-value theorem are satisfied on the given interval and find all values of c that satisfy the conclusion of the theorem.

5. $f(x) = x^2 + x$; [-4, 6].

6. $f(x) = x^3 + x - 4$; [-1, 2].

7. $f(x) = \sqrt{x + 1}$; [0, 3].

8. $f(x) = x + \dfrac{1}{x}$; [3, 4].

9. Let $f(x) = \tan x$.
  (a) Show that there is no point c in $(0, \pi)$ such that $f'(c) = 0$ even though $f(0) = f(\pi) = 0$.
  (b) Explain why the result in part (a) does not violate Rolle's theorem.

10. Let $f(x) = x^{2/3}$, $a = -1$, and $b = 8$.
  (a) Show that there is no point c in (a, b) such that

$$f'(c) = \frac{f(b) - f(a)}{b - a}$$

  (b) Explain why the result in part (a) does not violate the mean-value theorem.

11. Use the mean-value theorem to prove that

$$|\sin x - \sin y| \le |x - y|$$

holds for all real values of x and y.

12. Use the mean-value theorem to prove that

$$|\tan x + \tan y| \ge |x + y|$$

holds for all real values of x and y in the interval $(-\pi/2, \pi/2)$.

13. Use Rolle's theorem to prove that
$$4x^3 + 9x^2 - 4x - 2 = 0$$
has a solution in the interval $(0, 1)$.

$$\left[Hint: \frac{d}{dx} [x^4 + 3x^3 - 2x^2 - 2x] \right.$$
$$\left. = 4x^3 + 9x^2 - 4x - 2. \right]$$

14. Use Rolle's theorem to prove that

$$x^3 + 4x - 1 = 0$$

does not have more than one real root.

15. Let f and g be continuous on $[a, b]$ and differentiable on (a, b). Prove: If $f(a) = g(a)$ and $f(b) = g(b)$, then there is a point c in (a, b) where $f'(c) = g'(c)$.

16. Prove: If f is continuous on $[a, b]$ and $f'(x) = 0$ for all x in (a, b), then f is constant on $[a, b]$.

17. Prove: If f and g are continuous on $[a, b]$ and $f'(x) = g'(x)$ for all x in (a, b), then f and g differ by a constant on $[a, b]$.

18. (a) Prove: If $f''(x) > 0$ for all x in (a, b), then $f'(x) = 0$ at most once in (a, b).
  (b) Give a geometric interpretation of the result in part (a).

19. Prove: If f is continuous on $[a, b]$ and differentiable on (a, b), and if $f(a) = f(b)$, then there is a point c in (a, b) where $f'(c) = 0$.

20. Let $s = f(t)$ be the position versus time curve for a particle moving in the positive direction along a coordinate line. Prove: If f satisfies the hypotheses of the mean-value theorem on a time interval $[a, b]$ then there will be an instant t_0 in (a, b) where the instantaneous velocity at time t_0 equals the average velocity over the interval $[a, b]$.

21. Complete the proof of Rolle's theorem by considering the case where $f(x) < 0$ at some point x in (a, b).

22. Use the mean-value theorem to prove:

$$1.5 < \sqrt{3} < 1.75$$

[Hint: Let $f(x) = \sqrt{x}$, $a = 3$, and
$b = 4$ in the mean-value theorem.]

23. Theorem 4.4.3 gives conditions under which a function is increasing or decreasing on an *open* interval. Prove the following result concerned with *closed* intervals: Let the function f be *continuous* on the closed interval $[a, b]$ and differentiable on the open interval (a, b):
  (a) if $f'(x) > 0$ for all x in (a, b), then f is increasing on $[a, b]$;
  (b) if $f'(x) < 0$ for all x in (a, b), then f is decreasing on $[a, b]$.

4.10 PROOFS OF KEY RESULTS USING THE MEAN-VALUE THEOREM (OPTIONAL)

In this section we will prove three results that were stated without proof in earlier sections: Theorem 4.4.3, Theorem 4.7.5 (The First Derivative Test), and Theorem 4.7.6 (The Second Derivative Test). As we shall see, the proofs of all these important results rest on the mean-value theorem.

4.4.3 THEOREM (a) *If $f'(x) > 0$ on an open interval (a, b), then f is increasing on (a, b).*
(b) *If $f'(x) < 0$ on an open interval (a, b), then f is decreasing on (a, b).*

Proof of (a). Let x_1 and x_2 be points in (a, b) such that $x_1 < x_2$. We must show that $f(x_1) < f(x_2)$. Since f is differentiable on (a, b), it is continuous on (a, b). Therefore, f is continuous on $[x_1, x_2]$ and differentiable on (x_1, x_2) since the interval $[x_1, x_2]$ is contained inside the interval (a, b). Thus we can apply the mean-value theorem over the interval $[x_1, x_2]$ and conclude that there is a number c in the interval (x_1, x_2) such that

$$\frac{f(x_2) - f(x_1)}{x_2 - x_1} = f'(c)$$

or equivalently

$$f(x_2) - f(x_1) = f'(c)(x_2 - x_1) \tag{1}$$

But $x_2 - x_1 > 0$ since $x_1 < x_2$, and $f'(c) > 0$ follows from the hypothesis; thus from (1)

$$f(x_2) - f(x_1) > 0$$

or equivalently

$$f(x_1) < f(x_2)$$

The proof of part (b) is similar and is left as an exercise. ▨

4.7.5 THEOREM *Suppose f is continuous at a critical point x_0.*
(a) *If $f'(x) > 0$ on an open interval extending left from x_0 and $f'(x) < 0$ on an open interval extending right from x_0, then f has a relative maximum at x_0.*
(b) *If $f'(x) < 0$ on an open interval extending left from x_0 and $f'(x) > 0$ on an open interval extending right from x_0, then f has a relative minimum at x_0.*

(c) *If $f'(x)$ has the same sign [either $f'(x) > 0$ or $f'(x) < 0$] on an open interval extending left from x_0 and on an open interval extending right from x_0, then f does not have a relative extremum at x_0.*

Proof of (a). In accordance with the hypothesis, assume that $f'(x) > 0$ on the interval (a, x_0) and $f'(x) < 0$ on the interval (x_0, b). To prove that f has a relative maximum at x_0, we will show that

$$f(x_0) \geq f(x) \tag{2}$$

for all x in (a, b). First let x be any point in the interval (a, x_0). Because f is differentiable on the interval (a, x_0) and because of the hypothesis that f is continuous at x_0, it follows that the hypotheses of the mean-value theorem are satisfied on the interval $[x, x_0]$. Thus, there is a point c in the interval (x, x_0) such that

$$\frac{f(x_0) - f(x)}{x_0 - x} = f'(c)$$

or equivalently

$$f(x_0) - f(x) = (x_0 - x)f'(c) \tag{3}$$

But $x_0 - x > 0$ since $x_0 > x$, and $f'(c) > 0$ holds since f' is positive everywhere on the interval (a, x_0). Thus from (3)

$$f(x_0) - f(x) > 0$$

which shows that (2) holds for all x in the interval (a, x_0).

Next, let x be any point in the interval (x_0, b). As above, the hypotheses of the mean-value theorem hold on the interval $[x_0, x]$, so there is a point c in the interval (x_0, x) such that

$$\frac{f(x) - f(x_0)}{x - x_0} = f'(c)$$

or equivalently

$$f(x) - f(x_0) = (x - x_0)f'(c) \tag{4}$$

But $x - x_0 > 0$ since $x > x_0$, and $f'(c) < 0$ holds since f' is negative everywhere on the interval (x_0, b). Thus from (4)

$$f(x) - f(x_0) < 0$$

which shows that (2) holds for all x in the interval (x_0, b).

Since (2) obviously holds when $x = x_0$, we have shown that it holds for all x in (a, b). ▌

The proofs of parts (b) and (c) are left as exercises.

4.7.6   THEOREM   *Suppose f is twice differentiable at a stationary point x_0.*
(a)   If $f''(x_0) > 0$ then f has a relative minimum at x_0.
(b)   If $f''(x_0) < 0$ then f has a relative maximum at x_0.

Proof of (a).   Using the definition of a derivative we can write

$$f''(x_0) = \lim_{h \to 0} \frac{f'(x_0 + h) - f'(x_0)}{h} \tag{5}$$

By hypothesis, $f''(x_0) > 0$ so that we can use $\epsilon = \frac{1}{2} f''(x_0)$ in the definition of a limit and deduce from (5) that there exists a $\delta > 0$ such that

$$\left| \frac{f'(x_0 + h) - f'(x_0)}{h} - f''(x_0) \right| < \frac{1}{2} f''(x_0) \tag{6}$$

whenever h satisfies

$$0 < |h| < \delta \tag{7}$$

To prove that f has a relative minimum at x_0, we will show that

$$f'(x) > 0 \qquad \text{for all } x \text{ in } (x_0, x_0 + \delta) \tag{8}$$

and

$$f'(x) < 0 \qquad \text{for all } x \text{ in } (x_0 - \delta, x_0) \tag{9}$$

It will then follow from the first derivative test that f has a relative minimum at x_0.

From (6) and (7) it follows that

$$\frac{1}{2} f''(x_0) < \frac{f'(x_0 + h) - f'(x_0)}{h} < \frac{3}{2} f''(x_0) \tag{10}$$

whenever h satisfies

$$0 < |h| < \delta$$

By hypothesis, x_0 is a stationary point for f, so that $f'(x_0) = 0$. From this, the fact that $f''(x_0) > 0$, and the left-hand inequality in (10) it follows that

$$0 < \frac{f'(x_0 + h)}{h} \tag{11}$$

whenever h satisfies (7).

To prove (8), let x be any point in $(x_0, x_0 + \delta)$. If we let

$$h = x - x_0 \tag{12}$$

then $h > 0$ and h satisfies (7), so that from (11)

$$0 < \frac{f'(x_0 + h)}{h}$$

or on multiplying by h

$$0 < f'(x_0 + h) \tag{13}$$

Substituting (12) in (13) yields

$$0 < f'(x)$$

which establishes (8). To prove (9) let x be any point in $(x_0 - \delta, x_0)$. If we let

$$h = x - x_0 \tag{14}$$

then $h < 0$ and h satisfies (7) so that from (11)

$$0 < \frac{f'(x_0 + h)}{h}$$

or on multiplying by h

$$f'(x_0 + h) < 0 \tag{15}$$

Substituting (14) in (15) yields

$$f'(x) < 0$$

which establishes (9). ▮

The proof of part (b) is similar and is left as an exercise.

▶ Exercise Set 4.10

1. Prove part (b) of Theorem 4.4.3.
2. Prove part (b) of Theorem 4.7.5.
3. Prove part (c) of Theorem 4.7.5.
4. Prove part (b) of Theorem 4.7.6.
5. A function f is called **nondecreasing** on a given interval if

$$f(x_2) \geq f(x_1) \text{ whenever } x_2 > x_1$$

where x_1 and x_2 are points in the interval. Similarly, f is called **nonincreasing** on the interval if

$$f(x_2) \leq f(x_1) \text{ whenever } x_2 > x_1$$

   (a) Sketch the graph of a function that is nondecreasing, yet is not increasing.
   (b) Sketch the graph of a function that is nonincreasing, yet is not decreasing.
   (c) Prove: If $f'(x) \geq 0$ on an open interval (a, b), then f is nondecreasing on (a, b).
   (d) Prove: If $f'(x) \leq 0$ on an open interval (a, b), then f is nonincreasing on (a, b).
[REMARK. Unfortunately, terminology is not always consistent in the mathematical literature. Some writers use the terms "strictly increasing" and "strictly decreasing" where we have used the terms increasing and decreasing. These writers then use the terms "increasing" and "decreasing" where we have used the terms nondecreasing and nonincreasing.]

6. Prove: if $f'(x) < g'(x)$ for all x in (a, b), then for all points x_1, x_2 in (a, b), where $x_2 > x_1$,

$$f(x_2) - f(x_1) < g(x_2) - g(x_1)$$

7. (a) Prove: If f is continuous on (a, b) and $f'(x) > 0$ except at a single point x_0 in (a, b), then f is increasing on (a, b).
   (b) Does this result remain true if the condition $f'(x) > 0$ fails to hold at any finite number of points in (a, b)? Justify your answer.
   (c) Use part (a) to show that $f(x) = x^3$ is increasing on $(-\infty, +\infty)$.

8. (a) Prove: If $f'(x_0) > 0$ and f' is continuous at x_0, then there is an open interval containing x_0 on which f is increasing.
   (b) Prove: If $f'(x_0) < 0$ and f' is continuous at x_0, then there is an open interval containing x_0 on which f is decreasing.

9. (a) Use Exercise 8(a) to prove: If $f''(x_0) > 0$ and f'' is continuous at x_0, then there is an open interval containing x_0 on which f is concave up.
   (b) Use Exercise 8(b) to prove: If $f''(x_0) < 0$ and f'' is continuous at x_0, then there is an open interval containing x_0 on which f is concave down.

10. Prove: If f is differentiable at each point of the interval $[a, b]$ and f' is continuous at a and b, and $f'(a) f'(b) < 0$, then there is a number c in (a, b) such that $f'(c) = 0$. [*Hint:* Use the results in Exercise 8.]

▶ REVIEW EXERCISES

1. For the hollow cylinder shown, assume that R and r are increasing at a rate of 2 meters/sec, and h is decreasing at a rate of 3 meters/sec. At what rate is the volume changing when $R = 7$ m, $r = 4$ m, and $h = 5$ m?

2. The vessel shown is filled at the rate of 4 ft^3/min. How fast is the fluid level rising when the level is 1 ft?

3. A ball is dropped from a point 10 ft away from a light at the top of a 48 ft pole as shown. When the ball has dropped 16 ft, its velocity (downward) is 32 ft/sec. At what rate is its shadow moving along the ground at that instant?

In Exercises 4–9, find the minimum value m and the maximum value M of f on the indicated interval (if they exist) and state where these extreme values occur.

4. $f(x) = 1/x$; $[-2, -1]$.

5. $f(x) = x^3 - x^4$; $[-1, \frac{3}{2}]$.

6. $f(x) = x^2(x - 2)^{1/3}$; $(0, 3]$.

7. $f(x) = 2x/(x^2 + 3)$; $(0, 2]$.

8. $f(x) = 2x^5 - 5x^4 + 7$; $(-1, 3)$.

9. $f(x) = -|x^2 - 2x|$; $[1, 3]$.

10. Find d^2y/dx^2 implicitly if (a) $y^3 + 3x^2 = 4y$ (b) $\sin y + \cos x = 1$.

11. Verify that $y = \cos x - 3 \sin x$ satisfies
$$y''' + y'' + y' + y = 0.$$

In Exercises 12–19, sketch the graph of f. Use symmetry, where possible, and show all relative extrema, inflection points, and asymptotes.

12. $f(x) = (x^2 - 3)^2$.

13. $f(x) = \dfrac{1}{1 + x^2}$.

14. $f(x) = \dfrac{2x}{1 + x}$.

15. $f(x) = \dfrac{x^3 - 2}{x}$.

16. $f(x) = (1 + x)^{2/3}(3 - x)^{1/3}$.

17. $f(x) = 2 \cos^2 x$, $0 \le x \le \pi$.

18. $f(x) = x - \tan x$, $0 \le x \le 2\pi$

19. $f(x) = \dfrac{3x}{(x + 8)^2}$.

20. Use implicit differentiation to show that a function defined implicitly by $\sin x + \cos y = 2y$ has a critical point whenever $\cos x = 0$. Then use either the first or second derivative test to classify these critical points as relative maxima or minima.

21. Find the equations of the tangent lines at all inflection points of the graph of
$$f(x) = x^4 - 6x^3 + 12x^2 - 8x + 3.$$

In Exercises 22–24, find all critical points and use the first derivative test to determine their nature.

22. $f(x) = x^{1/3}(x - 7)^2$.

23. $f(x) = 2 \sin x - \cos 2x$, $0 \le x \le 2\pi$.

24. $f(x) = 3x - (x - 1)^{3/2}$.

In Exercises 25–27, find all critical points and use the second derivative test to determine their nature (if possible).

25. $f(x) = x^{-1/2} + \frac{1}{9}x^{1/2}$.

26. $f(x) = x^2 + 8/x$.

27. $f(x) = \sin^2 x - \cos x$, $0 \le x \le 2\pi$.

28. Find two nonnegative numbers whose sum is 20 and such that: (a) the sum of their squares is a maximum, and (b) the product of the square of one and the cube of the other is a maximum.

29. Find the dimensions of the rectangle of maximum area that can be inscribed in the ellipse $(x/4)^2 + (y/3)^2 = 1$.

30. Find the coordinates of the point on the curve $2y^2 = 5(x + 1)$ that is nearest to the origin. [Note that all points $P(x, y)$ on the curve satisfy $x \geq -1$.]

31. A church window consists of a blue semicircular section surmounting a clear rectangular section as shown. The blue glass lets through half as much light per unit area as the clear glass. Find the radius r of the window that admits the most light if the perimeter of the entire window is to be P feet.

32. The cost c (in dollars per hour) to run an ocean liner at a constant speed v (in miles per hour) is given by $c = a + bv^n$ where a, b, and n are positive constants with $n > 1$. Find the speed needed to make the cheapest 3000 mile run.

33. A soup can in the shape of a right circular cylinder of radius r and height h is to have a prescribed volume V. The top and bottom are cut from squares as shown. If the shaded corners are wasted, but there is no other waste, find the ratio r/h for the can requiring the least material (including waste).

34. If a calculator factory produces x calculators per day, the total daily cost (in dollars) is $0.25x^2 + 35x + 25$. If they are sold for $50 - \frac{1}{2}x$ dollars each, find the value of x that maximizes the daily profit.

In Exercises 35–37, determine if all hypotheses of Rolle's theorem are satisfied on the stated interval. If not, state which hypotheses fail; if so, find all values of c guaranteed in the conclusion of the theorem.

35. $f(x) = \sqrt{4 - x^2}$ on $[-2, 2]$.

36. $f(x) = x^{2/3} - 1$ on $[-1, 1]$.

37. $f(x) = \sin(x^2)$ on $[0, \sqrt{\pi}]$.

In Exercises 38–41, determine if all hypotheses of the mean value theorem are satisfied on the stated interval. If not, state which hypotheses fail; if so, find all values of c guaranteed in the conclusion of the theorem.

38. $f(x) = |x - 1|$ on $[-2, 2]$.

39. $f(x) = \sqrt{x}$ on $[0, 4]$.

40. $f(x) = \dfrac{x + 1}{x - 1}$ on $[2, 3]$.

41. $f(x) = \begin{cases} 3 - x^2 & \text{if } x \leq 1 \\ 2/x & \text{if } x > 1 \end{cases}$ on $[0, 2]$

5 integration

5.1 INTRODUCTION

In this chapter we will study the second major problem of calculus:

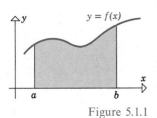

Figure 5.1.1

THE AREA PROBLEM. Given a function f, which is continuous and nonnegative on an interval $[a, b]$, find the area between the graph of f and the interval $[a, b]$ on the x-axis (Figure 5.1.1).

The key to solving the area problem rests on a fundamental relationship between areas and derivatives, which we will now derive. Consider a continuous curve $y = f(x)$ lying above the x-axis, and let $[a, x]$ be an interval on the x-axis with a fixed left-hand endpoint a and a variable right-hand endpoint. As we vary the right-hand endpoint, the area $A(x)$ between the graph of f and the interval $[a, x]$ will change (Figure 5.1.2).

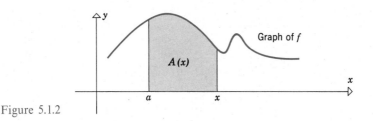

Figure 5.1.2

Provided the derivative $A'(x)$ exists, the instantaneous rate at which $A(x)$ changes with x is given by

$$A'(x) = \lim_{h \to 0} \frac{A(x + h) - A(x)}{h} \tag{1}$$

Consider this limit in the case where h is positive. Since $A(x + h)$ is the area between a and $x + h$ and $A(x)$ is the area between a and x, it follows that

$$A(x + h) - A(x)$$

is the area between x and $x + h$ (Figure 5.1.3).

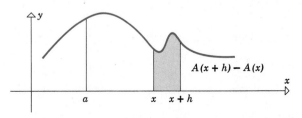

Figure 5.1.3

Since f is continuous, the extreme value theorem states that f has a maximum value and a minimum value on the interval $[x, x + h]$. If the minimum value occurs at x_1 and the maximum value occurs at x_2 then it is geometrically evident (Figure 5.1.4) that the area

$$A(x + h) - A(x)$$

between x and $x + h$ satisfies the inequalities

$$f(x_1) \cdot h \leq A(x + h) - A(x) \leq f(x_2) \cdot h \tag{2}$$

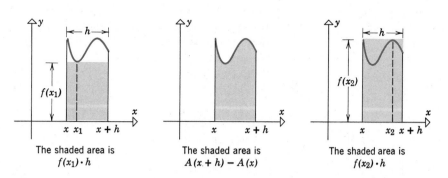

Figure 5.1.4

Since h is assumed positive, (2) can be rewritten as

$$f(x_1) \leq \frac{A(x + h) - A(x)}{h} \leq f(x_2) \tag{3}$$

Although (3) was derived under the assumption that h is positive, it is not hard to show that (3) holds for negative h as well (Exercise 11).

Because x_1 and x_2 lie between x and $x + h$, it follows that x_1 and x_2 must approach x as h approaches 0. Thus, by the continuity of f

$$\lim_{h \to 0} f(x_1) = f(x) \qquad \text{and} \qquad \lim_{h \to 0} f(x_2) = f(x)$$

From these results and the pinching theorem (Theorem 3.3.1) it follows from (3) that

$$\lim_{h \to 0} \frac{A(x + h) - A(x)}{h} = f(x)$$

In other words, $A(x)$ is differentiable at x, and the derivative is

$$A'(x) = f(x) \tag{4}$$

The following example illustrates how this fundamental result can be used to find areas.

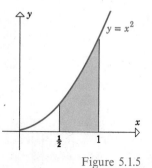

Figure 5.1.5

▶ Example 1

Find the area under the graph of $f(x) = x^2$ over the interval $[\frac{1}{2}, 1]$ (Figure 5.1.5).

Solution. Let $A(x)$ denote the area under the graph of $f(x) = x^2$ over the interval $[\frac{1}{2}, x]$. We are interested in finding the value $A(1)$. Since $f(x) = x^2$, it follows from (4) that

$$A'(x) = x^2 \tag{5}$$

Thus, to find $A(x)$ we must look for a function whose derivative is x^2. This is called an ***antidifferentiation*** problem because we are trying to find $A(x)$ by "undoing" a differentiation.

By simply guessing we see that

$$A(x) = \frac{1}{3}x^3$$

is one solution to (5). But this is not the only solution, since

$$A(x) = \frac{1}{3}x^3 + C \tag{6}$$

also satisfies (5) for any constant C. However, there is one additional condition that we can use to single out the appropriate choice for $A(x)$. If $x = \frac{1}{2}$, then the interval $[\frac{1}{2}, x]$ reduces to a single point and consequently the area under the graph of f above this interval is zero. In other words,

$$A(x) = 0 \quad \text{if} \quad x = \tfrac{1}{2}$$

It therefore follows from (6) with $x = \frac{1}{2}$ that

$$A\left(\frac{1}{2}\right) = \frac{1}{24} + C = 0$$

or

$$C = -\frac{1}{24}$$

Substituting this value in (6) gives

$$A(x) = \frac{1}{3}x^3 - \frac{1}{24}$$

Thus, the area under the graph of $f(x) = x^2$ over the interval $[\frac{1}{2}, 1]$ is

$$A(1) = \frac{1}{3} - \frac{1}{24} = \frac{7}{24} \qquad \blacktriangleleft$$

This example shows how certain areas, which are impossible to obtain using formulas from plane geometry, can be determined using calculus. However, the solution hinged on our ability to *guess* at a function $A(x)$ satisfying

$$A'(x) = x^2$$

But it is not always easy to find a function with a specified derivative. For example, at this stage we know no function $A(x)$ satisfying

$$A'(x) = \frac{1}{x}$$

It is one of the objectives of this chapter to develop some systematic methods for finding functions whose derivatives are specified.

Let us summarize the procedure used in the above example.

Calculating Areas.

If f is continuous on $[a, b]$ and if $f(x) \geq 0$ for all x in $[a, b]$, then the area between the graph of f and the interval $[a, b]$ can be obtained as follows:

Step 1. Let $A(x)$ represent the area between the graph of f and the interval $[a, x]$.

Step 2. Find $A(x)$ by solving the equation $A'(x) = f(x)$ subject to the condition $A(a) = 0$.

Step 3. Evaluate $A(x)$ at $x = b$ to find the area between the graph of f and the interval $[a, b]$.

In this section we have used the term "area" freely, assuming its meaning to be evident. However, in order to prove results about areas it is necessary to define the term precisely. We will consider this problem in Section 5.6.

► Exercise Set 5.1

In Exercises 1–6 use the method of Example 1 to find the area between the graph of f and the given interval.

1. $f(x) = 2x$; $[1, 2]$.

2. $f(x) = \cos x$; $[0, \pi/2]$.

3. $f(x) = \sec^2 x$; $[-\pi/4, \pi/4]$.

4. $f(x) = x^3 + 2$; $[3, 6]$.

5. $f(x) = \sqrt{x}$; $[1, 4]$.

6. $f(x) = 1/x^2$; $[2, 5]$.

7. (a) Use the method of Example 1 to find the area of the triangular region between the line $y = x$ and the interval $[0, 1]$.

    (b) Check your result by using the formula for the area of a triangle.

8. (a) Use the method of Example 1 to find the area of the trapezoid bounded above by the line $y = 4x$, below by the x-axis and on the sides by the lines $x = a$ and $x = b$, where $0 < a < b$.

    (b) Check your answer by using the formula for the area of a trapezoid.

9. Intuitively, is the area between the graph of f and the closed interval $[a, b]$ the same as the area between the graph of f and the open interval (a, b)? Explain.

10. (a) Use the method of Example 1 to find the area between the graph of $f(x) = 1/x^2$ and the interval $[1, b]$.

    (b) Let $A(b)$ denote the area obtained in part (a), and compute $\lim_{b \to +\infty} A(b)$.

    (c) What is the geometric meaning of the limit in part (b)?

11. Show that (3) holds for negative values of h.

5.2 ANTIDERIVATIVES; THE INDEFINITE INTEGRAL

The problem of finding a function whose derivative is specified arose in the last section. In this section we will discuss this problem.

5.2.1 DEFINITION   A function F is called an **antiderivative** of a function f if the derivative of F is f.

▶ Example 1

The functions

$$\frac{1}{3}x^3, \qquad \frac{1}{3}x^3 + 2, \qquad \frac{1}{3}x^3 - \pi$$

are antiderivatives of

$$f(x) = x^2$$

since

$$\frac{d}{dx}\left[\frac{1}{3}x^3\right] = \frac{d}{dx}\left[\frac{1}{3}x^3 + 2\right] = \frac{d}{dx}\left[\frac{1}{3}x^3 - \pi\right] = x^2 \qquad ◀$$

This example shows that a function can have many antiderivatives. In fact, if $F(x)$ is any antiderivative of $f(x)$ and C is any constant, then

$$F(x) + C$$

is also an antiderivative of $f(x)$ since

$$\frac{d}{dx}[F(x) + C] = \frac{d}{dx}[F(x)] + \frac{d}{dx}[C] = f(x) + 0 = f(x)$$

It is reasonable to ask if there are other antiderivatives of f that cannot be obtained by adding a constant to F. Provided we consider only values of x in an interval I, the answer is *no*. To see this, let $G(x)$ be any other antiderivative of $f(x)$; then

$$\frac{d}{dx}[F(x)] = \frac{d}{dx}[G(x)] = f(x)$$

so that by Corollary 4.9.4, F and G differ only by some constant C on I, that is,

$$G(x) = F(x) + C$$

for x in I. The following theorem summarizes these observations.

5.2.2 THEOREM *If $F(x)$ is any antiderivative of $f(x)$, then for any value of C*

$$F(x) + C$$

is also an antiderivative of $f(x)$; moreover on any interval, every antiderivative of $f(x)$ is expressible in the form $F(x)$ plus a constant.

▶ Example 2

For all x,

$$\frac{d}{dx}[\sin x] = \cos x$$

so that $F(x) = \sin x$ is an antiderivative of $f(x) = \cos x$ on the interval $(-\infty, +\infty)$; thus every antiderivative of $\cos x$ on this interval is expressible in the form

$$\sin x + C$$

where C is a constant. ◀

The process of finding antiderivatives is called *antidifferentiation* or *integration*. If

$$\frac{d}{dx}[F(x)] = f(x)$$

then the functions of the form $F(x) + C$ are the antiderivatives of $f(x)$. We denote this by writing

$$\int f(x)\,dx = F(x) + C \tag{1}$$

The symbol $\int$ is called an *integral sign*, and statement (1) is read, "the indefinite integral of $f(x)$ equals $F(x)$ plus C." The adjective "indefinite" is used because the right side of (1) is not a definite function, but rather a whole set of possible functions; the constant C is called the *constant of integration*.

REMARK. The symbol dx in the differentiation operation

$$\frac{d}{dx}[\quad]$$

and in the antidifferentiation operation*

$$\int[\quad]\,dx$$

serves to identify the independent variable. For example, the symbol

$$\int f(t)\,dt$$

denotes a function of t whose derivative with respect to t is $f(t)$.

The equation

$$\frac{d}{dx}[f(x)] = f'(x)$$

is equivalent to

$$\int f'(x)\,dx = f(x) + C \tag{2}$$

▶ Example 3

$$\frac{d}{dx}[x^3] = 3x^2 \qquad \text{is equivalent to} \qquad \int 3x^2\,dx = x^3 + C$$

$$\frac{d}{dt}[\tan t] = \sec^2 t \qquad \text{is equivalent to} \qquad \int \sec^2 t\,dt = \tan t + C$$

$$\frac{d}{du}[u^{3/2}] = \frac{3}{2}u^{1/2} \qquad \text{is equivalent to} \qquad \int \frac{3}{2}u^{1/2}\,du = u^{3/2} + C \blacktriangleleft$$

---

*This notation was devised by Leibniz. In his early papers Leibniz used the notation "omn." (an abbreviation for the Latin word "omnes") to denote integration. Then on October 29, 1675 he wrote, "It will be useful to write $\int$ for omn., thus $\int \ell$ for omn. ℓ . . . ." Two or three weeks later he refined the notation further and wrote $\int[\quad]\,dx$ rather than $\int$ alone. This notation is so useful and so powerful that its development by Leibniz must be regarded as a major milestone in the history of mathematics and science.

Antidifferentiation is guesswork. By looking only at the derivative of a function we try to guess the function itself. It simplifies this guessing process if we keep in mind that every differentiation formula produces a companion integration formula. For example, the differentiation formula

$$\frac{d}{dx}[\sin x] = \cos x$$

gives rise to the integration formula

$$\int \cos x \, dx = \sin x + C$$

Similarly,

$$\frac{d}{dx}[-\cos x] = \sin x \qquad \text{yields} \qquad \int \sin x \, dx = -\cos x + C$$

and

$$\frac{d}{dx}\left[\frac{x^{r+1}}{r+1}\right] = x^r \qquad \text{yields} \qquad \int x^r \, dx = \frac{x^{r+1}}{r+1} + C \qquad (r \neq -1)$$

Some basic integration formulas are given in Table 5.2.1.

Table 5.2.1

DIFFERENTIATION FORMULA	INTEGRATION FORMULA
1. $\dfrac{d}{dx}[x] = 1$	$\int 1 \, dx = x + C$
2. $\dfrac{d}{dx}\left[\dfrac{x^{r+1}}{r+1}\right] = x^r \ (r \neq -1)$	$\int x^r \, dx = \dfrac{x^{r+1}}{r+1} + C \ (r \neq -1)$
3. $\dfrac{d}{dx}[\sin x] = \cos x$	$\int \cos x \, dx = \sin x + C$
4. $\dfrac{d}{dx}[-\cos x] = \sin x$	$\int \sin x \, dx = -\cos x + C$
5. $\dfrac{d}{dx}[\tan x] = \sec^2 x$	$\int \sec^2 x \, dx = \tan x + C$
6. $\dfrac{d}{dx}[-\cot x] = \csc^2 x$	$\int \csc^2 x \, dx = -\cot x + C$
7. $\dfrac{d}{dx}[\sec x] = \sec x \tan x$	$\int \sec x \tan x \, dx = \sec x + C$
8. $\dfrac{d}{dx}[-\csc x] = \csc x \cot x$	$\int \csc x \cot x \, dx = -\csc x + C$

▶ Example 4

From the second integration formula in Table 5.2.1 we obtain

$$\int x^2 \, dx = \frac{x^3}{3} + C \qquad (r = 2)$$

$$\int x^3 \, dx = \frac{x^4}{4} + C \qquad (r = 3)$$

$$\int \frac{1}{x^5} \, dx = \int x^{-5} \, dx = \frac{x^{-5+1}}{-5+1} = -\frac{1}{4x^4} + C \qquad (r = -5)$$

$$\int \sqrt{x} \, dx = \int x^{\frac{1}{2}} \, dx = \frac{x^{\frac{1}{2}+1}}{\frac{1}{2}+1} + C = \frac{2}{3}x^{\frac{3}{2}} + C$$

$$= \frac{2}{3}(\sqrt{x})^3 + C \qquad (r = \tfrac{1}{2}) \qquad ◀$$

▶ Example 5

Evaluate

$$\int \frac{\cos x}{\sin^2 x} \, dx$$

Solution.

$$\int \frac{\cos x}{\sin^2 x} \, dx = \int \frac{1}{\sin x} \frac{\cos x}{\sin x} \, dx$$

$$= \int \csc x \cot x \, dx$$

$$= -\csc x + C \quad \text{[Formula 8 in Table 5.2.1]} \qquad ◀$$

If we differentiate an antiderivative of $f(x)$, we obtain $f(x)$ back again. Thus,

$$\frac{d}{dx}\left[\int f(x) \, dx\right] = f(x) \tag{3}$$

This result is helpful for proving the following basic properties of antiderivatives.

5.2.3 THEOREM *(a) A constant factor can be moved through an integral sign; that is,*

$$\int cf(x)\,dx = c\int f(x)\,dx$$

(b) An antiderivative of a sum is the sum of the antiderivatives; that is,

$$\int [f(x) + g(x)]\,dx = \int f(x)\,dx + \int g(x)\,dx$$

Proof. To prove formula *(a)* we must show that $c\int f(x)\,dx$ is an antiderivative of $cf(x)$ and to prove *(b)* we must show that $\int f(x)\,dx + \int g(x)\,dx$ is an antiderivative of $f(x) + g(x)$. But, these conclusions follow from (3):

$$\frac{d}{dx}\left[c\int f(x)\,dx \right] = c\frac{d}{dx}\left[\int f(x)\,dx \right] = cf(x)$$

$$\frac{d}{dx}\left[\int f(x)\,dx + \int g(x)\,dx \right] = \frac{d}{dx}\left[\int f(x)\,dx \right] + \frac{d}{dx}\left[\int g(x)\,dx \right]$$

$$= f(x) + g(x) \quad\blacksquare$$

▶ Example 6

Evaluate

(a) $\displaystyle\int 4\cos x\,dx$

(b) $\displaystyle\int (x + x^2)\,dx$

Solution (a).

$$\int 4\cos x\,dx = 4\int \cos x\,dx \qquad\qquad\text{[Theorem 5.2.3a]}$$

$$= 4\,(\sin x + C) \qquad\qquad\text{[Table 5.2.1]}$$

$$= 4\sin x + 4C$$

Since C is an arbitrary constant, so is $4C$. However, no purpose is served by keeping the arbitrary constant in this form, so we will replace it by a single letter, say $K = 4C$, and write

$$\int 4 \cos x \, dx = 4 \sin x + K$$

Solution (b).

$$\int (x + x^2) \, dx = \int x \, dx + \int x^2 \, dx \qquad \text{[Theorem 5.2.3}b\text{]}$$

$$= \left[\frac{x^2}{2} + C_1 \right] + \left[\frac{x^3}{3} + C_2 \right] \qquad \text{[Table 5.2.1]}$$

$$= \frac{x^2}{2} + \frac{x^3}{3} + C_1 + C_2$$

Since C_1 and C_2 are arbitrary constants, so is $C_1 + C_2$. If we denote this arbitrary constant by the single letter C, we obtain

$$\int (x + x^2) \, dx = \frac{x^2}{2} + \frac{x^3}{3} + C \qquad \blacktriangleleft$$

Part (*b*) of Theorem 5.2.3 can be extended to include more than two terms. More precisely,

$$\int [f_1(x) + f_2(x) + \cdots + f_n(x)] \, dx$$

$$= \int f_1(x) \, dx + \int f_2(x) \, dx + \cdots + \int f_n(x) \, dx$$

In addition, we have left it as an exercise to show:

$$\int [f(x) - g(x)] \, dx = \int f(x) \, dx - \int g(x) \, dx$$

▶ Example 7

$$\int (3x^6 - 2x^2 + 7x + 1) \, dx = 3 \int x^6 \, dx - 2 \int x^2 \, dx + 7 \int x \, dx + \int 1 \, dx$$

$$= \frac{3x^7}{7} - \frac{2x^3}{3} + \frac{7x^2}{2} + x + C \qquad \blacktriangleleft$$

REMARK. The function $f(x)$ in the expression $\int f(x)\,dx$ is called the **integrand**. Sometimes, for compactness of notation, the dx is incorporated into the integrand. For example,

$$\int 1\,dx \quad \text{can be written as} \quad \int dx$$

and

$$\int \frac{1}{x^2}\,dx \quad \text{can be written as} \quad \int \frac{dx}{x^2}.$$

▶ Exercise Set 5.2

In Exercises 1–22, evaluate the integrals and check your results by differentiating the answers.

1. $\displaystyle\int x^8\,dx.$

2. $\displaystyle\int \frac{1}{x^6}\,dx.$

3. $\displaystyle\int x^{5/7}\,dx.$

4. $\displaystyle\int \sqrt[3]{x^2}\,dx.$

5. $\displaystyle\int \frac{4}{\sqrt{t}}\,dt.$

6. $\displaystyle\int \frac{dx}{x^{-2}}.$

7. $\displaystyle\int x^3\sqrt{x}\,dx.$

8. $\displaystyle\int (u^3 - 2u + 7)\,du.$

9. $\displaystyle\int x(1 + x^3)\,dx.$

10. $\displaystyle\int (2 + y^2)^2\,dy.$

11. $\displaystyle\int [4\sin x + 2\cos x]\,dx.$

12. $\displaystyle\int \left[\frac{1}{t^2} - \cos t\right]\,dt.$

13. $\displaystyle\int [4\sec^2 x + \csc x \cot x]\,dx.$

14. $\displaystyle\int [\sqrt{\theta} - \csc^2 \theta]\,d\theta.$

15. $\displaystyle\int \sec x(\tan x + \cos x)\,dx.$

16. $\displaystyle\int \frac{dy}{\csc y}.$

17. $\displaystyle\int \frac{\sin x}{\cos^2 x}\,dx.$

18. $\displaystyle\int \frac{\sin 2x}{\cos x}\,dx.$

19. $\displaystyle\int [1 + \sin^2 \theta \csc \theta]\,d\theta.$

20. $\displaystyle\int \left[\phi + \frac{2}{\sin^2 \phi}\right]\,d\phi.$

21. $\displaystyle\int \frac{x^5 + 2x^2 - 1}{x^4}\,dx.$

22. $\displaystyle\int \frac{x^2 \sin x + 2\sin x}{2 + x^2}\,dx.$

23. Find an antiderivative, F, of $f(x) = \sqrt[3]{x}$ that satisfies $F(1) = 2$.

24. Find a function f such that

$$f'(x) + \sin x = 0$$

and

$$f(0) = 2$$

25. Find the general form of a function whose second derivative is $\sqrt{x}$. [*Hint:* Solve the equation $f''(x) = \sqrt{x}$ for $f(x)$ by integrating both sides twice.]

26. Find a function f such that

$$f''(x) = x + \cos x$$

and such that $f(0) = 1$ and $f'(0) = 2$.

[*Hint:* Integrate both sides of the equation twice.]

27. Prove: $\int [f(x) - g(x)]\, dx = \int f(x)\, dx - \int g(x)\, dx.$

28. (a) Show that

$$F(x) = \begin{cases} x, & x > 0 \\ -x, & x < 0 \end{cases}$$

and

$$F_1(x) = \begin{cases} x + 2, & x > 0 \\ -x + 3, & x < 0 \end{cases}$$

are both antiderivatives of

$$f(x) = \begin{cases} 1, & x > 0 \\ -1, & x < 0 \end{cases}$$

but that $F_1(x) \neq F(x)$ plus a constant.

(b) Does this violate Theorem 5.2.2? Explain.

5.3 INTEGRATION BY u-SUBSTITUTION

In this section we will discuss a technique, called **u-substitution,** which can often be used to transform complicated integration problems into simpler ones. The method hinges on the following formula in which u stands for a differentiable function of x.

$$\int \left[f(u)\, \frac{du}{dx} \right] dx = \int f(u)\, du \tag{1}$$

To justify this formula, let F be an antiderivative of f, so that

$$\frac{d}{du}[F(u)] = f(u)$$

or, equivalently,

$$\int f(u)\, du = F(u) + C \tag{2}$$

If u is a differentiable function of x, the chain rule implies that

$$\frac{d}{dx}[F(u)] = \frac{d}{du}[F(u)] \cdot \frac{du}{dx} = f(u)\, \frac{du}{dx}$$

or, equivalently,

$$\int \left[f(u)\, \frac{du}{dx} \right] dx = F(u) + C \tag{3}$$

Formula (1) follows from (2) and (3).

Suppose we are interested in evaluating

$$\int h(x)\, dx$$

It follows from (1) that if we can express this integral in the form

$$\int h(x)\, dx = \int f(g(x))g'(x)\, dx$$

then the substitution $u = g(x)$ and $du/dx = g'(x)$ will yield

$$\int h(x)\, dx = \int \left[f(u) \frac{du}{dx} \right] du = \int f(u)\, du$$

With a "good" choice of $u = g(x)$, the integral on the right will be easier to evaluate than the original.

In practice, this substitution process is carried out as follows:

Step 1. Make a choice for u, say $u = g(x)$.
Step 2. Compute $du/dx = g'(x)$.
Step 3. Make the substitution

$$u = g(x), \qquad du = g'(x)\, dx$$

At this stage, the *entire* integral must be in terms of u; no x's should remain. If this is not the case, try a different choice of u.

Step 4. Evaluate the resulting integral.
Step 5. Replace u by $g(x)$, so the final answer is in terms of x.

▶ Example 1

Evaluate

$$\int (x^2 + 1)^{50} \cdot 2x\, dx$$

Solution. If we let

$$u = x^2 + 1$$

then $du/dx = 2x$, so $du = 2x\, dx$

Therefore

$$\int (x^2 + 1)^{50} \cdot 2x \, dx = \int u^{50} \, du$$

$$= \frac{u^{51}}{51} + C$$

$$= \frac{(x^2 + 1)^{51}}{51} + C \qquad \blacktriangleleft$$

▶ Example 2

Evaluate

$$\int \sin^2 x \cos x \, dx$$

Solution. If we let

$$u = \sin x$$

then $du/dx = \cos x$, so $du = \cos x \, dx$

Thus

$$\int \sin^2 x \cos x \, dx = \int u^2 \, du$$

$$= \frac{u^3}{3} + C$$

$$= \frac{\sin^3 x}{3} + C \qquad \blacktriangleleft$$

▶ Example 3

Evaluate

$$\int \frac{\cos \sqrt{x}}{2\sqrt{x}} \, dx$$

Solution. If we let

$$u = \sqrt{x}$$

then $\dfrac{du}{dx} = \dfrac{1}{2\sqrt{x}}$, so $du = \dfrac{1}{2\sqrt{x}} \, dx$

Thus

$$\int \frac{\cos \sqrt{x}}{2\sqrt{x}} = \int \cos u \, du$$

$$= \sin u + C$$
$$= \sin \sqrt{x} + C \qquad \blacktriangleleft$$

▶ Example 4

Evaluate

$$\int 3x^2 \sqrt{x^3 + 1} \, dx$$

Solution. If we let

$$u = x^3 + 1$$

then $du/dx = 3x^2$, so $du = 3x^2 \, dx$

Thus

$$\int 3x^2 \sqrt{x^3 + 1} \, dx = \int \sqrt{u} \, du = \int u^{1/2} \, du$$

$$= \frac{u^{3/2}}{3/2} + C$$

$$= \frac{2}{3}(x^3 + 1)^{3/2} + C \qquad \blacktriangleleft$$

▶ Example 5

The easiest u-substitutions occur when the integrand is the derivative of a known function, except for a constant added to or subtracted from the independent variable. For example,

$$\int \sin (x + 9) \, dx = \int \sin u \, du \qquad \left[\begin{array}{l} u = x + 9 \\ du = 1 \cdot dx = dx \end{array} \right]$$

$$= -\cos u + C$$
$$= -\cos (x + 9) + C$$

$$\int (x-8)^{23}\, dx = \int u^{23}\, du \qquad \left[\begin{array}{l} u = x-8 \\ du = 1\cdot dx = dx \end{array}\right]$$

$$= \frac{u^{24}}{24} + C$$

$$= \frac{(x-8)^{24}}{24} + C \qquad \blacktriangleleft$$

▶ Example 6

Another easy *u*-substitution occurs when the integrand is the derivative of a known function, except for a constant that multiplies or divides the independent variable. For example,

$$\int \cos 5x\, dx = \int (\cos u)\cdot \frac{1}{5}\, du \qquad \left[\begin{array}{l} u = 5x \\ du = 5\, dx \text{ or } dx = \tfrac{1}{5}\, du \end{array}\right]$$

$$= \frac{1}{5}\int \cos u\, du$$

$$= \frac{1}{5}\sin u + C$$

$$= \frac{1}{5}\sin 5x + C \qquad \blacktriangleleft$$

▶ Example 7

$$\int \frac{dx}{(\tfrac{1}{3}x-8)^5} = \int \frac{3\, du}{u^5} \qquad \left[\begin{array}{l} u = \tfrac{1}{3}x - 8 \\ du = \tfrac{1}{3}\, dx \text{ or } dx = 3\, du \end{array}\right]$$

$$= 3\int u^{-5}\, du$$

$$= -\frac{3}{4}u^{-4} + C$$

$$= -\frac{3}{4}\left(\frac{1}{3}x - 8\right)^{-4} + C \qquad \blacktriangleleft$$

▶ Example 8

With the help of Theorem 5.2.3, a complicated integral can sometimes be computed by expressing it as a sum of simpler integrals. For example,

$$\int (x + \sec^2 \pi x)\, dx = \int x\, dx + \int \sec^2 \pi x\, dx \quad \left[\begin{array}{l} u = \pi x \\ du = \pi dx \text{ or } dx = \tfrac{1}{\pi} du \end{array} \right]$$

$$= \frac{x^2}{2} + \int \sec^2 \pi x\, dx$$

$$= \frac{x^2}{2} + \frac{1}{\pi} \int \sec^2 u\, du$$

$$= \frac{x^2}{2} + \frac{1}{\pi} \tan u + C$$

$$= \frac{x^2}{2} + \frac{1}{\pi} \tan \pi x + C \qquad \blacktriangleleft$$

▶ Example 9

Evaluate

$$\int t^4 \sqrt[3]{3 - 5t^5}\, dt$$

Solution. After some possible false starts most readers would eventually hit on the following substitution:

$$\int t^4 \sqrt[3]{3 - 5t^5}\, dt = -\frac{1}{25} \int \sqrt[3]{u}\, du \quad \left[\begin{array}{l} u = 3 - 5t^5 \\ du = -25t^4\, dt \text{ or } -\tfrac{1}{25} du = t^4\, dt \end{array} \right]$$

$$= -\frac{1}{25} \int u^{1/3}\, du$$

$$= -\frac{1}{25} \frac{u^{4/3}}{4/3} + C$$

$$= -\frac{3}{100} (3 - 5t^5)^{4/3} + C \qquad \blacktriangleleft$$

▶ Example 10

Evaluate

$$\int x^2 \sqrt{x - 1}\, dx$$

Solution. Let

$$u = x - 1 \tag{4}$$

so that

$$du = dx$$

From (4)

$$x^2 = (u + 1)^2 = u^2 + 2u + 1$$

so that

$$\int x^2 \sqrt{x - 1} \, dx = \int (u^2 + 2u + 1) \sqrt{u} \, du$$

$$= \int (u^{5/2} + 2u^{3/2} + u^{1/2}) \, du$$

$$= \frac{2}{7} u^{7/2} + \frac{4}{5} u^{5/2} + \frac{2}{3} u^{3/2} + C$$

$$= \frac{2}{7} (x - 1)^{7/2} + \frac{4}{5} (x - 1)^{5/2} + \frac{2}{3} (x - 1)^{3/2} + C \quad \blacktriangleleft$$

REMARK. If you find integration by substitution hard, don't despair—it is not an easy topic. Work lots of problems.

▶ Exercise Set 5.3

1. Evaluate the integrals by making the indicated substitutions.

(a) $\int 2x(x^2 + 1)^{23} \, dx$; $u = x^2 + 1$

(b) $\int \cos^3 x \sin x \, dx$; $u = \cos x$

(c) $\int \frac{1}{\sqrt{x}} \sin \sqrt{x} \, dx$; $u = \sqrt{x}$

(d) $\int \frac{3x \, dx}{\sqrt{4x^2 + 5}}$; $u = 4x^2 + 5$.

2. Evaluate the integrals by making the indicated substitutions.

(a) $\int \sec^2 (4x + 1) \, dx$; $u = 4x + 1$

(b) $\int y \sqrt{1 + 2y^2} \, dy$; $u = 1 + 2y^2$

(c) $\int \sqrt{\sin \pi \theta} \cos \pi \theta \, d\theta$; $u = \sin \pi \theta$

(d) $\int (2x + 7)(x^2 + 7x + 3)^{4/5} \, dx$;

$$u = x^2 + 7x + 3.$$

3. Evaluate the integrals by making the indicated substitutions.

(a) $\int \cot x \csc^2 x \, dx$; $u = \csc x$

(b) $\int (1 + \sin t)^9 \cos t \, dt$; $u = 1 + \sin t$

(c) $\int x^2 \sqrt{1 + x} \, dx$; $u = 1 + x$

(d) $\int [\csc (\sin x)]^2 \cos x \, dx$; $u = \sin x$.

In Exercises 4–25, evaluate the integrals.

4. $\int 4x^5 \, dx$.

5. $\int (1 + \cos x) \, dx$.

6. $\int \sin (x + 4) \, dx$.

7. $\int (x - 2)^{17} \, dx$.

8. $\int (7t^5 - t^3 + 2t + 1) \, dt$.

9. $\int 5 \cos (\pi x - 3) \, dx$.

10. $\int \left(\frac{1}{x^2} + \frac{1}{x^3} \right) dx$.

11. $\int \frac{ds}{\sqrt{3s + 1}}$.

12. $\int x \sqrt{7x^2 + 12} \, dx$.

13. $\int \sqrt[n]{a + bx} \, dx$ $(b \neq 0)$.

14. $\int 6x^2(x^3 + 1)^{19} \, dx$.

15. $\int \sin^6 x \cos x \, dx$.

16. $\int x^{2/3} \sin (x^{5/3}) \, dx$.

17. $\int \frac{x^2 \, dx}{\sqrt{x^3 + 1}}$.

18. $\int \csc^2 (\pi \theta + 1) \, d\theta$.

19. $\int \sqrt{\tan x} \sec^2 x \, dx$.

20. $\int (3u^2 + 1)(u^3 + u - 2)^{3/4} \, du$.

21. $\int \tan^2 x \sec^2 x \, dx$.

22. $\int \tan x \sec^3 x \, dx$.

23. $\int [\sin (\sin \theta)] \cos \theta \, d\theta$.

24. $\int \frac{\sin v \, dv}{(5 - \cos v)^{12}}$.

25. $\int \sin^n (a + bx) \cos (a + bx) \, dx$ $(n > 0, b \neq 0)$.

In Exercises 26–34, evaluate the integrals. These are a little trickier than those in the preceding exercises.

26. $\int \frac{t^2 - 1}{\sqrt{t}} \, dt$.

27. $\int x \sqrt[3]{(x^2 + 5)^2} \, dx$.

28. $\int (x^2 - 6x + 9)^{3/5} \, dx$.

29. $\int \frac{y \, dy}{\sqrt{y + 1}}$.

30. $\int x^2 \sqrt{2 - x} \, dx$.

31. $\int \sin^3 x \, dx$.

32. $\int \frac{1 + \cos \theta}{\sin^2 \theta} \, d\theta$.

33. $\int \tan^2 x \, dx.$

34. $\int (x^2 - 3)^3 \, dx.$

35. Evaluate the integral in Example 10 by making the substitution $u = \sqrt{x - 1}$. Check your answer against the one obtained in the example.

36. (a) Evaluate $\int \sin x \cos x \, dx$ by two methods: first, let $u = \sin x$; then, let $u = \cos x$.

(b) Explain why the two apparently different answers obtained in part (a) are really equivalent.

5.4   RECTILINEAR MOTION (AN APPLICATION OF THE INDEFINITE INTEGRAL)

In Section 2.3 we studied the simplest kind of motion, a particle moving in one direction along a straight line. In this section we will consider a more complicated kind of motion, a particle moving back and forth on a line; this is called **rectilinear motion.** Examples are: a piston moving up and down in a cylinder, a rock tossed straight up and returning to earth straight down, back and forth vibrations of a spring, and so forth.

To study the motion of a particle along a line, it is usually desirable to coordinatize the line by selecting an arbitrary origin and positive direction. As the particle moves along the coordinate line its coordinate s will vary as a function of time t. This function, denoted by $s(t)$, is called the **position function** of the particle. The rate at which the particle's coordinate changes with time is called the *velocity* of the particle, and the magnitude or absolute value of the velocity is called the *speed* of the particle. More precisely:

5.4.1   DEFINITION   If $s(t)$ is the position functon of a particle moving on a coordinate line, then the **instantaneous velocity** at time t is defined by

$$v(t) = s'(t) = \frac{ds}{dt}$$

and the **instantaneous speed** at time t is defined by

$$\text{speed at time } t = |v(t)| = |s'(t)| = \left| \frac{ds}{dt} \right|$$

$s(t)$ increasing
$v(t) = s'(t) > 0$

(a)

$s(t)$ decreasing
$v(t) = s'(t) < 0$

(b)

Figure 5.4.1

If $v(t) > 0$ at a given time t, then the coordinate s of the particle is increasing at that instant, which means that the particle is moving in the positive direction along the line; similarly if $v(t) < 0$, then s is decreasing at time t and the particle is moving in the negative direction (Figures 5.4.1a, b). The speed of a particle is always nonnegative; it tells how fast the particle is moving, but provides no information about the direction of motion.

▶ Example 1

Let

$$s = t^4 - 8t^2$$

be the position function of a particle moving on a coordinate line, where t is in seconds and s is in feet. The instantaneous velocity and instantaneous speed of the particle are given by

$$v(t) = \frac{ds}{dt} = 4t^3 - 16t$$

$$|v(t)| = \left|\frac{ds}{dt}\right| = |4t^3 - 16t|$$

Thus, when $t = 1$ sec the particle's velocity is $v(1) = -12$ ft/sec, which means the particle is moving in the negative direction with a speed of $|v(1)| = 12$ ft/sec. When $t = 3$ sec the velocity is $v(3) = 60$ ft/sec, which means that the particle is moving in the positive direction with a speed of $|v(3)| = 60$ ft/sec. When $t = 2$ sec the velocity is $v(2) = 0$ ft/sec, which means that the particle has momentarily stopped. ◀

If a particle moves along a coordinate line, the rate at which its velocity changes with time is called the *acceleration* of the particle; more precisely:

5.4.2 DEFINITION If $v(t)$ is the velocity at time t of a particle moving on a coordinate line, then its **instantaneous acceleration** at time t is defined by

$$a(t) = v'(t) = \frac{dv}{dt}$$

Since $v(t) = s'(t) = ds/dt$, the formula in this definition can also be written

$$a(t) = s''(t) = \frac{d^2s}{dt^2}$$

Since acceleration is the rate at which velocity changes with time, it is expressed in units of velocity per unit of time. For example, if t is in seconds and s is in feet, then velocity units are feet per second (ft/sec), and acceleration units are feet per second per second, which would be written

(ft/sec)/sec   or   ft/sec^2

▶ Example 2

Let

$$s = t^3 - 6t^2$$

where t is in seconds and s is in feet. Then

$$v(t) = \frac{ds}{dt} = 3t^2 - 12t$$

and

$$a(t) = \frac{dv}{dt} = 6t - 12$$

At $t = 1$, the acceleration is $a(1) = -6$ ft/sec², which means that the velocity is decreasing at the rate of 6 ft/sec². At $t = 2$ the acceleration is $a(2) = 0$ ft/sec², which means that the velocity is not changing at that instant. At $t = 4$ the acceleration is $a(4) = 12$ ft/sec², which means that the velocity is increasing at the rate of 12 ft/sec².   ◀

REMARK. We leave it as an exercise to show that the speed of a particle is increasing if its velocity and acceleration have the same sign and the speed is decreasing if they have opposite signs (Exercise 18). Thus, a particle moving in the positive direction $[v(t) > 0]$ is speeding up when its acceleration is positive and slowing down when its acceleration is negative, whereas a particle moving in the negative direction $[v(t) < 0]$ is speeding up when its acceleration is negative and slowing down when its acceleration is positive.

▶ Example 3

The position function of a particle moving on a coordinate line is given by $s(t) = 2t^3 - 21t^2 + 60t + 3$, where s is in feet and t is in seconds. Describe the motion of the particle for $t \geq 0$.

Solution. The velocity and acceleration at time t are

$$v(t) = \frac{ds}{dt} = 6t^2 - 42t + 60 = 6(t - 2)(t - 5)$$

$$a(t) = \frac{dv}{dt} = 12t - 42 = 12\left(t - \frac{7}{2}\right)$$

At each instant we can determine the direction of motion from the sign of $v(t)$ and whether the particle is speeding up or slowing down from the signs of $v(t)$ and $a(t)$ together (Figures 5.4.2a and 5.4.2b). The motion of the particle is described schematically by the curved line in Figure 5.4.3. At time $t = 0$ the particle is at the point $s(0) = 3$ moving right with velocity $v(0) = 60$ ft/sec, but slowing down with acceleration $a(0) = -42$ ft/sec². The particle continues moving right until time $t = 2$ when it stops at the point $s(2) = 55$, reverses direction, and begins to speed up with an acceleraton of $a(2) = -18$ ft/sec². At time $t = \frac{7}{2}$ the particle begins to slow down, but continues moving left until time $t = 5$ when it stops at the point $s(5) = 28$, reverses direction again, and begins to speed up with acceleration $a(5) = 18$ ft/sec². The particle then continues moving right thereafter with increasing speed. ◀

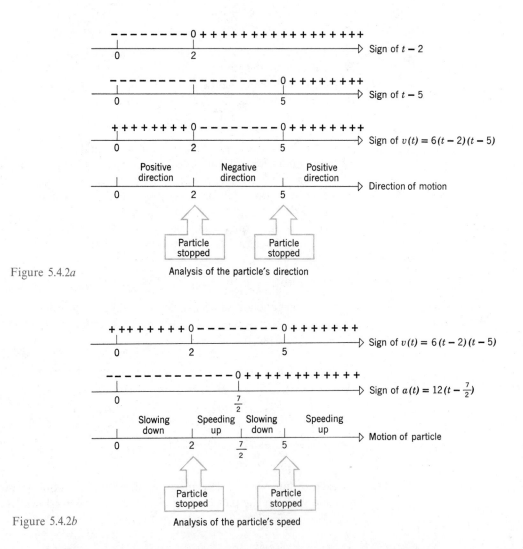

Figure 5.4.2a

Analysis of the particle's direction

Figure 5.4.2b

Analysis of the particle's speed

Figure 5.4.3

REMARK. The curved line in Figure 5.4.3 is descriptive only. The actual path of the particle is back and forth on the axis.

The formulas

$$v(t) = \frac{ds}{dt}$$

$$a(t) = \frac{dv}{dt}$$

can be written in the alternative forms,

$$s(t) = \int v(t)\, dt$$

$$v(t) = \int a(t)\, dt$$

Thus, if we have sufficient information to determine the constants of integration, the velocity can be determined from the acceleration, and the position can be determined from the velocity.

▶ Example 4

Find the position function of a particle moving with velocity $v(t) = \cos \pi t$ along a straight line, assuming the particle is at $s = 4$ when $t = 0$.

Solution. The position function is

$$s(t) = \int v(t)\, dt = \int \cos \pi t\, dt = \frac{1}{\pi} \sin \pi t + C$$

Since $s = 4$ when $t = 0$, it follows that

$$4 = s(0) = \frac{1}{\pi} \sin 0 + C = C$$

Thus

$$s(t) = \frac{1}{\pi} \sin \pi t + 4$$ ◀

It is a fact of physics that an object moving on a vertical line near the earth's surface and subject only to the force of gravity moves with constant acceleration.* This constant, denoted by the letter g, is approximately 32 ft/sec² or 9.8 m/sec², depending on whether distance is measured in feet or meters.

From this simple principle it is possible to describe completely the vertical motion of a particle moving freely near the earth's surface, provided the initial position and velocity of the particle are known. To see how, assume the line of motion has been coordinatized with the origin at the surface of the earth and the positive direction upward (Figure 5.4.4). Moreover, let us assume that time is measured in seconds, distance is measured in feet or meters and that:

the position at time $t = 0$ is s_0
the velocity at time $t = 0$ is v_0

Figure 5.4.4

where the initial displacement s_0 and initial velocity v_0 are known. Recall that a particle speeds up when its velocity and acceleration have the same sign and slows down when they have opposite signs. Because we have chosen the positive direction to be up, and because a particle moving up (positive velocity) slows down under the force of gravity and a particle moving down (negative velocity) speeds up under the force of gravity, it follows that the acceleration a of the particle must always be negative, that is,

$$a(t) = -g$$

Since the acceleration of the particle is constant, we obtain

$$v(t) = \int a(t)\, dt = \int -g\, dt = -gt + C_1 \tag{1}$$

To determine the constant of integration we use the fact that the velocity is v_0 when $t = 0$. Thus

$$v_0 = v(0) = -g \cdot 0 + C_1 = C_1$$

Substituting this in (1) yields

$$v(t) = -gt + v_0 \tag{2}$$

---

*Strictly speaking, the acceleration changes with the distance from the earth's center. However, near the surface of the earth this distance is approximately constant.

Since v_0 is a constant, it follows that

$$s(t) = \int v(t)\, dt = \int (-gt + v_0)\, dt = -\frac{1}{2}gt^2 + v_0 t + C_2 \qquad (3)$$

To determine the constant C_2 we use the fact that the position is s_0 when $t = 0$. Thus

$$s_0 = s(0) = -\frac{1}{2}g \cdot 0 + v_0 \cdot 0 + C_2 = C_2$$

Substituting this in (3) yields

$$s(t) = -\frac{1}{2}gt^2 + v_0 t + s_0 \qquad (4)$$

▶ Example 5

A rock, initially at rest, is dropped from a height of 400 ft. Assuming gravity is the only force acting, how long does it take for the rock to hit the ground and what is its speed at the time of impact?

Solution. Since distance is in feet, we take $g = 32$ ft/sec². Initially, we have $s_0 = 400$ and $v_0 = 0$, so that from (4)

$$s(t) = -16t^2 + 400$$

Impact occurs when $s(t) = 0$. Solving this equation for t we obtain

$$-16t^2 + 400 = 0$$
$$t^2 = 25$$
$$t = \pm 5$$

Since the rock is released at $t = 0$, the time of impact is positive, so that we can discard the negative solution and conclude that it takes 5 sec for the rock to hit the ground. Substituting $t = 5$ and $v_0 = 0$ in (2), we obtain the velocity at time of impact,

$$v(5) = -32(5) + 0 = -160 \text{ ft/sec}$$

Thus, the speed at impact is

$$|v(5)| = 160 \text{ ft/sec} \qquad\qquad\qquad ◀$$

► Example 6

A ball is thrown directly upward from a point 8 meters above the ground with an initial velocity of 49 meters/sec. Assuming gravity is the only force acting on the ball after its release, how high will the ball travel?

Solution. Since distance is in meters, we take $g = 9.8$ m/sec². Initially, $s_0 = 8$ and $v_0 = 49$, so that from (2) and (4)

$$v(t) = -9.8t + 49$$
$$s(t) = -4.9t^2 + 49t + 8$$

The ball will rise until $v(t) = 0$, that is, until $-9.8t + 49 = 0$ or $t = 5$. At this instant the height will be

$$s(5) = -4.9(5)^2 + 49(5) + 8 = 130.5 \text{ meters} \qquad ◄$$

► Exercise Set 5.4

In Exercises 1–3, $s(t)$, $v(t)$, and $a(t)$ are the position, velocity, and acceleration functions for a particle moving on a coordinate line; distance is measured in feet and time in seconds.

1. (a) Let $s = t^3 - 6t^2$. Make a table showing the position, velocity, speed, and acceleration at times $t = 1$, $t = 2$, $t = 3$, $t = 4$, and $t = 5$.
   (b) At each of these times specify the direction of motion, if any, and whether the particle is speeding up, slowing down, or neither.

2. In each part use the given information to find the position, velocity, speed, and acceleration at time $t = 1$.
   (a) $v = \sin \frac{1}{2}\pi t$; $s = 0$ when $t = 0$.
   (b) $a = -3t$; $s = 1$ and $v = 0$ when $t = 0$.

3. Let $s = 5t^2 - 22t$.
   (a) Find the maximum speed of the particle during the time interval $1 \le t \le 3$.
   (b) When, during the time interval $1 \le t \le 3$, is the particle farthest from the origin? What is its position at that instant?

In Exercises 4–9, describe the motion of the particle for $t \ge 0$ (as in Example 3) and make a sketch as in Figure 5.4.3.

4. $s = -3t + 2$.

5. $s = 1 + 6t - t^2$.

6. $s = t^3 - 6t^2 + 9t + 1$.

7. $s = t^3 - 9t^2 + 24t$.

8. $s = t + \dfrac{9}{t + 1}$.

9. $s = \begin{cases} \cos t, & 0 \le t \le 2\pi \\ 1 & , & t > 2\pi. \end{cases}$

10. A car traveling 60 mi/hr along a straight road decelerates at a constant rate of 10 ft/sec².
    (a) How long will it take until the speed is 45 mi/hr? [*Note:* 60 mi/hr = 88 ft/sec.]
    (b) How far will the car travel before coming to a stop?

In Exercises 11–14, assume that the only force acting is the earth's gravity, and apply formulas (2) and (4) to solve the problems.

11. A projectile is launched vertically upward from ground level with an initial velocity of 112 ft/sec.
    (a) Find the velocity at $t = 3$ and $t = 5$.
    (b) How high will the projectile rise?
    (c) Find the speed of the projectile when it hits the ground.

12. A projectile fired downward from a height of 112 ft reaches the ground in 2 sec. What is its initial velocity?

13. A projectile fired upward from ground level is to reach a height of 1000 ft. What must its initial velocity be?

14. A man drops a stone from a bridge. How high is the bridge if
   (a) the stone hits the water 4 sec later?
   (b) the sound of the splash reaches the man 4 sec later? (Take 1080 ft/sec as the speed of sound.)

15. The position function of a particle moving on a coordinate line is shown in Figure 5.4.5.
   (a) Is the particle moving faster at time t_0 or at time t_1?
   (b) Is the particle speeding up or slowing down at time t_0?
   (c) Is the particle speeding up or slowing down at time t_1?

Figure 5.4.5

16. For a particle moving on a coordinate line, the *average velocity* v_{ave} and *average acceleration* a_{ave} over a time interval $[t_0, t_1]$ are defined by:

$$v_{ave} = \frac{\text{change in position}}{\text{time elapsed}} = \frac{s(t_1) - s(t_0)}{t_1 - t_0}$$

$$a_{ave} = \frac{\text{change in velocity}}{\text{time elapsed}} = \frac{v(t_1) - v(t_0)}{t_1 - t_0}$$

   (a) Interpret v_{ave} geometrically on the graph of $s(t)$.
   (b) Interpret a_{ave} geometrically on the graph of $v(t)$.
   (c) Find the average velocity and acceleration over the time interval $2 \leq t \leq 4$ for a particle with position function $s(t) = t^3 - 3t^2$.

17. (a) Show that the average velocity v_{ave} over $[t_0, t_1]$ approaches the instantaneous velocity $v(t_0)$ as t_1 approaches t_0. [See Exercise 16 for definitions of v_{ave} and a_{ave}.]
   (b) Show that the average acceleration a_{ave} over $[t_0, t_1]$ approaches the instantaneous acceleration $a(t_0)$ as t_1 approaches t_0.

18. Let $r(t) = |v(t)|$ be the speed of a particle moving on a coordinate line. Prove:
   (a) $r'(t_0) > 0$ if $a(t_0) > 0$ and $v(t_0) > 0$ or $a(t_0) < 0$ and $v(t_0) < 0$.
   (b) $r'(t_0) < 0$ if $a(t_0) > 0$ and $v(t_0) < 0$ or $a(t_0) < 0$ and $v(t_0) > 0$.
   (To paraphrase: The particle is speeding up if velocity and acceleration have the same sign, and slowing down if they have opposite signs.)

5.5 SIGMA NOTATION

In this section we digress from the main theme of this chapter to introduce a notation that can be used to write lengthy sums in a compact form. This material will be helpful when we discuss areas in the next section. The notation uses the uppercase Greek letter Σ (sigma) and is called *sigma notation* or *summation notation.*

To illustrate the idea, consider the sum

$$1^2 + 2^2 + 3^2 + 4^2 + 5^2$$

Each term in the sum is of the form k^2, where k is one of the integers from 1 to 5. In sigma notation this sum can be written

$$\sum_{k=1}^{5} k^2$$

which is read, "the summation of k^2, where k runs from 1 to 5." The notation tells us to form the sum of the terms that result when we substitute successive integers for k in the expression k^2, starting with $k = 1$ and ending with $k = 5$.

More generally, if $f(k)$ is a function of k, and a and b are integers, then

$$\sum_{k=a}^{b} f(k) \tag{1}$$

denotes the sum of the terms that result when we substitute successive integers for k, starting with $k = a$ and ending with $k = b$.

▶ Example 1

$$\sum_{k=4}^{8} k^3 = 4^3 + 5^3 + 6^3 + 7^3 + 8^3$$

$$\sum_{k=1}^{5} 2k = 2 \cdot 1 + 2 \cdot 2 + 2 \cdot 3 + 2 \cdot 4 + 2 \cdot 5 = 2 + 4 + 6 + 8 + 10$$

$$\sum_{k=0}^{5} (2k + 1) = 1 + 3 + 5 + 7 + 9 + 11$$

$$\sum_{k=0}^{5} (-1)^k (2k + 1) = 1 - 3 + 5 - 7 + 9 - 11$$

$$\sum_{k=-3}^{1} k^3 = (-3)^3 + (-2)^3 + (-1)^3 + 0^3 + 1^3 = -27 - 8 - 1 + 0 + 1$$

$$\sum_{k=1}^{3} k \sin\left(\frac{k\pi}{5}\right) = \sin\frac{\pi}{5} + 2\sin\frac{2\pi}{5} + 3\sin\frac{3\pi}{5} \qquad ◀$$

The numbers a and b in (1) are called, respectively, the **lower** and **upper** **limits of summation;** and the letter k is called the **index of summation.** It is not essential to use k as the index of summation; any letter will do. For example,

$$\sum_{i=1}^{6} \frac{1}{i}, \qquad \sum_{j=1}^{6} \frac{1}{j}, \qquad \text{and} \qquad \sum_{n=1}^{6} \frac{1}{n}$$

all denote the sum

$$1 + \frac{1}{2} + \frac{1}{3} + \frac{1}{4} + \frac{1}{5} + \frac{1}{6}$$

If the upper and lower limits of summation are the same, then the "sum" in (1) reduces to one term. For example,

$$\sum_{k=2}^{2} k^3 = 2^3 \qquad \text{and} \qquad \sum_{i=1}^{1} \frac{1}{i+2} = \frac{1}{1+2} = \frac{1}{3}$$

In the sums

$$\sum_{i=1}^{5} 2, \qquad \sum_{k=3}^{6} 7, \qquad \text{and} \qquad \sum_{j=0}^{2} x^3$$

the expression to the right of the Σ sign does not involve the index of summation. In such cases we take all the terms in the sum to be the same, with one term for each allowable value of the summation index. Thus,

$$\sum_{i=1}^{5} 2 = 2 + 2 + 2 + 2 + 2$$

$$\sum_{k=3}^{6} 7 = 7 + 7 + 7 + 7$$

$$\sum_{j=0}^{2} x^3 = x^3 + x^3 + x^3$$

A sum can be written in various ways with sigma notation, depending on the limits of summation. For example, the sum of the first five positive even integers can be written as

$$\sum_{k=1}^{5} 2k = 2 + 4 + 6 + 8 + 10$$

or

$$\sum_{k=0}^{4} (2k + 2) = 2 + 4 + 6 + 8 + 10$$

or

$$\sum_{k=2}^{6} (2k - 2) = 2 + 4 + 6 + 8 + 10$$

On occasion we will want to change the sigma notation for a given sum to a sigma notation with different summation limits. The following example illustrates a method for doing this.

▶ Example 2

Express

$$\sum_{k=3}^{7} 5^{k-2}$$

in sigma notation so that the lower limit of summation is 0 rather than 3.

Solution. If we define a new summation index j by means of the formula

$$j = k - 3 \tag{2}$$

then j runs from 0 up to 4 as k runs from 3 up to 7. From (2), $k = j + 3$, so that

$$\sum_{k=3}^{7} 5^{k-2} = \sum_{j=0}^{4} 5^{(j+3)-2} = \sum_{j=0}^{4} 5^{j+1}$$

As a check, the reader can verify that

$$\sum_{j=0}^{4} 5^{j+1} \quad \text{and} \quad \sum_{k=3}^{7} 5^{k-2}$$

both denote the sum $5 + 5^2 + 5^3 + 5^4 + 5^5$. ◀

REMARK.   In this solution the summation index was changed from k to j. If it is desirable to keep the same symbol for the summation index, we can change the j back to k *at the very end* and express the final result as

$$\sum_{k=0}^{4} 5^{k+1} \quad \text{instead of} \quad \sum_{j=0}^{4} 5^{j+1}$$

When we want to represent a general sum we will use letters with subscripts. For example, a general sum with five terms might be written as

$$a_1 + a_2 + a_3 + a_4 + a_5$$

or in sigma notation as

$$\sum_{k=1}^{5} a_k, \quad \sum_{j=1}^{5} a_j, \quad \text{or} \quad \sum_{m=1}^{5} a_m$$

A general sum with n terms might be written as

$$b_1 + b_2 + \cdots + b_n$$

or in sigma notation as

$$\sum_{k=1}^{n} b_k, \quad \sum_{j=1}^{n} b_j, \quad \text{or} \quad \sum_{m=1}^{n} b_m$$

The following algebraic properties of sigma notation help to manipulate sums.

5.5.1   THEOREM   (a) $\displaystyle\sum_{k=1}^{n} (a_k + b_k) = \sum_{k=1}^{n} a_k + \sum_{k=1}^{n} b_k$

(b) $\displaystyle\sum_{k=1}^{n} (a_k - b_k) = \sum_{k=1}^{n} a_k - \sum_{k=1}^{n} b_k$

(c) $\displaystyle\sum_{k=1}^{n} ca_k = c \sum_{k=1}^{n} a_k$

We will prove parts (a) and (c) and leave (b) as an exercise.

Proof of (a). $\displaystyle\sum_{k=1}^{n} (a_k + b_k) = (a_1 + b_1) + (a_2 + b_2) + \cdots + (a_n + b_n)$

$$= (a_1 + a_2 + \cdots + a_n) + (b_1 + b_2 + \cdots + b_n)$$

$$= \sum_{k=1}^{n} a_k + \sum_{k=1}^{n} b_k$$

Proof of (c). $\displaystyle\sum_{k=1}^{n} ca_k = ca_1 + ca_2 + \cdots + ca_n$

$$= c(a_1 + a_2 + \cdots + a_n)$$

$$= c \sum_{k=1}^{n} a_k \quad \blacksquare$$

REMARK. Loosely phrased, this theorem states: sigma of a sum equals the sum of the sigmas; sigma of a difference equals the difference of the sigmas; and a constant factor can be moved through a sigma sign.

The following formulas will be used in our later work.

5.5.2 THEOREM (*a*) $\displaystyle\sum_{k=1}^{n} k = 1 + 2 + 3 + \cdots + n = \frac{n(n + 1)}{2}$

(*b*) $\displaystyle\sum_{k=1}^{n} k^2 = 1^2 + 2^2 + 3^2 + \cdots + n^2 = \frac{n(n + 1)(2n + 1)}{6}$

(*c*) $\displaystyle\sum_{k=1}^{n} k^3 = 1^3 + 2^3 + 3^3 + \cdots + n^3 = \left[\frac{n(n + 1)}{2}\right]^2$

We will prove parts (*a*) and (*b*) and leave part (*c*) as an exercise.

Proof of (a). If we write the terms of

$$\sum_{k=1}^{n} k = 1 + 2 + 3 + \cdots + (n - 2) + (n - 1) + n \tag{3}$$

in the opposite order, we obtain

$$\sum_{k=1}^{n} k = n + (n - 1) + (n - 2) + \cdots + 3 + 2 + 1 \tag{4}$$

Adding (3) and (4) term by term yields

$$2 \sum_{k=1}^{n} k = \underbrace{(n + 1) + (n + 1) + (n + 1) + \cdots + (n + 1)}_{n \text{ terms}} = n(n + 1)$$

Thus

$$\sum_{k=1}^{n} k = \frac{n(n + 1)}{2}$$

Proof of (b). This proof begins with a trick. Since

$$(k + 1)^3 - k^3 = k^3 + 3k^2 + 3k + 1 - k^3$$
$$= 3k^2 + 3k + 1$$

we obtain

$$\sum_{k=1}^{n} [(k + 1)^3 - k^3] = \sum_{k=1}^{n} (3k^2 + 3k + 1) \tag{5}$$

Writing out the left side of (5) yields

$$[2^3 - 1^3] + [3^3 - 2^3] + [4^3 - 3^3] + \cdots + [(n + 1)^3 - n^3] \tag{6}$$

Observe that each term in (6) cancels part of the next term, so that the entire sum collapses like a folding telescope, leaving only $-1^3 + (n + 1)^3$. Thus (5) can be rewritten as

$$-1 + (n + 1)^3 = \sum_{k=1}^{n} (3k^2 + 3k + 1) \tag{7}$$

or, from Theorem 5.5.1,

$$-1 + (n + 1)^3 = 3 \sum_{k=1}^{n} k^2 + 3 \sum_{k=1}^{n} k + \sum_{k=1}^{n} 1 \tag{8}$$

But

$$\sum_{k=1}^{n} 1 = \underbrace{1 + 1 + \cdots + 1}_{n \text{ terms}} = n$$

and by part (*a*) of this theorem

$$\sum_{k=1}^{n} k = \frac{n(n + 1)}{2}$$

Thus (8) can be written

$$-1 + (n + 1)^3 = 3 \sum_{k=1}^{n} k^2 + 3 \frac{n(n + 1)}{2} + n$$

Therefore

$$\sum_{k=1}^{n} k^2 = \frac{1}{3} \left[(n + 1)^3 - 3 \frac{n(n + 1)}{2} - (n + 1) \right]$$

$$= \frac{n + 1}{6} [2(n + 1)^2 - 3n - 2]$$

$$= \frac{n + 1}{6} (2n^2 + n)$$

$$= \frac{n(n + 1)(2n + 1)}{6}$$

▶ Example 3

Evaluate

$$\sum_{k=1}^{30} k(k + 1)$$

Solution.

$$\sum_{k=1}^{30} k(k + 1) = \sum_{k=1}^{30} (k^2 + k)$$

$$= \sum_{k=1}^{30} k^2 + \sum_{k=1}^{30} k$$

$$= \frac{30(31)(61)}{6} + \frac{30(31)}{2} \qquad \text{[Theorem 5.5.2}a,b\text{]}$$

$$= 9920 \qquad\qquad ◀$$

REMARK. In formulas like

$$\sum_{k=1}^{n} k^2 = \frac{n(n + 1)(2n + 1)}{6}$$

or

$$1^2 + 2^2 + \cdots + n^2 = \frac{n(n+1)(2n+1)}{6}$$

the left side of the equality is said to express the sum in **open form** and the right side is said to express it in **closed form;** the open form just indicates the terms to be added, while the closed form gives their total.

▶ Example 4

Express

$$\sum_{k=1}^{n} (3+k)^2$$

in closed form.

Solution.

$$\sum_{k=1}^{n} (3+k)^2 = \sum_{k=1}^{n} (9 + 6k + k^2) = \sum_{k=1}^{n} 9 + 6 \sum_{k=1}^{n} k + \sum_{k=1}^{n} k^2$$

$$= 9n + 6\frac{n(n+1)}{2} + \frac{n(n+1)(2n+1)}{6}$$

$$= \frac{1}{3}n^3 + \frac{7}{2}n^2 + \frac{73}{6}n$$ ◀

▶ Exercise Set 5.5

1. Evaluate

(a) $\displaystyle\sum_{k=1}^{3} k^3$

(b) $\displaystyle\sum_{j=2}^{6} (3j - 1)$

(c) $\displaystyle\sum_{i=-4}^{1} (i^2 - i)$

(d) $\displaystyle\sum_{n=0}^{5} 1.$

2. Evaluate

(a) $\displaystyle\sum_{k=1}^{4} k \sin\frac{k\pi}{2}$

(b) $\displaystyle\sum_{j=0}^{5} (-1)^j$

(c) $\displaystyle\sum_{i=7}^{20} \pi$

(d) $\displaystyle\sum_{m=3}^{5} 2^{m+1}.$

3. Express in sigma notation, but do not evaluate.
   (a) $1 + 2 + 3 + \cdots + 10$
   (b) $3 \cdot 1 + 3 \cdot 2 + 3 \cdot 3 + \cdots + 3 \cdot 20$
   (c) $1 \cdot 2 + 2 \cdot 3 + 3 \cdot 4 + \cdots + 49 \cdot 50$
   (d) $1 + 2 + 2^2 + 2^3 + 2^4$
   (e) $2 + 4 + 6 + 8 + \cdots + 20$
   (f) $1 + 3 + 5 + 7 + \cdots + 15.$

4. In (a)–(f), express in sigma notation, but do not evaluate.

   (a) $1 - \dfrac{1}{2} + \dfrac{1}{3} - \dfrac{1}{4} + \dfrac{1}{5}$

   (b) $-1 + \dfrac{1}{2} - \dfrac{1}{3} + \dfrac{1}{4} - \dfrac{1}{5}$

(c) $1 + \cos\frac{\pi}{7} + \cos\frac{2\pi}{7} + \cos\frac{3\pi}{7}$

(d) $2 + 4 + 8 + 16 + 32$

(e) $\frac{1}{2} + \frac{2}{3} + \frac{3}{4} + \frac{4}{5} + \frac{5}{6}$

(f) $15 + 24 + 35 + \cdots + (n^2 - 1)$.

5. Express in sigma notation.

(a) $a_1 - a_2 + a_3 - a_4 + a_5$

(b) $-b_0 + b_1 - b_2 + b_3 - b_4 + b_5$

(c) $a_0 + a_1 x + a_2 x^2 + \cdots + a_n x^n$

(d) $a^5 + a^4 b + a^3 b^2 + a^2 b^3 + ab^4 + b^5$.

6. Evaluate

(a) $\sum_{k=1}^{100} k$

(c) $\sum_{k=1}^{20} k^2$

(b) $\sum_{k=1}^{100} (7k + 1)$

(d) $\sum_{k=4}^{20} k^2$.

7. Evaluate $\sum_{k=1}^{30} k(k - 2)(k + 2)$.

8. When each term of a sum cancels part of the next term, leaving only portions of the first and last terms at the end, the sum is said to *telescope*. Evaluate the following telescoping sums.

(a) $\sum_{k=1}^{50} \left(\frac{1}{k} - \frac{1}{k+1} \right)$

(b) $\sum_{k=5}^{17} (3^k - 3^{k-1})$

(c) $\sum_{k=1}^{n} (a_k - a_{k+1})$

(d) $\sum_{k=1}^{n} (a_k - a_{k-1})$.

9. Evaluate

(a) $\sum_{k=1}^{n} n$

(b) $\sum_{i=0}^{0} (-3)$

(c) $\sum_{k=1}^{n} kx$

(d) $\sum_{k=m}^{n} c \ (n \geq m)$.

10. Express $1 + 2 + 2^2 + 2^3 + 2^4 + 2^5$ in sigma notation with:

(a) $j = 0$ as the lower limit of summation

(b) $j = 1$ as the lower limit of summation

(c) $j = 2$ as the lower limit of summation.

11. Express

$$\sum_{k=4}^{18} k(k - 3)$$

in sigma notation with:

(a) $k = 0$ as the lower limit of summation

(b) $k = 5$ as the lower limit of summation.

12. Express

$$\sum_{k=5}^{9} k2^{k+4}$$

in sigma notation with:

(a) $k = 1$ as the lower limit of summation

(b) $k = 13$ as the upper limit of summation.

13. Simplify

$$\sum_{k=11}^{28} (k - 10) \sin\left(\frac{\pi}{k - 10} \right)$$

by changing the limits of summation.

14. Which of the following are valid identities?

(a) $\sum_{i=1}^{n} a_i b_i = \sum_{i=1}^{n} a_i \sum_{i=1}^{n} b_i$

(b) $\sum_{i=1}^{n} \frac{a_i}{b_i} = \sum_{i=1}^{n} a_i \Bigg/ \sum_{i=1}^{n} b_i$

(c) $\sum_{i=1}^{n} a_i^2 = \left(\sum_{i=1}^{n} a_i \right)^2$.

15. By writing out the sums, determine if the following are valid identities.

(a) $\int \left[\sum_{i=1}^{n} f_i(x) \right] dx = \sum_{i=1}^{n} \left[\int f_i(x) \, dx \right]$

(b) $\frac{d}{dx} \left[\sum_{i=1}^{n} f_i(x) \right] = \sum_{i=1}^{n} \left[\frac{d}{dx} [f_i(x)] \right]$.

16. (a) Evaluate

$$\sum_{k=0}^{n} ar^k - r \sum_{k=0}^{n} ar^k$$

(b) Use the result in part (a) to prove that

$$\sum_{k=0}^{n} ar^k = a + ar + ar^2 + \cdots + ar^n$$

$$= \frac{a - ar^{n+1}}{1 - r} \qquad (r \neq 1)$$

(A sum of this form is called a **geometric sum.**)

17. Use Exercise 16 to evaluate

(a) $\sum_{k=1}^{20} 3^k$

(c) $\sum_{k=0}^{100} (-1)^{k+1} \frac{1}{2^k}.$

(b) $\sum_{k=5}^{30} 2^k$

18. Express the following sums in closed form:

(a) $\sum_{k=1}^{n} (k - \sin^k \theta)$

(b) $\sum_{k=1}^{m-3} k^3$

[*Hint:* Exercise 16 will help in part (a).]

19. Evaluate

$$\sum_{i=1}^{4} \left(\sum_{j=1}^{5} (i + j) \right)$$

20. Let $\bar{x}$ denote the arithmetic average of the n numbers $x_1, x_2, \ldots, x_n$. Use Theorem 5.5.1 to prove

$$\sum_{i=1}^{n} (x_i - \bar{x}) = 0$$

21. Prove part (b) of Theorem 5.5.1.

22. Prove part (c) of Theorem 5.5.2. [*Hint:* Begin with the difference $(k + 1)^4 - k^4$ and follow the steps used to prove part (b) of the theorem.]

5.6 AREAS AS LIMITS

Figure 5.6.1

Until now we have used the term "area" freely, assuming that its meaning is intuitively clear. However, we must ultimately replace this intuitive concept of area with a precise mathematical definition. It is natural to hope that we might be able to devise a definition that would specify the area of an arbitrary region in the plane (Figure 5.6.1). Surprisingly, this is impossible! It became clear in the late nineteenth century that there are regions in the plane of such complexity that any attempt to assign them areas would ultimately lead to mathematical inconsistencies. While we will not be concerned with such complicated regions, it is important to know that they exist. In this section we will show how to obtain areas of certain regions using limits. The ideas we develop here will form the basis for a precise definition of area in the next section.

5.6.1 PROBLEM Find the area of a region R bounded below by the x-axis, on the sides by the lines $x = a$ and $x = b$, and above by a curve $y = f(x)$, where f is continuous on $[a, b]$ and $f(x) \geq 0$ for all x in $[a, b]$ (Figure 5.6.2).

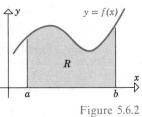

Figure 5.6.2

In Section 5.1 we showed how to obtain the area of this type of region using antiderivatives. However, we may also proceed as follows. Choose an arbitrary positive integer n and divide the interval $[a, b]$ into n subintervals of width $(b - a)/n$ by introducing points

$$x_1, x_2, \ldots, x_{n-1}$$

equally spaced between a and b (Figure 5.6.3). Next, draw vertical lines through the points $a, x_1, x_2, \ldots, x_{n-1}, b$ to divide the region R into n strips of uniform width. If we approximate each of these strips by a rectangle *inscribed*

Figure 5.6.3

under the curve $y = f(x)$ (Figure 5.6.3), then the union of these rectangles will form a region R_n, which we can view as an approximation to the entire region R. The area of this approximating region can be calculated by adding the areas of its component rectangles. Moreover, if we allow n to increase, the widths of the rectangles will get smaller and smaller so that the approximation of R by R_n will get better as the smaller rectangles fill in more of the gaps under the curve (Figure 5.6.4). Thus we can define the *exact* area of R as the limit of the areas of the approximating regions as n goes to plus infinity; that is

$$\text{area } (R) = \lim_{n \to +\infty} [\text{area } (R_n)] \tag{1}$$

Figure 5.6.4

REMARK. There is a difference between writing $\lim\limits_{n \to +\infty}$ and writing $\lim\limits_{x \to +\infty}$, where n represents a positive integer, and x has no such restriction. Later we will study limits of the type $\lim\limits_{n \to +\infty}$ in detail, but for now let it suffice to say that the computational techniques we have used for limits of the type $\lim\limits_{x \to +\infty}$ will also work for $\lim\limits_{n \to +\infty}$; in fact, if $\lim\limits_{x \to +\infty} f(x) = L$, then $\lim\limits_{n \to +\infty} f(n)$ is also L.

For computational purposes, (1) can be expressed in a more useful form. If we denote the heights of the inscribed rectangles by $h_1, h_2, \ldots, h_n$ and use the fact that each rectangle has a base of length $(b - a)/n$, then

$$\text{area } (R_n) = h_1 \cdot \frac{b - a}{n} + h_2 \cdot \frac{b - a}{n} + \cdots + h_n \cdot \frac{b - a}{n} \qquad (2)$$

Because f is assumed to be continuous on $[a, b]$, it follows from the extreme value theorem that f assumes a minimum value on each of the n closed subintervals

$$[a, x_1], [x_1, x_2], \ldots, [x_{n-1}, b]$$

If these minimum values occur at the points $c_1, c_2, \ldots, c_n$, then the heights of the inscribed rectangles are

$$h_1 = f(c_1), \ h_2 = f(c_2), \ldots, h_n = f(c_n)$$

(Figure 5.6.5), so (2) can be written

$$\text{area } (R_n) = f(c_1) \cdot \frac{b - a}{n} + f(c_2) \cdot \frac{b - a}{n} + \cdots + f(c_n) \cdot \frac{b - a}{n} \qquad (3)$$

The height of each inscribed rectangle is the minimum value of $f(x)$ on the subinterval

Figure 5.6.5

Finally, it will be helpful to write

$$\Delta x = \frac{b - a}{n}$$

for the base dimension of the rectangles, so that (3) becomes

$$\text{area } (R_n) = f(c_1) \cdot \Delta x + f(c_2) \cdot \Delta x + \cdots + f(c_n) \cdot \Delta x$$

or, in sigma notation,

$$\text{area } (R_n) = \sum_{k=1}^{n} f(c_k) \cdot \Delta x$$

With this notation (1) becomes

$$A = \lim_{n \to +\infty} \sum_{k=1}^{n} f(c_k) \cdot \Delta x \qquad (4)$$

▶ Example 1

Use inscribed rectangles to find the area under the line $y = x$ over the interval [1, 2].

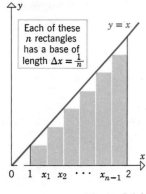

Figure 5.6.6

Solution. If we subdivide the interval [1, 2] into n equal parts (Figure 5.6.6), then each part will have length

$$\Delta x = \frac{b - a}{n} = \frac{2 - 1}{n} = \frac{1}{n}$$

and the points of subdivision will be

$$x_1 = 1 + \Delta x = 1 + \frac{1}{n}$$

$$x_2 = 1 + 2 \Delta x = 1 + \frac{2}{n}$$

$$x_3 = 1 + 3 \Delta x = 1 + \frac{3}{n}$$

$$\vdots$$

$$x_{n-1} = 1 + (n - 1) \Delta x = 1 + \frac{n - 1}{n}$$

Because $y = f(x) = x$ is increasing, the minimum value for $f(x)$ on each subinterval occurs at the left endpoint (Figure 5.6.6), so

$$c_1 = 1$$

$$c_2 = x_1 = 1 + \frac{1}{n}$$

$$c_3 = x_2 = 1 + \frac{2}{n}$$

$$c_4 = x_3 = 1 + \frac{3}{n}$$

$$\vdots$$

$$c_n = x_{n-1} = 1 + \frac{n-1}{n}$$

Thus, the inscribed rectangles have areas

$$f(c_1) \cdot \Delta x = c_1 \cdot \Delta x = 1 \cdot \frac{1}{n}$$

$$f(c_2) \cdot \Delta x = c_2 \cdot \Delta x = \left(1 + \frac{1}{n}\right) \cdot \frac{1}{n}$$

$$f(c_3) \cdot \Delta x = c_3 \cdot \Delta x = \left(1 + \frac{2}{n}\right) \cdot \frac{1}{n}$$

$$f(c_4) \cdot \Delta x = c_4 \cdot \Delta x = \left(1 + \frac{3}{n}\right) \cdot \frac{1}{n}$$

$$\vdots$$

$$f(c_n) \cdot \Delta x = c_n \cdot \Delta x = \left(1 + \frac{n-1}{n}\right) \cdot \frac{1}{n}$$

and the sum is

$$\sum_{k=1}^{n} f(c_k) \cdot \Delta x$$

$$= \left[1 + \left(1 + \frac{1}{n}\right) + \left(1 + \frac{2}{n}\right) + \left(1 + \frac{3}{n}\right) + \cdots + \left(1 + \frac{n-1}{n}\right)\right] \cdot \frac{1}{n}$$

$$= \left[n + \left(\frac{1}{n} + \frac{2}{n} + \frac{3}{n} + \cdots + \frac{n-1}{n}\right)\right] \cdot \frac{1}{n}$$

$$= 1 + \frac{1}{n^2}[1 + 2 + 3 + \cdots + (n-1)]$$

$$= 1 + \frac{1}{n^2} \cdot \frac{(n-1)(n)}{2} \qquad \text{[Theorem 5.5.2}a \text{ with } n-1 \text{ substituted for } n]$$

$$= \frac{3}{2} - \frac{1}{2n}$$

Thus, from (4) the area is

$$A = \lim_{n \to +\infty} \sum_{k=1}^{n} f(c_k) \cdot \Delta x$$

$$= \lim_{n \to +\infty} \left(\frac{3}{2} - \frac{1}{2n} \right)$$

$$= \frac{3}{2} - 0 = \frac{3}{2}$$

The region whose area we have computed is a trapezoid with height $h = 1$ and bases $b_1 = 1$ and $b_2 = 2$. From plane geometry, the area of this trapezoid is $A = \frac{1}{2}h(b_1 + b_2) = \frac{1}{2}(1)(1 + 2) = \frac{3}{2}$, which agrees with result obtained here. ◄

▶ Example 2

Use inscribed rectangles to find the area under the curve $y = 9 - x^2$ over the interval $[0, 3]$.

Solution. If we divide the interval $[0, 3]$ into n subintervals of equal length, then each subinterval will have length

$$\Delta x = \frac{b - a}{n} = \frac{3 - 0}{n} = \frac{3}{n}$$

and the points of subdivision will be

$$x_1 = 0 + \Delta x = \frac{3}{n}$$

$$x_2 = 0 + 2\,\Delta x = 2\left(\frac{3}{n} \right)$$

$$x_3 = 0 + 3\,\Delta x = 3\left(\frac{3}{n} \right)$$

$$\vdots$$

$$x_{n-1} = 0 + (n - 1)\,\Delta x = (n - 1)\left(\frac{3}{n} \right)$$

(Figure 5.6.7.) Because $y = f(x) = 9 - x^2$ is decreasing on $[0, 3]$, the minimum value for $f(x)$ on each subinterval occurs at the right endpoint (Figure 5.6.7) so that

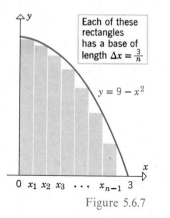

Each of these rectangles has a base of length $\Delta x = \frac{3}{n}$

$y = 9 - x^2$

$0 \quad x_1 \quad x_2 \quad x_3 \quad \cdots \quad x_{n-1} \quad 3$

Figure 5.6.7

$$c_1 = x_1 = \frac{3}{n}$$

$$c_2 = x_2 = 2 \cdot \frac{3}{n}$$

$$c_3 = x_3 = 3 \cdot \frac{3}{n}$$

$$\vdots$$

$$c_{n-1} = x_{n-1} = (n-1) \cdot \frac{3}{n}$$

$$c_n = 3$$

In short, $c_k = k \cdot (3/n)$ for $k = 1, 2, \ldots, n$. Thus, the kth inscribed rectangle has area

$$f(c_k) \cdot \Delta x = (9 - c_k^2) \cdot \Delta x = \left[9 - k^2\left(\frac{9}{n^2}\right)\right] \cdot \frac{3}{n} = \frac{27}{n} - \frac{27k^2}{n^3}$$

and the sum of these areas is

$$\sum_{k=1}^{n} f(c_k) \cdot \Delta x = \sum_{k=1}^{n} \left(\frac{27}{n} - \frac{27k^2}{n^3}\right)$$

$$= \sum_{k=1}^{n} \frac{27}{n} - \sum_{k=1}^{n} \frac{27k^2}{n^3}$$

$$= n \cdot \frac{27}{n} - \frac{27}{n^3} \sum_{k=1}^{n} k^2$$

$$= 27 - \frac{27}{n^3} \cdot \frac{n(n+1)(2n+1)}{6} \qquad \text{[Theorem 5.5.2}b\text{]}$$

Thus, from (4)

$$A = \lim_{n \to +\infty} \sum_{k=1}^{n} f(c_k) \cdot \Delta x$$

$$= \lim_{n \to +\infty} \left[27 - \frac{27}{n^3} \cdot \frac{n(n+1)(2n+1)}{6}\right]$$

$$= 27 - \lim_{n \to +\infty} \frac{27}{6}\left(1 + \frac{1}{n}\right)\left(2 + \frac{1}{n}\right)$$

$$= 27 - \frac{27}{6}(1)(2) = 18$$

◀

REMARK. As a check, the reader may want to verify this result by using the method of Section 5.1 to calculate the area.

It may already have occurred to the reader that we could have used *circumscribed* rather than inscribed rectangles in these examples. Indeed, if the *maximum* values of $f(x)$ in the various subintervals occur at the points

$$d_1, d_2, \ldots, d_n$$

then the sum

$$\sum_{k=1}^{n} f(d_k) \cdot \Delta x$$

is an approximation by circumscribed rectangles to the area under the curve (Figure 5.6.8) and the exact area A is

$$A = \lim_{n \to +\infty} \sum_{k=1}^{n} f(d_k) \cdot \Delta x \tag{5}$$

since the error (represented by the overlap) diminishes as the rectangles get thinner.

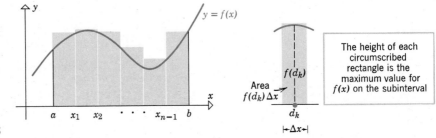

Figure 5.6.8

▶ Example 3

Use circumscribed rectangles to find the area under the line $y = x$ over the interval $[1, 2]$.

Solution. As in Example 1, the points

$$x_1 = 1 + \frac{1}{n}$$

$$x_2 = 1 + \frac{2}{n}$$

$$x_3 = 1 + \frac{3}{n}$$

$$\vdots$$

$$x_{n-1} = 1 + \frac{n-1}{n}$$

divide the interval $[1, 2]$ into n subintervals of length $\Delta x = 1/n$. Because $y = f(x) = x$ is increasing, the maximum value for $f(x)$ on each subinterval occurs at the right-hand endpoint (Figure 5.6.9), so that $d_1 = x_1 = 1 + 1/n$, $d_2 = x_2 = 1 + 2/n$, $d_3 = x_3 = 1 + 3/n, \ldots, d_n = 1 + n/n = 2$.

Figure 5.6.9

Thus, the kth circumscribed rectangle has area

$$f(d_k) \cdot \Delta x = d_k \cdot \Delta x = \left(1 + \frac{k}{n}\right) \cdot \Delta x = \left(1 + \frac{k}{n}\right) \cdot \frac{1}{n}$$

and the sum of the areas of the rectangles is

$$\sum_{k=1}^{n} f(d_k) \cdot \Delta x = \sum_{k=1}^{n} \left[\left(1 + \frac{k}{n}\right) \cdot \frac{1}{n}\right]$$

$$= \frac{1}{n} \sum_{k=1}^{n} 1 + \frac{1}{n^2} \sum_{k=1}^{n} k$$

$$= \frac{1}{n} \cdot n + \frac{1}{n^2} \left[\frac{n(n+1)}{2}\right] \qquad \text{[Theorem 5.5.2a]}$$

$$= \frac{3}{2} + \frac{1}{2n}$$

From (5), the area under the curve is

$$A = \lim_{n \to +\infty} \sum_{k=1}^{n} f(d_k) \cdot \Delta x = \lim_{n \to +\infty} \left(\frac{3}{2} + \frac{1}{2n}\right) = \frac{3}{2}$$

This result agrees with that obtained in Example 1 using inscribed rectangles. ◄

▶ Exercise Set 5.6

In Exercises 1–4, divide the interval $[a, b]$ into $n = 4$ subintervals of equal length and compute: (a) the sum of the areas of the inscribed rectangles, $\sum_{k=1}^{4} f(c_k) \cdot \Delta x$;

(b) the sum of the areas of the circumscribed rectangles, $\sum_{k=1}^{4} f(d_k) \cdot \Delta x$.

1. $y = 3x + 1$; $a = 2$, $b = 6$.

2. $y = 1/x$; $a = 1$, $b = 9$. *increasing*

3. $y = \cos x$; $a = -\dfrac{\pi}{2}$, $b = \dfrac{\pi}{2}$.

4. $y = 2x - x^2$; $a = 1$, $b = 2$.

In Exercises 5–10, use inscribed rectangles to find the area under the curve $y = f(x)$ over the interval $[a, b]$.

5. $y = \frac{1}{2}x$; $a = 1$, $b = 4$.

6. $y = -x + 5$; $a = 0$, $b = 5$.

7. $y = x^2$; $a = 0$, $b = 1$.

8. $y = 4 - \frac{1}{4}x^2$; $a = 0$, $b = 3$.

9. $y = x^3$; $a = 2$, $b = 6$.

10. $y = 1 - x^3$; $a = -3$, $b = -1$.

11. Use circumscribed rectangles to find the area of the region in Exercise 5.

12. Use circumscribed rectangles to find the area of the region in Exercise 6.

13. Use circumscribed rectangles to find the area of the region in Exercise 7.

14. Use inscribed rectangles to find the area under $y = mx$ over the interval $[a, b]$, where $m > 0$, and $a \geq 0$.

15. (a) Show that the area under $y = x^3$ over the interval $[0, b]$ is $b^4/4$.
    (b) Find a formula for the area under $y = x^3$ over the interval $[a, b]$, where $a \geq 0$.

16. Find the area between the curve $y = \sqrt{x}$ and the interval $0 \leq y \leq 1$ on the y-axis.

17. Assuming that $a > 0$, find the area under $y = x$ over the interval $[a, b]$ four different ways:
    (a) using the method of Section 5.1 (antidifferentiation)
    (b) using inscribed rectangles
    (c) using circumscribed rectangles
    (d) using an appropriate area formula from plane geometry.

5.7 THE DEFINITE INTEGRAL

In mathematics and science there are a variety of concepts, such as length, volume, density, probability, work, and others, whose properties are remarkably similar to properties of area. In this section we will introduce the concept of a *definite integral*, which is the unifying thread relating these diverse ideas.

In the last section we gave two equivalent ways of obtaining the area under a continuous curve $y = f(x)$ over an interval $[a, b]$:

$$A = \lim_{n \to +\infty} \sum_{k=1}^{n} f(c_k)\,\Delta x \qquad \text{(inscribed rectangles)} \qquad (1)$$

and

$$A = \lim_{n \to +\infty} \sum_{k=1}^{n} f(d_k)\,\Delta x \qquad \text{(circumscribed rectangles)} \qquad (2)$$

However, these are not the only possible formulas for the area A. Instead of choosing the height of the rectangle on the kth subinterval to be the minimum value or the maximum value of f, we can construct a rectangle of some intermediate height. For this purpose, let us select an *arbitrary* point in each subinterval; call these points

$$x_1^*, x_2^*, \ldots, x_n^*$$

Since $f(c_k)$ and $f(d_k)$ are respectively the smallest and largest values of f on the kth subinterval, it follows that

$$f(c_k) \leq f(x_k^*) \leq f(d_k)$$

and

$$f(c_k) \cdot \Delta x \leq f(x_k^*) \cdot \Delta x \leq f(d_k) \cdot \Delta x$$

so that

$$\sum_{k=1}^{n} f(c_k) \, \Delta x \leq \sum_{k=1}^{n} f(x_k^*) \, \Delta x \leq \sum_{k=1}^{n} f(d_k) \, \Delta x \tag{3}$$

As $n \to +\infty$, the two outside sums approach A [see (1) and (2)], so the inside sum must also approach A, since it is "pinched" between the outer sums. Thus, for all possible choices of $x_1^*, x_2^*, \ldots, x_n^*$ we obtain

$$A = \lim_{n \to +\infty} \sum_{k=1}^{n} f(x_k^*) \, \Delta x$$

While rectangles with equal widths are computationally convenient, they are not essential; we can just as well express the area A as the limit of a sum of rectangular areas with different widths (Figure 5.7.1).

Figure 5.7.1

However, there is a complication here that did not occur before. For rectangles of equal width,

$$\Delta x = \frac{b - a}{n}$$

it follows that $\Delta x \to 0$ as $n \to +\infty$. In other words we can ensure that the width of *every* rectangle decreases to zero by letting the number of rectangles increase to infinity. For rectangles whose widths vary in size this is not the case. For example, suppose we were to continually divide only the left half of the interval $[a, b]$ into more and more subintervals, but always leave the right half of the interval alone (Figure 5.7.2).

Figure 5.7.2

With this construction the number of rectangles increases to infinity, but the sum of the areas of the rectangles does not approach the area under the curve because there is always an "error" on the right half of the interval. To remedy this problem, we need to ensure that the widths of *all* rectangles decrease to zero as the number of rectangles increases to infinity.

Let us suppose that the interval $[a, b]$ is divided into n subintervals whose widths are

$$\Delta x_1, \Delta x_2, \ldots, \Delta x_n$$

and let us denote the largest of these by the symbol

$$\max \Delta x_k$$

(read, "the maximum of the Δx_k's.")

If x_k^* is an arbitrary point in the kth subinterval, then

$$f(x_k^*) \, \Delta x_k$$

is the area of a rectangle of height $f(x_k^*)$ and width Δx_k, so

$$\sum_{k=1}^{n} f(x_k^*) \, \Delta x_k \tag{4}$$

is the sum of the shaded rectangular areas in Figure 5.7.3. If we now increase n in such a way that

$$\max \Delta x_k \to 0$$

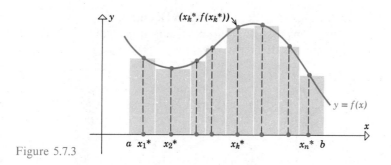

Figure 5.7.3

then the width of *every* rectangle tends to zero because no width exceeds the maximum. Thus (4) approaches the exact area under the curve as $\max \Delta x_k \to 0$. We denote this by writing

$$A = \lim_{\max \Delta x_k \to 0} \sum_{k=1}^{n} f(x_k^*) \Delta x_k$$

This limit is different from those studied in Chapter 2; we will define it precisely in terms of ϵ's and δ's at the end of this section.

The preceding informal discussion suggests the following definition.

5.7.1 DEFINITION
Area
If the function f is continuous on $[a, b]$ and if $f(x) \geq 0$ for all x in $[a, b]$, then the **area** under the curve $y = f(x)$ over the interval $[a, b]$ is defined by

$$A = \lim_{\max \Delta x_k \to 0} \sum_{k=1}^{n} f(x_k^*) \Delta x_k$$

where $x_1^*, x_2^*, \ldots, x_n^*$ are arbitrary points in successive subintervals.

REMARK. It is proved in advanced calculus that the limit in this definition exists, and that the area A is a nonnegative quantity.

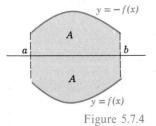

Figure 5.7.4

Definition 5.7.1 applies only to *nonnegative* continuous functions. However, if f is continuous and $f(x) \leq 0$ for all x in $[a, b]$, then $-f$ is continuous and $-f(x) \geq 0$ for all x in $[a, b]$ (Figure 5.7.4). We define the area A above $y = f(x)$ under the interval $[a, b]$ to be the same as the area under $y = -f(x)$ over the interval $[a, b]$.

The limiting process used in Definition 5.7.1 to define area is of fundamental importance in a host of physical and mathematical applications that have no immediate connection with area. In many such applications the function f may have discontinuities and may assume *both* positive and negative values on $[a, b]$. To pave the way for these applications, we will retrace the three steps leading to Definition 5.7.1, but now we will drop the assumptions that f is continuous and nonnegative on $[a, b]$; we will assume only that f is defined on $[a, b]$.

Step 1. Divide the interval $[a, b]$ into n subintervals by choosing arbitrary points $x_1, x_2, \ldots, x_{n-1}$ satisfying

$$a < x_1 < x_2 < \cdots < x_{n-1} < b$$

The points $a, x_1, x_2, \ldots, x_{n-1}, b$ are said to form a **partition** of the interval. Let

$$\Delta x_1, \Delta x_2, \ldots, \Delta x_n$$

denote the lengths of successive subintervals formed by the partition. The largest of these lengths is called the **norm** or **mesh size** of the partition and is denoted by

$$\max \Delta x_k$$

Step 2. Choose arbitrary points

$$x_1^*, x_2^*, \ldots, x_n^*$$

in successive subintervals and form the sum

$$f(x_1^*)\, \Delta x_1 + f(x_2^*)\, \Delta x_2 + \cdots + f(x_n^*)\, \Delta x_n$$

$$= \sum_{k=1}^{n} f(x_k^*)\, \Delta x_k$$

This is called a **Riemann* sum.** (See page 327.)

Step 3. Increase n, so that the mesh size of the partition approaches zero, and form the limit

$$\lim_{\max \Delta x_k \to 0} \sum_{k=1}^{n} f(x_k^*)\, \Delta x_k$$

if it exists.

The limit in Step 3 is of such importance that there is some notation and terminology associated with it.

5.7.2 DEFINITION If a function f is defined on a closed interval $[a, b]$, then the **definite integral** of f from a to b, denoted by $\int_a^b f(x)\, dx$, is defined to be

$$\int_a^b f(x)\, dx = \lim_{\max \Delta x_k \to 0} \sum_{k=1}^n f(x_k^*)\, \Delta x_k \tag{5}$$

provided the limit exists.

The numbers a and b in the symbol $\int_a^b f(x)\, dx$ are called respectively the **lower** and **upper limits of integration,** and $f(x)$ is called the **integrand.** The reason for the use of an integral sign in the above definition will become clear in the next section where we will establish a relationship between the definite integral defined by (5) and the indefinite integral studied earlier.

Because the definite integral is defined as a limit, the integral may or may not exist depending on the nature of the integrand f. If the integral does exist, then f is said to be **Riemann integrable** or simply **integrable** on $[a, b]$. The problem of determining precisely which functions are integrable is beyond the scope of this text. However, the following theorem, stated without proof, is useful to know.

5.7.3 THEOREM (a) *If f is continuous on $[a, b]$, then f is integrable on $[a, b]$.*

(b) *If f has only finitely many points of discontinuity on $[a, b]$ and if there is a positive number M such that $-M \le f(x) \le M$ for all x in $[a, b]$, then f is integrable on $[a, b]$.*

---

*GEORG FRIEDRICH BERNHARD RIEMANN (1826–1866) German mathematician. Bernard Riemann, as he is commonly known, was the son of a Protestant minister. He received his elementary education from his father and showed brilliance in arithmetic at an early age. In 1846 he enrolled at Göttingen University to study theology and philology, but he soon transferred to mathematics. He studied physics under W. E. Weber and mathematics under Karl Friedrich Gauss, whom some people consider to be the greatest mathematician who ever lived. In 1851 Riemann received his Ph.D. under Gauss, after which he remained at Göttingen to teach. In 1862, one month after his marriage, Riemann suffered an attack of pleuritis, and for the remainder of his life was a sick and dying man. He finally succumbed to tuberculosis in 1866 at age 39.

An interesting story surrounds Riemann's work in geometry. For his introductory lecture prior to becoming an associate professor, Riemann submitted three possible topics to Gauss. Gauss surprised Riemann by choosing the topic Riemann liked the least, the foundations of geometry. The lecture was like a scene from a movie. The old and failing Gauss, a giant in his day, watching intently as his brilliant and youthful protege skillfully pieced together portions of the old man's own work into a complete and beautiful system. Gauss is said to have gasped with delight as the lecture neared its end, and on the way home he marveled at his student's brilliance. Gauss died shortly thereafter. The results presented by Riemann that day eventually evolved into a fundamental tool that Einstein used some 50 years later to develop relativity theory.

In addition to his work in geometry, Riemann made major contributions to the theory of complex functions and mathematical physics. The notion of the definite integral, as it is presented in most basic calculus courses, is due to Riemann. Riemann's early death was a great loss to mathematics, for his mathematical work was brilliant and of fundamental importance.

REMARK. In part (*b*), the condition that there exists a positive number M such that $-M \leq f(x) \leq M$ for all x in $[a, b]$ is described by stating that f is **bounded** on $[a, b]$. Geometrically, this means that over the interval $[a, b]$ the entire graph lies between two horizontal lines, $y = -M$ and $y = M$. A function that approaches $+\infty$ or $-\infty$ somewhere on the interval $[a, b]$ does not satisfy this condition. It can be proved that every Riemann integrable function on $[a, b]$ must be bounded on $[a, b]$.

The function f graphed in Figure 5.7.5 is integrable on $[a, b]$, since it is bounded on the interval $[a, b]$ and has only finitely many points of discontinuity on $[a, b]$.

Figure 5.7.5

f is bounded on $[a, b]$ and has finitely many discontinuities

The following theorem follows from Definitions 5.7.1 and 5.7.2.

5.7.4 THEOREM *If f is nonnegative and continuous on $[a, b]$, then*

$$A = \begin{bmatrix} \text{area under} \\ y = f(x) \\ \text{over } [a, b] \end{bmatrix} = \int_a^b f(x)\, dx \tag{6}$$

▶ Example 1

In Example 2 of Section 5.6 we showed the area under the curve $y = 9 - x^2$ over the interval $[0, 3]$ to be 18. Thus,

$$\int_0^3 (9 - x^2)\, dx = 18 \qquad \blacktriangleleft$$

If f is continuous and assumes both positive and negative values on $[a, b]$, then the definite integral can be interpreted as a *difference* of areas. To see why, consider a typical Riemann sum

$$\sum_{k=1}^n f(x_k^*)\, \Delta x_k$$

If $f(x_k^*)$ is nonnegative, then the term $f(x_k^*)\, \Delta x_k$ represents the area, A_k, of a rectangle with height $f(x_k^*)$ and base Δx_k. On the other hand, if $f(x_k^*)$ is negative, then $f(x_k^*)\,\Delta x_k$ is *not* the area of a rectangle, but rather the negative, $-A_k$, of such an area. For example, in Figure 5.7.6

$$\sum_{k=1}^{6} f(x_k^*)\,\Delta x_k = f(x_1^*)\,\Delta x_1 + f(x_2^*)\,\Delta x_2 + \cdots + f(x_6^*)\,\Delta x_6$$

$$= A_1 + A_2 - A_3 - A_4 + A_5 + A_6$$
$$= (A_1 + A_2 + A_5 + A_6) - (A_3 + A_4)$$

Figure 5.7.6

It follows that the definite integral

$$\int_a^b f(x)\,dx = \lim_{\max \Delta x_k \to 0} \sum_{k=1}^{n} f(x_k^*)\,\Delta x_k$$

can be interpreted as a difference of areas: the area above the x-axis between $y = f(x)$ and $[a, b]$ minus the area below the x-axis between $y = f(x)$ and $[a, b]$. As an illustration, the Riemann sums in Figure 5.7.7 tend toward $A_{\mathrm{I}} - A_{\mathrm{II}} + A_{\mathrm{III}}$ as a limit, so that

$$\int_a^b f(x)\,dx = (A_I + A_{III}) - A_{II} = \begin{bmatrix} \text{area above} \\ x\text{-axis} \end{bmatrix} - \begin{bmatrix} \text{area below} \\ x\text{-axis} \end{bmatrix}$$

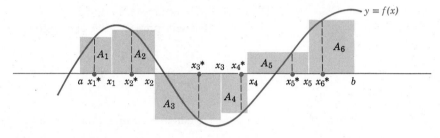

Figure 5.7.7

In the case where f is continuous and $f(x) \geq 0$ on $[a, b]$, the portion of area below the x-axis is zero so that $\int_a^b f(x)\,dx$ represents the area under $y = f(x)$ over $[a, b]$. On the other hand, if f is continuous and $f(x) \leq 0$ on $[a, b]$, then the portion of area above the x-axis is zero so that $\int_a^b f(x)\,dx$ represents the negative of the area above $y = f(x)$ under $[a, b]$. This agrees with our earlier results.

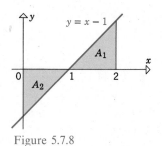

Figure 5.7.8

▶ Example 2

Referring to Figure 5.7.8, the area A_1 is equal to the area A_2 so that

$$\int_0^2 (x - 1)\, dx = A_1 - A_2 = 0$$

◀

OPTIONAL

The limits in Definitions 5.7.1 and 5.7.2 are different from those discussed in Chapter 2. Loosely phrased, the expression

$$\lim_{\max \Delta x_k \to 0} \sum_{k=1}^n f(x_k^*)\, \Delta x_k = L$$

is intended to convey the idea that we can force the Riemann sums to be as close as we please to L, regardless of how $x_1^*, x_2^*, \ldots, x_n^*$ are chosen, by making the mesh size of the partition sufficiently small. This idea is captured in the following definition.

5.7.5 DEFINITION We will write

$$\lim_{\max \Delta x_k \to 0} \sum_{k=1}^n f(x_k^*)\, \Delta x_k = L \tag{7}$$

if given any number $\epsilon > 0$ there is a number $\delta > 0$ such that

$$\left| \sum_{k=1}^n f(x_k^*)\, \Delta x_k - L \right| < \epsilon$$

whenever

$$\max \Delta x_k < \delta$$

and regardless of how $x_1^*, x_2^*, \ldots, x_n^*$ are chosen.

It can be shown that when a number L satisfying (7) exists, it is unique (Exercise 15). We denote this number by $\int_a^b f(x)\, dx$ and write

$$\lim_{\max \Delta x_k \to 0} \sum_{k=1}^n f(x_k^*)\, \Delta x_k = \int_a^b f(x)\, dx \tag{8}$$

REMARK. Some writers use the symbol $\|\Delta\|$ rather than max Δx_k for the mesh size of the partition, in which case (8) would be written

$$\lim_{\|\Delta\| \to 0} \sum_{k=1}^{n} f(x_k^*) \, \Delta x_k = \int_a^b f(x) \, dx$$

▶ Exercise Set 5.7

In Exercises 1–4, find the value of:

(a) $\displaystyle\sum_{k=1}^{n} f(x_k^*) \, \Delta x_k$ (b) max Δx_k.

1. $f(x) = x + 1$; $a = 0$, $b = 4$; $x_1 = 1$, $x_2 = 2$; $x_1^* = \frac{1}{3}$, $x_2^* = \frac{3}{2}$, $x_3^* = 3$.

2. $f(x) = \cos x$; $a = 0$, $b = 2\pi$; $x_1 = \pi/2$, $x_2 = 5\pi/4$, $x_3 = 7\pi/4$; $x_1^* = \pi/4$, $x_2^* = \pi$, $x_3^* = 3\pi/2$, $x_4^* = 7\pi/4$.

3. $f(x) = 4 - x^2$; $a = -3$, $b = 4$; $x_1 = -2$, $x_2 = 0$, $x_3 = 1$; $x_1^* = -\frac{5}{2}$, $x_2^* = -1$, $x_3^* = \frac{1}{4}$, $x_4^* = 3$.

4. $f(x) = x^3$; $a = -3$, $b = 3$; $x_1 = -1$, $x_2 = 0$, $x_3 = 1$; $x_1^* = -2$, $x_2^* = 0$, $x_3^* = 0$, $x_4^* = 2$.

5. Verify that relationship (3) holds when

$$f(x) = \frac{1}{x}; \quad a = 1, \quad b = 9; \quad x_1 = 3, \quad x_2 = 5,$$

$$x_3 = 7; \quad x_1^* = 2, \quad x_2^* = 4, \quad x_3^* = 6, \quad x_4^* = 8.$$

6. Using the given values of a and b, express the following limits as definite integrals. (Do not evaluate the integrals.)

(a) $\displaystyle\lim_{\text{max } \Delta x_k \to 0} \sum_{k=1}^{n} (x_k^*)^3 \, \Delta x_k$; $a = 1$, $b = 2$

(b) $\displaystyle\lim_{\text{max } \Delta x_k \to 0} \sum_{k=1}^{n} 4x_k^*(1 - 3x_k^*) \, \Delta x_k$; $a = -3$, $b = 3$

(c) $\displaystyle\lim_{\text{max } \Delta x_k \to 0} \sum_{k=1}^{n} (\sin^2 x_k^*) \, \Delta x_k$; $a = 0$, $b = \pi/2$.

7. Express the following definite integrals as limits. (Do not evaluate.)

(a) $\displaystyle\int_1^2 2x \, dx$ (b) $\displaystyle\int_{-\pi/2}^{\pi/2} (1 + \cos x) \, dx$

(c) $\displaystyle\int_0^1 \frac{x}{x+1} \, dx$

8. Evaluate the following definite integrals by using appropriate area formulas from plane geometry.

(a) $\displaystyle\int_0^3 x \, dx$ (b) $\displaystyle\int_{-2}^{-1} x \, dx$

(c) $\displaystyle\int_{-1}^4 x \, dx$ (d) $\displaystyle\int_{-5}^5 x \, dx$.

9. Let

$$f(x) = \begin{cases} -1, & x \geq 0 \\ -x, & x < 0 \end{cases}$$

Evaluate the following definite integrals by using appropriate area formulas from plane geometry.

(a) $\displaystyle\int_4^5 f(x) \, dx$ (b) $\displaystyle\int_{-3}^{-2} f(x) \, dx$

(c) $\displaystyle\int_{-5}^2 f(x) \, dx$ (d) $\displaystyle\int_{-k}^k f(x) \, dx$ $(k > 0)$.

10. Evaluate $\displaystyle\int_0^2 (1 - x^2) \, dx$.

[Hint: Find the areas above and below the x-axis using antiderivatives.]

11. Use the hint in Exercise 10 to evaluate

$$\int_{-2}^{1} x^3 \, dx$$

12. Prove that the function

$$f(x) = \begin{cases} 1 & \text{if } x \text{ is rational} \\ 0 & \text{if } x \text{ is irrational} \end{cases}$$

is not integrable on any closed interval $[a, b]$.

13. Prove that the function

$$f(x) = \begin{cases} \dfrac{1}{x^2}, & x > 0 \\ 0, & x = 0 \end{cases}$$

is not integrable on the interval $[0, 1]$.

14. Use Theorem 5.7.3 to show that

$$f(x) = \begin{cases} \sin \dfrac{1}{x}, & x \neq 0 \\ 0, & x = 0 \end{cases}$$

is integrable on the interval $[-1, 1]$.

Exercises 15–17 are for readers who have read the optional material at the end of this section.

15. (Optional) Prove: If f is integrable on $[a, b]$, then the value of $\int_a^b f(x) \, dx$ is unique. [*Hint:* Show that if

$$\lim_{\max \Delta x_k \to 0} \sum_{k=1}^{n} f(x_k^*) \, \Delta x_k = L_1$$

and

$$\lim_{\max \Delta x_k \to 0} \sum_{k=1}^{n} f(x_k^*) \, \Delta x_k = L_2$$

then $L_1 = L_2$.]

16. (Optional) Prove: If f is integrable on $[a, b]$ and c is any constant, then cf is integrable on $[a, b]$ and

$$\int_a^b cf(x) \, dx = c \int_a^b f(x) \, dx$$

17. (Optional) Prove: If f and g are integrable on $[a, b]$, then $f + g$ is integrable on $[a, b]$ and

$$\int_a^b [f(x) + g(x)] \, dx = \int_a^b f(x) \, dx + \int_a^b g(x) \, dx$$

5.8 THE FUNDAMENTAL THEOREM OF CALCULUS

In Section 5.1 we calculated areas under curves by antidifferentiation. In this section we will prove a result, called the *Fundamental Theorem of Calculus,* which shows that the relationship between areas and antiderivatives is a special case of a more general relationship between definite integrals and antiderivatives.

5.8.1 THEOREM

The Fundamental Theorem of Calculus

If f is continuous on $[a, b]$ and if F is an antiderivative of f on $[a, b]$, then

$$\int_a^b f(x) \, dx = F(b) - F(a) \tag{1}$$

Proof. Let $x_1, x_2, \ldots, x_{n-1}$ be any points in $[a, b]$ such that

$$a < x_1 < x_2 < \cdots < x_{n-1} < b$$

These points divide $[a, b]$ into n subintervals

$$[a, x_1], [x_1, x_2], \ldots, [x_{n-1}, b] \tag{2}$$

whose lengths, as usual, we denote by

$$\Delta x_1, \Delta x_2, \ldots, \Delta x_n \tag{3}$$

Now we write $F(b) - F(a)$ as a telescoping sum:

$$F(b) - F(a) = [F(x_1) - F(a)] + [F(x_2) - F(x_1)] + [F(x_3) - F(x_2)]$$
$$+ \cdots + [F(b) - F(x_{n-1})] \tag{4}$$

By hypothesis,

$$F'(x) = f(x) \tag{5}$$

for all x in $[a, b]$, so F satisfies the hypotheses of the mean-value theorem (Theorem 4.9.2) on each subinterval in (2). Hence (4) can be rewritten

$$F(b) - F(a) = F'(x_1^*)(x_1 - a) + F'(x_2^*)(x_2 - x_1) + F'(x_3^*)(x_3 - x_2)$$
$$+ \cdots + F'(x_n^*)(b - x_{n-1}) \tag{6}$$

where

$$x_1^*, x_2^*, \ldots, x_n^*$$

are points in successive subintervals.

Using (3) and (5) we can rewrite (6) as

$$F(b) - F(a) = f(x_1^*)\,\Delta x_1 + f(x_2^*)\,\Delta x_2 + f(x_3^*)\,\Delta x_3 + \cdots + f(x_n^*)\,\Delta x_n$$

or in sigma notation,

$$F(b) - F(a) = \sum_{k=1}^{n} f(x_k^*)\,\Delta x_k \tag{7}$$

Let us now increase n in such a way that max $\Delta x_k \to 0$. Since f is assumed to be continuous, the right side of (7) will approach $\int_a^b f(x)\,dx$, by Theorem 5.7.3a. However, the left side of (7) is a constant that is independent of n; thus,

$$F(b) - F(a) = \lim_{\max \Delta x_k \to 0} \sum_{k=1}^{n} f(x_k^*)\,\Delta x_k = \int_a^b f(x)\,dx$$

The difference $F(b) - F(a)$ is commonly denoted by $F(x)]_a^b$ so that (1) can be written

$$\int_a^b f(x)\, dx = F(x) \Big]_a^b \tag{8}$$

REMARK. The symbols $[F(x)]_a^b$, $[F(x)]_{x=a}^{x=b}$, and $F(x)]_{x=a}^b$ are also used to denote $F(b) - F(a)$.

▶ Example 1

Evaluate

$$\int_1^2 x\, dx$$

Solution. The function

$$F(x) = \frac{1}{2}x^2$$

is an antiderivative of $f(x) = x$; thus from (8)

$$\int_1^2 x\, dx = \frac{1}{2}x^2 \Big]_1^2 = \frac{1}{2}(2)^2 - \frac{1}{2}(1)^2 = 2 - \frac{1}{2} = \frac{3}{2} \qquad ◀$$

When applying the fundamental theorem of calculus, it does not matter which antiderivative of f is used, for if F is any antiderivative of f on $[a, b]$, then all others have the form

$$F(x) + C$$

(Theorem 5.2.2). Thus

$$\begin{aligned}
[F(x) + C]_a^b &= [F(b) + C] - [F(a) + C] \\
&= F(b) - F(a) \\
&= F(x)]_a^b \\
&= \int_a^b f(x)\, dx
\end{aligned} \tag{9}$$

which shows that all antiderivatives of f on $[a, b]$ yield the same value for $\int_a^b f(x)\, dx$.

Since

$$\int f(x)\, dx = F(x) + C$$

it follows from (9) that

$$\int_a^b f(x)\, dx = \left[\int f(x)\, dx \right]_a^b \tag{10}$$

which relates the definite and indefinite integrals of f.

▶ Example 2

Use the fundamental theorem of calculus to find the area under the curve $y = \cos x$ over the interval $[0, \pi/2]$.

Solution. The area is

$$A = \int_0^{\pi/2} \cos x\, dx$$

$$= \left[\int \cos x\, dx \right]_0^{\pi/2} \qquad \text{[from (10)]}$$

$$= \sin x \, \Big]_0^{\pi/2}$$

$$= \sin \frac{\pi}{2} - \sin 0$$

$$= 1 \qquad\qquad\qquad\qquad\qquad\qquad ◀$$

REMARK. In the third equality of this example we took the constant of integration for the indefinite integral to be $C = 0$. This is justified because we can select any antiderivative of f on $[a, b]$, in particular the one for which $C = 0$.

▶ Example 3

Evaluate

$$\int_0^2 x^2 \sqrt{x^3 + 1}\, dx$$

Solution. We first evaluate the indefinite integral

$$\int x^2 \sqrt{x^3 + 1}\, dx$$

by a *u*-substitution. Let

$$u = x^3 + 1 \qquad \text{so that} \qquad \frac{du}{dx} = 3x^2 \qquad \text{or} \qquad \frac{1}{3}\,du = x^2\,dx$$

Making this substitution gives

$$\int x^2 \sqrt{x^3 + 1}\,dx = \int \frac{1}{3}\sqrt{u}\,du$$

$$= \frac{2}{9}u^{3/2} + C$$

$$= \frac{2}{9}(x^3 + 1)^{3/2} + C$$

Therefore,

$$\int_0^2 x^2 \sqrt{x^3 + 1}\,dx = \frac{2}{9}(x^3 + 1)^{3/2}\Big]_0^2$$

$$= \frac{2}{9}(27) - \frac{2}{9}$$

$$= \frac{52}{9} \qquad\blacktriangleleft$$

The following properties of definite integrals follow from (10) and the corresponding properties for indefinite integrals. We will omit the proofs.

5.8.2   THEOREM   *If f and g are integrable on [a, b] and if c is a constant, then cf, f + g, and f − g are integrable on [a, b] and*

(a) $\displaystyle\int_a^b cf(x)\,dx = c\int_a^b f(x)\,dx$

(b) $\displaystyle\int_a^b [f(x) + g(x)]\,dx = \int_a^b f(x)\,dx + \int_a^b g(x)\,dx$

(c) $\displaystyle\int_a^b [f(x) - g(x)]\,dx = \int_a^b f(x)\,dx - \int_a^b g(x)\,dx$

Part (*b*) of Theorem 5.8.2 can be extended to include more than two functions. More precisely,

$$\int_a^b [f_1(x) + f_2(x) + \cdots + f_n(x)]\,dx \tag{11}$$

$$= \int_a^b f_1(x)\,dx + \int_a^b f_2(x)\,dx + \cdots + \int_a^b f_n(x)\,dx$$

REMARK. The bracket notation $F(x)]_a^b$ has some properties in common with the definite integral. We leave it as an exercise to show:

$$[F(x) + G(x)]_a^b = F(x)]_a^b + G(x)]_a^b$$
$$[F(x) - G(x)]_a^b = F(x)]_a^b - G(x)]_a^b$$
$$[cF(x)]_a^b = c[F(x)]_a^b$$

▶ Example 4

Evaluate

$$\int_0^3 (x^3 - 4x + 1)\, dx$$

Solution.

$$\int_0^3 (x^3 - 4x + 1)\, dx = \int_0^3 x^3\, dx - 4\int_0^3 x\, dx + \int_0^3 dx$$
$$= \left[\frac{x^4}{4} - 4 \cdot \frac{x^2}{2} + x \right]_0^3$$
$$= \left(\frac{81}{4} - 18 + 3 \right) - (0)$$
$$= \frac{21}{4} \qquad ◀$$

The following theorem is useful when it is necessary to compare the sizes of definite integrals.

5.8.3 THEOREM   (a)   *If f is continuous and f(x) ≥ 0 for all x in [a, b], then*

$$\int_a^b f(x)\, dx \geq 0$$

(b)   *If f and g are continuous and f(x) ≥ g(x) for all x in [a, b], then*

$$\int_a^b f(x)\, dx \geq \int_a^b g(x)\, dx$$

Proof of (a). Since f is continuous and nonnegative on $[a, b]$, the integral $\int_a^b f(x)\, dx$ represents the area under $y = f(x)$ over $[a, b]$. Thus the integral is nonnegative by the remark following Definition 5.7.1.

Proof of (b). It follows from the hypotheses that $f - g$ is continuous and $f(x) - g(x) \geq 0$ for all x in $[a, b]$. Thus by part (a),

$$\int_a^b [f(x) - g(x)]\, dx = \int_a^b f(x)\, dx - \int_a^b g(x)\, dx \geq 0$$

from which it follows that

$$\int_a^b f(x)\, dx \geq \int_a^b g(x)\, dx \quad \blacksquare$$

In the case where $f(x) \geq g(x) \geq 0$ for all x in $[a, b]$, part (b) of this theorem states that the area under the "higher" curve $y = f(x)$ is at least as large as the area under the "lower" curve $y = g(x)$.

Sometimes it is convenient to use a letter other than x for the variable of integration in a definite integral. For example:

$$\int_a^b f(t)\, dt, \qquad \int_a^b f(u)\, du, \qquad \int_a^b f(y)\, dy$$

It is an important fact that:

> *The value of a definite integral is unaffected if we change the letter used for the variable of integration, but do not change the limits of integration.*

As an illustration, let us compare the steps in the evaluation of $\int_1^3 x^2\, dx$, $\int_1^3 t^2\, dt$ and $\int_1^3 u^2\, du$. These integrals have the same limits of integration, but different variables of integration.

$$\int_1^3 x^2\, dx = \frac{x^3}{3}\bigg]_{x=1}^{x=3} = \frac{27}{3} - \frac{1}{3} = \frac{26}{3}$$

$$\int_1^3 t^2\, dt = \frac{t^3}{3}\bigg]_{t=1}^{t=3} = \frac{27}{3} - \frac{1}{3} = \frac{26}{3}$$

$$\int_1^3 u^2\, du = \frac{u^3}{3}\bigg]_{u=1}^{u=3} = \frac{27}{3} - \frac{1}{3} = \frac{26}{3}$$

Because the letter used for the variable of integration has no effect on the final value of the definite integral, it is sometimes called a ***dummy variable.***

► Exercise Set 5.8

In Exercises 1–20, evaluate the definite integrals using the fundamental theorem of calculus.

1. $\displaystyle\int_{2}^{3} x^3 \, dx.$

2. $\displaystyle\int_{1}^{2} \frac{1}{x^6} \, dx.$

3. $\displaystyle\int_{-1}^{2} x(1 + x^3) \, dx.$

4. $\displaystyle\int_{-3}^{0} (x^2 - 4x + 7) \, dx.$

5. $\displaystyle\int_{4}^{9} 2y \sqrt{y} \, dy.$

6. $\displaystyle\int_{1}^{8} (5x^{2/3} - 4x^{-2}) \, dx.$

7. $\displaystyle\int_{-\pi/2}^{\pi/2} \sin\theta \, d\theta.$

8. $\displaystyle\int_{0}^{\pi/4} \sec^2\theta \, d\theta.$

9. $\displaystyle\int_{-\pi/4}^{\pi/4} \frac{\sin 2x}{\cos x} \, dx.$

10. $\displaystyle\int_{0}^{1} (x - \sec x \tan x) \, dx.$

11. $\displaystyle\int_{\pi/6}^{\pi/2} \left(x + \frac{2}{\sin^2 x}\right) dx.$

12. $\displaystyle\int_{0}^{1} \sqrt[3]{a + bx} \, dx \ (b \neq 0).$

13. $\displaystyle\int_{0}^{1} \frac{du}{\sqrt{3u + 1}}.$

14. $\displaystyle\int_{-1}^{1} \pi \cos(x + 2) \, dx.$

15. $\displaystyle\int_{-3\pi/4}^{-\pi/4} \sin x \cos x \, dx.$

16. $\displaystyle\int_{0}^{\pi/4} \sqrt{\tan x} \sec^2 x \, dx.$

17. $\displaystyle\int_{-1}^{1} \frac{x^2 \, dx}{\sqrt{x^3 + 9}}.$

18. $\displaystyle\int_{-1}^{0} 6t^2(t^3 + 1)^{19} \, dt.$

19. $\displaystyle\int_{a}^{4a} (a^{1/2} - x^{1/2}) \, dx \ (a \text{ a positive constant}).$

20. $\displaystyle\int_{0}^{2\pi/t} t^2 \sin tx \, dx \ (t \text{ a positive constant}).$

21. Find the area under the curve $y = x^2 + 1$ over the interval $[0, 3]$. Make a sketch of the region.

22. Find the area above the x-axis, but below the curve $y = (1 - x)(x - 2)$. Make a sketch of the region.

23. Find the area under the curve $y = 3 \sin 2x$ over the interval $[0, \pi/8]$. Make a sketch of the region.

24. Find the area below the interval $[-2, -1]$, but above the curve $y = x^3$. Make a sketch of the region.

25. Find the total area between the curve $y = x^2 - 3x - 10$ and the interval $[-3, 8]$. Make a sketch of the region. [*Hint:* Find the portion of area above the interval and the portion of area below the interval separately.]

26. Express Equation (11) in sigma notation.

27. Prove: If f is continuous and $f(x) \leq 0$ for all x in $[a, b]$, then $\int_{a}^{b} f(x) \, dx \leq 0.$

28. Prove:
   (a) $[F(x) + G(x)]_{a}^{b} = F(x)]_{a}^{b} + G(x)]_{a}^{b}$
   (b) $[F(x) - G(x)]_{a}^{b} = F(x)]_{a}^{b} - G(x)]_{a}^{b}$
   (c) $[cF(x)]_{a}^{b} = c[F(x)]_{a}^{b}.$

29. Assuming the functions involved are all integrable on $[a, b]$, which of the following are always valid?

   (a) $\displaystyle\int_{a}^{b} f(x)g(x) \, dx = \int_{a}^{b} f(x) \, dx \int_{a}^{b} g(x) \, dx$

   (b) $\displaystyle\int_{a}^{b} [c_1 f(x) + c_2 g(x)] \, dx$
   $$= c_1 \int_{a}^{b} f(x) \, dx + c_2 \int_{a}^{b} g(x) \, dx$$
   $(c_1, c_2 \text{ constant})$

   (c) $\displaystyle\int_{a}^{b} \left(\sum_{k=1}^{n} c_k f_k(x)\right) dx = \sum_{k=1}^{n} \left[c_k \int_{a}^{b} f_k(x) \, dx\right]$
   $(c_1, c_2, \ldots, c_k \text{ constant})$

   (d) $\displaystyle\int_{a}^{b} [f(x)]^n \, dx = \left[\int_{a}^{b} f(x) \, dx\right]^n$

   (e) $\displaystyle\int_{a}^{b} \sqrt{f(x)} \, dx = \sqrt{\int_{a}^{b} f(x) \, dx}.$

5.9 PROPERTIES OF THE DEFINITE INTEGRAL; DISTANCE TRAVELED IN RECTILINEAR MOTION

In this section we discuss additional properties of the definite integral and apply some of our results to rectilinear motion.

Whenever we speak of the "closed interval $[a, b]$" it is assumed that $a < b$. Thus our definition of the definite integral

$$\int_a^b f(x)\, dx$$

(Definition 5.7.2) does not allow for the possibilities $a = b$ or $b < a$. The following definition extends the notion of the definite integral to include these cases.

5.9.1 DEFINITION (a) If a is in the domain of f, we define

$$\int_a^a f(x)\, dx = 0$$

(b) If $b < a$ and if f is integrable on $[b, a]$, then we define

$$\int_a^b f(x)\, dx = -\int_b^a f(x)\, dx$$

REMARK. Geometrically, part (a) states that there is zero area between a curve $y = f(x)$ and a single point a on the x-axis.

▶ Example 1

(a) $\displaystyle\int_1^1 x^2\, dx = 0$

(b) $\displaystyle\int_4^0 x\, dx = -\int_0^4 x\, dx = -\frac{x^2}{2}\Big]_0^4 = -8$ ◀

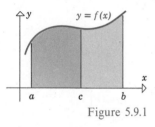

Figure 5.9.1

If f is continuous and nonnegative on $[a, b]$ and if c is a point between a and b, then it is evident that the area under $y = f(x)$ over the interval $[a, b]$ can be split into two parts, the area under curve from a to c plus the area under the curve from c to b (Figure 5.9.1), that is,

$$\int_a^b f(x)\, dx = \int_a^c f(x)\, dx + \int_c^b f(x)\, dx$$

This is a special case of the following theorem about definite integrals.

5.9.2 THEOREM *If f is integrable on a closed interval containing the three points a, b, and c, then*

$$\int_a^b f(x)\, dx = \int_a^c f(x)\, dx + \int_c^b f(x)\, dx \tag{1}$$

no matter how the points are ordered.

We omit the proof.

Theorem 5.9.2 is helpful when the formula for the integrand changes between the limits of integration.

▶ Example 2

Evaluate

$$\int_0^6 f(x)\, dx$$

if

$$f(x) = \begin{cases} x^2 & , & x \le 2 \\ 3x - 2, & x \ge 2 \end{cases}$$

Solution.

$$\int_0^6 f(x)\, dx = \int_0^2 f(x)\, dx + \int_2^6 f(x)\, dx$$

$$= \int_0^2 x^2\, dx + \int_2^6 (3x - 2)\, dx$$

$$= \frac{x^3}{3}\bigg]_0^2 + \left[\frac{3x^2}{2} - 2x\right]_2^6$$

$$= \left(\frac{8}{3} - 0\right) + (42 - 2)$$

$$= \frac{128}{3} \qquad \blacktriangleleft$$

▶ Example 3

Evaluate

$$\int_{-1}^2 |x|\, dx$$

Solution. Since $|x| = x$ when $x \geq 0$ and $|x| = -x$ when $x < 0$,

$$
\begin{aligned}
\int_{-1}^{2} |x|\, dx &= \int_{-1}^{0} |x|\, dx + \int_{0}^{2} |x|\, dx \\
&= \int_{-1}^{0} (-x)\, dx + \int_{0}^{2} x\, dx \\
&= -\frac{x^2}{2} \Big]_{-1}^{0} + \frac{x^2}{2} \Big]_{0}^{2} \\
&= \frac{1}{2} + 2 \\
&= \frac{5}{2}
\end{aligned}
$$

◀

In Section 5.4 we showed that for a particle with rectilinear motion the position function $s(t)$ and the velocity function $v(t)$ are related by

$$ s(t) = \int v(t)\, dt $$

It follows from this and the fundamental theorem of calculus that

$$ \int_{t_1}^{t_2} v(t)\, dt = s(t) \Big]_{t_1}^{t_2} = s(t_2) - s(t_1) \tag{2} $$

Since $s(t_1)$ is the position of the particle at time t_1 and $s(t_2)$ is the position at time t_2, the difference, $s(t_2) - s(t_1)$, is the change in position or **displacement** of the particle during the time interval $[t_1, t_2]$. A positive displacement means that the particle is farther to the right at time t_2 than at time t_1, and a negative displacement means that the particle is farther to the left. In the case where $v(t) \geq 0$ throughout the time interval $[t_1, t_2]$, the particle moves in the positive direction only; thus the displacement $s(t_2) - s(t_1)$ is the same as the distance traveled by the particle (Figure 5.9.2). In the case where $v(t) \leq 0$

Figure 5.9.2   Distance traveled $s(t_1)$ $s(t_2)$ Figure 5.9.3   Distance traveled $s(t_2)$ $s(t_1)$

throughout the time interval $[t_1, t_2]$, the particle moves in the negative direction only; thus the displacement $s(t_2) - s(t_1)$ is the negative of the distance traveled by the particle (Figure 5.9.3). In the case where $v(t)$ assumes both positive and negative values during the time interval $[t_1, t_2]$, the particle moves back and forth and the displacement is the distance traveled in the positive direction minus the distance traveled in the negative direction. If we want to find the total distance traveled in this case (distance traveled in the positive direction *plus* the distance traveled in the negative direction), we must integrate the absolute value of the velocity function, that is,

$$\begin{bmatrix} \text{total distance} \\ \text{traveled during} \\ \text{time interval} \\ [t_1, t_2] \end{bmatrix} = \int_{t_1}^{t_2} |v(t)| \, dt \tag{3}$$

▶ Example 4

A particle moves on a coordinate line so that its velocity at time t is $v(t) = t^2 - 2t$ meters per second. Find

(a) the displacement of the particle during the time interval $0 \leq t \leq 3$;

(b) the distance traveled by the particle during the time interval $0 \leq t \leq 3$.

Solution (a). From (2) the displacement is

$$\int_0^3 v(t) \, dt = \int_0^3 (t^2 - 2t) \, dt = \left[\frac{t^3}{3} - t^2 \right]_0^3 = 0$$

Thus the particle is at the same position at time $t = 3$ as at $t = 0$.

Solution (b). The velocity can be written as $v(t) = t^2 - 2t = t(t - 2)$ from which it is evident that $v(t) \leq 0$ for $0 \leq t \leq 2$ and $v(t) \geq 0$ for $2 \leq t \leq 3$. Thus it follows from (3) that the distance traveled is

$$\int_0^3 |v(t)| \, dt = \int_0^2 -v(t) \, dt + \int_2^3 v(t) \, dt$$

$$= \int_0^2 -(t^2 - 2t) \, dt + \int_2^3 (t^2 - 2t) \, dt$$

$$= -\left[\frac{t^3}{3} - t^2 \right]_0^2 + \left[\frac{t^3}{3} - t^2 \right]_2^3$$

$$= \frac{4}{3} + \frac{4}{3}$$

$$= \frac{8}{3} \text{ meters} \qquad ◀$$

In Section 5.3 we studied the technique of u-substitution, whereby an indefinite integral of the form

$$\int f(g(x))g'(x)\, dx$$

can be transformed into an equivalent but simpler indefinite integral,

$$\int f(u)\, du$$

by making the substitution

$$u = g(x) \qquad \text{and} \qquad du = g'(x)\, dx$$

The technique of u-substitution can also be used to evaluate a definite integral of the form

$$\int_a^b f(g(x))g'(x)\, dx \tag{4}$$

One way is to evaluate the indefinite integral

$$\int f(g(x))g'(x)\, dx$$

by u-substitution and then apply the fundamental theorem of calculus

$$\int_a^b f(g(x))g'(x)\, dx = \left[\int f(g(x))g'(x)\, dx \right]_a^b$$

This was the method used in Example 3 of the last section.
Another possibility is to make the substitution

$$u = g(x) \qquad \text{and} \qquad du = g'(x)\, dx$$

directly into the definite integral (4). However, to do this we must change the x-limits of integration to corresponding u-limits of integration. Since $u = g(x)$, it follows that

$$u = g(a) \qquad \text{when} \qquad x = a$$

and

$$u = g(b) \qquad \text{when} \qquad x = b$$

Thus, if (4) is expressed in terms of u, we obtain

$$\int_a^b f(g(x))g'(x)\,dx = \int_{g(a)}^{g(b)} f(u)\,du$$

The following examples illustrate this idea.

▶ Example 5

Evaluate

$$\int_0^2 2x(x^2 + 1)^3\,dx$$

. *Solution.* Let

$$u = x^2 + 1 \qquad \text{so that} \qquad du = 2x\,dx$$

Since $u = 1$ when $x = 0$, and $u = 5$ when $x = 2$, we obtain

$$\int_0^2 2x(x^2 + 1)^3\,dx = \int_1^5 u^3\,du = \frac{u^4}{4}\bigg]_1^5 = \frac{625}{4} - \frac{1}{4} = 156$$

As a check, let us compare this solution to the solution that results when we first evaluate the indefinite integral by u-substitution and then apply the fundamental theorem of calculus. We obtain

$$\int 2x(x^2 + 1)^3\,dx = \int u^3\,du = \frac{u^4}{4} + C = \frac{(x^2 + 1)^4}{4} + C$$

Thus

$$\int_0^2 2x(x^2 + 1)^3\,dx = \left[\int 2x(x^2 + 1)^3\,dx\right]_0^2 = \frac{(x^2 + 1)^4}{4}\bigg]_0^2$$

$$= \frac{625}{4} - \frac{1}{4} = 156$$

which agrees with our previous result. ◀

▶ Example 6

Evaluate

$$\int_0^{\pi/4} \cos(\pi - x)\,dx$$

Solution. Let

$$u = \pi - x \qquad \text{so that} \qquad du = -dx$$

Since $u = \pi$ when $x = 0$ and $u = 3\pi/4$ when $x = \pi/4$,

$$\int_0^{\pi/4} \cos(\pi - x)\,dx = \int_\pi^{3\pi/4} \cos u\,(-du)$$

$$= -\int_\pi^{3\pi/4} \cos u\,du$$

$$= \int_{3\pi/4}^{\pi} \cos u\,du \qquad \text{[Definition 5.9.1b]}$$

$$= \sin u \Big]_{3\pi/4}^{\pi}$$

$$= \sin \pi - \sin(3\pi/4)$$

$$= 0 - \tfrac{1}{2}\sqrt{2} = -\tfrac{1}{2}\sqrt{2} \qquad\blacktriangleleft$$

The following theorem justifies the method of *u*-substitution illustrated in these examples.

5.9.3 THEOREM *If g' is continuous on [a, b] and f is continuous and has an antiderivative on an interval containing the values of g(x) for $a \le x \le b$, then*

$$\int_a^b f(g(x))g'(x)\,dx = \int_{g(a)}^{g(b)} f(u)\,du$$

provided the integrals exist.

Proof. Let $u = g(x)$ and let F be an antiderivative of f on an interval containing the values of $g(x)$ for $a \le x \le b$. Then by the chain rule

$$\frac{d}{dx} F(g(x)) = \frac{d}{dx} F(u) = \frac{dF}{du}\frac{du}{dx} = f(u)\frac{du}{dx} = f(g(x))g'(x)$$

for each x in $[a, b]$. Thus $F(g(x))$ is an antiderivative of $f(g(x))g'(x)$ on $[a, b]$. Therefore, by the fundamental theorem of calculus (Theorem 5.8.1)

$$\int_a^b f(g(x))g'(x)\,dx = F(g(x)) \Big]_a^b$$

$$= F(g(b)) - F(g(a))$$

$$= \int_{g(a)}^{g(b)} f(u)\,du \quad\blacksquare$$

▶ Exercise Set 5.9

1. In each part express the integral in terms of the variable u, but do not evaluate.

(a) $\int_0^2 (x + 1)^7 \, dx; \ u = x + 1$

(b) $\int_{-1}^2 x\sqrt{8 - x^2} \, dx; \ u = 8 - x^2$

(c) $\int_{-1}^1 \sin(\pi\theta) \, d\theta; \ \theta = u/\pi$

(d) $\int_0^{\pi/4} \tan^2 x \sec^2 x \, dx; \ u = \tan x$

(e) $\int_0^1 x^3 \sqrt{x^2 + 3} \, dx; \ u = x^2 + 3$

(f) $\int_5^3 (x + 2)(x - 3)^{20} \, dx; \ u = x - 3.$

In Exercises 2–5, use Theorem 5.9.2 to evaluate the integrals.

2. $\int_1^5 |x - 2| \, dx.$

3. $\int_0^{3\pi/4} |\cos x| \, dx.$

4. $\int_{-2}^3 f(x) \, dx,$ where $f(x) = \begin{cases} -x, & x \geq 0 \\ x^2, & x < 0. \end{cases}$

5. $\int_0^9 g(x) \, dx,$ where $g(x) = \begin{cases} 1, & 0 \leq x < 1 \\ x^3, & 1 \leq x < 4 \\ \sqrt{x}, & 4 \leq x \leq 9. \end{cases}$

In Exercises 6–9, evaluate the integrals two ways; first by a u-substitution in the definite integral and then by a u-substitution in the corresponding indefinite integral.

6. $\int_1^2 (4 - 3x)^8 \, dx.$

7. $\int_0^8 x\sqrt{1 + x} \, dx.$

8. $\int_0^{\pi/6} 2\cos 3x \, dx.$

9. $\int_{-2}^{-1} \dfrac{x}{(x^2 + 2)^3} \, dx.$

In Exercises 10–17, evaluate the integrals by any method.

10. $\int_{-1}^4 \dfrac{x \, dx}{\sqrt{5 + x}}.$

11. $\int_0^1 \dfrac{y^2 \, dy}{\sqrt{4 - 3y}}.$

12. $\int_{-\pi/4}^{\pi} \sin\theta \cos\theta \, d\theta.$

13. $\int_0^{\sqrt{\pi}} 5x \cos(x^2) \, dx.$

14. $\int_{-2}^2 \sqrt{2 + |x|} \, dx.$

15. $\int_1^3 \dfrac{x + 2}{\sqrt{x^2 + 4x + 7}} \, dx.$

16. $\int_1^2 \dfrac{dx}{x^2 - 6x + 9}.$

17. $\int_{\pi^2}^{4\pi^2} \dfrac{1}{\sqrt{x}} \sin \sqrt{x} \, dx.$

In Exercises 18–23, $v(t)$ is the velocity (meters/sec) of a particle moving on a coordinate line. Find the displacement and the distance traveled by the particle during the given time interval.

18. $v(t) = 2t - 4; \ 0 \leq t \leq 4.$

19. $v(t) = t^2 + t - 2; \ 0 \leq t \leq 2.$

20. $v(t) = |t - 3|; \ 0 \leq t \leq 5.$

21. $v(t) = \cos t; \ 0 \leq t \leq \pi.$

22. $v(t) = 3\sin t; \ \pi/4 \leq t \leq \pi.$

23. $v(t) = t^3 - 3t^2 + 2t; \ 0 \leq t \leq 3.$

In Exercises 24–27, $a(t)$ is the acceleration (meters/sec^2) of a particle moving on a coordinate line, and v_0 is its velocity at time $t = 0$. Find the displacement and the distance traveled by the particle during the given time interval.

24. $a(t) = -2; \ v_0 = 3; \ 1 \leq t \leq 4.$

25. $a(t) = t - 2; \ v_0 = 0; \ 1 \leq t \leq 5.$

26. $a(t) = \sin t; \ v_0 = 1; \ \dfrac{\pi}{4} \leq t \leq \dfrac{\pi}{2}.$

27. $a(t) = \dfrac{1}{\sqrt{5t + 1}}; \ v_0 = 2; \ 0 \leq t \leq 3.$

28. Prove: If m and n are positive integers, then

$$\int_0^1 x^m(1-x)^n\,dx = \int_0^1 x^n(1-x)^m\,dx$$

[*Hint:* Do not evaluate; use a substitution.]

29. Prove: If n is a positive integer then

$$\int_0^{\pi/2} \sin^n x\,dx = \int_0^{\pi/2} \cos^n x\,dx.$$

[*Hint:* Do not evaluate; use a trigonometric identity and a substitution.]

30. (a) Prove: If $f(-x) = -f(x)$ for all x in $[-a, a]$ and if f is integrable on $[-a, a]$, then $\int_{-a}^a f(x)\,dx = 0$.

  (b) Give a geometric explanation of the result in part (a).

5.10 MEAN-VALUE THEOREM FOR INTEGRALS; AVERAGE VALUE

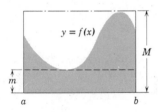

Figure 5.10.1

In this section we will obtain a fundamental property of definite integrals and define the "average value" of a function.

Let f be a continuous nonnegative function on $[a, b]$, and let m and M be the minimum and maximum values of $f(x)$ on this interval. Consider the rectangles of heights m and M over the interval $[a, b]$ (Figure 5.10.1). It is clear geometrically that the area

$$A = \int_a^b f(x)\,dx$$

is at least as large as the area of the rectangle of height m, and no larger than the area of the rectangle of height M. It seems reasonable, therefore, that somewhere between m and M there is an appropriate height $f(x^*)$ such that the rectangle of this height over $[a, b]$ has precisely area A; that is,

$$\int_a^b f(x)\,dx = f(x^*)(b-a)$$

Figure 5.10.2

(Figure 5.10.2). This conjecture is correct; it is a special case of the following result.

5.10.1 THEOREM

The Mean-Value Theorem for Integrals

If f is continuous on a closed interval $[a, b]$, then there is at least one number x^* in $[a, b]$ such that

$$\int_a^b f(x)\,dx = f(x^*)(b-a) \tag{1}$$

Proof.   By the extreme value theorem (4.2.4), f assumes a maximum value M and a minimum value m on $[a, b]$. Thus, for all x in $[a, b]$,

$$m \leq f(x) \leq M \tag{2}$$

and from Theorem 5.8.3(*b*)

$$\int_a^b m \, dx \leq \int_a^b f(x) \, dx \leq \int_a^b M \, dx$$

or

$$m(b - a) \leq \int_a^b f(x) \, dx \leq M(b - a)$$

or

$$m \leq \frac{1}{b - a} \int_a^b f(x) \, dx \leq M \tag{3}$$

Since (3) states that

$$\frac{1}{b - a} \int_a^b f(x) \, dx \tag{4}$$

is a number between m and M, and since $f(x)$ assumes the values m and M on $[a, b]$, it follows from the intermediate value theorem (3.7.8) that $f(x)$ must assume the value (4) at some point x^* in $[a, b]$, that is,

$$\frac{1}{b - a} \int_a^b f(x) \, dx = f(x^*)$$

or

$$\int_a^b f(x) \, dx = f(x^*)(b - a)$$

▶ Example 1

Since $f(x) = x^2$ is continuous on the interval $[1, 4]$, the mean value theorem for integrals guarantees that there is a number x^* in $[1, 4]$ such that

$$\int_1^4 x^2 \, dx = f(x^*)(4 - 1) = (x^*)^2(4 - 1) = 3(x^*)^2$$

But

$$\int_{1}^{4} x^2 \, dx = \frac{x^3}{3}\bigg]_{1}^{4} = 21$$

so that

$$3(x^*)^2 = 21 \qquad \text{or} \qquad (x^*)^2 = 7$$

Thus $x^* = \sqrt{7} \approx 2.65$ is the number in the interval $[1, 4]$ whose existence is guaranteed by the mean value theorem for integrals. ◀

The number $f(x^*)$ in Theorem 5.10.1 is closely related to the familiar notion of an *arithmetic average*. To see this, divide the interval $[a, b]$ into n subintervals of equal length

$$\Delta x = \frac{b - a}{n} \tag{5}$$

and choose arbitrary points $x_1^*, x_2^*, \ldots, x_n^*$ in successive subintervals. Then the arithmetic average of the numbers $f(x_1^*), f(x_2^*), \ldots, f(x_n^*)$ is

$$\text{ave} = \frac{1}{n}[f(x_1^*) + f(x_2^*) + \cdots + f(x_n^*)]$$

or from (5)

$$\text{ave} = \frac{1}{b - a}[f(x_1^*) \, \Delta x + f(x_2^*) \, \Delta x + \cdots + f(x_n^*) \, \Delta x]$$

$$= \frac{1}{b - a} \sum_{k=1}^{n} f(x_k^*) \, \Delta x$$

Taking the limit as $n \to +\infty$ yields

$$\lim_{n \to +\infty} \frac{1}{b - a} \sum_{k=1}^{n} f(x_k^*) \, \Delta x = \frac{1}{b - a} \int_{a}^{b} f(x) \, dx \tag{6}$$

Since this equation describes what happens when we average "more and more" values of $f(x)$, we are led to the following definition.

5.10.2 DEFINITION If f is integrable on $[a, b]$, then the *average value* (or *mean value*) of f on $[a, b]$ is defined to be

$$f_{\text{ave}} = \frac{1}{b - a} \int_{a}^{b} f(x) \, dx$$

▶ Example 2

Find the average value of $f(x) = x^2$ on the interval $[1, 4]$.

Solution.

$$f_{\text{ave}} = \frac{1}{b-a}\int_a^b f(x)\,dx = \frac{1}{4-1}\int_1^4 x^2\,dx = \frac{1}{3}(21) = 7 \qquad \blacktriangleleft$$

REMARK. In light of Definition 5.10.2, the quantity $f(x^*)$ in the mean-value theorem for integrals (5.10.1) is just the average value of f over $[a, b]$. Thus the mean-value theorem for integrals can be stated in the form

$$\int_a^b f(x)\,dx = (b-a)f_{\text{ave}}$$

▶ Exercise Set 5.10

1. (a) Find f_{ave} of $f(x) = x^2$ over $[0, 2]$.
   (b) Find a point x^* in $[0, 2]$ such that $f(x^*) = f_{\text{ave}}$.
   (c) Sketch the graph of $f(x) = x^2$ over $[0, 2]$ and construct a rectangle over the interval whose area is the same as the area under the graph of f over the interval.

In Exercises 2–5, find the average value of $f(x)$ over the interval and find all values of x^* described in the mean-value theorem for integrals.

2. $f(x) = x^3$; $[1, 2]$.

3. $f(x) = \sqrt{x}$; $[0, 4]$.

4. $f(x) = \sin x$; $[-\pi, \pi]$.

5. $f(x) = \alpha x + \beta$; $[x_0, x_1]$.

6. Let $s(t)$ be the position function of a particle moving on a coordinate line. In Section 5.4 (Exercise 16) we defined the average velocity v_{ave} over a time interval $[t_0, t_1]$ by

$$v_{\text{ave}} = \frac{\text{change in position}}{\text{time elapsed}} = \frac{s(t_1) - s(t_0)}{t_1 - t_0}$$

(a) Show that

$$v_{\text{ave}} = \frac{1}{t_1 - t_0}\int_{t_0}^{t_1} v(t)\,dt$$

so that v_{ave} can also be interpreted as the average value of the velocity function over the time interval $[t_0, t_1]$.

(b) Use the result in (a) to compute the average velocity over the time interval $[0, 5]$ for a particle moving on a coordinate line with velocity function $v(t) = 32t$.

7. For a particle moving on a coordinate line, show that the average acceleration a_{ave} over a time interval $[t_0, t_1]$ (as defined in Exercise 16 of Section 5.4) can be written as

$$a_{\text{ave}} = \frac{1}{t_1 - t_0}\int_{t_0}^{t_1} a(t)\,dt$$

which is the average of the acceleration function over the time interval $[t_0, t_1]$.

8. Consider a particle moving on a coordinate line. Use Exercises 6 and 7 to find:
   (a) v_{ave} for $1 \le t \le 4$ if $v(t) = 3t^3 + 2$
   (b) a_{ave} for $2 \le t \le 9$ if $a(t) = t^{1/2}$
   (c) v_{ave} for $t_0 \le t \le t_1$ if $v(t) = 32t + v_0$
                                (v_0 constant).

9. Water is run at a constant rate of $1 \text{ ft}^3/\text{min}$ to fill a cylindrical tank of radius 3 ft and height 5 ft. Assuming the tank is initially empty, find the average force on the bottom over the time period required to fill the tank (density of water $= 62.4 \text{ lb/ft}^3$).

10. Prove: If $f(x) = k$ is constant on $[a, b]$ then $f_{ave} = k$ on $[a, b]$.

11. (a) Prove: If f is integrable on $[a, b]$ then
$$\int_a^b [f(x) - f_{ave}]\, dx = 0$$
(b) Does there exist a constant $c \ne f_{ave}$ such that
$$\int_a^b [f(x) - c]\, dx = 0?$$

▶ REVIEW EXERCISES

In Exercises 1–10, evaluate the integrals and check your results by differentiation.

1. $\int \left[\dfrac{1}{x^3} + \dfrac{1}{\sqrt{x}} - 5 \sin x \right] dx.$

2. $\int \dfrac{2t^4 - t + 2}{t^3}\, dt.$

3. $\int \dfrac{(\sqrt{x} + 2)^8}{\sqrt{x}}\, dx.$

4. $\int x^3 \cos(2x^4 - 1)\, dx.$

5. $\int \dfrac{x \sin \sqrt{2x^2 - 5}}{\sqrt{2x^2 - 5}}\, dx.$

6. $\int \sqrt{\cos \theta}\, \sin(2\theta)\, d\theta.$

7. $\int \sqrt{x}(3 + \sqrt[3]{x^4})\, dx.$

8. $\int \dfrac{x^{1/3}\, dx}{x^{8/3} + 2x^{4/3} + 1}.$

9. $\int \sec^2(\sin 5t) \cos 5t\, dt.$

10. $\int \dfrac{\cot^2 x}{\sin^2 x}\, dx.$

11. Evaluate $\int y(y^2 + 2)^2\, dy$ two ways: (a) by multiplying out and integrating term by term; and (b) by using the substitution $u = y^2 + 2$. Show that your answers differ by a constant.

In Exercises 12–17, evaluate the definite integral by making the indicated substitution and changing the x-limits of integration to u-limits.

12. $\int_1^0 \sqrt[5]{1 - 2x}\, dx, \; u = 1 - 2x.$

13. $\int_0^{\pi/2} \sin^4 x \cos x\, dx, \; u = \sin x.$

14. $\int_0^{-3} \dfrac{x\, dx}{\sqrt{x^2 + 16}}, \; u = x^2 + 16.$

15. $\int_2^5 \dfrac{x - 2}{\sqrt{x - 1}}\, dx, \; u = x - 1.$

16. $\int_{\pi/6}^{\pi/4} \dfrac{\sin 2x\, dx}{\sqrt{1 - \frac{3}{2}\cos 2x}}, \; u = 1 - \frac{3}{2}\cos 2x$

17. $\int_0^1 \dfrac{1}{\sqrt{x}} \cos\left(\dfrac{\pi \sqrt{x}}{2} \right) dx, \; u = \dfrac{\pi \sqrt{x}}{2}.$

In Exercises 18 and 19, evaluate $\int_{-2}^2 f(x)\, dx.$

18. $f(x) = \begin{cases} x^3 & \text{for } x \ge 0 \\ -x & \text{for } x < 0 \end{cases}$

19. $f(x) = |2x - 1|.$

20. A car, initially at rest 10 ft from a pole, moves in a straight line away from the pole with a constant acceleration of a ft/sec^2. Express the distance $s(t)$ between the car and the pole in terms of a and t and then find a if the car travels 40 ft in the first 2 sec.

21. A projectile is fired upward from ground level. Find the initial velocity if it hits the ground 8 sec later. How high does it go?

22. A particle moving along a straight line is accelerating at a constant rate of 3 m/sec^2. Find the initial velocity if the particle moves 40 m in the first 4 sec.

23. Evaluate:

(a) $\displaystyle\sum_{i=3}^{6} 5$

(b) $\displaystyle\sum_{i=n}^{n+3} 2$

(c) $\displaystyle\sum_{i=n}^{n+3} n$

(d) $\displaystyle\sum_{k=1}^{3} \left(\frac{k-1}{k+3}\right)$

(e) $\displaystyle\sum_{k=2}^{4} \frac{6}{k^2}$

(f) $\displaystyle\sum_{n=4}^{4} (2n+1)$

(g) $\displaystyle\sum_{k=0}^{4} \sin(k\pi/4)$

(h) $\displaystyle\sum_{k=1}^{4} \sin^k(\pi/4)$.

24. Express in sigma notation and evaluate.
(a) $3 \cdot 1 + 4 \cdot 2 + 5 \cdot 3 + \cdots + 102 \cdot 100$
(b) $200 + 198 + \cdots + 4 + 2$.

25. Express in sigma notation, first starting with $k=1$, and then with $k=2$. Do not evaluate.

(a) $\dfrac{1}{4} - \dfrac{4}{9} + \dfrac{9}{16} - \cdots - \dfrac{64}{81} + \dfrac{81}{100}$

(b) $\dfrac{\pi^2}{1} - \dfrac{\pi^3}{2} + \dfrac{\pi^4}{3} - \cdots + \dfrac{\pi^{12}}{11}$.

In Exercises 26–29, use the partition of $[a, b]$ into n subintervals of equal length, and find a closed form for the sum of the areas of: (a) the inscribed rectangles and (b) the circumscribed rectangles. (c) Use your answer to either (a) or (b) to find the area under the curve $y = f(x)$ over the interval $[a, b]$. (Check your answer by integration.)

26. $f(x) = 6 - 2x$; $a = 1$, $b = 3$.

27. $f(x) = 16 - x^2$; $a = 0$, $b = 4$.

28. $f(x) = x^2 + 2$; $a = 1$, $b = 4$.

29. $f(x) = 6$; $a = -1$, $b = 1$.

30. (a) Prove: If $f(-x) = f(x)$ for all x in $[-a, a]$ and f is integrable on $[-a, a]$, then $\int_{-a}^{a} f(x)\, dx = 2\int_{0}^{a} f(x)\, dx$.
(b) Give a geometric explanation of the result in part (a).

31. In each part evaluate the definite integral geometrically, that is, by interpreting it as a difference of areas and using geometric reasoning to find the areas.

(a) $\displaystyle\int_{0}^{2} (2 - 4x)\, dx$

(b) $\displaystyle\int_{0}^{-3} \sqrt{9 - t^2}\, dt$

(c) $\displaystyle\int_{-4}^{2} \frac{|x|}{x}\, dx$

(d) $\displaystyle\int_{-\pi/3}^{\pi/3} \sin x\, dx$

(e) $\displaystyle\int_{0}^{\pi} \cos x\, dx$

(f) $\displaystyle\int_{-10}^{-5} 6\, dx$

(g) $\displaystyle\int_{-5}^{4} f(x)\, dx$ if $f(x) = \begin{cases} -4, & x \le -3 \\ 2, & -3 < x < 0 \\ 1, & x \ge 0. \end{cases}$

32. Given that
$$\int_{-1}^{0} f(x)\, dx = 3, \quad \int_{-1}^{2} f(x)\, dx = -1, \text{ and}$$
$$\int_{-1}^{2} g(x)\, dx = 2,$$
evaluate the following:

(a) $\displaystyle\int_{0}^{-1} 8f(x)\, dx$

(b) $\displaystyle\int_{-1}^{2} f(y)\, dy$

(c) $\displaystyle\int_{2}^{2} g(x)\, dx$

(d) $\displaystyle\int_{0}^{2} f(x)\, dx$

(e) $\displaystyle\int_{-1}^{2} [f(x) - 2g(x)]\, dx$.

33. Given that
$$\int_{1}^{5} P(x)\, dx = -1, \quad \int_{3}^{5} P(x)\, dx = 3, \text{ and}$$
$$\int_{3}^{5} Q(x)\, dx = 4,$$
evaluate the following:

(a) $\displaystyle\int_{3}^{5} [2P(x) + Q(x)]\, dx$

(b) $\displaystyle\int_{5}^{1} P(t)\, dt$

(c) $\displaystyle\int_{-3}^{-5} Q(-x)\, dx$

(d) $\displaystyle\int_{3}^{1} P(x)\, dx$.

34. Suppose that f is continuous and $x^2 \le f(x) \le 6$ for all x in $[-1, 2]$. Find values of A and B such that $A \le \int_{-1}^{2} f(x)\, dx \le B$.

In Exercises 35–38, find: (a) the average value of $f(x)$ over the indicated interval and (b) all values of x^* described in the mean value theorem for integrals.

35. $f(x) = 3x^2$; $[-2, -1]$.

36. $f(x) = \dfrac{x}{\sqrt{x^2 + 9}}$; $[0, 4]$.

37. $f(x) = 2 + |x|$; $[-3, 1]$.

38. $f(x) = \sin^2 x$; $[0, \pi]$
[*Hint:* $\sin^2 x = \frac{1}{2}(1 - \cos 2x)$.]

6 applications of the definite integral

6.1 AREA BETWEEN TWO CURVES

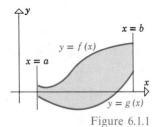

Figure 6.1.1

Let f and g be continuous functions such that

$$f(x) \geq g(x) \qquad \text{for} \qquad a \leq x \leq b \tag{1}$$

Geometrically, (1) states that the curve $y = f(x)$ does not dip below the curve $y = g(x)$ over the closed interval $[a, b]$. We will be interested in finding the area A of the region bounded above by $y = f(x)$, below by $y = g(x)$, and on the sides by the vertical lines $x = a$ and $x = b$ (Figure 6.1.1).

If g is nonnegative on $[a, b]$, then we recognize this area to be

$$A = [\text{area under } f] - [\text{area under } g]$$

$$= \int_a^b f(x)\, dx - \int_a^b g(x)\, dx = \int_a^b [f(x) - g(x)]\, dx$$

(Figure 6.1.2*a*). If g has negative values, let us translate the graphs of f and g upward until both are above the x-axis. To do this, let m be the minimum value of $g(x)$ on $[a, b]$. (There is such a number by virtue of the Extreme Value Theorem 4.2.4.) Since $g(x) \geq m$ it follows that $g(x) - m \geq 0$, so the functions $f - m$ and $g - m$ are nonnegative on $[a, b]$ (Figures 6.1.2*b* and 6.1.2*c*). If we accept that the area of a region is unchanged by translation,

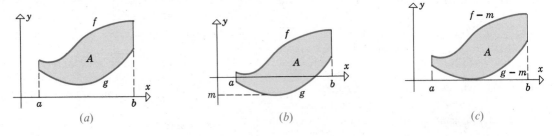

(*a*)   (*b*)   (*c*)

Figure 6.1.2

then the area A between f and g is the same as the area between $f - m$ and $g - m$. Thus

$$A = [\text{area under } f - m] - [\text{area under } g - m]$$

$$= \int_a^b [f(x) - m]\, dx - \int_a^b [g(x) - m]\, dx = \int_a^b [f(x) - g(x)]\, dx$$

which is the same formula we obtained above. These results suggest the following definition.

6.1.1 DEFINITION If f and g are continuous functions and if $f(x) \geq g(x)$ for all x in $[a, b]$, then the area of the region bounded above by $y = f(x)$, below by $y = g(x)$, on the left by the line $x = a$, and on the right by the line $x = b$ is

$$A = \int_a^b [f(x) - g(x)]\, dx \qquad (2)$$

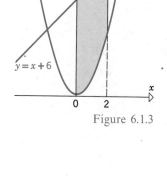

Figure 6.1.3

▶ **Example 1**

Find the area of the region bounded above by $y = x + 6$, bounded below by $y = x^2$, and bounded on the sides by the lines $x = 0$ and $x = 2$ (Figure 6.1.3).

Solution. From Formula (2) with $f(x) = x + 6$ and $g(x) = x^2$, the area is

$$A = \int_0^2 [(x + 6) - x^2]\, dx$$

$$= \left[\frac{x^2}{2} + 6x - \frac{x^3}{3} \right]_0^2 = \frac{34}{3} - 0 = \frac{34}{3} \qquad \blacktriangleleft$$

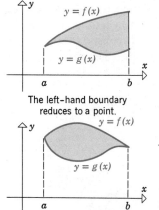

The left–hand boundary
reduces to a point.

Both side boundaries
reduce to points.

Figure 6.1.4

Sometimes the upper curve $y = f(x)$ intersects the lower curve $y = g(x)$ at either the left-hand boundary $x = a$, the right-hand boundary $x = b$, or both. When this happens the side of the region reduces to a point, rather than a vertical line segment (Figure 6.1.4).

▶ **Example 2**

Find the area of the region enclosed between the curves $y = x^2$ and $y = x + 6$.

Solution. A sketch of the region (Figure 6.1.5) shows that the lower boundary is $y = x^2$ and the upper boundary is $y = x + 6$. At the endpoints of the region, the upper and lower boundaries have the same y-coordinates; thus to find the endpoints we equate

$$y = x^2 \qquad \text{and} \qquad y = x + 6 \qquad (3)$$

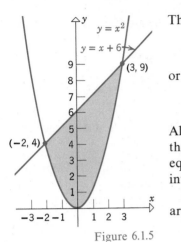

Figure 6.1.5

This yields

$$x^2 = x + 6$$

or

$$x^2 - x - 6 = 0, \qquad (x + 2)(x - 3) = 0, \qquad x = -2, \ x = 3$$

Although the y-coordinates of the endpoints are not essential to our solution, they may be obtained from (3) by substituting $x = -2$ and $x = 3$ in either equation. This yields $y = 4$ and $y = 9$, so the upper and lower boundaries intersect at $(-2, 4)$ and $(3, 9)$.

From (2) with $f(x) = x + 6$, $g(x) = x^2$, $a = -2$, and $b = 3$, we obtain the area

$$A = \int_{-2}^{3} [(x + 6) - x^2] \, dx$$

$$= \left[\frac{x^2}{2} + 6x - \frac{x^3}{3} \right]_{-2}^{3} = \frac{27}{2} - \left(-\frac{22}{3}\right) = \frac{125}{6} \qquad \blacktriangleleft$$

► **Example 3**

Find the area of the region enclosed by the curves $y = \sqrt{x}$, $y = -x + 6$ and $y = 1$.

Figure 6.1.6

Solution. The region is shown in Figure 6.1.6. By solving simultaneously, we find that $y = \sqrt{x}$ and $y = -x + 6$ intersect at $(4, 2)$; $y = 1$ and $y = \sqrt{x}$ intersect at $(1, 1)$; and $y = 1$ and $y = -x + 6$ intersect at $(5, 1)$.

As an aid to setting up the integral for the area between two curves, it is sometimes helpful to draw some line segments connecting the lower and upper boundaries. For example, the line segments in Figure 6.1.7 make it clear that upper boundary for the region consists of two parts: $y = \sqrt{x}$ when $1 \le x \le 4$ and $y = -x + 6$ when $4 \le x \le 5$. Thus it is necessary to divide the region into parts, R_1 and R_2, as shown in Figure 6.1.7, and find the area of each part separately.

From (2) with $f(x) = \sqrt{x}$, $g(x) = 1$, $a = 1$, and $b = 4$, the area of region R_1 is

Figure 6.1.7

$$A_1 = \int_{1}^{4} (\sqrt{x} - 1) \, dx$$

$$= \left[\frac{2}{3} x^{3/2} - x \right]_{1}^{4}$$

$$= \left(\frac{16}{3} - 4 \right) - \left(\frac{2}{3} - 1 \right) = \frac{5}{3}$$

From (2) with $f(x) = -x + 6$, $g(x) = 1$, $a = 4$, and $b = 5$, the area of region R_2 is

$$A_2 = \int_4^5 [(-x + 6) - 1]\,dx$$

$$= \int_4^5 (5 - x)\,dx$$

$$= \left[5x - \frac{x^2}{2}\right]_4^5$$

$$= \left(25 - \frac{25}{2}\right) - (20 - 8)$$

$$= \frac{1}{2}$$

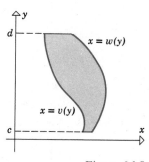

$x = w(y)$

$x = v(y)$

Figure 6.1.8

Thus the area of the entire region is

$$A = A_1 + A_2 = \frac{5}{3} + \frac{1}{2} = \frac{13}{6} \qquad \blacktriangleleft$$

Sometimes it is necessary to find the area of a region bounded above and below by horizontal lines and bounded on the left and right by the graphs of two functions of y (Figure 6.1.8).

The following definition is similar to 6.1.1, but with the roles of x and y reversed.

6.1.2 DEFINITION   If w and v are continuous functions and if $w(y) \geq v(y)$ for all y in $[c, d]$, then the area of the region bounded on the left and right by the curves $x = v(y)$, $x = w(y)$ and above and below by the lines $y = d$, $y = c$, is

$$A = \int_c^d [w(y) - v(y)]\,dy \qquad (4)$$

▶ Example 4

Find the area of the region in Example 3 using (4).

Solution. Figure 6.1.9 shows the region and a horizontal line extending between the left and right boundaries. (Note that we have rewritten the boundary curves $y = \sqrt{x}$ and $y = -x + 6$ as $x = y^2$ and $x = 6 - y$ so that x is expressed in terms of y.) The left boundary of the region is $x = v(y) = y^2$, the right boundary is $x = w(y) = 6 - y$, and the upper and lower boundaries are $y = d = 2$ and $y = c = 1$. Thus, from formula (4) the area of the region is

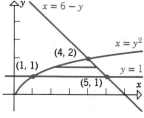

$x = 6 - y$

$x = y^2$

(4, 2)

(1, 1)

$y = 1$

(5, 1)

Figure 6.1.9

$$A = \int_1^2 [(6 - y) - y^2]\,dy$$

$$= \left[6y - \frac{y^2}{2} - \frac{y^3}{3} \right]_1^2$$

$$= \left(12 - 2 - \frac{8}{3} \right) - \left(6 - \frac{1}{2} - \frac{1}{3} \right) = \frac{13}{6}$$

which agrees with the answer obtained in Example 3. However, this solution is simpler than the solution in Example 3 since it is not necessary to break the region into two parts. ◄

▶ Exercise Set 6.1

1. Find the area of the region enclosed by the curves $y = x^2$ and $y = 4x$ by integrating
   (a) along the x-axis
   (b) along the y-axis.

2. Find the area of the region enclosed by the curves $y^2 = 4x$ and $y = 2x - 4$ by integrating
   (a) along the x-axis
   (b) along the y-axis.

In Exercises 3–12, sketch the region enclosed by the curves and find its area.

3. $y = x^2$, $y = \sqrt{x}$, $x = 1/4$, $x = 1$.

4. $x = \sin y$, $x = 0$, $y = \pi/4$, $y = 3\pi/4$.

5. $y = \sec^2 x$, $y = 2$, $x = -\frac{\pi}{4}$, $x = \frac{\pi}{4}$.

6. $y^2 = x$, $y = x - 2$.

7. $y = x^2 + 4$, $x + y = 6$.

8. $y = x^3$, $y = -x$, $y = 8$.

9. $y^2 = -x$, $y = x - 6$, $y = -1$, $y = 4$.

10. $y = x$, $y = 4x$, $y = -x + 2$.

11. $x = y^3 - y$, $x = 0$.

12. $y = x^3 - 2x^2$, $y = 2x^2 - 3x$, $x = 0$, $x = 3$.

13. Find a horizontal line $y = k$ that divides the area between $y = x^2$ and $y = 9$ into two equal parts.

14. Find the area enclosed between the curve $x^{1/2} + y^{1/2} = a^{1/2}$ and the coordinate axes.

15. Suppose that f and g are integrable on $[a, b]$, but neither $f(x) \geq g(x)$ nor $g(x) \geq f(x)$ holds for all x in $[a, b]$. (That is, the curves $y = f(x)$ and $y = g(x)$ are intertwined.)
    (a) What is the geometric significance of the integral
    $$\int_a^b [f(x) - g(x)]\,dx?$$
    (b) What is the geometric significance of the integral
    $$\int_a^b |f(x) - g(x)|\,dx?$$

6.2 VOLUMES BY SLICING; DISKS AND WASHERS

In this section we will use definite integrals to find volumes of three-dimensional solids.

A right circular cylinder (Figure 6.2.1, top) can be generated by moving a plane circular disk along a line perpendicular to the disk. In general, we define a *right cylinder* to be any solid that can be generated by moving a plane region along a line or *axis* perpendicular to the region. All of the solids in Figure 6.2.1 are right cylinders. Observe that all cross sections of a right cylinder taken perpendicular to the axis are identical in size and shape.

If a right cylinder is generated by moving a plane region of area A through a distance h (Figure 6.2.2), then the volume V of the cylinder is *defined* to be

$$V = A \cdot h$$

that is, the volume is the cross-sectional area times the height.

Volumes of solids that are neither right cylinders nor composed of finitely many right cylinders can be obtained by a technique called "slicing." To illustrate the idea, suppose that a solid S extends along the x-axis and is bounded on the left and right by planes perpendicular to the x-axis at $x = a$ and $x = b$ (Figure 6.2.3). Because S is not assumed to be a right cylinder, its cross sections perpendicular to the x-axis can vary from point to point; we will denote by $A(x)$ the area of the cross section at x (Figure 6.2.3).

Figure 6.2.1

Figure 6.2.3

Let us divide the interval $[a, b]$ into n subintervals with widths

$$\Delta x_1, \Delta x_2, \ldots, \Delta x_n$$

by inserting points

$$x_1, x_2, \ldots, x_{n-1}$$

Volume $= A \cdot h$

Figure 6.2.2

between a and b, and let us pass a plane perpendicular to the x-axis through each of these points. As illustrated in Figure 6.2.4, these planes cut the solid S into n slices

$$S_1, S_2, \ldots, S_n$$

Figure 6.2.4

The cross section here has area $A(x_k{}^*)$

S_k

x

$x_k{}^*$

Δx_k

Figure 6.2.5

Consider a typical slice S_k. In general, this slice will not be a right cylinder, since its cross section can vary. However, if the slice is very thin the cross section will not vary much. Therefore, if we choose an arbitrary point x_k^* in the kth subinterval, each cross section of slice S_k will be approximately the same as the cross section at x_k^*, and we can approximate slice S_k by a right cylinder of thickness (height) Δx_k and cross-sectional area $A(x_k^*)$ (Figure 6.2.5).

Thus the volume V_k of slice S_k is approximately the volume of this cylinder, that is,

$$V_k \approx A(x_k^*)\,\Delta x_k$$

and the volume V of the entire solid is approximately

$$V = V_1 + V_2 + \cdots + V_n \approx \sum_{k=1}^{n} A(x_k^*)\,\Delta x_k \tag{1}$$

If we now increase the number of slices in such a way that $\max \Delta x_k \to 0$ then the slices will become thinner and thinner and our approximations will get better and better. Thus, intuition suggests that approximation (1) will approach the exact value of the volume V as $\max \Delta x_k \to 0$, that is,

$$V = \lim_{\max \Delta x_k \to 0} \sum_{k=1}^{n} A(x_k^*)\,\Delta x_k \tag{2}$$

Since the right side of (2) is just the definite integral

$$\int_a^b A(x)\,dx$$

we are led to the following definition.

6.2.1 DEFINITION
Volumes by Cross Sections Perpendicular to the x-Axis

Let S be a solid bounded by two parallel planes perpendicular to the x-axis at $x = a$ and $x = b$. If, for each x in $[a, b]$, the cross-sectional area of S perpendicular to the x-axis is $A(x)$, then the volume of the solid is

$$V = \int_a^b A(x)\,dx \tag{3}$$

provided $A(x)$ is integrable.

There is a corresponding result for cross sections perpendicular to the y-axis.

6.2.2 DEFINITION
Volumes by Cross Sections
Perpendicular to the y-Axis

Let S be a solid bounded by two parallel planes perpendicular to the y-axis at $y = c$ and $y = d$. If, for each y in $[c, d]$, the cross-sectional area of S perpendicular to the y-axis is $A(y)$, then the volume of the solid is

$$V = \int_c^d A(y)\, dy \tag{4}$$

provided $A(y)$ is integrable.

▶ Example 1

Derive the formula for the volume of a right pyramid whose altitude is h and whose base is a square with sides of length a.

Solution. As illustrated in Figure 6.2.6a, we introduce a rectangular coordinate system so that the y-axis passes through the apex, and the x-axis passes through the base and is parallel to a side of the base.

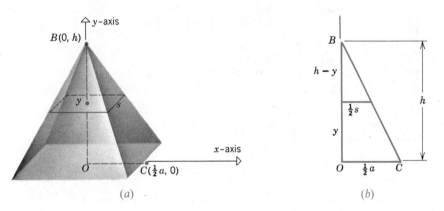

Figure 6.2.6   (a)   (b)

At any point y in the interval $[0, h]$ on the y-axis the cross section perpendicular to the y-axis is a square. If s denotes the length of a side of this square then by similar triangles (Figure 6.2.6b)

$$\frac{\frac{1}{2}s}{\frac{1}{2}a} = \frac{h - y}{h} \qquad \text{or} \qquad s = \frac{a}{h}(h - y)$$

Thus, the area $A(y)$ of the cross section at y is

$$A(y) = s^2 = \frac{a^2}{h^2}(h - y)^2$$

and by (4) the volume is

$$V = \int_0^h A(y)\, dy = \int_0^h \frac{a^2}{h^2}(h - y)^2\, dy = \frac{a^2}{h^2} \int_0^h (h - y)^2\, dy$$

$$= \frac{a^2}{h^2} \left[-\frac{1}{3}(h - y)^3 \right]_0^h$$

$$= \frac{a^2}{h^2} \left[0 + \frac{1}{3}h^3 \right] = \frac{1}{3}a^2 h \qquad \blacktriangleleft$$

VOLUMES OF SOLIDS OF REVOLUTION

Let f be nonnegative and continuous on $[a, b]$, and let R be the region bounded above by the graph of f, below by the x-axis, and on the sides by the lines $x = a$ and $x = b$ (Figure 6.2.7a). When this region is revolved about the x-axis, it generates a solid having circular cross sections (Figure 6.2.7b). Since the cross section at x has radius $f(x)$, the cross-sectional area is

$$A(x) = \pi [f(x)]^2$$

Therefore, from Definition 6.2.1, the volume of the solid is

$$V = \int_a^b \pi [f(x)]^2\, dx \qquad (5)$$

(a)

Because the cross sections are circular or disk shaped, the application of this formula is called the **method of disks.**

▶ Example 2

Find the volume of the solid obtained when the region under the curve $y = \sqrt{x}$ over the interval $[1, 4]$ is revolved about the x-axis (Figure 6.2.8).

Solution. From (5), the volume is

(b)

Figure 6.2.7

$$V = \int_a^b \pi [f(x)]^2\, dx = \int_1^4 \pi x\, dx = \frac{\pi x^2}{2} \Big]_1^4 = 8\pi - \frac{\pi}{2} = \frac{15\pi}{2} \qquad \blacktriangleleft$$

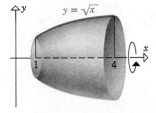

▶ Example 3

Derive the formula for the volume of a sphere of radius r.

Solution. As indicated in Figure 6.2.9, a sphere of radius r can be generated by revolving the upper half of the circle

Figure 6.2.8 $\qquad x^2 + y^2 = r^2$

Figure 6.2.9

about the *x*-axis. Since the upper half of this circle is the graph of $y = f(x) = \sqrt{r^2 - x^2}$, it follows from (5) that the volume of the sphere is

$$V = \int_a^b \pi[f(x)]^2\,dx = \int_{-r}^r \pi(r^2 - x^2)\,dx = \pi\left[r^2x - \frac{x^3}{3}\right]_{-r}^r = \frac{4}{3}\pi r^3 \blacktriangleleft$$

We will now consider more general solids of revolution. Suppose that *f* and *g* are nonnegative continuous functions such that

$$g(x) \le f(x) \qquad \text{for} \qquad a \le x \le b$$

and let *R* be the region enclosed between the graphs of these functions and the lines $x = a$ and $x = b$ (Figure 6.2.10*a*). When this region is revolved

Figure 6.2.10

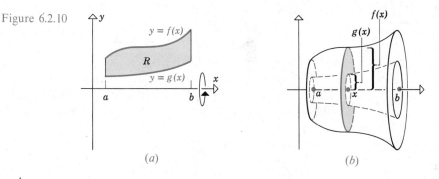

(*a*)   (*b*)

about the *x*-axis, it generates a solid having annular or washer-shaped cross sections (Figure 6.2.10*b*). Since the cross section at *x* has inner radius $g(x)$ and outer radius $f(x)$, its area is

$$A(x) = \pi[f(x)]^2 - \pi[g(x)]^2 = \pi([f(x)]^2 - [g(x)]^2)$$

Therefore, from Definition 6.2.1, the volume of the solid is

$$V = \int_a^b \pi([f(x)]^2 - [g(x)]^2)\,dx \qquad (6)$$

The application of this formula is called the ***method of washers.***

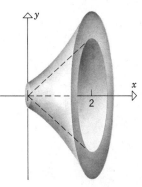

Figure 6.2.11

▶ Example 4

Find the volume of the solid generated when the region between the graphs of $f(x) = \frac{1}{2} + x^2$ and $g(x) = x$ over the interval [0, 2] is revolved about the *x*-axis (Figure 6.2.11).

(a)

(b)

Figure 6.2.12

Solution. From (6) the volume is

$$V = \int_a^b \pi([f(x)]^2 - [g(x)]^2) \, dx = \int_0^2 \pi([\tfrac{1}{2} + x^2]^2 - x^2) \, dx$$

$$= \int_0^2 \pi\left(\frac{1}{4} + x^4\right) dx = \pi\left[\frac{x}{4} + \frac{x^5}{5}\right]_0^2 = \frac{69\pi}{10} \qquad \blacktriangleleft$$

The methods of disks and washers have analogs for regions revolved about the y-axis. If the region of Figure 6.2.12 is revolved about the y-axis, the cross sections of the resulting solid taken perpendicular to the y-axis are disks, and it follows from Definition 6.2.2 that the volume is

$$V = \int_c^d \pi[u(y)]^2 \, dy \qquad (7)$$

(Verify.)

Also, if the region of Figure 6.2.13a is revolved about the y-axis, the cross sections taken perpendicular to the y-axis are washers, and it follows from Definition 6.2.2 that the volume of the solid in Figure 6.2.13b is

$$V = \int_c^d \pi([u(y)]^2 - [v(y)]^2) \, dy \qquad (8)$$

(Verify.)

▶ Example 5

Find the volume of the solid generated when the region enclosed by $y = \sqrt{x}$, $y = 2$, and $x = 0$ is revolved about the y-axis (Figure 6.2.14).

(a)

(b)

Figure 6.2.13

Cross section at y is
a circular disk

Figure 6.2.14

Solution. The cross sections taken perpendicular to the y-axis are disks, so we will apply (7). But first we must rewrite $y = \sqrt{x}$ as $x = y^2$. Thus, from (7) with $u(y) = y^2$, the volume is

$$V = \int_c^d \pi[u(y)]^2 \, dy = \int_0^2 \pi y^4 \, dy = \left. \frac{\pi y^5}{5} \right|_0^2 = \frac{32\pi}{5} \qquad \blacktriangleleft$$

▶ Exercise Set 6.2

1. Find the volume of the solid that results when the region under the curve $y = x^2$ between $x = 0$ and $x = 2$ is revolved about:
(a) the x-axis   (b) the y-axis.

2. Find the volume of the solid that results when the region under the curve $y = \sec x$ between $x = \pi/4$ and $x = \pi/3$ is revolved about the x-axis.

3. Find the volume of the solid that results when the region under the curve $y = 1 + x^3$ between $x = 1$ and $x = 2$ is revolved about
(a) the x-axis   (b) the y-axis.

4. Find the volume of the solid that results when the region enclosed between $y = x^2$ and $y = 9$ is revolved about the x-axis.

5. Find the volume of the solid that results when the region enclosed by $y = \sqrt{x}$, $y = 0$, and $x = 9$ is revolved about the line $x = 9$.

6. Find the volume of the solid that results when the region enclosed between $y = x^2 + 1$ and $y = x + 3$ is revolved about the x-axis.

7. Find the volume of the solid that results when the region above the x-axis and below the curve

$$\frac{x^2}{a^2} + \frac{y^2}{b^2} = 1 \qquad (a > 0, b > 0)$$

is revolved about the x-axis.

8. Let R_1, R_2, R_3, and R_4 be the regions indicated in Figure 6.2.15. Express the following volumes as definite integrals:
(a) The volume of the solid generated when R_1 is revolved about the x-axis.
(b) The volume of the solid generated when R_2 is revolved about the x-axis.
(c) The volume of the solid generated when R_3 is revolved about the y-axis.
(d) The volume of solid generated when R_4 is revolved about the y-axis.

Figure 6.2.15

9. Derive the formula for the volume of a right circular cone with radius r and height h.

10. A *general cylinder* is any solid that can be generated by translating a plane region along a line passing through but not contained in the region. Derive a formula for the volume of a general cylinder with base area A and height h (Figure 6.2.16).

Figure 6.2.16

11. A nose cone for a space reentry vehicle is designed so that a cross section, taken x feet from the tip and perpendicular to the axis of symmetry is a circle of radius $\frac{1}{4}x^2$ ft. Find the volume of the nose cone given that its length is 20 ft.

12. (a) A vat, shaped like a hemisphere of radius r ft, is filled with a fluid to a depth of h ft. Find the volume of the fluid.
    (b) If fluid enters a hemispherical vat of radius 10 ft at a rate of $\frac{1}{2}$ ft^3/min, how fast will the fluid be rising when the depth is 5 ft?

13. The base of a certain solid is the circle $x^2 + y^2 = 9$ and each cross section perpendicular to the x-axis is an equilateral triangle with one side across the base. Find the volume of the solid.

14. A certain solid is 1 ft high and a horizontal cross section taken x ft above the bottom of the solid is an annulus of inner radius x^2 and outer radius $\sqrt{x}$. Find the volume of the solid.

15. The base of a certain solid is the region enclosed by $y = \sin x$, $x = \pi/4$, and $x = 3\pi/4$; and every cross section perpendicular to the x-axis is a square with one side across the base. Find the volume of the solid. [*Hint:* To help with the integration, use the identity $\sin^2 x = \frac{1}{2}(1 - \cos 2x)$.]

16. A hole of radius $r/2$ is drilled through the center of a sphere of radius r. Find the volume of the remaining solid.

17. A wedge is cut from a right circular cylinder of radius r by two planes, one perpendicular to the axis of the cylinder and the other making an angle θ with the first. Find the volume of the wedge. [*Hint:* Introduce coordinate axes as shown in Figure 6.2.17.]

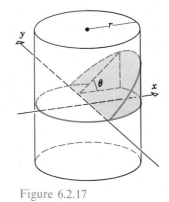

Figure 6.2.17

18. Two right circular cylinders of radius r have axes that intersect at right angles. Find the volume of the solid common to the two cylinders. [*Hint:* one eighth of the solid is sketched in Figure 6.2.18.]

Figure 6.2.18

6.3 VOLUMES BY CYLINDRICAL SHELLS

The methods discussed so far for computing volumes of solids depend on our ability to compute cross-sectional areas of the solid. In this section we will develop an alternative technique for computing volumes that can sometimes be applied when it is inconvenient or difficult to determine the cross-sectional areas or if the integration is too hard. This technique, called the *method of cylindrical shells,* applies only to *solids of revolution,* that is, solids generated by revolving a plane region about a line.

A *cylindrical shell* is a solid enclosed by two concentric right circular cylinders (Figure 6.3.1). The volume V of a cylindrical shell having inner radius r_1, outer radius r_2, and height h can be written

$$V = [\text{area of cross section}] \cdot [\text{height}]$$
$$= (\pi r_2{}^2 - \pi r_1{}^2)h$$
$$= \pi(r_2 + r_1)(r_2 - r_1)h$$
$$= 2\pi\left(\frac{r_2 + r_1}{2}\right)h(r_2 - r_1)$$

Figure 6.3.1

Thus,

$$V = 2\pi \cdot [\text{average radius}] \cdot [\text{height}] \cdot [\text{thickness}] \tag{1}$$

The method of cylindrical shells, which we will now discuss, makes use of this formula.

Let R be a plane region bounded above by a continuous curve $y = f(x)$, bounded below by the x-axis, and bounded on the left and right, respectively, by the lines $x = a$ and $x = b$; let S be the solid generated by revolving the region R about the y-axis (Figure 6.3.2.)

Figure 6.3.2

To find the volume of S, let us divide the interval $[a, b]$ into n subintervals with widths

$$\Delta x_1, \Delta x_2, \ldots, \Delta x_n$$

by inserting points

$$x_1, x_2, \ldots, x_{n-1}$$

between a and b, and let us draw a vertical line through each of these points to divide the region R into n strips $R_1, R_2, \ldots, R_n$ (Figure 6.3.3a). These strips, when revolved about the y-axis, generate solids $S_1, S_2, \ldots, S_n$.

Figure 6.3.3   (a)   (b)

As illustrated in Figure 6.3.3b, these solids are nested one inside the other and together form the entire solid S. Thus the volume of the solid S can be obtained by adding together the volumes of solids $S_1, S_2, \ldots, S_n$:

$$V(S) = V(S_1) + V(S_2) + \cdots + V(S_n) \qquad (2)$$

Consider a typical strip R_k and the solid S_k that it generates (Figure 6.3.4). Although solid S_k resembles a cylindrical shell, it will not, in general, be a

Figure 6.3.4

cylindrical shell because it can have a curved upper surface. However, if the interval width

$$\Delta x_k = x_k - x_{k-1}$$

is small we can obtain a good approximation to the region R_k by a rectangle of width Δx_k and height $f(x_k^*)$, where

$$x_k^* = \frac{x_k + x_{k-1}}{2}$$

is the midpoint of the interval $[x_{k-1}, x_k]$ (Figure 6.3.5a). This rectangle, when revolved about the y-axis, generates a cylindrical shell, which is a good approximation to the solid S_k (Figure 6.3.5b).

Figure 6.3.5   (a)   (b)

This approximating cylindrical shell has thickness Δx_k, height $f(x_k^*)$, and average radius x_k^*, so that by (1)

$$V(S_k) \approx \text{volume of approximating cylindrical shell} = 2\pi x_k^* f(x_k^*)\,\Delta x_k$$

Thus, from (2), the volume of the entire solid S is approximately

$$V(S) = \sum_{k=1}^{n} V(S_k) \approx \sum_{k=1}^{n} 2\pi x_k^* f(x_k^*)\,\Delta x_k \tag{3}$$

If we now divide $[a, b]$ into more and more subintervals in such a way that max $\Delta x_k \to 0$, then intuition suggests that our approximations will tend to get better and (3) will approach the exact value of the volume, that is,

$$V(S) = \lim_{\max \Delta x_k \to 0} \sum_{k=1}^{n} 2\pi x_k^* f(x_k^*)\,\Delta x_k \tag{4}$$

Since the right side of (4) is just the definite integral

$$\int_a^b 2\pi x f(x)\,dx$$

we are led to the following definition.

6.3.1   DEFINITION

Volumes by Cylindrical Shells about the y-Axis

Let R be a plane region bounded above by a continuous curve $y = f(x)$, below by the x-axis, and on the left and right, respectively, by the lines $x = a$ and $x = b$. Then the volume of the solid generated by revolving R about the y-axis is given by

$$V = \int_a^b 2\pi x f(x)\, dx \tag{5}$$

▶ Example 1

Use cylindrical shells to find the volume of the solid generated when the region enclosed between $y = \sqrt{x}$, $x = 1$, $x = 4$, and the x-axis is revolved about the y-axis (Figure 6.3.6).

Cut--away view of the solid

Figure 6.3.6

Solution. Since $f(x) = \sqrt{x}$, $a = 1$, and $b = 4$, Formula (5) yields

$$V = \int_1^4 2\pi x \sqrt{x}\, dx = 2\pi \int_1^4 x^{3/2}\, dx$$

$$= \left[2\pi \cdot \frac{2}{5} x^{5/2} \right]_1^4$$

$$= \frac{4\pi}{5}[32 - 1]$$

$$= \frac{124\pi}{5} \qquad\qquad\qquad ◀$$

There is a way of thinking about (5) that is sometimes useful. At each point x in $[a, b]$ the vertical line through x cuts the region R in a line segment that we can view as the vertical "cross-section" of R at x (Figure 6.3.7a). When the region R is revolved about the y-axis, the vertical cross section at x generates

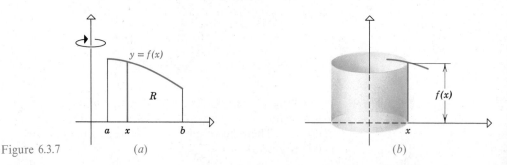

Figure 6.3.7   (a)   (b)

the *surface* of a right circular cylinder having height $f(x)$ and radius x (Figure 6.3.7*b*). The area of this surface is

$$2\pi x f(x)$$

(see Figure 4.8.1), which is precisely the integrand in (5). Thus (5) states:

6.3.2 DEFINITION   The volume V by cylindrical shells is the integral of the surface area generated by an arbitrary cross section of R taken parallel to the axis about which R is revolved.

Formulation 6.3.2 is helpful for adapting the method of cylindrical shells to problems that are not exactly of the form considered in 6.3.1.

▶ Example 2

Use cylindrical shells to find the volume of the solid generated when the region R in the first quadrant enclosed between $y = x$ and $y = x^2$ is revolved about the y-axis (Figure 6.3.8).

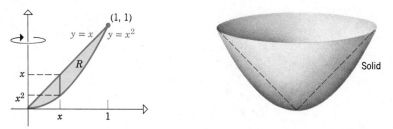

Figure 6.3.8

Solution. At each x in $[0, 1]$ the cross section of R parallel to the y-axis generates a cylindrical surface of height $x - x^2$ and radius x. Since the area of this surface is

$$2\pi x(x - x^2)$$

the volume of the solid is

$$V = \int_0^1 2\pi x(x - x^2)\,dx$$

$$= 2\pi \int_0^1 (x^2 - x^3)\,dx$$

$$= 2\pi \left[\frac{x^3}{3} - \frac{x^4}{4}\right]_0^1$$

$$= 2\pi \left[\frac{1}{3} - \frac{1}{4}\right] = \frac{\pi}{6} \qquad ◀$$

► Example 3

Use cylindrical shells to find the volume of the solid generated when the region R under $y = x^2$ over the interval $[0, 2]$ is revolved about the x-axis (Figure 6.3.9).

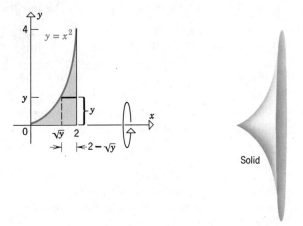

Solid

Figure 6.3.9

Solution. At each y in the interval $0 \le y \le 4$, the cross section of R parallel to the x-axis generates a cylindrical surface of height $2 - \sqrt{y}$ and radius y. Since the area of this surface is

$$2\pi y(2 - \sqrt{y})$$

the volume of the solid is

$$V = \int_0^4 2\pi y(2 - \sqrt{y})\, dy$$

$$= 2\pi \int_0^4 (2y - y^{3/2})\, dy$$

$$= 2\pi \left[y^2 - \frac{2}{5} y^{5/2} \right]_0^4$$

$$= \frac{32\pi}{5} \qquad\blacktriangleleft$$

The volume in Example 3 can also be obtained by the method of disks [Formula (5) of Section 6.2]. The computations are

$$V = \int_0^2 \pi (x^2)^2\, dx = \int_0^2 \pi x^4\, dx = \left. \frac{\pi x^5}{5} \right]_0^2 = \frac{32}{5}\pi$$

▶ Exercise Set 6.3

In Exercises 1–6, use cylindrical shells to find the volume of the solid generated when the region enclosed by the given curves is revolved about the *y*-axis.

1. $y = x^3$, $x = 1$, $y = 0$.

2. $y = \sqrt{x}$, $x = 4$, $x = 9$, $y = 0$.

3. $x^2 + y^3 = 4$, $x = 0$, $x = 4$, $y = 0$.

4. $y = \cos(x^2)$, $x = 0$ $x = \frac{1}{2}\sqrt{\pi}$, $y = 0$.

5. $y = 2x - 1$, $y = -2x + 3$, $x = 2$.

6. $x = y^2$, $y = x^2$.

In Exercises 7–10, use cylindrical shells to find the volume of the solid generated when the region enclosed by the given curves is revolved about the *x*-axis.

7. $y^2 = x$, $y = 1$, $x = 0$.

8. $x = 2y$, $y = 2$, $y = 3$, $x = 0$.

9. $y = x^2$, $x = 1$, $y = 0$.

10. $xy = 4$, $x + y = 5$.

11. (a) Use cylindrical shells to find the volume of the solid generated when the region under the curve $y = x^3 - 3x^2 + 2x$ over $[0, 1]$ is revolved about the *y*-axis.

(b) For this problem, is the method of cylindrical shells easier or harder than the method of slicing discussed in the last section? Explain.

12. Use cylindrical shells to find the volume of the solid generated when the region enclosed by

$$y = \frac{1}{x^3}, \; x = 1, \; x = 2, \; y = 0$$

is revolved about the line $x = -1$.

13. Use cylindrical shells to find the volume of the solid generated when the region enclosed by

$$y = x^3, \; y = 1, \; x = 0$$

is revolved about the line $y = 1$.

14. Let R_1 and R_2 be regions of the form shown in Figure 6.3.10. Use cylindrical shells to find a formula for the volume of the solid which results when:

(a) region R_1 is revolved about the *y*-axis

(b) region R_2 is revolved about the *x*-axis.

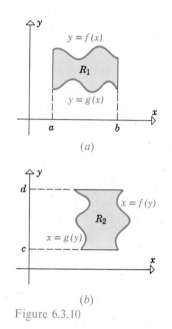

(*a*)

(*b*)

Figure 6.3.10

15. Use cylindrical shells to find the volume of the cone generated when the triangle with vertices $(0, 0)$, $(0, r)$, $(h, 0)$, where $r > 0$ and $h > 0$ is revolved about the *x*-axis.

16. The region enclosed between the curve $y^2 = kx$ and the line $x = \frac{1}{4}k$ is revolved about the line $x = \frac{1}{2}k$. Use cylindrical shells to find the volume of the resulting solid. Assume $k > 0$.

17. A round hole of radius a is drilled through the center of a solid sphere of radius r. Use cylindrical shells to find the volume of the portion removed. (Assume $r > a$.)

18. Use cylindrical shells to find the volume of the torus obtained by revolving the circle

$$x^2 + y^2 = a^2$$

about the line $x = b$, where $b > a$ [*Hint:* It may help in the integration to think of an integral as an area.]

6.4 LENGTH OF A PLANE CURVE

In this section we will use definite integrals to find arc lengths of plane curves. To start, we will consider only curves that are graphs of functions. In a later section, we will extend our results to more general curves.

If f' is continuous on an interval, we shall say that $y = f(x)$ is a **smooth curve** (or f is a **smooth function**) on that interval. We will restrict our discussion of arc length to smooth curves in order to eliminate some complications that would otherwise occur.

6.4.1 PROBLEM Suppose f is smooth on the interval $[a, b]$. Find the arc length L of the curve $y = f(x)$ over the interval $[a, b]$ (Figure 6.4.1).

Figure 6.4.1

In order to solve Problem 6.4.1 we must first define the term "arc length" precisely. For motivation, consider the graph of a smooth curve $y = f(x)$ over an interval $[a, b]$. As shown in Figure 6.4.2, we divide the interval $[a, b]$ into n subintervals with widths

$$\Delta x_1, \Delta x_2, \ldots, \Delta x_n$$

by inserting points

$$x_1, x_2, \ldots, x_{n-1}$$

between a and b, and let $P_0, P_1, \ldots, P_n$ be the points on the curve whose x-coordinates are $a, x_1, x_2, \ldots, x_{n-1}, b$. If we now join these points with straight-line segments we obtain a *polygonal path* that we can regard as an approximation to the curve $y = f(x)$. Intuition suggests that the length of the approximating polygonal path will approach the length of the curve if we increase the number of points in such a way that the lengths of the line segments approach zero.

Figure 6.4.2

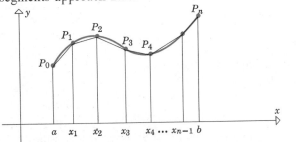

To examine this idea more closely, let us isolate a typical subinterval, say the kth (Figure 6.4.3). As suggested by this figure, the length L_k of the kth line segment in the polygonal path is given by

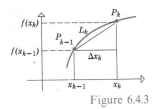

Figure 6.4.3

$$L_k = \sqrt{(\Delta x_k)^2 + [f(x_k) - f(x_{k-1})]^2} \tag{1}$$

By the mean-value theorem (4.9.2) there is a point x_k^* between x_{k-1} and x_k such that

$$\frac{f(x_k) - f(x_{k-1})}{x_k - x_{k-1}} = f'(x_k^*)$$

or

$$f(x_k) - f(x_{k-1}) = f'(x_k^*)\,\Delta x_k$$

Thus (1) can be rewritten as

$$L_k = \sqrt{1 + [f'(x_k^*)]^2}\,\Delta x_k$$

which means that the length of the *entire* polygonal path is

$$\sum_{k=1}^{n} L_k = \sum_{k=1}^{n} \sqrt{1 + [f'(x_k^*)]^2}\,\Delta x_k$$

If we now increase the number of subintervals in such a way that $\max \Delta x_k \to 0$ then the length of the polygonal path will approach the arc length L of the curve $y = f(x)$ over $[a, b]$. Thus

$$L = \lim_{\max \Delta x_k \to 0} \sum_{k=1}^{n} \sqrt{1 + [f'(x_k^*)]^2}\,\Delta x_k \qquad (2)$$

Since the right side of (2) is just the definite integral

$$\int_a^b \sqrt{1 + [f'(x)]^2}\,dx$$

we are led to the following definition.

6.4.2 DEFINITION If f is a smooth function on $[a, b]$, then the **arc length** L of the curve $y = f(x)$ from $x = a$ to $x = b$ is defined to be

$$L = \int_a^b \sqrt{1 + [f'(x)]^2}\ dx \qquad (3)$$

Similarly, for a curve expressed in the form $x = g(y)$, where g' is continuous on $[c, d]$, the arc length L from $y = c$ to $y = d$ is given by

$$L = \int_c^d \sqrt{1 + [g'(y)]^2}\,dy \qquad (4)$$

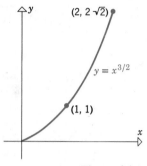

(2, 2 √2)

$y = x^{3/2}$

(1, 1)

Figure 6.4.4

▶ Example 1

Find the arc length of the curve $y = x^{3/2}$ from $(1, 1)$ to $(2, 2\sqrt{2})$ (Figure 6.4.4) using:

(a) formula (3)   (b) formula (4)

Solution (a). Since $f(x) = x^{3/2}$,

$$f'(x) = \frac{3}{2}x^{1/2}$$

Thus from (3), the arc length from $x = 1$ to $x = 2$ is

$$L = \int_1^2 \sqrt{1 + \frac{9}{4}x}\ dx$$

To evaluate this integral we make the u-substitution,

$$u = 1 + \frac{9}{4}x, \qquad du = \frac{9}{4}dx$$

and change the x-limits $(x = 1, x = 2)$ to the corresponding u-limits $(u = \frac{13}{4}, u = \frac{22}{4})$:

$$L = \frac{4}{9}\int_{13/4}^{22/4} u^{1/2}\ du = \frac{8}{27}u^{3/2}\Big]_{13/4}^{22/4}$$

$$= \frac{8}{27}\left[\left(\frac{22}{4}\right)^{3/2} - \left(\frac{13}{4}\right)^{3/2}\right]$$

$$= \frac{22\sqrt{22} - 13\sqrt{13}}{27}$$

Solution (b). Solving $y = x^{3/2}$ for x in terms of y, we obtain $x = y^{2/3}$. Hence $g(y) = y^{2/3}$ and

$$g'(y) = \frac{2}{3}y^{-1/3}$$

Thus from (4), the arc length from $y = 1$ to $y = 2\sqrt{2}$ is

$$L = \int_1^{2\sqrt{2}} \sqrt{1 + \frac{4}{9}y^{-2/3}}\ dy$$

$$= \frac{1}{3}\int_1^{2\sqrt{2}} y^{-1/3}\sqrt{9y^{2/3} + 4}\ dy$$

To evaluate this integral we make the u-substitution

$$u = 9y^{2/3} + 4, \qquad du = 6y^{-1/3}\,dy$$

and change the y limits ($y = 1, y = 2\sqrt{2}$) to the corresponding u limits ($u = 13, u = 22$). This gives

$$L = \frac{1}{18}\int_{13}^{22} u^{1/2}\,du = \frac{1}{27}u^{3/2}\Big]_{13}^{22}$$

$$= \frac{1}{27}[(22)^{3/2} - (13)^{3/2}]$$

$$= \frac{22\sqrt{22} - 13\sqrt{13}}{27}$$

This result agrees with that in part (a); however the integration here is more tedious. In problems where there is a choice between using (3) or (4), it is sometimes worthwhile to determine which one leads to the simpler integral. ◄

▶ Exercise Set 6.4

───

1. Find the arc length of the curve $y = 2x$ from $(1, 2)$ to $(2, 4)$ using
   (a) formula (3)
   (b) formula (4)
   (c) the theorem of Pythagoras.

2. Find the arc length of the curve $y = mx + b$ from $x = k_1$ to $x = k_2$ ($m \neq 0, k_2 > k_1$) using
   (a) formula (3)
   (b) formula (4)
   (c) the theorem of Pythagoras.

3. Find the arc length of the curve $y = 3x^{3/2} - 1$ from $x = 0$ to $x = 1$.

4. Find the arc length of the curve $y = \dfrac{x^4}{16} + \dfrac{1}{2x^2}$ from $x = 2$ to $x = 3$.

5. Find the arc length of the curve $24xy = y^4 + 48$ from $y = 2$ to $y = 4$.

6. Find the arc length of the curve $x = \frac{1}{8}y^4 + \frac{1}{4}y^{-2}$ from $y = 1$ to $y = 4$.

7. Consider the curve $y = x^{2/3}$.
   (a) Sketch the portion of the curve between $x = -1$ and $x = 8$.

   (b) Explain why (3) cannot be used to find the arc length of the curve sketched in (a).
   (c) Find the arc length of the curve sketched in (a).

8. Find the arc length of the curve $x^{2/3} + y^{2/3} = a^{2/3}$ in the second quadrant from $x = -a$ to $x = -\frac{1}{8}a$ ($a > 0$).

9. Let $y = f(x)$ be a smooth curve and suppose $f'(x) \geq 0$ on the closed interval $[a, b]$.
   (a) Prove: There are numbers m and M such that $m \leq f'(x) \leq M$ for all x in $[a, b]$.
   (b) Prove: The arc length L of $y = f(x)$ over the closed interval $[a, b]$ satisfies the inequalities

$$(b - a)\,\sqrt{1 + m^2} \leq L \leq (b - a)\,\sqrt{1 + M^2}$$

10. Use the result of Exercise 9 to show that the arc length L of $y = \sin x$ over the interval $0 \leq x \leq \frac{\pi}{4}$ satisfies

$$\frac{\pi}{4}\sqrt{\frac{3}{2}} \leq L \leq \frac{\pi}{4}\sqrt{2}$$

6.5   AREA OF A SURFACE OF REVOLUTION

In this section we will apply the definite integral to the problem of finding the area of a surface of revolution. For now we limit the discussion to the following situation:

6.5.1 PROBLEM Let f be a smooth, nonnegative function on $[a, b]$. Find the area of the surface generated by revolving the portion of the curve $y = f(x)$ between $x = a$ and $x = b$ about the x-axis (Figure 6.5.1).

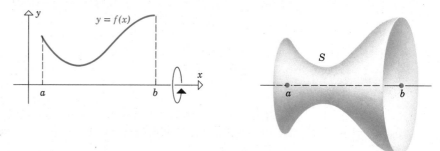

Figure 6.5.1

In order to solve this problem we must first define the term "surface area" precisely. For motivation, let us approximate the curve $y = f(x)$ by a polygonal path of straight line segments connecting the points on the curve that have x-coordinates

$$a, x_1, x_2, \ldots, x_{n-1}, b$$

(Figure 6.5.2a). As usual, let

$$\Delta x_1, \Delta x_2, \ldots, \Delta x_n$$

be the lengths of the subintervals determined by these x-coordinates. If these widths are small, then the surface generated by revolving the polygonal path about the x-axis will have approximately the same area as the surface generated by revolving the curve $y = f(x)$ about the x-axis (Figure 6.5.2b).

Figure 6.5.2   (a)   (b)

Figure 6.5.3

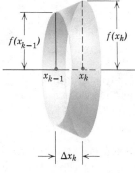

Figure 6.5.4

Observe that the surface generated by the polygonal path is made of parts, each of which is a frustum of a cone. Thus the lateral area of each of these parts can be obtained from the formula

$$S = \pi(r_1 + r_2)l \tag{1}$$

for the lateral area S of a frustum of slant height l and base radii r_1 and r_2 (Figure 6.5.3). A derivation of this formula is discussed in Exercise 15. Intuition suggests that the area of the approximating surface will approach the area of the desired surface if we increase the number of subdivisions in such a way that the lengths of the line segments in the polygonal path approach zero.

To examine this idea more closely, let us isolate a typical section of the approximating surface, say the kth (Figure 6.5.4). Applying (1), we obtain as the lateral area S_k of this kth section

$$S_k = \pi[f(x_{k-1}) + f(x_k)]\sqrt{(\Delta x_k)^2 + [f(x_k) - f(x_{k-1})]^2} \tag{2}$$

By the mean-value theorem (4.9.2) there is a point x_k^* between x_{k-1} and x_k such that

$$\frac{f(x_k) - f(x_{k-1})}{x_k - x_{k-1}} = f'(x_k^*)$$

or

$$f(x_k) - f(x_{k-1}) = f'(x_k^*)\,\Delta x_k$$

Thus, (2) can be rewritten as

$$S_k = \pi[f(x_{k-1}) + f(x_k)]\sqrt{1 + [f'(x_k^*)]^2}\,\Delta x_k \tag{3}$$

Since the arithmetic average of two numbers lies between those numbers, $\frac{1}{2}[f(x_{k-1}) + f(x_k)]$ is between $f(x_{k-1})$ and $f(x_k)$. Since f is continuous on the interval $[x_{k-1}, x_k]$, the intermediate value theorem (3.7.8) implies that there exists a point x_k^{**} in this interval such that

$$\frac{1}{2}[f(x_{k-1}) + f(x_k)] = f(x_k^{**})$$

Thus (3) can be rewritten

$$S_k = 2\pi f(x_k^{**})\sqrt{1 + [f'(x_k^*)]^2}\,\Delta x_k$$

so that the area of the *entire* polygonal surface is

$$\sum_{k=1}^{n} S_k = \sum_{k=1}^{n} 2\pi f(x_k^{**}) \sqrt{1 + [f'(x_k^*)]^2}\, \Delta x_k$$

If we now increase the number of subintervals in such a way that $\max \Delta x_k \to 0$, then the area of the approximating polygonal surface will approach the exact surface area S. Thus

$$S = \lim_{\max \Delta x_k \to 0} \sum_{k=1}^{n} 2\pi f(x_k^{**}) \sqrt{1 + [f'(x_k^*)]^2}\, \Delta x_k \tag{4}$$

If it were the case that $x_k^{**} = x_k^*$, then the right side of (4) would be the definite integral

$$\int_a^b 2\pi f(x) \sqrt{1 + [f'(x)]^2}\, dx \tag{5}$$

However, it is proved in advanced calculus that this is true even if $x_k^* \neq x_k^{**}$ because of the continuity of f and f'. In light of (4) and (5) we are led to the following definition.

6.5.2 DEFINITION   Let f be a smooth, nonnegative function on $[a, b]$. Then the **surface area** S generated by revolving the portion of the curve $y = f(x)$ between $x = a$ and $x = b$ about the x-axis is

$$S = \int_a^b 2\pi f(x) \sqrt{1 + [f'(x)]^2}\, dx \tag{6}$$

For a curve expressed in the form $x = g(y)$, where g' is continuous on $[c, d]$, and $g(y) \geq 0$ for $c \leq y \leq d$, the surface area S generated by revolving the portion of the curve from $y = c$ to $y = d$ about the y-axis is given by

$$S = \int_c^d 2\pi g(y) \sqrt{1 + [g'(y)]^2}\, dy \tag{7}$$

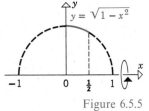

Figure 6.5.5

▶ Example 1

Find the surface area generated by revolving the curve

$$y = \sqrt{1 - x^2}, \qquad 0 \leq x \leq \tfrac{1}{2}$$

about the x-axis (Figure 6.5.5).

Solution.   Since $f(x) = \sqrt{1 - x^2}$,

$$f'(x) = -\frac{x}{\sqrt{1 - x^2}}$$

Thus (6) yields

$$S = \int_0^{1/2} 2\pi \sqrt{1 - x^2} \sqrt{1 + \frac{x^2}{1 - x^2}}\, dx$$

$$= \int_0^{1/2} 2\pi\, dx$$

$$= 2\pi x \Big]_0^{1/2} = \pi$$ ◀

▶ Example 2

Find the surface area generated by revolving the curve

$$y = \sqrt[3]{3x}, \qquad 0 \le y \le 2$$

about the y-axis (Figure 6.5.6).

Figure 6.5.6

Solution. We will apply (7) after rewriting $y = \sqrt[3]{3x}$ as $x = g(y) = \frac{1}{3}y^3$. Thus

$$g'(y) = y^2$$

so that from (7) we obtain

$$S = \int_0^2 2\pi \left(\frac{1}{3}y^3\right) \sqrt{1 + y^4}\, dy$$

$$= \frac{2\pi}{3} \int_0^2 y^3 \sqrt{1 + y^4}\, dy \tag{8}$$

The u-substitution

$$u = 1 + y^4, \qquad du = 4y^3\, dy$$

yields

$$\int y^3 \sqrt{1 + y^4}\, dy = \frac{1}{4} \int \sqrt{u}\, du = \frac{1}{4} \cdot \frac{2}{3} u^{3/2} + C$$

$$= \frac{1}{6}(1 + y^4)^{3/2} + C$$

Thus, from (8),

$$S = \frac{2\pi}{3} \left[\frac{1}{6}(1 + y^4)^{3/2}\right]_0^2 = \frac{\pi}{9}(17^{3/2} - 1)$$ ◀

▶ Exercise Set 6.5

In Exercises 1–6, find the area of the surface generated by revolving the given curve about the x-axis.

1. $y = 7x$, $0 \leq x \leq 1$.

2. $y = \sqrt{x}$, $1 \leq x \leq 4$.

3. $y = \sqrt{4 - x^2}$, $-1 \leq x \leq 1$.

4. $x = \sqrt[3]{y}$, $1 \leq y \leq 8$.

5. $y = \sqrt{x} - \frac{1}{3}x^{3/2}$, $1 \leq x \leq 3$.

6. $y = \frac{1}{3}x^3 + \frac{1}{4}x^{-1}$, $1 \leq x \leq 2$.

In Exercises 7–12, find the area of the surface generated by revolving the given curve about the y-axis.

7. $x = 9y + 1$, $0 \leq y \leq 2$.

8. $x = y^3$, $0 \leq y \leq 1$.

9. $x = \sqrt{9 - y^2}$, $-2 \leq y \leq 2$.

10. $x = 2\sqrt{1 - y}$, $-1 \leq y \leq 0$.

11. $8xy^2 = 2y^6 + 1$, $1 \leq y \leq 2$.

12. $x = |y - 11|$, $0 \leq y \leq 2$.

13. The lateral area S of a right circular cone with height h and base radius r is $S = \pi r \sqrt{r^2 + h^2}$. Obtain this result using (6).

14. Assume $f(x) \geq 0$ for $a \leq x \leq b$. Derive a formula for the surface area generated when the curve $y = f(x)$, $a \leq x \leq b$ is revolved about the line $y = -k$ ($k > 0$).

15. (*Formula for Surface Area of a Frustum*)
    (a) If a cone of slant height l and base radius r is cut along a lateral edge and laid flat, it becomes a sector of a circle of radius l (Figure 6.5.7).

Figure 6.5.7

Use the formula $\frac{1}{2}l^2\theta$ for the area of a sector with radius l and central angle θ (in radians) to show that the lateral surface area of the cone is $\pi r l$.
    (b) Use the result in (a) to obtain Formula (1) for the lateral surface area of a frustum.

16. Let $y = f(x)$ be a smooth curve on the interval $[a, b]$ and assume $f(x) \geq 0$ for $a \leq x \leq b$. By the extreme-value theorem (4.2.4), the function f has a maximum value K and a minimum value k on $[a, b]$. Prove: If L is the arc length of the curve $y = f(x)$ between $x = a$ and $x = b$ and if S is the area of the surface that is generated by revolving this curve about the x-axis, then

$$2\pi k L \leq S \leq 2\pi K L$$

17. Let $y = f(x)$ be a smooth curve on $[a, b]$ and assume $f(x) \geq 0$ for $a \leq x \leq b$. Let A be the area under the curve $y = f(x)$ between $x = a$ and $x = b$ and let S be the area of the surface obtained when this section of curve is revolved about the x-axis.
    (a) Prove: $2\pi A \leq S$.
    (b) For what functions f is $2\pi A = S$?

6.6 WORK

If an object moves a distance d along a line while subjected to a *constant* force F applied in the direction of motion, then physicists and engineers define the **work** W done by the force F to be

$$W = F \cdot d$$
[work] = [force] · [distance] $\hspace{3em}$ (1)

If force is measured in pounds and distance in feet then the unit of work is foot-pounds. In the metric system force is measured in **dynes** (the force required to give a mass of 1 gm an acceleration of 1 cm/sec^2) or **newtons** (the force required to give a mass of 1 kg an acceleration of 1 m/sec^2). In this system common units of work are newton-meters and dyne-centimeters.

▶ Example 1

An object moves 5 ft, while subjected to a constant force of 100 lb along its direction of motion. The work done is

$$W = F \cdot d = 100 \cdot 5 = 500 \text{ ft-lb} \hspace{3em} ◀$$

Calculus is needed when it is required to calculate the work done by a *variable* force. We shall consider the following problem.

6.6.1 PROBLEM Suppose an object moves in the positive direction along a coordinate line while subject to a force $F(x)$, in the direction of motion, whose magnitude depends on the coordinate x. Find the work done by the force when the object moves over an interval $[a, b]$.

For example, Figure 6.6.1 shows a block subjected to the force of a compressed spring. As the block moves from a to b the spring expands and the force it applies diminishes. Thus, the force $F(x)$ applied by the spring varies with x.

Figure 6.6.1

Before we can solve Problem 6.6.1 we must define precisely what is meant by the work done by a variable force. To motivate this definition, let us divide the interval $[a, b]$ into n subintervals with widths

$$\Delta x_1, \Delta x_2, \ldots, \Delta x_n$$

by inserting points

$$x_1, x_2, \ldots, x_{n-1}$$

between a and b. If we denote by W_k the work done when the object moves across the kth subinterval, then total work W done when the object moves across the entire interval $[a, b]$ will be

$$W = W_1 + W_2 + \cdots + W_n$$

Let us estimate W_k. If the kth subinterval is short and if $F(x)$ is continuous, then the force will not vary much over this subinterval; it will be almost constant. We can approximate this nearly constant force by $F(x_k^*)$ where x_k^* is any point in the kth subinterval. Thus from (1), the work done over the kth subinterval is approximately

$$W_k \approx F(x_k^*) \, \Delta x_k$$

and the work done over the entire interval $[a, b]$ is approximately

$$W = \sum_{k=1}^{n} W_k \approx \sum_{k=1}^{n} F(x_k^*) \, \Delta x_k$$

If we now increase the number of subintervals in such a way that $\max \Delta x_k \to 0$, then intuition suggests that our approximations will tend to get better; hence

$$W = \lim_{\max \Delta x_k \to 0} \sum_{k=1}^{n} F(x_k^*) \, \Delta x_k$$

Since the limit on the right is just the definite integral

$$\int_a^b F(x) \, dx$$

the following definition is suggested.

6.6.2 DEFINITION If an object moves in the positive direction over the interval $[a, b]$ while subjected to a variable force $F(x)$ in the direction of motion, then the **work** done by the force is

$$W = \int_a^b F(x) \, dx \qquad (2)$$

Hooke's law [Robert Hooke (1635–1703)-English physicist] states that a spring stretched x units beyond its equilibrium position pulls back with a force

$$F(x) = kx$$

where k is a constant (called the **spring constant**). The value of k depends on such factors as the thickness of the spring, the material used in its composition, and the units of force and distance.

▶ Example 2

A spring whose natural length is 24 in. exerts a force of 5 lb when stretched 10 in. beyond its natural length.
(a) Find the spring constant k.
(b) How much work is required to stretch the spring from its natural length to a length of 42 in.?

Solution (a). From Hooke's law,

$$F(x) = kx$$

From the data, $F(x) = 5$ lb when $x = 10$ in. so that

$$5 = k \cdot 10$$

Thus the spring constant is

$$k = \frac{5}{10} = \frac{1}{2}$$

This means that the force $F(x)$ required to stretch the spring x in. is

$$F(x) = \frac{1}{2}x \tag{3}$$

Solution (b). Place the spring along a coordinate line as shown in Figure 6.6.2.

Figure 6.6.2

We want to find the work W required to stretch the spring over the interval from $x = 0$ to $x = 18$. From (2) and (3) the work W required is

$$W = \int_a^b F(x)\, dx = \int_0^{18} \frac{1}{2} x\, dx$$

$$= \frac{x^2}{4} \Bigg]_0^{18}$$

$$= 81 \text{ in-lb}$$ ◀

► Example 3

A cylindrical water tank of radius 10 ft and height 30 ft is half filled with water. How much work is required to pump all the water over the upper rim of the tank?

Solution. Introduce a coordinate line as shown in Figure 6.6.3. Imagine the water to be divided into n thin layers with thicknesses

$$\Delta x_1, \Delta x_2, \ldots, \Delta x_n$$

Figure 6.6.3

The force required to move the kth layer equals the weight of the layer, which can be found by multiplying its volume by the density of water (62.4 lb/ft³). Since the kth layer is a cylinder of radius $r = 10$ ft and height Δx_k, the force required to move it is

$$\begin{bmatrix} \text{Force to move} \\ \text{the } k\text{th layer} \end{bmatrix} = (\pi r^2 \,\Delta x_k) \cdot [\text{density of water}]$$
$$= (\pi (10)^2 \,\Delta x_k)(62.4)$$
$$= 6240\pi \,\Delta x_k$$

Because the kth layer has a finite thickness, the upper and lower surfaces are at different distances from the origin. However, if the layer is thin, the difference in these distances is small, and we can reasonably assume that the entire layer is concentrated at a single distance x_k^* from the origin (Figure 6.6.3). With this assumption, the work W_k required to pump the kth layer over the rim will be approximately

$$W_k \approx \underbrace{(30 - x_k^*)}_{\text{distance}} \cdot \underbrace{6240\pi \, \Delta x_k}_{\text{force}}$$

and the work W required to pump all n layers will be approximately

$$W = \sum_{k=1}^{n} W_k \approx \sum_{k=1}^{n} (30 - x_k^*)(6240\pi) \, \Delta x_k$$

To find the *exact* value of the work we take the limit as $\max \Delta x_k \to 0$. This yields

$$W = \lim_{\max \Delta x_k \to 0} \sum_{k=1}^{n} (30 - x_k^*)(6240\pi) \, \Delta x_k$$

$$= \int_0^{15} (30 - x)(6240\pi) \, dx$$

$$= 6240\pi \left(30x - \frac{x^2}{2} \right) \Big]_0^{15}$$

$$= 2,106,000 \pi \ \text{ft-lb} \qquad \blacktriangleleft$$

▶ Exercise Set 6.6

1. Find the work done when
   (a) a constant force of 30 lb along the x-axis moves an object from $x = -2$ to $x = 5$ in.
   (b) a variable force of $F(x) = 1/x^2$ lb along the x-axis moves an object from $x = 1$ to $x = 6$ in.

2. A spring whose natural length is 15 in. exerts a force of 45 lb when stretched to a length of 20 in.
   (a) Find the spring constant.
   (b) Find the work done in stretching the spring 3 in. beyond its natural length.
   (c) Find the work done in stretching the spring from a length of 20 in. to a length of 25 in.

3. A spring exerts a force of $\frac{1}{2}$ ton when stretched 5 ft beyond its natural length. How much work is required to stretch the spring 6 ft beyond its natural length?

4. Assume a force of 6 N (newtons) is required to compress a spring from a natural length of 4 m (meters) to a length of $3\frac{1}{2}$ m. Find the work required to compress the spring from its natural length to a length of 2 m. (Hooke's law applies to compression as well as extension.)

5. Assume 10 ft-lb of work is required to stretch a spring 1 ft beyond its natural length. What is the spring constant?

6. A cylindrical tank of radius 5 ft and height 9 ft is two-thirds filled with water. Find the work required to pump all the water over the upper rim.

7. Solve Problem 6 assuming that the tank is two-thirds filled with a liquid of density ρ lb/ft^3.

8. A cone-shaped water reservoir is 20 ft in diameter across the top and 15 ft deep. If the reservoir is filled to a depth of 10 ft, how much work is required to pump all the water to the top of the reservoir?

9. A swimming pool is built in the shape of a rectangular parallelepiped 10 ft deep, 15 ft wide, and 20 ft long.
   (a) If the pool is filled 1 ft below the top, how much work is required to pump all the water into a drain at the top edge of the pool?
   (b) If a one-horsepower motor can do 550 ft-lb of work per second, what size motor is required to empty the pool in one hour?

10. A water tower in the shape of a hemisphere of radius 10 ft is to be filled by pumping water over the top edge from a lake 200 ft below the top. How much work is required to fill the tank? [*Hint:* Solve without calculus.]

11. A 100 ft length of steel chain weighing 15 lb/ft is dangling from a pulley. How much work is required to wind the chain onto the pulley?

12. A rocket weighing 3 tons is filled with 40 tons of liquid fuel. In the initial part of the flight fuel is burned off at a constant rate of 2 tons per 1000 ft of vertical height. How much work is done in lifting the rocket to 3000 ft?

13. **(Satellite Problem)** The weight of an object is the force exerted on it by the earth's gravity. Thus, if a person weighs 100 lb on the surface of the earth, the earth's gravity is pulling on that person with a force of 100 lb. It is a fundamental law of physics that the force the earth exerts on an object varies inversely as the square of its distance from the earth's center. Thus an object's weight $F(x)$ is related to its distance x from the earth's center by a formula of the form

$$F(x) = k/x^2$$

where k is a constant of proportionality depending on the mass of the object and the units of force and distance.
   (a) Assuming that the earth is a sphere of radius 4000 miles, find the constant k in the formula above if the satellite weighs 6000 lb on the earth's surface.
   (b) How much work must be performed to lift this satellite to an orbital position 1000 miles above the earth's surface?

14. **(Coulomb's law)** It follows from Coulomb's law in physics that two like electrostatic charges repel each other with a force inversely proportional to the square of the distance between them. Suppose that two charges A and B repel with a force of k pounds when they are positioned at points $A(-a, 0)$ and $B(a, 0)$. Find the work W required to move charge A along the x-axis to the origin if charge B remains stationary.

6.7 LIQUID PRESSURE AND FORCE

In this section we use the definite integral to calculate the force exerted by a liquid on a submerged surface.

If a flat surface of area A is submerged horizontally at a depth h in a container of fluid, then the weight of the fluid above exerts a force F on the surface given by

$$F = \rho h A \qquad (1)$$

where ρ is the density of the fluid (weight per unit volume). It is a physical fact that the force F in (1) does not depend on the shape or size of the container. Thus, if the three containers in Figure 6.7.1 have bases of the same area and are filled with a fluid to the same depth, h, then each container will have the same fluid force on its base.

Figure 6.7.1

The units in (1) must be compatible. For example, if height is in feet and force in pounds, then the density ρ must be expressed in pounds per cubic foot. The density of water is approximately 62.4 lb/ft^3.

Pressure is defined to be force per unit area. Thus from (1), the pressure p exerted at each point of a flat surface of area A submerged horizontally at a depth h is given by

$$p = \frac{F}{A} = \rho h \tag{2}$$

▶ Example 1

If a flat circular plate of radius $r = 2$ ft is submerged horizontally in water so the top surface is at a depth of 3 ft, then the force on the top surface of the plate is

$$F = \rho h A = \rho h(\pi r^2) = (62.4)(3)(4\pi) = 748.8\pi \text{ lb}$$

and the pressure at each point on the plate surface is

$$p = \rho h = (62.4)(3) = 187.2 \text{ lb/ft}^2 \qquad \blacktriangleleft$$

Pascal's* principle in physics states that *fluid pressure is the same in all directions.* Thus, if a flat surface is submerged vertically (or at any angle at

*BLAISE PASCAL (1623–1662) French mathematician and scientist. Pascal's mother died when he was three years old and his father, a highly educated magistrate, personally provided the boy's early education. Although Pascal showed an inclination for science and mathematics, his father refused to tutor him in those subjects until he mastered Latin and Greek. Pascal's sister and primary biographer claimed that he independently discovered the first thirty-two propositions of Euclid without ever reading a book on geometry. (However, it is generally agreed that the story is apocryphal.) Nevertheless, the precocious Pascal published a highly respected essay on conic sections by the time he was sixteen years old. Descartes, who read the essay, thought it so brilliant that he could not believe that it was written by such a young man. By age 18 his health began to fail and until his death he was in frequent pain. However, his creativity was unimpaired.

Pascal's contributions to physics include the discovery that air pressure decreases with altitude and the principle of fluid pressure that bears his name. However, the originality of his work is questioned by some historians. Pascal made major contributions to a branch of mathematics called "projective geometry," and he helped to develop probability theory through a series of letters with Fermat.

In 1646, Pascal's health problems resulted in a deep emotional crisis that led him to become increasingly concerned with religious matters. Although born a Catholic, he converted to a religious doctrine called Jansenism, and spent most of his final years writing on religion and philosophy.

all), the pressure at a point of depth h is exactly the same as the pressure at a point of depth h on a horizontal surface (Figure 6.7.2). However, (1) cannot be applied to obtain the total force acting on a flat surface that is not submerged horizontally, because the depth h can vary from point to point on the surface. It is in such problems that calculus comes into play.

By Pascal's Principle the fluid pressure at points A, B and C is the same.

Figure 6.7.2

Suppose we want to compute the total fluid force against a submerged vertical surface. As shown in Figure 6.7.3, let us introduce a vertical x-axis whose positive direction is downward and whose origin is at any convenient point.

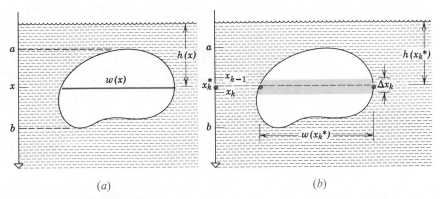

Figure 6.7.3   (a)   (b)

As indicated in Figure 6.7.3a, let us assume that the submerged portion of the plate extends from $x = a$ to $x = b$ on the x-axis. Also, suppose that a point x on the axis lies $h(x)$ units below the surface and that the cross section of the plate at x has width $w(x)$.

Next, divide the interval $[a, b]$ into n subintervals with lengths

$$\Delta x_1, \Delta x_2, \ldots, \Delta x_n$$

and in each subinterval choose an arbitrary point x_k^*. Following a familiar pattern, we approximate the section of plate along the kth subinterval by a rectangle of length $w(x_k^*)$ and width Δx_k^* (Figure 6.7.3b). Because the upper and lower edges of this rectangle are at different depths, (1) cannot be used to calculate the force on this rectangle. However, if Δx_k is small, the difference in depth between the upper and lower edges is small, and we can reasonably assume that the entire rectangle is concentrated at a single depth $h(x_k^*)$ below the surface. With this assumption, (1) can be used to *approximate* the force F_k on the kth rectangle. We obtain

$$F_k \approx \underbrace{\rho h(x_k^*)}_{\text{depth}} \cdot \underbrace{w(x_k^*)\,\Delta x_k}_{\text{area of rectangle}}$$

Thus the total force F on the plate is approximately

$$F = \sum_{k=1}^{n} F_k \approx \sum_{k=1}^{n} \rho h(x_k^*) w(x_k^*)\,\Delta x_k$$

To find the *exact* value of the force we take the limit as $\max \Delta x_k \to 0$. This yields

$$F = \lim_{\max \Delta x_k \to 0} \sum_{k=1}^{n} \rho h(x_k^*) w(x_k^*)\,\Delta x_k = \int_a^b \rho h(x) w(x)\,dx$$

which suggests the following definition.

6.7.1 DEFINITION   Assume that a plate is immersed vertically in a liquid of density ρ and that the submerged portion extends from $x = a$ to $x = b$ on a vertical x-axis. For $a \le x \le b$, let $w(x)$ be the width of the plate at x and let $h(x)$ be the depth of the point x. Then the total *fluid force* on the plate is

$$F = \int_a^b \rho h(x) w(x)\,dx \tag{3}$$

▶ **Example 2**

The face of a dam is a vertical rectangle of height 100 ft and width 200 ft (Figure 6.7.4). Find the total fluid force exerted on the face when the water surface is level with the top of the dam.

Figure 6.7.4

Solution. Introduce an x-axis with origin at the water surface as shown in Figure 6.7.4. At a point x on this axis, the width of the dam in feet is $w(x) = 200$ and the depth in feet is $h(x) = x$. Thus from (3) with $\rho = 62.4$ lb/ft^3 (the density of water) we obtain as the total force on the face

$$F = \int_0^{100} (62.4)(x)(200)\, dx$$

$$= 12{,}480 \int_0^{100} x\, dx$$

$$= 12{,}480 \frac{x^2}{2} \Big]_0^{100}$$

$$= 62{,}400{,}000 \text{ lb} \qquad \blacktriangleleft$$

▶ Example 3

A plate in the form of an isosceles triangle with base 10 ft and altitude 4 ft is submerged vertically in oil as shown in Figure 6.7.5a. Find the fluid force F against a surface of the plate if the oil has density $\rho = 30$ lb/ft^3.

Solution. Introduce an x-axis as shown in Figure 6.7.5b. By similar triangles,

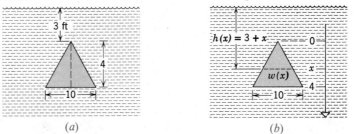

Figure 6.7.5   (a)   (b)

the width of the plate at a depth of $h(x) = 3 + x$ ft satisfies

$$\frac{w(x)}{10} = \frac{x}{4}$$

so

$$w(x) = \frac{5}{2}x$$

Thus, it follows from (3) that the force on the plate is

$$F = \int_a^b \rho h(x) w(x)\, dx = \int_0^4 (30)(3 + x)\left(\frac{5}{2}x\right) dx$$

$$= 75 \int_0^4 (3x + x^2)\, dx$$

$$= 75 \left[\frac{3x^2}{2} + \frac{x^3}{3}\right]_0^4 = 3400 \text{ lb} \qquad \blacktriangleleft$$

▶ Exercise Set 6.7

1. A flat square plate with 3 ft sides and negligible thickness is submerged horizontally in a liquid. Find the force and pressure on a surface of the plate if
    (a) the liquid is water and the plate is at a depth of 5 ft.
    (b) the liquid has density 40 lb/ft^3 and the plate is at a depth of 10 ft.

In Exercises 2–7, the flat surfaces shown are submerged vertically in water. Find the fluid force against the suface.

2.

3.

4.

5.

6.

7.

8. An oil tank is shaped like a right circular cylinder of diameter 4 ft. Find the total fluid force against one end when the axis is horizontal and the tank is half filled with oil of density 50 lb/ft^3.

9. A square plate of side a ft is dipped in a liquid of density ρ lb/ft^3. Find the fluid force on the plate if a vertex is at the surface and a diagonal is perpendicular to the surface.

10. Figure 6.7.6 shows a dam whose face is an inclined rectangle. Find the fluid force on the face when the water is level with the top of this dam.

Figure 6.7.6

11. Figure 6.7.7 shows a rectangular swimming pool whose bottom is an inclined plane. Find the fluid force on the bottom when the pool is filled to the top.

Figure 6.7.7

12. An observation window on a submarine is a square with 2 ft sides. Using ρ_0 for the density of sea water, find the fluid force on the window when the submarine has descended vertically so that top of the window is at a depth of h feet.

13. (a) Show: If the submarine in Exercise 12 descends vertically at a constant rate, then the fluid force on the window increases at a constant rate.

(b) At what rate is the force on the window increasing if the submarine is descending vertically at 20 ft/min?

14. Find the fluid force on the lateral surface of a right circular cylinder of base radius r and altitude h when the cylinder is submerged so its top is level with the surface of the water.

▶ REVIEW EXERCISES

In Exercises 1–3, set up, but do not evaluate an integral or sum of integrals that gives the area of the region R. (Set up the integral with respect to x or y as directed.)

1. R is the region in the first quadrant enclosed by $y = x^2$, $y = 2 + x$, and $x = 0$.
   (a) Integrate with respect to x.
   (b) Integrate with respect to y.

2. R is enclosed by $x = 4y - y^2$ and $y = \frac{1}{2}x$.
   (a) Integrate with respect to x.
   (b) Integrate with respect to y.

3. R is enclosed by $x = 9$ and $x = y^2$.
   (a) Integrate with respect to x.
   (b) Integrate with respect to y.

In Exercises 4–9, set up, but do not evaluate an integral or sum of integrals that gives the stated volume. (Set up the integral with respect to x or y as directed.)

4. The volume generated by revolving the region in Exercise 1 about the x-axis.
   (a) Integrate with respect to x.
   (b) Integrate with respect to y.

5. The volume generated by revolving the region in Exercise 1 about the y-axis.
   (a) Integrate with respect to x.
   (b) Integrate with respect to y.

6. The volume generated by revolving the region in Exercise 2 about the x-axis.
   (a) Integrate with respect to x.
   (b) Integrate with respect to y.

7. The volume generated by revolving the region in Exercise 2 about the y-axis.
   (a) Integrate with respect to x.
   (b) Integrate with respect to y.

8. The volume generated by revolving the region in Exercise 3 about the x-axis.
   (a) Integrate with respect to x.
   (b) Integrate with respect to y.

9. The volume generated by revolving the region in Exercise 3 about the y-axis.
   (a) Integrate with respect to x.
   (b) Integrate with respect to y.

In Exercises 10–11, find:
   (a) the area of the region described;
   (b) the volume generated by revolving the region about the indicated line.

10. The region in the first quadrant enclosed by $y = \sin x$, $y = \cos x$, and $x = 0$; revolved about the x-axis.
    [*Hint:* $\cos^2 x - \sin^2 x = \cos 2x$]

11. The region enclosed by the x-axis, the y-axis and $x = \sqrt{4 - y}$; revolved about the y-axis.

12. Set up a sum of definite integrals that represents the total shaded area between the curves $y = f(x)$ and $y = g(x)$ below.

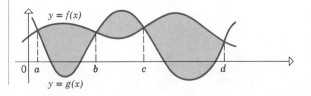

13. Find the *total* area bounded between $y = x^3$ and $y = x$ over the interval $[-1, 2]$. (See previous exercise.)

14. Find the volume of the solid whose base is the region bounded between the curves $y = x$ and $y = x^2$, and whose cross sections perpendicular to the x-axis are squares.

15. Find the volume of the solid whose base is the triangular region with vertices $(0, 0)$, $(a, 0)$, and $(0, b)$, where $a > 0$ and $b > 0$, and whose cross sections perpendicular to the y-axis are semicircles.

In Exercises 16–17, find the volume generated by revolving the region described about the axis indicated.

16. The region bounded above by the curve $y = \cos x^2$, on the left by the y-axis, and below by the x-axis; revolved about the y-axis.

17. The region bounded by $y = \sqrt{x}$, $x = 4$, and $y = 0$; revolved about: (a) the line $x = 4$, (b) the line $y = 2$.

18. A football has the shape of the solid generated by revolving the region bounded between the x-axis and the parabola $y = 4R(x^2 - \frac{1}{4}L^2)/L^2$ about the x-axis. Find its volume.

In Exercises 19–22, find the arc length of the indicated curve.

19. $8y^2 = x^3$ between $(0, 0)$ and $(2, 1)$.

20. $y = \frac{1}{3}(x^2 + 2)^{3/2}$, $0 \le x \le 3$.

21. $y = \frac{1}{10}x^5 + \frac{1}{6}x^{-3}$, $1 \le x \le 2$.

22. $y = \frac{1}{3}x^3 + \frac{1}{4x}$, $1 \le x \le 2$.

In Exercises 23–28, find the area of the surface generated by revolving the given curve about the indicated axis.

23. $y = x^3$ between $(1, 1)$ and $(2, 8)$; x-axis.

24. $y^2 = 12x$ between $(0, 0)$ and $(3, 6)$; x-axis.

25. $y = \frac{2}{3}x^{3/2} - \frac{1}{2}x^{1/2}$ between $(0, 0)$ and $(9, 33/2)$; y-axis.

26. The curve in Exercise 25 revolved about the line $x = 9$.

27. $3y = \sqrt{x}(3 - x)$ between $(0, 0)$ and $(3, 0)$; y-axis.

28. $y = \sqrt{2x - x^2}$ between $(\frac{1}{2}, \sqrt{3}/2)$ and $(1, 1)$; x-axis.

29. The natural length of a spring is 6 in. If a force of 2 lb is needed to hold it at a length of 10 in., find the work done in stretching it from 8 in. to 10 in.

30. Find the spring constant if 180 in.-lbs of work are required to stretch a spring 3 in. from its natural length.

31. A 250-lb weight is suspended from a ledge by a uniform 40 ft cable weighing 30 lb. How much work is required to bring the weight up to the ledge?

32. A tank in the shape of a right circular cone has a 6-ft diameter at the top and a height of 5 ft. It is filled with a liquid of density 64 lb/ft³. How much work can be done by the liquid if it runs out of the bottom of the tank?

33. A vessel has the shape obtained by revolving about the y-axis the part of the parabola $y = 2(x^2 - 4)$ lying below the x-axis. If x and y are in feet, how much work is required to pump all the water in the full vessel to a point 4 ft above its top?

34. Two like magnetic poles repel each other with a force $F = k/x^2$ newtons. Express the work needed to move them along a line from D meters apart to $D/3$ meters apart.

In Exercises 35–36, the flat surface shown is submerged vertically in a liquid of density ρ lb/ft³. Find the fluid force against the surface.

35.

36.

(Semicircle)

7 logarithm and exponential functions

7.1 INTRODUCTION

There are many simple functions whose integrals we cannot yet evaluate. For example, if $n \neq -1$ we know that

$$\int x^n \, dx = \frac{x^{n+1}}{n+1} + C$$

However, if $n = -1$ this formula does not apply. At this stage we have no solution to the simple-looking integration problem

$$\int \frac{1}{x} \, dx = ?$$

This problem is more difficult than it appears since there is no antiderivative for $1/x$ expressible in terms of familiar functions from algebra or trigonometry. Similarly, such basic integration problems as

$$\int \tan x \, dx \qquad \text{and} \qquad \int \sec x \, dx$$

have no elementary solutions. In this chapter we will create new functions to help us handle such problems.

7.2 THE NATURAL LOGARITHM; THE SECOND FUNDAMENTAL THEOREM OF CALCULUS

In this section we discuss one of the most useful theoretical applications of the definite integral, the construction of functions with specified derivatives.

We will use this work to obtain a new function of fundamental importance, called the **natural logarithm.**

In our subsequent work we will need to consider definite integrals of the form

$$\int_a^x \underline{\hspace{3cm}}$$

where the upper limit x is allowed to vary. For such integrals we will use a letter different from x (often t) for the variable of integration; thus we would write

$$\int_a^x f(t)\,dt \tag{1}$$

rather than

$$\int_a^x f(x)\,dx \tag{2}$$

We do this because x is used in two different ways in (2), as a variable of integration and as a limit of integration. This can cause errors.

▶ Example 1

Evaluate

$$\int_2^x t^2\,dt$$

Solution.

$$\int_2^x t^2\,dt = \frac{t^3}{3}\Bigg]_{t=2}^{t=x} = \frac{x^3}{3} - \frac{8}{3} \qquad\qquad ◀$$

As this example shows, an expression of the form

$$\int_a^x f(t)\,dt$$

represents a function of x—the variable t does not enter into the final result. The following major theorem tells us how to differentiate functions of this type.

7.2.1 THEOREM

Second Fundamental Theorem of Calculus

Let f be continuous on an open interval I, and let a be any point in I. If F is defined by

$$F(x) = \int_a^x f(t)\, dt \tag{3}$$

then at each point x in the interval I

$$F'(x) = f(x)$$

Proof. From the definition of derivative,

$$F'(x) = \lim_{h \to 0} \frac{F(x+h) - F(x)}{h}$$

$$= \lim_{h \to 0} \frac{1}{h} \left[\int_a^{x+h} f(t)\, dt - \int_a^x f(t)\, dt \right]$$

$$= \lim_{h \to 0} \frac{1}{h} \left[\int_a^{x+h} f(t)\, dt + \int_x^a f(t)\, dt \right]$$

$$= \lim_{h \to 0} \frac{1}{h} \int_x^{x+h} f(t)\, dt \qquad \text{[Theorem 5.9.2]}$$

Applying the mean-value theorem for integrals (5.10.1) to the last expression, we obtain

$$F'(x) = \lim_{h \to 0} \frac{1}{h} [f(t^*) \cdot h] = \lim_{h \to 0} f(t^*) \tag{4}$$

where t^* is some number between x and $x + h$. Because t^* is between x and $x + h$, it follows that $t^* \to x$ as $h \to 0$. Thus $f(t^*) \to f(x)$ as $h \to 0$, since f is assumed continuous at x. Therefore, from (4)

$$F'(x) = f(x) \quad \blacksquare$$

The second fundamental theorem of calculus ensures that if f is continuous on an open interval, then f has an antiderivative on that interval, namely, $\int_a^x f(t)\, dt$. However, since we can add an arbitrary constant to an antiderivative of f and still have an antiderivative of f, it follows that a function, continuous on an open interval, has infinitely many antiderivatives on the interval. Note, however, that (3) singles out that antiderivative of f whose value at $x = a$ is zero since

$$F(a) = \int_a^a f(t)\, dt = 0$$

▶ Example 2

The function $f(x) = x^2$ is continuous on $(-\infty, +\infty)$. Thus, by the second fundamental theorem of calculus and the remarks above, the function

$$F(x) = \int_a^x f(t)\, dt = \int_2^x t^2\, dt$$

must be that antiderivative of $f(x) = x^2$ on $(-\infty, +\infty)$ having value zero at $x = 2$. This is indeed the case since

$$F(x) = \frac{x^3}{3} - \frac{8}{3} \tag{5}$$

(see Example 1) so that

$$F'(x) = x^2 \qquad \text{and} \qquad F(2) = \frac{8}{3} - \frac{8}{3} = 0 \qquad \blacktriangleleft$$

▶ Example 3

Since $f(x) = 1/x$ is continuous on the interval $(0, +\infty)$,

$$F(x) = \int_1^x \frac{1}{t}\, dt$$

is the antiderivative of $1/x$ on $(0, +\infty)$ whose value at 1 is zero. That is,

$$F'(x) = \frac{1}{x} \qquad \text{and} \qquad F(1) = 0 \qquad \blacktriangleleft$$

In Example 1 we were able to "carry out" the integration

$$F(x) = \int_2^x t^2\, dt = \frac{x^3}{3} - \frac{8}{3}$$

and express $F(x)$ as a familiar function. However, we cannot "carry out" the integration

$$F(x) = \int_1^x \frac{1}{t}\, dt \tag{6}$$

so that we are forced to leave this antiderivative of $1/x$ in integral form. This raises an interesting computational question. How do we evaluate the function $F(x)$ in (6)? For example, how would we evaluate

$$F(2) = \int_1^2 \frac{1}{t}\, dt?$$

Since we cannot carry out the integration, we must approximate the value of the definite integral. This can be done using Riemann sums or other numerical techniques that we will study in later sections.

The function defined by (6) is of fundamental importance in mathematics. It is called the ***natural logarithm*** of x (for reasons that will become clear as we progress) and is denoted by the symbol ***ln*** (read, "ell-en").

7.2.2 DEFINITION
The Natural Logarithm Function

$$\ln x = \int_1^x \frac{1}{t}\, dt, \qquad x > 0 \tag{7}$$

It follows from this definition that $\ln x$ is that antiderivative of $1/x$ on $(0, +\infty)$ whose value is zero when $x = 1$. Thus

$$\frac{d}{dx}[\ln x] = \frac{1}{x}, \qquad x > 0 \tag{8}$$

$$\ln 1 = \int_1^1 \frac{1}{t}\, dt = 0$$

If $u(x) > 0$ and if the function u is differentiable at x, the chain rule implies that

$$\frac{d}{dx}[\ln u] = \frac{d}{du}[\ln u] \cdot \frac{du}{dx}$$

Since

$$\frac{d}{du}[\ln u] = \frac{1}{u}$$

we obtain

$$\frac{d}{dx}[\ln u] = \frac{1}{u} \cdot \frac{du}{dx} \tag{9}$$

▶ Example 4

Find

$$\frac{d}{dx}[\ln (x^2)]$$

Solution. From (9) with $u = x^2$,

$$\frac{d}{dx}[\ln (x^2)] = \frac{1}{x^2} \cdot \frac{d}{dx}[x^2] = \frac{1}{x^2} \cdot 2x = \frac{2}{x} \qquad ◀$$

▶ Example 5

Find

$$\frac{d}{dx}[\ln|x|]$$

Solution. We consider the cases where $x > 0$ and $x < 0$ separately. [The case where $x = 0$ is excluded since ln 0 is undefined; see (7).] If $x > 0$, then $|x| = x$ so that

$$\frac{d}{dx}[\ln|x|] = \frac{d}{dx}[\ln x] = \frac{1}{x}$$

If $x < 0$, then $|x| = -x$ so that from (9) with $u = -x$,

$$\frac{d}{dx}[\ln|x|] = \frac{d}{dx}[\ln(-x)] = \frac{1}{(-x)} \cdot \frac{d}{dx}[-x] = \frac{1}{x}$$

Thus

$$\frac{d}{dx}[\ln|x|] = \frac{1}{x} \tag{10}$$

if $x \neq 0$. ◀

Formula (10) states that the function $\ln|x|$ is an antiderivative of $1/x$ everywhere except at $x = 0$. In contrast, the function $\ln x$ is an antiderivative for $1/x$ only for $x > 0$.

The companion integration formula for (10) is

$$\int \frac{1}{x} dx = \ln|x| + C \tag{11}$$

▶ Example 6

From (10) and the chain rule,

$$\frac{d}{dx}[\ln|\sin x|] = \frac{1}{\sin x} \cdot \frac{d}{dx}[\sin x]$$

$$= \frac{\cos x}{\sin x}$$

$$= \cot x \qquad\qquad ◀$$

▶ Example 7

Evaluate

$$\int \frac{3x^2}{x^3 + 5} \, dx$$

Solution. Make the substitution

$$u = x^3 + 5 \qquad du = 3x^2 \, dx$$

so that

$$\int \frac{3x^2}{x^3 + 5} \, dx = \int \frac{1}{u} \, du$$

$$= \ln |u| + C \qquad \text{[Formula (11)]}$$
$$= \ln |x^3 + 5| + C \qquad\qquad\blacktriangleleft$$

REMARK. This example illustrates an important point. Any integral of the form

$$\int \frac{g'(x)}{g(x)} \, dx$$

(where the numerator of the integrand is the derivative of the denominator) can be evaluated by the *u*-substitution $u = g(x)$, $du = g'(x) \, dx$ since this substitution yields

$$\int \frac{g'(x)}{g(x)} \, dx = \int \frac{du}{u} = \ln |u| + C = \ln |g(x)| + C$$

▶ Example 8

Evaluate

$$\int \tan x \, dx$$

Solution.

$$\int \tan x \, dx = \int \frac{\sin x}{\cos x} \, dx$$

$$= -\int \frac{1}{u} \, du \qquad \begin{bmatrix} u = \cos x \\ du = -\sin x \, dx \end{bmatrix}$$

$$= -\ln |u| + C$$
$$= -\ln |\cos x| + C \qquad\qquad\blacktriangleleft$$

▶ **Exercise Set 7.2**

In Exercises 1–14, find dy/dx.

1. $y = \ln 2x.$

2. $y = \ln (x^3).$

3. $y = (\ln x)^2.$

4. $y = \ln (\sin x).$

5. $y = \ln |\tan x|.$

6. $y = \ln (2 + \sqrt{x}).$

7. $y = \ln \left(\dfrac{x}{1 + x^2} \right).$

8. $y = \ln (\ln x).$

9. $y = \ln |x^3 - 7x^2 - 3|.$

10. $y = x^3 \ln x.$

11. $y = \sqrt{\ln x}.$

12. $y = \cos (\ln x).$

13. $y = \dfrac{x^2}{1 + \ln x}.$

14. $y = \ln \left| \dfrac{1 - \cos \pi x}{1 + \cos \pi x} \right|.$

15. Find dy/dx by implicit differentiation if
$y + \ln xy = 1.$

16. Find dy/dx by implicit differentiation if
$y = \ln (x \tan y).$

In Exercises 17–28, evaluate the indefinite integrals.

17. $\displaystyle \int \frac{dx}{2x}.$

18. $\displaystyle \int \frac{5x^4}{x^5 + 1} \, dx.$

19. $\displaystyle \int \frac{x^2}{x^3 - 4} \, dx.$

20. $\displaystyle \int \frac{t + 1}{t} \, dt.$

21. $\displaystyle \int \frac{\sec^2 x}{\tan x} \, dx.$

22. $\displaystyle \int \cot x \, dx.$

23. $\displaystyle \int \frac{\sin 3\theta}{1 + \cos 3\theta} \, d\theta.$

24. $\displaystyle \int \frac{dx}{x \ln x}.$

25. $\displaystyle \int \frac{x^3}{x^2 + 1} \, dx.$

26. $\displaystyle \int \frac{1}{x} \cos (\ln x) \, dx.$

27. $\displaystyle \int \frac{1}{y} (\ln y)^3 \, dy.$

28. $\displaystyle \int \frac{dx}{\sqrt{x}(1 - 2\sqrt{x})}.$

29. Define $F(x)$ by

$$F(x) = \int_1^x (t^3 + 1) \, dt$$

   (a) Use the second fundamental theorem of calculus to find $F'(x)$.

   (b) Check the result in (a) by first integrating and then differentiating.

30. The formulas

$$\int \frac{dx}{x} = \ln x + C \quad \text{and} \quad \int \frac{dx}{x} = \ln |x| + C$$

are both correct. What advantage does the second formula have over the first?

31. Express the following as integrals:

   (a) An antiderivative for $1/(1 + x^2)$ on the interval $(-\infty, +\infty)$ whose value at $x = 1$ is 0.

   (b) An antiderivative for $1/(x - 1)$ on the interval $(1, +\infty)$ whose value at $x = 2$ is 0.

   (c) An antiderivative for $1/(x - 1)$ on the interval $(-\infty, 1)$ whose value at $x = -3$ is 0.

32. (a) Over what open interval does the formula

$$F(x) = \int_1^x \frac{1}{t^2 - 9} \, dt$$

represent an antiderivative of

$$f(x) = \frac{1}{x^2 - 9} ?$$

   (b) Find a point where the graph of F crosses the x-axis.

33. Prove: If f is continuous on an open interval I and b is any point in I, then at each point in I

$$\frac{d}{dx} \int_x^b f(t) \, dt = -f(x)$$

34. Use the second fundamental theorem of calculus and the chain rule to show that

$$\frac{d}{dx} \int_a^{g(x)} f(t) \, dt = f(g(x)) g'(x)$$

35. Apply the result in Exercise 34 to perform the following differentiations:

   (a) $\dfrac{d}{dx} \displaystyle\int_1^{x^3} \frac{1}{t} \, dt$

   (b) $\dfrac{d}{dx} \displaystyle\int_3^{\sin x} \frac{1}{1 + t^2} \, dt.$

36. Use Exercise 34 to show that

$$\frac{d}{dx} \int_{h(x)}^{g(x)} f(t)\, dt = f(g(x))g'(x) - f(h(x))h'(x)$$

[*Hint:* Theorem 5.9.2.]

37. Use the result in Exercise 36 to perform the following differentiations:

(a) $\dfrac{d}{dx} \int_{x^2}^{x^3} \sin^2 t\, dt$ (b) $\dfrac{d}{dx} \int_{\ln x}^{x} \dfrac{1}{1+t}\, dt.$

38. Prove: If f is continuous on an open interval I and a is any point in I, then

$$F(x) = \int_{a}^{x} f(t)\, dt$$

is continuous on I.

7.3 PROPERTIES OF THE NATURAL LOGARITHM

In this section we develop basic properties of the logarithm function and obtain the graph of $y = \ln x$.

Our first theorem explains why the function $\ln x$ has the word *logarithm* in its name. The properties of $\ln x$ stated in this theorem are identical to properties of $\log_{10} x$ (logarithm to the base 10), which is studied in high school.

7.3.1 THEOREM *For any positive numbers a and b and any positive rational number r,*

(a) $\ln ab = \ln a + \ln b$

(b) $\ln \dfrac{a}{b} = \ln a - \ln b$

(c) $\ln a^r = r \ln a$

(d) $\ln \dfrac{1}{b} = -\ln b$

Our proof of this theorem uses Corollary 4.9.4, which states that two functions with the same derivative on an interval differ by a constant on that interval.

Proof of (a).   Consider the function

$$f(x) = \ln ax$$

Treating a as constant and differentiating with respect to x, we obtain

$$f'(x) = \frac{d}{dx}[\ln ax] = \frac{1}{ax} \cdot \frac{d}{dx}[ax] = \frac{1}{ax} \cdot a = \frac{1}{x}$$

which shows that $\ln ax$ and $\ln x$ have the same derivative on $(0, +\infty)$. Thus, by Corollary 4.9.4, there is a constant k such that

$$\ln ax - \ln x = k \qquad (1)$$

If we let $x = 1$ in this equation and use the fact that $\ln 1 = 0$, we obtain

$$\ln a = k$$

so that (1) can be written

$$\ln ax - \ln x = \ln a$$

In particular, if $x = b$ we obtain

$$\ln ab - \ln b = \ln a$$

or

$$\ln ab = \ln a + \ln b$$

Proofs of (b) *and* (d). Using part (a) we can write

$$0 = \ln 1 = \ln \left(b \cdot \frac{1}{b} \right) = \ln b + \ln \frac{1}{b}$$

so that

$$\ln \frac{1}{b} = -\ln b$$

which proves (d). To prove (b) we write

$$\ln \frac{a}{b} = \ln \left(a \cdot \frac{1}{b} \right) = \ln a + \ln \frac{1}{b} = \ln a - \ln b$$

Proof of (c). The functions $\ln x^r$ and $r \ln x$

have the same derivative on $(0, +\infty)$, since

$$\frac{d}{dx} [\ln x^r] = \frac{1}{x^r} \cdot \frac{d}{dx} [x^r] = \frac{1}{x^r} \cdot rx^{r-1} = \frac{r}{x}$$

and

$$\frac{d}{dx} [r \ln x] = r \frac{d}{dx} [\ln x] = \frac{r}{x}$$

Thus there is a constant k such that

$$\ln x^r - r \ln x = k \tag{2}$$

If we let $x = 1$ in this equation and use the fact that $\ln 1 = 0$, we obtain $k = 0$ so that (2) can be written

$$\ln x^r - r \ln x = 0 \qquad \text{or} \qquad \ln x^r = r \ln x$$

Letting $x = a$ completes the proof. ▮

▶ Example 1

Property (c) in the above theorem yields the following useful result:

$$\ln \sqrt[n]{a} = \ln a^{1/n} = \frac{1}{n} \ln a$$ ◀

The values of $\ln x$ have useful interpretations as areas. If $x > 1$ then

$$\ln x = \int_1^x \frac{1}{t} \, dt$$

can be viewed as the area under the curve $y = 1/t$ between the points 1 and x on the t-axis (Figure 7.3.1a). On the other hand, if $0 < x < 1$, then by writing

$$\ln x = -\int_x^1 \frac{1}{t} \, dt$$

we can view $\ln x$ as the *negative* of the area under the curve $y = 1/t$ between the points x and 1 on the t-axis (Figure 7.3.1b). Therefore

$$\ln x > 0 \qquad \text{if} \qquad x > 1$$
$$\ln x < 0 \qquad \text{if} \qquad 0 < x < 1$$
$$\ln x = 0 \qquad \text{if} \qquad x = 1$$

Since $\ln x$ is differentiable for $x > 0$, it is continuous on the interval $(0, +\infty)$. The first derivative

$$\frac{d}{dx}[\ln x] = \frac{1}{x}$$

(a)

(b)

Figure 7.3.1

is positive for all x in $(0, +\infty)$, so that $\ln x$ is *increasing* on this interval, and the second derivative

$$\frac{d^2}{dx^2} [\ln x] = \frac{d}{dx} \left[\frac{1}{x} \right] = -\frac{1}{x^2}$$

is negative for all x in $(0, +\infty)$, so that the graph of $\ln x$ is *concave down* on this interval.

To complete the picture of the graph, we must investigate the behavior of $\ln x$ as $x \to 0^+$ and $x \to +\infty$. For this purpose we will need an estimate of $\ln 2$. From the mean-value theorem for integrals (5.10.1) we obtain

$$\ln 2 = \int_1^2 \frac{1}{t} \, dt = (2 - 1) \cdot \frac{1}{t^*} = \frac{1}{t^*} \tag{3}$$

for some t^* satisfying

$$1 \le t^* \le 2 \tag{4}$$

Taking reciprocals in (4) yields

$$\frac{1}{2} \le \frac{1}{t^*} \le 1$$

so that from (3),

$$\frac{1}{2} \le \ln 2 \le 1 \tag{5}$$

We will now use this estimate to deduce the following results.

7.3.2 THEOREM   *(a)* $\lim\limits_{x \to +\infty} \ln x = +\infty$

                         (b) $\lim\limits_{x \to 0^+} \ln x = -\infty$

Proof of (a). We will show that for any positive number N (no matter how large) the values of $\ln x$ eventually exceed N as $x \to +\infty$. It will then follow that $\lim\limits_{x \to +\infty} \ln x = +\infty$.

Since $\ln x$ is an increasing function, it follows from (5) that for $x > 2^{2N}$ we must have

$$\ln x > \ln 2^{2N} = 2N \ln 2 \ge 2N \left(\frac{1}{2} \right) = N$$

Therefore $\lim\limits_{x \to +\infty} \ln x = +\infty$.

Proof of (b). If we let

$$v = \frac{1}{x}$$

then $v \to +\infty$ as $x \to 0^+$, so that from part (a) of this theorem and Theorem 7.3.1d we obtain

$$\lim_{x \to 0^+} \ln x = \lim_{x \to 0^+} \left(-\ln \frac{1}{x}\right) = \lim_{v \to +\infty} (-\ln v)$$

$$= -\lim_{v \to +\infty} \ln v = -\infty \quad \blacksquare$$

In light of this theorem and the fact that $y = \ln x$ is increasing and concave down on $(0, +\infty)$ the graph of $\ln x$ is as shown in Figure 7.3.2. The graph crosses the x-axis at $x = 1$ since $\ln 1 = 0$. A more accurate picture can be obtained by plotting some points and sketching some tangent lines. For this purpose we use the approximation,

Figure 7.3.2

$$\ln 2 \approx 0.6931$$

which is accurate to four decimal places. (We will show how to obtain this approximation later.) Using this value of $\ln 2$ and applying Theorem 7.3.1 we can estimate the natural logarithm of any power of 2. For example,

$$\ln 8 = \ln 2^3 = 3 \ln 2 \approx 2.0793$$
$$\ln 4 = \ln 2^2 = 2 \ln 2 \approx 1.3862$$
$$\ln \tfrac{1}{4} = \ln 2^{-2} = -2 \ln 2 \approx -1.3862$$
$$\ln \tfrac{1}{8} = \ln 2^{-3} = -3 \ln 2 \approx -2.0793$$
$$\ln \tfrac{1}{2} = -\ln 2 \approx -0.6931$$

In Figure 7.3.3 we have plotted these points and drawn some tangent lines to obtain a more accurate sketch of the curve $y = \ln x$. The slopes of the tangent lines were obtained from the derivative of $\ln x$.

When possible, the properties of logarithms in Theorem 7.3.1 should be used to convert products, quotients, and exponents into sums, differences, and constant multiples *before* differentiating a function involving natural logarithms.

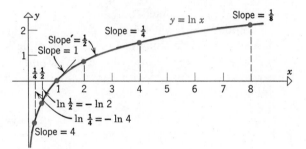

Figure 7.3.3

▶ Example 2

$$\frac{d}{dx}\left[\ln\left(\frac{x^2 \sin x}{\sqrt{1+x}}\right)\right] = \frac{d}{dx}\left[2\ln x + \ln(\sin x) - \frac{1}{2}\ln(1+x)\right]$$

$$= \frac{2}{x} + \frac{\cos x}{\sin x} - \frac{1}{2(1+x)}$$

$$= \frac{2}{x} + \cot x - \frac{1}{2+2x} \qquad ◀$$

We conclude this section by discussing a technique called *logarithmic differentiation.*

▶ Example 3

The derivative of

$$y = \frac{x^2 \sqrt[3]{7x-14}}{(1+x^2)^4} \qquad (6)$$

is messy to calculate directly. However, if we first take ln of both sides and use properties of the natural logarithm, we can write

$$\ln y = 2\ln x + \frac{1}{3}\ln(7x-14) - 4\ln(1+x^2)$$

Differentiating both sides with respect to x yields

$$\frac{1}{y}\frac{dy}{dx} = \frac{2}{x} + \frac{7/3}{7x-14} - \frac{8x}{1+x^2}$$

Thus, on solving for dy/dx and using (6) we obtain

$$\frac{dy}{dx} = \frac{x^2 \sqrt[3]{7x-14}}{(1+x^2)^4}\left[\frac{2}{x} + \frac{1}{3x-6} - \frac{8x}{1+x^2}\right] \qquad ◀$$

Since $\ln y$ is defined only for $y > 0$, results obtained by logarithmic differentiation are valid only at points where this condition is satisfied. The derivative formula obtained in the last example is valid for $x > 2$ since $y > 0$ for such x [see (6)]. As the next example shows, logarithmic differentiation can sometimes be used with functions having negative values by first taking absolute values.

▶ Example 4

The derivative of

$$y = \frac{x \sqrt[3]{x - 5}}{1 + \sin^3 x}$$

can be obtained by writing

$$\ln |y| = \ln \left| \frac{x \sqrt[3]{x - 5}}{1 + \sin^3 x} \right|$$

$$\ln |y| = \ln |x| + \frac{1}{3} \ln |x - 5| - \ln |1 + \sin^3 x|$$

$$\frac{1}{y} \frac{dy}{dx} = \frac{1}{x} + \frac{1}{3(x - 5)} - \frac{3 \sin^2 x \cos x}{1 + \sin^3 x}$$

$$\frac{dy}{dx} = \frac{x \sqrt[3]{x - 5}}{1 + \sin^3 x} \left[\frac{1}{x} + \frac{1}{3(x - 5)} - \frac{3 \sin^2 x \cos x}{1 + \sin^3 x} \right] \qquad ◀$$

▶ Exercise Set 7.3

1. Let $r = \ln 2$ and $s = \ln 3$. Express the following in terms of r and s.
    (a) $\ln 6$ (b) $\ln 1.5$
    (c) $\ln \frac{1}{2}$ (d) $\ln 9$
    (e) $\ln \sqrt[5]{3}$ (f) $\ln \frac{1}{36}$.

2. Assume that $\ln x_0 = 1$. Solve the following equations for x in terms of x_0.
    (a) $\ln x = -1$ (b) $\ln x = 2$
    (c) $\ln \sqrt[3]{x} = -\frac{1}{2}$.

3. (a) On graph paper, make an accurate sketch of the curve $y = \ln x$ by first plotting the points where $x = \frac{1}{9}, \frac{1}{3}, 1, 3, 9$, then sketching tangent lines at these points, and finally drawing a smooth curve through the points. [Use $\ln 3 \approx 1.1$.]
    (b) Use your sketch to estimate $\ln 2$.
    (c) Use your sketch to estimate x_0 such that $\ln x_0 = 1$.

In Exercises 4–7, draw the graph. Avoid point plotting.

4. $y = \ln |x|$.

5. $y = \ln \frac{1}{x}$.

6. $y = \ln \sqrt{x}$.

7. $y = \ln (x - 1)$.

In Exercises 8–13, use Theorem 7.3.1 to help perform the indicated differentiations and integrations.

8. $\dfrac{d}{dx}\left[\ln\sqrt{\dfrac{x-1}{x+1}}\,\right].$

9. $\dfrac{d}{dx}[\ln(\sqrt{x}\,\sqrt[3]{x}+3\,\sqrt[5]{(3x-2)}\,)].$

10. $\dfrac{d}{dx}\left[\ln\left(\dfrac{\sqrt{x}\,\sqrt[3]{x}+1}{\sin x \sec x}\right)\right].$

11. $\displaystyle\int \dfrac{1}{x}\ln(x^3)\,dx.$

12. $\displaystyle\int \dfrac{dx}{x\ln\sqrt{x}}.$

13. $\displaystyle\int \dfrac{\ln(1/x)}{x}\,dx.$

14. Use an area interpretation of $\ln x$ to obtain (5).

In Exercises 15–18, obtain dy/dx by logarithmic differentiation.

15. $y = x\sqrt[3]{1+x^2}.$

16. $y = \sqrt[5]{\dfrac{x-1}{x+1}}.$

17. $y = \dfrac{(x^2-8)^{1/3}\sqrt{x^3+1}}{x^6-7x+5}.$

18. $y = \dfrac{\sin x \cos x \tan^3 x}{\sqrt{x}}.$

19. (a) Make an appropriate u-substitution to show that
$$\int_a^{ab} \frac{1}{t}\,dt = \int_1^b \frac{1}{t}\,dt.$$

 (b) Use this equality to obtain another proof of $\ln(ab) = \ln a + \ln b.$

20. If the interval $[1, x]$ on the t-axis is divided into n equal subintervals and A_1 and A_2 denote, respectively, the total areas of the inscribed and circumscribed rectangles shown in Figure 7.3.4, then
$$A_1 \le \int_1^x \frac{1}{t}\,dt \le A_2$$

Use this to solve the following problems:
 (a) Find upper and lower estimates for $\ln 2$ using $n = 4$ subintervals.

 (b) Find upper and lower estimates for $\ln 2$ using $n = 10$ subintervals.
 (c) Compare your answers to the value listed in Appendix 3, Table 3.

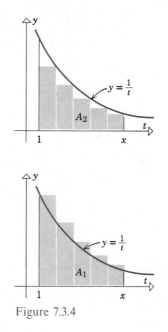

Figure 7.3.4

21. Prove: For all $x > 0$
$$1 - \frac{1}{x} \le \ln x \le x - 1$$
[*Hint*: Review the derivation of (5).]

22. Use the result of Exercise 21 to prove that
$$\lim_{x\to 0} \frac{\ln(x+1)}{x} = 1$$

23. (a) Use the result of Exercise 21 to prove that $0 < \ln x < x$ when $x > 1.$
 (b) Use the result in (a) to prove that
$$0 < \frac{\ln x}{x} < \frac{2}{\sqrt{x}}$$
when $x > 1.$ [*Hint*: $\ln x = 2\ln\sqrt{x}.$]
 (c) Use the result in (b) to prove that
$$\lim_{x\to +\infty} \frac{\ln x}{x} = 0$$

(d) Use the result in (c) to prove that

$$\lim_{x \to 0^+} x \ln x = 0$$

[*Remark:* This problem was suggested by an article published by D. S. Greenstein in the *American Mathematical Monthly*, vol. 72 (1965), p. 767).]

24. *Boyle's law* in physics states that under appropriate conditions the pressure p exerted by a gas is related to its volume v by $pv = c$, where c is a constant depending on the units and various physical factors. Prove: When such a gas increases in volume from v_0 to v_1 the average pressure it exerts is

$$p_{\text{ave}} = \frac{c}{v_1 - v_0} \ln \frac{v_1}{v_0}$$

7.4 THE NUMBER e; THE FUNCTIONS a^x AND e^x

In this section we use the natural logarithm to extend the concept of an exponent. Our objective is to give a definition of a^k in which k can be irrational. This will give meaning to such expressions as

$$2^\pi, \qquad 5^{-\sqrt{2}}, \qquad \text{and} \qquad \sqrt{3}^{-\sin(\pi/8)}$$

Figure 7.4.1

We also define a new function of fundamental importance, called the *exponential function*.

Because $\ln x$ is a continuous function that varies from $-\infty$ to $+\infty$ as x varies over the interval $(0, +\infty)$, it follows that the curve $y = \ln x$ must cross each horizontal line $y = c$ at a point to the right of the y-axis (Figure 7.4.1). Moreover, because $\ln x$ is an increasing function, it cannot cross any horizontal line $y = c$ more than once; otherwise there would be distinct values of x, say $x_1 < x_2$, such that $\ln x_1 = \ln x_2 = c$, which contradicts the fact that $\ln x$ is increasing. Thus we have the following result.

7.4.1 THEOREM *For every real number c there is a unique positive real number x such that*

$$\ln x = c$$

It follows as a special case that there is a unique positive real number x such that $\ln x = 1$. This number is of special importance and is denoted by e. Thus,

$$\ln e = 1$$

Figure 7.4.2

(Figure 7.4.2). To obtain some information about the size of e, consider the inequalities

$$\frac{1}{2} \leq \ln 2 \leq 1 \tag{1}$$

obtained in Section 7.3. Multiplying through by 2 and writing $2 \ln 2 = \ln 2^2 = \ln 4$ yields

$$1 \leq \ln 4 \leq 2 \tag{2}$$

Since $\ln e = 1$, the right inequality in (1) and the left inequality in (2) yield

$$\ln 2 \leq \ln e \leq \ln 4$$

Therefore,

$$2 \leq e \leq 4$$

since $\ln$ is an increasing function. In later sections we will show how to estimate e to any degree of accuracy. However, for now we note that its value to 5 decimal places is

$$e \approx 2.71828 \ldots$$

A table of powers of e appears in Appendix 3, Table 2.

If a is a positive real number and r is a rational number, then Theorem 7.3.1c states that

$$\ln a^r = r \ln a$$

This equation gives us a way of defining a^k when k is an *arbitrary* real number.

7.4.2   DEFINITION   If a is any positive real number and k is any real number, then a^k is defined to be that real number whose natural logarithm has value $k \ln a$; that is,

$$\ln a^k = k \ln a \tag{3}$$

▶ Example 1

By definition, 2^π is that real number whose natural logarithm is

$$\begin{aligned}
\ln 2^\pi &= \pi \ln 2 \\
&\approx (3.142)(.6931) \quad \text{(Appendix 3, Table 3)} \\
&\approx 2.1778
\end{aligned}$$

From Appendix 3, Table 3, we see that the value of 2^π is between 8.8 and 8.9.

◀

The following theorem shows that the familiar laws for rational exponents are valid for all real exponents.

7.4.3 THEOREM *If a and b are positive, then for all real numbers k and l the following laws of exponents hold.*

(a) $a^0 = 1$ (e) $\dfrac{a^k}{a^l} = a^{k-l}$

(b) $a^1 = a$ (f) $a^k \cdot b^k = (ab)^k$

(c) $a^k \cdot a^l = a^{k+l}$ (g) $\dfrac{a^k}{b^k} = \left(\dfrac{a}{b}\right)^k$

(d) $a^{-l} = \dfrac{1}{a^l}$ (h) $(a^k)^l = a^{kl}$

Proof. We will prove (c) and leave the rest as exercises. We can write

$$
\begin{aligned}
\ln(a^k \cdot a^l) &= \ln a^k + \ln a^l \\
&= k \ln a + l \ln a \\
&= (k + l) \ln a \\
&= \ln(a^{k+l})
\end{aligned}
$$

Thus $a^k \cdot a^l = a^{k+l}$ since the two sides have the same natural logarithm.

If a is a positive real constant, then the formula

$$ f(x) = a^x \tag{4} $$

defines a function of x on the interval $-\infty < x < +\infty$. Some examples are

$$ 2^x, \qquad \pi^x, \qquad \text{and} \qquad e^x $$

The function $f(x) = e^x$, called the ***exponential function,*** has special importance because of the following theorem, which states that the natural logarithm and exponential functions undo the effect of one another.

7.4.4 THEOREM (a) $\ln e^x = x$ for $-\infty < x < +\infty$
                  (b) $e^{\ln x} = x$ for $x > 0$

Proof of (a).

$$ \ln e^x = x \ln e = x $$

Proof of (b). Let $y = e^{\ln x}$. Then

$$ \ln y = \ln(e^{\ln x}) = (\ln x)(\ln e) = \ln x $$

Since y and x have the same natural logarithm, it follows that $y = x$ or, equivalently, $e^{\ln x} = x$. ▪

The following corollary gives a useful computational formula for a^x in terms of the natural logarithm and exponential functions.

7.4.5 COROLLARY   *If a is a positive real number and k is any real number, then*

$$a^k = e^{k \ln a}$$

Proof.   From (3)

$$\ln a^k = k \ln a$$

so that

$$e^{\ln a^k} = e^{k \ln a}$$

or by Theorem 7.4.4b,

$$a^k = e^{k \ln a} \quad ▪$$

▶ Example 2

From this corollary and the tables in Appendix 3 we can approximate $3^{\sqrt{2}}$ as follows:

$$3^{\sqrt{2}} = e^{\sqrt{2}\,\ln 3}$$
$$\approx e^{(1.4)(1.1)} \qquad \text{[Appendix 3, Table 3]}$$
$$\approx e^{1.5}$$
$$\approx 4.5 \qquad \text{[Appendix 3, Table 2]} \qquad\qquad ◀$$

Later, we will see that a function of the form $f(x) = a^x$ $(a > 0)$ is differentiable everywhere. If we accept this for now we can obtain the derivative of a^x as follows. Let $y = a^x$ so that $\ln y = \ln a^x = x \ln a$; thus

$$\frac{d}{dx}[\ln y] = \frac{d}{dx}[x \ln a]$$

Implicit differentiation yields

$$\frac{1}{y}\frac{dy}{dx} = \ln a$$

so

$$\frac{dy}{dx} = y \ln a = a^x \ln a$$

or

$$\frac{d}{dx}[a^x] = a^x \ln a \qquad (5)$$

▶ Example 3

$$\frac{d}{dx}[\pi^x] = \pi^x \ln \pi \qquad \blacktriangleleft$$

Since $\ln e = 1$, it follows as a special case of (5) that

$$\frac{d}{dx}[e^x] = e^x \qquad (6)$$

If u is a differentiable function of x, then the chain rule yields the following generalizations of (5) and (6):

$$\frac{d}{dx}[a^u] = a^u \ln a \cdot \frac{du}{dx} \qquad (7)$$

$$\frac{d}{dx}[e^u] = e^u \cdot \frac{du}{dx} \qquad (8)$$

▶ Example 4

From (7) with $a = 2$ and $u = \sin x$

$$\frac{d}{dx}[2^{\sin x}] = (2^{\sin x})(\ln 2) \cdot \frac{d}{dx}[\sin x] = (2^{\sin x})(\ln 2)(\cos x)$$

and from (8) with $u = x^3$

$$\frac{d}{dx}[e^{x^3}] = e^{x^3} \cdot \frac{d}{dx}[x^3] = 3x^2 e^{x^3} \qquad \blacktriangleleft$$

The following theorem extends a familiar derivative formula so it applies to arbitrary exponents.

7.4.6   THEOREM   *For any real exponent* r

$$\frac{d}{dx}[x^r] = rx^{r-1}$$

Proof.   From Corollary 7.4.5

$$x^r = e^{r \ln x}$$

so that

$$\frac{d}{dx}[x^r] = \frac{d}{dx}[e^{r \ln x}] = e^{r \ln x} \cdot \frac{d}{dx}[r \ln x]$$

$$= x^r \cdot \frac{r}{x}$$

$$= rx^{r-1} \quad \blacksquare$$

▶ Example 5

$$\frac{d}{dx}[x^\pi] = \pi x^{\pi - 1} \qquad\qquad ◀$$

REMARK.   It is important to distinguish between the derivative formulas for a^x (variable exponent) and x^a (constant exponent). Compare, for example, the derivative of x^π in Example 5 and the derivative of π^x in Example 3.

Associated with derivatives (5) and (6) are the companion integration formulas

$$\int a^x\, dx = \frac{a^x}{\ln a} + C \tag{9}$$

and

$$\int e^x\, dx = e^x + C \tag{10}$$

▶ Example 6

$$\int 2^x\, dx = \frac{2^x}{\ln 2} + C \qquad\qquad ◀$$

▶ Example 7

To evaluate

$$\int e^{5x} \, dx$$

let $u = 5x$ so that $du = 5 \, dx$ or $dx = \frac{1}{5} \, du$, which yields

$$\int e^{5x} \, dx = \int e^{u} \left(\frac{1}{5} \, du \right) = \frac{1}{5} \int e^{u} \, du = \frac{1}{5} e^{u} + C$$

$$= \frac{1}{5} e^{5x} + C \qquad \blacktriangleleft$$

▶ Example 8

$$\int e^{-x} \, dx = -\int e^{u} \, du \qquad \begin{bmatrix} u = -x \\ du = -dx \end{bmatrix}$$

$$= -e^{u} + C$$

$$= -e^{-x} + C \qquad \blacktriangleleft$$

▶ Example 9

$$\int x^{2} e^{x^{3}} \, dx = \int \frac{1}{3} e^{u} \, du \qquad \begin{bmatrix} u = x^{3} \\ du = 3x^{2} \, dx \end{bmatrix}$$

$$= \frac{1}{3} e^{u} + C$$

$$= \frac{1}{3} e^{x^{3}} + C \qquad \blacktriangleleft$$

▶ Example 10

Evaluate

$$\int_{0}^{\ln 3} e^{x} (1 + e^{x})^{1/2} \, dx$$

Solution. Make the *u*-substitution

$$u = 1 + e^{x}, \qquad du = e^{x} \, dx$$

and change the x-limits ($x = 0, x = \ln 3$) to the u-limits ($u = 1 + e^0 = 2$, $u = 1 + e^{\ln 3} = 1 + 3 = 4$).

$$\int_0^{\ln 3} e^x (1 + e^x)^{1/2} \, dx = \int_2^4 u^{1/2} \, du = \frac{2}{3} u^{3/2} \Big]_2^4 = \frac{2}{3} [4^{3/2} - 2^{3/2}]$$

$$= \frac{16}{3} - \frac{4\sqrt{2}}{3} \qquad \blacktriangleleft$$

REMARK.   For ease of printing, the exponential function is often denoted by $\exp x$ rather than e^x. For example, in this notation the statements

$$e^{x_1 + x_2} = e^{x_1} e^{x_2}$$

$$e^{\ln x} = x$$

$$\ln (e^x) = x$$

would be written as

$$\exp (x_1 + x_2) = \exp x_1 \cdot \exp x_2$$

$$\exp (\ln x) = x$$

$$\ln (\exp x) = x$$

▶ Exercise Set 7.4

1. Simplify the expression and state the values of x for which your simplification is valid.
    (a) $e^{-\ln x}$ (b) $e^{\ln x^2}$
    (c) $\ln (e^{-x^2})$ (d) $\ln (1/e^x)$
    (e) $\exp (3 \ln x)$ (f) $\ln (x e^x)$
    (g) $\ln (e^{x - \sqrt[3]{x}})$ (h) $e^{x - \ln x}$.

2. Solve for x:
    (a) $\ln (x^2) = 5$ (b) $e^{-4x} = 3$
    (c) $\ln (\ln x) = 0$ (d) $e^x + e^{-x} = 2$.

3. Use the method of Example 2 to approximate the following. (Round off each computation to two significant digits.)
    (a) $2^{1.7}$ (b) $5^{\sqrt{3}}$ (c) 3^{π}

In Exercises 4–13, find dy/dx.
4. $y = e^{7x}$.
5. $y = e^{-5x^2}$.
6. $y = e^{1/x}$.
7. $y = x^3 e^x$.
8. $y = \sin (e^x)$.
9. $y = \dfrac{e^x - e^{-x}}{e^x + e^{-x}}$.

10. $y = \dfrac{e^x}{\ln x}$.
11. $y = e^{x \tan x}$.
12. $y = \exp (\sqrt{1 + 5x^3})$.
13. $y = e^{(x - e^{3x})}$.
14. Find $f'(x)$ by formula (7) and then by logarithmic differentiation.
    (a) $f(x) = 2^x$
    (b) $f(x) = \pi^{\sin x}$
    (c) $f(x) = (\sqrt{2})^{x \ln x}$
15. (a) Explain why formula (5) cannot be used to find $(d/dx)[x^x]$.
    (b) Find this derivative.

In Exercises 16–18, find dy/dx by implicit differentiation.
16. $e^x - e^y = e^{xy}$.
17. $e^y = \ln (x^2 - 3y)$.
18. $\sin (e^{x+y}) = x$.

In Exercises 19–34, evaluate the integrals.
19. $\displaystyle\int e^{7x} \, dx$.

20. $\int \dfrac{dx}{e^x}.$

21. $\int e^{\sin x} \cos x \, dx.$

22. $\int x^3 e^{x^4} \, dx.$

23. $\int \dfrac{e^x}{1 + e^x} \, dx.$

24. $\int \sqrt{e^x} \, dx.$

25. $\int e^{2t} \sqrt{1 + e^{2t}} \, dt.$

26. $\int (x + 3) \exp (x^2 + 6x) \, dx.$

27. $\int \dfrac{e^{\sqrt{y}}}{\sqrt{y}} \, dy.$

28. $\int 2^{5x} \, dx.$

29. $\int \pi^{\sin x} \cos x \, dx.$

30. $\int_0^{\ln 2} e^{-3x} \, dx.$

31. $\int_0^{\ln 5} e^x(3 - 4e^x) \, dx.$

32. $\int_1^{\sqrt{2}} x 4^{-x^2} \, dx.$

33. $\int_1^2 (3 - e^x) \, dx.$

34. $\int_0^e \dfrac{dx}{x + e}.$

35. Prove: If $f'(x) = f(x)$ for all x in $(-\infty, +\infty)$, then $f(x)$ has the form $f(x) = ke^x$ for some constant k. [*Hint:* Let $g(x) = e^{-x}f(x)$ and consider $g'(x)$.]

36. A particle moves along the x-axis so that its x-coordinate at time t is given by $x = ae^{kt} + be^{-kt}$. Show that its acceleration is proportional to x.

37. Let $f(x) = ce^{kx}$. Find a formula for the nth derivative, $f^{(n)}(x)$.

38. Find a point on the graph of $y = e^{3x}$ at which the tangent line passes through the origin.

39. Find the derivatives of
    (a) $f(x) = (x^3 - 2x)^{\ln x}$
    (b) $f(x) = x^e$
    (c) $f(x) = x^{(e^x)}.$

40. Show that
    (a) $y = xe^{-x}$ satisfies the equation
    $xy' = (1 - x)y$
    (b) $y = xe^{-x^2/2}$ satisfies the equation
    $xy' = (1 - x^2)y.$

41. The equilibrium constant k of a balanced chemical reaction changes with the absolute temperature T according to the law

$$k = k_0 \exp \left(-\frac{q}{2} \frac{T - T_0}{T_0 T}\right)$$

where k_0, q, and T_0 are constants. Find the rate of change of k with respect to T.

Exercises 42–46, depend on the following definition.

DEFINITION. If a is any positive real number except 1, and if x and y satisfy

$$a^y = x$$

then we call y the **logarithm to the base a of x,** and write

$$y = \log_a x$$

42. Evaluate the following logarithms:
    (a) $\log_2 8$
    (b) $\log_5 \dfrac{1}{25}$
    (c) $\log_9 3$
    (d) $\log_8 32$
    (e) $\log_{10} 0.001$
    (f) $\log_{0.5} \dfrac{1}{8}.$

43. Prove: For all $x > 0$,
    (a) $\log_e x = \ln x$
    (b) $\log_a x = \dfrac{\ln x}{\ln a}.$

44. Use part (b) of Exercise 43 and properties of the natural logarithm to prove that
    (a) $\log_a cd = \log_a c + \log_a d$
    (b) $\log_a \dfrac{c}{d} = \log_a c - \log_a d$
    (c) $\log_a c^r = r \log_a c$
    (d) $\log_a \dfrac{1}{c} = -\log_a c.$

45. Use part (b) of Exercise 43 to obtain the derivative formula

$$\frac{d}{dx}[\log_a x] = \frac{1}{x \ln a}$$

[*Remark:* This formula is simplest when $a = e$ since $\ln e = 1$. For this reason natural logarithms are preferred over logarithms to the base 10 in calculus.]

46. Use part (b) of Exercise 43 and the table of natural logarithms in the appendix to evaluate the following. (Express your answer to two decimal places.)
    (a) $\log_2 6$ (b) $\log_{10} 4$ (c) $\log_{1.3} 2.8.$

47. Prove parts (*a*) and (*d*) of Theorem 7.4.3.

48. Prove parts (*b*) and (*h*) of Theorem 7.4.3.

49. Evaluate

$$\int \frac{e^{2x}}{e^x + 3} \, dx$$

7.5 ADDITIONAL PROPERTIES OF e^x

In this section we obtain the graph of $y = e^x$ and derive some additional properties of the exponential function.

The graph of $y = e^x$ is easiest to obtain by first writing this equation in the alternate form

$$\ln y = \ln e^x$$

or

$$x = \ln y$$

Figure 7.5.1

To obtain the graph of this equation we need some preliminary results. It is easy to prove (Exercise 24) that the points (a, b) and (b, a) are symmetrically positioned about the line $y = x$; that is, the line $y = x$ is the perpendicular bisector of the line segment joining these points (Figure 7.5.1). Thus, interchanging the x- and y-coordinates of a point causes the point to be reflected about the line $y = x$. Consequently, interchanging the x and y variables in an equation causes the graph of that equation to be reflected about this line. It follows, therefore, that the graph of the curve $x = \ln y$ (or equivalently $y = e^x$) is the reflection of the curve $y = \ln x$ about the line $y = x$ (Figure 7.5.2).

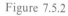
Figure 7.5.2

In Section 8.1 we will prove that $f(x) = e^x$ is everywhere differentiable and consequently everywhere continuous. However, the continuity should be intuitively clear since the graph of $y = e^x$ is the reflection of the continuous curve $y = \ln x$.

The following properties are evident from the graph of e^x (proofs are discussed in the exercises):

$$e^x > 0 \text{ for all } x$$

$$\lim_{x \to +\infty} e^x = +\infty$$

$$\lim_{x \to -\infty} e^x = 0$$

$y = e^{-x}$ $y = e^x$

(0, 1)

Figure 7.5.3

▶ Example 1

Sketch the graph of $y = e^{-x}$.

Solution. The effect of replacing x by $-x$ in the equation $y = e^x$ is to reflect the graph of this equation about the y-axis. Thus, the graph of $y = e^{-x}$ is as shown in Figure 7.5.3. Note that

$$\lim_{x \to +\infty} e^{-x} = \lim_{x \to +\infty} \frac{1}{e^x} = 0$$

and

$$\lim_{x \to -\infty} e^{-x} = \lim_{x \to -\infty} \frac{1}{e^x} = +\infty$$

▶ Example 2

Sketch the graph of $f(x) = xe^x$.

Solution.

0 + + + +

−1

Sign of $(x + 1)e^x$

0 + + + + + + + + +

−2

Sign of $(x + 2)e^x$

Figure 7.5.4

$$f'(x) = xe^x + e^x = (x + 1)e^x$$

$$f''(x) = \frac{d}{dx}[(x + 1)e^x] = (x + 1)e^x + e^x = (x + 2)e^x$$

Since $e^x > 0$ for all x, the signs of $f'(x)$ and $f''(x)$ are the same as the signs of $x + 1$ and $x + 2$, respectively (Figure 7.5.4). The information these signs convey is summarized in Figure 7.5.5a. Because the curve is concave up at $x = -1$, there is a relative minimum at this stationary point. The value of f at this point is

$$f(-1) = (-1)e^{-1} = -\frac{1}{e} \approx -\frac{1}{2.72} \approx -0.37$$

At the inflection point $x = -2$ the value of f is

$$f(-2) = (-2)e^{-2} = -\frac{2}{e^2} \approx -\frac{2}{7.39} \approx -0.27$$

Finally, $f(x) = xe^x$ is positive if $x > 0$, negative if $x < 0$ and $f(0) = 0$. Piecing together our information leads to the graph in Figure 7.5.5b. ◄

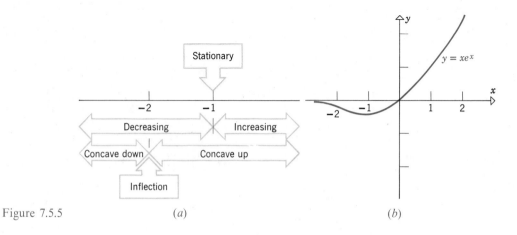

Figure 7.5.5   (a)   (b)

► Exercise Set 7.5

In Exercises 1–6, sketch the graph of f and label the relative extreme points and points of inflection.

1. $f(x) = 4e^{3x}$.

2. $f(x) = x - e^x$.

3. $f(x) = e^x + e^{-x}$.

4. $f(x) = x^2 e^x$.

5. $f(x) = e^{-x^2}$.

6. $f(x) = e^{1/x}$.

7. Sketch the graphs of e^x, $-e^x$, and $e^x \cos x$, all in the same figure, and label any points of intersection.

8. Sketch the following curves:

(a) $y = \dfrac{\ln x}{x}$

(b) $y = x \ln x$.

[*Remark:* The limits in Exercise 23 of Section 7.3 may be helpful.]

In Exercises 9–12, find the limits.

9. (a) $\displaystyle\lim_{x \to +\infty} 3e^{4x}$

(b) $\displaystyle\lim_{x \to -\infty} 3e^{4x}$.

10. (a) $\displaystyle\lim_{x \to +\infty} e^{-x^2}$

(b) $\displaystyle\lim_{x \to -\infty} e^{-x^2}$.

11. (a) $\displaystyle\lim_{x \to +\infty} \frac{e^x + e^{-x}}{e^x - e^{-x}}$

(b) $\displaystyle\lim_{x \to -\infty} \frac{e^x + e^{-x}}{e^x - e^{-x}}$.

12. (a) $\displaystyle\lim_{x \to +\infty} \frac{2 + e^x}{1 + 3e^x}$.

(b) $\displaystyle\lim_{x \to +\infty} e^{(-e^x)}$.

13. Let $f(x) = e^{|x|}$.
   (a) Is f continuous at $x = 0$?
   (b) Is f differentiable at $x = 0$?
   (c) Sketch the graph of f.

14. One of the fundamental functions of mathematical statistics is

$$f(x) = \frac{1}{\sqrt{2\pi}\sigma} \exp\left[-\frac{1}{2}\left(\frac{x-\mu}{\sigma}\right)^2\right]$$

where μ and σ are constants such that $\sigma > 0$ and $-\infty < \mu < +\infty$

(a) Locate the inflection points and relative extreme points.

(b) Find $\lim\limits_{x \to +\infty} f(x)$ and $\lim\limits_{x \to -\infty} f(x)$.

(c) Sketch the graph of f.

15. By writing the derivative

$$\frac{d}{dx}[e^x]\bigg|_{x=0} = 1$$

as a limit, prove that

$$\lim_{h \to 0} \frac{e^h - 1}{h} = 1.$$

16. Use the method of Exercise 15 to find the following limits.

(a) $\lim\limits_{x \to 0} \dfrac{e^{2x} - e^x}{x}$ (b) $\lim\limits_{x \to 0} \dfrac{1 - e^{-x}}{x}$

(c) $\lim\limits_{x \to a} \dfrac{e^x - e^a}{x - a}$ (d) $\lim\limits_{x \to +\infty} x(e^{1/x} - 1)$.

17. It was proved in Exercise 23 of Section 7.3 that $0 < \ln x < x$ for $x > 1$.

(a) Use this result to prove that $\ln x < 2\sqrt{x}$ for $x > 1$.

(b) Use the result in (a) to obtain

$$\frac{1}{e^x} < \frac{x}{e^x} < e^{2\sqrt{x} - x}$$

for $x > 1$.

(c) Use the result in (b) to show that

$$\lim_{x \to +\infty} \frac{x}{e^x} = 0$$

(d) Sketch the graph of $f(x) = x/e^x$.

18. Use the result in part (c) of Exercise 17, where needed, to find the following limits.

(a) $\lim\limits_{x \to +\infty} \dfrac{x^2}{e^{2x}}$ (b) $\lim\limits_{x \to 0^+} \dfrac{e^x}{x}$

(c) $\lim\limits_{x \to -\infty} xe^x$ (d) $\lim\limits_{x \to +\infty} xe^x$.

19. Let c_1, c_2, k_1, and k_2 be positive constants. In each part make a sketch that shows the relative positions of the curves $y = c_1 e^{k_1 x}$ and $y = c_2 e^{k_2 x}$ when:

(a) $c_1 = c_2$ and $k_1 < k_2$
(b) $c_1 < c_2$ and $k_1 = k_2$
(c) $c_1 < c_2$ and $k_2 < k_1$.

20. Repeat Exercise 19 for the curves $y = c_1 e^{-k_1 x}$ and $y = c_2 e^{-k_2 x}$

21. (a) Sketch the curve $y = 2^x$.
(b) Sketch the curve $y = a^x$ assuming $a > 1$.
(c) Make a sketch that shows the relative positions of the curves $y = a_1^x$ and $y = a_2^x$ if $a_2 > a_1 > 1$.

22. Find the volume of the solid generated when the region bounded by $y = e^x$, $x = 0$, $x = \ln 3$, and $y = 0$ is revolved about the x-axis.

23. Find the area of the region enclosed by the curve $y = e^{-x}$ and the line through the points $(0, 1)$ and $(1, 1/e)$.

24. Prove that the line $y = x$ is the perpendicular bisector of the line segment joining (a, b) and (b, a).

25. Use the mean-value theorem (4.9.2) to prove that
(a) $e^x \geq 1 + x$ if $x \geq 0$
(b) $e^{-x} \geq 1 - x$ if $x \geq 0$.

26. Prove that $\lim\limits_{x \to +\infty} e^x = +\infty$ by showing that for any $N > 0$, there is a point x_0 such that $e^x > N$ whenever $x > x_0$.

27. Prove that $\lim\limits_{x \to -\infty} e^x = 0$ by showing that for any $\epsilon > 0$ there is a point x_0 such that $0 < e^x < \epsilon$ whenever $x < x_0$.

7.6 THE HYPERBOLIC FUNCTIONS

In this section we will study certain combinations of e^x and e^{-x}, called **hyperbolic functions.** These functions have numerous engineering applications and arise naturally in many mathematical problems. It will become evident as we progress that the hyperbolic functions have many properties in common with the trigonometric functions. This similarity is reflected in the names of the hyperbolic functions.

7.6.1 DEFINITION   The **hyperbolic sine** and **hyperbolic cosine** functions, denoted by **sinh** and **cosh,** respectively, are defined by

$$\sinh x = \frac{e^x - e^{-x}}{2}$$

$$\cosh x = \frac{e^x + e^{-x}}{2}$$

(a)

Figure 7.6.1

REMARK.   sinh rhymes with "cinch" and cosh rhymes with "gosh."

The graph of $\cosh x = \frac{1}{2}e^x + \frac{1}{2}e^{-x}$ can be obtained by separately graphing $\frac{1}{2}e^x$ and $\frac{1}{2}e^{-x}$ and then adding the y-coordinates together at each point (Figure 7.6.1a). This graphing technique, called **addition of ordinates,** can also be used to graph $\sinh x = \frac{1}{2}e^x - \frac{1}{2}e^{-x}$ (Figure 7.6.1b).

As an illustration of how hyperbolic functions occur in physical problems, consider a homogeneous flexible cable hanging suspended between two points (for example, an electrical transmission line suspended between two poles). The cable forms a curve called a **catenary** (from the Latin "catena" meaning chain.) If a coordinate system is introduced in such a way that the low point of the cable occurs on the y-axis at $(0, a)$ where $a > 0$, then it can be shown using principles of physics that the equation of the curve formed by the cable is

$$y = a \cosh\left(\frac{x}{a}\right),$$

where a depends on the tension and physical properties of the cable (Figure 7.6.2).

Figure 7.6.2

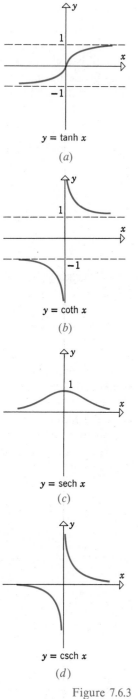

$y = \tanh x$

(a)

$y = \coth x$

(b)

$y = \text{sech } x$

(c)

$y = \text{csch } x$

(d)

Figure 7.6.3

The remaining hyperbolic functions, *hyperbolic tangent, hyperbolic cotangent, hyperbolic secant,* and *hyperbolic cosecant* are defined in terms of sinh and cosh as follows.

$$\tanh x = \frac{\sinh x}{\cosh x} = \frac{e^x - e^{-x}}{e^x + e^{-x}}$$

$$\coth x = \frac{\cosh x}{\sinh x} = \frac{e^x + e^{-x}}{e^x - e^{-x}}$$

$$\text{sech } x = \frac{1}{\cosh x} = \frac{2}{e^x + e^{-x}}$$

$$\text{csch } x = \frac{1}{\sinh x} = \frac{2}{e^x - e^{-x}}$$

The graphs of these functions are shown in Figure 7.6.3 (see Exercises 31–34).

The hyperbolic functions satisfy various identities similar to the identities for the trigonometric functions. The most fundamental of these is

$$\cosh^2 x - \sinh^2 x = 1 \tag{1}$$

which can be proved by writing

$$\cosh^2 x - \sinh^2 x = \left(\frac{e^x + e^{-x}}{2}\right)^2 - \left(\frac{e^x - e^{-x}}{2}\right)^2$$

$$= \frac{1}{4}(e^{2x} + 2e^0 + e^{-2x}) - \frac{1}{4}(e^{2x} - 2e^0 + e^{-2x})$$

$$= 1$$

If we divide (1) by $\cosh^2 x$ we obtain

$$1 - \tanh^2 x = \text{sech}^2 x \tag{2}$$

and if we divide (1) by $\sinh^2 x$ we obtain

$$\coth^2 x - 1 = \text{csch}^2 x \tag{3}$$

The addition formulas for sinh and cosh are

$$\sinh (x + y) = \sinh x \cosh y + \cosh x \sinh y \tag{4a}$$

$$\cosh (x + y) = \cosh x \cosh y + \sinh x \sinh y \tag{4b}$$

These are easier to prove than the corresponding identities for the trigono-metric functions. For example, to prove (4a) we use the relations

$$\cosh x + \sinh x = e^x \tag{5a}$$
$$\cosh x - \sinh x = e^{-x} \tag{5b}$$

which follow immediately from Definition 7.6.1. From (5a) and (5b) it follows that

$$\sinh (x + y) = \frac{e^{(x+y)} - e^{-(x+y)}}{2} = \frac{e^x e^y - e^{-x} e^{-y}}{2}$$

$$= \frac{1}{2}[(\cosh x + \sinh x)(\cosh y + \sinh y)$$
$$- (\cosh x - \sinh x)(\cosh y - \sinh y)]$$
$$= \sinh x \cosh y + \cosh x \sinh y$$

which proves (4a). The proof of (4b) is similar.

If we let $x = y$ in (4a) and (4b) we obtain the following analogs of the trigonometric double angle formulas:

$$\sinh 2x = 2 \sinh x \cosh x \tag{6a}$$
$$\cosh 2x = \cosh^2 x + \sinh^2 x \tag{6b}$$

By using (1) and (6b) the reader should be able to show that

$$\cosh 2x = 2 \sinh^2 x + 1 \tag{7a}$$
$$\cosh 2x = 2 \cosh^2 x - 1 \tag{7b}$$

To obtain subtraction formulas for sinh and cosh we will use the addition formulas and the relations

$$\cosh (-x) = \cosh x \tag{8a}$$
$$\sinh (-x) = -\sinh x \tag{8b}$$

which follow immediately from Definition 7.6.1. If we replace y by $-y$ in (4a) and (4b) and then apply (8a) and (8b) we obtain

$$\sinh (x - y) = \sinh x \cosh y - \cosh x \sinh y \tag{9a}$$
$$\cosh (x - y) = \cosh x \cosh y - \sinh x \sinh y \tag{9b}$$

Derivative formulas for $\sinh x$ and $\cosh x$ follow easily from Definition 7.6.1. For example,

$$\frac{d}{dx}[\cosh x] = \frac{d}{dx}\left[\frac{e^x + e^{-x}}{2}\right]$$

$$= \frac{1}{2}\left(\frac{d}{dx}[e^x] + \frac{d}{dx}[e^{-x}]\right)$$

$$= \frac{e^x - e^{-x}}{2}$$

$$= \sinh x$$

Similarly,

$$\frac{d}{dx}[\sinh x] = \cosh x$$

The derivatives of the remaining hyperbolic functions can be obtained by first expressing them in terms of $\sinh$ and $\cosh$. For example,

$$\frac{d}{dx}[\tanh x] = \frac{d}{dx}\left[\frac{\sinh x}{\cosh x}\right]$$

$$= \frac{\cosh x \dfrac{d}{dx}[\sinh x] - \sinh x \dfrac{d}{dx}[\cosh x]}{\cosh^2 x}$$

$$= \frac{\cosh^2 x - \sinh^2 x}{\cosh^2 x} = \frac{1}{\cosh^2 x}$$

$$= \operatorname{sech}^2 x$$

Derivations of differentiation formulas for the remaining hyperbolic functions are exercises. For reference we summarize the results.

$$\frac{d}{dx}[\sinh x] = \cosh x$$

$$\frac{d}{dx}[\cosh x] = \sinh x$$

$$\frac{d}{dx}[\tanh x] = \operatorname{sech}^2 x$$

$$\frac{d}{dx}[\coth x] = -\operatorname{csch}^2 x$$

$$\frac{d}{dx}[\operatorname{sech} x] = -\operatorname{sech} x \tanh x$$

$$\frac{d}{dx}[\operatorname{csch} x] = -\operatorname{csch} x \coth x$$

REMARK. Except for a difference in the pattern of signs, these formulas are similar to the formulas for the derivatives of the trigonometric functions.

These differentiation formulas produce the following companion integration formulas.

$$\int \sinh x \, dx = \cosh x + C$$

$$\int \cosh x \, dx = \sinh x + C$$

$$\int \operatorname{sech}^2 x \, dx = \tanh x + C$$

$$\int \operatorname{csch}^2 x \, dx = -\coth x + C$$

$$\int \operatorname{sech} x \tanh x \, dx = -\operatorname{sech} x + C$$

$$\int \operatorname{csch} x \coth x \, dx = -\operatorname{csch} x + C$$

▶ Example 1

$$\frac{d}{dx}[\cosh (x^3)] = \sinh (x^3) \cdot \frac{d}{dx}[x^3] = 3x^2 \sinh (x^3)$$

$$\frac{d}{dx}[\ln (\tanh x)] = \frac{1}{\tanh x} \cdot \frac{d}{dx}[\tanh x] = \frac{\operatorname{sech}^2 x}{\tanh x} \qquad ◀$$

▶ Example 2

$$\int \sinh^5 x \cosh x \, dx = \frac{1}{6} \sinh^6 x + C \qquad \begin{bmatrix} u = \sinh \ x \\ du = \cosh x \, dx \end{bmatrix}$$

$$\int \tanh x \, dx = \int \frac{\sinh x}{\cosh x} \, dx$$

$$= \ln |\cosh x| + C \qquad \begin{bmatrix} u = \cosh x \\ du = \sinh x \, dx \end{bmatrix}$$

$$= \ln (\cosh x) + C$$

We are justified in writing $|\cosh x| = \cosh x$ because $\cosh x$ is positive for all x. In fact, $\cosh x \geq 1$ for all x. This is geometrically clear from Figure 7.6.1a and can be proved using (1) (Exercise 33). ◄

The trigonometric functions are sometimes called the **circular functions** because of their relationship to the unit circle $x^2 + y^2 = 1$. (See Figure 8 in Appendix 1.) For the hyperbolic functions there is an analogous relationship to the curve $x^2 - y^2 = 1$, called the **unit hyperbola.** We will pursue this idea in later sections.

▶ Exercise Set 7.6

1. In each part, a value for one of the hyperbolic functions is given at x_0. Find the values of the remaining five hyperbolic functions at x_0.

   (a) $\sinh x_0 = -2$

   (b) $\cosh x_0 = \dfrac{5}{4}, \ x_0 < 0$

   (c) $\tanh x_0 = -\dfrac{4}{5}$

   (d) $\coth x_0 = 2$

   (e) $\operatorname{sech} x_0 = \dfrac{15}{17}, \ x_0 > 0$

   (f) $\operatorname{csch} x_0 = -1$.

In Exercises 2–14, establish the identities.

2. $\cosh (x + y) = \cosh x \cosh y + \sinh x \sinh y$.

3. $\cosh 2x = 2 \sinh^2 x + 1$.

4. $\cosh 2x = 2 \cosh^2 x - 1$.

5. $\cosh (-x) = \cosh x$.

6. $\sinh (-x) = -\sinh x$.

7. $\tanh (x + y) = \dfrac{\tanh x + \tanh y}{1 + \tanh x \tanh y}$.

8. $\tanh (x - y) = \dfrac{\tanh x - \tanh y}{1 - \tanh x \tanh y}$.

9. $\tanh 2x = \dfrac{2 \tanh x}{1 + \tanh^2 x}$.

10. $\cosh \dfrac{1}{2} x = \sqrt{\dfrac{1}{2}(\cosh x + 1)}$.

11. $\sinh \dfrac{1}{2} x = \pm \sqrt{\dfrac{1}{2}(\cosh x - 1)}$.

12. $\sinh x + \sinh y = 2 \sinh \left(\dfrac{x + y}{2}\right) \cosh \left(\dfrac{x - y}{2}\right)$.

13. $\cosh x + \cosh y = 2 \cosh \left(\dfrac{x + y}{2}\right) \cosh \left(\dfrac{x - y}{2}\right)$.

14. $\cosh 3x = 4 \cosh^3 x - 3 \cosh x$.

15. Derive the differentiation formula for
   (a) $\sinh x$ (b) $\coth x$
   (c) $\operatorname{sech} x$ (d) $\operatorname{csch} x$.

In Exercises 16–21, find dy/dx.

16. $y = \cosh (x^4)$.   17. $y = \sinh (4x - 8)$.

18. $y = \ln (\tanh 2x)$.   19. $y = \coth (\ln x)$.

20. $y = \operatorname{sech} (e^{2x})$.   21. $y = \operatorname{csch} (1/x)$.

In Exercises 22–29, evaluate the integral.

22. $\displaystyle\int \cosh (2x - 3)\, dx$.   23. $\displaystyle\int \sinh^6 x \cosh x\, dx$.

24. $\displaystyle\int \operatorname{csch}^2 (3x)\, dx$.   25. $\displaystyle\int \sqrt{\tanh x}\, \operatorname{sech}^2 x\, dx$.

26. $\displaystyle\int \coth^2 x \operatorname{csch}^2 x\, dx$.   27. $\displaystyle\int \tanh x\, dx$.

28. $\displaystyle\int \dfrac{e^x - e^{-x}}{e^x + e^{-x}}\, dx$.   29. $\displaystyle\int \tanh x \operatorname{sech}^3 x\, dx$.

30. Prove the following facts about $\tanh x$:
   (a) $-1 < \tanh x < 1$ for all x
   (b) $\displaystyle\lim_{x \to +\infty} \tanh x = 1$
   (c) $\displaystyle\lim_{x \to -\infty} \tanh x = -1$.

31. Determine the intervals on which $y = \tanh x$ is positive, negative, increasing, decreasing, concave up, and concave down. Then use these results and Exercise 30 to obtain the graph of $y = \tanh x$.

32. Use the results found in Exercises 30 and 31 to help obtain the graph of $y = \coth x$.

33. (a) Prove: $\cosh x \geq 1$ for all x.
(b) Prove: $0 < \operatorname{sech} x \leq 1$ for all x.
(c) Obtain the graph of $y = \operatorname{sech} x$.

34. (a) Show that $\lim\limits_{x \to +\infty} \operatorname{csch} x = \lim\limits_{x \to -\infty} \operatorname{csch} x = 0$.
(b) Show that $\lim\limits_{x \to 0^+} \operatorname{csch} x = +\infty$ and
$\lim\limits_{x \to 0^-} \operatorname{csch} x = -\infty$.
(c) Obtain the graph of $y = \operatorname{csch} x$.

35. Prove: $(\sinh x + \cosh x)^n = \sinh nx + \cosh nx$.

36. (a) It is sometimes said that $\sinh x$ and $\cosh x$ "behave like $\frac{1}{2}e^x$" for x large and positive. Give an informal justification of this statement.
(b) Make a sketch showing the graphs of $y = \sinh x$, $y = \cosh x$, and $y = \frac{1}{2}e^x$, all in the same figure.

37. Find the arc length of the catenary
$y = a \cosh (x/a)$ between $x = 0$ and $x = x_1$
$(x_1 > 0)$.

7.7 FIRST-ORDER DIFFERENTIAL EQUATIONS AND APPLICATIONS

Many problems in science and engineering require the determination of functions satisfying equations containing derivatives. These equations are called *differential equations.* Some examples are

$$\frac{dy}{dx} = 3y \tag{1}$$

$$\frac{d^2y}{dx^2} - 6\frac{dy}{dx} + 8y = 0 \tag{2}$$

$$y' - y = e^{2x} \tag{3}$$

$$\frac{d^3y}{dt^3} - t\frac{dy}{dt} + (t^2 - 1)y = e^t \tag{4}$$

In the first three equations, $y = y(x)$ is an unknown function of x and in the last equation $y = y(t)$ is an unknown function of t. The *order* of a differential equation is the order of the highest derivative that appears in the equation. Thus (1) and (3) are first-order equations, (2) is second order, and (4) is third order.

A function $y = y(x)$ is a *solution* of a differential equation if the equation is satisfied when $y(x)$ and its derivatives are substituted. For example,

$$y = e^{2x} \tag{5}$$

is a solution of the equation

$$\frac{dy}{dx} - y = e^{2x} \tag{6}$$

since

$$\frac{dy}{dx} - y = 2e^{2x} - e^{2x} = e^{2x}$$

Similarly,

$$y = e^x + e^{2x} \tag{7}$$

is a solution. More generally,

$$y = Ce^x + e^{2x} \tag{8}$$

is a solution for any constant C (verify). This solution is of special importance since it can be proved that *all* solutions of (6) can be obtained by substituting values for the arbitrary constant C. For example, $C = 0$ yields solution (5) and $C = 1$ yields solution (7). A solution of a differential equation from which all other solutions can be derived by substituting values for arbitrary constants is called the **general solution** of the equation. Usually, the general solution of an nth-order equation contains n arbitrary constants. Thus (8), which is the general solution of the first-order equation (6), contains one arbitrary constant.

Often, solutions of differential equations are expressed as implicitly defined functions. For example,

$$\ln y = xy + C \tag{9}$$

defines a solution of

$$\frac{dy}{dx} = \frac{y^2}{1 - xy} \tag{10}$$

for any value of the constant C since implicit differentiation of (9) yields

$$\frac{1}{y}\frac{dy}{dx} = x\frac{dy}{dx} + y$$

or

$$\frac{dy}{dx} - xy\frac{dy}{dx} = y^2$$

from which (10) follows.

In this section we will discuss some important classes of first-order differential equations. However, we are only touching the surface of a vast and important field. More complete studies of differential equations are available in numerous textbooks devoted to the subject.

A first-order differential equation is called **separable** if it is expressible in the form

$$\frac{dy}{dx} = \frac{g(x)}{h(y)}$$

To solve such an equation we rewrite it in the differential form

$$h(y)\, dy = g(x)\, dx \tag{11}$$

and integrate both sides, thereby obtaining the general solution

$$\int h(y)\, dy = \int g(x)\, dx + C$$

where C is an arbitrary constant.

In (11), the x and y variables are "separated" from each other, hence the term *separable* differential equation.

▶ Example 1

Solve the equation

$$\frac{dy}{dx} = \frac{x}{y^2}$$

Solution. Changing to differential form and integrating yields

$$y^2\, dy = x\, dx$$
$$\int y^2\, dy = \int x\, dx$$
$$\frac{y^3}{3} = \frac{x^2}{2} + C$$

This expresses the solution implicitly. If desired, we can solve explicitly for y to obtain

$$y = \left(\frac{3}{2}x^2 + 3C\right)^{1/3}$$

or

$$y = \left(\frac{3}{2}x^2 + K\right)^{1/3}$$

where $K(=3C)$ is an arbitrary constant. ◀

▶ Example 2

Solve the equation

$$x(y - 1)\frac{dy}{dx} = y$$

Solution. Separating the variables, and integrating yields

$$x(y - 1)\,dy = y\,dx$$

$$\frac{y - 1}{y}\,dy = \frac{dx}{x}$$

$$\int \frac{y - 1}{y}\,dy = \int \frac{dx}{x}$$

$$\int \left(1 - \frac{1}{y}\right)dy = \int \frac{dx}{x}$$

$$y - \ln|y| = \ln|x| + C$$

or, using properties of logarithms,

$$y = \ln|xy| + C \qquad \blacktriangleleft$$

When a physical problem leads to a differential equation, there are usually conditions in the problem that determine specific values for the arbitrary constants in the general solution of the equation. For a first-order equation, a condition that specifies the value of the unknown function $y(x)$ at some point $x = x_0$ is called an **initial condition.** A first-order differential equation together with one initial condition constitutes a **first-order initial-value problem.**

▶ Example 3

Solve the initial-value problem

$$\frac{dy}{dx} = -4xy^2, \qquad y(0) = 1$$

Solution. We first solve the differential equation:

$$\frac{dy}{y^2} = -4x\,dx$$

$$\int \frac{dy}{y^2} = \int -4x\,dx$$

$$-\frac{1}{y} = -2x^2 + C_1$$

or on multiplying by -1, taking reciprocals, and writing C in place of $-C_1$,

$$y = \frac{1}{2x^2 + C} \tag{12}$$

The initial condition, $y(0) = 1$, requires that $y = 1$ when $x = 0$. Substituting these values in (12) yields $C = 1$. Thus, the solution of the initial-value problem is

$$y = \frac{1}{2x^2 + 1}$$

◀

Not every first-order differential equation is separable. For example it is impossible to separate the variables in the equation

$$\frac{dy}{dx} + x^2 y = e^x$$

However, this equation can be solved by a different method that we will now consider.

A first-order differential equation is called *linear* if it is expressible in the form

$$\frac{dy}{dx} + p(x)y = q(x) \tag{13}$$

where the functions $p(x)$ and $q(x)$ may be constant. Some examples are

$$\frac{dy}{dx} + x^2 y = e^x \qquad\qquad [p(x) = x^2,\ q(x) = e^x]$$

$$y' - 3e^x y = 0 \qquad\qquad [p(x) = -3e^x,\ q(x) = 0]$$

$$\frac{dy}{dx} + 5y = 2 \qquad\qquad [p(x) = 5,\ q(x) = 2]$$

$$\frac{dy}{dx} + (\sin x)y + x^3 = 0 \qquad [p(x) = \sin x,\ q(x) = -x^3]$$

One procedure for solving (13) is based on the observation that if we define $\rho = \rho(x)$ by

$$\rho = e^{\int p(x)\,dx}$$

then

$$\frac{d\rho}{dx} = e^{\int p(x)\,dx} \cdot \frac{d}{dx}\int p(x)\,dx = \rho p(x)$$

Thus

$$\frac{d}{dx}(\rho y) = \rho\frac{dy}{dx} + \frac{d\rho}{dx}y = \rho\frac{dy}{dx} + \rho p(x)y \tag{14}$$

If (13) is multiplied through by ρ, it becomes

$$\rho\frac{dy}{dx} + \rho p(x)y = \rho q(x)$$

or from (14),

$$\frac{d}{dx}(\rho y) = \rho q(x)$$

This equation can be solved by integrating both sides to obtain

$$\rho y = \int \rho q(x)\, dx + C$$

or

$$y = \frac{1}{\rho}\left[\int \rho q(x)\, dx + C\right]$$

To summarize, (13) can be solved in three steps:

Step 1. Calculate

$$\rho = e^{\int p(x)\, dx}$$

This is called the **_integrating factor._** Since any ρ will suffice, we can take the constant of integration to be zero in this step.

Step 2. Multiply both sides of (13) by ρ and express the result as

$$\frac{d}{dx}(\rho y) = \rho q(x)$$

Step 3. Integrate both sides of the equation obtained in Step 2 and then solve for y. Be sure to include a constant of integration in this step.

▶ Example 4

Solve the equation

$$\frac{dy}{dx} - 4xy = x \tag{15}$$

Solution. Since $p(x) = -4x$, the integrating factor is

$$\rho = e^{\int (-4x)\, dx} = e^{-2x^2}$$

If we multiply (15) by ρ we obtain

$$\frac{d}{dx}(e^{-2x^2}y) = xe^{-2x^2}$$

Integrating both sides of this equation yields

$$e^{-2x^2}y = \int xe^{-2x^2}\, dx = -\frac{1}{4}e^{-2x^2} + C$$

and then multiplying both sides by e^{2x^2}, yields

$$y = -\frac{1}{4} + Ce^{2x^2} \qquad\qquad ◀$$

▶ Example 5

Solve the equation

$$x\frac{dy}{dx} - y = x \qquad (x > 0)$$

Solution. To put the equation in form (13), we divide through by x to obtain

$$\frac{dy}{dx} - \frac{1}{x}y = 1 \tag{16}$$

Since $p(x) = -1/x$, the integrating factor is

$$\rho = e^{\int -(1/x)\, dx} = e^{-\ln |x|} = \frac{1}{|x|} = \frac{1}{x}$$

(The absolute value was dropped because of the assumption that $x > 0$.)
If we multiply (16) by ρ we obtain

$$\frac{d}{dx}\left(\frac{1}{x}y\right) = \frac{1}{x}$$

Integrating both sides of this equation yields

$$\frac{1}{x}y = \int \frac{1}{x}\,dx = \ln x + C$$

or

$$y = x \ln x + Cx \qquad \blacktriangleleft$$

We conclude this section with some applications of first-order differential equations.

▶ Example 6 *(Geometry)*

Find a curve in the xy-plane that passes through $(0, 3)$ and whose tangent at a point (x, y) has slope $2x/y^2$.

Solution. Since the slope of the tangent is dy/dx, we have

$$\frac{dy}{dx} = \frac{2x}{y^2} \qquad (17)$$

and, since the curve passes through $(0, 3)$, we have the initial condition

$$y(0) = 3 \qquad (18)$$

Equation (17) is separable and can be written as

$$y^2\,dy = 2x\,dx$$

so

$$\int y^2\,dy = \int 2x\,dx$$

or

$$\frac{1}{3}y^3 = x^2 + C$$

From the initial condition, (18), it follows that $C = 9$, and so the curve has the equation

$$\frac{1}{3}y^3 = x^2 + 9$$

or

$$y = (3x^2 + 27)^{1/3} \qquad\blacktriangleleft$$

▶ Example 7 (*Mixing Problems*)

At time $t = 0$, a tank contains 4 lb of salt dissolved in 100 gal of water. Suppose that brine containing 2 lb of salt per gallon of water is allowed to enter the tank at a rate of 5 gal/min and that the mixed solution is drained from the tank at the same rate. Find a formula for the amount of salt in the tank after 10 minutes.

Solution. Let $y(t)$ be the amount of salt (in pounds) at time t. We are interested in finding $y(10)$, the amount of salt at time $t = 10$. We will begin by finding an expression for dy/dt, the rate of change of the amount of salt in the tank at time t. Clearly,

$$\frac{dy}{dt} = \text{rate in} - \text{rate out} \qquad (19)$$

where *rate in* is the rate at which salt enters the tank and *rate out* is the rate at which salt leaves the tank. But

$$\text{rate in} = (2 \text{ lb/gal}) \cdot (5 \text{ gal/min}) = 10 \text{ lb/min}$$

At time t, the mixture contains $y(t)$ pounds of salt in 100 gallons of water; thus the concentration of salt at time t is $y(t)/100$ pounds per gallon and

$$\text{rate out} = \left(\frac{y(t)}{100} \text{ lb/gal}\right) \cdot (5 \text{ gal/min}) = \frac{y(t)}{20} \text{ lb/min}$$

Therefore, (19) can be written

$$\frac{dy}{dt} = 10 - \frac{y}{20}$$

or

$$\frac{dy}{dt} + \frac{y}{20} = 10 \qquad (20)$$

which is a first-order linear differential equation. We also have the initial condition

$$y(0) = 4 \tag{21}$$

since the tank contains four pounds of salt at time $t = 0$.

Multiplying both sides of (20) by the integrating factor

$$\rho = e^{\int (1/20)\, dt} = e^{t/20}$$

yields

$$\frac{d}{dt}(e^{t/20}y) = 10e^{t/20}$$

so

$$e^{t/20}y = \int 10e^{t/20}\, dt = 200e^{t/20} + C$$

or

$$y(t) = 200 + Ce^{-t/20}$$

From the initial condition, (21), it follows that

$$4 = 200 + C$$

or $C = -196$, so

$$y(t) = 200 - 196e^{-t/20}$$

Thus, after 10 minutes ($t = 10$) the amount of salt in the tank is

$$y(10) = 200 - 196e^{-0.5} \approx 81.1 \text{ lb} \qquad \blacktriangleleft$$

Many quantities increase or decrease with time in proportion to the amount of the quantity present. Some examples are human population, bacteria in a culture, drug concentration in the bloodstream, radioactivity, and the values of certain kinds of investments. We will show how differential equations can be used to study the growth and decay of such quantities.

7.7.1 DEFINITION   A quantity is said to have an ***exponential growth (decay) model*** if at each instant of time its rate of increase (decrease) is proportional to the amount of the quantity present.

Consider a quantity with an exponential growth or decay model and denote by $y(t)$ the amount of the quantity present at time t. We assume that $y(t) > 0$ for all t. Since the rate of change of $y(t)$ is proportional to the amount present, it follows that $y(t)$ satisfies

$$\frac{dy}{dt} = ky \tag{22}$$

where k is a constant of proportionality.

The constant k in (22) is called the **growth constant** if $k > 0$ and the **decay constant** if $k < 0$. If $k > 0$, then $dy/dt > 0$, so that y is increasing with time (a growth model). If $k < 0$, then $dy/dt < 0$, so that y is decreasing with time (a decay model).

REMARK. In problems of exponential growth and decay, the constant k is often expressed as a percentage. Thus, a growth constant of 3% means $k = 0.03$ and a decay constant of -500% means $k = -5$. We also note that the constant k is often called the **growth rate.** Strictly speaking, this is not correct; as indicated by (22), the growth rate dy/dt is not k, but rather ky. However, the description of k as a growth rate is so common that we will use this standard terminology in this section.

If we are given a quantity with an exponential growth or decay model, and if we know the amount y_0 present at some initial time $t = 0$, then we can find the amount present at any time t by solving the initial-value problem

$$\frac{dy}{dt} = ky, \qquad y(0) = y_0 \tag{23}$$

The differential equation is both separable and linear and thus can be solved by either of the methods we have studied in this section. We will treat it as a linear equation. Rewriting the differential equation as

$$\frac{dy}{dt} - ky = 0$$

and multiplying through by the integrating factor

$$\rho = e^{\int -k \, dt} = e^{-kt}$$

yields

$$\frac{d}{dt}(e^{-kt}y) = 0$$

After integrating,

$$e^{-kt}y = C$$

or

$$y = Ce^{kt}$$

y(t) = population at time t

$y = y(t)$

Figure 7.7.1

From the initial condition, $y(0) = y_0$, it follows that $C = y_0$; thus the solution of (23) is

$$y(t) = y_0 e^{kt} \tag{24}$$

Exponential models have proved useful in studies of population growth. Although populations (e.g., people, bacteria, and flowers) grow in discrete steps, we can apply the results of this section if we are willing to approximate the population graph by a continuous curve $y = y(t)$ (Figure 7.7.1).

▶ Example 8 *(Population Growth)*

According to United Nations data, the world population at the beginning of 1975 was approximately 4 billion and growing at a rate of about 2% per year. Assuming an exponential growth model, estimate the world population at the beginning of the year 2000.

Solution. Let

$t =$ time (in years) elapsed from the beginning of 1975
$y =$ world population in billions

Since the beginning of 1975 corresponds to $t = 0$, it follows from the given data that

$$y_0 = y(0) = 4 \text{ (billion)}$$

Since the growth rate is 2% ($k = 0.02$), it follows from (24) that the world population at time t will be

$$y(t) = y_0 e^{kt} = 4e^{0.02t} \tag{25}$$

Since the beginning of the year 2000 corresponds to $t = 25$ ($2000 - 1975 = 25$), it follows from (25) that the world population by the year 2000 will be

$$y(25) = 4e^{0.02(25)} = 4e^{0.5}$$

or from Appendix 3, Table 2,

$$y(25) = 4(1.6487) = 6.5948 \text{ (billion)}$$

which is a population of approximately 6.6 billion.   ◄

If a quantity has an exponential growth model, then the time required for it to double in size is called the ***doubling time.*** Similarly, if a quantity has an exponential decay model, then the time required for it to reduce in value by half is called the ***halving time.*** As it turns out, doubling and halving times depend only on the growth rate and not on the amount present initially. To see why, suppose y has an exponential growth model so that

$$y = y_0 e^{kt} \qquad (k > 0)$$

At any fixed time t_1 let

$$y_1 = y_0 e^{kt_1} \tag{26}$$

be the value of y and let T denote the amount of time required for y to double in size. Thus, at time $t_1 + T$ the value of y will be $2y_1$ so that

$$2y_1 = y_0 e^{k(t_1+T)} = y_0 e^{kt_1} e^{kT}$$

or, from (26),

$$2y_1 = y_1 e^{kt}$$

Thus

$$2 = e^{kT} \qquad \text{and} \qquad \ln 2 = kT$$

Therefore the doubling time T is

$$T = \frac{1}{k} \ln 2 \tag{27}$$

which does not depend on y_0 or t_1. We leave it as an exercise to show that the halving time for a quantity with an exponential decay model ($k < 0$) is

$$T = -\frac{1}{k} \ln 2 \tag{28}$$

▶ Example 9

It follows from (27) that at the current 2% annual growth rate, the doubling time for the world population is

$$T = \frac{1}{0.02} \ln 2$$

$$= \frac{1}{0.02}(0.6931) \qquad \text{[Appendix 3, Table 3]}$$

$$= 34.655$$

or approximately 35 years. Thus, with a continued 2% annual growth rate the population of 4 billion in 1975 will double to 8 billion by the year 2010 and will double again to 16 billion by 2045. ◀

Radioactive elements continually undergo a process of disintegration called **radioactive decay.** It is a physical fact that at each instant of time the rate of decay is proportional to the amount of the element present. Consequently, the amount of any radioactive element has an exponential decay model. For radioactive elements, halving time is called **half-life.**

▶ Example 10 (*Radioactive Decay*)

The radioactive element carbon-14 has a half-life of 5750 years. If 100 grams of this element are present initially, how much will be left after 1000 years?

Solution. From (28) the decay constant is

$$k = -\frac{1}{T} \ln 2$$

$$= -\frac{1}{5750}(0.6931) \qquad \text{[Appendix 3, Table 3]}$$

$$\approx -0.00012$$

Thus, if we take $t = 0$ to be the present time, then $y_0 = y(0) = 100$ so that (24) implies that the amount of carbon-14 after 1000 years will be

$$y(1000) = 100e^{-0.00012(1000)}$$

$$= 100e^{-0.12}$$

$$\approx 100\,(0.88692)$$

$$= 88.692$$

Thus, about 88.69 grams of carbon-14 will remain. ◀

▶ Exercise Set 7.7

In Exercises 1–6, solve the given separable differential equation. Where convenient, express the solution explicitly as a function of x.

1. $\dfrac{dy}{dx} = \dfrac{y}{x}$.

2. $\dfrac{dy}{dx} = \dfrac{x^3}{(1 + x^4)y}$.

3. $\sqrt{1 + x^2}\, y' + x(1 + y) = 0$.

4. $x \tan y - \dfrac{dy}{dx} \sec x = 0$.

5. $e^{-y} \sin x - y' \cos^2 x = 0$.

6. $\dfrac{dy}{dx} = 1 - y + x^2 - yx^2$.

In Exercises 7–12, solve the given first-order linear differential equation. Where convenient, express the solution explicitly as a function of x.

7. $\dfrac{dy}{dx} + 3y = e^{-2x}$.

8. $\dfrac{dy}{dx} - \dfrac{5}{x}y = x \quad (x > 0)$.

9. $y' + y = \cos(e^x)$.

10. $2\dfrac{dy}{dx} + 4y = 1$.

11. $x^2 y' + 3xy + 2x^5 = 0 \quad (x > 0)$.

12. $\dfrac{dy}{dx} + y - \dfrac{1}{1 + e^x} = 0$.

In Exercises 13–18, solve the initial-value problems.

13. $\dfrac{dy}{dx} - xy = x, \quad y(0) = 3$.

14. $2y\dfrac{dy}{dx} = 3x^2(x^3 + 1)^{-1/2}, \quad y(2) = 1$.

15. $\dfrac{dy}{dt} + y = 2, \quad y(0) = 1$.

16. $y' - xe^y = 2e^y, \quad y(0) = 0$.

17. $y^2 t\dfrac{dy}{dt} - t + 1 = 0, \quad y(1) = 3$.

18. $y' \cosh x + y \sinh x = \cosh^2 x, \quad y(0) = \dfrac{1}{4}$.

19. Find a curve in the xy-plane that passes through the point $(1, 2)$ and whose tangent at a point (x, y) has slope $2 - (y/x)$.

20. Find a curve in the xy-plane that passes through the point $(1, 1)$ and whose normal at a point (x, y) has slope $-2y(3x^2)$.

21. The line normal to a curve at any point (x, y) passes through $(3, 0)$. Given that the curve passes through $(3, 2)$ find its equation.

22. A function f with positive values has the property that its value at every point is twice the slope of the tangent at the point. Given that $f(0) = 2$, find f.

In Exercises 23–40, use a hand-held calculator or the tables in the appendix where needed.

23. At time $t = 0$, a tank contains 25 lb of salt dissolved in 50 gal of water. Then brine containing 4 lb of salt per gallon of water is allowed to enter the tank at a rate of 2 gal/min and the mixed solution is drained from the tank at the same rate.
   (a) How much salt is in the tank at an arbitrary time t?
   (b) How much salt is in the tank after 25 minutes?

24. A tank initially contains 200 gallons of pure water. Then at time $t = 0$ brine containing 5 lb of salt per gallon of water is allowed to enter the tank at a rate of 10 gal/min and the mixed solution is drained from the tank at the same rate.
   (a) How much salt is in the tank at an arbitrary time t?
   (b) How much salt is in the tank after 30 minutes?

25. A tank with a 1000-gal capacity initially contains 500 gal of brine containing 50 lb of salt. At time $t = 0$, pure water is added at a rate of 20 gal/min and the mixed solution is drained off at a rate of 10 gal/min. How much salt is in the tank when it reaches the point of overflowing?

26. The number of bacteria in a certain culture grows exponentially at a rate of 1% per hour. Assuming that 10,000 bacteria are present initially, find
   (a) the number of bacteria present at any time t,
   (b) the number of bacteria present after 5 hours,
   (c) the time required for the number of bacteria to reach 45,000.

27. Polonium-210 is a radioactive element with a half-life of 140 days. Assume a sample weighs 10 mg initially.
    (a) Find a formula for the amount that will remain after t days.
    (b) How much will remain after 10 weeks?

28. In a certain chemical reaction a substance decomposes at a rate proportional to the amount present. Tests show that under appropriate conditions 15,000 grams will reduce to 5000 grams in 10 hours.
    (a) Find a formula for the amount that will remain from a 15,000 g sample after t hours.
    (b) How long will it take for 50% of an initial sample of y_0 grams to decompose?

29. One hundred fruit flies are placed in a breeding container that can support a population of at most 5000 flies. If the population grows exponentially at a rate of 2% per day, how long will it take for the container to reach capacity?

30. In 1960 the American scientist W. F. Libby won the Nobel prize for his discovery of carbon dating, a method for determining the age of certain fossils. Carbon dating is based on the fact that nitrogen is converted to radioactive carbon-14 by cosmic radiation in the upper atmosphere. This radioactive carbon is absorbed by plant and animal tissue through the life processes while the plant or animal lives. However, when the plant or animal dies the absorption process stops and the amount of carbon-14 decreases through radioactive decay. Suppose that tests on a fossil show that 70% of its carbon-14 has decayed. Estimate the age of the fossil, assuming a half-life of 5750 years for carbon-14.

31. Forty percent of a radioactive substance decays in 5 years. Find the half-life of the substance.

32. Assume that if the temperature is constant, then the atmospheric pressure p varies with the altitude h (above sea level) in such a way that

$$\frac{dp}{dh} = kp$$

where k is a constant.

(a) Find a formula for p in terms of k, h, and the atmospheric pressure p_0 at sea level.
(b) Given that p measures 15 lb per square inch at sea level and 12 lb per square inch at 5000 ft above sea level, find the pressure at 10,000 ft (assuming temperature is constant).

33. The town of Grayrock had a population of 10,000 in 1960 and 12,000 in 1970.
    (a) Assuming an exponential growth model, estimate the population in 1980.
    (b) What is the doubling time for the town's population?

34. Prove: If a quantity A has an exponential growth or decay model and A has values A_1 and A_2 at times t_1 and t_2, respectively, then the growth rate k is

$$k = \frac{1}{t_1 - t_2} \ln\left(\frac{A_1}{A_2}\right)$$

35. Newton's law of cooling states that the rate at which an object cools is proportional to the difference in temperature between the object and the surrounding medium. Show that if C is the constant temperature of a surrounding medium, then the temperature $T(t)$ at time t of a cooling object is given by

$$T(t) = (T_0 - C)e^{kt} + C$$

where T_0 is the temperature of the object at $t = 0$ and k is a negative constant.

36. A liquid with initial temperature 200° is surrounded by a body of air at a constant temperature of 80°. If the liquid cools to 120° in 30 minutes, what will the temperature be after 1 hour? (Use the result of Exercise 35.)

37. Suppose P dollars is invested at an annual interest rate of $r \times 100\%$. If the accumulated interest is credited to the account at the end of the year, then the interest is said to be *compounded annually;* if it is credited at the end of each six-month period then it is said to be *compounded semi-annually;* and if it is credited at the end of each three-month period, then it is said to be *compounded quarterly.* The more frequently the interest is compounded,

the better it is for the investor since more of the interest is itself earning interest.

(a) Show that if interest is compounded n times a year at equally spaced intervals, then the value A of the investment after t years is

$$A = P\left(1 + \frac{r}{n}\right)^{nt}$$

(b) One can imagine interest to be compounded each day, each hour, each minute, and so forth. Carried to the limit one can conceive of interest compounded at each instant of time; this is called *continuous compounding*. Thus, from (a), the value A of P dollars after t years when invested at an annual rate of $r \times 100\%$, compounded continuously is

$$A = \lim_{n \to +\infty} P\left(1 + \frac{r}{n}\right)^{nt}$$

In Section 10.3 we will show that $\lim_{x \to 0}(1 + x)^{1/x} = e$. Use this result to prove that

$$A = Pe^{rt}$$

(c) Use the result in (b) to show that money invested at continuous compound interest increases at a rate proportional to the amount present.

38. (a) If $1000 is invested at 8% per year compounded continuously (Exercise 37) what will the investment be worth after 5 years?

(b) If it is desired that an investment at 8% per year compounded continuously should have a value of $10,000 after 10 years, how much should be invested now?

(c) How long does it take for an investment at 8% per year compounded continuously to double in value?

39. Derive Formula (28) for halving time.

40. Let a quantity have an exponential growth model with growth rate k. How long does it take for the quantity to triple in size?

▶ REVIEW EXERCISES

1. If $r = \ln 2$ and $s = \ln 3$, express the following in terms of r and s:
   (a) $\ln(1/12)$ (b) $\ln(9/\sqrt{8})$
   (c) $\ln(\sqrt[4]{8/3})$.

2. Simplify.
   (a) $e^{2 - \ln x}$ (b) $\exp(\ln x^2 - 2\ln y)$
   (c) $\ln[x^3 \exp(-x^2)]$.

3. Solve for x in terms of $\ln 3$ and $\ln 5$.
   (a) $25^x = 3^{1-x}$ (b) $\sinh x = \frac{1}{4}\cosh x$.

4. Express the following as a rational function of x:
   $3\ln\left(e^{2x}(e^x)^3\right) + 2\exp(\ln 1)$.

5. If $\sinh x = -3/5$, find:
   (a) $\cosh x$ (b) $\tanh x$ (c) $\sinh(2x)$.

6. Use appendix tables to approximate the following quantities to two decimal places.
   (a) 2^e (b) $(\sqrt{2})^\pi$

In Exercises 7–28, find dy/dx. When appropriate, use implicit or logarithmic differentiation.

7. $y = 1/\sqrt{e^x}$.

8. $y = 1/e^{\sqrt{x}}$.

9. $y = x/\ln x$.

10. $y = e^x \ln(1/x)$.

11. $y = x/e^{\ln x}$.

12. $y = \ln\sqrt{x^2 + 2x}$.

13. $y = \ln(10^x/\sin x)$.

14. $y = \cos(e^{-2x})$.

15. $y = e^{\tan x} e^{4 \ln x}$.

16. $y = \ln\left|\dfrac{a+x}{a-x}\right|$.

17. $y = \ln |x + \sqrt{x^2 + a^2}|$.

18. $y = \ln |\tan 3x + \sec 3x|$.

19. $y = [\exp (x^2)]^3$.

20. $y = \ln (x^3/\sqrt{5 + \sin x})$.

21. $y = \sqrt{\ln (\sqrt{x})}$.

22. $y = e^{5x} + (5x)^e$.

23. $y = \pi^x x^\pi$.

24. $y = 4(e^x)^3/\sqrt{\exp (5x)}$.

25. $y = \sinh [\tanh (5x)]$.

26. $x^4 + e^{xy} - y^2 = 20$.

27. $y = e^{3x}(1 + e^{-x})^2$

28. $y = (\cosh x)^{x^3}$.

29. Show that $y = e^{ax} \sin bx$ satisfies $y'' - 2ay' + (a^2 + b^2)y = 0$ for any real constants a and b.

30. Assuming that u and v are differentiable functions of x, use logarithmic differentiation to derive a formula for $d(u^v)/dx$ in terms of u, v, du/dx, and dv/dx. What formula results when the base u is constant? When the exponent v is constant?

31. Find dy if

    (a) $y = e^{-x}$,   (b) $y = \ln (1 + x)$,

    (c) $y = 2^{x^2}$.

32. If $y = 3^{2x}5^{7x}$, show that dy/dx is proportional to y.

33. Use the chain rule and the second fundamental theorem of calculus to find the indicated derivative.

    (a) $\dfrac{d}{dx}\left(\displaystyle\int_0^{\ln x} \dfrac{dt}{\sqrt{4 + e^t}}\right)$

    (b) $\dfrac{d}{dx}\left(\displaystyle\int_1^{e^{5x}} \sqrt{\ln u + u}\; du\right)$.

[*Hint:* See Exercise 34, Section 7.2.]

34. Suppose $F(-\pi) = 0$ and that $y = F(x)$ satisfies $dy/dx = (3 \sin^2 2x + 2 \cos^2 3x)^{1/2}$. Express $F(x)$ as an integral.

35. If $y = Ce^{kt}$ (C, k constant) and $Y = \ln y$, show that the graph of Y versus t is a straight line.

36. Evaluate $\int e^{2x}(4 + e^{2x})\, dx$ two ways: (a) using the substitution $u = 4 + e^{2x}$, and (b) changing the

integrand to a sum. Verify that the antiderivatives in (a) and (b) differ by a constant.

In Exercises 37–52, evaluate the indicated integral.

37. $\displaystyle\int \frac{e^x}{1 + e^x}\, dx$.

38. $\displaystyle\int \frac{1 + e^x}{e^x}\, dx$.

39. $\displaystyle\int x^e\, dx$.

40. $\displaystyle\int \frac{x^2}{5 - 2x^3}\, dx$.

41. $\displaystyle\int \frac{4x^2 - 3x}{x^3}\, dx$.

42. $\displaystyle\int \frac{(\ln x^2)^2}{x}\, dx$.

43. $\displaystyle\int \frac{\sec x \tan x}{2 \sec x - 1}\, dx$.

44. $\displaystyle\int \frac{e^{5x}}{3 + e^{5x}}\, dx$.

45. $\displaystyle\int (\cos 2x) \exp (\sin 2x)\, dx$.

46. $\displaystyle\int \tanh (3x + 1)\, dx$.

47. $\displaystyle\int \operatorname{sech}^2 x \tanh x\, dx$.

48. $\displaystyle\int e^{2x}(4 + e^{-3x})\, dx$

49. $\displaystyle\int_e^{e^2} \frac{dx}{x \ln x}$.

50. $\displaystyle\int_0^1 \frac{dx}{\sqrt{e^x}}$.

51. $\displaystyle\int_0^{\pi/4} \frac{2\tan x}{\cos^2 x}\, dx$.

52. $\displaystyle\int_1^4 \frac{dx}{\sqrt{xe^{\sqrt{x}}}}$.

53. Show that $\int e^{kx}\, dx = (e^{kx}/k) + C$ for any nonzero k.

54. In each part find the indicated limit by interpreting the expression as a derivative and evaluating the derivative.

    (a) $\displaystyle\lim_{h \to 0} \frac{10^h - 1}{h}$

    (b) $\displaystyle\lim_{h \to 0} \frac{e^{(3+h)^2} - e^9}{h}$

    (c) $\displaystyle\lim_{h \to 0} \frac{\ln (e^2 + h) - 2}{h}$

    (d) $\displaystyle\lim_{x \to 1} \frac{2^x - 2}{x - 1}$.

55. Sketch $y = x^3 e^{-x}$, showing all relative extrema and inflection points. You may assume that $\lim_{x \to +\infty} x^3 e^{-x} = 0$.

56. Use the second derivative test to determine the nature of the critical points of $f(x) = x^2 e^{-x}$.

57. Show that if $y = f(x)$ satisfies $y' = e^y + 2y + x$, then every critical point of y is a relative minimum.

58. Sketch the curve $y = e^{-x/2} \sin 2x$ on $[-\pi/2, 3\pi/2]$. Show all x-intercepts and all points where $y = \pm e^{-x/2}$.

59. Let A denote the area under the curve $y = e^{-2x}$ over the interval $[0, b]$. Express A as a function of b, and then find the value of b for which $A = \frac{1}{4}$. What is the limiting value of A as $b \to +\infty$?

60. Let R be the region bounded by the curve $y = 4(2x - 1)^{-1/2}$, the x-axis, and the lines $x = 1$ and $x = 3$. Find the volume of the solid generated by revolving R about the x-axis.

61. Find the surface area generated when the curve $y = \cosh x$, $0 \le x \le 1$, is revolved about the x-axis.

[*Hint:* Equation (7b) of Section 7.6 may help in evaluating the integral.]

62. Suppose that a crystal dissolves at a rate proportional to the amount *un*dissolved. If 9 grams are undissolved initially and 6 grams remain undissolved after 1 minute, how many grams remain undissolved after 3 minutes?

63. The population of the United States was 205 million in 1970. Assuming an annual population increase of 1.8%, find: (a) the population in the year 2000, (b) the year in which the population will reach one billion.

8 inverse trigonometric and hyperbolic functions

8.1 INVERSE FUNCTIONS

The functions $f(x) = x^{1/3}$ and $g(x) = x^3$ have the property that each cancels out the effect of the other in the sense that

$$f(g(x)) = f(x^3) = (x^3)^{1/3} = x$$
$$g(f(x)) = g(x^{1/3}) = (x^{1/3})^3 = x$$

Similarly, the functions $f(x) = \ln x$ and $g(x) = e^x$ cancel the effect of one another since

$$f(g(x)) = f(e^x) = \ln (e^x) = x$$
$$g(f(x)) = g(\ln x) = e^{\ln x} = x$$

In this section we will study pairs of functions like these and in subsequent sections we will apply our results.

8.1.1 DEFINITION If two functions f and g satisfy

$$f(g(x)) = x \text{ for every } x \text{ in the domain of } g$$
$$g(f(x)) = x \text{ for every } x \text{ in the domain of } f$$

then we say that ***f is an inverse of g*** and ***g is an inverse of f***. We also say that ***f and g are inverses of one another.***

▶ Example 1

The functions $x^{1/3}$ and x^3 are inverses of one another, as are e^x and $\ln x$.

◀

It can be shown that a function cannot have two different inverses. Thus, if f has an inverse we are entitled to talk about *the* inverse of f. The inverse of f is commonly denoted by f^{-1} (read, "f inverse"); thus

$$f(f^{-1}(x)) = x \quad \text{and} \quad f^{-1}(f(x)) = x$$

WARNING.   The symbol f^{-1} does not mean $1/f$.

Given a function f, we can ask:

1.   Does f have an inverse?
2.   If so, how do we find it?

Let us address the second question first. Assume that f has an inverse, and let us apply f^{-1} to both sides of the equation

$$y = f(x) \tag{1}$$

We obtain

$$f^{-1}(y) = f^{-1}(f(x))$$

or

$$f^{-1}(y) = x \tag{2}$$

Since (2) expresses x as a function of y, the net effect of applying f^{-1} to both sides of (1) is to "solve" $y = f(x)$ for x in terms of y. Moreover, the left side of (2) is the inverse of f, expressed with y as the independent variable. This suggests the following procedure for finding inverses.

How to find the inverse of f.

   Step 1.   Write down the equation $y = f(x)$.
   Step 2.   Solve for x as a function of y.
   Step 3.   The resulting equation is $x = f^{-1}(y)$.

▶ Example 2

Find the inverse of $f(x) = 4x - 5$.

Solution.   Following the procedure above we obtain

$$y = 4x - 5$$

or

$$x = \frac{1}{4}(y + 5)$$

Thus $f^{-1}(y) = \frac{1}{4}(y + 5)$ is the inverse of f, expressed with y as the independent variable. For many purposes it is desirable to have the formulas for f and f^{-1} expressed in terms of the same variable. Thus, if we use x rather than y in the formula for f^{-1} we obtain

$$f^{-1}(x) = \frac{1}{4}(x + 5)$$

As a check, we verify that f and f^{-1} are in fact inverses:

$$f(f^{-1}(x)) = f\left(\frac{1}{4}(x + 5)\right) = 4\left[\frac{1}{4}(x + 5)\right] - 5 = x$$

$$f^{-1}(f(x)) = f^{-1}(4x - 5) = \frac{1}{4}[(4x - 5) + 5] = x \quad \blacktriangleleft$$

The procedure for finding inverses can break down for two reasons: the function f may not have an inverse; or the function f may have an inverse, but the equation $y = f(x)$ may be too complicated to solve for x in terms of y. Thus it becomes important to know what kinds of functions have inverses and to be able to study properties of an inverse even if a formula for it cannot be obtained. To attack these problems we will first investigate how the graphs of inverse functions are related.

In Section 7.5 we saw that the graphs of the inverse functions $f(x) = \ln x$ and $f^{-1}(x) = e^x$ are reflections of one another about the line $y = x$ (Figure 7.5.2). This is not accidental; the graphs of any two inverse functions are related in this way. To prove this, we must show that for each point (a, b) on the graph of f, the reflection (b, a) about $y = x$ lies on the graph of f^{-1}, and conversely. We argue as follows: Let (a, b) be a point on the curve $y = f(x)$. Thus

$$b = f(a)$$
$$f^{-1}(b) = f^{-1}(f(a))$$
$$f^{-1}(b) = a$$

The last equation implies that (b, a) lies on the curve $y = f^{-1}(x)$. By a similar argument, if (a, b) lies on the curve $y = f^{-1}(x)$, then (b, a) lies on the curve $y = f(x)$.

▶ Example 3

In Figure 8.1.1 we have sketched the graphs of the inverse functions $f(x) = x^{1/3}$ and $f^{-1}(x) = x^3$ as well as the inverse functions $f(x) = 4x - 5$ and $f^{-1}(x) = \frac{1}{4}(x + 5)$ obtained in Example 2. $\quad \blacktriangleleft$

Figure 8.1.1

(a)   (b)

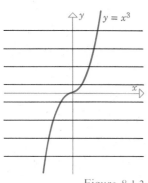

(a)

(b)

Figure 8.1.2

Let us turn now to the problem of determining conditions which ensure that a function f has an inverse. We begin with a simple example of a function with no inverse, $f(x) = x^2$. To see that no inverse exists, we need only observe that if f had an inverse then the curve $y = f^{-1}(x)$ would be the reflection about $y = x$ of the curve $y = x^2$ (Figure 8.1.2a). But the reflected curve (Figure 8.1.2b) cannot be the graph of any function, since it does not pass the vertical line test (see Theorem 2.1.4). Thus $f(x) = x^2$ has no inverse.

Figure 8.1.2 suggests that the reflection about $y = x$ of a curve $y = f(x)$ will fail to pass the vertical line test if there is some *horizontal* line that cuts the curve $y = f(x)$ more than once. Thus, for a function f to have an inverse its graph cannot be cut more than once by any horizontal line. A function f with this property is said to be *one-to-one* (sometimes written 1-1). This idea can also be expressed as follows.

8.1.2 DEFINITION   A function is said to be ***one-to-one*** if it does not take on the same value at two distinct points in its domain.

▶ Example 4

The function $f(x) = x^3$ is one-to-one since two different numbers cannot have the same cube, that is, if $x_1 \neq x_2$ then $x_1{}^3 \neq x_2{}^3$. Geometrically, no horizontal line cuts the graph of $f(x) = x^3$ more than once (Figure 8.1.3). ◀

▶ Example 5

Figure 8.1.3

The function $f(x) = \sin x$ is not one-to-one since distinct values of x may produce the same value for $\sin x$; for example, $\pi \neq 0$, yet $\sin \pi = \sin 0 = 0$.

Geometrically, there are horizontal lines that cut the graph of $f(x) = \sin x$ more than once (Figure 8.1.4). Similarly, $f(x) = x^2$ is not one-to-one since $f(2) = f(-2) = 4$. ◄

Figure 8.1.4

► Example 6

If f is an increasing function on an interval and if $x_1 < x_2$ are distinct points in the interval, then $f(x_1) < f(x_2)$ (Definition 4.4.1), so that $f(x_1) \neq f(x_2)$. Thus f is one-to-one on the interval. Similarly, if f is decreasing on an interval, it is a one-to-one function on that interval. ◄

REMARK. For a one-to-one function f, the value of the independent variable x is uniquely determined once the value for $f(x)$ is specified. For example, if $f(x) = x^3$ then knowing $x^3 = 8$ tells us that $x = 2$. For functions that are not one-to-one, the value of $f(x)$ need not determine x uniquely. For example, if $f(x) = x^2 = 4$, then $x = 2$ and $x = -2$ are both possibilities for x.

8.1.3 THEOREM *If f is one-to-one then f has an inverse, and conversely if f has an inverse then f is one-to-one.*

Proof. Assume f to be one-to-one and *define* a function g as follows. Take the domain of g to be the range of f; that is, the domain of g is the set of all possible values of $f(x)$. Thus, if x is any number in the domain of g, there is a number x' in the domain of f such that

$$x = f(x') \tag{3}$$

Moreover, by the remark preceding this theorem, the number x' is uniquely determined. Define g by the formula

$$g(x) = x' \tag{4}$$

Using (4) and then (3) gives

$$f(g(x)) = f(x') = x \tag{5}$$

On the other hand, using (3) and then (4) gives

$$g(f(x')) = g(x) = x' \tag{6}$$

From (5) and (6), the functions f and g are inverses since they cancel the effect of one another.

For the converse, assume that f has an inverse, and let x_1 and x_2 be distinct numbers in the domain of f. To prove that f is one-to-one we must show that $f(x_1) \neq f(x_2)$. But if

$$f(x_1) = f(x_2)$$

then

$$f^{-1}(f(x_1)) = f^{-1}(f(x_2))$$

from which it follows that

$$x_1 = x_2$$

in contradiction of the assumption that $x_1 \neq x_2$. Thus $f(x_1) \neq f(x_2)$.

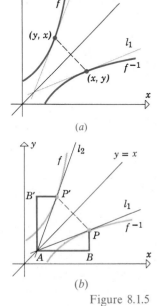

(a)

(b)

Figure 8.1.5

We conclude this section by investigating the relationship between the derivative of a function and the derivative of its inverse. The result we want can be motivated by Figure 8.1.5. In Figure 8.1.5a we show graphs of two inverse functions and tangent lines l_1 and l_2 at the points (x, y) and (y, x). Referring to Figure 8.1.5b, the slopes m_1 and m_2 of tangents l_1 and l_2 are

$$m_1 = \frac{\overline{BP}}{\overline{AB}} \quad \text{and} \quad m_2 = \frac{\overline{AB'}}{\overline{B'P'}} \tag{7}$$

But triangles ABP and $AB'P'$ are congruent (verify) so that $\overline{BP} = \overline{B'P'}$ and $\overline{AB} = \overline{AB'}$. From these equalities and (7) it follows that

$$m_1 = \frac{1}{m_2} \tag{8}$$

Since $m_1 = (f^{-1})'(x)$ and $m_2 = f'(y)$, (8) yields

$$(f^{-1})'(x) = \frac{1}{f'(y)} \tag{9}$$

But (x, y) is on the graph of f^{-1}, so $y = f^{-1}(x)$. Consequently, (9) can also be written

$$(f^{-1})'(x) = \frac{1}{f'(f^{-1}(x))}$$

Observe that these formulas break down if $f'(y) = 0$. However, this is to be expected since $f'(y) = 0$ indicates a horizontal tangent for f at y, which reflects about $y = x$ into a vertical tangent for f^{-1} at x.

The following theorem summarizes these informal results.

8.1.4 THEOREM *Suppose that f has an inverse and is differentiable on an open interval I. If $f^{-1}(x)$ is a point in I at which f' is nonzero, then f^{-1} is differentiable at x and*

$$(f^{-1})'(x) = \frac{1}{f'(f^{-1}(x))} \tag{10}$$

REMARK. If we let

$$y = f^{-1}(x) \qquad \text{so that} \qquad x = f(y)$$

then

$$\frac{dy}{dx} = (f^{-1})' \qquad \text{and} \qquad \frac{dx}{dy} = f'$$

so that (10) can be written in more compact form as

$$\frac{dy}{dx} = \frac{1}{dx/dy} \tag{11}$$

▶ Example 7

In Section 7.4 we obtained the derivative formula for e^x by *assuming* this function to be differentiable. Using Theorem 8.1.4 we can now prove the differentiability of e^x as follows: Let

$$f(x) = \ln x$$

so that

$$f^{-1}(x) = e^x$$

If x is a point in $(-\infty, +\infty)$ then $f^{-1}(x) = e^x$ is a point in $(0, +\infty)$, since e^x assumes only positive values. Moreover, f' is nonzero on $(0, +\infty)$ since

$$f'(x) = \frac{1}{x} > 0$$

In particular, f' is nonzero at the point $f^{-1}(x) = e^x$. Thus, f^{-1} is differentiable at x by Theorem 8.1.4. Since x is any point in $(-\infty, +\infty)$, we have shown that e^x is differentiable everywhere. ◀

As a point of interest, the familiar derivative formula for e^x follows from (10) with $f(x) = \ln x$ and $f^{-1}(x) = e^x$, since

$$\frac{d}{dx}[e^x] = (f^{-1})'(x) = \frac{1}{f'(f^{-1}(x))} = \frac{1}{f'(e^x)} = \frac{1}{1/e^x} = e^x$$

▶ **Example 8**

Let $y = \sqrt[3]{x - 2}$. Use (11) to find dy/dx.

Solution.

$$y = \sqrt[3]{x - 2}$$
$$x = y^3 + 2$$
$$\frac{dx}{dy} = 3y^2$$
$$\frac{dy}{dx} = \frac{1}{dx/dy} = \frac{1}{3y^2} = \frac{1}{3(\sqrt[3]{x - 2})^2}$$

This agrees with the result that would be obtained by differentiating directly. (Verify.) ◀

The method of Example 8 is especially useful when direct calculation of dy/dx is difficult or impossible. We will encounter such examples in the next few sections.

▶ **Exercise Set 8.1**

1. In (a)–(d), determine if f and g are inverse functions.

    (a) $f(x) = 4x$, $g(x) = \frac{1}{4}x$

    (b) $f(x) = 3x + 1$, $g(x) = 3x - 1$

    (c) $f(x) = \ln(x + 1)$, $g(x) = e^x - 1$

    (d) $f(x) = x^4$, $g(x) = \sqrt[4]{x}$.

    In Exercises 2–7, find f^{-1}.

    2. $f(x) = 6x$.

    3. $f(x) = 7x - 6$.

4. $f(x) = \dfrac{x+1}{x-1}$.

5. $f(x) = 3x^3 - 5$.

6. $f(x) = \sqrt[5]{4x+2}$.

7. $f(x) = \sqrt{e^x + 1}$.

8. Which of the following functions are one-to-one on the stated interval?

   (a) $f(x) = 1 - x$ on $(-\infty, +\infty)$

   (b) $f(x) = x^2 - 2x + 1$ on $(-\infty, +\infty)$

   (c) $f(x) = x + \dfrac{1}{x}$ on $(0, +\infty)$

   (d) $f(x) = \tan x$ on $\left(-\dfrac{\pi}{2}, \dfrac{\pi}{2}\right)$.

In Exercises 9–12, find dy/dx using (11). (See Example 8.) Check your work by differentiating directly.

9. $y = 8x + 5$.

10. $y = \sqrt[5]{x+5}$.

11. $y = \dfrac{x}{x-1}$.

12. $y = e^{(x-1)/3}$.

13. (a) Show that $f(x) = (3-x)/(1-x)$ is its own inverse.

   (b) What does the result in (a) tell you about the graph of f?

14. Prove that the graph of

$$f(x) = \frac{ax+b}{cx+d}$$

is symmetric about the line $y = x$ if $d = -a$.

15. (a) Show that $f(x) = x^3 - 3x^2 + 2x$ is not one-to-one on $(-\infty, +\infty)$.

   (b) Find the largest value of k such that f is one-to-one on the interval $(-k, k)$.

16. A line of nonzero slope m intersects the x-axis at $(x_0, 0)$. Find an equation for the reflection of this line about $y = x$.

17. Let $f(x) = \int_1^x \sqrt[3]{1+t^2}\, dt$.

   (a) Without integrating, show that f is one-to-one on the interval $(-\infty, +\infty)$.

   (b) Find $\dfrac{d}{dx}[f^{-1}(x)]\Big|_{x=0}$

18. (a) Prove: If f and g are one-to-one, then so is the composition $f \circ g$.

   (b) Prove: If f and g are one-to-one, then

$$(f \circ g)^{-1} = g^{-1} \circ f^{-1}$$

19. Sketch the graph of a function that is one-to-one on $(-\infty, +\infty)$, yet not monotone on $(-\infty, +\infty)$.

20. Prove: A one-to-one function f cannot have two different inverses.

21. Prove: If $a > 0$, then a^x is a differentiable function. [*Hint:* Use Example 7 and Corollary 7.4.5.]

8.2 INVERSE TRIGONOMETRIC FUNCTIONS

Since the six basic trigonometric functions are periodic, they are not one-to-one, and consequently have no inverses. However, we will show in this section that it is possible to restrict the domains of the trigonometric functions in such a way that the restricted functions do have inverses.

In Figure 8.2.1 we have sketched the graph of

$$f(x) = \sin x, \qquad -\frac{\pi}{2} \le x \le \frac{\pi}{2} \tag{1}$$

Although $\sin x$ is defined on the entire interval $(-\infty, \infty)$, the function $f(x)$ in (1) is defined only on the interval $[-\pi/2, \pi/2]$. Since no horizontal line cuts the graph of $f(x) = \sin x$, $-\pi/2 \le x \le \pi/2$ more than once, this function

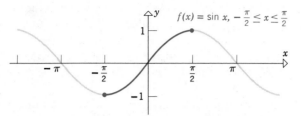

Figure 8.2.1

has an inverse. This inverse is called the *inverse sine,* or sometimes the *arcsine function,* and is denoted by

$$\sin^{-1} x \qquad \text{or} \qquad \arcsin x$$

The graph of $\sin^{-1} x$ is obtained by reflecting the graph of $\sin x$, $-\pi/2 \le x \le \pi/2$ about the line $y = x$ (Figure 8.2.2). As noted in the previous section, the -1 in $\sin^{-1} x$ is not an exponent, but rather a notation for the inverse function. The reciprocal of $\sin x$ would be written as $(\sin x)^{-1}$, or better, $\csc x$.

By definition, the domain of $\sin^{-1} x$ is the interval $[-1, 1]$ and

Figure 8.2.2

$$\sin^{-1}(\sin x) = x \qquad \text{if} \qquad -\frac{\pi}{2} \le x \le \frac{\pi}{2}$$

$$\sin(\sin^{-1} x) = x \qquad \text{if} \qquad -1 \le x \le 1$$

Moreover,

$$y = \sin x \qquad \text{and} \qquad x = \sin^{-1} y \qquad\qquad (2)$$

are equivalent statements if $-\pi/2 \le x \le \pi/2$. Geometrically, if $x = \sin^{-1} y$, then x is an angle whose sine is y.

▶ Example 1

Find

(a) $\sin^{-1}\left(\dfrac{1}{2}\right)$ (b) $\sin^{-1}\left(-\dfrac{1}{\sqrt{2}}\right)$

Solution (a). Let $x = \sin^{-1}(\frac{1}{2})$ so that from (2), $\frac{1}{2} = \sin x$, where $-\pi/2 \le x \le \pi/2$. Thus $x = \pi/6$. (Note that $\frac{1}{2} = \sin x$ has infinitely many solutions; however $x = \pi/6$ is the only solution that satisfies the requirement, $-\pi/2 \le x \le \pi/2$.)

Solution (b). Let $x = \sin^{-1}(-1/\sqrt{2})$, so that from (2) $\sin x = -1/\sqrt{2}$, where $-\pi/2 \le x \le \pi/2$. Thus $x = -\pi/4$. ◀

REMARK. To define $\sin^{-1} x$ we restricted the domain of $\sin x$ to the interval $[-\pi/2, \pi/2]$. This yielded a one-to-one function. There are other ways to restrict the domain of $\sin x$ to obtain one-to-one functions; for example we might have required that $3\pi/2 \le x \le 5\pi/2$ or $-5\pi/2 \le x \le -3\pi/2$. However, the choice $-\pi/2 \le x \le \pi/2$ is customary.

The remaining inverse trigonometric functions, $\cos^{-1} x$, $\tan^{-1} x$, $\cot^{-1} x$, $\sec^{-1} x$, and $\csc^{-1} x$ are also defined as inverses of the corresponding trigonometric functions, with appropriate restrictions on their domains. The definitions are as follows:

8.2.1 DEFINITION

$\sin^{-1} x$	is the inverse of	$\sin x,\ -\pi/2 \le x \le \pi/2$
$\cos^{-1} x$	is the inverse of	$\cos x,\ \ 0 \le x \le \pi$
$\tan^{-1} x$	is the inverse of	$\tan x,\ -\pi/2 < x < \pi/2$
$\cot^{-1} x$	is the inverse of	$\cot x,\ \ 0 < x < \pi$
$\sec^{-1} x$	is the inverse of	$\sec x,\ \begin{cases} 0 \le x < \pi/2 \\ \pi \le x < 3\pi/2 \end{cases}$
$\csc^{-1} x$	is the inverse of	$\csc x,\ \begin{cases} 0 < x \le \pi/2 \\ \pi < x \le 3\pi/2 \end{cases}$

By reflecting the graphs of the restricted trigonometric functions in this definition about $y = x$, the reader should be able to obtain the graphs of the inverses shown in Figure 8.2.3. The missing graphs are left as exercises.

Figure 8.2.3

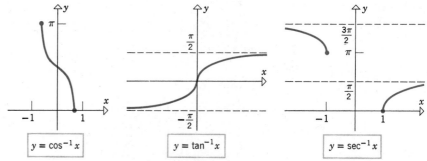

$y = \cos^{-1} x$ $y = \tan^{-1} x$ $y = \sec^{-1} x$

$\sin \alpha = \cos \beta$

Figure 8.2.4

In Definition 8.2.1, the restrictions on the domains were selected not only to produce one-to-one functions, but also to ensure that certain "natural" identities hold for the inverse trigonometric functions. For example, if α and β are acute complementary angles, then from basic trigonometry, $\sin \alpha$ and $\cos \beta$ are equal (Figure 8.2.4). Let us write $x = \sin \alpha = \cos \beta$ so that

$$\alpha = \sin^{-1} x \quad \text{and} \quad \beta = \cos^{-1} x$$

since $\alpha + \beta = \pi/2$ we obtain the identity

$$\sin^{-1} x + \cos^{-1} x = \frac{\pi}{2} \tag{3}$$

(Because α and β were assumed acute, this derivation is only valid for non-negative x; for a derivation valid for all x in $[-1, 1]$ see Exercise 21.) Similarly, we can obtain the identities

$$\tan^{-1} x + \cot^{-1} x = \frac{\pi}{2} \tag{4}$$

$$\sec^{-1} x + \csc^{-1} x = \frac{\pi}{2} \tag{5}$$

The proofs of the following identities are left as exercises. Identities (6) and (8) are evident from Figures 8.2.2 and 8.2.3.

$$\sin^{-1}(-x) = -\sin^{-1} x \tag{6}$$

$$\cos^{-1}(-x) = \pi - \cos^{-1} x \tag{7}$$

$$\tan^{-1}(-x) = -\tan^{-1} x \tag{8}$$

REMARK.   There is no universal agreement among mathematicians about the definitions of $\sec^{-1} x$ and $\csc^{-1} x$. Some writers define $\sec^{-1} x$ and $\csc^{-1} x$ to be the inverses of $\sec x, 0 \leq x \leq \pi$ and $\csc x, -\pi/2 \leq x \leq \pi/2$ respectively. With these definitions one obtains some useful identities that are not valid with our definitions. However, our definitions produce simpler derivative formulas for $\sec^{-1} x$ and $\csc^{-1} x$ and simplify the solution of certain integration problems to be studied in the next chapter.

▶ Exercise Set 8.2

1. Find the exact value of
    (a) $\sin^{-1}(-1)$ (b) $\cos^{-1}(-1)$
    (c) $\tan^{-1}(-1)$ (d) $\cot^{-1}(1)$
    (e) $\sec^{-1}(1)$ (f) $\csc^{-1}(1)$.

2. Find the exact value of
    (a) $\sin^{-1}(\frac{1}{2}\sqrt{3})$ (b) $\cos^{-1}(\frac{1}{2})$
    (c) $\tan^{-1}(1)$ (d) $\cot^{-1}(-1)$
    (e) $\sec^{-1}(-2)$ (f) $\csc^{-1}(-2)$.

3. Given that $\theta = \sin^{-1}(-\frac{1}{2}\sqrt{3})$, find the exact values of $\cos\theta$, $\tan\theta$, $\cot\theta$, $\sec\theta$, and $\csc\theta$.

4. Given that $\theta = \cos^{-1}(\frac{1}{2})$, find the exact values of $\sin\theta$, $\tan\theta$, $\cot\theta$, $\sec\theta$, and $\csc\theta$.

5. Given that $\theta = \tan^{-1}(\frac{4}{3})$, find the exact values of $\sin\theta$, $\cos\theta$, $\cot\theta$, $\sec\theta$, and $\csc\theta$.

6. State the domains of the six inverse trigonometric functions.

7. Find the exact value of

   (a) $\sin^{-1}\left(\sin\frac{\pi}{7}\right)$ (b) $\sin^{-1}(\sin\pi)$

   (c) $\sin^{-1}\left(\sin\frac{5\pi}{7}\right)$.

8. Find the exact value of

   (a) $\cos^{-1}\left(\cos\frac{\pi}{7}\right)$ (b) $\cos^{-1}(\cos\pi)$

   (c) $\cos^{-1}\left(\cos\frac{12\pi}{7}\right)$.

9. For which values of x is it true that
   (a) $\cos^{-1}(\cos x) = x$ (b) $\cos(\cos^{-1}x) = x$
   (c) $\tan^{-1}(\tan x) = x$ (d) $\tan(\tan^{-1}x) = x$
   (e) $\csc^{-1}(\csc x) = x$ (f) $\csc(\csc^{-1}x) = x$.

In Exercises 10–15, find the exact value of the given quantity.

10. $\sec\left[\sin^{-1}\left(-\frac{3}{4}\right)\right]$.   **11.** $\sin\left[2\cos^{-1}\left(\frac{3}{5}\right)\right]$.

12. $\tan^{-1}\left[\sin\left(-\frac{\pi}{2}\right)\right]$.   **13.** $\sin^{-1}\left[\cot\left(\frac{\pi}{4}\right)\right]$.

14. $\sin\left[\sin^{-1}\left(\frac{2}{3}\right) + \cos^{-1}\left(\frac{1}{3}\right)\right]$.

15. $\tan\left[2\sec^{-1}\left(\frac{3}{2}\right)\right]$.

16. Sketch the graphs of $\cot^{-1}x$ and $\csc^{-1}x$.

17. Sketch the graphs of
   (a) $y = \sin^{-1}2x$ (b) $y = \tan^{-1}\frac{1}{2}x$.

18. Sketch the graphs of
   (a) $y = \cos^{-1}\frac{1}{3}x$ (b) $y = 2\cot^{-1}2x$.

19. Prove:
   (a) $\sin^{-1}(-x) = -\sin^{-1}x$
   (b) $\tan^{-1}(-x) = -\tan^{-1}x$.

20. Prove: $\cos^{-1}(-x) = \pi - \cos^{-1}x$.

21. In the text we proved that

$$\sin^{-1}x + \cos^{-1}x = \frac{\pi}{2}$$

for $0 \leq x \leq 1$. Use this result together with Exercises 19(a) and 20 to prove this identity for $-1 \leq x \leq 1$.

22. A camera is positioned x feet from the base of a missile launching pad. If a missile of length a feet is launched vertically, show that when the base of the missile is b feet above the camera lens, the angle θ subtended at the lens by the missile is

$$\theta = \cot^{-1}\frac{x}{a+b} - \cot^{-1}\frac{x}{b}$$

Camera x Launch pad

8.3 DERIVATIVES AND INTEGRALS INVOLVING INVERSE TRIGONOMETRIC FUNCTIONS

In this section we obtain the derivatives of the inverse trigonometric functions and derive some new integration formulas.

We begin with some techniques for simplifying expressions involving inverse trigonometric functions. Below we will encounter the expression

$$\sec^2 (\tan^{-1} x)$$

This can be simplified by writing sec in terms of tan to take advantage of the simplification $\tan (\tan^{-1} x) = x$. Thus, if we let $\theta = \tan^{-1} x$ in the identity

$$\sec^2 \theta = 1 + \tan^2 \theta$$

we obtain

$$\sec^2 (\tan^{-1} x) = 1 + \tan^2 (\tan^{-1} x)$$

or

$$\sec^2 (\tan^{-1} x) = 1 + x^2 \tag{1}$$

We will also encounter the expression

$$\cos (\sin^{-1} x)$$

This time we will write cos in terms of sin to take advantage of the simplification $\sin (\sin^{-1} x) = x$. Thus, if we start with

$$\cos^2 \theta = 1 - \sin^2 \theta$$

or

$$|\cos \theta| = \sqrt{1 - \sin^2 \theta}$$

and let

$$\theta = \sin^{-1} x$$

we obtain

$$|\cos (\sin^{-1} x)| = \sqrt{1 - \sin^2 (\sin^{-1} x)}$$

or

$$|\cos (\sin^{-1} x)| = \sqrt{1 - x^2}$$

Since $-\pi/2 \leq \sin^{-1} x \leq \pi/2$, it follows that $\cos (\sin^{-1} x)$ is nonnegative; thus we can drop the absolute value sign and write

$$\cos (\sin^{-1} x) = \sqrt{1 - x^2} \tag{2}$$

The reader may wish to check for mastery of this technique by proving

$$\tan(\sec^{-1} x) = \sqrt{x^2 - 1} \tag{3}$$

We can now derive the following differentiation formulas.

8.3.1 THEOREM

$$\frac{d}{dx}[\sin^{-1} x] = \frac{1}{\sqrt{1 - x^2}} \tag{4}$$

$$\frac{d}{dx}[\cos^{-1} x] = -\frac{1}{\sqrt{1 - x^2}} \tag{5}$$

$$\frac{d}{dx}[\tan^{-1} x] = \frac{1}{1 + x^2} \tag{6}$$

$$\frac{d}{dx}[\cot^{-1} x] = -\frac{1}{1 + x^2} \tag{7}$$

$$\frac{d}{dx}[\sec^{-1} x] = \frac{1}{x\sqrt{x^2 - 1}} \tag{8}$$

$$\frac{d}{dx}[\csc^{-1} x] = -\frac{1}{x\sqrt{x^2 - 1}} \tag{9}$$

Proof. To obtain (4), let

$$y = \sin^{-1} x$$

so

$$x = \sin y, \qquad -\frac{\pi}{2} \le y \le \frac{\pi}{2}$$

Thus, from (11) of Section 8.1 and (2) above,

$$\frac{d}{dx}[\sin^{-1} x] = \frac{dy}{dx} = \frac{1}{dx/dy} = \frac{1}{\cos y} = \frac{1}{\cos(\sin^{-1} x)} = \frac{1}{\sqrt{1 - x^2}}$$

Formula (6) is obtained similarly; let

$$y = \tan^{-1} x$$

so

$$x = \tan y, \qquad -\frac{\pi}{2} < y < \frac{\pi}{2}$$

Thus, from (11) of Section 8.1 and (1) above,

$$\frac{d}{dx}[\tan^{-1} x] = \frac{dy}{dx} = \frac{1}{dx/dy} = \frac{1}{\sec^2 y} = \frac{1}{\sec^2 (\tan^{-1} x)} = \frac{1}{1 + x^2}$$

To obtain (8), let

$$y = \sec^{-1} x$$

so

$$x = \sec y, \qquad 0 \leq y < \frac{\pi}{2} \quad \text{or} \quad \pi \leq y < \frac{3\pi}{2}$$

Thus, from (11) of Section 8.1 and (3) above,

$$\frac{d}{dx}[\sec^{-1} x] = \frac{dy}{dx} = \frac{1}{dx/dy} = \frac{1}{\sec y \tan y} = \frac{1}{\sec (\sec^{-1} x) \tan (\sec^{-1} x)}$$

$$= \frac{1}{x \sqrt{x^2 - 1}}$$

The remaining formulas can all be obtained from the derivatives of $\sin^{-1}$, $\tan^{-1}$, and $\sec^{-1}$ by using appropriate identities. For example, using identity (3) of Section 8.2 yields

$$\frac{d}{dx}[\cos^{-1} x] = \frac{d}{dx}\left[\frac{\pi}{2} - \sin^{-1} x\right] = -\frac{d}{dx}[\sin^{-1} x] = -\frac{1}{\sqrt{1 - x^2}}$$

Similarly, the derivatives of $\cot^{-1} x$ and $\csc^{-1} x$ follow from identities (4) and (5) of Section 8.2. ▨

▶ Example 1

Find dy/dx if $y = \sin^{-1}(x^3)$.

Solution. From (4) and the chain rule,

$$\frac{dy}{dx} = \frac{1}{\sqrt{1 - (x^3)^2}}(3x^2) = \frac{3x^2}{\sqrt{1 - x^6}}$$ ◀

▶ Example 2

Find dy/dx if $y = \sec^{-1}(e^x)$.

Solution. From (8) and the chain rule,

$$\frac{dy}{dx} = \frac{1}{e^x \sqrt{(e^x)^2 - 1}}(e^x) = \frac{1}{\sqrt{e^{2x} - 1}}$$ ◀

Differentiation formulas (4)–(9) yield useful integration formulas. Those most commonly needed are:

$$\int \frac{dx}{\sqrt{1 - x^2}} = \sin^{-1} x + C \tag{10}$$

$$\int \frac{dx}{1 + x^2} = \tan^{-1} x + C \tag{11}$$

$$\int \frac{dx}{x\sqrt{x^2 - 1}} = \sec^{-1} x + C \tag{12}$$

▶ Example 3

Evaluate

$$\int \frac{dx}{1 + 3x^2}$$

Solution. Substituting

$$u = \sqrt{3}\, x, \qquad du = \sqrt{3}\, dx$$

yields

$$\int \frac{dx}{1 + 3x^2} = \frac{1}{\sqrt{3}} \int \frac{du}{1 + u^2} = \frac{1}{\sqrt{3}} \tan^{-1} u + C$$

$$= \frac{1}{\sqrt{3}} \tan^{-1}(\sqrt{3}\, x) + C \qquad ◀$$

▶ Example 4

Evaluate

$$\int \frac{e^x}{\sqrt{1 - e^{2x}}}\, dx$$

Solution. Substituting

$$u = e^x, \qquad du = e^x\, dx$$

yields

$$\int \frac{e^x}{\sqrt{1 - e^{2x}}} \, dx = \int \frac{du}{\sqrt{1 - u^2}} = \sin^{-1} u + C = \sin^{-1}(e^x) + C \quad \blacktriangleleft$$

▶ **Example 5**

Evaluate

$$\int \frac{dx}{a^2 + x^2}$$

where $a \neq 0$ is a constant.

Solution. If we had a 1 in place of a^2, we could use (11). Thus, we look for a u-substitution that will enable us to replace the a^2 with a 1. If we let

$$x = au, \qquad dx = a \, du$$

then

$$\int \frac{dx}{a^2 + x^2} = \int \frac{a \, du}{a^2 + a^2 u^2} = \frac{1}{a} \int \frac{du}{1 + u^2}$$

$$= \frac{1}{a} \tan^{-1} u + C$$

$$= \frac{1}{a} \tan^{-1} \frac{x}{a} + C \quad \blacktriangleleft$$

The method of Example 5 leads to the following generalizations of (10), (11), and (12).

$$\int \frac{dx}{\sqrt{a^2 - x^2}} = \sin^{-1} \frac{x}{a} + C \qquad (a > 0) \tag{13}$$

$$\int \frac{dx}{a^2 + x^2} = \frac{1}{a} \tan^{-1} \frac{x}{a} + C \qquad (a \neq 0) \tag{14}$$

$$\int \frac{dx}{x\sqrt{x^2 - a^2}} = \frac{1}{a} \sec^{-1} \frac{x}{a} + C \qquad (a > 0) \tag{15}$$

▶ Example 6

Evaluate

$$\int \frac{dx}{\sqrt{2 - x^2}}$$

Solution. Applying (13) with $a = \sqrt{2}$ yields

$$\int \frac{dx}{\sqrt{2 - x^2}} = \sin^{-1} \frac{x}{\sqrt{2}} + C$$ ◀

▶ Exercise Set 8.3

In Exercises 1–9, find dy/dx.

1. (a) $y = \sin^{-1}(\frac{1}{3}x)$ (b) $y = \cos^{-1}(2x + 1)$.

2. (a) $y = \tan^{-1}(x^2)$ (b) $y = \cot^{-1}(\sqrt{x})$.

3. (a) $y = \sec^{-1}(x^7)$ (b) $y = \csc^{-1}(e^x)$.

4. (a) $y = (\tan x)^{-1}$ (b) $y = \dfrac{1}{\tan^{-1} x}$.

5. (a) $y = \sin^{-1}\left(\dfrac{1}{x}\right)$ (b) $y = \cos^{-1}(\cos x)$.

6. (a) $y = \ln(\cos^{-1} x)$ (b) $y = \sqrt{\cot^{-1} x}$.

7. (a) $y = e^x \sec^{-1} x$

     (b) $y = x^2(\sin^{-1} x)^3$.

8. (a) $y = \sin^{-1} x + \cos^{-1} x$

     (b) $y = \sec^{-1} x + \csc^{-1} x$.

9. (a) $y = \tan^{-1}\left(\dfrac{1 - x}{1 + x}\right)$

     (b) $y = (1 + x \csc^{-1} x)^{10}$.

In Exercises 10–11, find dy/dx by implicit differentiation.

10. $x^3 + x \tan^{-1} y = e^y$.

11. $\sin^{-1}(xy) = \cos^{-1}(x - y)$.

In Exercises 12–23, evaluate the integral.

12. $\displaystyle\int_0^{1/\sqrt{2}} \frac{dx}{\sqrt{1 - x^2}}$.   **13.** $\displaystyle\int_{-1}^1 \frac{dx}{1 + x^2}$.

14. $\displaystyle\int_{\sqrt{2}}^2 \frac{dx}{x\sqrt{x^2 - 1}}$.   **15.** $\displaystyle\int_{-\sqrt{2}}^{-2/\sqrt{3}} \frac{dx}{x\sqrt{x^2 - 1}}$.

16. $\displaystyle\int \frac{dx}{\sqrt{1 - 4x^2}}$.   **17.** $\displaystyle\int \frac{dx}{1 + 16x^2}$.

18. $\displaystyle\int \frac{dx}{x\sqrt{9x^2 - 1}}$.   **19.** $\displaystyle\int \frac{e^x}{1 + e^{2x}} dx$.

20. $\displaystyle\int \frac{t}{t^4 + 1} dt$.   **21.** $\displaystyle\int \frac{\sec^2 x \, dx}{\sqrt{1 - \tan^2 x}}$.

22. $\displaystyle\int \frac{\sin \theta}{\cos^2 \theta + 1} d\theta$.   **23.** $\displaystyle\int \frac{dx}{x\sqrt{1 - (\ln x)^2}}$.

24. Derive integration formulas (13) and (15).

In Exercises 25 and 26, use (13), (14), and (15) to evaluate the integrals.

25. (a) $\displaystyle\int \frac{dx}{\sqrt{9 - x^2}}$ (b) $\displaystyle\int \frac{dx}{5 + x^2}$

     (c) $\displaystyle\int \frac{dx}{x\sqrt{x^2 - \pi}}$.

26. (a) $\displaystyle\int \frac{e^x}{4 + e^{2x}} dx$ (b) $\displaystyle\int \frac{dx}{\sqrt{9 - 4x^2}}$

     (c) $\displaystyle\int \frac{dy}{y\sqrt{5y^2 - 3}}$.

27. In the beginning of this section we derived the identity $\cos(\sin^{-1} x) = \sqrt{1 - x^2}$ [see (2)]. If we view $\sin^{-1} x$ as an angle having x for its sine, then this result can be remembered from the right triangle.

This triangle suggests the identity

$$\tan(\sin^{-1} x) = \frac{x}{\sqrt{1 - x^2}}, \quad 0 < x < 1$$

Prove this result.

28. Using the idea suggested in Exercise 27, find the length of the unmarked side of each triangle below and then use the triangles to complete the indicated identities.

(a) $\sin(\cos^{-1} x) =$?   for $0 < x < 1$
(b) $\tan(\cos^{-1} x) =$?   for $0 < x < 1$
(c) $\csc(\tan^{-1} x) =$?   for $x > 0$
(d) $\sin(\tan^{-1} x) =$?   for $x > 0$
(e) $\cos(\tan^{-1} x) =$?   for $x > 0$
(f) $\tan(\cot^{-1} x) =$?   for $x > 0$
(g) $\sin(\sec^{-1} x) =$?   for $x > 1$
(h) $\cot(\csc^{-1} x) =$?   for $x > 1$.

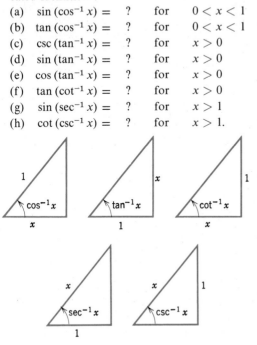

29. Find the volume of the solid generated when the region bounded by $x = 2$, $x = -2$, $y = 0$, and $y = 1/\sqrt{4 + x^2}$ is revolved about the x-axis.

30. In Exercise 22 of Section 8.2, how far from the launching pad should the camera be positioned to maximize the angle θ subtended at the lens by the missile?

31. Given points $A(2, 1)$ and $B(5, 4)$, find the point P in the interval $[2, 5]$ on the x-axis that maximizes angle APB.

32. An aircraft is flying at an altitude of 4 miles at a speed of 800 mi/hr in a direction away from a tracking station on the ground. How fast is the angle of elevation changing when the aircraft is above a point 10 miles from the station?

33. A lighthouse is located 3 miles off a straight shore. If the light revolves at 2 revolutions per minute, how fast is the beam moving along the coastline at a point 2 miles down the coast?

34. A 25-ft ladder leans against a vertical wall. If the bottom of the ladder slides away from the base of the wall at the rate of 4 ft/sec, how fast is the angle between the ladder and the wall changing when the top of the ladder is 20 ft above the ground?

35. The lower edge of a painting, 10 ft in height, is 2 ft above an observer's eye level. Assuming the best view is obtained when the angle subtended at the observer's eye by the painting is maximum, how far from the wall should the observer stand?

8.4 INVERSE HYPERBOLIC FUNCTIONS

In this section we discuss inverses of hyperbolic functions. The primary interest in these functions results from their usefulness for obtaining integration formulas.

Because the hyperbolic sine function is increasing on the interval $(-\infty, +\infty)$ (Figure 7.6.1b), it follows that $\sinh x$ is one-to-one on this interval and consequently has an inverse function, $\sinh^{-1} x$. Thus for all real values of x

$$\sinh^{-1} (\sinh x) = x$$
$$\sinh (\sinh^{-1} x) = x$$

Moreover,

$$y = \sinh x \quad \text{and} \quad x = \sinh^{-1} y$$

$y = \sinh^{-1} x$

Figure 8.4.1

are equivalent statements. The graph of $y = \sinh^{-1} x$ can be obtained by reflecting the graph of $y = \sinh x$ about the line $y = x$ (Figure 8.4.1).

Because $\sinh x$ is expressible in terms of e^x, one might suspect that $\sinh^{-1} x$ is expressible in terms of natural logarithms. To see that this is so, we write $y = \sinh^{-1} x$ in the alternate form

$$x = \sinh y = \frac{e^y - e^{-y}}{2}$$

so that

$$e^y - 2x - e^{-y} = 0$$

or, on multiplying through by e^y

$$e^{2y} - 2xe^y - 1 = 0$$

Applying the quadratic formula yields

$$e^y = \frac{2x \pm \sqrt{4x^2 + 4}}{2} = x \pm \sqrt{x^2 + 1}$$

Since $e^y > 0$, the solution involving the minus sign is extraneous and must be discarded. Thus

$$e^y = x + \sqrt{x^2 + 1}$$

Taking natural logarithms yields

$$y = \ln(x + \sqrt{x^2 + 1})$$

or

$$\sinh^{-1} x = \ln(x + \sqrt{x^2 + 1}) \tag{1}$$

This formula and a table of natural logarithms can be used to evaluate $\sinh^{-1} x$. Moreover, we can use it to obtain the derivative formula for $\sinh^{-1} x$ as follows: from (1),

$$\frac{d}{dx}[\sinh^{-1} x] = \frac{d}{dx}[\ln(x + \sqrt{x^2 + 1})]$$

$$= \frac{1}{x + \sqrt{x^2 + 1}}\left(1 + \frac{x}{\sqrt{x^2 + 1}}\right)$$

$$= \frac{\sqrt{x^2 + 1} + x}{(x + \sqrt{x^2 + 1})(\sqrt{x^2 + 1})}$$

or, on cancelling

$$\frac{d}{dx}[\sinh^{-1} x] = \frac{1}{\sqrt{x^2 + 1}} \tag{2}$$

Derivatives of the remaining inverse hyperbolic functions are obtained in a similar manner with the exception of $\cosh^{-1}$ and sech^{-1}. As evidenced by Figures 7.6.1a and 7.6.3, $\cosh x$ and $\mathrm{sech}\, x$ are not one-to-one, so that we must restrict the domains of these functions appropriately before we can obtain inverses. For example, if we let

$$f(x) = \cosh x, \; x \geq 0$$

then f is one-to-one (Figure 8.4.2a) and consequently has an inverse. We call this inverse $\cosh^{-1} x$. The graph of $y = \cosh^{-1} x$ is the reflection of the graph of f about the line $y = x$ (Figure 8.4.2b). The domain of $\cosh^{-1}$ is the interval $[1, +\infty)$ and

$$\cosh^{-1}(\cosh x) = x \quad \text{if} \quad x \geq 0$$
$$\cosh(\cosh^{-1} x) = x \quad \text{if} \quad x \geq 1$$

Moreover, if $x \geq 0$ then the statements

$$y = \cosh x \quad \text{and} \quad x = \cosh^{-1} y$$

are equivalent.

By following the derivations (1) and (2), the reader should be able to show that

$$\cosh^{-1} x = \ln(x + \sqrt{x^2 - 1})$$

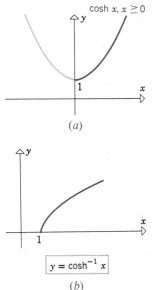

cosh x, $x \geq 0$

(a)

$y = \cosh^{-1} x$

(b)

Figure 8.4.2

and

$$\frac{d}{dx}[\cosh^{-1} x] = \frac{1}{\sqrt{x^2 - 1}}$$

if $x > 1$.

For reference, we summarize the definitions, logarithmic expressions, and derivatives of the inverse hyperbolic functions.

8.4.1 DEFINITION

$\sinh^{-1} x$	is the inverse of	$\sinh x$
$\cosh^{-1} x$	is the inverse of	$\cosh x,\ x \geq 0$
$\tanh^{-1} x$	is the inverse of	$\tanh x$
$\coth^{-1} x$	is the inverse of	$\coth x$
$\operatorname{sech}^{-1} x$	is the inverse of	$\operatorname{sech} x,\ x \geq 0$
$\operatorname{csch}^{-1} x$	is the inverse of	$\operatorname{csch} x$

8.4.2 THEOREM

$$\sinh^{-1} x = \ln\left(x + \sqrt{x^2 + 1}\right) \qquad -\infty < x < +\infty$$

$$\cosh^{-1} x = \ln\left(x + \sqrt{x^2 - 1}\right) \qquad x \geq 1$$

$$\tanh^{-1} x = \frac{1}{2}\ln\frac{1 + x}{1 - x} \qquad -1 < x < 1$$

$$\coth^{-1} x = \frac{1}{2}\ln\frac{x + 1}{x - 1} \qquad |x| > 1$$

$$\operatorname{sech}^{-1} x = \ln\left(\frac{1 + \sqrt{1 - x^2}}{x}\right) \qquad 0 < x \leq 1$$

$$\operatorname{csch}^{-1} x = \ln\left(\frac{1}{x} + \frac{\sqrt{1 + x^2}}{|x|}\right) \qquad x \neq 0$$

8.4.3 THEOREM

$$\frac{d}{dx}[\sinh^{-1} x] = \frac{1}{\sqrt{1 + x^2}}$$

$$\frac{d}{dx}[\cosh^{-1} x] = \frac{1}{\sqrt{x^2 - 1}}$$

$$\frac{d}{dx}[\tanh^{-1} x] = \frac{1}{1 - x^2} \qquad (|x| < 1)$$

$$\frac{d}{dx}[\coth^{-1} x] = \frac{1}{1 - x^2} \qquad (|x| > 1)$$

$$\frac{d}{dx}[\operatorname{sech}^{-1} x] = -\frac{1}{x\sqrt{1 - x^2}}$$

$$\frac{d}{dx}[\operatorname{csch}^{-1} x] = -\frac{1}{|x|\sqrt{1 + x^2}}$$

REMARK. Care must be exercised in differentiating $\tanh^{-1} x$ and $\coth^{-1} x$. In both cases the expression for the derivative is

$$\frac{1}{1 - x^2} \tag{3}$$

however, the domain of $\tanh^{-1} x$ is the interval $-1 < x < 1$ (verify this) so that the derivative formula for $\tanh^{-1} x$ applies only when $|x| < 1$. Similarly the derivative formula for $\coth^{-1} x$ applies only on the domain of $\coth^{-1} x$, that is, when $|x| > 1$. This distinction becomes important when we integrate (3); for then we must write

$$\int \frac{1}{1 - x^2}\, dx = \begin{cases} \tanh^{-1} x + C & \text{if } |x| < 1 \\ \coth^{-1} x + C & \text{if } |x| > 1 \end{cases}$$

The following integration formulas follow from 8.4.3.

8.4.4 THEOREM

$$\int \frac{dx}{\sqrt{1 + x^2}} = \sinh^{-1} x + C$$

$$\int \frac{dx}{\sqrt{x^2 - 1}} = \cosh^{-1} x + C$$

$$\int \frac{dx}{1 - x^2} = \begin{cases} \tanh^{-1} x + C & \text{if } |x| < 1 \\ \coth^{-1} x + C & \text{if } |x| > 1 \end{cases}$$

$$\int \frac{dx}{x\sqrt{1 - x^2}} = -\operatorname{sech}^{-1} |x| + C$$

$$\int \frac{dx}{x\sqrt{1 + x^2}} = -\operatorname{csch}^{-1} |x| + C$$

The first three integration formulas follow immediately from the corresponding differentiation formulas. The last two require additional work (see Exercises 25 and 26). We also leave it as an exercise to show that the third formula can be expressed in the alternative form:

$$\int \frac{dx}{1 - x^2} = \frac{1}{2} \ln \left| \frac{1 + x}{1 - x} \right| + C \tag{4}$$

▶ Exercise Set 8.4

1. (a) Prove: $\cosh^{-1} x = \ln (x + \sqrt{x^2 - 1})$, $x \geq 1$.
    (b) Use (a) to obtain the derivative of $\cosh^{-1} x$.

2. (a) Prove: $\tanh^{-1} x = \dfrac{1}{2} \ln \dfrac{1 + x}{1 - x}$, $-1 < x < 1$.
    (b) Use (a) to obtain the derivative of $\tanh^{-1} x$.

3. (a) Prove: $\operatorname{sech}^{-1} x = \cosh^{-1} \left(\dfrac{1}{x}\right)$, $0 < x \leq 1$

    $\coth^{-1} x = \tanh^{-1} \left(\dfrac{1}{x}\right)$, $|x| > 1$

    $\operatorname{csch}^{-1} x = \sinh^{-1} \left(\dfrac{1}{x}\right)$, $x \neq 0$

    (b) Use (a) and the derivatives of $\sinh^{-1} x$, $\cosh^{-1} x$, and $\tanh^{-1} x$ to find the derivatives of $\operatorname{sech}^{-1} x$, $\coth^{-1} x$, and $\operatorname{csch}^{-1} x$.
    (c) Use (a) and the logarithmic expressions for $\sinh^{-1} x$, $\cosh^{-1} x$, and $\tanh^{-1} x$ to derive the logarithmic expressions for $\operatorname{sech}^{-1} x$, $\coth^{-1} x$, and $\operatorname{csch}^{-1} x$.

4. Without referring to the text, state the domains of the six inverse hyperbolic functions.

In Exercises 5–13, find dy/dx.

5. (a) $y = \sinh^{-1} (\frac{1}{3}x)$
    (b) $y = \cosh^{-1} (2x + 1)$.

6. (a) $y = \tanh^{-1} (x^2)$
    (b) $y = \coth^{-1} (\sqrt{x})$.

7. (a) $y = \operatorname{sech}^{-1} (x^7)$
    (b) $y = \operatorname{csch}^{-1} (e^x)$.

8. (a) $y = (\tanh^{-1} x)^{-1}$
    (b) $y = \dfrac{1}{\tanh^{-1} x}$.

9. (a) $y = \sinh^{-1} \left(\dfrac{1}{x}\right)$
    (b) $y = \cosh^{-1} (\cosh x)$.

10. (a) $y = \ln (\cosh^{-1} x)$
    (b) $y = \sqrt{\coth^{-1} x}$.

11. (a) $y = e^x \operatorname{sech}^{-1} x$
    (b) $y = x^2 (\sinh^{-1} x)^3$.

12. (a) $y = \sinh^{-1} (\tanh x)$
    (b) $y = \cosh^{-1} (\sinh^{-1} x)$.

13. (a) $y = \tanh^{-1} \left(\dfrac{1 - x}{1 + x}\right)$
    (b) $y = (1 + x \operatorname{csch}^{-1} x)^{10}$.

In Exercises 14–19, evaluate the integral.

14. $\displaystyle\int \dfrac{dx}{\sqrt{1 + 9x^2}}$.

15. $\displaystyle\int \dfrac{dx}{\sqrt{x^2 - 2}}$.

16. $\displaystyle\int \dfrac{dx}{\sqrt{9x^2 - 25}}$.

17. $\displaystyle\int \dfrac{dx}{\sqrt{1 - e^{2x}}}$.

18. $\displaystyle\int \dfrac{\sin \theta \, d\theta}{\sqrt{1 + \cos^2 \theta}}$.

19. $\displaystyle\int \dfrac{dx}{x \sqrt{1 + x^6}}$.

In Exercises 20–22, use the natural logarithm table in Appendix 3 to obtain a numerical value for the integral.

20. $\displaystyle\int_0^{1/2} \dfrac{dx}{1 - x^2}$.

21. $\displaystyle\int_2^3 \dfrac{dx}{1 - x^2}$.

22. $\displaystyle\int_0^{\sqrt{3}} \dfrac{dt}{\sqrt{t^2 + 1}}$.

23. Sketch the graphs of
    (a) $\tanh^{-1} x$ (b) $\operatorname{csch}^{-1} x$.

24. Sketch the graphs of
    (a) $\coth^{-1} x$ (b) $\operatorname{sech}^{-1} x$.

25. Show that
    $$\dfrac{d}{dx} [\operatorname{sech}^{-1} |x|] = -\dfrac{1}{x \sqrt{1 - x^2}}.$$

26. Show that
    $$\dfrac{d}{dx} [\operatorname{csch}^{-1} |x|] = -\dfrac{1}{x \sqrt{1 + x^2}}.$$

27. Derive integration formula (4).

28. Let a be a positive constant. Derive an integration formula for:
    (a) $\displaystyle\int \dfrac{du}{\sqrt{a^2 + u^2}}$ (b) $\displaystyle\int \dfrac{du}{\sqrt{u^2 - a^2}}$
    (c) $\displaystyle\int \dfrac{du}{a^2 - u^2}$.

    [Hint: See Example 5 of Section 8.3.]

29. Our derivation of the differentiation formula for $\sinh^{-1} x$ used the logarithmic expression for this function. Show that the derivative can also be obtained by the method we used in Section 8.3 to obtain the derivative formula for $\sin^{-1} x$.

30. Find

(a) $\lim\limits_{x \to +\infty} \sinh^{-1} x$ (b) $\lim\limits_{x \to +\infty} \coth^{-1} x$

(c) $\lim\limits_{x \to 0^+} \operatorname{csch}^{-1} x$ (d) $\lim\limits_{x \to +\infty} (\cosh^{-1} x - \ln x)$.

▶ REVIEW EXERCISES

1. In each part determine if f and g are inverse functions.

(a) $f(x) = mx$ $g(x) = 1/(mx)$

(b) $f(x) = 3/(x + 1)$ $g(x) = (3 - x)/x$

(c) $f(x) = x^3 - 8$ $g(x) = \sqrt[3]{x} + 2$

(d) $f(x) = x^3 - 1$ $g(x) = \sqrt[3]{x + 1}$

(e) $f(x) = \sqrt{e^x}$ $g(x) = 2 \ln x$.

In Exercises 2–6, find $f^{-1}(x)$ if it exists.

2. $f(x) = 8x^3 - 1$.

3. $f(x) = x^2 - 2x + 1$.

4. $f(x) = x^2 - 2x + 1$, $x \geq 1$.

5. $f(x) = (e^x)^2 + 1$.

6. $f(x) = \exp(x^2) + 1$.

7. Let $f(x) = (ax + b)/(cx + d)$.

What conditions on a, b, c, d guarantee that f^{-1} exists? Find $f^{-1}(x)$.

8. Show that

$$f(x) = (x + 2)/(x - 1)$$

is its own inverse.

9. Find the largest open interval containing the origin on which f is one-to-one.

(a) $f(x) = |2x - 5|$ (b) $f(x) = x^2 + 4x$

(c) $f(x) = \cos(x - 2\pi/3)$.

In Exercises 10–13, find dy/dx using (11) of Section 8.1. Check your work by differentiating directly. [*Hint:* Example 8, Section 8.1.]

10. $y = x^3 - 8$.   **12.** $y = mx + b$ $(m \neq 0)$.

11. $y = 3/(x + 1)$.   **13.** $y = \sqrt{e^x}$.

In Exercises 14–17, find the exact value of the expression.

14. (a) $\sec^{-1}(-1)$ (b) $\cos^{-1}(-\sqrt{3}/2)$

(c) $\tan[\sec^{-1}(5/4)]$ (d) $\sec[\sin^{-1}(2/3)]$.

15. (a) $\cos^{-1}(-1/2)$ (b) $\cot^{-1}[\cot(3/4)]$

(c) $\cos[\sin^{-1}(4/5)]$ (d) $\cos[\sin^{-1}(-4/5)]$.

16. (a) $\tan^{-1}(-1)$ (b) $\csc^{-1}(-2/\sqrt{3})$

(c) $\cos^{-1}[\cos(-\pi/3)]$ (d) $\sin[-\sec^{-1}(2/\sqrt{3})]$.

17. (a) $\sin^{-1}(1/\sqrt{2})$ (b) $\sin^{-1}[\sin(5\pi/4)]$

(c) $\tan(\sec^{-1} 5)$ (d) $\tan^{-1}[\cot(\pi/6)]$.

18. Use a double angle formula to convert the given expression to an algebraic function of x.

(a) $\sin(2 \csc^{-1} x)$, $|x| \geq 1$

(b) $\cos(2 \sin^{-1} x)$, $|x| \leq 1$

(c) $\sin(2 \tan^{-1} x)$.

19. Express as a rational number:

(a) $\cos[\cos^{-1}(4/5) + \sin^{-1}(5/13)]$

(b) $\sin[\sin^{-1}(4/5) + \cos^{-1}(5/13)]$

(c) $\tan[\tan^{-1}(1/3) + \tan^{-1}(2)]$.

20. If $u = \operatorname{csch}^{-1}(-5/12)$, find $\coth u$, $\sinh u$, $\cosh u$, and $\sinh(2u)$.

21. If $u = \tanh^{-1}(-3/5)$, find $\cosh u$, $\sinh u$, and $\cosh(2u)$.

In Exercises 22–23, sketch the graph of f.

22. (a) $f(x) = 3 \sin^{-1}(x/2)$

(b) $f(x) = \cos^{-1} x - \pi/2$.

23. (a) $f(x) = 2 \tan^{-1}(-3x)$

(b) $f(x) = \cos^{-1} x + \sin^{-1} x$.

In Exercises 24–37, find dy/dx, using implicit or logarithmic differentiation where convenient.

24. $y = \sin^{-1}(e^x) + 2 \tan^{-1}(3x)$.

25. $y = \dfrac{1}{\sec^{-1} x^2}$.

26. $y = x \sin^{-1} x + \sqrt{1 - x^2}$.

27. $y = \cosh^{-1}(\sec x)$.

28. $\tan^{-1} y = \sin^{-1} x$.

29. $y = x \tanh^{-1}(\ln x)$.

30. $y = \tan^{-1}\left(\dfrac{2x}{1 - x^2}\right)$.

31. $y = \sqrt{\sin^{-1} 3x}$.

32. $y = (\sin^{-1} 2x)^{-1}$.

33. $y = \exp(\sec^{-1} x)$.

34. $y = (\tan^{-1} x)/\ln x$.

35. $y = \pi^{\sin^{-1} x}$.

36. $y = (\sinh^{-1} x)^{\pi}$.

37. $y = \tanh^{-1}\left(\dfrac{1}{\coth x}\right)$.

38. Let $f(x) = \tan^{-1} x + \tan^{-1}(1/x)$.
    (a) By considering $f'(x)$, show that $f(x) = C_1$ on $(-\infty, 0)$ and $f(x) = C_2$ on $(0, +\infty)$, where C_1 and C_2 are constants.
    (b) Find C_1 and C_2 in part (a).

39. Show that $y = \tan^{-1} x$ satisfies
    $y'' = -2 \sin y \cos^3 y$.

In Exercises 40–49, evaluate the integral.

40. $\displaystyle\int \frac{dx}{\sqrt{9 - 4x^2}}$.

41. $\displaystyle\int \frac{e^x \, dx}{1 - e^{2x}}$.

42. $\displaystyle\int \frac{\cot x \, dx}{\sqrt{1 - \sin^2 x}}$.

43. $\displaystyle\int \frac{dx}{x \sqrt{(\ln x)^2 - 1}}$.

44. $\displaystyle\int \frac{dx}{e^x \sqrt{1 - e^{-2x}}}$.

45. $\displaystyle\int_{2\sqrt{3}/9}^{2/3} \frac{dx}{3x \sqrt{9x^2 - 1}}$.

46. $\displaystyle\int_0^{\sqrt{2}} \frac{x \, dx}{4 + x^4}$.

47. $\displaystyle\int \frac{dx}{x^{1/2} + x^{3/2}}$.

48. $\displaystyle\int \frac{x^2 \, dx}{\sqrt{1 + x^6}}$.

49. $\displaystyle\int_0^{1/4} \frac{dx}{\sqrt{x} \sqrt{1 - x}}$.

50. The hypotenuse of a right triangle is growing at a rate of a cm/sec and one leg is decreasing at a rate of b cm/sec. How fast is the acute angle between the hypotenuse and the other leg changing at the instant when both legs are 1 cm?

51. Find the value of x at which $f(x) = \tan^{-1}(2x) - \tan^{-1} x$ assumes its maximum value on the interval $[0, +\infty)$.

52. Find the area between the curve $y = 1/(9 + x^2)$, the x-axis, and the lines $x = \pm \sqrt{3}$.

53. Suppose that a computer can calculate values of $\tan^{-1} x$, but no other inverse trigonometric functions. How can you use it to evaluate $\sin^{-1} x$?

9 techniques of integration

9.1 A BRIEF REVIEW

Although we can now integrate a wide variety of functions, there remain many important kinds of integrals that we cannot yet evaluate. The purpose of this chapter is to develop some additional techniques of integration and also to systematize the procedure of integration. We begin with a review of the basic integration formulas we have encountered thus far.

CONSTANTS, POWERS, EXPONENTIALS (See Sections 5.2, 7.2 and 7.4)

1. $\int dx = x + C$

2. $\int a\, dx = a \int dx = ax + C$

3. $\int x^r\, dx = \dfrac{x^{r+1}}{r+1} + C, \quad r \neq -1$

4. $\int \dfrac{dx}{x} = \ln |x| + C$

5. $\int e^x\, dx = e^x + C$

6. $\int a^x\, dx = \dfrac{a^x}{\ln a} + C$

TRIGONOMETRIC FUNCTIONS (See Sections 5.2 and 7.2)

7. $\int \sin x\, dx = -\cos x + C$

8. $\int \cos x\, dx = \sin x + C$

9. $\int \sec^2 x\, dx = \tan x + C$

10. $\int \csc^2 x\, dx = -\cot x + C$

11. $\int \sec x \tan x\, dx = \sec x + C$

12. $\int \csc x \cot x\, dx = -\csc x + C$

13. $\int \tan x\, dx = -\ln |\cos x| + C$

14. $\int \cot x\, dx = \ln |\sin x| + C$

477

HYPERBOLIC   (See Section 7.6)
FUNCTIONS

15. $\displaystyle\int \sinh x \, dx = \cosh x + C$ **18.** $\displaystyle\int \text{csch}^2 x \, dx = -\coth x + C$

16. $\displaystyle\int \cosh x \, dx = \sinh x + C$ **19.** $\displaystyle\int \text{sech } x \tanh x \, dx = -\text{sech } x + C$

17. $\displaystyle\int \text{sech}^2 x \, du = \tanh x + C$ **20.** $\displaystyle\int \text{csch } x \coth x \, dx = -\text{csch } x + C$

ALGEBRAIC   (See Sections 8.3 and 8.4)
FUNCTIONS

21. $\displaystyle\int \frac{dx}{\sqrt{1-x^2}} = \sin^{-1} x + C$

22. $\displaystyle\int \frac{dx}{1+x^2} = \tan^{-1} x + C$

23. $\displaystyle\int \frac{dx}{x\sqrt{x^2-1}} = \sec^{-1} x + C$

24. $\displaystyle\int \frac{dx}{\sqrt{1+x^2}} = \sinh^{-1} x + C$ **27.** $\displaystyle\int \frac{dx}{x\sqrt{1-x^2}} = -\text{sech}^{-1} |x| + C$

25. $\displaystyle\int \frac{dx}{\sqrt{x^2-1}} = \cosh^{-1} x + C$ **28.** $\displaystyle\int \frac{dx}{x\sqrt{1+x^2}} = -\text{csch}^{-1} |x| + C$

26. $\displaystyle\int \frac{dx}{1-x^2} = \begin{cases} \tanh^{-1} x + C & \text{if } |x| < 1 \\ \coth^{-1} x + C & \text{if } |x| > 1 \end{cases}$

$$= \frac{1}{2} \ln \left| \frac{1+x}{1-x} \right| + C$$

▶ Exercise Set 9.1

With the text closed, complete the following integration
formulas.

Constants, Powers, Exponentials

1. $\displaystyle\int dx =$ **2.** $\displaystyle\int a \, dx =$

3. $\displaystyle\int x^r \, dx =$ **4.** $\displaystyle\int \frac{dx}{x} =$

5. $\displaystyle\int e^x \, dx =$ **6.** $\displaystyle\int a^x \, dx =$

Trigonometric Functions

7. $\displaystyle\int \sin x \, dx =$ **8.** $\displaystyle\int \cos x \, dx =$

9. $\displaystyle\int \sec^2 x \, dx =$ **10.** $\displaystyle\int \csc^2 x \, dx =$

11. $\displaystyle\int \sec x \tan x \, dx =$ **12.** $\displaystyle\int \csc x \cot x \, dx =$

13. $\displaystyle\int \tan x\, dx =$ **14.** $\displaystyle\int \cot x\, dx =$

Hyperbolic Functions

15. $\displaystyle\int \sinh x\, dx =$ **16.** $\displaystyle\int \cosh x\, dx =$

17. $\displaystyle\int \operatorname{sech}^2 x\, dx =$ **18.** $\displaystyle\int \operatorname{csch}^2 x\, dx =$

19. $\displaystyle\int \operatorname{sech} x \tanh x\, dx =$ **20.** $\displaystyle\int \operatorname{csch} x \coth x\, dx =$

Algebraic Functions

21. $\displaystyle\int \frac{dx}{\sqrt{1-x^2}} =$ **22.** $\displaystyle\int \frac{dx}{1+x^2} =$

23. $\displaystyle\int \frac{dx}{x\sqrt{x^2-1}} =$ **24.** $\displaystyle\int \frac{dx}{\sqrt{1+x^2}} =$

25. $\displaystyle\int \frac{dx}{\sqrt{x^2-1}} =$ **26.** $\displaystyle\int \frac{dx}{1-x^2} =$

27. $\displaystyle\int \frac{dx}{x\sqrt{1-x^2}} =$ **28.** $\displaystyle\int \frac{dx}{x\sqrt{1+x^2}} =$

9.2 INTEGRATION BY PARTS

In this section we develop a technique that will help us to evaluate a wide variety of integrals that do not fit any of the basic integration formulas.

If f and g are differentiable functions, then by the rule for differentiating products

$$\frac{d}{dx}[f(x)g(x)] = f(x)g'(x) + g(x)f'(x)$$

Integrating both sides we obtain

$$\int \frac{d}{dx}[f(x)g(x)]\, dx = \int f(x)g'(x)\, dx + \int g(x)f'(x)\, dx$$

or

$$f(x)g(x) + C = \int f(x)g'(x)\, dx + \int g(x)f'(x)\, dx$$

or

$$\int f(x)g'(x)\, dx = f(x)g(x) - \int g(x)f'(x)\, dx + C$$

Since the integral on the right will produce another constant of integration, there is no need to keep the C in this last equation; thus we obtain

$$\int f(x)g'(x)\, dx = f(x)g(x) - \int f'(x)g(x)\, dx \qquad (1)$$

which is called the formula for *integration by parts.* By using this formula we can sometimes reduce a hard integration problem to an easier one.

In practice it is usual to rewrite (1) by letting

$$u = f(x), \qquad du = f'(x)\, dx$$
$$v = g(x), \qquad dv = g'(x)\, dx$$

This yields the following formula.

> **Integration by Parts for Indefinite Integrals**
> $$\int u\, dv = uv - \int v\, du \qquad \text{(2a)}$$

For definite integrals the corresponding formula is

> **Integration by Parts for Definite Integrals**
> $$\int_a^b u\, dv = uv \Big]_a^b - \int_a^b v\, du \qquad \text{(2b)}$$

▶ Example 1

Evaluate

$$\int xe^x\, dx$$

Solution. To apply (2a) we must write the integral in the form

$$\int u\, dv$$

One way to do this is to let

$$u = x \qquad \text{and} \qquad dv = e^x\, dx$$

so that

$$du = dx \qquad \text{and} \qquad v = \int e^x\, dx = e^x$$

Thus from (2a)

$$\int \underset{u}{x} \underset{dv}{e^x\, dx} = \underset{u}{x}\,\underset{v}{e^x} - \int \underset{v}{e^x}\,\underset{du}{dx}$$

or

$$\int xe^x\, dx = xe^x - e^x + C \qquad \blacktriangleleft$$

REMARK. In the calculation of v from dv above, we omitted the constant of integration and wrote $v = \int e^x\, dx = e^x$. Had we written $v = \int e^x\, dx = e^x + C_1$, the constant C_1 would have eventually cancelled out (Exercise 26a). This is always the case in integration by parts (Exercise 26b), so we will usually omit the constant when calculating v from dv.

To use integration by parts successfully, the choice of u and dv must be made so that the new integral is easier than the original. For example, had we decided above to let

$$u = e^x \qquad dv = x\, dx$$

$$du = e^x\, dx \qquad v = \int x\, dx = \frac{x^2}{2}$$

then we would have obtained

$$\int xe^x\, dx = \int u\, dv = uv - \int v\, du = \frac{x^2}{2}e^x - \frac{1}{2}\int x^2 e^x\, dx$$

For this choice of u and dv the new integral is actually more complicated than the original.

The next example shows that it is sometimes necessary to use integration by parts more than once in the same problem.

▶ Example 2

Evaluate

$$\int x^2 e^{-x}\, dx$$

Solution. Let

$$u = x^2 \qquad dv = e^{-x}\, dx$$

$$du = 2x\, dx \qquad v = \int e^{-x}\, dx = -e^{-x}$$

so that

$$\int x^2 e^{-x}\, dx = \int u\, dv = uv - \int v\, du = -x^2 e^{-x} + 2 \int x e^{-x}\, dx \qquad (3)$$

The last integral is similar to the original except that we have replaced x^2 by x. Another integration by parts applied to $\int x e^{-x}\, dx$ will complete the problem. We let

$$u = x \qquad dv = e^{-x}\, dx$$

$$du = dx \qquad v = \int e^{-x}\, dx = -e^{-x}$$

so that

$$\int x e^{-x}\, dx = \int u\, dv = uv - \int v\, du$$

$$= -x e^{-x} + \int e^{-x}\, dx$$

$$= -x e^{-x} - e^{-x} + C_1$$

Substituting in (3) we obtain

$$\int x^2 e^{-x}\, dx = -x^2 e^{-x} + 2(-x e^{-x} - e^{-x} + C_1)$$
$$= -x^2 e^{-x} - 2x e^{-x} - 2e^{-x} + 2C_1$$
$$= -(x^2 + 2x + 2)e^{-x} + C$$

where $C = 2C_1$. ◄

▶ Example 3

Evaluate

$$\int \ln x\, dx$$

Solution. Let

$$u = \ln x \qquad dv = dx$$

$$du = \frac{1}{x} dx \qquad v = \int dx = x$$

so that

$$\int \ln x \, dx = \int u \, dv = uv - \int v \, du$$

$$= x \ln x - \int x \left(\frac{1}{x} \right) dx$$

$$= x \ln x - \int dx = x \ln x - x + C \qquad \blacktriangleleft$$

The next example illustrates how integration by parts can be used to integrate the inverse trigonometric functions.

▶ Example 4

Evaluate

$$\int_0^1 \tan^{-1} x \, dx$$

Solution. Let

$$u = \tan^{-1} x \qquad dv = dx$$

$$du = \frac{1}{1 + x^2} dx \qquad v = \int dx = x$$

Thus

$$\int_0^1 \tan^{-1} x \, dx = \int_0^1 u \, dv = uv \Big]_0^1 - \int_0^1 v \, du$$

$$= x \tan^{-1} x \Big]_0^1 - \int_0^1 \frac{x}{1 + x^2} dx$$

But

$$\int_0^1 \frac{x}{1 + x^2} dx = \frac{1}{2} \int_0^1 \frac{2x}{1 + x^2} dx = \frac{1}{2} \ln (1 + x^2) \Big]_0^1 = \frac{1}{2} \ln 2$$

so

$$\int_0^1 \tan^{-1} x \, dx = x \tan^{-1} x \Big]_0^1 - \frac{1}{2} \ln 2 = \left(\frac{\pi}{4} - 0\right) - \frac{1}{2} \ln 2$$

$$= \frac{\pi}{4} - \ln \sqrt{2} \qquad \blacktriangleleft$$

▶ Example 5

Evaluate

$$\int e^x \cos x \, dx$$

Solution. Let

$$u = e^x \qquad dv = \cos x \, dx$$

$$du = e^x \, dx \qquad v = \int \cos x \, dx = \sin x$$

Thus

$$\int e^x \cos x \, dx = \int u \, dv = uv - \int v \, du$$

$$= e^x \sin x - \int e^x \sin x \, dx \qquad (4)$$

Since the integral $\int e^x \sin x \, dx$ is similar to the original integral $\int e^x \cos x \, dx$ it seems that nothing has been accomplished. However, let us integrate this new integral by parts; we let

$$u = e^x \qquad dv = \sin x \, dx$$

$$du = e^x \, dx \qquad v = \int \sin x \, dx = -\cos x$$

Thus,

$$\int e^x \sin x \, dx = \int u \, dv = uv - \int v \, du$$

$$= -e^x \cos x + \int e^x \cos x \, dx$$

Substituting in (4) yields

$$\int e^x \cos x\, dx = e^x \sin x - \left[-e^x \cos x + \int e^x \cos x\, dx \right]$$

or

$$\int e^x \cos x\, dx = e^x \sin x + e^x \cos x - \int e^x \cos x\, dx$$

which is an equation we can solve for the unknown integral. We obtain

$$2 \int e^x \cos x\, dx = e^x \sin x + e^x \cos x$$

or

$$\int e^x \cos x\, dx = \frac{1}{2} e^x \sin x + \frac{1}{2} e^x \cos x + C \qquad \blacktriangleleft$$

If n is a positive integer, then

$$\int \sin^n x\, dx \qquad \text{and} \qquad \int \cos^n x\, dx$$

can be evaluated by using **reduction formulas.** These are formulas that express the given integral in terms of a similar integral involving a *lower* power. For example, we can obtain a reduction formula for $\int \cos^n x\, dx$ by writing $\cos^n x$ as $\cos^{n-1} x \cdot \cos x$ and letting

$$u = \cos^{n-1} x$$
$$dv = \cos x\, dx$$
$$du = (n - 1) \cos^{n-2} x(-\sin x)\, dx = -(n - 1) \cos^{n-2} x \sin x\, dx$$
$$v = \int \cos x\, dx = \sin x$$

so that

$$\int \cos^n x\, dx = \int \cos^{n-1} x \cos x\, dx = \int u\, dv$$

$$= uv - \int v\, du$$

$$= \cos^{n-1} x \sin x + (n - 1) \int \sin^2 x \cos^{n-2} x\, dx$$

$$= \cos^{n-1} x \sin x + (n - 1) \int (1 - \cos^2 x) \cos^{n-2} x\, dx$$

$$= \cos^{n-1} x \sin x + (n - 1) \int \cos^{n-2} x\, dx - (n - 1) \int \cos^n x\, dx$$

Transposing the last integral on the right to the left side yields

$$n \int \cos^n x \, dx = \cos^{n-1} x \sin x + (n-1) \int \cos^{n-2} x \, dx$$

or

$$\int \cos^n x \, dx = \frac{1}{n} \cos^{n-1} x \sin x + \frac{n-1}{n} \int \cos^{n-2} x \, dx \qquad (5)$$

This reduction formula reduces the exponent by 2. Thus if we apply it repeatedly we can eventually express $\int \cos^n x \, dx$ in terms of

$$\int \cos x \, dx = \sin x + C$$

if n is odd, or

$$\int \cos^0 x \, dx = \int dx = x + C$$

if n is even.

▶ Example 6

Evaluate

$$(a) \quad \int \cos^3 x \, dx \qquad (b) \quad \int \cos^4 x \, dx$$

Solution (a). From (5) with $n = 3$

$$\int \cos^3 x \, dx = \frac{1}{3} \cos^2 x \sin x + \frac{2}{3} \int \cos x \, dx$$

$$= \frac{1}{3} \cos^2 x \sin x + \frac{2}{3} \sin x + C$$

Solution (b). From (5) with $n = 4$

$$\int \cos^4 x \, dx = \frac{1}{4} \cos^3 x \sin x + \frac{3}{4} \int \cos^2 x \, dx$$

and from (5) with $n = 2$

$$\int \cos^2 x \, dx = \frac{1}{2} \cos x \sin x + \frac{1}{2} \int dx = \frac{1}{2} \cos x \sin x + \frac{1}{2} x + C_1$$

so that

$$\int \cos^4 x \, dx = \frac{1}{4} \cos^3 x \sin x + \frac{3}{4} \left(\frac{1}{2} \cos x \sin x + \frac{1}{2} x + C_1 \right)$$

$$= \frac{1}{4} \cos^3 x \sin x + \frac{3}{8} \cos x \sin x + \frac{3}{8} x + C$$

where $C = \frac{3}{4} C_1$ ◄

We leave it as an exercise to derive the following companion formula to (5).

$$\int \sin^n x \, dx = -\frac{1}{n} \sin^{n-1} x \cos x + \frac{n-1}{n} \int \sin^{n-2} x \, dx \qquad (6)$$

▶ Exercise Set 9.2

In Exercises 1–20, evaluate the integral.

1. $\int x e^{-x} \, dx$.

2. $\int x \ln x \, dx$.

3. $\int x^2 e^x \, dx$.

4. $\int \sin^{-1} x \, dx$.

5. $\int e^x \sin x \, dx$.

6. $\int x^2 \ln x \, dx$.

7. $\int x^2 \cos x \, dx$.

8. $\int x^3 e^{-x} \, dx$.

9. $\int x \sec^2 x \, dx$.

10. $\int x \tan^{-1} x \, dx$.

11. $\int x \sin (3x + 1) \, dx$.

12. $\int x \sinh x \, dx$.

13. $\int e^{2x} \cos 3x \, dx$.

14. $\int e^{ax} \sin bx \, dx$.

15. $\int \sin (\ln x) \, dx$.

16. $\int \frac{x^3 \, dx}{\sqrt{1 - x^2}}$.

17. $\int_0^1 \frac{x^3}{\sqrt{x^2 + 1}} \, dx$.

18. $\int_0^2 x e^{2x} \, dx$.

19. $\int_0^{\pi/2} x \sin 4x \, dx$.

20. $\int_1^2 x \sec^{-1} x \, dx$.

21. Use reduction formula (6) to evaluate

(a) $\int \sin^3 x \, dx$ (b) $\int_0^{\pi/4} \sin^4 x \, dx$.

22. Use reduction formula (5) to evaluate

(a) $\int \cos^5 x \, dx$ (b) $\int_0^{\pi/2} \cos^6 x \, dx$.

23. Derive reduction formula (6).

In Exercises 24–25, derive the reduction formula in part (a) and use it to evaluate the integral in part (b).

24. (a) $\int \sec^n x \, dx$

$$= \frac{\sec^{n-2} x \tan x}{n - 1} + \frac{n - 2}{n - 1} \int \sec^{n-2} x \, dx$$

(b) $\int \sec^4 x \, dx$.

25. (a) $\displaystyle\int x^n e^x \, dx = x^n e^x - n \int x^{n-1} e^x \, dx$

(b) $\displaystyle\int x^3 e^x \, dx.$

26. (a) In Example 1, let

$$u = x \qquad dv = e^x \, dx$$

$$du = dx \qquad v = \int e^x \, dx = e^x + C_1$$

and show that the constant C_1 cancels out, thus leading to the same solution we obtained by omitting C_1.

(b) Show that

$$uv - \int v \, du = u(v + C_1) - \int (v + C_1) \, du$$

thereby justifying the omission of the constant of integration when calculating v in integration by parts.

9.3 INTEGRATING POWERS OF SINE AND COSINE

Integrals of the form

$$\int \sin^m x \cos^n x \, dx$$

where m and n are nonnegative integers are evaluated in various ways, depending on the values of m and n. The cases

$$\int \sin^m x \, dx \qquad (n = 0)$$

and

$$\int \cos^n x \, dx \qquad (m = 0)$$

can be treated using reduction formulas (5) and (6) of the previous section.

▶ Example 1

Show that

$$\int \sin^2 x \, dx = \frac{1}{2}x - \frac{1}{4}\sin 2x + C \tag{1}$$

$$\int \cos^2 x \, dx = \frac{1}{2}x + \frac{1}{4}\sin 2x + C \tag{2}$$

Solution. Letting $n = 2$ in reduction formulas (5) and (6) of Section 9.2 yields

$$\int \sin^2 x \, dx = -\frac{1}{2} \sin x \cos x + \frac{1}{2} x + C$$

$$= \frac{1}{2} x - \frac{1}{4} \sin 2x + C$$

and

$$\int \cos^2 x \, dx = \frac{1}{2} \cos x \sin x + \frac{1}{2} x + C$$

$$= \frac{1}{2} x + \frac{1}{4} \sin 2x + C$$

Alternate Solution. Apply the identities

$$\sin^2 x = \frac{1}{2}(1 - \cos 2x) \tag{3a}$$

$$\cos^2 x = \frac{1}{2}(1 + \cos 2x) \tag{3b}$$

which follow from the double-angle formulas $\cos 2x = 2 \cos^2 x - 1$ and $\cos 2x = 1 - 2 \sin^2 x$. We obtain

$$\int \sin^2 x \, dx = \frac{1}{2} \int (1 - \cos 2x) \, dx = \frac{1}{2} x - \frac{1}{4} \sin 2x + C$$

$$\int \cos^2 x \, dx = \frac{1}{2} \int (1 + \cos 2x) \, dx = \frac{1}{2} x + \frac{1}{4} \sin 2x + C \qquad \blacktriangleleft$$

▶ Example 2

The formulas

$$\int \sin^4 x \, dx = \frac{3}{8} x - \frac{1}{4} \sin 2x + \frac{1}{32} \sin 4x + C \tag{4}$$

$$\int \cos^4 x \, dx = \frac{3}{8} x + \frac{1}{4} \sin 2x + \frac{1}{32} \sin 4x + C \tag{5}$$

can be obtained from the reduction formulas and basic trigonometric identities. However, a more direct approach is as follows. From (3b)

$$\int \cos^4 x \, dx = \int (\cos^2 x)^2 \, dx = \int \left[\frac{1}{2}(1 + \cos 2x) \right]^2 dx$$

$$= \frac{1}{4} \int (1 + 2 \cos 2x + \cos^2 2x) \, dx$$

To finish, we apply (3b) again and write

$$\cos^2 2x = \frac{1}{2}(1 + \cos 4x) = \frac{1}{2} + \frac{1}{2}\cos 4x$$

which gives

$$\int \cos^4 x \, dx = \frac{1}{4} \int \left(\frac{3}{2} + 2 \cos 2x + \frac{1}{2}\cos 4x \right) dx$$

$$= \frac{3}{8}x + \frac{1}{4}\sin 2x + \frac{1}{32}\sin 4x + C$$

A similar procedure, using (3a), will yield (4).   ◀

▶ Example 3

Show that

$$\int \sin^3 x \, dx = -\cos x + \frac{1}{3}\cos^3 x + C \qquad (6)$$

$$\int \cos^3 x \, dx = \sin x - \frac{1}{3}\sin^3 x + C \qquad (7)$$

Solution. In the exercises we ask the reader to derive these results from reduction formulas. An alternate approach is as follows.

$$\int \sin^3 x \, dx = \int \sin^2 x \sin x \, dx$$

$$= \int (1 - \cos^2 x) \sin x \, dx$$

$$= \int \sin x \, dx - \int \cos^2 x \sin x \, dx$$

$$= -\cos x + \frac{1}{3}\cos^3 x + C$$

To integrate, let
$u = \cos x, \ du = -\sin x \, dx$

$$\int \cos^3 x \, dx = \int \cos^2 x \cos x \, dx$$

$$= \int (1 - \sin^2 x) \cos x \, dx$$

$$= \int \cos x \, dx - \int \sin^2 x \cos x \, dx \quad \boxed{\begin{array}{l} \text{To integrate, let} \\ u = \sin x, \; du = \cos x \, dx \end{array}}$$

$$= \sin x - \frac{1}{3}\sin^3 x + C \qquad \blacktriangleleft$$

If m and n are both positive integers, then the integral

$$\int \sin^m x \cos^n x \, dx$$

can be evaluated by one of three procedures, depending on whether m and n are odd or even. The procedures are outlined in the following table.

Table 9.3.1

CASE	PROCEDURE	RELEVANT IDENTITIES
n odd	Substitute $u = \sin x$	$\cos^2 x = 1 - \sin^2 x$
m odd	Substitute $u = \cos x$	$\sin^2 x = 1 - \cos^2 x$
m and n even	Use identities to reduce the powers on sin and cos.	$\begin{cases} \sin^2 x = \frac{1}{2}(1 - \cos 2x) \\ \cos^2 x = \frac{1}{2}(1 + \cos 2x) \end{cases}$

▶ Example 4

Evaluate

$$\int \sin^4 x \cos^5 x \, dx$$

Solution. Since $n = 5$ is odd, we will follow the first procedure in Table 9.3.1 and make the substitution

$$u = \sin x \qquad du = \cos x \, dx$$

To form the du, we will first split off a factor of $\cos x$.

$$\int \sin^4 x \cos^5 x \, dx = \int \sin^4 x \cos^4 x \cos x \, dx$$

$$= \int \sin^4 x (1 - \sin^2 x)^2 \cos x \, dx$$

$$= \int u^4 (1 - u^2)^2 \, du$$

$$= \int (u^4 - 2u^6 + u^8) \, du$$

$$= \frac{1}{5} u^5 - \frac{2}{7} u^7 + \frac{1}{9} u^9 + C$$

$$= \frac{1}{5} \sin^5 x - \frac{2}{7} \sin^7 x + \frac{1}{9} \sin^9 x + C \qquad \blacktriangleleft$$

▶ Example 5

Evaluate

$$\int \sin^3 x \cos^2 x \, dx$$

Solution. Since $m = 3$ is odd, we will follow the second procedure in Table 9.3.1 and make the substitution

$$u = \cos x \qquad du = -\sin x \, dx$$

To form the *du*, we will split off a factor of sin *x*.

$$\int \sin^3 x \cos^2 x \, dx = \int \sin^2 x \cos^2 x \sin x \, dx$$

$$= \int (1 - \cos^2 x) \cos^2 x \sin x \, dx$$

$$= -\int (1 - u^2) u^2 \, du$$

$$= \int (u^4 - u^2) \, du$$

$$= \frac{1}{5} u^5 - \frac{1}{3} u^3 + C$$

$$= \frac{1}{5} \cos^5 x - \frac{1}{3} \cos^3 x + C \qquad \blacktriangleleft$$

▶ Example 6

Evaluate

$$\int \sin^4 x \cos^4 x \, dx$$

Solution. Since $m = n = 4$, we will follow the third procedure in Table 9.3.1.

$$\int \sin^4 x \cos^4 x \, dx = \int (\sin^2 x)^2 (\cos^2 x)^2 \, dx$$

$$= \int (\tfrac{1}{2}[1 - \cos 2x])^2 (\tfrac{1}{2}[1 + \cos 2x])^2 \, dx$$

$$= \frac{1}{16} \int (1 - \cos^2 2x)^2 \, dx$$

$$= \frac{1}{16} \int \sin^4 2x \, dx$$

To finish, we will let $u = 2x$, $du = 2 \, dx$ and then use (4).

$$\int \sin^4 x \cos^4 x \, dx = \frac{1}{32} \int \sin^4 u \, du$$

$$= \frac{1}{32} \left(\frac{3}{8} u - \frac{1}{4} \sin 2u + \frac{1}{32} \sin 4u \right) + C$$

$$= \frac{3}{128} x - \frac{1}{128} \sin 4x + \frac{1}{1024} \sin 8x + C \qquad ◀$$

Integrals of the form

$$\int \sin mx \cos nx \, dx \quad \int \sin mx \sin nx \, dx \quad \int \cos mx \cos nx \, dx$$

can be found using the product formulas from trigonometry (see 19a, 19b, and 19c of the Unit I trigonometry review in Appendix 1).

▶ Example 7

Evaluate

$$\int \sin 7x \cos 3x \, dx$$

Solution. Since

$$\sin 7x \cos 3x = \frac{1}{2}(\sin 4x + \sin 10x)$$

we can write

$$\int \sin 7x \cos 3x \, dx = \frac{1}{2} \int (\sin 4x + \sin 10x) \, dx$$

$$= -\frac{1}{8} \cos 4x - \frac{1}{20} \cos 10x + C \qquad \blacktriangleleft$$

▶ Exercise Set 9.3

In Exercises 1–30, perform the indicated integration.

1. $\int \cos^5 x \sin x \, dx.$

2. $\int \sin^4 3x \cos 3x \, dx.$

22. $\int \sin ax \cos bx \, dx \qquad (a > 0, \ b > 0, \ a \neq b).$

3. $\int \sin ax \cos ax \, dx \qquad (a \neq 0).$

23. $\int \frac{\sin x}{\cos^8 x} \, dx.$

24. $\int \sqrt{\cos \theta} \sin \theta \, d\theta.$

4. $\int \cos^2 3x \, dx.$

5. $\int \sin^2 5\theta \, d\theta.$

25. $\int_0^{\pi/4} \cos^3 x \, dx.$

26. $\int_{-\pi}^{\pi} \cos^2 5\theta \, d\theta.$

6. $\int \cos^3 at \, dt \quad (a \neq 0).$

7. $\int \cos^4\left(\frac{x}{4}\right) dx.$

27. $\int_0^{\pi/3} \sin^4 3x \cos^3 3x \, dx.$

8. $\int \sin^5 x \, dx.$

9. $\int \cos^5 \theta \, d\theta.$

28. $\int_0^{\pi/2} \sin^2 \frac{x}{2} \cos^2 \frac{x}{2} \, dx.$

10. $\int \sin^3 x \cos^3 x \, dx.$

11. $\int \sin^2 2t \cos^3 2t \, dt.$

29. $\int_0^{\pi/6} \sin 2x \cos 4x \, dx.$

12. $\int \sin^4 x \cos^5 x \, dx.$

13. $\int \cos^4 x \sin^3 x \, dx.$

30. $\int_0^{2\pi} \sin^2 kx \, dx \qquad (k \neq 0).$

14. $\int \sin^3 2x \cos^2 2x \, dx.$

15. $\int \sin^5 \theta \cos^4 \theta \, d\theta.$

31. Let m, n be distinct nonnegative integers. Prove:

(a) $\int_0^{2\pi} \sin mx \cos nx \, dx = 0$

16. $\int \cos^{1/5} x \sin x \, dx.$

17. $\int \sin^2 x \cos^2 x \, dx.$

(b) $\int_0^{2\pi} \cos mx \cos nx \, dx = 0$

18. $\int \sin^2 x \cos^4 x \, dx.$

19. $\int \sin x \cos 2x \, dx.$

20. $\int \sin 3\theta \cos 2\theta \, d\theta.$

21. $\int \sin x \cos\left(\frac{x}{2}\right) dx.$

(c) $\int_0^{2\pi} \sin mx \sin nx \, dx = 0.$

32. The region bounded below by the x-axis and above by the portion of $y = \sin x$ from $x = 0$ to $x = \pi$ is revolved about the x-axis. Find the volume of the resulting solid.

33. Derive (6) and (7) using reduction formulas and appropriate trigonometric identities.

9.4 INTEGRATING POWERS OF SECANT AND TANGENT

In this section we discuss methods for evaluating integrals of the form

$$\int \tan^m x \sec^n x \, dx$$

where m and n are nonnegative integers. The simplest integrals of this form are

$$\int \tan x \, dx \qquad (m = 1, n = 0)$$

and

$$\int \sec x \, dx \qquad (m = 0, n = 1)$$

The first integral is evaluated by writing

$$\int \tan x \, dx = \int \frac{\sin x}{\cos x} \, dx$$

from which it follows ($u = \cos x$, $du = -\sin x \, dx$) that

$$\int \tan x = -\ln |\cos x| + C \tag{1}$$

The second integral requires a trick. We write

$$
\begin{aligned}
\int \sec x \, dx &= \int \sec x \left(\frac{\sec x + \tan x}{\sec x + \tan x} \right) dx \\
&= \int \frac{\sec^2 x + \sec x \tan x}{\sec x + \tan x} \, dx \\
&= \int \frac{du}{u} \qquad\qquad \begin{bmatrix} u = \sec x + \tan x \\ du = (\sec^2 x + \sec x \tan x) \, dx \end{bmatrix} \\
&= \ln |u| + C
\end{aligned}
$$

from which it follows that

$$\int \sec x \, dx = \ln |\sec x + \tan x| + C \qquad (2)$$

Higher powers of secant and tangent can be evaluated using the reduction formulas

$$\int \sec^n x \, dx = \frac{\sec^{n-2} x \tan x}{n-1} + \frac{n-2}{n-1} \int \sec^{n-2} x \, dx \qquad (3)$$

$$\int \tan^m x \, dx = \frac{\tan^{m-1} x}{m-1} - \int \tan^{m-2} x \, dx \qquad (4)$$

Formula (3) was discussed in Exercise 24a of Section 9.2, and Formula (4) can be obtained from the identity

$$1 + \tan^2 x = \sec^2 x$$

by writing

$$\int \tan^m x \, dx = \int \tan^{m-2} x \tan^2 x \, dx$$

$$= \int \tan^{m-2} x \, (\sec^2 x - 1) \, dx$$

$$= \int \tan^{m-2} x \sec^2 x \, dx - \int \tan^{m-2} x \, dx$$

$$= \frac{\tan^{m-1} x}{m-1} - \int \tan^{m-2} x \, dx$$

▶ Example 1

Evaluate

$$\int \sec^3 dx$$

Solution. From (3) with $n = 3$,

$$\int \sec^3 dx = \frac{\sec x \tan x}{2} + \frac{1}{2} \int \sec x \, dx$$

$$= \frac{1}{2} \sec x \tan x + \frac{1}{2} \ln |\sec x + \tan x| + C \qquad ◀$$

▶ Example 2

Evaluate

$$\int \tan^5 x \, dx$$

Solution. We will use Formula (4) twice.

$$\int \tan^5 x \, dx = \frac{\tan^4 x}{4} - \int \tan^3 x \, dx$$

$$= \frac{\tan^4 x}{4} - \left[\frac{\tan^2 x}{2} - \int \tan x \, dx\right]$$

$$= \frac{1}{4} \tan^4 x - \frac{1}{2} \tan^2 x - \ln|\cos x| + C \qquad ◀$$

If m and n are positive integers, then the integral

$$\int \tan^m x \, \sec^n x \, dx$$

can be evaluated by one of the three procedures in Table 9.4.1.

Table 9.4.1

CASE	PROCEDURE	RELEVANT IDENTITIES
n even	Substitute $u = \tan x$	$\sec^2 x = \tan^2 x + 1$
m odd	Substitute $u = \sec x$	$\tan^2 x = \sec^2 x - 1$
m even and n odd	Reduce to powers of $\sec x$ alone.	$\tan^2 x = \sec^2 x - 1$

▶ Example 3

Evaluate

$$\int \tan^2 x \, \sec^4 x \, dx$$

Solution. Since $n = 4$ is even, we will follow the first procedure in Table 9.4.1 and make the substitution

$$u = \tan x, \qquad du = \sec^2 x \, dx$$

To form the *du*, we will split off a factor of $\sec^2 x$.

$$\int \tan^2 x \sec^4 x \, dx = \int \tan^2 x \sec^2 x \sec^2 x \, dx$$

$$= \int \tan^2 x \, (\tan^2 x + 1) \sec^2 x \, dx$$

$$= \int u^2(u^2 + 1) \, du$$

$$= \frac{1}{5}u^5 + \frac{1}{3}u^3 + C = \frac{1}{5}\tan^5 x + \frac{1}{3}\tan^3 x + C$$

◀

▶ Example 4

Evaluate

$$\int \tan^3 x \sec^3 x \, dx$$

Solution. Since $m = 3$ is odd, we will follow the second procedure in Table 9.4.1 and make the substitution

$$u = \sec x, \qquad du = \sec x \tan x \, dx$$

To form the *du*, we will split off the product $\sec x \tan x$.

$$\int \tan^3 x \sec^3 x \, dx = \int \tan^2 x \sec^2 x \, (\sec x \tan x) \, dx$$

$$= \int (\sec^2 x - 1) \sec^2 x \, (\sec x \tan x) \, dx$$

$$= \int (u^2 - 1)u^2 \, du$$

$$= \frac{1}{5}u^5 - \frac{1}{3}u^3 + C = \frac{1}{5}\sec^5 x - \frac{1}{3}\sec^3 x + C$$

◀

▶ Example 5

Evaluate

$$\int \tan^2 x \sec x \, dx$$

Solution. Since $m = 2$ and $n = 1$, we will follow the third procedure in Table 9.4.1.

$$\int \tan^2 x \sec x \, dx = \int (\sec^2 x - 1) \sec x \, dx$$

$$= \int \sec^3 x \, dx - \int \sec x \, dx$$

See Example 1

$$= \frac{1}{2} \sec x \tan x + \frac{1}{2} \ln |\sec x + \tan x|$$

$$- \ln |\sec x + \tan x| + C$$

$$= \frac{1}{2} \sec x \tan x - \frac{1}{2} \ln |\sec x + \tan x| + C \qquad \blacktriangleleft$$

▶ Example 6

Instead of using reduction formula (3), the first procedure of Table 9.4.1 can be used to integrate an even power of sec x. For example,

$$\int \sec^6 x \, dx = \int \sec^4 x \sec^2 x \, dx$$

$$= \int (\sec^2 x)^2 \sec^2 x \, dx$$

$$= \int (\tan^2 x + 1)^2 \sec^2 x \, dx$$

$$= \int (u^2 + 1)^2 \, du \qquad (u = \tan x)$$

$$= \int (u^4 + 2u^2 + 1) \, du$$

$$= \frac{1}{5} u^5 + \frac{2}{3} u^3 + u + C$$

$$= \frac{1}{5} \tan^5 x + \frac{2}{3} \tan^3 x + \tan x + C \qquad \blacktriangleleft$$

REMARK.   With the aid of the identity

$$1 + \cot^2 x = \csc^2 x$$

some of the techniques in this section can be adapted to treat integrals of the form

$$\int \cot^m x \csc^n x \, dx$$

▶ Exercise Set 9.4

In Exercises 1–30, perform the indicated integration.

1. $\int \sec^2 (3x + 1)\, dx$.

2. $\int \tan 5x\, dx$.

3. $\int \tan^2 x \sec^2 x\, dx$.

4. $\int \tan^5 x \sec^4 x\, dx$.

5. $\int \tan^3 4x \sec^4 4x\, dx$.

6. $\int \tan^4 \theta \sec^4 \theta\, d\theta$.

7. $\int \sec^5 x \tan^3 x\, dx$.

8. $\int \tan^5 \theta \sec \theta\, d\theta$.

9. $\int \tan^4 x \sec x\, dx$.

10. $\int \tan^2 \frac{x}{2} \sec^3 \frac{x}{2}\, dx$.

11. $\int \tan 2t \sec^3 2t\, dt$.

12. $\int \tan x \sec^5 x\, dx$.

13. $\int \sec^4 x\, dx$.

14. $\int \sec^5 x\, dx$.

15. $\int \sec^6 (\pi x)\, dx$.

16. $\int \tan^3 4x\, dx$.

17. $\int \tan^4 x\, dx$.

18. $\int \tan^7 \theta\, d\theta$.

19. $\int x \tan^2 (x^2) \sec^2 (x^2)\, dx$.

20. $\int \tan^2 (1 - 2x) \sec (1 - 2x)\, dx$.

21. $\int \cot^3 x \csc^3 x\, dx$.

22. $\int \cot^2 3t \sec 3t\, dt$.

23. $\int \cot^3 x\, dx$.

24. $\int \csc^4 x\, dx$.

25. $\int \sqrt{\tan x} \sec^4 x\, dx$.

26. $\int \tan x \sec^{3/2} x\, dx$.

27. $\int_0^{\pi/6} \tan^2 2x\, dx$.

28. $\int_0^{\pi/6} \sec^3 \theta \tan \theta\, d\theta$.

29. $\int_0^{\pi/2} \tan^5 \frac{x}{2}\, dx$.

30. $\int_{\pi/4}^{\pi/2} \csc^3 x \cot x\, dx$.

31. (a) Show that

$$\int \csc x\, dx = -\ln |\csc x + \cot x| + C$$

(b) Show that the result in (a) can also be written

$$\int \csc x\, dx = \ln |\csc x - \cot x| + C$$

and

$$\int \csc x\, dx = \ln |\tan \tfrac{1}{2}x| + C$$

32. Rewrite $\sin x + \cos x$ in the form

$$A \sin (x + \phi)$$

and use your result together with Exercise 31 to evaluate

$$\int \frac{dx}{\sin x + \cos x}.$$

33. Use the method of Exercise 32 to evaluate

$$\int \frac{dx}{a \sin x + b \cos x} \qquad (a, b \text{ not both zero})$$

9.5 TRIGONOMETRIC SUBSTITUTIONS

Integrals containing expressions of the form

$$\sqrt{a^2 - x^2}, \quad \sqrt{x^2 + a^2}, \quad \text{and} \quad \sqrt{x^2 - a^2}$$

$(a > 0)$ can often be evaluated by making appropriate substitutions involving trigonometric functions. The idea is to choose a substitution that will eliminate the radical. For example, substituting

$$x = a \sin \theta, \qquad -\frac{\pi}{2} \leq \theta \leq \frac{\pi}{2} \tag{1}$$

in

$$\sqrt{a^2 - x^2}$$

yields

$$\begin{aligned}
\sqrt{a^2 - x^2} = \sqrt{a^2 - a^2 \sin^2 \theta} &= \sqrt{a^2 (1 - \sin^2 \theta)} \\
&= a \sqrt{\cos^2 \theta} \\
&= a \, |\cos \theta| \\
&= a \cos \theta \qquad \left(\cos \theta \geq 0 \text{ since } -\frac{\pi}{2} \leq \theta \leq \frac{\pi}{2} \right)
\end{aligned}$$

Thus we have eliminated the radical by this substitution. The purpose of the restriction $-\pi/2 \leq \theta \leq \pi/2$ in (1) is twofold. First it enables us to rewrite (1) as

$$\theta = \sin^{-1}\left(\frac{x}{a}\right)$$

if desired (see Section 8.2), and second this restriction enables us to replace $|\cos \theta|$ by the simpler expression $\cos \theta$ in the resulting calculations.

▶ Example 1

Evaluate

$$\int \frac{dx}{x^2 \sqrt{4 - x^2}}$$

Solution. To eliminate the radical we make the substitution

$$x = 2 \sin \theta, \qquad -\frac{\pi}{2} \leq \theta \leq \frac{\pi}{2}$$

so that

$$\frac{dx}{d\theta} = 2 \cos \theta \qquad \text{or} \qquad dx = 2 \cos \theta \, d\theta$$

This yields

$$\int \frac{dx}{x^2 \sqrt{4 - x^2}} = \int \frac{2 \cos \theta \, d\theta}{(2 \sin \theta)^2 \sqrt{4 - 4 \sin^2 \theta}}$$

$$= \int \frac{2 \cos \theta \, d\theta}{(2 \sin \theta)^2 (2 \cos \theta)}$$

$$= \frac{1}{4} \int \frac{d\theta}{\sin^2 \theta}$$

$$= \frac{1}{4} \int \csc^2 \theta \, d\theta = -\frac{1}{4} \cot \theta + C$$

$x = 2 \sin \theta$

Figure 9.5.1

$$\cot \theta = \frac{\sqrt{4 - x^2}}{x}$$

To complete the solution we must express $\cot \theta$ in terms of x. This can be done using trigonometric identities (Exercise 25) or more simply by displaying the substitution $x = 2 \sin \theta$ ($\sin \theta = x/2$) as in Figure 9.5.1. From the figure we obtain

so that

$$\int \frac{dx}{x^2 \sqrt{4 - x^2}} = -\frac{1}{4} \cot \theta + C = -\frac{1}{4} \frac{\sqrt{4 - x^2}}{x} + C \qquad \blacktriangleleft$$

To eliminate the radical in $\sqrt{x^2 + a^2}$ $(a > 0)$, we can make the substitution

$$x = a \tan \theta, \qquad -\frac{\pi}{2} < \theta < \frac{\pi}{2} \qquad (2)$$

to obtain

$$\sqrt{x^2 + a^2} = \sqrt{a^2 \tan^2 \theta + a^2} = \sqrt{a^2 (1 + \tan^2 \theta)}$$

$$= \sqrt{a^2 \sec^2 \theta}$$

$$= a \, |\sec \theta|$$

$$= a \sec \theta \qquad \left(\sec \theta > 0 \text{ since } -\frac{\pi}{2} < \theta < \frac{\pi}{2} \right)$$

As before, the restriction on θ in (2) enables us to write

$$\theta = \tan^{-1} \left(\frac{x}{a} \right)$$

if needed, and also eliminates the absolute value sign in the resulting computation.

▶ Example 2

Evaluate

$$\int \frac{dx}{\sqrt{x^2 + a^2}}$$

Solution. To eliminate the radical we make the substitution

$$x = a \tan \theta, \qquad -\frac{\pi}{2} < \theta < \frac{\pi}{2}$$

so that

$$\frac{dx}{d\theta} = a \sec^2 \theta \qquad \text{or} \qquad dx = a \sec^2 \theta \, d\theta$$

This yields

$$\int \frac{dx}{\sqrt{x^2 + a^2}} = \int \frac{a \sec^2 d\theta}{\sqrt{a^2 \tan^2 \theta + a^2}}$$

$$= \int \frac{a \sec^2 \theta \, d\theta}{a \sec \theta}$$

$$= \int \sec \theta \, d\theta$$

$$= \ln |\sec \theta + \tan \theta| + C$$

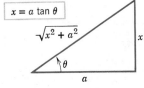

Figure 9.5.2

To express the solution in terms of x, we use Figure 9.5.2, which yields

$$\int \frac{dx}{\sqrt{x^2 + a^2}} = \ln \left| \frac{\sqrt{x^2 + a^2}}{a} + \frac{x}{a} \right| + C$$

or if we prefer we can rewrite the expression on the right as

$$\ln |\sqrt{x^2 + a^2} + x| - \ln a + C$$

and combine the constant $\ln a$ with the constant of integration to obtain

$$\int \frac{dx}{\sqrt{x^2 + a^2}} = \ln |\sqrt{x^2 + a^2} + x| + C'$$

Moreover, $\sqrt{x^2 + a^2} + x > 0$ for all x, so that we can drop the absolute value sign and write

$$\int \frac{dx}{\sqrt{x^2 + a^2}} = \ln\left(\sqrt{x^2 + a^2} + x\right) + C' \tag{3}$$

◀

The integral in the last example can also be evaluated by making the substitution

$$x = au \qquad \text{or} \qquad u = \frac{x}{a}$$

so that

$$dx = a\, du$$

which yields

$$\int \frac{dx}{\sqrt{x^2 + a^2}} = \int \frac{a\, du}{\sqrt{a^2 u^2 + a^2}}$$

$$= \int \frac{du}{\sqrt{u^2 + 1}} = \sinh^{-1} u + C$$

or

$$\int \frac{dx}{\sqrt{x^2 + a^2}} = \sinh^{-1}\left(\frac{x}{a}\right) + C \tag{4}$$

Using the logarithmic expression for $\sinh^{-1}$ in Theorem 8.4.2, the reader can show that (3) and (4) are equivalent.

To eliminate the radical in $\sqrt{x^2 - a^2}$, we can make the substitution

$$x = a \sec\theta, \qquad 0 \le \theta < \frac{\pi}{2} \ \text{ or } \ \pi \le \theta < \frac{3\pi}{2} \tag{5}$$

to obtain

$$\sqrt{x^2 - a^2} = \sqrt{a^2 \sec^2\theta - a^2} = \sqrt{a^2(\sec^2\theta - 1)}$$
$$= \sqrt{a^2 \tan^2\theta}$$
$$= a\,|\tan\theta|$$
$$= a \tan\theta$$

The removal of the absolute value sign is justified because the restriction on θ in (5) implies that $\tan \theta \geq 0$. This restriction also enables us to write

$$\theta = \sec^{-1}\left(\frac{x}{a}\right)$$

when needed (Definition 8.2.1).

▶ Example 3

Evaluate

$$\int \frac{\sqrt{x^2 - 25}}{x} \, dx$$

Solution. To eliminate the radical, we make the substitution

$$x = 5 \sec \theta, \qquad 0 \leq \theta < \frac{\pi}{2} \text{ or } \pi \leq \theta < \frac{3\pi}{2}$$

so that

$$\frac{dx}{d\theta} = 5 \sec \theta \tan \theta \qquad \text{or} \qquad dx = 5 \sec \theta \tan \theta \, d\theta$$

Thus,

$$\int \frac{\sqrt{x^2 - 25}}{x} \, dx = \int \frac{\sqrt{25 \sec^2 \theta - 25}}{5 \sec \theta} (5 \sec \theta \tan \theta) \, d\theta$$

$$= \int \frac{5 \tan \theta}{5 \sec \theta} (5 \sec \theta \tan \theta) \, d\theta$$

$$= 5 \int \tan^2 \theta \, d\theta$$

$$= 5 \int (\sec^2 \theta - 1) \, d\theta$$

$$= 5 \tan \theta - 5\theta + C$$

But $\theta = \sec^{-1}(x/5)$ and from Figure 9.5.3,

$$\tan \theta = \frac{\sqrt{x^2 - 25}}{5}$$

$x = 5 \sec \theta$

Figure 9.5.3

so that

$$\int \frac{\sqrt{x^2 - 25}}{x} dx = \sqrt{x^2 - 25} - 5 \sec^{-1}\left(\frac{x}{5}\right) + C \qquad \blacktriangleleft$$

The integral in the next example will arise frequently in later sections.

▶ Example 4

Evaluate

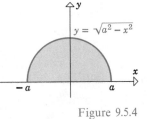

$$\int_{-a}^{a} \sqrt{a^2 - x^2} \, dx \qquad (a > 0)$$

Figure 9.5.4

Solution. This integral can be evaluated by the substitution $x = a \sin\theta$, but the computations are tedious. A better approach is to observe that the integral represents the area of a semicircle of radius a (Figure 9.5.4). Thus

$$\int_{-a}^{a} \sqrt{a^2 - x^2} \, dx = \frac{1}{2}\pi a^2 \qquad \blacktriangleleft$$

▶ Exercise Set 9.5

In Exercises 1–22, perform the indicated integration.

1. $\int \frac{x^2}{\sqrt{9 - x^2}} dx.$

2. $\int \frac{x}{\sqrt{9 + x^2}} dx.$

3. $\int \frac{x}{\sqrt{4 - x^2}} dx.$

4. $\int \frac{dx}{x^2\sqrt{x^2 - 16}}.$

5. $\int \frac{dx}{x^2\sqrt{4x^2 - 9}}.$

6. $\int \frac{\sqrt{1 + t^2}}{t} dt.$

7. $\int \frac{dx}{(1 - x^2)^{3/2}}.$

8. $\int \frac{dx}{x^2\sqrt{x^2 + 25}}.$

9. $\int \frac{x^2}{1 + x^2} dx.$

10. $\int \frac{dx}{1 + 2x^2 + x^4}.$

11. $\int \frac{dx}{x^2\sqrt{9 - 4x^2}}.$

12. $\int \frac{x^2}{\sqrt{x^2 - 25}} dx.$

13. $\int \frac{dx}{(9x^2 - 1)^{3/2}}.$

14. $\int \frac{\cos\theta}{\sqrt{2 - \sin^2\theta}} d\theta.$

15. $\int e^x \sqrt{1 - e^{2x}} \, dx.$

16. $\int_0^{1/3} \frac{dx}{(4 - 9x^2)^2}.$

17. $\int_0^4 x^3 \sqrt{16 - x^2} \, dx.$

18. $\int_{\sqrt{2}}^2 \frac{\sqrt{2x^2 - 4}}{x} dx.$

19. $\int_{\sqrt{2}}^2 \frac{dx}{x^2\sqrt{x^2 - 1}}.$

20. $\int_{-1/\sqrt{2}}^{1/\sqrt{2}} (1 - 2x^2)^{3/2} \, dx.$

21. $\int_{-1}^1 \frac{dx}{x^2 + 2x + 5}$

$$\begin{bmatrix} Hint: \text{Complete} \\ \text{the square.} \end{bmatrix}$$

22. $\displaystyle\int_0^3 \frac{x^3}{(3 + x^2)^{5/2}}\, dx.$

In cases where the trigonometric substitutions $x = a \sec\theta$ and $x = a\tan\theta$ lead to difficult integrals, it is sometimes possible to use the **hyperbolic substitutions:**

$x = a \sinh u$ for integrals involving $\sqrt{x^2 + a^2}$

$x = a \cosh u$ for integrals involving $\sqrt{x^2 - a^2}$

These substitutions are useful because in each case the hyperbolic identity

$$a^2 \cosh^2 u - a^2 \sinh^2 u = a^2$$

removes the radical.

23. (a) Evaluate

$$\int \frac{dx}{\sqrt{x^2 + 9}}$$

using the hyperbolic substitution suggested above.

   (b) Evaluate the integral in (a) by a trigonometric substitution and show that the results in (a) and (b) agree.

24. Follow the directions of Exercise 23 for the integral

$$\int \sqrt{x^2 - 1}\, dx, \qquad x \geq 1$$

25. For the substitution $x = a \sin\theta$, $-\pi/2 \leq \theta \leq \pi/2$, the triangle

suggests the relationships

(a) $\cos\theta = \dfrac{1}{a}\sqrt{a^2 - x^2}$

(b) $\tan\theta = \dfrac{x}{\sqrt{a^2 - x^2}}$

Use trigonometric identities to prove that these results are correct.

9.6 INTEGRALS INVOLVING $ax^2 + bx + c$

In this section we discuss integrals whose integrands contain a quadratic expression $ax^2 + bx + c$ that cannot be factored into real linear factors. Some cases where factorization is possible are treated in the next section. If $b = 0$, then methods discussed in previous sections will frequently work. For cases where $b \neq 0$, the integral can often be evaluated by first completing the square as follows:

$$ax^2 + bx + c = a\left(x^2 + \frac{b}{a}x\right) + c$$

$$= a\left(x^2 + \frac{b}{a}x + \frac{b^2}{4a^2}\right) + c - \frac{b^2}{4a}$$

$$= a\left(x + \frac{b}{2a}\right)^2 + c - \frac{b^2}{4a}$$

At this point, the substitution

$$u = x + \frac{b}{2a}$$

will reduce the original expression $ax^2 + bx + c$ to the simpler form $au^2 + d$, where $d = c - (b^2/4a)$.

▶ Example 1

Evaluate

$$\int \frac{dx}{x^2 - 2x + 5}$$

Solution. Completing the square yields

$$x^2 - 2x + 5 = (x^2 - 2x + 1) + 5 - 1 = (x - 1)^2 + 4$$

Thus

$$\int \frac{dx}{x^2 - 2x + 5} = \int \frac{dx}{(x - 1)^2 + 4}$$

$$= \int \frac{du}{u^2 + 4} \qquad \begin{bmatrix} u = x - 1 \\ du = dx \end{bmatrix}$$

$$= \frac{1}{2} \tan^{-1} \frac{u}{2} + C \qquad \begin{bmatrix} \text{Formula (14)} \\ \text{of Section 8.3} \end{bmatrix}$$

$$= \frac{1}{2} \tan^{-1} \left(\frac{x - 1}{2} \right) + C \qquad ◀$$

▶ Example 2

Evaluate

$$\int \frac{dx}{\sqrt{5 - 4x - 2x^2}}$$

Solution. Completing the square yields

$$5 - 4x - 2x^2 = 5 - 2(x^2 + 2x)$$

$$= 5 - 2(x^2 + 2x + 1) + 2$$

$$= 5 - 2(x + 1)^2 + 2 = 7 - 2(x + 1)^2$$

Thus

$$\int \frac{dx}{\sqrt{5 - 4x - 2x^2}} = \int \frac{dx}{\sqrt{7 - 2(x + 1)^2}}$$

$$= \int \frac{du}{\sqrt{7 - 2u^2}} \qquad \begin{bmatrix} u = x + 1 \\ du = dx \end{bmatrix}$$

$$= \frac{1}{\sqrt{2}} \int \frac{du}{\sqrt{(7/2) - u^2}}$$

$$= \frac{1}{\sqrt{2}} \sin^{-1}\left(\frac{u}{\sqrt{7/2}} \right) + C \qquad \begin{bmatrix} \text{Formula (13),} \\ \text{Section 8.3} \\ \text{with } a = \sqrt{7/2}. \end{bmatrix}$$

$$= \frac{1}{\sqrt{2}} \sin^{-1}(\sqrt{\tfrac{2}{7}}u) + C$$

$$= \frac{1}{\sqrt{2}} \sin^{-1}(\sqrt{\tfrac{2}{7}}(x + 1)) + C \qquad \blacktriangleleft$$

▶ Example 3

Evaluate

$$\int \frac{x}{x^2 - 4x + 8} dx$$

Solution.   Completing the square yields

$$x^2 - 4x + 8 = (x^2 - 4x + 4) + 8 - 4 = (x - 2)^2 + 4$$

Thus the substitution

$$u = x - 2 \qquad du = dx$$

yields

$$\int \frac{x}{x^2 - 4x + 8} dx = \int \frac{x}{(x - 2)^2 + 4} dx = \int \frac{u + 2}{u^2 + 4} du$$

$$= \int \frac{u}{u^2 + 4} du + 2 \int \frac{du}{u^2 + 4}$$

$$= \frac{1}{2} \int \frac{2u}{u^2 + 4} du + 2 \int \frac{du}{u^2 + 4}$$

$$= \frac{1}{2} \ln (u^2 + 4) + 2\left(\frac{1}{2}\right) \tan^{-1} \frac{u}{2} + C$$

$$= \frac{1}{2} \ln [(x - 2)^2 + 4] + \tan^{-1}\left(\frac{x - 2}{2}\right) + C \qquad \blacktriangleleft$$

▶ Exercise Set 9.6

Evaluate the integrals in Exercises 1–13.

1. $\displaystyle\int \frac{dx}{x^2 - 4x + 13}$.

2. $\displaystyle\int \frac{dx}{\sqrt{2x - x^2}}$.

9. $\displaystyle\int \frac{dx}{2x^2 + 4x + 7}$.

3. $\displaystyle\int \frac{dx}{\sqrt{8 + 2x - x^2}}$.

4. $\displaystyle\int \frac{dx}{16x^2 + 16x + 5}$.

10. $\displaystyle\int \frac{\cos\theta}{\sin^2\theta - 6\sin\theta + 12}\, d\theta$.

5. $\displaystyle\int \frac{dx}{\sqrt{x^2 - 6x + 10}}$.

6. $\displaystyle\int \frac{x}{x^2 + 6x + 10}\, dx$.

11. $\displaystyle\int \frac{2x + 5}{x^2 + 2x + 5}\, dx$.

12. $\displaystyle\int \frac{2x + 3}{4x^2 + 4x + 5}\, dx$.

7. $\displaystyle\int \sqrt{3 - 2x - x^2}\, dx$.

8. $\displaystyle\int \frac{e^x}{\sqrt{1 + e^x + e^{2x}}}\, dx$.

13. $\displaystyle\int \frac{x + 3}{\sqrt{x^2 + 2x + 2}}\, dx$.

9.7 INTEGRATING RATIONAL FUNCTIONS; PARTIAL FRACTIONS

Recall that a *rational function* is the quotient of two polynomials. Some examples are:

$$\frac{3x}{x^2 + 1}, \quad \frac{6x^2 - x + 2}{(x - 1)(x^2 + 4)}, \quad \frac{\frac{1}{2}x^3 + 2}{x^3 - \frac{1}{3}}, \quad \frac{7x^4 - 8x^2 + 2x + 1}{3x^3 + 2x - 5} \tag{1}$$

In this section we will discuss a method for integrating rational functions.

In algebra we learn to combine two or more fractions into a single fraction; for example,

$$\frac{1}{x} + \frac{3}{x - 1} + \frac{2}{x + 2} = \frac{(x - 1)(x + 2) + 3x(x + 2) + 2x(x - 1)}{x(x - 1)(x + 2)}$$

$$= \frac{6x^2 + 5x - 2}{x^3 + x^2 - 2x} \tag{2}$$

However, the left side of (2) is easier to integrate than the right. Thus it would be helpful if we knew how to obtain the left side of the equation starting with the right. The procedure for doing this is called *partial fraction decomposition.*

A rational function

$$\frac{P(x)}{Q(x)} \tag{3}$$

is called **proper** if the degree of $P(x)$ is less than the degree of $Q(x)$; otherwise it is **improper.** Thus in (1), the first two rational functions are proper and the last two are improper. It can be proved that any proper rational function is expressible as a sum of terms (called **partial fractions**) having the form:

$$\frac{A}{(ax + b)^k} \qquad \text{or} \qquad \frac{Bx + C}{(ax^2 + bx + c)^k}$$

The exact number of terms of each type depends on how the denominator $Q(x)$ in (3) factors. In theory, a polynomial $Q(x)$ with real coefficients can always be factored into a product of linear and quadratic factors with real coefficients. For example, the polynomial

$$Q(x) = x^3 - 3x^2 + x - 3$$

factors into

$$Q(x) = (x - 3)(x^2 + 1)$$

The quadratic factor $x^2 + 1$ cannot be further decomposed into linear factors without using imaginary numbers $[x^2 + 1 = (x - i)(x + i)]$. Such quadratic factors are said to be **irreducible.**

The first step in the partial fraction decomposition of $P(x)/Q(x)$ is to completely factor the denominator $Q(x)$ into linear and irreducible quadratic factors and then collect all repeated factors so that $Q(x)$ is expressed as a product of *distinct* factors of the form

$$(ax + b)^m \qquad \text{and} \qquad (ax^2 + bx + c)^m$$

where $ax^2 + bx + c$ is irreducible. Once this is done, the structure of the partial fraction decomposition of $P(x)/Q(x)$ is determined as follows:

Linear Factors

For each factor of the form $(ax + b)^m$, introduce the m terms

$$\frac{A_1}{ax + b} + \frac{A_2}{(ax + b)^2} + \cdots + \frac{A_m}{(ax + b)^m}$$

where $A_1, A_2, \ldots, A_m$ are constants to be determined.

Irreducible Quadratic Factors
For each factor of the form $(ax^2 + bx + c)^m$, introduce m terms

$$\frac{A_1 x + B_1}{ax^2 + bx + c} + \frac{A_2 x + B_2}{(ax^2 + bx + c)^2} + \cdots + \frac{A_m x + B_m}{(ax^2 + bx + c)^m}.$$

where $A_1, A_2, \ldots, A_m, B_1, B_2, \ldots, B_m$ are constants to be determined.

▶ Example 1

Evaluate

$$\int \frac{dx}{x^2 + x - 2}$$

Solution. The integrand can be written

$$\frac{1}{x^2 + x - 2} = \frac{1}{(x - 1)(x + 2)}$$

According to the rule above for linear factors, the factor $x - 1$ introduces one term (since $m = 1$):

$$\frac{A}{x - 1}$$

and the factor $x + 2$ introduces one term:

$$\frac{B}{x + 2}$$

so that the partial fraction decomposition is

$$\frac{1}{(x - 1)(x + 2)} = \frac{A}{x - 1} + \frac{B}{x + 2} \tag{4}$$

where A and B are constants to be determined so that (4) becomes an *identity*. To find these constants we multiply both sides of (4) by $(x - 1)(x + 2)$ to obtain

$$1 = A(x + 2) + B(x - 1) \tag{5}$$

Next we substitute values of x to make the various terms zero. Setting $x = -2$ in (5) yields

$$1 = -3B \qquad \text{or} \qquad B = -\frac{1}{3}$$

and setting $x = 1$ in (5) yields

$$1 = 3A \qquad \text{or} \qquad A = \frac{1}{3}$$

Thus (4) becomes

$$\frac{1}{(x-1)(x+2)} = \frac{1/3}{x-1} + \frac{-1/3}{x+2}$$

and

$$\int \frac{dx}{(x-1)(x+2)} = \frac{1}{3}\int \frac{dx}{x-1} - \frac{1}{3}\int \frac{dx}{x+2}$$

$$= \frac{1}{3}\ln|x-1| - \frac{1}{3}\ln|x+2| + C$$

$$= \frac{1}{3}\ln\left|\frac{x-1}{x+2}\right| + C$$

Alternate Solution. The constants A and B in (5) can also be determined by collecting like terms:

$$1 = (A + B)x + (2A - B) \tag{6}$$

and then equating corresponding coefficients on both sides:

$$A + B = 0$$
$$2A - B = 1$$

and then solving these equations simultaneously to obtain $A = \frac{1}{3}$, $B = -\frac{1}{3}$. This method is justified because (6) is an *identity* holding for all x, and two polynomials are equal for all x if and only if their corresponding coefficients are equal (Exercise 24). ◀

▶ Example 2

Evaluate

$$\int \frac{2x+4}{x^3 - 2x^2}\,dx$$

Solution. The integrand can be written

$$\frac{2x + 4}{x^3 - 2x^2} = \frac{2x + 4}{x^2(x - 2)}$$

Although x^2 is a quadratic factor, it is *not* irreducible since $x^2 = xx$. Thus by the rule for linear factors, x^2 introduces two terms (since $m = 2$):

$$\frac{A}{x} + \frac{B}{x^2}$$

and the factor $x - 2$ introduces one term (since $m = 1$):

$$\frac{C}{x - 2}$$

so the partial fraction decomposition is

$$\frac{2x + 4}{x^2(x - 2)} = \frac{A}{x} + \frac{B}{x^2} + \frac{C}{x - 2} \tag{7}$$

Multiplying by $x^2(x - 2)$ yields

$$2x + 4 = Ax(x - 2) + B(x - 2) + Cx^2 \tag{8}$$

To determine A, B, and C we will follow the alternate method used in Example 1. Multiplying and collecting like terms in (8) yields

$$2x + 4 = (A + C)x^2 + (-2A + B)x - 2B$$

Equating corresponding coefficients gives

$$\begin{aligned} A \quad\quad + C &= 0 \\ -2A + B \quad\quad &= 2 \\ -2B \quad\quad &= 4 \end{aligned}$$

and solving this system yields

$$A = -2, \quad B = -2, \quad C = 2$$

so (7) becomes

$$\frac{2x + 4}{x^2(x - 2)} = \frac{-2}{x} + \frac{-2}{x^2} + \frac{2}{x - 2}$$

Thus

$$\int \frac{2x+4}{x^2(x-2)}\,dx = -2\int \frac{dx}{x} - 2\int \frac{dx}{x^2} + 2\int \frac{dx}{x-2}$$

$$= -2\ln|x| + \frac{2}{x} + 2\ln|x-2| + C$$

$$= 2\ln\left|\frac{x-2}{x}\right| + \frac{2}{x} + C \qquad \blacktriangleleft$$

▶ Example 3

Evaluate

$$\int \frac{x^2+x-2}{3x^3-x^2+3x-1}\,dx$$

Solution. The denominator in the integrand can be factored by grouping:

$$\frac{x^2+x-2}{3x^3-x^2+3x-1} = \frac{x^2+x-2}{x^2(3x-1)+(3x-1)} = \frac{x^2+x-2}{(3x-1)(x^2+1)}$$

By the rule for linear factors, the factor $3x-1$ introduces one term:

$$\frac{A}{3x-1}$$

and by the rule for irreducible quadratic factors, the factor x^2+1 introduces one term:

$$\frac{Bx+C}{x^2+1}$$

Thus the partial fraction decomposition is

$$\frac{x^2+x-2}{(3x-1)(x^2+1)} = \frac{A}{3x-1} + \frac{Bx+C}{x^2+1} \qquad (9)$$

Multiplying by $(3x-1)(x^2+1)$ yields

$$x^2+x-2 = A(x^2+1) + (Bx+C)(3x-1)$$

To determine A, B, and C, we multiply out and collect like terms:

$$x^2+x-2 = (A+3B)x^2 + (-B+3C)x + (A-C)$$

Equating corresponding coefficients gives

$$A + 3B \quad = \quad 1$$
$$- \ B + 3C = \quad 1$$
$$A \quad\quad - \ C = -2$$

To solve this system, subtract the third equation from the first to eliminate A. Then use the resulting equation together with the second equation to solve for B and C. Finally, determine A from the first or third equation. This yields (verify):

$$A = -\frac{7}{5}, \qquad B = \frac{4}{5}, \qquad C = \frac{3}{5}$$

Thus (9) becomes

$$\frac{x^2 + x - 2}{(3x - 1)(x^2 + 1)} = \frac{-\frac{7}{5}}{3x - 1} + \frac{\frac{4}{5}x + \frac{3}{5}}{x^2 + 1}$$

and

$$\int \frac{x^2 + x - 2}{(3x - 1)(x^2 + 1)} \, dx$$

$$= -\frac{7}{5} \int \frac{dx}{3x - 1} + \frac{4}{5} \int \frac{x}{x^2 + 1} \, dx + \frac{3}{5} \int \frac{dx}{x^2 + 1}$$

$$= -\frac{7}{15} \ln |3x - 1| + \frac{2}{5} \ln (x^2 + 1) + \frac{3}{5} \tan^{-1} x + C \quad \blacktriangleleft$$

▶ Example 4

Evaluate

$$\int \frac{3x^4 + 4x^3 + 16x^2 + 20x + 9}{(x + 2)(x^2 + 3)^2} \, dx$$

Solution. By the rule for linear factors, the factor $x + 2$ introduces one term:

$$\frac{A}{x + 2}$$

and by the rule for irreducible quadratic factors, the factor $(x^2 + 3)^2$ introduces two terms (since $m = 2$):

$$\frac{Bx + C}{x^2 + 3} + \frac{Dx + E}{(x^2 + 3)^2}$$

Thus the partial fraction decomposition of the integrand is

$$\frac{3x^4 + 4x^3 + 16x^2 + 20x + 9}{(x + 2)(x^2 + 3)^2} = \frac{A}{x + 2} + \frac{Bx + C}{x^2 + 3} + \frac{Dx + E}{(x^2 + 3)^2} \quad (10)$$

Multiplying by $(x + 2)(x^2 + 3)^2$ yields

$$3x^4 + 4x^3 + 16x^2 + 20x + 9$$
$$= A(x^2 + 3)^2 + (Bx + C)(x^2 + 3)(x + 2) + (Dx + E)(x + 2) \quad (11)$$

Multiplying out and collecting terms on the right side of (11) yields

$$(A + B)x^4 + (2B + C)x^3 + (6A + 3B + 2C + D)x^2$$
$$+ (6B + 3C + 2D + E)x + (9A + 6C + 2E)$$

and equating corresponding coefficients with the left side of (11) yields

$$\begin{aligned} A + B &= 3 \\ 2B + C &= 4 \\ 6A + 3B + 2C + D &= 16 \\ 6B + 3C + 2D + E &= 20 \\ 9A + 6C + 2E &= 9 \end{aligned} \quad (12)$$

This system of five equations in five unknowns is tedious to solve. However, we can reduce the work considerably by substituting $x = -2$ in (11) to obtain

$$49 = 49A \quad \text{or} \quad A = 1$$

Substituting this value in the first equation of (12) gives $B = 2$ and substituting this value of B in the second equation of (11) gives $C = 0$; continuing in this way we are led to the values (verify):

$$A = 1, \quad B = 2, \quad C = 0, \quad D = 4, \quad E = 0$$

Thus, (10) becomes

$$\frac{3x^4 + 4x^3 + 16x^2 + 20x + 9}{(x + 2)(x^2 + 3)^2} = \frac{1}{x + 2} + \frac{2x}{x^2 + 3} + \frac{4x}{(x^2 + 3)^2}$$

and

$$\int \frac{3x^4 + 4x^3 + 16x^2 + 20x + 9}{(x + 2)(x^2 + 3)^2} dx$$

$$= \int \frac{dx}{x + 2} + \int \frac{2x}{x^2 + 3} dx + 4 \int \frac{x}{(x^2 + 3)^2} dx$$

$$= \ln |x + 2| + \ln (x^2 + 3) - \frac{2}{x^2 + 3} + C$$

(The third integral on the right was evaluated by the substitution $u = x^2 + 3$.) ◀

As noted in the beginning of this section, partial fraction decomposition only applies to *proper* rational functions. However, the next example shows that improper rational functions can be integrated by first performing a long division, and then working with the remainder term.

▶ Example 5

Evaluate

$$\int \frac{3x^4 + 3x^3 - 5x^2 + x - 1}{x^2 + x - 2} dx$$

Solution. Since the integrand is an improper rational function, we cannot use a partial fraction decomposition directly. However, if we perform the long division

$$
\begin{array}{r}
3x^2 + 1 \\
x^2 + x - 2 \overline{\smash{\big)}\ 3x^4 + 3x^3 - 5x^2 + x - 1} \\
\underline{3x^4 + 3x^3 - 6x^2} \\
x^2 + x - 1 \\
\underline{x^2 + x - 2} \\
1
\end{array}
$$

we can write the integrand as the quotient plus the remainder over the divisor, that is

$$\frac{3x^4 + 3x^3 - 5x^2 + x - 1}{x^2 + x - 2} = (3x^2 + 1) + \frac{1}{x^2 + x - 2}$$

Thus

$$\int \frac{3x^4 + 3x^3 - 5x^2 + x - 1}{x^2 + x - 2} dx = \int (3x^2 + 1) dx + \int \frac{dx}{x^2 + x - 2}$$

The second integral on the right now involves a proper rational function and can thus be evaluated by a partial fraction decomposition. Using the result of Example 1 we obtain

$$\int \frac{3x^4 + 3x^3 - 5x^2 + x - 1}{x^2 + x - 2}\,dx = x^3 + x + \frac{1}{3}\ln\left|\frac{x-1}{x+2}\right| + C \quad \blacktriangleleft$$

▶ **Exercise Set 9.7**

In Exercises 1–18, perform the integrations.

1. $\displaystyle\int \frac{x}{x^2 - 5x + 6}\,dx.$

2. $\displaystyle\int \frac{5x - 4}{x^2 - 4x}\,dx.$

12. $\displaystyle\int \frac{dx}{x(x^2 + x + 1)}.$

13. $\displaystyle\int \frac{dx}{x^4 - 16}.$

3. $\displaystyle\int \frac{11x + 17}{2x^2 + 7x - 4}\,dx.$

4. $\displaystyle\int \frac{x^2}{x^2 - 3x + 2}\,dx.$

14. $\displaystyle\int \frac{x^3 + x^2 + x + 2}{(x^2 + 1)(x^2 + 2)}\,dx.$

5. $\displaystyle\int \frac{3x^2 - 10}{x^2 - 4x + 4}\,dx.$

6. $\displaystyle\int \frac{2x^2 + 4x - 8}{x^3 - 4x}\,dx.$

15. $\displaystyle\int \frac{x^2 + 1}{(x^2 + 2x + 3)^2}\,dx.$

7. $\displaystyle\int \frac{dx}{(x - 2)^3}.$

8. $\displaystyle\int \frac{3x^2 - x + 1}{x^3 - x^2}\,dx.$

16. $\displaystyle\int \frac{x^5 + x^4 + 4x^3 + 4x^2 + 4x + 4}{(x^2 + 2)^3}\,dx.$

9. $\displaystyle\int \frac{2x^2 + 3}{x(x - 1)^2}\,dx.$

10. $\displaystyle\int \frac{e^t}{e^{2t} - 4}\,dt.$

17. $\displaystyle\int \frac{2x^2 - 1}{4x^3 - x^2 + 4x - 1}\,dx.$

11. $\displaystyle\int \frac{\cos\theta}{\sin^2\theta + 4\sin\theta - 5}\,d\theta.$

18. $\displaystyle\int \frac{dx}{1 + e^x}.$

FACTORING POLYNOMIALS—OPTIONAL DISCUSSION AND EXERCISES

The method of partial fractions depends on our ability to carry out the necessary factorization. This is not always easy to do. The following results, usually proved in algebra courses, are helpful to know.

9.7.1 THEOREM *Factor Theorem*   *If $p(x)$ is a polynomial and r is a solution of the equation $p(x) = 0$, then $x - r$ is a factor of $p(x)$.*

9.7.2 THEOREM   *Let*

$$p(x) = a_0 x^n + a_1 x^{n-1} + \cdots + a_{n-1}x + a_n$$

be a polynomial with integer coefficients.

(a) If r is an integer solution of $p(x) = 0$, then the constant term a_n is an integer multiple of r.

(b) *If c/d is a rational solution of $p(x) = 0$, and if c/d is expressed in lowest terms, then the constant term a_n is an integer multiple of c; and the leading coefficient a_0 is an integer multiple of d.*

▶ Example 6

For the equation

$$x^3 + x^2 - 10x + 8 = 0$$

the only possible integer solutions are

$$\pm 1, \ \pm 2, \ \pm 4, \ \pm 8$$

By substitution, or by using synthetic division, the reader can show that 1, 2, -4 are solutions and the rest are not. It follows that

$$x^3 + x^2 - 10x + 8 = (x - 1)(x - 2)(x + 4) \qquad ◀$$

▶ Example 7

For the equation

$$2x^3 + x^2 + 8x + 4 = 0$$

the only possible numerators for rational solutions are $\pm 1, \ \pm 2, \ \pm 4$ and the only possible denominators are $\pm 1, \ \pm 2$; thus the only possible rational solutions are

$$\pm 1, \ \pm 2, \ \pm 4, \ \pm \tfrac{1}{2}$$

By substitution, or by using synthetic division, the reader can show that $-\tfrac{1}{2}$ is a solution but the rest are not. It follows by division that

$$2x^3 + x^2 + 8x + 4 = (x + \tfrac{1}{2})(2x^2 + 8) = 2(x + \tfrac{1}{2})(x^2 + 4) \qquad ◀$$

In Exercises 19–22, use Theorems 9.7.1 and 9.7.2.

19. Find all rational solutions, if any, and use your results to factor the polynomial into a product of linear and irreducible quadratic factors.
   (a) $x^3 - 6x^2 + 11x - 6 = 0$
   (b) $x^3 - 3x^2 + x - 20 = 0$
   (c) $x^4 - 5x^3 + 7x^2 - 5x + 6 = 0$

20. Find all rational solutions, if any, and use your results to factor the polynomial into a product of linear and irreducible quadratic factors.
   (a) $8x^3 + 4x^2 - 2x - 1 = 0$
   (b) $6x^4 - 7x^3 + 6x^2 - 1 = 0$
   (c) $9x^4 - 56x^3 + 57x^2 + 98x - 24 = 0$

21. Evaluate

$$\int \frac{dx}{x^4 - 3x^3 - 7x^2 + 27x - 18}$$

22. Evaluate

$$\int \frac{dx}{16x^3 - 4x^2 + 4x - 1}$$

23. Use Theorem 9.7.2 to prove:
(a) $\sqrt{2}$ is irrational.

(b) If a is a positive integer, then $\sqrt{a}$ is either an integer or is irrational.

24. (a) Prove: $a_0 x^n + a_1 x^{n-1} + \cdots + a_n \equiv 0$ if and only if $a_0 = a_1 = \cdots = a_n = 0$.

(b) Use the result in (a) to prove:

$$a_0 x^n + a_1 x^{n-1} + \cdots + a_n \equiv$$
$$b_0 x^n + b_1 x^{n-1} + \cdots + b_n$$

if and only if $a_0 = b_0, a_1 = b_1, \ldots, a_n = b_n$.

9.8 MISCELLANEOUS SUBSTITUTIONS (OPTIONAL)

In this section we consider some integrals that do not fit in any of the categories previously studied.

INTEGRALS
INVOLVING
RATIONAL
EXPONENTS

Integrals involving rational powers of x can often be simplified by substituting

$$u = x^{1/n}$$

where n is the least common multiple of the denominators of the exponents. The effect of this substitution is to replace fractional exponents with integer exponents, which are easier to work with.

▶ Example 1

Evaluate

$$\int \frac{\sqrt{x}}{1 + \sqrt[3]{x}}\, dx$$

Solution. The integrand involves $x^{1/2}$ and $x^{1/3}$, so we make the substitution

$$u = x^{1/6}$$

or

$$x = u^6 \quad \text{and} \quad dx = 6u^5\, du$$

Thus

$$\int \frac{\sqrt{x}}{1 + \sqrt[3]{x}}\, dx = \int \frac{(u^6)^{1/2}}{1 + (u^6)^{1/3}}(6u^5)\, du = 6\int \frac{u^8}{1 + u^2}\, du$$

By long division

$$\frac{u^8}{1 + u^2} = u^6 - u^4 + u^2 - 1 + \frac{1}{1 + u^2}$$

Thus

$$\int \frac{\sqrt{x}}{1 + \sqrt[3]{x}}\, dx = 6\int \left(u^6 - u^4 + u^2 - 1 + \frac{1}{1 + u^2}\right) du$$

$$= \frac{6}{7}u^7 - \frac{6}{5}u^5 + 2u^3 - 6u + 6\tan^{-1} u + C$$

$$= \frac{6}{7}x^{7/6} - \frac{6}{5}x^{5/6} + 2x^{1/2} - 6x^{1/6} + 6\tan^{-1}(x^{1/6}) + C \qquad \blacktriangleleft$$

▶ Example 2

Evaluate

$$\int \frac{dx}{2 + 2\sqrt{x}}$$

Solution. The integrand contains $\sqrt{x} = x^{1/2}$, so we make the substitution

$$u = x^{1/2}$$

or

$$x = u^2 \qquad \text{and} \qquad dx = 2u\, du$$

This yields

$$\int \frac{dx}{2 + 2\sqrt{x}} = \int \frac{2u}{2 + 2u}\, du = \int \left(1 - \frac{1}{1 + u}\right) du$$

$$= u - \ln|1 + u| + C = \sqrt{x} - \ln|1 + \sqrt{x}| + C \qquad \blacktriangleleft$$

The following example illustrates a variation on the above idea.

▶ Example 3

Evaluate

$$\int \sqrt{1 + e^x}\, dx$$

Solution. The substitution

$$u^2 = 1 + e^x$$

will eliminate the square root. To express dx in terms of du, it is helpful to solve this equation for x and then differentiate. We obtain

$$e^x = u^2 - 1$$
$$x = \ln(u^2 - 1)$$
$$\frac{dx}{du} = \frac{2u}{u^2 - 1}, \qquad dx = \frac{2u}{u^2 - 1}\, du$$

Thus

$$
\begin{aligned}
\int \sqrt{1 + e^x}\, dx &= \int u\left(\frac{2u}{u^2 - 1}\right) du \\[2mm]
&= \int \frac{2u^2}{u^2 - 1}\, du \\[2mm]
&= \int \left(2 + \frac{2}{u^2 - 1}\right) du && \text{(long division)} \\[2mm]
&= 2u + \int \left(\frac{1}{u - 1} - \frac{1}{u + 1}\right) du && \text{(partial fractions)} \\[2mm]
&= 2u + \ln|u - 1| - \ln|u + 1| + C \\[2mm]
&= 2u + \ln\left|\frac{u - 1}{u + 1}\right| + C \\[2mm]
&= 2\sqrt{1 + e^x} + \ln\left|\frac{\sqrt{1 + e^x} - 1}{\sqrt{1 + e^x} + 1}\right| + C \qquad ◀
\end{aligned}
$$

INTEGRALS
CONTAINING
RATIONAL
EXPRESSIONS IN SIN x
AND COS x

Functions like

$$\frac{\sin x + 3\cos^2 x}{\cos x + 4\sin x}, \qquad \frac{\sin x}{1 + \cos x - \cos^2 x}, \qquad \frac{3\sin^5 x}{1 + 4\sin x}$$

are called *rational expressions in sin x and cos x* since they would become rational functions of x if we were to replace $\sin x$ and $\cos x$ by the letter x.

If an integrand is a rational expression in $\sin x$ and $\cos x$, then the substitution

$$u = \tan\left(\frac{x}{2}\right), \qquad -\pi < x < \pi \tag{1}$$

will transform the integrand into a rational function of u (which can then be integrated by methods already discussed.) To see this, observe that

$$\cos\left(\frac{x}{2}\right) = \frac{1}{\sec(x/2)} = \frac{1}{\sqrt{1 + \tan^2(x/2)}} = \frac{1}{\sqrt{1 + u^2}}$$

$$\sin\left(\frac{x}{2}\right) = \tan\left(\frac{x}{2}\right)\cos\left(\frac{x}{2}\right) = u\left(\frac{1}{\sqrt{1 + u^2}}\right)$$

Therefore,

$$\sin x = 2\sin\left(\frac{x}{2}\right)\cos\left(\frac{x}{2}\right) = 2u\left(\frac{1}{\sqrt{1 + u^2}}\right)\left(\frac{1}{\sqrt{1 + u^2}}\right)$$

or

$$\sin x = \frac{2u}{1 + u^2} \tag{2}$$

Also,

$$\cos x = 1 - 2\sin^2\left(\frac{x}{2}\right) = 1 - \frac{2u^2}{1 + u^2}$$

or

$$\cos x = \frac{1 - u^2}{1 + u^2} \tag{3}$$

Moreover, from (1)

$$\frac{x}{2} = \tan^{-1} u$$

so that

$$\frac{dx}{du} = \frac{2}{1 + u^2}$$

or

$$dx = \frac{2}{1 + u^2}\, du \qquad\qquad (4)$$

It follows from (2), (3), and (4) that substitution (1) will yield an integrand that is a rational function of u.

▶ Example 4

Evaluate

$$\int \frac{dx}{1 + \sin x}$$

Solution. Making substitution (1), it follows from (2) and (4) above that

$$\int \frac{dx}{1 + \sin x} = \int \frac{1}{1 + \left(\dfrac{2u}{1 + u^2}\right)} \left(\frac{2}{1 + u^2}\right) du$$

$$= \int \frac{2}{1 + 2u + u^2}\, du$$

$$= \int \frac{2}{(1 + u)^2}\, du$$

$$= -\frac{2}{1 + u} + C$$

$$= -\frac{2}{1 + \tan\left(\dfrac{x}{2}\right)} + C \qquad\qquad ◀$$

REMARK. While substitution (1) is a useful tool, it can lead to cumbersome partial fraction decompositions. Consequently, this method should be used only after looking for simpler methods.

▶ Exercise Set 9.8

In Exercises 1–14, perform the integrations.

1. $\displaystyle\int \frac{dx}{\sqrt{x} + \sqrt[3]{x}}.$

2. $\displaystyle\int \frac{dx}{x - x^{3/5}}.$

5. $\displaystyle\int \frac{dt}{t^{1/2} - t^{1/3}}.$

6. $\displaystyle\int \frac{1 + \sqrt{x}}{1 - \sqrt{x}}\, dx.$

3. $\displaystyle\int \frac{dv}{v(1 - v^{1/4})}.$

4. $\displaystyle\int \frac{x^{2/3}}{x + 1}\, dx.$

7. $\displaystyle\int \frac{x^3}{\sqrt{1 + x^2}}\, dx.$

8. $\displaystyle\int \frac{x}{(x + 3)^{1/5}}\, dx.$

9. $\displaystyle\int \frac{dx}{1 + \sin x + \cos x}$.   10. $\displaystyle\int \frac{dx}{2 + \sin x}$.

11. $\displaystyle\int_{\pi/2}^{\pi} \frac{d\theta}{1 - \cos \theta}$.   12. $\displaystyle\int \frac{dx}{4 \sin x - 3 \cos x}$.

13. $\displaystyle\int \frac{\cos x}{2 - \cos x}\, dx$.   14. $\displaystyle\int \frac{dx}{\sin x + \tan x}$.

15. (a) Use the substitution $u = \tan (x/2)$ to show that

$$\int \sec x \, dx = \ln \left| \frac{1 + \tan \frac{1}{2}x}{1 - \tan \frac{1}{2}x} \right| + C$$

(b) Use the result in (a) to show that

$$\int \sec x \, dx = \ln \left| \tan \left(\frac{\pi}{4} + \frac{x}{2} \right) \right| + C$$

(c) Show that this agrees with (2) of Section 9.4.

16. (a) Use the substitution $u = \tan (x/2)$ to show that

$$\int \csc x \, dx = \frac{1}{2} \ln \left[\frac{1 - \cos x}{1 + \cos x} \right] + C$$

(b) Show that this agrees with Exercise 31a of Section 9.4.

17. Develop a substitution that can be used to integrate rational functions of $\sinh x$ and $\cosh x$ and use your substitution to evaluate

$$\int \frac{dx}{2 \cosh x + \sinh x}$$

without expressing the integrand in terms of e^x and e^{-x}.

9.9 NUMERICAL INTEGRATION; SIMPSON'S RULE

If an antiderivative of f cannot be found, then the integral

$$\int_a^b f(x) \, dx$$

cannot be evaluated by the fundamental theorem of calculus. In such cases, the value of the integral can be approximated using methods we will study in this section.

Suppose we want to approximate

$$\int_a^b f(x) \, dx$$

where f is continuous on $[a, b]$. Using the definition of a definite integral (Definition 5.7.2), we can write this integral as

$$\int_a^b f(x) \, dx = \lim_{\max \Delta x_k \to 0} \sum_{k=1}^{n} f(x_k^*) \, \Delta x_k \qquad (1)$$

If we divide the interval $[a, b]$ into n subintervals of equal length, then

$$\Delta x_1 = \Delta x_2 = \cdots = \Delta x_n = \frac{b - a}{n}$$

so that we can rewrite (1) as

$$\int_a^b f(x)\,dx = \lim_{n \to +\infty} \sum_{k=1}^{n} f(x_k^*)\left(\frac{b-a}{n}\right)$$

Thus we expect the following approximation to be good when n is large:

$$\int_a^b f(x)\,dx \approx \sum_{k=1}^{n} f(x_k^*)\left(\frac{b-a}{n}\right) = \frac{b-a}{n}\sum_{k=1}^{n} f(x_k^*) \tag{2}$$

where x_k^* denotes an arbitrary point in the kth subinterval. Expression (2) yields various approximation formulas, depending on how the points $x_1^*, x_2^*, \ldots, x_n^*$ are chosen. For example, let

$$y_0, y_1, \ldots, y_n$$

be the values of f at the endpoints of the subintervals (Figure 9.9.1).

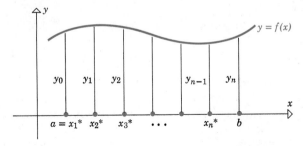

Figure 9.9.1

If we choose $x_1^*, x_2^*, \ldots, x_n^*$ to be the left-hand endpoints of the subintervals, then

$$y_0 = f(x_1^*), \; y_1 = f(x_2^*), \ldots, y_{n-1} = f(x_n^*)$$

so that (2) yields:

Left-hand Endpoint Approximation

$$\int_a^b f(x)\,dx \approx \left(\frac{b-a}{n}\right)[y_0 + y_1 + \cdots + y_{n-1}] \tag{3}$$

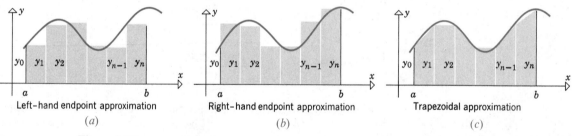

Left-hand endpoint approximation
(a)

Right-hand endpoint approximation
(b)

Trapezoidal approximation
(c)

Figure 9.9.2

(Figure 9.9.2a). Similarly, if $x_1^*, x_2^*, \ldots, x_n^*$ are the right-hand endpoints of the subintervals, then (2) yields

Right-hand Endpoint Approximation

$$\int_a^b f(x)\,dx \approx \left(\frac{b-a}{n}\right)[y_1 + y_2 + \cdots + y_n] \qquad (4)$$

(Figure 9.9.2b). Averaging the left-hand and right-hand endpoint approximations, we obtain:

Trapezoidal Approximation

$$\int_a^b f(x)\,dx \approx \left(\frac{b-a}{2n}\right)[y_0 + 2y_1 + \cdots + 2y_{n-1} + y_n] \qquad (5)$$

The name **_trapezoidal approximation_** can be explained by considering the case where $f(x) \geq 0$ on $[a, b]$, so that $\int_a^b f(x)\,dx$ represents the area under $f(x)$ over $[a, b]$. Geometrically, the trapezoidal approximation formula is what results if we approximate this area by the sum of the trapezoidal areas shown in Figure 9.9.2c (Exercise 8). By comparison, the left-hand and right-hand approximation formulas result when we approximate the area using the rectangles shown in Figures 9.9.2a and b.

▶ Example 1

Below we have approximated

$$\int_1^2 \frac{1}{x}\,dx$$

by three methods, left-hand endpoint approximation, right-hand endpoint approximation, and trapezoidal approximation. In each case we used $n = 10$ subdivisions of the interval $[1, 2]$ so that the subintervals have width

$$\frac{b-a}{n} = \frac{2-1}{10} = \frac{1}{10} = 0.1$$

Left-Hand Endpoint Approximation

i	Endpoint x_i	$y_i = f(x_i) = 1/x_i$
0	1.0	1.0000
1	1.1	0.9091
2	1.2	0.8333
3	1.3	0.7692
4	1.4	0.7143
5	1.5	0.6667
6	1.6	0.6250
7	1.7	0.5882
8	1.8	0.5556
9	1.9	0.5263
		Total
		7.1877

Right-Hand Endpoint Approximation

i	Endpoint x_i	$y_i = f(x_i) = 1/x_i$
1	1.1	0.9091
2	1.2	0.8333
3	1.3	0.7692
4	1.4	0.7143
5	1.5	0.6667
6	1.6	0.6250
7	1.7	0.5882
8	1.8	0.5556
9	1.9	0.5263
10	2.0	0.5000
		Total
		6.6877

From (3)

$$\int_1^2 \frac{1}{x}\, dx \approx (0.1)(7.1877)$$
$$= 0.71877$$

From (4)

$$\int_1^2 \frac{1}{x}\, dx \approx (0.1)(6.6877)$$
$$= 0.66877$$

Trapezoidal Approximation

i	Endpoint x_i	$y_i = f(x_i) = 1/x_i$	Multiplier w_i	$w_i y_i$
0	1.0	1.0000	1	1.0000
1	1.1	0.9091	2	1.8182
2	1.2	0.8333	2	1.6666
3	1.3	0.7692	2	1.5384
4	1.4	0.7143	2	1.4286
5	1.5	0.6667	2	1.3334
6	1.6	0.6250	2	1.2500
7	1.7	0.5882	2	1.1764
8	1.8	0.5556	2	1.1112
9	1.9	0.5263	2	1.0526
10	2.0	0.5000	1	0.5000
				Total
				13.8754

Since

$$\frac{b - a}{2n} = \frac{2 - 1}{2(10)} = \frac{1}{20} = 0.05$$

we obtain from (5)

$$\int_1^2 \frac{1}{x}\,dx \approx (0.05)(13.8754) = 0.69377 \qquad \blacktriangleleft$$

REMARK. Because $1/x$ is decreasing on the interval $[1, 2]$, we would expect the left-hand endpoint approximation to be too large and the right-hand endpoint approximation to be too small. This is in fact the case since,

$$\int_1^2 \frac{1}{x}\,dx = \ln 2 \approx 0.6931$$

to four-decimal-place accuracy.

Intuition suggests that we might improve on the endpoint approximations and the trapezoidal approximation by replacing the rectangular and trapezoidal strips of Figure 9.9.2 with strips having curved upper boundaries chosen to match the shape of the curve $y = f(x)$. This is the idea behind **Simpson's* rule,** which uses curves of the form

$$y = ax^2 + bx + c$$

---

*THOMAS SIMPSON (1710–1761) English mathematician. Simpson was the son of a weaver. He was trained to follow in his father's footsteps and had little formal education in his early life. His interest in science and mathematics was aroused in 1724, when he witnessed an eclipse of the sun and received two books from a peddler, one on astrology and the other on arithmetic. Simpson quickly absorbed their contents and soon became a successful local fortune teller. His improved financial situation enabled him to give up weaving and marry his landlady, an older woman. Then in 1733 some mysterious "unfortunate incident" forced him to move. He settled in Derby, where he taught in an evening school and worked at weaving during the day. In 1736 he moved to London and published his first mathematical work in a periodical called the *Ladies' Diary* (of which he later became the editor). In 1737 he published a successful calculus textbook that enabled him to give up weaving completely and concentrate on textbook writing and teaching. His fortunes improved further in 1740 when one Robert Heath accused him of plagiarism. The publicity was marvelous, and Simpson proceeded to dash off a succession of best selling textbooks, *Algebra* (ten editions plus translations), *Geometry* (twelve editions plus translations), *Trigonometry* (five editions plus translations) and numerous others.

It is interesting to note that Simpson did not discover the rule that bears his name. It was a well-known result by Simpson's time.

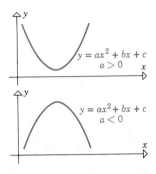

Figure 9.9.3

to approximate sections of the curve $y = f(x)$. As noted at the end of Section 2.2, these curves are parabolas with axes of symmetry parallel to the y-axis (Figure 9.9.3).

To simplify the description of Simpson's rule we will assume $f(x) \geq 0$ on $[a, b]$ so we can interpret $\int_a^b f(x)\, dx$ as an area. However, the method is valid without this assumption.

The heart of Simpson's rule is the formula

$$A = \frac{h}{3}[Y_0 + 4Y_1 + Y_2] \tag{6}$$

which gives the area under the curve

$$y = ax^2 + bx + c$$

over an arbitrary interval of width $2h$. In this formula Y_0, Y_1, and Y_2 represent the y values at the left-hand endpoint, the midpoint, and the right-hand endpoint of the interval (Figure 9.9.4).

To derive formula (6), denote the midpoint of the interval by m. Since the interval has length $2h$, the left-hand endpoint will be $m - h$ and the right-hand endpoint $m + h$. Thus the area A under $y = ax^2 + bx + c$ over this interval will be

Figure 9.9.4

$$
\begin{aligned}
A &= \int_{m-h}^{m+h} (ax^2 + bx + c)\, dx \\
&= \left. \frac{a}{3}x^3 + \frac{b}{2}x^2 + cx \right]_{m-h}^{m+h} \\
&= \frac{a}{3}[(m + h)^3 - (m - h)^3] + \frac{b}{2}[(m + h)^2 - (m - h)^2] \\
&\quad\quad + c[(m + h) - (m - h)]
\end{aligned}
$$

or on simplifying

$$A = \frac{h}{3}[a(6m^2 + 2h^2) + b(6m) + 6c] \tag{7}$$

But the values of $y = ax^2 + bx + c$ at the left-hand endpoint, the midpoint, and the right-hand endpoint are, respectively,

$$
\begin{aligned}
Y_0 &= a(m - h)^2 + b(m - h) + c \\
Y_1 &= am^2 + bm + c \\
Y_2 &= a(m + h)^2 + b(m + h) + c
\end{aligned}
$$

from which it follows that

$$Y_0 + 4Y_1 + Y_2 = a(6m^2 + 2h^2) + b(6m) + 6c \tag{8}$$

Thus (6) follows from (7) and (8).

Simpson's rule is obtained by dividing the interval $[a, b]$ into an *even* number of equal subintervals of width h and applying formula (6) to approximate the area under $y = f(x)$ over successive pairs of subintervals. The sum of these approximations then serves as an estimate of $\int_a^b f(x)\,dx$. More precisely, let $[a, b]$ be divided into n subintervals of width h (n even) and let

$$y_0, y_1, \ldots, y_n$$

be the values of $y = f(x)$ at the interval endpoints

$$a = x_0, x_1, \ldots, x_n = b$$

By (6) the area under $y = f(x)$ over the first two subintervals is approximately

$$\frac{h}{3}[y_0 + 4y_1 + y_2]$$

and the area over the second pair of subintervals is approximately

$$\frac{h}{3}[y_2 + 4y_3 + y_4]$$

and the area over the last pair of subintervals is approximately

$$\frac{h}{3}[y_{n-2} + 4y_{n-1} + y_n]$$

Adding all the approximations, collecting terms, and replacing h by $(b - a)/n$ yields:

Simpson's Rule

$$\int_a^b f(x)\,dx \approx \left(\frac{b - a}{3n}\right)[y_0 + 4y_1 + 2y_2 + 4y_3 + 2y_4 + \cdots$$
$$+ 2y_{n-2} + 4y_{n-1} + y_n]$$

▶ Example 2

Below we have approximated

$$\int_1^2 \frac{1}{x} \, dx$$

by Simpson's rule using $n = 10$ subdivisions so that the subintervals have width

$$\frac{b - a}{n} = \frac{2 - 1}{10} = \frac{1}{10} = 0.1$$

Simpson's Rule

i	Endpoint x_i	$y_i = f(x_i) = 1/x_i$	Multiplier w_i	$w_i y_i$
0	1.0	1.0000	1	1.0000
1	1.1	0.9091	4	3.6364
2	1.2	0.8333	2	1.6666
3	1.3	0.7692	4	3.0768
4	1.4	0.7143	2	1.4286
5	1.5	0.6667	4	2.6668
6	1.6	0.6250	2	1.2500
7	1.7	0.5882	4	2.3528
8	1.8	0.5556	2	1.1112
9	1.9	0.5263	4	2.1052
10	2.0	0.5000	1	0.5000
				Total
				20.7944

Since

$$\frac{b - a}{3n} = \frac{2 - 1}{3(10)} = \frac{1}{30}$$

we obtain from Simpson's rule

$$\int_1^2 \frac{1}{x} \, dx \approx \frac{1}{30}(20.7944) \approx 0.69315$$

To six decimal places, the value of $\ln 2 = \int_1^2 (1/x) \, dx$ is 0.693147 so that with $n = 10$ Simpson's rule is more accurate than the methods of Example 1.

◀

With all the methods studied in this section, there are two sources of error, the *intrinsic* or *trunction error* due to the approximation formula, and the *roundoff* error introduced in the calculations. In general, increasing n reduces the truncation error but increases the roundoff error, since more computations are required for larger n. In practical applications, it is important to know how large n must be taken to ensure a specified degree of accuracy (see Exercise 10). Such problems are studied in a branch of mathematics called *numerical analysis.*

▶ Exercise Set 9.9

In Exercises 1–6, use the given value of n to approximate the integral by:
(a) the left-hand endpoint rule
(b) the right-hand endpoint rule
(c) the trapezoidal rule
(d) Simpson's rule.

Round off all computations to four decimal places and use the appendix tables of trigonometric functions and exponentials where needed.

1. $\displaystyle\int_0^1 \frac{dx}{x+1}$, $\quad n = 6$.
2. $\displaystyle\int_0^4 x^2\, dx$, $\quad n = 8$.

3. $\displaystyle\int_0^\pi \sin x\, dx$, $\quad n = 4$.
4. $\displaystyle\int_0^1 \frac{dx}{\sqrt{1+x^2}}$, $\quad n = 4$.

5. $\displaystyle\int_0^2 e^{-x^2}\, dx$, $\quad n = 4$.
6. $\displaystyle\int_1^4 \sqrt{1+x^3}\, dx$, $\quad n = 6$.

7. Use the relationship

$$\frac{\pi}{4} = \tan^{-1}(1) = \int_0^1 \frac{dx}{1+x^2}$$

and Simpson's rule with $n = 10$ to estimate π. Round off all computations to four decimal places.

8. Derive the trapezoidal rule by summing the areas of the trapezoids in Figure 9.9.2c.

9. If, in (2), x_k^* is taken as the midpoint of the kth subinterval, we obtain the **midpoint rule** for approximating $\int_a^b f(x)\, dx$. Use the midpoint rule with $n = 10$ to approximate

$$\int_1^2 \frac{1}{x}\, dx$$

(Compare your results to those obtained in Examples 1 and 2.)

10. Let K be any number such that $|f''(x)| \le K$, and let M be any number such that $|f^{(4)}(x)| \le M$ for all x in $[a, b]$. It is shown in advanced calculus that the absolute value of the error resulting from Simpson's rule with n subintervals is at most

$$\frac{M(b-a)^5}{180n^4}$$

and the absolute value of the error from the trapezoidal rule is at most

$$\frac{K(b-a)^3}{12n^2}$$

(a) In Examples 1 and 2, $\int_1^2 (1/x)\, dx$ was estimated by the trapezoidal rule and Simpson's rule using $n = 10$. Show that the absolute value of the error for the trapezoidal rule is at most

$$\frac{2}{12 \times 10^2} \approx 0.00167$$

and the absolute value of the error for Simpson's rule is at most

$$\frac{24}{180 \times 10^4} \approx 0.0000133$$

(b) How large should n be taken to ensure that the absolute value of the error in the approximation of $\int_1^2 (1/x)\, dx$ by Simpson's rule is at most 0.0002?

(c) Answer (b) for the trapezoidal rule.

► REVIEW EXERCISES

In Exercises 1–64, evaluate the integrals.

1. $\displaystyle\int x \cos 2x \, dx.$

2. $\displaystyle\int x \cos x^2 \, dx.$

3. $\displaystyle\int \tan^3 x \sec x \, dx.$

4. $\displaystyle\int \sin^3 x \cos^2 x \, dx.$

5. $\displaystyle\int \tan^2 3t \sec^2 3t \, dt.$

6. $\displaystyle\int \cot 2x \csc^3 2x \, dx.$

7. $\displaystyle\int \frac{\sin^2 x \, dx}{1 + \cos x}.$

8. $\displaystyle\int \frac{\sin 2x \, dx}{\cos x \,(1 + \cos x)}.$

9. $\displaystyle\int x^2 \cos^2 x \, dx.$

10. $\displaystyle\int \sin^2 2x \cos^2 2x \, dx.$

11. $\displaystyle\int \sec^5 x \sin x \, dx.$

12. $\displaystyle\int \tan^5 2x \, dx.$

13. $\displaystyle\int \sin^4 x \cos^2 x \, dx.$

14. $\displaystyle\int \frac{dx}{\sec^4 x}.$

15. $\displaystyle\int_0^{\pi/4} \sin 5x \sin 3x \, dx.$

16. $\displaystyle\int_{-\pi/10}^{0} \sin 2x \cos 3x \, dx.$

17. $\displaystyle\int_0^1 \sin^2 \pi x \, dx.$

18. $\displaystyle\int_0^{\pi/3} \sin^3 3x \, dx.$

19. $\displaystyle\int_0^{\sqrt{\pi/2}} x \sec^2 (x^2) \, dx.$

20. $\displaystyle\int_0^{\pi/4} \frac{\sec^2 x \, dx}{\sqrt{1 + 3 \tan x}}.$

21. $\displaystyle\int \frac{\sin (\cot^{-1} x) \, dx}{1 + x^2}.$

22. $\displaystyle\int \frac{e^{\tan 3x} \, dx}{\cos^2 3x}.$

23. $\displaystyle\int e^x \sec (e^x) \, dx.$

24. $\displaystyle\int x \sec^2 3x \, dx.$

25. $\displaystyle\int \frac{e^{2x}}{\sqrt{e^{2x} + 1}} \, dx.$

26. $\displaystyle\int \frac{e^x}{\sqrt{e^{2x} + 1}} \, dx.$

27. $\displaystyle\int e^{3x} \sin 2x \, dx.$

28. $\displaystyle\int \ln (a^2 + x^2) \, dx.$

29. $\displaystyle\int_1^2 \sin^{-1} (x/2) \, dx.$

30. $\displaystyle\int (\ln x)^2 \, dx.$

31. $\displaystyle\int (\ln x)/x^2 \, dx.$

32. $\displaystyle\int x^3 e^{-x^2} \, dx.$

33. $\displaystyle\int \frac{x \, dx}{\sqrt{x^2 - 9}}.$

34. $\displaystyle\int_1^2 \frac{\sqrt{4x^2 - 1}}{x} \, dx.$

35. $\displaystyle\int_1^3 \frac{\sqrt{9 - x^2}}{x} \, dx.$

36. $\displaystyle\int_{-\sqrt{8}}^{-\sqrt{8/3}} \frac{ds}{s\sqrt{s^2 + 8}}.$

37. $\displaystyle\int \frac{x^2 \, dx}{\sqrt{2x + 3}}.$

38. $\displaystyle\int \frac{dx}{\sqrt{x^2 + 16}}.$

39. $\displaystyle\int \frac{dt}{\sqrt{3 - 4t - 4t^2}}.$

40. $\displaystyle\int_4^7 \frac{dx}{\sqrt{x^2 - 2x - 8}}.$

41. $\displaystyle\int \frac{dx}{x^2 \sqrt{a^2 - x^2}}.$

42. $\displaystyle\int \frac{x^3}{(x^2 + 4)^{1/3}} \, dx.$

43. $\displaystyle\int \sqrt{a^2 - x^2} \, dx.$

44. $\displaystyle\int x \sqrt{a^2 - x^2} \, dx.$

45. $\displaystyle\int \frac{x - 2}{\sqrt{4x - x^2}} \, dx.$

46. $\displaystyle\int_1^3 \frac{dx}{x^2 - 2x + 5}.$

47. $\displaystyle\int \frac{dx}{2x^2 + 3x + 1}.$

48. $\displaystyle\int \frac{dx}{(x^2 + 4)^2}.$

49. $\displaystyle\int \frac{x + 1}{x^3 + x^2 - 6x} \, dx.$

50. $\displaystyle\int \frac{x^3 + 1}{x - 2} \, dx.$

51. $\displaystyle\int \frac{x^2 - 1}{x^3 - 3x} \, dx.$

52. $\displaystyle\int \frac{x - 3}{x^3 - 1} \, dx.$

53. $\displaystyle\int \frac{2x^2 + 5}{x^4 - 1} \, dx.$

54. $\displaystyle\int \frac{x^4 - x^3 - x - 1}{x^3 - x^2} \, dx.$

55. $\displaystyle\int \frac{dx}{(x^2 + 4)(x - 3)}.$

56. $\displaystyle\int \frac{x \, dx}{(x + 1)^3}.$

57. $\displaystyle\int \frac{3x^2 + 12x + 2}{(x^2 + 4)^2} \, dx.$

58. $\displaystyle\int \frac{(4x + 2) \, dx}{x^4 + 2x^3 + x^2}.$

59. $\displaystyle\int \frac{x \, dx}{x^2 + 2x + 5}.$

60. $\displaystyle\int \frac{6x \, dx}{(x^2 + 9)^3}.$

61. $\displaystyle\int \frac{dx}{\sqrt{3 - 2x^2}}.$

62. $\displaystyle\int \frac{1 + t}{\sqrt{t}} \, dt.$

63. $\displaystyle\int \frac{\sqrt{t} \, dt}{1 + t}.$

64. $\displaystyle\int \frac{\sqrt{1 - x^2}}{x^2} \, dx.$

Exercises 65–70 relate to the optional Section 9.8. Evaluate the integrals.

65. $\displaystyle\int_0^1 \frac{x^{2/3}}{1 + x^{1/3}}\,dx.$

66. $\displaystyle\int \frac{dx}{x^{1/2} + x^{1/4}}.$

67. $\displaystyle\int \frac{dx}{1 - \tan x}.$

68. $\displaystyle\int \frac{dx}{3\cos x + 5}.$

69. $\displaystyle\int \frac{dx}{\sin x - \tan x}.$

70. $\displaystyle\int_0^{\pi/3} \frac{dx}{5\sec x - 3}.$

71. Use partial fractions to show that

$$\int \frac{dx}{x^2 - a^2} = \frac{1}{2a} \ln \left| \frac{x - a}{x + a} \right| + C \quad (a \neq 0).$$

72. Find the arc length of: (a) the parabola $y = x^2/2$ from $(0, 0)$ to $(2, 2)$. (b) the curve $y = \ln(\sec x)$ from $(0, 0)$ to $(\pi/4, \frac{1}{2}\ln 2)$.

73. Let R be the region bounded by the curve $y = 1/(4 + x^2)$ and the lines $x = 0$, $y = 0$, and $x = 2$. Find: (a) the area of R, (b) the volume of the solid obtained by revolving R about the x-axis, (c) the volume of the solid obtained by revolving R about the y-axis.

74. Derive the following reduction formulas for $a \neq 0$:

(a) $\displaystyle\int x^n e^{ax}\,dx = \frac{x^n e^{ax}}{a} - \frac{n}{a}\int x^{n-1}e^{ax}\,dx.$

(b) $\displaystyle\int x^n \sin ax\,dx =$

$$\frac{-x^n \cos ax}{a} + \frac{n}{a}\int x^{n-1}\cos ax\,dx.$$

$$\int x^n \cos ax\,dx =$$

$$\frac{x^n \sin ax}{a} - \frac{n}{a}\int x^{n-1}\sin ax\,dx.$$

(c) $\displaystyle\int \sin^n ax \cos^m ax\,dx = -\frac{\sin^{n-1} ax \cos^{m+1} ax}{a(m + n)}$

$$+ \frac{n - 1}{m + n}\int \sin^{n-2} ax \cos^m ax\,dx$$

$$= \frac{\sin^{n+1} ax \cos^{m-1} ax}{a(m + n)} + \frac{m - 1}{m + n}\int \sin^n ax \cos^{m-2} ax\,dx$$

75. Use Exercise 74 to evaluate the following integrals.

(a) $\displaystyle\int x^3 e^{2x}\,dx.$

(b) $\displaystyle\int_0^{\pi/10} x^2 \sin 5x\,dx$

(c) $\displaystyle\int \sin^2 x \cos^4 x\,dx.$

76. Evaluate the following integrals assuming that $a \neq 0$.

(a) $\displaystyle\int x^n \ln ax\,dx \quad (n \neq -1)$

(b) $\displaystyle\int \sec^n ax \tan ax\,dx \quad (n \geq 1).$

77. Find $\int (\sin^3 \theta / \cos^5 \theta)\,d\theta$ two ways: (a) letting $u = \cos \theta$ and (b) expressing the integrand in terms of $\sec \theta$ and $\tan \theta$. Show that your answers differ by a constant.

In Exercises 78–81, approximate the integral using the given value of n and (a) the trapezoidal rule, (b) Simpson's rule. Use a calculator and express the answer to four decimal places.

78. $\displaystyle\int_0^1 \sqrt{x}\,dx, \; n = 4.$

79. $\displaystyle\int_{-4}^2 e^{-x}\,dx, \; n = 6.$

80. $\displaystyle\int_0^4 \sinh x\,dx, \; n = 4.$

81. $\displaystyle\int_4^{5.2} \ln x\,dx, \; n = 6.$

In Exercises 82 and 83, use Simpson's rule with $n = 10$ to approximate the given integral. Use a calculator and express the answer to five decimal places.

82. $\displaystyle\int_0^2 \cos(\sinh x)\,dx.$

83. $\displaystyle\int_1^2 \sin(\ln x)\,dx.$

10

improper integrals; L'Hôpital's rule

10.1 IMPROPER INTEGRALS

In the definition of $\int_a^b f(x)\,dx$, it is assumed that the interval $[a, b]$ is *finite*. Moreover, in order for the limit

$$\lim_{\max \Delta x_k \to 0} \sum_{k=1}^{n} f(x_k^*)\,\Delta x_k = \int_a^b f(x)\,dx$$

to exist the function f must be *bounded* on the interval $[a, b]$. (See the remark following Theorem 5.7.3.) For example, the integral

$$\int_0^3 \frac{1}{(x-2)^{2/3}}\,dx$$

does not exist, since the integrand approaches $+\infty$ at the point $x = 2$ in the interval of integration.

In this section we will extend the concept of the definite integral to include:

(a) integrals with infinite intervals of integration;
(b) integrals in which the integrand becomes infinite within the interval of integration.

These are called **improper integrals.**

If f is continuous on the interval $[a, +\infty)$, then we define the improper integral $\int_a^{+\infty} f(x)\,dx$ as the limit:

$$\int_a^{+\infty} f(x)\,dx = \lim_{l \to +\infty} \int_a^l f(x)\,dx \tag{1}$$

537

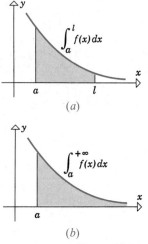

$$\int_a^l f(x)\,dx$$

(a)

$$\int_a^{+\infty} f(x)\,dx$$

(b)

Figure 10.1.1

If this limit exists, the improper integral is said to **converge,** and the value of the limit is the value assigned to the integral. If the limit does not exist, then the improper integral is said to **diverge,** in which case it is not assigned a value.

If f is nonnegative and continuous on $[a, +\infty)$, then (1) has an important geometric interpretation. For each value of $l > a$, the definite integral $\int_a^l f(x)\,dx$ represents the area under the curve $y = f(x)$ over the finite interval $[a, l]$ (Figure 10.1.1a). As we let $l \to +\infty$ this area tends toward the area under $y = f(x)$ over the entire interval $[a, +\infty)$ (Figure 10.1.1b). Thus $\int_a^{+\infty} f(x)\,dx$ can be regarded as the area under $y = f(x)$ over the interval $[a, +\infty)$.

▶ Example 1

Find

$$\int_1^{+\infty} \frac{dx}{x^2}$$

Solution. We begin by replacing the infinite upper limit with a finite upper limit l.

$$\int_1^l \frac{dx}{x^2} = -\frac{1}{x}\Big]_1^l = -\frac{1}{l} - (-1) = 1 - \frac{1}{l}$$

Thus

$$\int_1^{+\infty} \frac{dx}{x^2} = \lim_{l \to +\infty} \int_1^l \frac{dx}{x^2} = \lim_{l \to +\infty} \left(1 - \frac{1}{l}\right) = 1 \qquad \blacktriangleleft$$

▶ Example 2

Evaluate

$$\int_1^{+\infty} \frac{dx}{x}$$

Solution.

$$\int_1^{+\infty} \frac{dx}{x} = \lim_{l \to +\infty} \int_1^l \frac{dx}{x} = \lim_{l \to +\infty} [\ln |x|]_1^l = \lim_{l \to +\infty} \ln |l| = +\infty$$

Thus the integral diverges. $\qquad \blacktriangleleft$

Figure 10.1.2

It is worthwhile to reflect on the results of Examples 1 and 2. Why is it that $\int_1^{+\infty} dx/x^2$ converges while $\int_1^{+\infty} dx/x$ diverges when, in fact, the graphs of $1/x$ and $1/x^2$ look similar over the interval $[1, +\infty)$ (Figure 10.1.2)? The explanation is that $1/x^2$ approaches zero more rapidly than $1/x$ as $x \to +\infty$, so that area accumulates under the curve $y = 1/x$ more rapidly than under the curve $y = 1/x^2$. When we calculate

$$\lim_{l \to +\infty} \int_1^l \frac{dx}{x} \quad \text{and} \quad \lim_{l \to +\infty} \int_1^l \frac{dx}{x^2}$$

the difference is enough that the first limit is infinite while the second is finite.

If f is continuous on the interval $(-\infty, b]$, then we define the improper integral $\int_{-\infty}^b f(x)\, dx$ to be the limit,

$$\int_{-\infty}^b f(x)\, dx = \lim_{l \to -\infty} \int_l^b f(x)\, dx$$

As before, the improper integral is said to **converge** if the limit exists and **diverge** if it does not. If f is nonnegative, then the integral represents the area under $y = f(x)$ over the interval $(-\infty, b]$.

▶ Example 3

Evaluate

$$\int_{-\infty}^0 e^x\, dx$$

Solution. We begin by replacing the infinite lower limit with a finite lower limit l.

$$\int_l^0 e^x\, dx = e^x \Big]_l^0 = e^0 - e^l = 1 - e^l$$

Thus

$$\int_{-\infty}^0 e^x\, dx = \lim_{l \to -\infty} \int_l^0 e^x\, dx = \lim_{l \to -\infty} (1 - e^l) = 1 \qquad ◀$$

If the two improper integrals $\int_{-\infty}^0 f(x)\, dx$ and $\int_0^{+\infty} f(x)\, dx$ both converge, then we say that $\int_{-\infty}^{+\infty} f(x)\, dx$ **converges** and we define:

$$\int_{-\infty}^{+\infty} f(x)\, dx = \int_{-\infty}^0 f(x)\, dx + \int_0^{+\infty} f(x)\, dx \qquad (2)$$

If either integral on the right side of (2) diverges, then we say that $\int_{-\infty}^{+\infty} f(x)\,dx$ **diverges.** If f is nonnegative, then the integral $\int_{-\infty}^{+\infty} f(x)\,dx$ represents the area under $y = f(x)$ over the interval $(-\infty, +\infty)$.

▶ Example 4

Evaluate

$$\int_{-\infty}^{+\infty} \frac{dx}{1 + x^2}$$

Solution.

$$\int_{0}^{+\infty} \frac{dx}{1 + x^2} = \lim_{l \to +\infty} \int_{0}^{l} \frac{dx}{1 + x^2} = \lim_{l \to +\infty} \left[\tan^{-1} x \right]_{0}^{l} = \lim_{l \to +\infty} (\tan^{-1} l) = \frac{\pi}{2}$$

$$\int_{-\infty}^{0} \frac{dx}{1 + x^2} = \lim_{l \to -\infty} \int_{l}^{0} \frac{dx}{1 + x^2} = \lim_{l \to -\infty} \tan^{-1} x \Big]_{l}^{0} = \lim_{l \to -\infty} (-\tan^{-1} l) =$$

$$-\left(-\frac{\pi}{2}\right) = \frac{\pi}{2}$$

Thus

$$\int_{-\infty}^{+\infty} \frac{dx}{1 + x^2} = \int_{-\infty}^{0} \frac{dx}{1 + x^2} + \int_{0}^{+\infty} \frac{dx}{1 + x^2} = \frac{\pi}{2} + \frac{\pi}{2} = \pi \qquad ◀$$

REMARK. In (2), the decision to split the integral $\int_{-\infty}^{+\infty} f(x)\,dx$ at $x = 0$ is arbitrary. We can just as well make the split at any other point $x = c$ without affecting the convergence, divergence, or the value of the integral; that is,

$$\int_{-\infty}^{+\infty} f(x)\,dx = \int_{-\infty}^{c} f(x)\,dx + \int_{c}^{+\infty} f(x)\,dx$$

(We omit the proof.)

Next, we consider improper integrals in which the integrand approaches $+\infty$ or $-\infty$ somewhere in the interval of integration. We start with the cases where the integrand becomes infinite at one of the endpoints of integration.

If f is continuous on the interval $[a, b)$, but $f(x) \to +\infty$ or $f(x) \to -\infty$ as x approaches b from the left, then we define the improper integral $\int_{a}^{b} f(x)\,dx$ as the limit:

$$\int_{a}^{b} f(x)\,dx = \lim_{l \to b^-} \int_{a}^{l} f(x)\,dx \qquad (3)$$

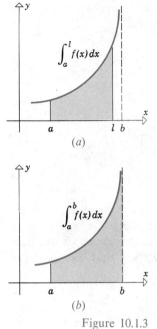

(a)

(b)

Figure 10.1.3

To visualize this definition geometrically, consider the case where f is non-negative on the interval $[a, b]$. For each number l satisfying $a \le l < b$, the integral $\int_a^l f(x)\,dx$ represents the area under $y = f(x)$ over the interval $[a, l]$ (Figure 10.1.3a). As we let l approach b from the left, these areas tend to fill out the entire area under $y = f(x)$ over the interval $[a, b)$ (Figure 10.1.3b). As before, the improper integral $\int_a^b f(x)\,dx$ is said to **converge** or **diverge** depending on whether the limit in (3) does or does not exist.

▶ Example 5

Evaluate

$$\int_0^1 \frac{dx}{\sqrt{1-x}}$$

Solution. The integral is improper because the integrand approaches $+\infty$ as x approaches the upper limit 1 from the left. From (3),

$$\int_0^1 \frac{dx}{\sqrt{1-x}} = \lim_{l \to 1^-} \int_0^l \frac{dx}{\sqrt{1-x}}$$
$$= \lim_{l \to 1^-} [-2\sqrt{1-x}]_0^l$$
$$= \lim_{l \to 1^-} [-2\sqrt{1-l} + 2]$$
$$= 2 \qquad \blacktriangleleft$$

If f is continuous on the interval $(a, b]$, but $f(x) \to +\infty$ or $f(x) \to -\infty$ as x approaches a from the right, then we define the improper integral $\int_a^b f(x)\,dx$ as:

$$\int_a^b f(x)\,dx = \lim_{l \to a^+} \int_l^b f(x)\,dx$$

The improper integral **converges** or **diverges** depending on whether the limit does or does not exist.

▶ Example 6

The integral

$$\int_1^2 \frac{dx}{1-x}$$

is improper because the integrand approaches $-\infty$ as x approaches the lower limit 1 from the right. The integral diverges because

$$\int_1^2 \frac{dx}{1-x} = \lim_{l \to 1^+} \int_l^2 \frac{dx}{1-x}$$

$$= \lim_{l \to 1^+} \left[-\ln|1-x|\right]_l^2$$

$$= \lim_{l \to 1^+} \left[-\ln|-1| + \ln|1-l|\right]$$

$$= \lim_{l \to 1^+} \ln|1-l|$$

$$= -\infty$$ ◀

Let f be continuous on the interval $[a, b]$ with the exception that at some point c satisfying $a < c < b$, $f(x)$ becomes infinite (tends to $+\infty$ or $-\infty$) as x approaches c from the left or the right. If the two improper integrals $\int_a^c f(x)\,dx$ and $\int_c^b f(x)\,dx$ both converge, then we say that the improper integral $\int_a^b f(x)\,dx$ **converges**, and we define:

$$\int_a^b f(x)\,dx = \int_a^c f(x)\,dx + \int_c^b f(x)\,dx \qquad (4)$$

If either integral on the right side of (4) diverges, then we say that $\int_a^b f(x)\,dx$ **diverges**.

▶ Example 7

Evaluate

$$\int_1^4 \frac{dx}{(x-2)^{2/3}}$$

Solution. The integrand approaches $+\infty$ as $x \to 2$, so we use (4) to write

$$\int_1^4 \frac{dx}{(x-2)^{2/3}} = \int_1^2 \frac{dx}{(x-2)^{2/3}} + \int_2^4 \frac{dx}{(x-2)^{2/3}} \qquad (5)$$

But

$$\int_1^2 \frac{dx}{(x-2)^{2/3}} = \lim_{l \to 2^-} \int_1^l \frac{dx}{(x-2)^{2/3}} = \lim_{l \to 2^-} [3(l-2)^{1/3} - 3(1-2)^{1/3}] = 3$$

$$\int_2^4 \frac{dx}{(x-2)^{2/3}} = \lim_{l \to 2^+} \int_l^4 \frac{dx}{(x-2)^{2/3}} = \lim_{l \to 2^+} [3(4-2)^{1/3} - 3(l-2)^{1/3}] = 3\sqrt[3]{2}$$

Thus from (5)

$$\int_1^4 \frac{dx}{(x-2)^{2/3}} = 3 + 3\sqrt[3]{2}$$ ◀

It is sometimes tempting to apply the fundamental theorem of calculus directly to an improper integral without taking the appropriate limits. To illustrate what can go wrong with this procedure, suppose we ignore the fact that the integral

$$\int_0^2 \frac{dx}{(x-1)^2}$$ (6)

is improper and write

$$\int_0^2 \frac{dx}{(x-1)^2} = -\frac{1}{x-1}\Big]_0^2 = -1 - (1) = -2$$

This result is clearly nonsense since the integrand is never negative and consequently the integral cannot be negative! To correctly evaluate (6) we should write

$$\int_0^2 \frac{dx}{(x-1)^2} = \int_0^1 \frac{dx}{(x-1)^2} + \int_1^2 \frac{dx}{(x-1)^2}$$ (7)

But

$$\int_0^1 \frac{dx}{(x-1)^2} = \lim_{l\to 1^-} \int_0^l \frac{dx}{(x-1)^2} = \lim_{l\to 1^-} \left[-\frac{1}{l-1} - 1 \right] = +\infty$$

so that (6) is a divergent integral.

▶ Exercise Set 10.1

In Exercises 1–20, evaluate the integrals that converge.

1. $\int_0^{+\infty} e^{-x} \, dx.$

2. $\int_1^{+\infty} \frac{dx}{x^3}.$

7. $\int_{-\infty}^0 \frac{dx}{(2x-1)^3}.$

8. $\int_{-\infty}^2 \frac{dx}{x^2+4}.$

3. $\int_1^{+\infty} \frac{dx}{\sqrt{x}}.$

4. $\int_{-1}^{+\infty} \frac{x}{1+x^2} \, dx.$

9. $\int_{-\infty}^{+\infty} x^3 \, dx.$

10. $\int_{-\infty}^{+\infty} \frac{x}{\sqrt{x^2+2}} \, dx.$

5. $\int_4^{+\infty} \frac{2}{x^2-1} \, dx.$

6. $\int_0^{+\infty} xe^{-x^2} \, dx.$

11. $\int_{-\infty}^{+\infty} \frac{x}{(x^2+3)^2} \, dx.$

12. $\int_0^8 \frac{dx}{\sqrt[3]{x}}.$

13. $\displaystyle\int_0^{\pi/2} \tan x\, dx.$

14. $\displaystyle\int_0^9 \frac{dx}{\sqrt{9-x}}.$

15. $\displaystyle\int_0^1 \frac{dx}{\sqrt{1-x^2}}.$

16. $\displaystyle\int_{-2}^2 \frac{dx}{x^2}.$

17. $\displaystyle\int_{-1}^8 x^{-1/3}\, dx.$

18. $\displaystyle\int_0^4 \frac{dx}{(x-2)^{2/3}}.$

19. $\displaystyle\int_0^{\pi/2} \frac{\cos\theta}{\sqrt{1-\sin\theta}}\, d\theta.$

20. $\displaystyle\int_1^{+\infty} \frac{dx}{x\sqrt{x^2-1}}.$

21. (a) It is possible for an improper integral to diverge without becoming infinite. Show that $\int_0^{+\infty} \cos x\, dx$ does this.

   (b) Evaluate $\int_0^{+\infty} e^{-x} \cos x\, dx.$

22. Show that $\displaystyle\int_1^{+\infty} \frac{dx}{x^p}$ converges if $p > 1$ and diverges if $p \le 1.$

23. Show that $\displaystyle\int_0^1 \frac{dx}{x^p}$ converges if $p < 1$ and diverges if $p \ge 1.$

24. Find the area of the region between the x-axis and the curve $y = 8/(x^2 - 4)$ for $x \ge 3.$

25. Find the area of the region bounded by $y = 0$, $x = 0$, $x = 4$, and $y = (1-x)^{-2}.$

26. (a) Show that
$$\int_{-1}^0 \frac{dx}{x} = -\infty \quad \text{and} \quad \int_0^1 \frac{dx}{x} = +\infty$$

   (b) Evaluate
$$\int_{-1}^1 \frac{dx}{x}$$

27. Let R be the region to the right of $x = 1$ that is bounded by the x-axis and the curve $y = 1/x$.

Show that the solid obtained by revolving R about the x-axis has a finite volume, but an infinite surface area. [It has been suggested that by filling this solid with paint and letting it seep through to the surface one could paint an infinite surface area with a finite amount of paint!]

28. For what values of p does $\int_0^{+\infty} e^{px}\, dx$ converge?

29. In electromagnetic theory, the magnetic potential at a point on the axis of a circular coil is given by
$$u = \frac{2\pi NIr}{k} \int_a^{+\infty} \frac{dx}{(r^2 + x^2)^{3/2}}$$
where N, I, r, k, and a are constants. Find u.

30. Sketch the region whose area is $\displaystyle\int_0^{+\infty} \frac{dx}{1+x^2}$, and use your sketch to show that
$$\int_0^{+\infty} \frac{dx}{1+x^2} = \int_0^1 \sqrt{\frac{1-y}{y}}\, dy$$

31. **(Satellite Problem).** In Exercise 13 of Section 6.6, we determined the work required to lift a 6000 lb satellite to a specified orbital position. The result in part (a) of that problem will be needed here.
   (a) Find a definite integral that represents the work required to lift a 6000 lb satellite to a position l miles above the earth's surface.
   (b) Find a definite integral that represents the work required to lift a 6000 lb satellite an "infinite distance" above the earth's surface. Evaluate the integral. [The result obtained here is sometimes called the work required to "escape" the earth's gravity.]

10.2 L'HÔPITAL'S RULE (INDETERMINATE FORMS OF TYPE 0/0)

In this section we develop an important new technique for finding limits of functions.

In each of the limits

$$\lim_{x\to 2} \frac{x^2-4}{x-2} \quad \text{and} \quad \lim_{x\to 0} \frac{\sin x}{x} \tag{1}$$

the numerator and denominator both approach zero. It is customary to describe such limits as ***indeterminate forms of type* 0/0.** The word "indeterminate" conveys the idea that the limit cannot be determined without additional work. In Example 8 of Section 2.5 we evaluated the first limit in (1) by canceling the common factor $x - 2$ from the numerator and denominator, and in Theorem 3.3.4 we resorted to a rather intricate geometric argument to obtain the second limit in (1). Because geometric arguments and the technique of canceling factors apply only to a limited range of problems, it is desirable to have a general method for handling indeterminate forms. This is provided by ***L'Hôpital's* rule,** which we now discuss.

10.2.1 THEOREM

L'Hôpital's Rule for Form 0/0

Let lim *stand for one of the limits* $\lim_{x \to a}$, $\lim_{x \to a^+}$, $\lim_{x \to a^-}$, $\lim_{x \to +\infty}$, *or* $\lim_{x \to -\infty}$, *and suppose that* $\lim f(x) = 0$ *and* $\lim g(x) = 0$. *If* $\lim [f'(x)/g'(x)]$ *has a finite value L, or if this limit is* $+\infty$ *or* $-\infty$, *then*

$$\lim \frac{f(x)}{g(x)} = \lim \frac{f'(x)}{g'(x)}$$

REMARK.   There are some hypotheses implicit in this theorem. For example, in the case where $x \to a$, the statement $\lim_{x \to a} [f'(x)/g'(x)] = L$ requires that f'/g' be defined in some open interval I containing a (except possibly at a). This implies that f and g are differentiable and $g'(x) \neq 0$ in I (except possibly at a). Similar hypotheses are implicit in the other cases.

In essence, L'Hôpital's rule enables us to replace one limit problem with another. Sometimes the new limit problem is simpler than the original. Before discussing the proof of L'Hôpital's rule, we will consider some examples.

---

*GUILLAUME FRANCOIS ANTOINE DE L'HÔPITAL (1661–1704) French mathematician. L'Hôpital, born to parents of the French high nobility, held the title of Marquis de Sainte-Mesme Comte d'Entrement. He showed mathematical talent quite early and at age 15 solved a difficult problem about cycloids posed by Pascal. As a young man he served briefly as a cavalry officer, but resigned because of nearsightedness. In his own time he gained fame as the author of the first textbook ever published on differential calculus, *Analyse des Infiniment Petits pour L'Intelligence des Lignes Courbes* (1696). L'Hôpital's rule appeared for the first time in that book. Actually, L'Hôpital's rule and most of the material in the calculus text were due to John Bernoulli who was L'Hôpital's teacher. L'Hôpital dropped his plans for a book on integral calculus when Leibniz informed him that he intended to write such a text.

L'Hôpital was apparently generous and personable and his many contacts with major mathematicians provided the vehicle for disseminating major discoveries in calculus throughout Europe.

▶ Example 1

In Theorem 3.3.4 we proved

$$\lim_{x \to 0} \frac{\sin x}{x} = 1$$

geometrically. We can verify this result using L'Hôpital's rule. Since

$$\lim_{x \to 0} \sin x = 0 \quad \text{and} \quad \lim_{x \to 0} x = 0$$

the limit

$$\lim_{x \to 0} \frac{\sin x}{x}$$

is an indeterminate form of type 0/0. Thus, by L'Hôpital's rule,

$$\lim_{x \to 0} \frac{\sin x}{x} = \lim_{x \to 0} \frac{\dfrac{d}{dx}[\sin x]}{\dfrac{d}{dx}[x]} = \lim_{x \to 0} \frac{\cos x}{1} = 1 \qquad ◀$$

REMARK. To be rigorous, the first equality in this calculation is not justified until the limit on the right is shown to exist. However, for simplicity we will usually arrange the computations as shown when applying L'Hôpital's rule.

▶ Example 2

Evaluate

$$\lim_{x \to \pi/2} \frac{1 - \sin x}{\cos x}$$

Solution. Since

$$\lim_{x \to \pi/2} (1 - \sin x) = \lim_{x \to \pi/2} \cos x = 0$$

the given limit is an indeterminate form of type 0/0. Thus by L'Hôpital's rule

$$\lim_{x \to \pi/2} \frac{1 - \sin x}{\cos x} = \lim_{x \to \pi/2} \frac{\dfrac{d}{dx}[1 - \sin x]}{\dfrac{d}{dx}[\cos x]} = \lim_{x \to \pi/2} \frac{-\cos x}{-\sin x} = \frac{0}{-1} = 0 \qquad ◀$$

▶ Example 3

Evaluate

$$\lim_{x \to 0} \frac{e^x - 1}{x^3}$$

Solution. Since

$$\lim_{x \to 0} (e^x - 1) = \lim_{x \to 0} x^3 = 0$$

the given limit is an indeterminate form of type 0/0. Thus by L'Hôpital's rule

$$\lim_{x \to 0} \frac{e^x - 1}{x^3} = \lim_{x \to 0} \frac{\dfrac{d}{dx}[e^x - 1]}{\dfrac{d}{dx}[x^3]} = \lim_{x \to 0} \frac{e^x}{3x^2} = +\infty \qquad ◀$$

As the following example shows, it is sometimes necessary to apply L'Hôpital's rule more than once in the same problem.

▶ Example 4

Evaluate

$$\lim_{x \to 0} \frac{1 - \cos x}{x^2}$$

Solution. Since

$$\lim_{x \to 0} (1 - \cos x) = \lim_{x \to 0} x^2 = 0$$

the given limit is an indeterminate form of type 0/0. Thus by L'Hôpital's rule

$$\lim_{x \to 0} \frac{1 - \cos x}{x^2} = \lim_{x \to 0} \frac{\sin x}{2x}$$

However, the new limit is also an indeterminate form of type 0/0, so we apply L'Hôpital's rule again. This yields

$$\lim_{x \to 0} \frac{1 - \cos x}{x^2} = \lim_{x \to 0} \frac{\sin x}{2x} = \lim_{x \to 0} \frac{\cos x}{2} = \frac{1}{2} \qquad ◀$$

WARNING. We warn the reader about two possible sources of errors. First, when applying L'Hôpital's rule to $\lim f(x)/g(x)$, the derivatives of $f(x)$ and

$g(x)$ are taken separately to yield the new limit, $\lim [f'(x)/g'(x)]$. Do not make the mistake of differentiating $f(x)/g(x)$ according to the quotient rule. Second, make sure that $\lim [f(x)/g(x)]$ is an indeterminate form; otherwise L'Hôpital's rule does not apply.

▶ Example 5

Evaluate

$$\lim_{x \to 0} \frac{e^x}{x^2}$$

Solution.

$$\lim_{x \to 0} e^x = 1 \qquad \text{and} \qquad \lim_{x \to 0} x^2 = 0$$

so the given problem is not an indeterminate form of type $0/0$. By inspection

$$\lim_{x \to 0} \frac{e^x}{x^2} = +\infty \qquad \qquad ◀$$

▶ Example 6

Evaluate

$$\lim_{x \to +\infty} \frac{x^{-4/3}}{\sin (1/x)}$$

Solution. Since

$$\lim_{x \to +\infty} x^{-4/3} = \lim_{x \to +\infty} \sin (1/x) = 0$$

the given limit is an indeterminate form of type $0/0$. Thus by L'Hôpital's rule

$$\lim_{x \to +\infty} \frac{x^{-4/3}}{\sin (1/x)} = \lim_{x \to +\infty} \frac{-\frac{4}{3}x^{-7/3}}{(-1/x^2)\cos (1/x)} = \lim_{x \to +\infty} \frac{\frac{4}{3}x^{-1/3}}{\cos (1/x)} = \frac{0}{1} = 0 \qquad ◀$$

▶ Example 7

Evaluate

$$\lim_{x \to 0^-} \frac{\tan x}{x^2}$$

Solution. Since

$$\lim_{x \to 0^-} \tan x = \lim_{x \to 0^-} x^2 = 0$$

the given limit is an indeterminate form of type 0/0. Thus, by L'Hôpital's rule

$$\lim_{x \to 0^-} \frac{\tan x}{x^2} = \lim_{x \to 0^-} \frac{\sec^2 x}{2x} = -\infty \qquad \blacktriangleleft$$

The proof of L'Hôpital's rule depends on the following result, called the **Extended Mean Value Theorem** or sometimes the **Cauchy* Mean Value Theorem.**

10.2.2 THEOREM
Extended Mean Value Theorem

Let the functions f and g be differentiable on (a, b) and continuous on [a, b]. If $g'(x) \neq 0$ for any x in (a, b) then there is at least one point c in (a, b) such that

$$\frac{f'(c)}{g'(c)} = \frac{f(b) - f(a)}{g(b) - g(a)} \qquad (2)$$

Proof. Observe first that $g(b) - g(a) \neq 0$, since otherwise it would follow from the mean value theorem (4.9.2) that $g'(x) = 0$ at some point x in (a, b),

---

*AUGUSTIN LOUIS CAUCHY (1789–1857) French mathematician. Cauchy's early education was acquired from his father, a barrister and master of the classics. Cauchy entered L'Ecole Polytechnique in 1805 to study engineering, but because of poor health, was advised to concentrate on mathematics. His major mathematical work began in 1811 with a series of brilliant solutions to some difficult outstanding problems.

In 1814 he wrote a treatise on integrals that was to become the basis for modern complex variable theory; in 1816 there followed a classic paper on wave propagation in liquids that won a prize from the French Academy; and in 1822 he wrote a paper that formed the basis of modern elasticity theory.

Cauchy's mathematical contributions for the next 35 years were brilliant and staggering in quantity, over 700 papers filling 26 modern volumes. Cauchy's work initiated the era of modern analysis. He brought to mathematics standards of precision and rigor undreamed of by Leibniz and Newton.

Cauchy's life was inextricably tied to the political upheavals of the time. A strong partisan of the Bourbons, he left his wife and children in 1830 to follow the Bourbon king Charles X into exile. For his loyalty he was made a baron by the ex-king. Cauchy eventually returned to France, but refused to accept a university position until the government waived its requirement that he take a loyalty oath.

It is difficult to get a clear picture of the man. Devoutly Catholic, he sponsored charitable work for unwed mothers, criminals, and relief for Ireland. Yet other aspects of his life cast him in an unfavorable light. The Norwegian mathematician Abel described him as, "mad, infinitely Catholic, and bigoted." Some writers praise his teaching, yet others say he rambled incoherently and, according to a report of the day, he once devoted an entire lecture to extracting the square root of seventeen to ten decimal places by a method well-known to his students. In any event, Cauchy is undeniably one of the greatest minds in the history of science.

contradicting our hypothesis. For convenience, introduce a new function F defined by

$$F(x) = [f(b) - f(a)]g(x) - [g(b) - g(a)]f(x) \qquad (3)$$

It follows from our assumptions about f and g, that F is continuous on $[a, b]$ and differentiable on (a, b). Moreover, $F(b) = F(a)$ (verify), so that the mean value theorem (4.9.2) implies that there is at least one point c in (a, b) where

$$F'(c) = 0$$

Thus from (3)

$$[f(b) - f(a)]g'(c) - [g(b) - g(a)]f'(c) = 0$$

or

$$\frac{f'(c)}{g'(c)} = \frac{f(b) - f(a)}{g(b) - g(a)}$$

Note that in the special case where $g(x) = x$, formula (2) reduces to

$$f'(c) = \frac{f(b) - f(a)}{b - a}$$

which is precisely the conclusion of the mean value theorem (4.9.2). Thus the extended mean value theorem is, in fact, an extension of the mean value theorem.

We will now prove one of the cases of L'Hôpital's rule.

Proof of Theorem 10.2.1 (L'Hôpital's Rule).
We will consider only the case where L is finite and lim is the two-sided limit $\lim\limits_{x \to a}$, with a finite.

By hypothesis

$$\lim_{x \to a} \frac{f'(x)}{g'(x)} = L \qquad (4)$$

As noted in the remark following the statement of Theorem 10.2.1, (4) implies that there is an interval $(a, r]$ extending to the right of a and an interval $[l, a)$ extending to the left of a on which $f'(x)$ and $g'(x)$ are defined and $g'(x) \neq 0$. For convenience, define two new functions F and G by

$$F(x) = \begin{cases} f(x), & x \neq a \\ 0, & x = a \end{cases} \qquad G(x) = \begin{cases} g(x), & x \neq a \\ 0, & x = a \end{cases}$$

Both F and G are continuous on $[l, r]$; they are continuous on the intervals $[l, a)$ and $(a, r]$ because on these intervals $F(x) = f(x)$, $G(x) = g(x)$, and f and g are continuous (since they are differentiable). The continuity of F and G at a follows from the fact that $\lim_{x \to a} f(x)/g(x)$ is an indeterminate form of type $0/0$, so

$$\lim_{x \to a} F(x) = \lim_{x \to a} f(x) = 0$$

$$\lim_{x \to a} G(x) = \lim_{x \to a} g(x) = 0$$

Thus, F and G satisfy the hypotheses of the extended mean value theorem (10.2.2) on $[l, a]$ and $[a, r]$. Moreover, the definitions of F and G imply that

$$F'(c) = f'(c) \quad \text{and} \quad G'(c) = g'(c) \tag{5}$$

at any point c (different from a) in one of these intervals. If we choose a point $x \neq a$ in one of the intervals $[l, a]$ or $[a, r]$ and apply the extended mean value theorem to $[a, x]$ (or $[x, a]$) we conclude that there is a number c between a and x such that

$$\frac{F(x) - F(a)}{G(x) - G(a)} = \frac{F'(c)}{G'(c)} \tag{6}$$

From (5) together with the fact that $F(a) = G(a) = 0$ and $F(x) = f(x)$, $G(x) = g(x)$, we can rewrite (6) as

$$\frac{f(x)}{g(x)} = \frac{f'(c)}{g'(c)}$$

Thus

$$\lim_{x \to a} \frac{f(x)}{g(x)} = \lim_{x \to a} \frac{f'(c)}{g'(c)} \tag{7}$$

Since c is between a and x it follows that $c \to a$ as $x \to a$. This fact together with (4) yields

$$\lim_{x \to a} \frac{f'(c)}{g'(c)} = \lim_{c \to a} \frac{f'(c)}{g'(c)} = L$$

Thus from (7)

$$\lim_{x \to a} \frac{f(x)}{g(x)} = L \quad ∎$$

▶ Exercise Set 10.2

In Exercises 1–20, find the limits.

1. $\lim\limits_{x \to 1} \dfrac{\ln x}{x - 1}$.

2. $\lim\limits_{x \to 0} \dfrac{\sin 2x}{\sin 5x}$.

3. $\lim\limits_{x \to 0} \dfrac{e^x - 1}{\sin x}$.

4. $\lim\limits_{x \to 3} \dfrac{x - 3}{3x^2 - 13x + 12}$.

5. $\lim\limits_{\theta \to 0} \dfrac{\tan \theta}{\theta}$.

6. $\lim\limits_{t \to 0} \dfrac{te^t}{1 - e^t}$.

7. $\lim\limits_{x \to 1} \dfrac{\ln x}{\tan \pi x}$.

8. $\lim\limits_{x \to c} \dfrac{x^{1/3} - c^{1/3}}{x - c}$.

9. $\lim\limits_{x \to \pi^+} \dfrac{\sin x}{x - \pi}$.

10. $\lim\limits_{x \to 0^+} \dfrac{\sin x}{x^2}$.

11. $\lim\limits_{x \to \frac{1}{2}\pi^-} \dfrac{\cos x}{\sqrt{\frac{1}{2}\pi - x}}$.

12. $\lim\limits_{x \to 0^+} \dfrac{1 - \cos x}{x^3}$.

13. $\lim\limits_{x \to 0} \dfrac{e^x + e^{-x} - 2}{1 - \cos 2x}$.

14. $\lim\limits_{x \to 0} \dfrac{2 \cosh x - 2}{1 - \cos 2x}$.

15. $\lim\limits_{x \to 0} \dfrac{x - \tan x}{\sin x - x}$.

16. $\lim\limits_{x \to \pi} \dfrac{\sin^2 x}{1 + \cos 3x}$.

17. $\lim\limits_{x \to +\infty} \dfrac{\frac{1}{2}\pi - \tan^{-1} x}{\ln\left(1 + \dfrac{1}{x^2}\right)}$.

18. $\lim\limits_{x \to 0^+} \dfrac{\tan x}{\tan 2x}$.

19. $\lim\limits_{x \to 0^+} \dfrac{\ln \cos x}{\ln \cos 3x}$.

20. $\lim\limits_{\theta \to 0} \dfrac{\sin^2 \theta - \sin (\theta^2)}{\theta^4}$.

21. (a) Find the error in the following calculation:

$$\lim_{x \to 1} \frac{x^3 - x^2 + x - 1}{x^3 - x^2} = \lim_{x \to 1} \frac{3x^2 - 2x + 1}{3x^2 - 2x}$$

$$= \lim_{x \to 1} \frac{6x - 2}{6x - 2} = 1$$

(b) Find the correct answer.

22. Find $\lim\limits_{x \to 1} \dfrac{x^4 - 4x^3 + 6x^2 - 4x + 1}{x^4 - 3x^3 + 3x^2 - x}$.

23. Find all values of k and l such that

$$\lim_{x \to 0} \frac{k + \cos lx}{x^2} = -4$$

24. Evaluate

$$\lim_{x \to 0} \frac{(2 + x) \ln (1 - x)}{(1 - e^x) \cos x}$$

25. Evaluate

$$\lim_{x \to +\infty} x \ln \left(\frac{x + 1}{x - 1}\right)$$

[*Hint:* Rewrite the problem as an indeterminate form of type 0/0.]

26. (a) Explain why L'Hôpital's rule does not apply to the problem

$$\lim_{x \to 0} \frac{x^2 \sin (1/x)}{\sin x}$$

(b) Find the limit.

10.3 OTHER INDETERMINATE FORMS
$$\left(\frac{\infty}{\infty},\ 0 \cdot \infty,\ 0^0,\ \infty^0,\ 1^\infty,\ \infty - \infty\right)$$

In this section we continue our study of indeterminate forms and introduce another version of L'Hôpital's rule.

We begin by introducing some new notation. In Figure 10.3.1 we have graphed portions of four functions that become "infinite at a." For the first function, $\lim\limits_{x \to a} f(x) = +\infty$ since both one-sided limits are $+\infty$; for the second

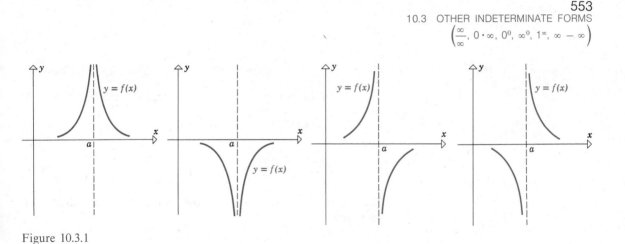

Figure 10.3.1

function, $\lim_{x \to a} f(x) = -\infty$ since both one-sided limits are $-\infty$; while for the last two functions, the one-sided limits are infinite, but opposite in sign. We will write

$$\lim_{x \to a} f(x) = \infty$$

to indicate that one of these four possibilities occurs. Also, we will write

$$\lim_{x \to +\infty} f(x) = \infty$$

to mean either $\lim_{x \to +\infty} f(x) = +\infty$ or $\lim_{x \to +\infty} f(x) = -\infty$. Similarly, for the symbol $\lim_{x \to -\infty} f(x) = \infty$.

An **indeterminate form of type** ∞/∞ is a limit $\lim f(x)/g(x)$

in which $\lim f(x) = \infty$ and $\lim g(x) = \infty$. Some examples are

$$\lim_{x \to 0^+} \frac{\ln x}{\csc x} \quad (\text{numerator} \to -\infty, \ \text{denominator} \to +\infty)$$

$$\lim_{x \to +\infty} \frac{x}{e^x} \quad (\text{numerator} \to +\infty, \ \text{denominator} \to +\infty)$$

$$\lim_{x \to 0} \frac{1 + \dfrac{1}{x}}{\cot x} \quad (\text{numerator} \to \infty, \ \text{denominator} \to \infty)$$

The following version of L'Hôpital's rule, usually proved in advanced courses, is applicable to problems like these.

10.3.1 THEOREM

L'Hôpital's Rule for Form $\frac{\infty}{\infty}$

Let $\lim$ *stand for one of the limits* $\lim\limits_{x \to a}$, $\lim\limits_{x \to a^+}$, $\lim\limits_{x \to a^-}$, $\lim\limits_{x \to +\infty}$, *or* $\lim\limits_{x \to -\infty}$, *and suppose that* $\lim f(x) = \infty$ *and* $\lim g(x) = \infty$. *If* $\lim \left[f'(x)/g'(x) \right]$ *has a finite value L, or if this limit is* $+\infty$ *or* $-\infty$, *then*

$$\lim \frac{f(x)}{g(x)} = \lim \frac{f'(x)}{g'(x)}$$

REMARK. The remark about hypotheses, which follows Theorem 10.2.1, is also applicable here.

► Example 1

Evaluate

$$\lim_{x \to +\infty} \frac{x}{e^x}$$

Solution.

$$\lim_{x \to +\infty} x = \lim_{x \to +\infty} e^x = +\infty$$

so that the given limit is an indeterminate form of type ∞/∞. Thus by L'Hôpital's rule

$$\lim_{x \to +\infty} \frac{x}{e^x} = \lim_{x \to +\infty} \frac{\dfrac{d}{dx}[x]}{\dfrac{d}{dx}[e^x]} = \lim_{x \to +\infty} \frac{1}{e^x} = 0 \qquad\qquad ◄$$

► Example 2

Evaluate

$$\lim_{x \to 0^+} \frac{\ln x}{\csc x}$$

Solution.

$$\lim_{x \to 0^+} \ln x = -\infty \qquad \text{and} \qquad \lim_{x \to 0^+} \csc x = +\infty$$

so that the given limit is an indeterminate form of type ∞/∞. Thus, by L'Hôpital's rule

$$\lim_{x\to0^+}\frac{\ln x}{\csc x}=\lim_{x\to0^+}\frac{1/x}{-\csc x\cot x}\qquad\qquad(1)$$

This last limit is again an indeterminate form of type ∞/∞. Moreover, any additional applications of L'Hôpital's rule will yield powers of $1/x$ in the numerator and expressions involving $\csc x$ and $\cot x$ in the denominator; thus repeated application of L'Hôpital's rule simply produces new indeterminate forms. We must try something else. The last limit in (1) can be rewritten

$$\lim_{x\to0^+}\left(-\frac{\sin x}{x}\tan x\right)=-\lim_{x\to0^+}\frac{\sin x}{x}\cdot\lim_{x\to0^+}\tan x=-(1)(0)=0$$

Thus

$$\lim_{x\to0^+}\frac{\ln x}{\csc x}=0 \qquad\qquad\qquad\blacktriangleleft$$

A limit, $\lim f(x)g(x)$, is called an **indeterminate form of type $0\cdot\infty$** if $\lim f(x)=0$ and $\lim g(x)=\infty$. Frequently, limit problems of this type can be converted to the form $0/0$ by writing

$$f(x)g(x)=\frac{f(x)}{1/g(x)}$$

or to the form ∞/∞ by writing

$$f(x)g(x)=\frac{g(x)}{1/f(x)}$$

They can then be treated by L'Hôpital's rule.

▶ Example 3

Evaluate

$$\lim_{x\to0^+}x\ln x$$

Solution.   Since

$$\lim_{x\to0^+}x=0 \qquad\text{and}\qquad \lim_{x\to0^+}\ln x=-\infty$$

the given problem is an indeterminate form of type $0 \cdot \infty$. We will convert the problem to the form ∞/∞ and apply L'Hôpital's rule:

$$\lim_{x \to 0^+} x \ln x = \lim_{x \to 0^+} \frac{\ln x}{1/x}$$

$$= \lim_{x \to 0^+} \frac{1/x}{-1/x^2}$$

$$= \lim_{x \to 0^+} (-x)$$

$$= 0 \qquad \blacktriangleleft$$

In this example we could have converted the problem to form $0/0$ by writing

$$x \ln x = \frac{x}{1/\ln x}$$

However, this is a less desirable first choice because of the relatively complicated derivative of $1/\ln x$.

WARNING. It is tempting to argue that an indeterminate form of type $0 \cdot \infty$ has value 0 since "zero times anything is zero." However this is fallacious since $0 \cdot \infty$ is not a product of numbers, but rather a statement about limits. The following example gives an indeterminate form of type $0 \cdot \infty$ whose value is not zero.

▶ Example 4

Evaluate

$$\lim_{x \to \pi/4} (1 - \tan x) \sec 2x$$

Solution. The given problem is an indeterminate form of type $0 \cdot \infty$. We convert it to type $0/0$ and apply L'Hôpital's rule as follows:

$$\lim_{x \to \pi/4} (1 - \tan x) \sec 2x = \lim_{x \to \pi/4} \frac{1 - \tan x}{1/\sec 2x}$$

$$= \lim_{x \to \pi/4} \frac{1 - \tan x}{\cos 2x}$$

$$= \lim_{x \to \pi/4} \frac{-\sec^2 x}{-2 \sin 2x}$$

$$= \frac{-2}{-2} = 1 \qquad \blacktriangleleft$$

Limits of the form

$$\lim_{x\to a} f(x)^{g(x)}$$

give rise to **indeterminate forms of the type 0^0, ∞^0, and 1^∞.** (At this stage the meaning of these symbols should be clear.) All three types are treated by first introducing a dependent variable

$$y = f(x)^{g(x)}$$

and then calculating

$$\lim_{x\to a} \ln y = \lim_{x\to a} (\ln f(x)^{g(x)}) = \lim_{x\to a} [g(x) \ln f(x)]$$

Once $\lim_{x\to a} \ln y$ is known it is a simple matter to determine $\lim_{x\to a} y = \lim_{x\to a} f(x)^{g(x)}$ as our examples will show.

▶ Example 5

Evaluate

$$\lim_{x\to 0} (1 + x)^{1/x}$$

Solution.   Since

$$\lim_{x\to 0} (1 + x) = 1 \qquad \text{and} \qquad \lim_{x\to 0} \frac{1}{x} = \infty$$

the given limit is an indeterminate form of type 1^∞. As discussed above, we introduce a dependent variable

$$y = (1 + x)^{1/x}$$

and take the natural logarithm of both sides

$$\ln y = \ln (1 + x)^{1/x} = \frac{1}{x} \ln (1 + x) = \frac{\ln (1 + x)}{x}$$

The limit

$$\lim_{x\to 0} \ln y = \lim_{x\to 0} \frac{\ln (1 + x)}{x}$$

is an indeterminate form of type $0/0$, so by L'Hôpital's rule,

$$\lim_{x \to 0} \ln y = \lim_{x \to 0} \frac{\ln (1 + x)}{x} = \lim_{x \to 0} \frac{1/(1 + x)}{1} = 1 \tag{2}$$

To complete the calculation, we use the known limit, $\lim_{x \to 0} \ln y = 1$, to find the unknown limit, $\lim_{x \to 0} y = \lim_{x \to 0} (1 + x)^{1/x}$. We argue as follows. As $x \to 0$ we have:

$\ln y \to 1 \qquad$ [from (2)]

so

$e^{\ln y} \to e^1 \qquad$ [since e^x is a continuous function]

and thus

$y \to e$

Therefore,

$$\lim_{x \to 0} (1 + x)^{1/x} = e \tag{3}$$

◄

The limit in the above example provides an effective way to approximate e. If we let $h = 1/x$, then it follows from (3) that

$$\lim_{h \to +\infty} \left(1 + \frac{1}{h}\right)^h = e \tag{4}$$

Thus $e \approx (1 + 1/h)^h$ is a good approximation for large values of h. Using a hand-held calculator we constructed Table 10.3.1 below. The table values compare favorably to the true value of e, which to five decimal places is $e \approx 2.71828$.

Table 10.3.1

h	1	10	100	1000	10,000	100,000
$\left(1 + \dfrac{1}{h}\right)^h$	2	2.5937	2.7048	2.7169	2.7181	2.7183

If a limit problem such as

$$\lim [f(x) - g(x)] \qquad \text{or} \qquad \lim [f(x) + g(x)]$$

559

10.3 OTHER INDETERMINATE FORMS
$$\left(\frac{\infty}{\infty}, 0 \cdot \infty, 0^0, \infty^0, 1^\infty, \infty - \infty\right)$$

leads to one of the expressions

$$(+\infty) - (+\infty), \quad (-\infty) - (-\infty)$$

$$(+\infty) + (-\infty), \quad (-\infty) + (+\infty)$$

then one term tends to make the expression large and the other tends to make it small, resulting in an indeterminate form. These are called *indeterminate forms of type* $\infty - \infty$. Such forms can sometimes be treated by combining the two terms into one and manipulating the result into one of the previous forms.

▶ Example 6

Evaluate

$$\lim_{x \to 0^+} \left(\frac{1}{x} - \frac{1}{\sin x}\right)$$

Solution. Since

$$\lim_{x \to 0^+} \frac{1}{x} = +\infty \quad \text{and} \quad \lim_{x \to 0^+} \frac{1}{\sin x} = +\infty$$

the given limit is an indeterminate form of type $\infty - \infty$. Combining terms yields

$$\lim_{x \to 0^+} \left(\frac{1}{x} - \frac{1}{\sin x}\right) = \lim_{x \to 0^+} \left(\frac{\sin x - x}{x \sin x}\right)$$

which is an indeterminate form of type $0/0$. Applying L'Hôpital's rule twice yields

$$\lim_{x \to 0^+} \left(\frac{\sin x - x}{x \sin x}\right) = \lim_{x \to 0^+} \frac{\cos x - 1}{\sin x + x \cos x}$$

$$= \lim_{x \to 0^+} \frac{-\sin x}{\cos x + \cos x - x \sin x}$$

$$= \frac{0}{2} = 0 \qquad ◀$$

▶ Example 7

Evaluate

$$\lim_{x \to 0^+} (\cot x - \ln x)$$

Solution. Since

$$\lim_{x \to 0^+} \cot x = +\infty \qquad \text{and} \qquad \lim_{x \to 0^+} \ln x = -\infty$$

we are dealing with a problem of the type $(+\infty) - (-\infty)$. This is not an indeterminate form. The first term tends to make the limit large and because of the subtraction, the second term also tends to make the limit large. Thus,

$$\lim_{x \to 0^+} (\cot x - \ln x) = +\infty \qquad \blacktriangleleft$$

► Exercise Set 10.3

In Exercises 1–24, find the limits.

1. $\lim\limits_{x \to +\infty} \dfrac{\ln x}{x}$.

2. $\lim\limits_{x \to +\infty} \dfrac{e^{3x}}{x^2}$.

3. $\lim\limits_{x \to 0^+} \dfrac{\cot x}{\ln x}$.

4. $\lim\limits_{x \to 0^+} \dfrac{1 - \ln x}{e^{1/x}}$.

5. $\lim\limits_{x \to +\infty} \dfrac{x \ln x}{x + \ln x}$.

6. $\lim\limits_{x \to +\infty} \dfrac{x^3 - 2x + 1}{4x^3 + 2}$.

7. $\lim\limits_{x \to +\infty} \dfrac{x^{100}}{e^x}$.

8. $\lim\limits_{x \to 0^+} \dfrac{\ln \sin x}{\ln \tan x}$.

9. $\lim\limits_{x \to +\infty} xe^{-x}$.

10. $\lim\limits_{x \to \pi^-} (x - \pi) \tan \tfrac{1}{2}x$.

11. $\lim\limits_{x \to +\infty} x \sin \dfrac{\pi}{x}$.

12. $\lim\limits_{x \to 0^+} \tan x \ln x$.

13. $\lim\limits_{x \to 0} (e^x + x)^{1/x}$.

14. $\lim\limits_{x \to 0^+} x^x$.

15. $\lim\limits_{x \to 0^+} x^{\sin x}$.

16. $\lim\limits_{x \to +\infty} (\ln x)^{1/x}$.

17. $\lim\limits_{x \to \frac{1}{2}\pi^-} (\tan x)^{\cos x}$.

18. $\lim\limits_{x \to 0^+} x^{1/\ln x}$.

19. $\lim\limits_{\theta \to 0} \left(\dfrac{1}{1 - \cos \theta} - \dfrac{2}{\sin^2 \theta} \right)$.

20. $\lim\limits_{x \to 0} \left(\dfrac{1}{x} - \dfrac{1}{e^x - 1} \right)$.

21. $\lim\limits_{x \to 0} (\cot x - \csc x)$.

22. $\lim\limits_{x \to +\infty} [\ln x - \ln (1 + x)]$.

23. $\lim\limits_{x \to 0^+} \dfrac{\cot x}{\cot 2x}$.

24. $\lim\limits_{x \to \frac{1}{2}\pi^-} \dfrac{4 \tan x}{1 + \sec x}$.

25. Show that for any positive integer n:

(a) $\lim\limits_{x \to +\infty} \dfrac{x^n}{e^x} = 0$ (b) $\lim\limits_{x \to +\infty} \dfrac{e^x}{x^n} = +\infty$.

26. Show that for any positive integer n:

(a) $\lim\limits_{x \to +\infty} \dfrac{\ln x}{x^n} = 0$ (b) $\lim\limits_{x \to +\infty} \dfrac{x^n}{\ln x} = +\infty$.

27. Limits of the type

$$\frac{0}{\infty}, \quad \frac{\infty}{0}, \quad 0^\infty, \quad \infty \cdot \infty, \quad +\infty + (+\infty),$$
$$+\infty - (-\infty), \quad -\infty + (-\infty), \quad -\infty - (+\infty)$$

are *not* indeterminate forms. Find the following limits by inspection.

(a) $\lim\limits_{x \to 0^+} \dfrac{x}{\ln x}$ (b) $\lim\limits_{x \to +\infty} \dfrac{x^3}{e^{-x}}$

(c) $\lim\limits_{x \to \frac{1}{2}\pi^-} (\cos x)^{\tan x}$.

(d) $\lim\limits_{x \to 0^+} (\ln x) \cot x$

(e) $\lim\limits_{x \to \frac{1}{2}\pi^-} \left(\dfrac{1}{\frac{1}{2}\pi - x} + \tan x \right)$.

(f) $\lim\limits_{x \to 0^+} \left(\dfrac{1}{x} - \ln x \right)$

(g) $\lim\limits_{x \to -\infty} (x + x^3)$ (h) $\lim\limits_{x \to +\infty} \left(\ln \left(\dfrac{1}{x} \right) - e^x \right)$.

28. Find $\lim\limits_{x \to +\infty} (e^x - x^2)$.

29. Sketch the graph of x^x, $x > 0$.

30. Sketch the graph of $(1/x)\tan x$, $-\pi/2 < x < \pi/2$.

31. Evaluate $\int_0^1 \ln x\, dx$ or else show that the integral diverges.

32. (a) Show that $\int_1^{+\infty} e^{t^2}\, dt = +\infty$.
[*Hint:* Don't try to carry out the integration. Instead, compare the sizes of $\int_1^l e^t\, dt$ and $\int_1^l e^{t^2}\, dt$.]

(b) Use L'Hôpital's rule to evaluate

$$\lim_{x \to +\infty} \frac{1}{x}\int_1^x e^{t^2}\, dt$$

▶ REVIEW EXERCISES

In Exercises 1–14, evaluate the integrals that converge.

1. $\displaystyle\int_{-\infty}^{+\infty} \frac{dx}{x^2 + 4}$.

2. $\displaystyle\int_4^6 \frac{dx}{4 - x}$.

3. $\displaystyle\int_0^1 \frac{x\, dx}{\sqrt{1 - x^2}}$.

4. $\displaystyle\int_0^5 \frac{dx}{\sqrt{25 - x^2}}$.

5. $\displaystyle\int_{-1}^1 \frac{dx}{\sqrt[3]{x^2}}$.

6. $\displaystyle\int_0^{\pi/2} \sec^2 x\, dx$.

7. $\displaystyle\int_0^{+\infty} xe^{-x^2}\, dx$.

8. $\displaystyle\int_{-\infty}^0 xe^x\, dx$.

9. $\displaystyle\int_0^{\pi/2} \tan x\, dx$.

10. $\displaystyle\int_0^{+\infty} \frac{dx}{x^5}$.

11. $\displaystyle\int_e^{+\infty} \frac{dx}{x(\ln x)^2}$.

12. $\displaystyle\int_0^1 \sqrt{x}\, \ln x\, dx$.

13. $\displaystyle\int_0^{+\infty} \frac{dx}{x^2 + 2x + 2}$.

14. $\displaystyle\int_0^4 \frac{e^{-\sqrt{x}}}{\sqrt{x}}\, dx$.

15. For what values of n does $\displaystyle\int_0^1 x^n \ln x\, dx$ converge? What is its value?

In Exercises 16–31, find the limit.

16. $\displaystyle\lim_{x \to 1} \frac{\ln x}{x - 1}$.

17. $\displaystyle\lim_{x \to 0} \frac{xe^{3x} - x}{1 - \cos 2x}$.

18. $\displaystyle\lim_{x \to +\infty} \frac{\ln(\ln x)}{\sqrt{x}}$.

19. $\displaystyle\lim_{x \to 0^+} x^2 e^{1/x}$.

20. $\displaystyle\lim_{x \to +\infty} (\sqrt{x^2 + x} - x)$.

21. $\displaystyle\lim_{x \to 0^-} x^2 e^{1/x}$.

22. $\displaystyle\lim_{x \to 0^-} (1 - x)^{2/x}$.

23. $\displaystyle\lim_{\theta \to 0} (\csc\theta - \cot\theta)$.

24. $\displaystyle\lim_{x \to 0} \frac{x - \tan^{-1} x}{x^4}$.

25. $\displaystyle\lim_{x \to 2} \frac{x - 1 - e^{x-2}}{1 - \cos 2\pi x}$.

26. $\displaystyle\lim_{x \to 0} \frac{9^x - 3^x}{x}$.

27. $\displaystyle\lim_{x \to 0} \frac{\int_0^x \sin t^2\, dt}{\sin x^2}$.

28. $\displaystyle\lim_{x \to +\infty} x^{1/x}$.

29. $\displaystyle\lim_{x \to +\infty} \frac{(\ln x)^3}{x}$.

30. $\displaystyle\lim_{x \to +\infty} \left(\frac{x}{x - 3}\right)^x$.

31. $\displaystyle\lim_{x \to 0^+} (1 + x)^{\ln x}$.

32. Find the area of the region in the first quadrant between the curves $y = e^{-x}$ and $y = (e^{-x})^2$.

In Exercises 33–34, find:
(a) the area of the region R;
(b) the volume obtained by revolving R about the x-axis.

33. The region R bounded between the x-axis and the curve $y = x^{-2/3}$, $x \geq 8$.

34. The region R bounded between the x-axis and the curve $y = x^{-1/3}$, $0 \leq x \leq 1$.

35. Find the total arc length of the graph of $f(x) = \sqrt{x - x^2} - \sin^{-1}\sqrt{x}$. [*Hint:* First find the domain of f.]

11 infinite series

11.1 SEQUENCES

In everyday language, we use the term "sequence" to suggest a succession of objects or events given in a specified order. *Informally* speaking, the term "sequence" in mathematics is used to describe an unending succession of numbers. Some possibilities are

$$1, 2, 3, 4, \ldots$$
$$2, 4, 6, 8, \ldots$$
$$1, \tfrac{1}{2}, \tfrac{1}{3}, \tfrac{1}{4}, \ldots$$
$$1, -1, 1, -1, \ldots$$

In each case, the dots are used to suggest that the sequence continues indefinitely, following the obvious pattern. The numbers in a sequence are called the *terms* of the sequence. The terms may be described according to the positions they occupy. Thus a sequence has a *first term,* a *second term,* a *third term,* and so forth. Because a sequence continues indefinitely, there is no last term.

The most common way to specify a sequence is to give a formula for the terms. To illustrate the idea, we have listed in the table below the terms in the sequence $2, 4, 6, 8, \ldots$ together with their term numbers:

term number	1	2	3	4	...
term	2	4	6	8	...

There is a clear relationship here; each term is twice its term number. Thus for each positive integer n the nth term in the sequence $2, 4, 6, 8, \ldots$ is given by the formula $2n$. This is denoted by writing

$$2, 4, 6, 8, \ldots, 2n, \ldots$$

or more compactly in **bracket notation** as

$$\{2n\}_{n=1}^{+\infty}$$

From the bracket notation, the terms in the sequence can be generated by successively substituting the integer values $n = 1, 2, 3, \ldots$ into the formula $2n$.

▶ Example 1

List the first five terms of the sequence $\{2^n\}_{n=1}^{+\infty}$.

Solution. Substituting $n = 1, 2, 3, 4, 5$ into the formula 2^n yields

$$2^1, 2^2, 2^3, 2^4, 2^5, \ldots$$

or, equivalently,

$$2, 4, 8, 16, 32, \ldots \qquad\qquad\qquad ◀$$

▶ Example 2

Express the following sequences in bracket notation

(a) $\dfrac{1}{2}, \dfrac{2}{3}, \dfrac{3}{4}, \dfrac{4}{5}, \ldots$

(b) $\dfrac{1}{2}, \dfrac{1}{4}, \dfrac{1}{8}, \dfrac{1}{16}, \ldots$

(c) $1, -1, 1, -1, \ldots$

(d) $\dfrac{1}{2}, -\dfrac{2}{3}, \dfrac{3}{4}, -\dfrac{4}{5}, \ldots$

(e) $1, 3, 5, 7, \ldots$

Solution.

(a) Begin by comparing terms and term numbers:

term number	1	2	3	4	...
term	$\dfrac{1}{2}$	$\dfrac{2}{3}$	$\dfrac{3}{4}$	$\dfrac{4}{5}$	...

In each term, the numerator is the same as the term number, and the denominator is one greater than the term number. Thus the nth term is $n/(n+1)$ and the sequence may be written

$$\left\{\frac{n}{n+1}\right\}_{n=1}^{+\infty}$$

(b)   Observe that the sequence can be rewritten as

$$\frac{1}{2}, \frac{1}{2^2}, \frac{1}{2^3}, \frac{1}{2^4}, \cdots$$

and construct a table comparing terms and term numbers:

term number	1	2	3	4	...
term	$\frac{1}{2}$	$\frac{1}{2^2}$	$\frac{1}{2^3}$	$\frac{1}{2^4}$	...

From the table we see that the nth term is $1/2^n$, so the sequence can be written as

$$\left\{\frac{1}{2^n}\right\}_{n=1}^{+\infty}$$

(c)   Observe first that $(-1)^r$ is either 1 or -1 according to whether r is an even integer or an odd integer. In the sequence $1, -1, 1, -1, \ldots$ the odd-numbered terms are 1's and the even-numbered terms are -1's. Thus a formula for the nth term can be obtained by raising -1 to a power that will be even when n is odd and odd when n is even. This is accomplished by the formula $(-1)^{n+1}$, so that the sequence can be written as

$$\{(-1)^{n+1}\}_{n=1}^{+\infty}$$

(d)   Combining the results in parts (a) and (c), we may write this sequence as

$$\left\{(-1)^{n+1}\frac{n}{n+1}\right\}_{n=1}^{+\infty}$$

(e)   Begin by comparing terms and term numbers:

term number	1	2	3	4	...
term	1	3	5	7	...

From the table we see that each term is one less than twice the term number. Thus the nth term is $2n - 1$ and the sequence can be written as

$$\{2n - 1\}_{n=1}^{+\infty}$$ ◀

Frequently we shall want to write down a sequence without specifying the numerical values of the terms. We do this by writing

$$a_1, a_2, \ldots, a_n, \ldots$$

or in bracket notation

$$\{a_n\}_{n=1}^{+\infty}$$

or sometimes simply

$$\{a_n\}$$

(There is nothing special about the letter a; any other letter may be used.)

In the beginning of this section we stated informally that a sequence is an unending succession of numbers. However, this is not a satisfactory mathematical definition since the word "succession" is itself an undefined term. The time has come to formally define the term "sequence." When we write a sequence such as

$$2, 4, 6, 8, \ldots, 2n, \ldots$$

in bracket notation

$$\{2n\}_{n=1}^{+\infty} \tag{1}$$

we are specifying a rule that tells us how to associate a numerical value, namely $2n$, with each positive integer n. Stated another way, $\{2n\}_{n=1}^{+\infty}$ may be regarded as a formula for a function whose independent variable n ranges over the positive integers. Indeed, we could rewrite (1) in functional notation as

$$f(n) = 2n, \qquad n = 1, 2, 3, \ldots$$

From this point of view, the notation

$$2, 4, 6, 8, \ldots, 2n, \ldots$$

represents a listing of the function values

$$f(1), f(2), f(3), \ldots, f(n), \ldots$$

This suggests the following definition.

11.1.1 DEFINITION A *sequence* (or *infinite sequence*) is a function whose domain is the set of positive integers.

Because sequences are functions, we may inquire about the graph of a sequence. For example, the graph of the sequence

$$\left\{\frac{1}{n}\right\}_{n=1}^{+\infty}$$

is the graph of the equation

$$y = \frac{1}{n}, \qquad n = 1, 2, 3, \dots$$

Because the right side of this equation is defined only for positive integer values of n, the graph consists of a succession of isolated points (Figure 11.1.1a). This is in marked distinction to the graph of

$$y = \frac{1}{x}, \qquad x \geq 1$$

which is a continuous curve (Figure 11.1.1b).

Figure 11.1.1   (a)   (b)

In Figure 11.1.2 we have sketched the graphs of four sequences:

$$\{n+1\}_{n=1}^{+\infty}, \qquad \{(-1)^{n+1}\}_{n=1}^{+\infty}, \qquad \left\{\frac{n}{n+1}\right\}_{n=1}^{+\infty}, \qquad \left\{1+\left(-\frac{1}{2}\right)^n\right\}_{n=1}^{+\infty}$$

Each of these sequences behaves differently as n gets larger and larger. In the sequence $\{n+1\}$, the terms grow without bound; in the sequence $\{(-1)^{n+1}\}$ the terms oscillate between 1 and -1; in the sequence $\left\{\frac{n}{n+1}\right\}$ the terms increase toward a "limit" of 1; and finally, in the sequence $\{1 + (-\frac{1}{2})^n\}$ the terms also tend toward a "limit" of 1, but do so in an oscillatory fashion.

Let us try to make the term "limit" more precise. To say that a sequence $\{a_n\}_{n=1}^{+\infty}$ approaches a limit L as n gets large is intended to mean that eventually the terms in the sequence become arbitrarily close to the number L. Thus, if we choose *any* positive number ϵ, the terms in the sequence will

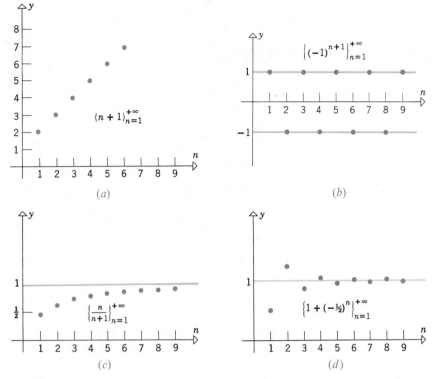

Figure 11.1.2

eventually be within ϵ units of L. Geometrically, this means that if we sketch the lines $y = L + \epsilon$ and $y = L - \epsilon$, the terms in the sequence will eventually be trapped within the band between these lines, and thus be within ϵ units of L (Figure 11.1.3).

Figure 11.1.3

The following definition expresses this idea precisely.

11.1.2   DEFINITION   A sequence $\{a_n\}_{n=1}^{+\infty}$ is said to have the **limit** L if given any $\epsilon > 0$ there is a positive integer N such that $|a_n - L| < \epsilon$ when $n \geq N$.

If a sequence $\{a_n\}_{n=1}^{+\infty}$ has a limit L, we say that the sequence **converges** to L and write

$$\lim_{n \to +\infty} a_n = L$$

A sequence that does not have a finite limit is said to **diverge.**

▶ Example 3

Figure 11.1.2 suggests that the sequences $\{n + 1\}_{n=1}^{+\infty}$ and $\{(-1)^{n+1}\}_{n=1}^{+\infty}$ diverge, while the sequences $\left\{\dfrac{n}{n+1}\right\}_{n=1}^{+\infty}$ and $\left\{1 + \left(-\dfrac{1}{2}\right)^n\right\}_{n=1}^{+\infty}$ converge to 1; that is,

$$\lim_{n \to +\infty} \frac{n}{n+1} = 1 \quad \text{and} \quad \lim_{n \to +\infty} \left(1 + \left(-\frac{1}{2}\right)^n\right) = 1 \qquad ◀$$

Many familiar properties of limits apply to limits of sequences.

11.1.3 THEOREM *Suppose that the sequences $\{a_n\}$ and $\{b_n\}$ converge to limits L_1 and L_2, respectively, and c is a constant. Then*

(a) $\displaystyle \lim_{n \to +\infty} c = c$

(b) $\displaystyle \lim_{n \to +\infty} ca_n = c \lim_{n \to +\infty} a_n = cL_1$

(c) $\displaystyle \lim_{n \to +\infty} (a_n + b_n) = \lim_{n \to +\infty} a_n + \lim_{n \to +\infty} b_n = L_1 + L_2$

(d) $\displaystyle \lim_{n \to +\infty} (a_n - b_n) = \lim_{n \to +\infty} a_n - \lim_{n \to +\infty} b_n = L_1 - L_2$

(e) $\displaystyle \lim_{n \to +\infty} (a_n b_n) = \lim_{n \to +\infty} a_n \cdot \lim_{n \to +\infty} b_n = L_1 L_2$

(f) $\displaystyle \lim_{n \to +\infty} \left(\frac{a_n}{b_n}\right) = \frac{\displaystyle\lim_{n \to +\infty} a_n}{\displaystyle\lim_{n \to +\infty} b_n} = \frac{L_1}{L_2}$ (if $L_2 \neq 0$)

(We omit the proof.)

▶ Example 4

In each part, determine if the given sequence converges or diverges. If it converges, find the limit.

(a) $\left\{\dfrac{n}{2n+1}\right\}_{n=1}^{+\infty}$ (b) $\left\{(-1)^{n+1} \dfrac{n}{2n+1}\right\}_{n=1}^{+\infty}$

(c) $\left\{(-1)^{n+1} \dfrac{1}{n}\right\}_{n=1}^{+\infty}$ (d) $\{8 - 2n\}_{n=1}^{+\infty}$ (e) $\left\{\dfrac{n}{e^n}\right\}_{n=1}^{+\infty}$

Solution.

(a) Dividing numerator and denominator by n yields

$$\lim_{n \to +\infty} \frac{n}{2n + 1} = \lim_{n \to +\infty} \frac{1}{\left(2 + \dfrac{1}{n}\right)} = \frac{\displaystyle\lim_{n \to +\infty} 1}{\displaystyle\lim_{n \to +\infty}\left(2 + \dfrac{1}{n}\right)} = \frac{\displaystyle\lim_{n \to +\infty} 1}{\displaystyle\lim_{n \to +\infty} 2 + \lim_{n \to +\infty} \dfrac{1}{n}}.$$

$$= \frac{1}{2 + 0} = \frac{1}{2}$$

Thus $\left\{\dfrac{n}{2n + 1}\right\}_{n=1}^{+\infty}$ converges to $\dfrac{1}{2}$.

(b) From part (a),

$$\lim_{n \to +\infty} \frac{n}{2n + 1} = \frac{1}{2}$$

Thus, since $(-1)^{n+1}$ oscillates between $+1$ and -1, the product $(-1)^{n+1} \dfrac{n}{2n + 1}$ oscillates between positive and negative values, with the odd-numbered terms approaching $\frac{1}{2}$ and the even-numbered terms approaching $-\frac{1}{2}$. Therefore the sequence $\left\{(-1)^{n+1} \dfrac{n}{2n + 1}\right\}_{n=1}^{+\infty}$ approaches no limit—it diverges.

(c) Since $\lim_{n \to +\infty} 1/n = 0$, the product $(-1)^{n+1} (1/n)$ oscillates between positive and negative values, with the odd-numbered terms approaching 0 through positive values and the even-numbered terms approaching 0 through negative values. Thus

$$\lim_{n \to +\infty} (-1)^{n+1} \frac{1}{n} = 0$$

so that the sequence converges to 0.

(d) $\lim_{n \to +\infty} (8 - 2n) = -\infty$ so the sequence $\{8 - 2n\}_{n=1}^{+\infty}$ diverges.

(e) We want to find $\lim_{n \to +\infty} n/e^n$, which is an indeterminate form of type ∞/∞. Unfortunately, we cannot apply L'Hôpital's rule directly since e^n and n are not differentiable functions (n assumes only integer values). However, we can apply L'Hôpital's rule to the related problem $\lim_{x \to +\infty} x/e^x$ to obtain

$$\lim_{x \to +\infty} \frac{x}{e^x} = \lim_{x \to +\infty} \frac{1}{e^x} = 0$$

We conclude from this that $\lim\limits_{n\to+\infty} n/e^n = 0$ since the values of n/e^n and x/e^x are the same when x is a positive integer. ◄

▶ Example 5

Show

$$\lim_{n\to+\infty} \sqrt[n]{n} = 1$$

Solution. With the aid of L'Hôpital's rule, $\lim\limits_{n\to+\infty} \dfrac{1}{n} \ln n = 0$. Thus

$$\lim_{n\to+\infty} \sqrt[n]{n} = \lim_{n\to+\infty} n^{1/n} = \lim_{n\to+\infty} e^{(1/n)\ln n} = e^0 = 1 \qquad ◄$$

▶ Exercise Set 11.1

In Exercises 1–14, show the first five terms of the sequence, determine if the sequence converges, and if so find the limit. (When writing out the terms of the sequence, you need not find numerical values; leave the terms in the first form you obtain.)

1. $\left\{\dfrac{n}{n+2}\right\}_{n=1}^{+\infty}$.

2. $\left\{\dfrac{n^2}{2n+1}\right\}_{n=1}^{+\infty}$.

3. $\{2\}_{n=1}^{+\infty}$.

4. $\left\{\ln\left(\dfrac{1}{n}\right)\right\}_{n=1}^{+\infty}$.

5. $\left\{\dfrac{\ln n}{n}\right\}_{n=1}^{+\infty}$.

6. $\left\{n \sin\dfrac{\pi}{n}\right\}_{n=1}^{+\infty}$.

7. $\{1 + (-1)^n\}_{n=1}^{+\infty}$.

8. $\left\{\dfrac{(-1)^{n+1}}{n^2}\right\}_{n=1}^{+\infty}$.

9. $\left\{(-1)^n \dfrac{2n^3}{n^3+1}\right\}_{n=1}^{+\infty}$.

10. $\left\{\dfrac{n}{2^n}\right\}_{n=1}^{+\infty}$.

11. $\left\{\dfrac{(n+1)(n+2)}{2n^2}\right\}_{n=1}^{+\infty}$.

12. $\left\{\dfrac{\pi^n}{4^n}\right\}_{n=1}^{+\infty}$.

13. $\{n^{1/n}\}_{n=1}^{+\infty}$.

14. $\left\{\left(1 - \dfrac{2}{n}\right)^n\right\}_{n=1}^{+\infty}$.

In Exercises 15–22, express the sequence in the notation $\{a_n\}_{n=1}^{+\infty}$, determine if the sequence converges, and if so find its limit.

15. $\dfrac{1}{2}, \dfrac{3}{4}, \dfrac{5}{6}, \dfrac{7}{8}, \ldots$

16. $0, \dfrac{1}{2^2}, \dfrac{2}{3^2}, \dfrac{3}{4^2}, \ldots$

17. $\dfrac{1}{3}, \dfrac{1}{9}, \dfrac{1}{27}, \dfrac{1}{81}, \ldots$

18. $-1, 2, -3, 4, -5, \ldots$

19. $\left(1 - \dfrac{1}{2}\right), \left(\dfrac{1}{2} - \dfrac{1}{3}\right), \left(\dfrac{1}{3} - \dfrac{1}{4}\right), \left(\dfrac{1}{4} - \dfrac{1}{5}\right), \ldots$

20. $3, \dfrac{3}{2}, \dfrac{3}{2^2}, \dfrac{3}{2^3}, \ldots$

21. $(\sqrt{2} - \sqrt{3}), (\sqrt{3} - \sqrt{4}), (\sqrt{4} - \sqrt{5}), \ldots$

22. $\dfrac{1}{3^5}, -\dfrac{1}{3^6}, \dfrac{1}{3^7}, -\dfrac{1}{3^8}, \ldots$

23. (a) Let $\{a_n\}$ be a sequence for which $a_1 = 3$ and $a_n = 2a_{n-1}$ when $n \geq 2$. Find the first eight terms.

(b) Let $\{a_n\}$ be a sequence for which $a_1 = 1$, $a_2 = 1$, and $a_n = a_{n-1} + a_{n-2}$ when $n \geq 3$. Find the first eight terms.

24. The nth term a_n of the sequence $1, 2, 1, 4, 1, 6, \ldots$ is best written in the form

$$a_n = \begin{cases} 1, & \text{if } n \text{ is odd} \\ n, & \text{if } n \text{ is even} \end{cases}$$

since it would be tedious to find one formula applicable to all terms. By considering even and odd terms separately, find a formula for the nth term of the given sequence.

(a) $1, \dfrac{1}{2^2}, 3, \dfrac{1}{2^4}, 5, \dfrac{1}{2^6}, \ldots$

(b) $1, \dfrac{1}{3}, \dfrac{1}{3}, \dfrac{1}{5}, \dfrac{1}{5}, \dfrac{1}{7}, \dfrac{1}{7}, \dfrac{1}{9}, \dfrac{1}{9}, \ldots$

25. Consider the sequence $\{a_n\}_{n=1}^{+\infty}$ where

$$a_n = \frac{1}{n^2} + \frac{2}{n^2} + \cdots + \frac{n}{n^2}$$

(a) Write out the first four terms of the sequence.
(b) Find the limit of the sequence.
[*Hint:* Sum up the terms in the formula for a_n.]

26. Follow the directions in Exercise 25 with

$$a_n = \frac{1^2}{n^3} + \frac{2^2}{n^3} + \cdots + \frac{n^2}{n^3}$$

27. If we accept the fact that the sequence $\{1/n\}_{n=1}^{+\infty}$ converges to the limit $L = 0$, then according to Definition 11.1.2, for every $\epsilon > 0$, there exists an integer N such that $|a_n - L| = |(1/n) - 0| < \epsilon$ when $n \ge N$. In each part, find the smallest possible value of N for the given value of ϵ.
(a) $\epsilon = 0.5$ (b) $\epsilon = 0.1$ (c) $\epsilon = 0.001$

28. If we accept the fact that the sequence $\left\{\dfrac{n}{n+1}\right\}_{n=1}^{+\infty}$

converges to the limit $L = 1$, then according to Definition 11.1.2, for every $\epsilon > 0$, there exists an integer N such that $|a_n - L| = \left|\dfrac{n}{n+1} - 1\right| < \epsilon$ when $n \ge N$. In each part, find the smallest value of N for the given value of ϵ.
(a) $\epsilon = 0.25$ (b) $\epsilon = 0.1$ (c) $\epsilon = 0.001$

29. Prove:

(a) The sequence $\left\{\dfrac{1}{n}\right\}_{n=1}^{+\infty}$ converges to 0.

(b) The sequence $\left\{\dfrac{n}{n+1}\right\}_{n=1}^{+\infty}$ converges to 1.

30. Consider the sequence $\{a_n\}_{n=1}^{+\infty}$ whose nth term is

$$a_n = \sum_{k=0}^{n-1} \frac{1}{1 + \dfrac{k}{n}} \cdot \frac{1}{n}$$

Show that $\lim\limits_{n \to +\infty} a_n = \ln 2$. [*Hint:* Interpret $\lim\limits_{n \to +\infty} a_n$ as a definite integral.]

31. (a) Show that a polygon with n equal sides inscribed in a circle of radius r, has perimeter $P_n = 2rn \sin(\pi/n)$.
(b) By finding the limit of the sequence $\{P_n\}_{n=1}^{+\infty}$, show that the perimeters approach the circumference of the circle as n increases.

11.2 MONOTONE SEQUENCES

Sometimes the critical information about a sequence is whether it converges or not, with the limit being of little or no importance. In this section we discuss results that are used to study convergence of sequences.

We begin with some terminology.

11.2.1 DEFINITION   A sequence $\{a_n\}$ is called

> ***increasing*** if $a_1 < a_2 < a_3 < \cdots < a_n < \cdots$
> ***nondecreasing*** if $a_1 \le a_2 \le a_3 \le \cdots \le a_n \le \cdots$
> ***decreasing*** if $a_1 > a_2 > a_3 > \cdots > a_n > \cdots$
> ***nonincreasing*** if $a_1 \ge a_2 \ge a_3 \ge \cdots \ge a_n \ge \cdots$

A sequence that is either nondecreasing or nonincreasing is called ***monotone,*** and a sequence that is increasing or decreasing is called ***strictly monotone.*** Observe that a strictly monotone sequence is monotone, but not conversely. (Why?)

▶ Example 1

$$\frac{1}{2}, \frac{2}{3}, \frac{3}{4}, \dots, \frac{n}{n+1}, \dots \quad \text{is increasing}$$

$$1, \frac{1}{2}, \frac{1}{3}, \dots, \frac{1}{n}, \dots \quad \text{is decreasing}$$

$$1, 1, 2, 2, 3, 3, \dots \quad \text{is nondecreasing}$$

$$1, 1, \frac{1}{2}, \frac{1}{2}, \frac{1}{3}, \frac{1}{3}, \dots \quad \text{is nonincreasing}$$

All four of these sequences are monotone, but the sequence

$$1, -\frac{1}{2}, \frac{1}{3}, -\frac{1}{4}, \dots, (-1)^{n+1}\frac{1}{n}, \dots$$

is not. The first and second sequences are strictly monotone. ◀

Frequently, one can *guess* whether a sequence is increasing, decreasing, nondecreasing, or nonincreasing after writing out some of the initial terms. However, to be certain that the guess is correct, the kind of analysis illustrated in the next few examples is needed.

▶ Example 2

Show that

$$\frac{1}{2}, \frac{2}{3}, \frac{3}{4}, \dots, \frac{n}{n+1}, \dots$$

is an increasing sequence.

Solution. Although it is intuitively clear that the sequence is increasing, intuition cannot replace a definitive proof. To prove that the sequence is increasing, let

$$f(x) = \frac{x}{x+1}$$

so the nth term in the given sequence is $a_n = f(n)$. The function f is increasing for $x \geq 1$ since

$$f'(x) = \frac{(x+1)(1) - x(1)}{(x+1)^2} = \frac{1}{(x+1)^2}$$

so $f'(x) > 0$ for $x \geq 1$. Thus,

$$a_n = f(n) < f(n+1) = a_{n+1}$$

which proves that the given sequence is increasing.

Alternate Solution. Since

$$a_n = \frac{n}{n+1}$$

we obtain on replacing n by $n+1$,

$$a_{n+1} = \frac{n+1}{(n+1)+1} = \frac{n+1}{n+2}$$

Thus, for $n \geq 1$

$$a_n - a_{n+1} = \frac{n}{n+1} - \frac{n+1}{n+2} = \frac{n^2 + 2n - n^2 - 2n - 1}{(n+1)(n+2)}$$

$$= -\frac{1}{(n+1)(n+2)} < 0$$

so that

$$a_n < a_{n+1}$$

for $n \geq 1$. This proves that the sequence is increasing. ◄

To prove that a sequence $\{a_n\}$ is decreasing, one can show either that $f'(x) < 0$ for $x \geq 1$ [where $f(n) = a_n$] or $a_n - a_{n+1} > 0$ for $n \geq 1$.

If a sequence $\{a_n\}$ has *positive* terms, then it is increasing if

$$\frac{a_{n+1}}{a_n} > 1 \tag{1a}$$

for $n \geq 1$, and it is decreasing if

$$\frac{a_{n+1}}{a_n} < 1 \tag{1b}$$

for $n \geq 1$.

▶ Example 3

Use (1a) to prove that the sequence in Example 2 is increasing.

Solution. As shown in the alternate solution of Example 2,

$$a_n = \frac{n}{n+1} \quad \text{and} \quad a_{n+1} = \frac{n+1}{n+2}$$

Thus

$$\frac{a_{n+1}}{a_n} = \frac{(n+1)/(n+2)}{n/(n+1)} = \frac{n+1}{n+2} \cdot \frac{n+1}{n} = \frac{n^2 + 2n + 1}{n^2 + 2n} \tag{2}$$

Since the numerator in (2) exceeds the denominator, the ratio exceeds 1, that is, $a_{n+1}/a_n > 1$ for $n \geq 1$. This proves that the sequence is increasing.

◀

In our subsequent work we will encounter sequences involving factorials. The reader will recall that if n is a positive integer, then $n!$ (n factorial) is the product of the first n positive integers, that is, $n! = 1 \cdot 2 \cdot 3 \cdots n$. Furthermore, it is agreed that $0! = 1$.

▶ Example 4

Show that the sequence

$$\frac{e}{2!}, \frac{e^2}{3!}, \frac{e^3}{4!}, \dots, \frac{e^n}{(n+1)!}, \dots$$

is decreasing.

Solution. We will use (1b). Since

$$a_n = \frac{e^n}{(n+1)!}$$

it follows on replacing n by $n+1$ that

$$a_{n+1} = \frac{e^{n+1}}{[(n+1)+1]!} = \frac{e^{n+1}}{(n+2)!}$$

Thus

$$\frac{a_{n+1}}{a_n} = \frac{e^{n+1}/(n+2)!}{e^n/(n+1)!} = \frac{e^{n+1}}{e^n} \cdot \frac{(n+1)!}{(n+2)!} = \frac{e}{n+2}$$

For $n \geq 1$, it follows that $n + 2 \geq 3 > e(\approx 2.718...)$, so

$$\frac{a_{n+1}}{a_n} = \frac{e}{n+2} < 1$$

for $n \geq 1$. This proves that the sequence is decreasing. ◄

The convergence properties of monotone sequences are especially simple; a monotone sequence either converges or it becomes infinite—divergence by oscillation cannot occur. This is the content of the following two theorems.

11.2.2 THEOREM *For a nondecreasing sequence $a_1 \leq a_2 \leq a_3 \leq \cdots \leq a_n \leq \cdots$ there are two possibilities:*

(a) *There is a constant M such that $a_n \leq M$ for all n, in which case the sequence converges to a limit L satisfying $L \leq M$.*

(b) *No such constant exists, in which case $\lim_{n \to +\infty} a_n = +\infty$.*

11.2.3 THEOREM *For a nonincreasing sequence $a_1 \geq a_2 \geq a_3 \geq \cdots \geq a_n \geq \cdots$ there are two possibilities:*

(a) *There is a constant M such that $a_n \geq M$ for all n, in which case the sequence converges to a limit L satisfying $L \geq M$.*

(b) *No such constant exists, in which case $\lim_{n \to +\infty} a_n = -\infty$.*

It should be noted that these results do not yield a method for evaluating limits; they tell us only whether a limit exists. To prove these theorems we need a preliminary result that takes us to the very foundations of the real number system. In this text we have not been concerned with a logical development of the real numbers. Our approach has been to accept the familiar properties of real numbers and to work with them. Indeed, we have not even attempted to define the term "real number." However, by the late nineteenth century, the study of limits and functions in calculus necessitated a precise axiomatic formulation of the real numbers in much the same way that Euclidean geometry is developed from axioms. While we will not attempt to pursue this development, we will have need for the following axiom about real numbers.

11.2.4 AXIOM

The Completeness Axiom

If S is a nonempty set of real numbers, and if there is some real number that is greater than or equal to every number in S, then there is a smallest real number that is greater than or equal to every number in S.

For example, let S be the set of numbers in the interval $(1, 3)$. It is true that there exists a number u greater than or equal to every number in S; some examples are $u = 10$, $u = 100$, and $u = 3.2$. The smallest number u that is greater than or equal to every number in S is $u = 3$.

There is an alternate phrasing of the completeness axiom that is useful to know. Let us call a number u an **upper bound** for a set S if u is greater than or equal to every number in S. Using this terminology the completeness axiom states:

11.2.5 AXIOM

The Completeness Axiom

(Alternate Form)

If a nonempty set S of real numbers has an upper bound, then S has a smallest upper bound.

The smallest upper bound of S is commonly called the **least upper bound** of S. Using the completeness axiom we can prove Theorem 11.2.2.

Proof of Theorem 11.2.2.

(a) Assume there exists a number M such that $a_n \leq M$ for $n = 1, 2, \ldots$ Then M is an upper bound for the set of terms in the sequence. By the completeness axiom there is a least upper bound for the terms, call it L. Now let ϵ be any positive number. Since L is the least upper bound for the terms, $L - \epsilon$ is not an upper bound for the terms, which means that there is at least one term a_N such that

$$a_N > L - \epsilon$$

Moreover, since $\{a_n\}$ is a nondecreasing sequence, we must have

$$a_n \geq a_N > L - \epsilon \qquad (3)$$

when $n \geq N$. But a_n cannot exceed L since L is an upper bound for the terms. This observation together with (3) tells us that $L \geq a_n > L - \epsilon$ for $n \geq N$, so that all terms from the Nth on are within ϵ units of L. This is exactly the requirement to have

$$\lim_{n \to +\infty} a_n = L$$

Observe that $L \leq M$ since M is an upper bound for the terms and L is the least upper bound. This proves (a).

(b) If there is no number M such that $a_n \leq M$ for $n = 1, 2, \ldots$, then no matter how large we choose M, there is a term a_N such that

$$a_N > M$$

and, since the sequence is nondecreasing,

$$a_n \geq a_N > M$$

when $n \geq N$. Thus the terms in the sequence become arbitrarily large as n increases. That is,

$$\lim_{n \to +\infty} a_n = +\infty \quad \blacksquare$$

The proof of Theorem 11.2.3 will be omitted since it is similar to 11.2.2.

▶ Example 5

As shown in Examples 2 and 3, the sequence $\left\{ \dfrac{n}{n+1} \right\}_{n=1}^{+\infty}$ is nondecreasing. Since

$$a_n = \frac{n}{n+1} < 1, \qquad n = 1, 2, \ldots$$

the terms in the sequence have $M = 1$ as an upper bound. By Theorem 11.2.2 the sequence must converge to a limit $L \leq M$. This is indeed the case since

$$\lim_{n \to +\infty} \frac{n}{n+1} = \lim_{n \to +\infty} \frac{1}{1 + \dfrac{1}{n}} = 1 \qquad \blacktriangleleft$$

Because the limit of a sequence $\{a_n\}$ describes the behavior of the terms as n gets *large,* one can alter or even delete a *finite* number of terms in a sequence without affecting either the convergence or the value of the limit. That is, the original and the modified sequence will both converge or both diverge, and in the case of convergence both will have the same limit. We omit the proof.

▶ Example 6

Show that the sequence

$$\left\{ \frac{5^n}{n!} \right\}_{n=1}^{+\infty}$$

converges.

Solution. It is tedious to determine convergence directly from the limit

$$\lim_{n \to +\infty} \frac{5^n}{n!}$$

Thus, we will proceed indirectly. If we let

$$a_n = \frac{5^n}{n!}$$

then

$$a_{n+1} = \frac{5^{n+1}}{(n+1)!}$$

so that

$$\frac{a_{n+1}}{a_n} = \frac{5^{n+1}/(n+1)!}{5^n/n!} = \frac{5^{n+1}}{5^n} \cdot \frac{n!}{(n+1)!} = \frac{5}{n+1}$$

For $n = 1, 2$, and 3, the value of a_{n+1}/a_n is greater than 1 so $a_{n+1} > a_n$. Thus

$$a_1 < a_2 < a_3 < a_4$$

For $n = 4$ the value of a_{n+1}/a_n is 1, so

$$a_4 = a_5$$

For $n \geq 5$ the value of a_{n+1}/a_n is less than 1, so

$$a_5 > a_6 > a_7 > a_8 > \cdots$$

Thus, if we discard the first four terms of the given sequence (which won't affect convergence) the resulting sequence will be decreasing. Moreover, each term in the sequence is positive, so that by Theorem 11.2.3, the sequence converges to some limit that is ≥ 0. ◀

▶ Exercise Set 11.2
_____

In Exercises 1–6, use differentiation to show that the sequence is strictly monotone and classify it as increasing or decreasing.

1. $\left\{\dfrac{n}{2n+1}\right\}_{n=1}^{+\infty}$.

2. $\left\{3 - \dfrac{1}{n}\right\}_{n=1}^{+\infty}$.

3. $\left\{\dfrac{1}{n + \ln n}\right\}_{n=1}^{+\infty}$.

4. $\{ne^{-2n}\}_{n=1}^{+\infty}$.

5. $\left\{\dfrac{\ln(n+2)}{n+2}\right\}_{n=1}^{+\infty}$.

6. $\{\tan^{-1} n\}_{n=1}^{+\infty}$.

In Exercises 7–12, determine if the given sequence $\{a_n\}$ is monotone by examining a_{n+1}/a_n. If so, classify it as increasing, decreasing, nonincreasing, or nondecreasing.

7. $\left\{\dfrac{n}{2n+1}\right\}_{n=1}^{+\infty}$.

8. $\left\{\dfrac{2^n}{n!}\right\}_{n=1}^{+\infty}$.

9. $\left\{\dfrac{n}{2^n}\right\}_{n=1}^{+\infty}$.

10. $\left\{\dfrac{2^n}{1+2^n}\right\}_{n=1}^{+\infty}$.

11. $\left\{\dfrac{10^n}{2^{n^2}}\right\}_{n=1}^{+\infty}$.

12. $\left\{\dfrac{n^n}{n!}\right\}_{n=1}^{+\infty}$.

In Exercises 13–18, show that the sequence is monotone and apply Theorem 11.2.2 or 11.2.3 to determine if the sequence converges.

13. $\left\{2 + \dfrac{1}{n}\right\}_{n=1}^{+\infty}$.

14. $\left\{4 - \dfrac{2}{n}\right\}_{n=1}^{+\infty}$.

15. $\left\{n - \dfrac{1}{n}\right\}_{n=1}^{+\infty}$.

16. $\left\{\dfrac{4n - 1}{5n + 2}\right\}_{n=1}^{+\infty}$.

17. $\left\{\dfrac{n}{5^n}\right\}_{n=1}^{+\infty}$.

18. $\left\{\cos \dfrac{\pi}{2n}\right\}_{n=1}^{+\infty}$.

19. Find the limit (if it exists) of the sequence.

(a) $1, -1, 1, \dfrac{1}{2}, \dfrac{1}{3}, \dfrac{1}{4}, \dfrac{1}{5}, \ldots$

(b) $-\dfrac{1}{2}, 0, 0, 0, 1, 2, 3, 4, \ldots$

20. (a) Is $\left\{\dfrac{100^n}{n!}\right\}_{n=1}^{+\infty}$ a monotone sequence?

(b) Is it a convergent sequence? Justify your answer.

21. Show that $\left\{\dfrac{3^n}{1 + 3^{2n}}\right\}_{n=1}^{+\infty}$ is a decreasing sequence.

22. Show that $\left\{\dfrac{1 \cdot 3 \cdot 5 \cdots (2n - 1)}{n!}\right\}_{n=1}^{+\infty}$ is an increasing sequence.

23. (a) Show that if $\{a_n\}_{n=1}^{+\infty}$ is a nonincreasing sequence, then $\{-a_n\}_{n=1}^{+\infty}$ is a nondecreasing sequence.

(b) Use part (a) and Theorem 11.2.2 to help prove Theorem 11.2.3.

11.3 INFINITE SERIES

The purpose of this section is to discuss sums

$$u_1 + u_2 + u_3 + \cdots + u_k + \cdots$$

that contain infinitely many terms. The most familiar examples of such sums occur in the decimal representation of real numbers. For example, when we write $\frac{1}{3}$ in the decimal form

$$\frac{1}{3} = 0.3333\ldots$$

we mean

$$\frac{1}{3} = 0.3 + 0.03 + 0.003 + 0.0003 + \cdots$$

$$= \frac{3}{10} + \frac{3}{10^2} + \frac{3}{10^3} + \frac{3}{10^4} + \cdots$$

Our first objective will be to define what is meant by a sum with infinitely many terms. Since it is impossible to "add up" infinitely many numbers, we will deal with infinite sums by means of a limiting process involving sequences.

11.3.1 DEFINITION   An *infinite series* is an expression of the form

$$u_1 + u_2 + u_3 + \cdots + u_k + \cdots$$

or in sigma notation

$$\sum_{k=1}^{\infty} u_k$$

The numbers $u_1, u_2, u_3, \ldots$ are called the *terms* of the series.

Informally speaking, the expression $\displaystyle\sum_{k=1}^{\infty} u_k$ directs us to obtain the "sum" of the terms $u_1, u_2, u_3, \ldots$ To carry out this summation process we proceed as follows: Let s_n denote the sum of the first n terms of the series. Thus

$$s_1 = u_1$$
$$s_2 = u_1 + u_2$$
$$s_3 = u_1 + u_2 + u_3$$
$$\vdots$$
$$s_n = u_1 + u_2 + u_3 + \cdots + u_n = \sum_{k=1}^{n} u_k$$

The number s_n is called the *nth partial sum* of the series and the sequence $\{s_n\}_{n=1}^{+\infty}$ is called the *sequence of partial sums.*

▶ Example 1

For the infinite series

$$\frac{3}{10} + \frac{3}{10^2} + \frac{3}{10^3} + \frac{3}{10^4} + \cdots$$

the partial sums are

$$s_1 = \frac{3}{10}$$
$$s_2 = \frac{3}{10} + \frac{3}{10^2} = \frac{33}{100}$$
$$s_3 = \frac{3}{10} + \frac{3}{10^2} + \frac{3}{10^3} = \frac{333}{1000}$$
$$s_4 = \frac{3}{10} + \frac{3}{10^2} + \frac{3}{10^3} + \frac{3}{10^4} = \frac{3333}{10000}$$
$$\vdots$$

◀

As n increases, the partial sum $s_n = u_1 + u_2 + \cdots + u_n$ includes more and more terms of the series. Thus, if s_n tends toward a limit as $n \to +\infty$, it is reasonable to view this limit as the sum of *all* the terms in the series. This suggests the following definition.

11.3.2 DEFINITION   Let $\{s_n\}$ be the sequence of partial sums of the series $\sum\limits_{k=1}^{\infty} u_k$. If the sequence $\{s_n\}$ converges to a limit S, then the series is said to **converge** and S is called the **sum** of the series. We denote this by writing

$$S = \sum_{k=1}^{\infty} u_k$$

If the sequence of partial sums of a series diverges, then the series is said to **diverge.** A divergent series has no sum.

▶ Example 2

If Definition 11.3.2 is to be reasonable, it should be the case that

$$\frac{1}{3} = \frac{3}{10} + \frac{3}{10^2} + \frac{3}{10^3} + \cdots + \frac{3}{10^k} + \cdots$$

Let us verify that this is indeed the case. The nth partial sum is

$$s_n = \frac{3}{10} + \frac{3}{10^2} + \cdots + \frac{3}{10^n} \tag{1}$$

The problem of calculating $\lim\limits_{n \to +\infty} s_n$ is complicated by the fact that the number of terms in (1) changes with n. For purposes of calculation, it is desirable to rewrite (1) in closed form (see Section 5.5). To do this, we multiply both sides of (1) by $\frac{1}{10}$ to obtain

$$\frac{1}{10} s_n = \frac{3}{10^2} + \frac{3}{10^3} + \cdots + \frac{3}{10^n} + \frac{3}{10^{n+1}} \tag{2}$$

and then subtract (2) from (1) to obtain

$$s_n - \frac{1}{10} s_n = \frac{3}{10} - \frac{3}{10^{n+1}}$$

or

$$\frac{9}{10} s_n = \frac{3}{10}\left(1 - \frac{1}{10^n}\right)$$

or

$$s_n = \frac{1}{3}\left(1 - \frac{1}{10^n}\right)$$ (3)

Since $1/10^n \to 0$ as $n \to +\infty$, it follows from (3) that $S = \lim\limits_{n \to +\infty} s_n = \frac{1}{3}$. Thus,

$$\frac{1}{3} = \frac{3}{10} + \frac{3}{10^2} + \frac{3}{10^3} + \cdots + \frac{3}{10^n} + \cdots$$ ◄

▶ Example 3

Determine if the series

$$1 - 1 + 1 - 1 + 1 - 1 + \cdots$$

converges or diverges. If it converges, find the sum.

Solution. The partial sums are $s_1 = 1, s_2 = 1 - 1 = 0, s_3 = 1 - 1 + 1 = 1$, $s_4 = 1 - 1 + 1 - 1 = 0$, and so forth. Thus the sequence of partial sums is

$$1, 0, 1, 0, 1, 0, \ldots$$

Since this is a divergent sequence, the given series diverges and consequently has no sum. ◄

The series in Examples 2 and 3 are examples of **geometric series.** A geometric series is one of the form

$$a + ar + ar^2 + ar^3 + \cdots + ar^{k-1} + \cdots \qquad (a \neq 0)$$

where each term is obtained by multiplying the previous one by a constant r. The multiplier r is called the **ratio** for the series. Some examples of geometric series are:

$$1 + 2 + 4 + 8 + \cdots + 2^{k-1} + \cdots \qquad (a = 1, r = 2)$$

$$3 + \frac{3}{10} + \frac{3}{10^2} + \frac{3}{10^3} + \cdots + \frac{3}{10^{k-1}} + \cdots \qquad \left(a = 3, r = \frac{1}{10}\right)$$

$$\frac{1}{2} - \frac{1}{4} + \frac{1}{8} - \frac{1}{16} + \cdots + (-1)^{k+1}\frac{1}{2^k} + \cdots \qquad \left(a = \frac{1}{2}, r = -\frac{1}{2}\right)$$

$$1 + 1 + 1 + \cdots + 1 + \cdots \qquad (a = 1, r = 1)$$

$$1 - 1 + 1 - 1 + \cdots + (-1)^{k+1} + \cdots \qquad (a = 1, r = -1)$$

The following theorem is the fundamental result on convergence of geometric series.

11.3.3 THEOREM   *A geometric series*

$$a + ar + ar^2 + \cdots + ar^{k-1} + \cdots \qquad (a \neq 0)$$

converges if $|r| < 1$ and diverges if $|r| \geq 1$. When the series converges the sum is

$$\frac{a}{1-r} = a + ar + ar^2 + \cdots + ar^{k-1} + \cdots$$

Proof. Let us treat the case $|r| = 1$ first. If $r = 1$, then the series is

$$a + a + a + \cdots + a + \cdots$$

so that the *n*th partial sum is $s_n = na$ and $\lim_{n \to +\infty} s_n = \lim_{n \to +\infty} na = \pm\infty$ (the sign depending on whether a is positive or negative). This proves divergence. If $r = -1$, the series is

$$a - a + a - a + \cdots$$

so the sequence of partial sums is

$$a, 0, a, 0, a, 0, \ldots$$

which diverges.

Now let us consider the case where $|r| \neq 1$. The *n*th partial sum of the series is

$$s_n = a + ar + ar^2 + \cdots + ar^{n-1} \tag{4}$$

Multiplying both sides of (4) by r yields

$$rs_n = ar + ar^2 + \cdots + ar^{n-1} + ar^n \tag{5}$$

and subtracting (5) from (4) gives

$$s_n - rs_n = a - ar^n$$

or

$$(1 - r)s_n = a - ar^n \tag{6}$$

Since $r \neq 1$ in the case we are considering, this can be rewritten as

$$s_n = \frac{a - ar^n}{1 - r} = \frac{a}{1 - r} - \frac{ar^n}{1 - r} \qquad (7)$$

If $|r| < 1$, then $\lim_{n \to +\infty} r^n = 0$, so that $\{s_n\}$ converges. From (7)

$$\lim_{n \to +\infty} s_n = \frac{a}{1 - r}$$

If $|r| > 1$, then either $r > 1$ or $r < -1$. In the case $r > 1$, $\lim_{n \to +\infty} r^n = +\infty$, and in the case $r < -1$, r^n oscillates between positive and negative values that grow in magnitude, so $\{s_n\}$ diverges. ▮

▶ **Example 4**

The series

$$5 + \frac{5}{4} + \frac{5}{4^2} + \cdots + \frac{5}{4^{k-1}} + \cdots$$

is a geometric series with $a = 5$ and $r = \frac{1}{4}$. Since $|r| = \frac{1}{4} < 1$, the series converges and the sum is

$$\frac{a}{1 - r} = \frac{5}{1 - \frac{1}{4}} = \frac{20}{3} \qquad ◀$$

▶ **Example 5**

Find the rational number represented by the repeating decimal

$$0.784784784\ldots$$

Solution. We can write

$$0.784784784\ldots = 0.784 + 0.000784 + 0.000000784 + \cdots$$

so the given decimal is the sum of a geometric series with $a = 0.784$ and $r = 0.001$. Thus

$$0.784784784\ldots = \frac{a}{1 - r} = \frac{0.784}{1 - 0.001} = \frac{0.784}{0.999} = \frac{784}{999} \qquad ◀$$

▶ Example 6

Determine if the series

$$\sum_{k=1}^{\infty} \frac{1}{k(k+1)} = \frac{1}{1\cdot 2} + \frac{1}{2\cdot 3} + \frac{1}{3\cdot 4} + \frac{1}{4\cdot 5} + \cdots$$

converges or diverges. If it converges, find the sum.

Solution. The *n*th partial sum of the series is

$$s_n = \sum_{k=1}^{n} \frac{1}{k(k+1)} = \frac{1}{1\cdot 2} + \frac{1}{2\cdot 3} + \frac{1}{3\cdot 4} + \cdots + \frac{1}{n(n+1)}$$

To calculate $\lim_{n\to +\infty} s_n$ we will rewrite s_n in closed form. This may be accomplished by using the method of partial fractions to obtain (verify):

$$\frac{1}{k(k+1)} = \frac{1}{k} - \frac{1}{k+1}$$

from which it follows that

$$s_n = \sum_{k=1}^{n} \left(\frac{1}{k} - \frac{1}{k+1}\right)$$
$$= \left(1 - \frac{1}{2}\right) + \left(\frac{1}{2} - \frac{1}{3}\right) + \left(\frac{1}{3} - \frac{1}{4}\right) + \cdots + \left(\frac{1}{n} - \frac{1}{n+1}\right)$$
$$= 1 + \left(-\frac{1}{2} + \frac{1}{2}\right) + \left(-\frac{1}{3} + \frac{1}{3}\right) + \cdots + \left(-\frac{1}{n} + \frac{1}{n}\right) - \frac{1}{n+1} \quad (8)$$

The sum in (8) is an example of a ***telescoping sum,*** which means that each term cancels part of the next term, thereby collapsing the sum (like a folding telescope) into only two terms. After the cancellation, (8) can be written as

$$s_n = 1 - \frac{1}{n+1}$$

so

$$\lim_{n\to +\infty} s_n = \lim_{n\to +\infty} \left(1 - \frac{1}{n+1}\right) = 1$$

and therefore

$$1 = \sum_{k=1}^{\infty} \frac{1}{k(k+1)}$$

◀

▶ Example 7

One of the most famous and important of all diverging series is the **harmonic series,**

$$\sum_{k=1}^{\infty} \frac{1}{k} = 1 + \frac{1}{2} + \frac{1}{3} + \frac{1}{4} + \frac{1}{5} + \cdots$$

which arises in connection with the overtones produced by a vibrating musical string. It is not immediately evident that this series diverges. However the divergence will become apparent when we examine the partial sums in detail. Because the terms in the series are all positive, the partial sums

$$s_1 = 1, \ s_2 = 1 + \frac{1}{2}, \ s_3 = 1 + \frac{1}{2} + \frac{1}{3}, \ s_4 = 1 + \frac{1}{2} + \frac{1}{3} + \frac{1}{4}, \ldots$$

form an increasing sequence

$$s_1 < s_2 < s_3 < \cdots < s_n < \cdots$$

Thus, by Theorem 11.2.2 we can prove divergence by demonstrating that there is no constant M that is greater than or equal to *every* partial sum. To this end, we will consider some selected partial sums, namely $s_2, s_4, s_8, s_{16}, s_{32}, \ldots$. Note that the subscripts are successive powers of 2, so that these are the partial sums of the form s_{2^n}. These partial sums satisfy the inequalities

$$s_2 = 1 + \frac{1}{2} > \frac{1}{2} + \frac{1}{2} = \frac{2}{2}$$

$$s_4 = s_2 + \frac{1}{3} + \frac{1}{4} > s_2 + \left(\frac{1}{4} + \frac{1}{4} \right) = s_2 + \frac{1}{2} > \frac{3}{2}$$

$$s_8 = s_4 + \frac{1}{5} + \frac{1}{6} + \frac{1}{7} + \frac{1}{8} > s_4 + \left(\frac{1}{8} + \frac{1}{8} + \frac{1}{8} + \frac{1}{8} \right) = s_4 + \frac{1}{2} > \frac{4}{2}$$

$$s_{16} = s_8 + \frac{1}{9} + \frac{1}{10} + \frac{1}{11} + \frac{1}{12} + \frac{1}{13} + \frac{1}{14} + \frac{1}{15} + \frac{1}{16}$$

$$> s_8 + \left(\frac{1}{16} + \frac{1}{16} + \frac{1}{16} + \frac{1}{16} + \frac{1}{16} + \frac{1}{16} + \frac{1}{16} + \frac{1}{16} \right) = s_8 + \frac{1}{2} > \frac{5}{2}$$

$$\vdots$$

$$s_{2^n} > \frac{n+1}{2}$$

Now if M is any constant, we can certainly find a positive integer n such that $(n+1)/2 > M$. But for this n

$$s_{2^n} > \frac{n+1}{2} > M$$

so that no constant M is greater than or equal to *every* partial sum of the harmonic series. This proves divergence. ◄

► Exercise Set 11.3

1. In each part, find the first four partial sums; find a closed form for the nth partial sum; determine if the series converges, and if so give the sum.

   (a) $\displaystyle\sum_{k=1}^{\infty} \frac{2}{5^{k-1}}$ (b) $\displaystyle\sum_{k=1}^{\infty} \frac{1}{(k+1)(k+2)}$

   (c) $\displaystyle\sum_{k=1}^{\infty} \frac{2^{k-1}}{4}$.

In Exercises 2–16, determine if the series converges or diverges. If it converges, find the sum.

2. $\displaystyle\sum_{k=1}^{\infty} \frac{1}{5^k}$.

3. $\displaystyle\sum_{k=1}^{\infty} \left(-\frac{3}{4}\right)^{k-1}$.

4. $\displaystyle\sum_{k=1}^{\infty} \left(\frac{2}{3}\right)^{k+2}$.

5. $\displaystyle\sum_{k=1}^{\infty} (-1)^{k-1} \frac{7}{6^{k-1}}$.

6. $\displaystyle\sum_{k=1}^{\infty} 4^{k-1}$.

7. $\displaystyle\sum_{k=1}^{\infty} \left(-\frac{3}{2}\right)^{k+1}$.

8. $\displaystyle\sum_{k=1}^{\infty} \left(\frac{1}{k+3} - \frac{1}{k+4}\right)$.

9. $\displaystyle\sum_{k=1}^{\infty} \frac{1}{(k+2)(k+3)}$.

10. $\displaystyle\sum_{k=1}^{\infty} \left(\frac{1}{2^k} - \frac{1}{2^{k+1}}\right)$.

11. $\displaystyle\sum_{k=1}^{\infty} \frac{1}{9k^2 + 3k - 2}$.

12. $\displaystyle\sum_{k=2}^{\infty} \frac{1}{k^2 - 1}$.

13. $\displaystyle\sum_{k=1}^{\infty} \frac{4^{k+2}}{7^{k-1}}$.

14. $\displaystyle\sum_{k=1}^{\infty} \left(\frac{e}{\pi}\right)^{k-1}$.

15. $\displaystyle\sum_{k=1}^{\infty} \left(-\frac{1}{2}\right)^k$.

16. $\displaystyle\sum_{k=3}^{\infty} \frac{5}{k-2}$.

In Exercises 17–22, express the repeating decimal as a fraction.

17. $0.4444\ldots$

18. $0.9999\ldots$

19. $5.373737\ldots$

20. $0.159159159\ldots$

21. $0.782178217821\ldots$

22. $0.451141414\ldots$

23. Find a closed form for the nth partial sum of the series

$$\ln\frac{1}{2} + \ln\frac{2}{3} + \ln\frac{3}{4} + \cdots + \ln\frac{n}{n+1} + \cdots$$

and determine if the series converges.

24. A ball is dropped from a height of 10 meters. Each time it strikes the ground it bounces vertically to a height that is three fourths of the previous height. Find the total distance the ball will travel if it is allowed to bounce indefinitely.

25. Show: $\displaystyle\sum_{k=2}^{\infty} \ln(1 - 1/k^2) = -\ln 2$.

26. Show: $\displaystyle\sum_{k=1}^{\infty} \frac{\sqrt{k+1} - \sqrt{k}}{\sqrt{k^2 + k}} = 1$.

27. Use geometric series to show:

   (a) $\displaystyle\sum_{k=0}^{\infty} (-1)^k x^k = \frac{1}{1+x}$ if $-1 < x < 1$

   (b) $\displaystyle\sum_{k=0}^{\infty} (x-3)^k = \frac{1}{4-x}$ if $2 < x < 4$

   (c) $\displaystyle\sum_{k=0}^{\infty} (-1)^k x^{2k} = \frac{1}{1+x^2}$ if $-1 < x < 1$.

28. Prove the following decimal equality assuming $a_n \neq 9$:
$$0.a_1 a_2 \ldots a_n 9999 \ldots = 0.a_1 a_2 \ldots (a_n + 1)0000 \ldots$$

11.4 CONVERGENCE; THE INTEGRAL TEST

In the previous section we found sums of series and investigated convergence by first writing the nth partial sum s_n in closed form and then examining the limit $\lim_{n \to +\infty} s_n$. Unfortunately, it is relatively rare that the nth partial sum of a series can be written in closed form; for most series, convergence or divergence is determined by using convergence tests, some of which we will introduce in this section. Once it is established that a series converges, the sum of the series can always be approximated to any degree of accuracy by a partial sum with sufficiently many terms.

Our first theorem states that the terms of an infinite series must tend toward zero if the series is to converge.

11.4.1 THEOREM   *If the series Σu_k converges, then $\lim_{k \to +\infty} u_k = 0$.*

Proof. The term u_k can be written

$$u_k = s_k - s_{k-1} \tag{1}$$

where s_k is the sum of the first k terms and s_{k-1} is the sum of the first $k - 1$ terms. If S denotes the sum of the series, then $\lim_{k \to +\infty} s_k = S$, and since $(k - 1) \to +\infty$ as $k \to +\infty$ we also have $\lim_{k \to +\infty} s_{k-1} = S$. Thus from (1)

$$\lim_{k \to +\infty} u_k = \lim_{k \to +\infty} (s_k - s_{k-1}) = S - S = 0 \quad \blacksquare$$

The following result is just an alternate phrasing of the above theorem and needs no additional proof.

11.4.2 THEOREM
The Divergence Test   *If $\lim_{k \to +\infty} u_k \neq 0$, then the series Σu_k diverges.*

▶ Example 1

The series

$$\sum_{k=1}^{\infty} \frac{k}{k+1} = \frac{1}{2} + \frac{2}{3} + \frac{3}{4} + \cdots + \frac{k}{k+1} + \cdots$$

diverges since

$$\lim_{k \to +\infty} \frac{k}{k+1} = \lim_{k \to +\infty} \frac{1}{1 + 1/k} = 1 \neq 0 \qquad \blacktriangleleft$$

WARNING. The converse of Theorem 11.4.1 is false. To prove that a series converges it does not suffice to show that $\lim_{k \to +\infty} u_k = 0$ since this property may hold for divergent as well as convergent series. For example, the kth term of the divergent harmonic series $1 + 1/2 + 1/3 + \cdots + 1/k + \cdots$ tends to zero as $k \to +\infty$, and the kth term of the convergent geometric series $1/2 + 1/2^2 + \cdots + 1/2^k + \cdots$ tends to zero as $k \to +\infty$.

For brevity, the proof of the following result is left for the exercises.

11.4.3 THEOREM *(a) If Σu_k and Σv_k are convergent series, then $\Sigma(u_k + v_k)$ and $\Sigma(u_k - v_k)$ are convergent series and the sums of these series are related by*

$$\sum_{k=1}^{\infty} (u_k + v_k) = \sum_{k=1}^{\infty} u_k + \sum_{k=1}^{\infty} v_k$$

$$\sum_{k=1}^{\infty} (u_k - v_k) = \sum_{k=1}^{\infty} u_k - \sum_{k=1}^{\infty} v_k$$

(b) If c is a nonzero constant then the series Σu_k and $\Sigma c u_k$ both converge or both diverge. In the case of convergence, the sums are related by

$$\sum_{k=1}^{\infty} c u_k = c \sum_{k=1}^{\infty} u_k$$

(c) Convergence or divergence is unaffected by deleting a finite number of terms from the beginning of a series; that is, for any positive integer K, the series

$$\sum_{k=1}^{\infty} u_k = u_1 + u_2 + u_3 + \cdots$$

and

$$\sum_{k=K}^{\infty} u_k = u_K + u_{K+1} + u_{K+2} + \cdots$$

both converge or both diverge.

▶ Example 2

Find the sum of the series

$$\sum_{k=1}^{\infty} \left(\frac{3}{4^k} - \frac{2}{5^{k-1}} \right)$$

Solution. The series

$$\sum_{k=1}^{\infty} \frac{3}{4^k} = \frac{3}{4} + \frac{3}{4^2} + \frac{3}{4^3} + \cdots$$

is a convergent geometric series ($a = \frac{3}{4}, r = \frac{1}{4}$) and the series

$$\sum_{k=1}^{\infty} \frac{2}{5^{k-1}} = 2 + \frac{2}{5} + \frac{2}{5^2} + \frac{2}{5^3} + \cdots$$

is also a convergent geometric series ($a = 2, r = \frac{1}{5}$). Thus from Theorems 11.4.3*a* and 11.3.3 the given series converges and

$$\sum_{k=1}^{\infty} \left(\frac{3}{4^k} - \frac{2}{5^{k-1}} \right) = \sum_{k=1}^{\infty} \frac{3}{4^k} - \sum_{k=1}^{\infty} \frac{2}{5^{k-1}}$$

$$= \frac{\frac{3}{4}}{1 - \frac{1}{4}} - \frac{2}{1 - \frac{1}{5}}$$

$$= 1 - \frac{5}{2} = -\frac{3}{2} \qquad \blacktriangleleft$$

▶ Example 3

The series

$$\sum_{k=1}^{\infty} \frac{5}{k} = 5 + \frac{5}{2} + \frac{5}{3} + \cdots + \frac{5}{k} + \cdots$$

diverges by part (*b*) of Theorem 11.4.3 since

$$\sum_{k=1}^{\infty} \frac{5}{k} = \sum_{k=1}^{\infty} 5 \left(\frac{1}{k} \right)$$

so each term is a constant times the corresponding term of the divergent harmonic series. ◀

▶ Example 4

The series

$$\sum_{k=10}^{\infty} \frac{1}{k} = \frac{1}{10} + \frac{1}{11} + \frac{1}{12} + \cdots$$

diverges by part (*c*) of Theorem 11.4.3, since this series results by deleting the first nine terms from the divergent harmonic series. ◀

If an infinite series $u_1 + u_2 + u_3 + \cdots + u_k + \cdots$ has *positive terms,* then the partial sums $s_1 = u_1,\ s_2 = u_1 + u_2,\ s_3 = u_1 + u_2 + u_3, \ldots$ form an increasing sequence, that is

$$s_1 < s_2 < s_3 < \cdots < s_n < \cdots$$

If there is a finite constant *M* such that $s_n \le M$ for all *n*, then according to Theorem 11.2.2, the sequence of partial sums will converge to a limit *S* satisfying $S \le M$. If no such constant exists, then $\lim\limits_{n \to +\infty} s_n = +\infty$. This yields the following theorem.

11.4.4 THEOREM *If Σu_k is a series with positive terms, and if there is a constant M such that*

$$s_n = u_1 + u_2 + \cdots + u_n \le M$$

for every n, then the series converges and the sum S satisfies $S \le M$. If no such M exists then the series diverges.

If we have a series with positive terms, say

$$\sum_{k=1}^{\infty} \frac{1}{k^2}$$

and if we form the improper integral

$$\int_1^{+\infty} \frac{1}{x^2}\, dx$$

whose integrand is obtained by replacing the summation index *k* by *x*, then there is a relationship between convergence of the series and convergence of the improper integral.

11.4.5 THEOREM
The Integral Test

Let Σu_k be a series with positive terms, and let f(x) be the function that results when k is replaced by x in the formula for u_k. If f is decreasing and continuous for $x \ge 1$, then

$$\sum_{k=1}^{\infty} u_k \qquad and \qquad \int_1^{+\infty} f(x)\, dx$$

both converge or both diverge.

We will defer the proof to the end of the section and proceed with some examples.

▶ Example 5

Determine if

$$\sum_{k=1}^{\infty} \frac{1}{k^2}$$

converges or diverges.

Solution. If we replace k by x in the formula for u_k, we obtain the function

$$f(x) = \frac{1}{x^2}$$

which satisfies the hypotheses of the integral test. (Verify.) Since

$$\int_1^{+\infty} \frac{1}{x^2}\, dx = \lim_{l \to +\infty} \int_1^l \frac{dx}{x^2} = \lim_{l \to +\infty} \left[-\frac{1}{x} \right]_1^l = \lim_{l \to +\infty} \left[1 - \frac{1}{l} \right] = 1$$

the integral converges and consequently the series converges. ◀

REMARK. In the above example, do *not* erroneously conclude that

$$\sum_{k=1}^{\infty} \frac{1}{k^2} = 1 \text{ from the fact that } \int_1^{+\infty} \frac{1}{x^2}\, dx = 1.$$ (To see that this is false, write out the series: $1 + 1/2^2 + 1/3^2 + \cdots$; the sum obviously exceeds 1.)

▶ Example 6

The integral test provides another way to demonstrate divergence of the harmonic series $\sum_{k=1}^{\infty} \frac{1}{k}$. If we replace k by x in the formula for u_k we obtain the function $f(x) = 1/x$ which satisfies the hypotheses of the integral test. (Verify.) Since

$$\int_1^{+\infty} \frac{1}{x}\, dx = \lim_{l \to +\infty} \int_1^l \frac{1}{x}\, dx = \lim_{l \to +\infty} [\ln l - \ln 1] = +\infty$$

the integral and the series diverge. ◀

▶ Example 7

Determine if the series

$$\frac{1}{e} + \frac{2}{e^4} + \frac{3}{e^9} + \cdots + \frac{k}{e^{k^2}} + \cdots$$

converges or diverges.

Solution. If we replace k by x in the formula for u_k, we obtain the function

$$f(x) = \frac{x}{e^{x^2}} = xe^{-x^2}$$

For $x \geq 1$, this function has positive values and is continuous. Moreover, for $x \geq 1$ the derivative

$$f'(x) = e^{-x^2} - 2x^2e^{-x^2} = e^{-x^2}(1 - 2x^2)$$

is negative, so that f is decreasing for $x \geq 1$. Thus, the hypotheses of the integral test are met. But

$$\int_1^{+\infty} xe^{-x^2}\,dx = \lim_{l \to +\infty} \int_1^l xe^{-x^2}\,dx$$

$$= \lim_{l \to +\infty} \left[-\frac{1}{2}e^{-x^2} \right]_1^l$$

$$= \left(-\frac{1}{2} \right) \lim_{l \to +\infty} \left[e^{-l^2} - e^{-1} \right]$$

$$= \frac{1}{2e}$$

Thus the improper integral and the series converge. ◀

The harmonic series and the series in Example 5 are special cases of a class of series called ***p-series*** or ***hyperharmonic series***. A p-series is an infinite series of the form

$$\sum_{k=1}^{\infty} \frac{1}{k^p} = 1 + \frac{1}{2^p} + \frac{1}{3^p} + \cdots + \frac{1}{k^p} + \cdots$$

where $p > 0$. Examples of p-series are

$$\sum_{k=1}^{\infty} \frac{1}{k} = 1 + \frac{1}{2} + \frac{1}{3} + \cdots + \frac{1}{k} + \cdots \qquad (p = 1)$$

$$\sum_{k=1}^{\infty} \frac{1}{k^2} = 1 + \frac{1}{2^2} + \frac{1}{3^2} + \cdots + \frac{1}{k^2} + \cdots \qquad (p = 2)$$

$$\sum_{k=1}^{\infty} \frac{1}{\sqrt{k}} = 1 + \frac{1}{\sqrt{2}} + \frac{1}{\sqrt{3}} + \cdots + \frac{1}{\sqrt{k}} + \cdots \qquad (p = \tfrac{1}{2})$$

The following theorem tells when a p-series converges.

11.4.6 THEOREM
Convergence of p-Series

$$\sum_{k=1}^{\infty} \frac{1}{k^p} = 1 + \frac{1}{2^p} + \frac{1}{3^p} + \cdots + \frac{1}{k^p} + \cdots$$

converges if $p > 1$ and diverges if $0 < p \leq 1$.

Proof. To establish this result when $p \neq 1$, we will use the integral test.

$$\int_{1}^{+\infty} \frac{1}{x^p} \, dx = \lim_{l \to +\infty} \int_{1}^{l} x^{-p} \, dx$$

$$= \lim_{l \to +\infty} \frac{x^{1-p}}{1-p} \Big]_{1}^{l}$$

$$= \lim_{l \to +\infty} \left[\frac{l^{1-p}}{1-p} - \frac{1}{1-p} \right]$$

For $p > 1$, $1 - p < 0$ and $l^{1-p} \to 0$ as $l \to +\infty$, so the integral and the series converge. For $0 < p < 1$, $1 - p > 0$ and $l^{1-p} \to +\infty$ as $l \to +\infty$, so the integral and the series diverge. The case $p = 1$ is the harmonic series, which was previously shown to diverge. ∎

▶ Example 8

$$1 + \frac{1}{\sqrt[3]{2}} + \frac{1}{\sqrt[3]{3}} + \cdots + \frac{1}{\sqrt[3]{k}} + \cdots$$

diverges since it is a p-series with $p = \tfrac{1}{3} < 1$. ◀

We conclude this section by proving Theorem 11.4.5 (The Integral Test).

Proof. Let $f(x)$ satisfy the hypotheses of the theorem. Since

$$f(1) = u_1, \, f(2) = u_2, \ldots, f(n) = u_n, \ldots$$

the rectangles shown in Figures 11.4.1a and 11.4.1b have areas $u_1, u_2, \ldots, u_n$ as indicated.

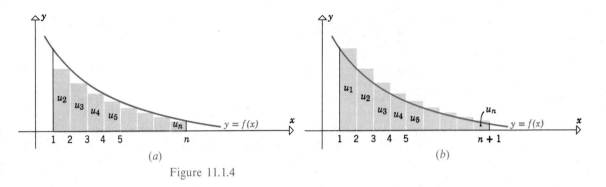

Figure 11.1.4

In Figure 11.4.1a, the total area of the rectangles is less than the area under the curve from $x = 1$ to $x = n$, so

$$u_2 + u_3 + \cdots + u_n < \int_1^n f(x)\,dx$$

Therefore

$$u_1 + u_2 + u_3 + \cdots + u_n < u_1 + \int_1^n f(x)\,dx \tag{2}$$

In Figure 11.4.1b, the total area of the rectangles is greater than the area under the curve from $x = 1$ to $x = n + 1$, so that

$$\int_1^{n+1} f(x)\,dx < u_1 + u_2 + \cdots + u_n \tag{3}$$

If we let $s_n = u_1 + u_2 + \cdots + u_n$ be the nth partial sum of the series, then (2) and (3) yield

$$\int_1^{n+1} f(x)\,dx < s_n < u_1 + \int_1^n f(x)\,dx \tag{4}$$

If the integral $\int_1^{+\infty} f(x)\,dx$ converges to a finite value L, then from the right-hand inequality in (4)

$$s_n < u_1 + \int_1^n f(x)\,dx < u_1 + \int_1^{+\infty} f(x)\,dx = u_1 + L$$

Thus each partial sum is less than the finite constant $u_1 + L$, and the series converges by Theorem 11.4.4. On the other hand, if the integral $\int_1^{+\infty} f(x)\,dx$ diverges, then

$$\lim_{n \to +\infty} \int_{1}^{n+1} f(x)\, dx = +\infty$$

so that from the left-hand inequality in (4), $\lim_{n \to +\infty} s_n = +\infty$. This means that the series also diverges. ∎

REMARK. If the summation index in a series Σu_k does not begin with $k = 1$, a variation of the integral test may still apply. It can be shown that

$$\sum_{k=K}^{\infty} u_k \quad \text{and} \quad \int_{K}^{+\infty} f(x)\, dx$$

both converge or both diverge provided the hypotheses of Theorem 11.4.5 hold for $x \geq K$.

▶ Exercise Set 11.4

In Exercises 1–4, use Theorem 11.4.3 to find the sum of the series.

1. $\displaystyle\sum_{k=1}^{\infty} \left[\frac{1}{2^k} + \frac{1}{4^k} \right].$

2. $\displaystyle\sum_{k=1}^{\infty} \left[\frac{1}{5^k} - \frac{1}{k(k+1)} \right].$

3. $\displaystyle\sum_{k=2}^{\infty} \left[\frac{1}{k^2 - 1} - \frac{7}{10^{k-1}} \right].$

4. $\displaystyle\sum_{k=1}^{\infty} \left[\frac{7}{3^k} + \frac{6}{(k+3)(k+4)} \right].$

In Exercises 5–26, determine if the series converges or diverges.

5. $\displaystyle\sum_{k=1}^{\infty} \frac{1}{k+6}.$

6. $\displaystyle\sum_{k=1}^{\infty} \frac{3}{5k}.$

7. $\displaystyle\sum_{k=1}^{\infty} \frac{1}{5k+2}.$

8. $\displaystyle\sum_{k=1}^{\infty} \frac{k}{1+k^2}.$

9. $\displaystyle\sum_{k=1}^{\infty} \frac{1}{1+9k^2}.$

10. $\displaystyle\sum_{k=1}^{\infty} \frac{1}{(4+2k)^{3/2}}.$

11. $\displaystyle\sum_{k=1}^{\infty} \frac{1}{\sqrt{k+5}}.$

12. $\displaystyle\sum_{k=1}^{\infty} \frac{1}{\sqrt[k]{e}}.$

13. $\displaystyle\sum_{k=1}^{\infty} \frac{1}{\sqrt[3]{2k-1}}.$

14. $\displaystyle\sum_{k=3}^{\infty} \frac{\ln k}{k}.$

15. $\displaystyle\sum_{k=1}^{\infty} \frac{k}{\ln (k+1)}.$

16. $\displaystyle\sum_{k=1}^{\infty} ke^{-k^2}.$

17. $\displaystyle\sum_{k=1}^{\infty} \frac{1}{(k+1)[\ln (k+1)]^2}.$

18. $\displaystyle\sum_{k=1}^{\infty} \frac{k^2 + 1}{k^2 + 3}.$

19. $\displaystyle\sum_{k=1}^{\infty} \left(1 + \frac{1}{k} \right)^k.$

20. $\displaystyle\sum_{k=1}^{\infty} \frac{1}{\sqrt{k^2 + 1}}.$

21. $\displaystyle\sum_{k=1}^{\infty} \frac{\tan^{-1} k}{1 + k^2}.$

22. $\displaystyle\sum_{k=1}^{\infty} \operatorname{sech}^2 k.$

23. $\displaystyle\sum_{k=5}^{\infty} 7k^{-p} \quad (p > 1).$

24. $\displaystyle\sum_{k=1}^{\infty} 7(k+5)^{-p} \quad (p \leq 1).$

25. $\displaystyle\sum_{k=1}^{\infty} k^2 \sin^2 \left(\frac{1}{k} \right).$

26. $\displaystyle\sum_{k=1}^{\infty} k^2 e^{-k^3}.$

27. Prove: $\displaystyle\sum_{k=2}^{\infty} \frac{1}{k(\ln k)^p}$ converges if $p > 1$ and diverges if $p \leq 1$.

28. Prove: $\displaystyle\sum_{k=3}^{\infty} \frac{1}{k(\ln k)[\ln (\ln k)]^p}$ converges if $p > 1$ and diverges if $p \leq 1$.

29. Prove: If Σu_k converges and Σv_k diverges, then $\Sigma(u_k + v_k)$ diverges and $\Sigma(u_k - v_k)$ diverges. [*Hint:* Assume $\Sigma(u_k + v_k)$ converges and use Theorem 11.4.3 to obtain a contradiction. Similarly for $\Sigma(u_k - v_k)$.]

30. Find examples to show that $\Sigma(u_k + v_k)$ and $\Sigma(u_k - v_k)$ may converge or may diverge if Σu_k and Σv_k both diverge.

31. With the help of Exercise 29, determine if the given series converges or diverges.

(a) $\displaystyle\sum_{k=1}^{\infty} \left[\left(\frac{2}{3}\right)^{k-1} + \frac{1}{k} \right]$

(b) $\displaystyle\sum_{k=1}^{\infty} \left[\frac{k^2}{1 + k^2} + \frac{1}{k(k + 1)} \right]$

(c) $\displaystyle\sum_{k=1}^{\infty} \left[\frac{1}{3k + 2} + \frac{1}{k^{3/2}} \right]$

(d) $\displaystyle\sum_{k=2}^{\infty} \left[\frac{1}{k(\ln k)^2} - \frac{1}{k^2} \right]$.

11.5 ADDITIONAL CONVERGENCE TESTS

In this section we will develop some additional convergence tests for series with *positive* terms.

Our first result is not only a practical test for convergence, but is also a theoretical tool that will be used to develop other convergence tests.

11.5.1 THEOREM
The Comparison Test

Let Σa_k and Σb_k be series with positive terms and suppose

$$a_1 \leq b_1, \ a_2 \leq b_2, \ a_3 \leq b_3, \ldots, a_k \leq b_k, \ldots$$

(a) *If the "bigger series" Σb_k converges, then the "smaller series" Σa_k also converges.*

(b) *On the other hand, if the "smaller series" Σa_k diverges, then the "bigger series" Σb_k also diverges.*

Proof of (a). Suppose that the series Σb_k converges and its sum is B. Then for all n

$$b_1 + b_2 + \cdots + b_n < \sum_{k=1}^{\infty} b_k = B$$

From our hypothesis it follows that

$$a_1 + a_2 + \cdots + a_n \leq b_1 + b_2 + \cdots + b_n$$

so that

$$a_1 + a_2 + \cdots + a_n < B$$

Thus each partial sum of the series Σa_k is less than B, so that Σa_k converges by Theorem 11.4.4.

Proof of (b). This part is really just an alternate phrasing of part (*a*). If Σa_k diverges, then Σb_k must diverge since convergence of Σb_k would imply convergence of Σa_k, contrary to hypothesis. ∎

Since the comparison test requires a little ingenuity to use, we will wait until the next section before applying it. For now, we will use the comparison test to develop some other tests that are easier to apply.

11.5.2 THEOREM

The Ratio Test

Let Σu_k be a series with positive terms and suppose

$$\lim_{k \to +\infty} \frac{u_{k+1}}{u_k} = \rho$$

(a) *If $\rho < 1$, the series converges.*
(b) *If $\rho > 1$ or $\rho = +\infty$, the series diverges.*
(c) *If $\rho = 1$, the series may converge or diverge, so that another test must be tried.*

Proof of (a). Assume $\rho < 1$, and let $r = \frac{1}{2}(1 + \rho)$. Thus $\rho < r < 1$, since r is the midpoint between 1 and ρ. It follows that the number

$$\epsilon = r - \rho \tag{1}$$

is positive. Since

$$\rho = \lim_{k \to +\infty} \frac{u_{k+1}}{u_k}$$

it follows that for k sufficiently large, say $k \geq K$, the ratios u_{k+1}/u_k are within ϵ units of ρ. Thus we will have

$$\frac{u_{k+1}}{u_k} < \rho + \epsilon \qquad \text{when} \quad k \geq K$$

or on substituting (1)

$$\frac{u_{k+1}}{u_k} < r \qquad \text{when} \quad k \geq K$$

that is,

$$u_{k+1} < r u_k \qquad \text{when} \quad k \geq K$$

This yields the inequalities

$$
\begin{aligned}
u_{K+1} &< ru_K \\
u_{K+2} &< ru_{K+1} < r^2 u_K \\
u_{K+3} &< ru_{K+2} < r^3 u_K \\
u_{K+4} &< ru_{K+3} < r^4 u_K \\
&\quad\vdots
\end{aligned}
\tag{2}
$$

But $|r| < 1$ (why?) so that

$$
ru_K + r^2 u_K + r^3 u_K + \cdots
$$

is a convergent geometric series. From the inequalities in (2) and the comparison test it follows that

$$
u_{K+1} + u_{K+2} + u_{K+3} + \cdots
$$

must also be a convergent series. Thus, $u_1 + u_2 + u_3 + \cdots + u_k + \cdots$ converges by Theorem 11.4.3c.

Proof of (b). Assume $\rho > 1$. Thus

$$
\epsilon = \rho - 1
\tag{3}
$$

is a positive number. Since

$$
\rho = \lim_{k \to +\infty} \frac{u_{k+1}}{u_k}
$$

it follows that for k sufficiently large, say $k \geq K$, the ratio u_{k+1}/u_k is within ϵ units of ρ. Thus,

$$
\frac{u_{k+1}}{u_k} > \rho - \epsilon \qquad \text{when} \quad k \geq K
$$

or on substituting (3)

$$
\frac{u_{k+1}}{u_k} > 1 \qquad \text{when} \quad k \geq K
$$

that is,

$$
u_{k+1} > u_k \qquad \text{when} \quad k \geq K
$$

This yields the inequalities

$$u_{K+1} > u_K$$
$$u_{K+2} > u_{K+1} > u_K$$
$$u_{K+3} > u_{K+2} > u_K \qquad (4)$$
$$u_{K+4} > u_{K+3} > u_K$$
$$\vdots$$

Since $u_K > 0$, it follows from (4) that $\lim\limits_{k \to +\infty} u_k \neq 0$, so $u_1 + u_2 + \cdots +$ $u_k + \cdots$ diverges by Theorem 11.4.2. The proof in the case where $\rho = +\infty$ is omitted.

Proof of (c). The series

$$\sum_{k=1}^{\infty} \frac{1}{k} \qquad \text{and} \qquad \sum_{k=1}^{\infty} \frac{1}{k^2}$$

both have $\rho = 1$ (verify). Since the first is the divergent harmonic series and the second is a convergent p-series, the ratio test does not distinguish between convergence and divergence when $\rho = 1$. ∎

▶ Example 1

The series

$$\sum_{k=1}^{\infty} \frac{1}{k!}$$

converges by the ratio test since

$$\rho = \lim_{k \to +\infty} \frac{u_{k+1}}{u_k} = \lim_{k \to +\infty} \frac{1/(k+1)!}{1/k!} = \lim_{k \to +\infty} \frac{k!}{(k+1)!} = \lim_{k \to +\infty} \frac{1}{k+1} = 0$$

so that $\rho < 1$. ◀

▶ Example 2

The series

$$\sum_{k=1}^{\infty} \frac{k}{2^k}$$

converges by the ratio test since

$$\rho = \lim_{k \to +\infty} \frac{u_{k+1}}{u_k} = \lim_{k \to +\infty} \frac{k+1}{2^{k+1}} \cdot \frac{2^k}{k} = \frac{1}{2} \lim_{k \to +\infty} \frac{k+1}{k} = \frac{1}{2}$$

so that $\rho < 1$. ◀

▶ Example 3

The series

$$\sum_{k=1}^{\infty} \frac{k^k}{k!}$$

diverges by the ratio test since

$$\rho = \lim_{k \to +\infty} \frac{u_{k+1}}{u_k} = \lim_{k \to +\infty} \frac{(k+1)^{k+1}}{(k+1)!} \cdot \frac{k!}{k^k}$$

$$= \lim_{k \to +\infty} \frac{(k+1)^k}{k^k}$$

$$= \lim_{k \to +\infty} \left(1 + \frac{1}{k}\right)^k = e \qquad \begin{bmatrix} \text{See Section 10.3} \\ \text{Formula (4).} \end{bmatrix}$$

Since $\rho = e > 1$, the series diverges. ◀

▶ Example 4

Determine whether the series

$$1 + \frac{1}{3} + \frac{1}{5} + \frac{1}{7} + \cdots + \frac{1}{2k-1} + \cdots$$

converges or diverges.

Solution. The ratio test is of no help since

$$\rho = \lim_{k \to +\infty} \frac{u_{k+1}}{u_k} = \lim_{k \to +\infty} \frac{1}{2(k+1)-1} \cdot \frac{2k-1}{1}$$

$$= \lim_{k \to +\infty} \frac{2k-1}{2k+1} = 1$$

However, the integral test proves that the series diverges since

$$\int_1^{+\infty} \frac{dx}{2x-1} = \lim_{l \to +\infty} \int_1^l \frac{dx}{2x-1} = \lim_{l \to +\infty} \frac{1}{2} \ln(2x-1) \Big]_1^l = +\infty \qquad ◀$$

▶ Example 5

The series

$$\frac{2!}{4} + \frac{4!}{4^2} + \frac{6!}{4^3} + \cdots + \frac{(2k)!}{4^k} + \cdots$$

diverges since

$$\rho = \lim_{k \to +\infty} \frac{u_{k+1}}{u_k} = \lim_{k \to +\infty} \frac{[2(k+1)]!}{4^{k+1}} \cdot \frac{4^k}{(2k)!}$$

$$= \lim_{k \to +\infty} \left(\frac{(2k+2)!}{(2k)!} \cdot \frac{1}{4} \right)$$

$$= \frac{1}{4} \lim_{k \to +\infty} (2k+1)(2k+2) = +\infty \qquad ◀$$

Sometimes the following result is easier to apply than the ratio test.

11.5.3 THEOREM
The Root Test

Let Σu_k be a series with positive terms and suppose

$$\rho = \lim_{k \to +\infty} \sqrt[k]{u_k} = \lim_{k \to +\infty} (u_k)^{1/k}$$

(a) *If $\rho < 1$, the series converges.*
(b) *If $\rho > 1$ or $\rho = +\infty$, the series diverges.*
(c) *If $\rho = 1$, the series may converge or diverge, so that another test must be tried.*

Since the proof of the root test is similar to the proof of the ratio test, we will omit it.

▶ Example 6

The series

$$\sum_{k=1}^{\infty} \left(\frac{4k-5}{2k+1} \right)^k$$

diverges by the root test since

$$\rho = \lim_{k \to +\infty} (u_k)^{1/k} = \lim_{k \to +\infty} \frac{4k-5}{2k+1} = 2 > 1 \qquad ◀$$

► Example 7

The series

$$\sum_{k=1}^{\infty} \frac{1}{(\ln (k + 1))^k}$$

converges by the root test, since

$$\lim_{k \to +\infty} (u_k)^{1/k} = \lim_{k \to +\infty} \frac{1}{\ln (k + 1)} = 0 < 1 \qquad \blacktriangleleft$$

We conclude this section with a remark about notation. Until now we have written most of our infinite series in the form

$$\sum_{k=1}^{\infty} u_k \qquad (5)$$

with the summation index beginning at 1. If the summation index begins at some other integer, it is always possible to rewrite the series in form (5). Thus, for example, the series

$$\sum_{k=0}^{\infty} \frac{2^k}{k!} = 1 + 2 + \frac{2^2}{2!} + \frac{2^3}{3!} + \cdots \qquad (6)$$

can be written as

$$\sum_{k=1}^{\infty} \frac{2^{k-1}}{(k - 1)!} = 1 + 2 + \frac{2^2}{2!} + \frac{2^2}{3!} + \cdots \qquad (7)$$

However, for purposes of applying the convergence tests, it is not necessary that the series have form (5). For example, we can apply the ratio test to (6) without converting to the more complicated form (7). Doing so yields

$$\rho = \lim_{k \to +\infty} \frac{u_{k+1}}{u_k} = \lim_{k \to +\infty} \frac{2^{k+1}}{(k + 1)!} \cdot \frac{k!}{2^k} = \lim_{k \to +\infty} \frac{2}{k + 1} = 0$$

which shows that the series converges since $\rho < 1$.

▶ Exercise Set 11.5

In Exercises 1–6, apply the ratio test. According to the test, does the series converge, does the series diverge, or are the results inconclusive?

1. $\displaystyle\sum_{k=1}^{\infty} \frac{3^k}{k!}$.

2. $\displaystyle\sum_{k=1}^{\infty} \frac{4^k}{k^2}$.

3. $\displaystyle\sum_{k=2}^{\infty} \frac{1}{5k}$.

4. $\displaystyle\sum_{k=1}^{\infty} k\left(\frac{1}{2}\right)^k$.

5. $\displaystyle\sum_{k=1}^{\infty} \frac{k!}{k^3}$.

6. $\displaystyle\sum_{k=1}^{\infty} \frac{k}{k^2+1}$.

In Exercises 7–10, apply the root test. According to the test, does the series converge, does the series diverge, or are the results inconclusive?

7. $\displaystyle\sum_{k=1}^{\infty} \left(\frac{3k+2}{2k-1}\right)^k$.

8. $\displaystyle\sum_{k=1}^{\infty} \left(\frac{k}{100}\right)^k$.

9. $\displaystyle\sum_{k=1}^{\infty} \frac{k}{5^k}$.

10. $\displaystyle\sum_{k=1}^{\infty} \frac{1}{k^2}$.

In Exercises 11–32, use any appropriate test to determine if the series converges.

11. $\displaystyle\sum_{k=1}^{\infty} \frac{2^k}{k^3}$.

12. $\displaystyle\sum_{k=1}^{\infty} \frac{1}{k^2}$.

13. $\displaystyle\sum_{k=0}^{\infty} \frac{7^k}{k!}$.

14. $\displaystyle\sum_{k=1}^{\infty} \frac{1}{2k+1}$.

15. $\displaystyle\sum_{k=1}^{\infty} \frac{k^2}{5^k}$.

16. $\displaystyle\sum_{k=1}^{\infty} \frac{k!\,10^k}{3^k}$.

17. $\displaystyle\sum_{k=1}^{\infty} k^{50}e^{-k}$.

18. $\displaystyle\sum_{k=1}^{\infty} \frac{k^2}{k^3+1}$.

19. $\displaystyle\sum_{k=1}^{\infty} k\left(\frac{2}{3}\right)^k$.

20. $\displaystyle\sum_{k=1}^{\infty} k^k$.

21. $\displaystyle\sum_{k=2}^{\infty} \frac{1}{k\ln k}$.

22. $\displaystyle\sum_{k=1}^{\infty} \frac{2^k}{k^3+1}$.

23. $\displaystyle\sum_{k=1}^{\infty} \left(\frac{4}{7k-1}\right)^k$.

24. $\displaystyle\sum_{k=1}^{\infty} \frac{(k!)^2 2^k}{(2k+2)!}$.

25. $\displaystyle\sum_{k=0}^{\infty} \frac{(k!)^2}{(2k)!}$.

26. $\displaystyle\sum_{k=1}^{\infty} \frac{1}{k^2+25}$.

27. $\displaystyle\sum_{k=1}^{\infty} \frac{1}{1+\sqrt{k}}$.

28. $\displaystyle\sum_{k=1}^{\infty} \frac{k^k}{k!}$.

29. $\displaystyle\sum_{k=1}^{\infty} \frac{\ln k}{e^k}$.

30. $\displaystyle\sum_{k=1}^{\infty} \frac{k!}{e^{k^2}}$.

31. $\displaystyle\sum_{k=0}^{\infty} \frac{(k+4)!}{4!k!4^k}$.

32. $\displaystyle\sum_{k=1}^{\infty} \left(\frac{k}{k+1}\right)^{k^2}$.

33. Determine if the following series converges:

$$1 + \frac{1\cdot 2}{1\cdot 3} + \frac{1\cdot 2\cdot 3}{1\cdot 3\cdot 5} + \frac{1\cdot 2\cdot 3\cdot 4}{1\cdot 3\cdot 5\cdot 7} + \cdots$$

34. For which positive values of α does $\displaystyle\sum_{k=1}^{\infty} \frac{\alpha^k}{k^\alpha}$ converge?

35. (a) Show $\displaystyle\lim_{k\to+\infty} (\ln k)^{1/k} = 1$ [*Hint:* Let $y = (\ln x)^{1/x}$ and find $\displaystyle\lim_{x\to+\infty} \ln y$.]

   (b) Use the result in (a) and the root test to show that $\displaystyle\sum_{k=1}^{\infty} \frac{\ln k}{3^k}$ converges.

   (c) Show that the series converges using the ratio test.

36. Prove: $\displaystyle\lim_{k\to+\infty} k!/k^k = 0$ [*Hint:* Exploit Theorem 11.4.1.]

37. Prove: $\displaystyle\lim_{k\to+\infty} \frac{a^k}{k!} = 0$ for every real number a. [See Hint to Exercise 36.]

11.6 APPLYING THE COMPARISON TEST

The comparison test requires some skill and experience to apply. One must first guess whether the given series Σu_k converges or diverges. If the guess is divergence, then one tries to prove it by producing a divergent series whose terms are "smaller" than the corresponding terms of Σu_k, while if the guess is convergence one tries to prove it by producing a convergent series whose terms are "bigger" than the corresponding terms of Σu_k. In this section we will illustrate these ideas through examples and also introduce a useful variation of the comparison test. Before starting, we remind the reader that the comparison test applies only to series with positive terms.

As noted above, the first step in applying the comparison test is to guess whether the series converges or diverges. To aid in this guessing process we have formulated some principles that sometimes *suggest* whether a series is likely to converge or diverge. We have called these "informal principles" because they are not intended as formal theorems. In fact, we will not guarantee that they *always* work. However, they work often enough to be useful as a starting point for the comparison test.

11.6.1 INFORMAL PRINCIPLE *Constants added or subtracted in the formula for u_k may usually be deleted without affecting the convergence or divergence behavior of the series.*

▶ Example 1

Use the above principle to help guess whether the following series converge or diverge.

(a) $\displaystyle\sum_{k=1}^{\infty} \frac{1}{2^k + 1}$ (b) $\displaystyle\sum_{k=5}^{\infty} \frac{1}{\sqrt{k} - 2}$ (c) $\displaystyle\sum_{k=1}^{\infty} \frac{1}{(k + \frac{1}{2})^3}$

Solution.
(a) Deleting the constant 1 suggests that

$$\sum_{k=1}^{\infty} \frac{1}{2^k + 1} \quad \text{behaves like} \quad \sum_{k=1}^{\infty} \frac{1}{2^k}$$

The modified series is a convergent geometric series, so the given series is likely to converge.

(b) Deleting the -2 suggests that

$$\sum_{k=5}^{\infty} \frac{1}{\sqrt{k} - 2} \quad \text{behaves like} \quad \sum_{k=5}^{\infty} \frac{1}{\sqrt{k}}$$

The modified series is a portion of a divergent p-series ($p = \frac{1}{2}$), so the given series is likely to diverge.

(c) Deleting the $\frac{1}{2}$ suggests that

$$\sum_{k=1}^{\infty} \frac{1}{(k + \frac{1}{2})^3} \qquad \text{behaves like} \qquad \sum_{k=1}^{\infty} \frac{1}{k^3}$$

The modified series is a convergent p-series ($p = 3$), so the given series is likely to converge. ◀

11.6.2 INFORMAL PRINCIPLE *If a polynomial in k appears as a factor in the numerator or denominator of u_k, all but the highest power of k in the polynomial may usually be deleted without affecting the convergence or divergence behavior of the series.*

▶ Example 2

Use the above principle to help guess whether the following series converge or diverge.

(a) $\displaystyle\sum_{k=1}^{\infty} \frac{1}{\sqrt{k^3 + 2k}}$ (b) $\displaystyle\sum_{k=1}^{\infty} \frac{6k^4 - 2k^3 + 1}{k^5 + k^2 - 2k}$

Solution.
(a) Deleting the term $2k$ suggests that

$$\sum_{k=1}^{\infty} \frac{1}{\sqrt{k^3 + 2k}} \qquad \text{behaves like} \qquad \sum_{k=1}^{\infty} \frac{1}{\sqrt{k^3}} = \sum_{k=1}^{\infty} \frac{1}{k^{3/2}}$$

Since the modified series is a convergent p-series ($p = \frac{3}{2}$), the given series is likely to converge.

(b) Deleting all but the highest powers of k in the numerator and also in the denominator suggests that

$$\sum_{k=1}^{\infty} \frac{6k^4 - 2k^3 + 1}{k^5 + k^2 - 2k} \qquad \text{behaves like} \qquad \sum_{k=1}^{\infty} \frac{6k^4}{k^5} = 6 \sum_{k=1}^{\infty} \frac{1}{k}$$

Since the modified series is a constant times the divergent harmonic series, the given series is likely to diverge. ◀

Once it is decided if a series is likely to converge or diverge, the second step in applying the comparison test is to produce a series with which the given series may be compared to substantiate the guess. Let us consider the case of convergence first. To prove Σa_k converges by the comparison test, we must find a *convergent* series Σb_k such that

$$a_k \leq b_k$$

for all k. Frequently, b_k is derived from the formula for a_k by either increasing the numerator of a_k, or decreasing the denominator of a_k, or both.

▶ Example 3

Use the comparison test to determine whether

$$\sum_{k=1}^{\infty} \frac{1}{2k^2 + k}$$

converges or diverges.

Solution. Using Principle 11.6.2, the series behaves like the series

$$\sum_{k=1}^{\infty} \frac{1}{2k^2} = \frac{1}{2} \sum_{k=1}^{\infty} \frac{1}{k^2}$$

which is a constant times a convergent *p*-series. Thus the given series is likely to converge. To prove the convergence, observe that when we discard the k from the denominator of $1/(2k^2 + k)$, the denominator decreases and the ratio increases, so that

$$\frac{1}{2k^2 + k} < \frac{1}{2k^2}$$

for $k = 1, 2, \dots$. Since

$$\sum_{k=1}^{\infty} \frac{1}{2k^2} = \frac{1}{2} \sum_{k=1}^{\infty} \frac{1}{k^2}$$

converges, so does $\displaystyle\sum_{k=1}^{\infty} \frac{1}{2k^2 + k}$ by the comparison test.   ◄

▶ Example 4

Use the comparison test to determine whether

$$\sum_{k=1}^{\infty} \frac{1}{2k^2 - k}$$

converges or diverges.

Solution. Using Principle 11.6.2, the series behaves like the convergent series

$$\sum_{k=1}^{\infty} \frac{1}{2k^2} = \frac{1}{2} \sum_{k=1}^{\infty} \frac{1}{k^2}$$

Thus the given series is likely to converge. However, if we discard the k from the denominator of $1/(2k^2 - k)$, the denominator increases and the ratio decreases, so that

$$\frac{1}{2k^2 - k} > \frac{1}{2k^2}$$

Unfortunately, this inequality is in the wrong direction to prove convergence of the given series. A different approach is needed; we must do something to decrease the denominator, not increase it. We accomplish this by replacing k by k^2 to obtain

$$\frac{1}{2k^2 - k} \leq \frac{1}{2k^2 - k^2} = \frac{1}{k^2}$$

Since $\sum_{k=1}^{\infty} \frac{1}{k^2}$ is a convergent p-series, the given series converges by the comparison test. ◀

To prove that a series Σa_k diverges by the comparison test, we must produce a divergent series Σb_k of positive terms such that $a_k \geq b_k$ for all k.

▶ Example 5

Use the comparison test to determine whether

$$\sum_{k=1}^{\infty} \frac{1}{k - \frac{1}{4}}$$

converges or diverges.

Solution. Using Principle 11.6.1, the series behaves like the divergent harmonic series

$$\sum_{k=1}^{\infty} \frac{1}{k}$$

Thus the given series is likely to diverge. Since

$$\frac{1}{k - \frac{1}{4}} > \frac{1}{k} \qquad \text{for } k = 1, 2, \ldots .$$

and since $\sum_{k=1}^{\infty} \frac{1}{k}$ diverges, the given series diverges by the comparison test. ◀

▶ Example 6

Use the comparison test to determine whether

$$\sum_{k=1}^{\infty} \frac{1}{\sqrt{k} + 5}$$

converges or diverges.

Solution. Using Principle 11.6.1, the series behaves like the divergent *p*-series

$$\sum_{k=1}^{\infty} \frac{1}{\sqrt{k}}$$

Thus the given series is likely to diverge. For $k \geq 25$ we have

$$\frac{1}{\sqrt{k} + 5} \geq \frac{1}{\sqrt{k} + \sqrt{k}} = \frac{1}{2\sqrt{k}}$$

and since

$$\sum_{k=25}^{\infty} \frac{1}{2\sqrt{k}}$$

diverges (why?) the series

$$\sum_{k=25}^{\infty} \frac{1}{\sqrt{k}+5}$$

diverges by the comparison test; consequently the given series diverges by Theorem 11.4.3c. ◀

The following result is often easier to apply than the comparison test.

11.6.3 THEOREM
The Limit Comparison Test

Let Σa_k and Σb_k be series with positive terms and suppose

$$\rho = \lim_{k \to +\infty} \frac{a_k}{b_k}$$

If ρ is finite and $\rho > 0$, then the series both converge or both diverge.

Proof. The proof will be complete if we can show that Σb_k converges when Σa_k converges and conversely. To this end, let

$$\epsilon = \frac{\rho}{2} \tag{1}$$

so that $\epsilon > 0$ because $\rho > 0$ by hypothesis. From the assumption that

$$\rho = \lim_{k \to +\infty} \frac{a_k}{b_k}$$

it follows that for k sufficiently large, say $k \geq K$, the ratio a_k/b_k will be within ϵ units of ρ. Thus

$$\rho - \epsilon < \frac{a_k}{b_k} < \rho + \epsilon \qquad \text{when} \qquad k \geq K$$

or on substituting (1)

$$\frac{1}{2}\rho < \frac{a_k}{b_k} < \frac{3}{2}\rho \qquad \text{when} \quad k \geq K$$

or

$$\frac{1}{2}\rho b_k < a_k < \frac{3}{2}\rho b_k \qquad \text{when} \quad k \geq K \tag{2}$$

If $\displaystyle\sum_{k=1}^{\infty} a_k$ converges, then $\displaystyle\sum_{k=K}^{\infty} a_k$ converges by Theorem 11.4.3c. It follows from the comparison test and the left-hand inequality in (2), that $\displaystyle\sum_{k=K}^{\infty} \frac{1}{2}\rho b_k$

converges. Thus $\displaystyle\sum_{k=K}^{\infty} b_k$ converges (Theorem 11.4.3b) and $\displaystyle\sum_{k=1}^{\infty} b_k$ converges (Theorem 11.4.3c).

Conversely, if $\displaystyle\sum_{k=1}^{\infty} b_k$ converges, then $\displaystyle\sum_{k=K}^{\infty} \frac{3}{2}\rho b_k$ converges (Theorem

11.4.3b,c), so that $\displaystyle\sum_{k=K}^{\infty} a_k$ converges by the comparison test and the right-hand

inequality in (2). Thus $\displaystyle\sum_{k=1}^{\infty} a_k$ converges. ▪

▶ Example 7

Use the limit comparison test to determine if

$$\sum_{k=1}^{\infty} \frac{3k^3 - 2k^2 + 4}{k^5 - k^3 + 2}$$

converges or diverges.

Solution. From Principle 11.6.2, the series behaves like

$$\sum_{k=1}^{\infty} \frac{3k^3}{k^5} = \sum_{k=1}^{\infty} \frac{3}{k^2} \tag{3}$$

which converges since it is a constant times a convergent *p*-series. Thus the given series is likely to converge. To substantiate this, we apply the limit comparison test to series (3) and the given series. We obtain

$$\rho = \lim_{k \to +\infty} \frac{\dfrac{3k^3 - 2k^2 + 4}{k^5 - k^3 + 2}}{\dfrac{3}{k^2}} = \lim_{k \to +\infty} \frac{3k^5 - 2k^4 + 4k^2}{3k^5 - 3k^3 + 6} = 1$$

Since $\rho > 0$, the given series converges because series (3) converges. ◀

Unlike the comparison test, the limit comparison test does not require any tricky manipulation of inequalities. However the test only applies when $0 < \rho < +\infty$. We note, though, that if $\rho = 0$ or $\rho = +\infty$, conclusions about convergence or divergence may be drawn in certain cases (see Exercise 42).

▶ Exercise Set 11.6

In Exercises 1–6, prove that the series converges by the comparison test.

1. $\displaystyle\sum_{k=1}^{\infty} \frac{1}{3^k + 5}$.

2. $\displaystyle\sum_{k=1}^{\infty} \frac{2}{k^4 + k}$.

3. $\displaystyle\sum_{k=1}^{\infty} \frac{1}{5k^2 - k}$.

4. $\displaystyle\sum_{k=1}^{\infty} \frac{k}{8k^3 + 2k^2 - 1}$.

5. $\displaystyle\sum_{k=1}^{\infty} \frac{2^k - 1}{3^k + 2k}$.

6. $\displaystyle\sum_{k=1}^{\infty} \frac{5 \sin^2 k}{k!}$.

In Exercises 7–12, prove that the series diverges by the comparison test.

7. $\displaystyle\sum_{k=1}^{\infty} \frac{3}{k - \frac{1}{4}}$.

8. $\displaystyle\sum_{k=1}^{\infty} \frac{1}{\sqrt{k + 8}}$.

9. $\displaystyle\sum_{k=1}^{\infty} \frac{9}{\sqrt{k} + 1}$.

10. $\displaystyle\sum_{k=2}^{\infty} \frac{k + 1}{k^2 - k}$.

11. $\displaystyle\sum_{k=1}^{\infty} \frac{k^{4/3}}{8k^2 + 5k + 1}$.

12. $\displaystyle\sum_{k=1}^{\infty} \frac{k^{-1/2}}{2 + \sin^2 k}$.

In Exercises 13–18, use the limit comparison test to determine if the series converges or diverges.

13. $\displaystyle\sum_{k=1}^{\infty} \frac{4k^2 - 2k + 6}{8k^7 + k - 8}$.

14. $\displaystyle\sum_{k=1}^{\infty} \frac{1}{9k + 6}$.

15. $\displaystyle\sum_{k=1}^{\infty} \frac{5}{3^k + 1}$.

16. $\displaystyle\sum_{k=1}^{\infty} \frac{k(k + 3)}{(k + 1)(k + 2)(k + 5)}$.

17. $\displaystyle\sum_{k=1}^{\infty} \frac{1}{\sqrt[3]{8k^2 - 3k}}$.

18. $\displaystyle\sum_{k=1}^{\infty} \frac{1}{(2k + 3)^{17}}$.

In Exercises 19–33, use any method to determine if the series converges or diverges. In some cases you may have to use tests from earlier sections.

19. $\displaystyle\sum_{k=1}^{\infty} \frac{1}{k^3 + 2k + 1}$.

20. $\displaystyle\sum_{k=1}^{\infty} \frac{1}{(3 + k)^{2/5}}$.

21. $\displaystyle\sum_{k=1}^{\infty} \frac{1}{9k - 2}$.

22. $\displaystyle\sum_{k=1}^{\infty} \frac{\ln k}{k}$.

23. $\displaystyle\sum_{k=1}^{\infty} \frac{\sqrt{k}}{k^3 + 1}$.

24. $\displaystyle\sum_{k=1}^{\infty} \frac{4}{2 + k3^k}$.

25. $\displaystyle\sum_{k=1}^{\infty} \frac{1}{\sqrt{k(k + 1)}}$.

26. $\displaystyle\sum_{k=1}^{\infty} \frac{2 + (-1)^k}{5^k}$.

27. $\displaystyle\sum_{k=1}^{\infty} \frac{2 + \sqrt{k}}{(k + 1)^3 - 1}$.

28. $\displaystyle\sum_{k=1}^{\infty} \frac{4 + |\cos k|}{k^3}$.

29. $\displaystyle\sum_{k=1}^{\infty} \frac{1}{4 + 2^{-k}}$.

30. $\displaystyle\sum_{k=1}^{\infty} \frac{\sqrt{k} \ln k}{k^3 + 1}$.

31. $\displaystyle\sum_{k=1}^{\infty} \frac{\tan^{-1} k}{k^2}$.

32. $\displaystyle\sum_{k=1}^{\infty} \frac{5^k + k}{k! + 3}$.

33. $\displaystyle\sum_{k=1}^{\infty} \frac{\ln k}{k\sqrt{k}}$.

34. Use the limit comparison test to show that $\displaystyle\sum_{k=1}^{\infty} \sin\left(\frac{\pi}{k}\right)$ diverges. [*Hint:* Compare with the series $\displaystyle\sum_{k=1}^{\infty} \frac{\pi}{k}$.]

35. Use the comparison test to determine if $\displaystyle\sum_{k=1}^{\infty} \frac{\ln k}{k^2}$ converges or diverges. [*Hint:* $\ln x < 2\sqrt{x}$ for $x > 1$ by Exercise 23 of Section 7.3.]

36. Determine if $\displaystyle\sum_{k=2}^{\infty} \frac{1}{(\ln k)^2}$ converges or diverges. [*Hint:* See Hint to Exercise 35.]

37. Let a, b, and p be positive constants. For which values of p does the series $\displaystyle\sum_{k=1}^{\infty} \frac{1}{(a + bk)^p}$ converge?

38. (a) Show that $k^k \geq k!$ and use this result to prove that the series $\sum\limits_{k=1}^{\infty} k^{-k}$ converges by the comparison test.

(b) Prove convergence using the root test.

39. Use the limit comparison test to investigate convergence of $\sum\limits_{k=1}^{\infty} \dfrac{(k+1)^2}{(k+2)!}$.

40. Use the limit comparison test to investigate convergence of the series $1 + \frac{1}{3} + \frac{1}{5} + \frac{1}{7} + \cdots$

41. Prove that $\sum\limits_{k=1}^{\infty} \dfrac{1}{k!}$ converges by comparison with a suitable geometric series.

42. Let Σa_k and Σb_k be series with positive terms. Prove:

(a) If $\lim\limits_{k \to +\infty} (a_k/b_k) = 0$ and Σb_k converges, then Σa_k converges.

(b) If $\lim\limits_{k \to +\infty} (a_k/b_k) = +\infty$ and Σb_k diverges, then Σa_k diverges.

11.7 ALTERNATING SERIES; CONDITIONAL CONVERGENCE

So far our emphasis has been on series with positive terms. In this section we discuss series containing negative terms.

Of special importance are series whose terms are alternately positive and negative. These are called **alternating series.** Such series have one of two possible forms:

$$a_1 - a_2 + a_3 - a_4 + \cdots + (-1)^{k+1} a_k + \cdots$$

or

$$-a_1 + a_2 - a_3 + a_4 - \cdots + (-1)^k a_k + \cdots$$

where the a_k's are all positive.

The following theorem is the key result on convergence of alternating series.

11.7.1 THEOREM
Alternating Series Test

An alternating series

$$a_1 - a_2 + a_3 - a_4 + \cdots + (-1)^{k+1} a_k + \cdots$$

or

$$-a_1 + a_2 - a_3 + a_4 - \cdots + (-1)^k a_k + \cdots$$

converges if the following two conditions are satisfied:

(a) $a_1 \geq a_2 \geq a_3 \geq \cdots \geq a_k \geq \cdots$

(b) $\lim\limits_{k \to +\infty} a_k = 0$

To prove this result, we will need the following fact about sequences:

If the even-numbered terms of a sequence tend toward a limit L, and if the odd-numbered terms of the sequence tend toward the same limit L, then the entire sequence tends toward the limit L.

This result should be intuitively obvious; we omit the proof.

Proof of 11.7.1. We will consider the case

$$a_1 - a_2 + a_3 - a_4 + \cdots + (-1)^{k+1} a_k + \cdots$$

The other case is left as an exercise. Consider the partial sums

$$
\begin{aligned}
s_2 &= a_1 - a_2 \\
s_4 &= (a_1 - a_2) + (a_3 - a_4) \\
s_6 &= (a_1 - a_2) + (a_3 - a_4) + (a_5 - a_6) \\
s_8 &= (a_1 - a_2) + (a_3 - a_4) + (a_5 - a_6) + (a_7 - a_8) \\
&\ \vdots
\end{aligned}
$$

By hypothesis (*a*), each of the differences appearing in the parentheses is nonnegative so that

$$s_2 \leq s_4 \leq s_6 \leq s_8 \leq \cdots$$

Moreover, the terms in this sequence are all less than or equal to a_1 since we can write

$$
\begin{aligned}
s_2 &= a_1 - a_2 \\
s_4 &= a_1 - (a_2 - a_3) - a_4 \\
s_6 &= a_1 - (a_2 - a_3) - (a_4 - a_5) - a_6 \\
s_8 &= a_1 - (a_2 - a_3) - (a_4 - a_5) - (a_6 - a_7) - a_8 \\
&\ \vdots
\end{aligned}
$$

Thus the sequence

$$s_2, s_4, s_6, s_8, \ldots s_{2n}, \ldots$$

converges to some limit S, by Theorem 11.2.2. That is,

$$\lim_{n \to +\infty} s_{2n} = S \tag{1}$$

Now consider the sequence

$$s_1, s_3, s_5, \ldots, s_{2n-1}, \ldots$$

We can write

$$s_{2n-1} = s_{2n} + a_{2n} \tag{2}$$

But $2n \to +\infty$ as $n \to +\infty$ so, $\lim\limits_{n \to +\infty} a_{2n} = 0$ by hypothesis (b). Thus, from (1) and (2)

$$\lim_{n \to +\infty} s_{2n-1} = \lim_{n \to +\infty} s_{2n} + \lim_{n \to +\infty} a_{2n} = S + 0 = S \tag{3}$$

From (1) and (3), the sequence $s_1, s_2, s_3, \ldots, s_n, \ldots$ converges to S, so the series converges. ∎

▶ Example 1

The series

$$1 - \frac{1}{2} + \frac{1}{3} - \frac{1}{4} + \cdots + (-1)^{k+1} \frac{1}{k} + \cdots$$

is called the **alternating harmonic series.** Since

$$a_k = \frac{1}{k} > \frac{1}{k+1} = a_{k+1}$$

and

$$\lim_{k \to +\infty} a_k = \lim_{k \to +\infty} \frac{1}{k} = 0$$

this series converges by the alternating series test. ◀

▶ Example 2

Determine whether the alternating series

$$\sum_{k=1}^{\infty} (-1)^{k+1} \frac{k+3}{k(k+1)}$$

converges or diverges.

Solution. Requirement (*b*) of the alternating series test is satisfied since

$$\lim_{k \to +\infty} a_k = \lim_{k \to +\infty} \frac{k + 3}{k(k + 1)} = \lim_{k \to +\infty} \frac{\dfrac{1}{k} + \dfrac{3}{k^2}}{1 + \dfrac{1}{k}} = 0$$

To see if requirement (*a*) is met, we must determine if the sequence

$$\{a_k\}_{k=1}^{+\infty} = \left\{ \frac{k + 3}{k(k + 1)} \right\}_{k=1}^{+\infty}$$

is nonincreasing. Since

$$\frac{a_{k+1}}{a_k} = \frac{k + 4}{(k + 1)(k + 2)} \cdot \frac{k(k + 1)}{k + 3} = \frac{k^2 + 4k}{k^2 + 5k + 6}$$

$$= \frac{k^2 + 4k}{(k^2 + 4k) + (k + 6)} < 1$$

we have $a_k > a_{k+1}$, so the series converges by the alternating series test. ◄

$s_2 \quad s_4 \ S \ s_5 \ s_3 \qquad s_1$

Figure 11.7.1

REMARK. If an alternating series violates condition (*b*) of the alternating series test, then the series must diverge by the divergence test (11.4.2). However if condition (*b*) is satisfied, but (*a*) is not, the series may either converge or diverge. The interested reader will find some nice examples in an article by R. Lariviere, "On a Convergence Test for Alternating Series," *Mathematics Magazine,* Vol. 29 (1956), p. 88.

Figure 11.7.1 provides some insight into the way in which an alternating series

$$a_1 - a_2 + a_3 - a_4 + \cdots + (-1)^{k+1} a_k + \cdots$$

converges to its sum S when the hypotheses of the alternating test are satisfied. In the figure we have plotted the successive partial sums on the *x*-axis. Because

$$a_1 \geq a_2 \geq a_3 \geq a_4 \geq \cdots$$

and

$$\lim_{k \to +\infty} a_k = 0$$

the successive partial sums oscillate in smaller and smaller steps, closing in on the sum S. It is of interest to note that the even-numbered partial sums are smaller than S and the odd-numbered partial sums are greater than S. Thus the sum S falls between any two successive partial sums; that is, for any positive integer n

$$s_n < S < s_{n+1} \quad \text{or} \quad s_{n+1} < S < s_n$$

depending on whether n is even or odd. In either case

$$|S - s_n| < |s_{n+1} - s_n| \tag{4}$$

But $s_{n+1} - s_n = \pm a_{n+1}$ (the sign depending on whether n is even or odd) so (4) yields

$$|S - s_n| < a_{n+1} \tag{5}$$

Since $|S - s_n|$ represents the magnitude of error that results when we approximate the sum of the entire series by the sum of the first n terms, (5) tells us that this error is less than the magnitude of the $(n + 1)$-st term in the series. The reader can check that (5) also holds for alternating series of the form

$$-a_1 + a_2 - a_3 + a_4 - \cdots + (-1)^k a_k + \cdots$$

In summary, we have the following result.

11.7.2 THEOREM *If an alternating series satisfies the hypotheses of the alternating series test, and if the sum S of the series is approximated by the nth partial sum s_n, then the absolute value of the error is less than a_{n+1}.*

▶ Example 3

As shown in Example 1, the alternating harmonic series

$$1 - \frac{1}{2} + \frac{1}{3} - \frac{1}{4} + \cdots + (-1)^{k+1} \frac{1}{k} + \cdots$$

satisfies the hypotheses of the alternating series test. If we approximate the sum of the series by

$$1 - \frac{1}{2} + \frac{1}{3} - \frac{1}{4} = \frac{7}{12}$$

then the absolute value of the error is less than $\frac{1}{5} = 0.2$, and if we approximate the sum by

$$1 - \frac{1}{2} + \frac{1}{3} - \frac{1}{4} + \frac{1}{5} - \frac{1}{6} + \frac{1}{7} = \frac{319}{420}$$

then the absolute value of the error is less than $\frac{1}{8} = 0.125$. ◄

The series

$$1 - \frac{1}{2} - \frac{1}{2^2} + \frac{1}{2^3} + \frac{1}{2^4} - \frac{1}{2^5} - \frac{1}{2^6} + \cdots$$

does not fit in any of the categories studied so far—it has mixed signs, but is not alternating. We will now develop some convergence tests that can be applied to such series.

11.7.3 DEFINITION A series

$$\sum_{k=1}^{\infty} u_k = u_1 + u_2 + \cdots + u_k + \cdots$$

is said to **converge absolutely** if the series of absolute values

$$\sum_{k=1}^{\infty} |u_k| = |u_1| + |u_2| + \cdots + |u_k| + \cdots$$

converges.

▶ Example 4

The series

$$1 - \frac{1}{2} - \frac{1}{2^2} + \frac{1}{2^3} + \frac{1}{2^4} - \frac{1}{2^5} - \frac{1}{2^6} + \cdots$$

converges absolutely since the series of absolute values

$$1 + \frac{1}{2} + \frac{1}{2^2} + \frac{1}{2^3} + \frac{1}{2^4} + \frac{1}{2^5} + \frac{1}{2^6} + \cdots$$

is a convergent geometric series. On the other hand, the alternating harmonic series

$$1 - \frac{1}{2} + \frac{1}{3} - \frac{1}{4} + \frac{1}{5} - \cdots$$

does not converge absolutely since the series of absolute values

$$1 + \frac{1}{2} + \frac{1}{3} + \frac{1}{4} + \frac{1}{5} + \cdots$$

diverges.   ◀

Absolute convergence is of importance because of the following theorem.

11.7.4 THEOREM *If the series*

$$\sum_{k=1}^{\infty} |u_k| = |u_1| + |u_2| + \cdots + |u_k| + \cdots$$

converges, then so does the series

$$\sum_{k=1}^{\infty} u_k = u_1 + u_2 + \cdots + u_k + \cdots$$

In other words, if a series converges absolutely, then it converges.

Proof. Our proof is based on a trick. We will show that the series

$$\sum_{k=1}^{\infty} (u_k + |u_k|) \tag{6}$$

converges. Since $\Sigma |u_k|$ is assumed to converge, it will then follow from Theorem 11.4.3a that Σu_k converges, since

$$\sum_{k=1}^{\infty} u_k = \sum_{k=1}^{\infty} [(u_k + |u_k|) - |u_k|]$$

For all k, the value of $u_k + |u_k|$ is either 0 or $2|u_k|$, depending on whether u_k is negative or not. Thus for all values of k

$$0 \leq u_k + |u_k| \leq 2|u_k| \tag{7}$$

But $\Sigma 2|u_k|$ is a convergent series since it is a constant times the convergent series $\Sigma |u_k|$. Thus from (7), series (6) converges by the comparison test. ▮

▶ Example 5

In Example 4 we showed that

$$1 - \frac{1}{2} - \frac{1}{2^2} + \frac{1}{2^3} + \frac{1}{2^4} - \frac{1}{2^5} - \frac{1}{2^6} + \cdots$$

converges absolutely. It follows from Theorem 11.7.4 that the series converges. ◀

▶ Example 6

Show that the series

$$\sum_{k=1}^{\infty} \frac{\cos k}{k^2}$$

converges.

Solution. Since $|\cos k| \leq 1$ for all k

$$\left| \frac{\cos k}{k^2} \right| \leq \frac{1}{k^2}$$

Thus

$$\sum_{k=1}^{\infty} \left| \frac{\cos k}{k^2} \right|$$

converges by the comparison test, and consequently

$$\sum_{k=1}^{\infty} \frac{\cos k}{k^2}$$

converges. ◀

If $\Sigma |u_k|$ *diverges,* no conclusion can be drawn about the convergence or divergence of Σu_k. For example, consider the two series

$$1 - \frac{1}{2} + \frac{1}{3} - \frac{1}{4} + \cdots + (-1)^{k+1} \frac{1}{k} + \cdots \tag{8}$$

$$-1 - \frac{1}{2} - \frac{1}{3} - \frac{1}{4} - \cdots - \frac{1}{k} - \cdots \tag{9}$$

Series (8), the alternating harmonic series, converges; while series (9), being a constant times the harmonic series, diverges. Yet in each case the series of absolute values is

$$1 + \frac{1}{2} + \frac{1}{3} + \cdots + \frac{1}{k} + \cdots$$

which diverges. A series such as (8), which is convergent, but not absolutely convergent, is called ***conditionally convergent.***

The following version of the ratio test is useful for investigating absolute convergence.

11.7.5 THEOREM

The Ratio Test for Absolute Convergence

Let Σu_k be a series with nonzero terms and suppose

$$\lim_{k \to +\infty} \frac{|u_{k+1}|}{|u_k|} = \rho$$

(a) *If $\rho < 1$, the series Σu_k converges absolutely.*
(b) *If $\rho > 1$ or if $\rho = +\infty$, then the series Σu_k diverges.*
(c) *If $\rho = 1$, no conclusion about convergence can be drawn.*

The proof is discussed in the exercises.

▶ Example 7

The series

$$\sum_{k=1}^{\infty} (-1)^k \frac{2^k}{k!}$$

converges absolutely since

$$\rho = \lim_{k \to +\infty} \frac{|u_{k+1}|}{|u_k|} = \lim_{k \to +\infty} \frac{2^{k+1}}{(k+1)!} \cdot \frac{k!}{2^k} = \lim_{k \to +\infty} \frac{2}{k+1} = 0 < 1 \qquad ◀$$

▶ Example 8

We proved earlier (Theorem 11.3.3) that a geometric series

$$a + ar + ar^2 + \cdots + ar^{k-1} + \cdots$$

converges if $|r| < 1$ and diverges if $|r| \geq 1$. However, a stronger statement can be made—the series converges *absolutely* if $|r| < 1$. This follows from Theorem 11.7.5 since

$$\rho = \lim_{k \to +\infty} \frac{|u_{k+1}|}{|u_k|} = \lim_{k \to +\infty} \frac{|ar^k|}{|ar^{k-1}|} = \lim_{k \to +\infty} |r| = |r|$$

so that $\rho < 1$ if $|r| < 1$. ◄

The following review is included as a ready reference to convergence tests.

Review of Convergence Tests

NAME	STATEMENT	COMMENTS
Divergence test (11.4.2)	If $\lim_{k \to +\infty} u_k \neq 0$, then $\sum u_k$ diverges.	If $\lim_{k \to +\infty} u_k = 0$, $\sum u_k$ may or may not converge.
Integral test (11.4.5)	Let $\sum u_k$ be a series with positive terms and let $f(x)$ be the function that results when k is replaced by x in the formula for u_k. If f is decreasing and continuous for $x \geq 1$, then $$\sum_{k=1}^{\infty} u_k \quad \text{and} \quad \int_{1}^{+\infty} f(x)\, dx$$ both converge or both diverge.	Use this test when $f(x)$ is easy to integrate.
Comparison test (11.5.1)	Let $\sum a_k$ and $\sum b_k$ be series with positive terms such that $$a_1 \leq b_1, a_2 \leq b_2, \ldots, a_k \leq b_k, \ldots$$ If $\sum b_k$ converges, then $\sum a_k$ converges; and if $\sum a_k$ diverges, then $\sum b_k$ diverges.	Use this test as a last resort. Other tests are often easier to apply.
Ratio test (11.5.2)	Let $\sum u_k$ be a series with positive terms and suppose $$\lim_{k \to +\infty} \frac{u_{k+1}}{u_k} = \rho$$ (a) Series converges if $\rho < 1$. (b) Series diverges if $\rho > 1$ or $\rho = +\infty$. (c) No conclusion if $\rho = 1$.	Try this test when u_k involves factorials or kth powers.

Review of Convergence Tests (Continued)

NAME	STATEMENT	COMMENTS
Root test (11.5.3)	Let $\sum u_k$ be a series with positive terms such that $$\rho = \lim_{k \to +\infty} \sqrt[k]{u_k}$$ (a) Series converges if $\rho < 1$. (b) Series diverges if $\rho > 1$ or $\rho = +\infty$. (c) No conclusion if $\rho = 1$.	Try this test when u_k involves kth powers.
Limit comparison test (11.6.3)	Let $\sum a_k$ and $\sum b_k$ be series with positive terms such that $$\rho = \lim_{k \to +\infty} \frac{a_k}{b_k}$$ If $0 < \rho < +\infty$, then both series converge or both diverge.	This is easier to apply than the comparison test, but still requires some skill in choosing the series $\sum b_k$ for comparison.
Alternating series test (11.7.1)	The series $$a_1 - a_2 + a_3 - a_4 + \cdots$$ and $$-a_1 + a_2 - a_3 + a_4 - \cdots$$ converge if (a) $a_1 \geq a_2 \geq a_3 \geq \cdots$ (b) $\lim_{k \to +\infty} a_k = 0$	This test applies only to alternating series.
Ratio test for absolute convergence (11.7.5)	Let $\sum u_k$ be a series with nonzero terms such that $$\rho = \lim_{k \to +\infty} \frac{|u_{k+1}|}{|u_k|}$$ (a) Series converges absolutely if $\rho < 1$. (b) Series diverges if $\rho > 1$ or $\rho = +\infty$. (c) No conclusion if $\rho = 1$.	The series need not have positive terms and need not be alternating to use this test.

▶ Exercise Set 11.7

In Exercises 1–6, use the alternating series test to determine if the series converges or diverges.

1. $\displaystyle\sum_{k=1}^{\infty} \frac{(-1)^{k+1}}{2k+1}$.

2. $\displaystyle\sum_{k=1}^{\infty} (-1)^{k+1} \frac{k}{3^k}$.

3. $\displaystyle\sum_{k=1}^{\infty} (-1)^{k+1} \frac{k+1}{3k+1}$.

4. $\displaystyle\sum_{k=1}^{\infty} (-1)^{k+1} \frac{k+4}{k^2+k}$.

5. $\displaystyle\sum_{k=1}^{\infty} (-1)^{k+1} e^{-k}$.

6. $\displaystyle\sum_{k=3}^{\infty} (-1)^k \frac{\ln k}{k}$.

In Exercises 7–12, use the ratio test for absolute convergence to determine if the series converges absolutely or diverges.

7. $\displaystyle\sum_{k=1}^{\infty} \left(-\frac{3}{5}\right)^k$.

8. $\displaystyle\sum_{k=1}^{\infty} (-1)^{k+1} \frac{2^k}{k!}$.

9. $\displaystyle\sum_{k=1}^{\infty} (-1)^{k+1} \frac{3^k}{k^2}$.

10. $\displaystyle\sum_{k=1}^{\infty} (-1)^k \left(\frac{k}{5^k}\right)$.

11. $\displaystyle\sum_{k=1}^{\infty} (-1)^k \left(\frac{k^3}{e^k}\right)$.

12. $\displaystyle\sum_{k=1}^{\infty} (-1)^{k+1} \frac{k^k}{k!}$.

In Exercises 13–30, classify the series as: absolutely convergent, conditionally convergent, or divergent.

13. $\displaystyle\sum_{k=1}^{\infty} \frac{(-1)^{k+1}}{3k}$.

14. $\displaystyle\sum_{k=1}^{\infty} \frac{(-1)^{k+1}}{k^{4/3}}$.

15. $\displaystyle\sum_{k=1}^{\infty} \frac{(-4)^k}{k^2}$.

16. $\displaystyle\sum_{k=1}^{\infty} \frac{(-1)^{k+1}}{k!}$.

17. $\displaystyle\sum_{k=1}^{\infty} \frac{\cos k\pi}{k}$.

18. $\displaystyle\sum_{k=3}^{\infty} \frac{(-1)^k \ln k}{k}$.

19. $\displaystyle\sum_{k=1}^{\infty} (-1)^{k+1} \left(\frac{k+2}{3k-1}\right)^k$.

20. $\displaystyle\sum_{k=1}^{\infty} \frac{(-1)^{k+1}}{k^2+1}$.

21. $\displaystyle\sum_{k=1}^{\infty} (-1)^{k+1} \frac{k+2}{k(k+3)}$.

22. $\displaystyle\sum_{k=1}^{\infty} \frac{(-1)^{k+1}k^2}{k^3+1}$.

23. $\displaystyle\sum_{k=1}^{\infty} \sin \frac{k\pi}{2}$.

24. $\displaystyle\sum_{k=1}^{\infty} \frac{\sin k}{k^3}$.

25. $\displaystyle\sum_{k=2}^{\infty} \frac{(-1)^k}{k \ln k}$.

26. $\displaystyle\sum_{k=1}^{\infty} \frac{(-1)^k}{\sqrt{k}(k+1)}$.

27. $\displaystyle\sum_{k=2}^{\infty} \left(-\frac{1}{\ln k}\right)^k$.

28. $\displaystyle\sum_{k=1}^{\infty} \frac{(-1)^{k+1}}{\sqrt{k+1} + \sqrt{k}}$.

29. $\displaystyle\sum_{k=2}^{\infty} \frac{(-1)^k(k^2+1)}{k^3+2}$.

30. $\displaystyle\sum_{k=1}^{\infty} \frac{k \cos k\pi}{k^2+1}$.

In Exercises 31–34, the given series satisfies the hypotheses of the alternating series test. For the stated value of n, estimate the error that results if the sum of the series is approximated by the nth partial sum.

31. $\displaystyle\sum_{k=1}^{\infty} \frac{(-1)^{k+1}}{k}$; $n = 7$.

32. $\displaystyle\sum_{k=1}^{\infty} \frac{(-1)^{k+1}}{k!}$; $n = 5$.

33. $\displaystyle\sum_{k=1}^{\infty} \frac{(-1)^{k+1}}{\sqrt{k}}$; $n = 99$.

34. $\displaystyle\sum_{k=1}^{\infty} \frac{(-1)^{k+1}}{(k+1)\ln(k+1)}$; $n = 3$.

In Exercises 35–38, the given series satisfies the hypotheses of the alternating series test. Find the smallest value of n for which the nth partial sum approximates the sum of the series to the stated accuracy.

35. $\displaystyle\sum_{k=1}^{\infty} \frac{(-1)^{k+1}}{k}$; $|\text{error}| < 0.0001$.

36. $\displaystyle\sum_{k=1}^{\infty} \frac{(-1)^{k+1}}{k!}$; $|\text{error}| < 0.00001$.

37. $\displaystyle\sum_{k=1}^{\infty} \frac{(-1)^{k+1}}{\sqrt{k}}$; $|\text{error}| < 0.005$.

38. $\displaystyle\sum_{k=1}^{\infty} \frac{(-1)^{k+1}}{(k+1)\ln(k+1)}$; $|\text{error}| < 0.1$.

39. Prove: If Σa_k converges absolutely, then Σa_k^2 converges.

40. Show that the converse of the result in Exercise 39 is false by finding a series for which Σa_k^2 converges, but $\Sigma |a_k|$ diverges.

41. Prove Theorem 11.7.1 for series of the form

$$-a_1 + a_2 - a_3 + a_4 - \cdots + (-1)^k a_k + \cdots$$

42. Prove Theorem 11.7.5. [*Hint:* Theorem 11.7.4 will help in part (*a*). For part (*b*), it may help to review the proof of Theorem 11.5.2.]

11.8 POWER SERIES

In previous sections we studied series with constant terms. In this section we will consider series whose terms involve variables. Such series are of fundamental importance in many branches of mathematics and the physical sciences.

If $c_0, c_1, c_2, \ldots$ are constants and x is a variable, then a series of the form

$$\sum_{k=0}^{\infty} c_k x^k = c_0 + c_1 x + c_2 x^2 + \cdots + c_k x^k + \cdots$$

is called a *power series* in x. Some examples are

$$\sum_{k=0}^{\infty} x^k = 1 + x + x^2 + x^3 + \cdots$$

$$\sum_{k=0}^{\infty} \frac{x^k}{k!} = 1 + x + \frac{x^2}{2!} + \frac{x^3}{3!} + \cdots$$

$$\sum_{k=0}^{\infty} (-1)^k \frac{x^{k+1}}{k+1} = x - \frac{x^2}{2} + \frac{x^3}{3} - \frac{x^4}{4} + \cdots$$

$$\sum_{k=0}^{\infty} (-1)^k \frac{x^{2k}}{(2k)!} = 1 - \frac{x^2}{2!} + \frac{x^4}{4!} - \frac{x^6}{6!} + \cdots$$

$$\sum_{k=0}^{\infty} (-1)^k \frac{x^{2k+1}}{(2k+1)!} = x - \frac{x^3}{3!} + \frac{x^5}{5!} - \frac{x^7}{7!} + \cdots$$

If a numerical value is substituted for x in a power series $\Sigma c_k x^k$, then we obtain a series of constants that may either converge or diverge. This leads to the following basic problem.

A Fundamental Problem. For what values of x does a given power series, $\Sigma c_k x^k$, converge?

The following theorem is the fundamental result on convergence of power series. The proof can be found in most advanced calculus texts.

11.8.1 THEOREM   *For any power series in x, exactly one of the following is true:*
(a)   *The series converges only for $x = 0$.*
(b)   *The series converges absolutely for all x.*
(c)   *The series converges absolutely for all x in some finite open interval $(-R, R)$, and diverges if $x < -R$ or $x > R$ (Figure 11.8.1). At the points $x = R$ and $x = -R$ the series may converge absolutely, converge conditionally, or diverge, depending on the particular series.*

Series diverges	Series converges absolutely	Series diverges

Figure 11.8.1 $-R$ 0 R

In case (c), where the power series converges absolutely for $|x| < R$ and diverges for $|x| > R$, we call R the **radius of convergence.** In case (a), where the series converges only for $x = 0$, we define the radius of convergence to be $R = 0$; and in case (b), where the series converges absolutely for all x, we define the radius of convergence to be $R = +\infty$. The set of all values of x for which a power series converges is called the **interval of convergence.**

▶ Example 1

Find the interval of convergence and radius of convergence of the power series

$$\sum_{k=0}^{\infty} x^k = 1 + x + x^2 + \cdots + x^k + \cdots$$

Solution. For every x, the given series is a geometric series with ratio $r = x$. Thus by Example 8 of Section 11.7, the series converges absolutely if $-1 < x < 1$ and diverges if $|x| \geq 1$. Therefore the interval of convergence is $(-1, 1)$ and the radius of convergence is $R = 1$.   ◀

▶ Example 2

Find the interval of convergence and radius of convergence of

$$\sum_{k=0}^{\infty} \frac{x^k}{k!}$$

Solution. We will apply the ratio test for absolute convergence (11.7.5). For every real number x,

$$\rho = \lim_{k \to +\infty} \left| \frac{u_{k+1}}{u_k} \right| = \lim_{k \to +\infty} \left| \frac{x^{k+1}}{(k+1)!} \cdot \frac{k!}{x^k} \right| = \lim_{k \to +\infty} \left| \frac{x}{k+1} \right| = 0$$

Since $\rho < 1$ for all x, the series converges absolutely for all x. Thus the interval of convergence is $(-\infty, +\infty)$ and the radius of convergence is $R = +\infty$. ◀

REMARK. There is a useful byproduct of Example 2. Since

$$\sum_{k=0}^{\infty} \frac{x^k}{k!}$$

converges for all x, Theorem 11.4.1 implies that for all values of x

$$\lim_{k \to +\infty} \frac{x^k}{k!} = 0 \tag{1}$$

We will need this result later.

▶ Example 3

Find the interval and radius of convergence of

$$\sum_{k=0}^{\infty} k!x^k$$

Solution. If $x = 0$, the series has only one nonzero term and therefore converges. If $x \neq 0$, the ratio test yields

$$\rho = \lim_{k \to +\infty} \left| \frac{u_{k+1}}{u_k} \right| = \lim_{k \to +\infty} \left| \frac{(k+1)!x^{k+1}}{k!x^k} \right| = \lim_{k \to +\infty} |(k+1)x| = +\infty$$

Therefore, the series converges if $x = 0$, but diverges for all other x. Consequently, the interval of convergence is the single point $x = 0$ and the radius of convergence is $R = 0$. ◀

▶ Example 4

Find the interval of convergence and radius of convergence of

$$\sum_{k=0}^{\infty} \frac{(-1)^k x^k}{3^k(k+1)}$$

Solution. Since $|(-1)^k| = |(-1)^{k+1}| = 1$ we obtain

$$\rho = \lim_{k \to +\infty} \left| \frac{u_{k+1}}{u_k} \right| = \lim_{k \to +\infty} \left| \frac{x^{k+1}}{3^{k+1}(k+2)} \cdot \frac{3^k(k+1)}{x^k} \right|$$

$$= \lim_{k \to +\infty} \left[\frac{|x|}{3} \cdot \left(\frac{k+1}{k+2} \right) \right]$$

$$= \frac{|x|}{3} \lim_{k \to +\infty} \left(\frac{1 + 1/k}{1 + 2/k} \right) = \frac{|x|}{3}$$

The ratio test for absolute convergence implies that the series converges absolutely if

$$|x| < 3$$

and diverges if

$$|x| > 3$$

The ratio test fails when

$$|x| = 3$$

so the cases $x = -3$ and $x = 3$ need separate analyses. Substituting $x = -3$ in the given series yields

$$\sum_{k=0}^{\infty} \frac{(-1)^k(-3)^k}{3^k(k+1)} = \sum_{k=0}^{\infty} \frac{(-1)^k(-1)^k 3^k}{3^k(k+1)} = \sum_{k=0}^{\infty} \frac{1}{k+1}$$

which is the divergent harmonic series $1 + \frac{1}{2} + \frac{1}{3} + \frac{1}{4} + \cdots$. Substituting $x = 3$ in the given series yields

$$\sum_{k=0}^{\infty} \frac{(-1)^k 3^k}{3^k(k+1)} = \sum_{k=0}^{\infty} \frac{(-1)^k}{k+1}$$

which is the conditionally convergent alternating harmonic series $1 - \frac{1}{2} + \frac{1}{3} - \frac{1}{4} + \cdots$. Thus the interval of convergence for the given series is $(-3, 3]$ and the radius of convergence is $R = 3$. ◄

▶ Example 5

Find the interval and radius of convergence of the series

$$\sum_{k=0}^{\infty} (-1)^k \frac{x^{2k}}{(2k)!}$$

Solution. Since $|(-1)^k| = |(-1)^{k+1}| = 1$, we have

$$\rho = \lim_{k \to +\infty} \left| \frac{u_{k+1}}{u_k} \right| = \lim_{k \to +\infty} \left| \frac{x^{2(k+1)}}{[2(k+1)]!} \cdot \frac{(2k)!}{x^{2k}} \right|$$

$$= \lim_{k \to +\infty} \left| \frac{x^{2k+2}}{(2k+2)!} \cdot \frac{(2k)!}{x^{2k}} \right|$$

$$= \lim_{k \to +\infty} \left| \frac{x^2}{(2k+2)(2k+1)} \right|$$

$$= x^2 \lim_{k \to +\infty} \frac{1}{(2k+2)(2k+1)}$$

$$= x^2 \cdot 0 = 0$$

Thus $\rho < 1$ for all x, which means that the interval of convergence is $(-\infty, +\infty)$ and the radius of convergence is $R = +\infty$. ◀

In addition to power series in x, we shall be interested in series of the form

$$\sum_{k=0}^{\infty} c_k(x-a)^k = c_0 + c_1(x-a) + c_2(x-a)^2 + \cdots + c_k(x-a)^k + \cdots$$

where $c_0, c_1, c_2, \ldots$ and a are constants. Such a series is called a **power series in $x - a$.** Some examples are

$$\sum_{k=0}^{\infty} \frac{(x-1)^k}{k+1} = 1 + \frac{(x-1)}{2} + \frac{(x-1)^2}{3} + \frac{(x-1)^3}{4} + \cdots$$

$$\sum_{k=0}^{\infty} \frac{(-1)^k(x+3)^k}{k!} = 1 - (x+3) + \frac{(x+3)^2}{2!} - \frac{(x+3)^3}{3!} + \cdots$$

The first is a power series in $x - 1$ ($a = 1$) and the second a power series in $x + 3$ ($a = -3$).

The convergence properties of a power series $\Sigma c_k(x-a)^k$ may be obtained from Theorem 11.8.1 by substituting $X = x - a$, to obtain

$$\sum_{k=0}^{\infty} a_k X^k$$

which is a power series in X. There are three possibilities for this series; it converges only when $X = 0$, or equivalently only when $x = a$; it converges absolutely for all values of X, or equivalently for all values of x; or finally, it converges absolutely for all X satisfying

$$-R < X < R \qquad \qquad (2)$$

and diverges when $X < -R$ or $X > R$. But (2) can be written

$$-R < x - a < R$$

or

$$a - R < x < a + R$$

Thus we are led to the following result.

11.8.2 THEOREM *For a power series $\Sigma c_k(x - a)^k$, exactly one of the following is true:*
(a) *The series converges only for $x = a$.*
(b) *The series converges absolutely for all x.*
(c) *The series converges absolutely for all x in some finite open interval $(a - R, a + R)$ and diverges if $x < a - R$ or $x > a + R$ (Figure 11.8.2). At the points $x = a - R$ and $x = a + R$, the series may converge absolutely, converge conditionally, or diverge, depending on the particular series.*

Series diverges	Series converges absolutely	Series diverges

Figure 11.8.2

$a - R \qquad a \qquad a + R$

In cases (a), (b), and (c) of Theorem 11.8.2 the series is said to have **radius of convergence,** 0, $+\infty$, and R respectively. The set of all values of x for which the series converges is called the **interval of convergence.**

▶ Example 6

Find the interval of convergence and radius of convergence of the series

$$\sum_{k=1}^{\infty} \frac{(x - 5)^k}{k^2}$$

Solution. We apply the ratio test for absolute convergence.

$$\rho = \lim_{k \to +\infty} \left| \frac{u_{k+1}}{u_k} \right| = \lim_{k \to +\infty} \left| \frac{(x - 5)^{k+1}}{(k + 1)^2} \cdot \frac{k^2}{(x - 5)^k} \right|$$

$$= \lim_{k \to +\infty} \left[|x - 5| \left(\frac{k}{k + 1} \right)^2 \right]$$

$$= |x - 5| \lim_{k \to +\infty} \left(\frac{1}{1 + 1/k} \right)^2 = |x - 5|$$

Thus the series converges absolutely if $|x - 5| < 1$, or $-1 < x - 5 < 1$, or $4 < x < 6$. The series diverges if $x < 4$ or $x > 6$.

To determine the convergence behavior at the endpoints $x = 4$ and $x = 6$, we substitute these values in the given series. If $x = 6$ the series becomes

$$\sum_{k=1}^{\infty} \frac{1^k}{k^2} = \sum_{k=1}^{\infty} \frac{1}{k^2} = 1 + \frac{1}{2^2} + \frac{1}{3^2} + \frac{1}{4^2} + \cdots$$

which is a convergent p-series ($p = 2$). If $x = 4$ the series becomes

$$\sum_{k=1}^{\infty} \frac{(-1)^k}{k^2} = -1 + \frac{1}{2^2} - \frac{1}{3^2} + \frac{1}{4^2} - \cdots$$

Since this series converges absolutely, the interval of convergence for the given series is $[4, 6]$. The radius of convergence is $R = 1$ (Figure 11.8.3).

◄

Figure 11.8.3

▶ Exercise Set 11.8

In Exercises 1–24, find the radius of convergence and the interval of convergence.

1. $\displaystyle\sum_{k=0}^{\infty} \frac{x^k}{k + 1}$.

2. $\displaystyle\sum_{k=0}^{\infty} 3^k x^k$.

3. $\displaystyle\sum_{k=0}^{\infty} \frac{(-1)^k x^k}{k!}$.

4. $\displaystyle\sum_{k=0}^{\infty} \frac{k!}{2^k} x^k$.

5. $\displaystyle\sum_{k=1}^{\infty} \frac{5^k}{k^2} x^k$.

6. $\displaystyle\sum_{k=2}^{\infty} \frac{x^k}{\ln k}$.

7. $\displaystyle\sum_{k=1}^{\infty} \frac{x^k}{k(k + 1)}$.

8. $\displaystyle\sum_{k=0}^{\infty} \frac{(-2)^k x^{k+1}}{k + 1}$.

9. $\displaystyle\sum_{k=1}^{\infty} (-1)^{k-1} \frac{x^k}{\sqrt{k}}$.

10. $\displaystyle\sum_{k=0}^{\infty} \frac{(-1)^k x^{2k}}{(2k)!}$.

11. $\displaystyle\sum_{k=0}^{\infty} (-1)^k \frac{x^{2k+1}}{(2k + 1)!}$.

12. $\displaystyle\sum_{k=1}^{\infty} (-1)^k \frac{x^{3k}}{k^{3/2}}$.

13. $\displaystyle\sum_{k=0}^{\infty} \frac{3^k}{k!} x^k$.

14. $\displaystyle\sum_{k=2}^{\infty} (-1)^{k+1} \frac{x^k}{k(\ln k)^2}$.

15. $\displaystyle\sum_{k=0}^{\infty} \frac{x^k}{1 + k^2}$.

16. $\displaystyle\sum_{k=0}^{\infty} \frac{(x - 3)^k}{2^k}$.

17. $\displaystyle\sum_{k=1}^{\infty} (-1)^{k+1} \frac{(x+1)^k}{k}$. **18.** $\displaystyle\sum_{k=0}^{\infty} (-1)^k \frac{(x-4)^k}{(k+1)^2}$.

19. $\displaystyle\sum_{k=0}^{\infty} \left(\frac{3}{4}\right)^k (x+5)^k$. **20.** $\displaystyle\sum_{k=1}^{\infty} \frac{(2k+1)!}{k^3}(x-2)^k$.

21. $\displaystyle\sum_{k=1}^{\infty} (-1)^k \frac{(x+1)^{2k+1}}{k^2+4}$. **22.** $\displaystyle\sum_{k=1}^{\infty} \frac{(\ln k)(x-3)^k}{k}$.

23. $\displaystyle\sum_{k=0}^{\infty} \frac{\pi^k (x-1)^{2k}}{(2k+1)!}$. **24.** $\displaystyle\sum_{k=0}^{\infty} \frac{(2x-3)^k}{4^{2k}}$.

25. Use the root test to find the interval of convergence of $\displaystyle\sum_{k=2}^{\infty} \frac{x^k}{(\ln k)^k}$.

26. Find the radius of convergence of

$$\sum_{k=1}^{\infty} (-1)^k \frac{1 \cdot 2 \cdot 3 \cdots k}{1 \cdot 3 \cdot 5 \cdots (2k-1)} x^{2k+1}$$

27. Find the interval of convergence of

$$\sum_{k=0}^{\infty} \frac{(x-a)^k}{b^k}$$

where $b > 0$.

28. Find the radius of convergence of the power series

$$\sum_{k=0}^{\infty} \frac{(pk)!}{(k!)^p} x^k, \text{ where } p \text{ is a positive integer.}$$

29. Find the radius of convergence of the power series

$$\sum_{k=0}^{\infty} \frac{(k+p)!}{k!(k+q)!} x^k, \text{ where } p \text{ and } q \text{ are positive integers.}$$

30. Prove: If $\displaystyle\lim_{k\to+\infty} |c_k|^{1/k} = L$, where $L \neq 0$, then the power series $\displaystyle\sum_{k=0}^{\infty} c_k x^k$ has radius of convergence $1/L$.

31. Prove: If the power series $\displaystyle\sum_{k=0}^{\infty} c_k x^k$ has radius of convergence R, then the series $\displaystyle\sum_{k=0}^{\infty} c_k x^{2k}$ has radius of convergence $\sqrt{R}$.

32. Prove: If the interval of convergence of $\displaystyle\sum_{k=0}^{\infty} c_k (x-a)^k$ is $(a-R, a+R]$, then the series converges conditionally at $a + R$.

11.9 TAYLOR AND MACLAURIN SERIES

One of the early applications of calculus was the calculation of values for functions like $\sin x$, $\ln x$, and e^x. The basic idea is to approximate the given function by a polynomial in such a way that the resulting error is within some specified tolerance. In this section we will study the approximation of functions by polynomials, and introduce an important class of power series.

Suppose we are interested in approximating a function f by a polynomial

$$p(x) = c_0 + c_1 x + \cdots + c_n x^n \tag{1}$$

over an interval centered at $x = 0$. Because $p(x)$ has $n + 1$ coefficients, it seems reasonable that we will be able to impose $n + 1$ conditions on this polynomial. We will assume that the first n derivatives of f exist at $x = 0$, and we will choose these $n + 1$ conditions to be:

$$f(0) = p(0), \ f'(0) = p'(0), \ f''(0) = p''(0), \ldots, f^{(n)}(0) = p^{(n)}(0) \qquad (2)$$

These conditions require that the value of $p(x)$ and its first n derivatives match the value of $f(x)$ and its first n derivatives at $x = 0$. By forcing this high degree of "match" at $x = 0$, it is reasonable to hope that $f(x)$ and $p(x)$ will remain close over some interval (possibly small) centered at $x = 0$. Since

$$p(x) = c_0 + c_1 x + c_2 x^2 + c_3 x^3 + \cdots + c_n x^n$$
$$p'(x) = c_1 + 2c_2 x + 3c_3 x^2 + \cdots + nc_n x^{n-1}$$
$$p''(x) = 2c_2 + 3 \cdot 2c_3 x + \cdots + n(n-1)c_n x^{n-2}$$
$$p'''(x) = 3 \cdot 2c_3 + \cdots + n(n-1)(n-2)c_n x^{n-3}$$
$$\vdots$$
$$p^{(n)}(x) = n(n-1)(n-2) \cdots (1)c_n$$

we obtain on substituting $x = 0$

$$p(0) = c_0$$
$$p'(0) = c_1$$
$$p''(0) = 2c_2 = 2!c_2$$
$$p'''(0) = 3 \cdot 2c_3 = 3!c_3$$
$$\vdots$$
$$p^{(n)}(0) = n(n-1)(n-2) \cdots (1)c_n = n!c_n$$

Thus, from (2),

$$f(0) = c_0$$
$$f'(0) = c_1$$
$$f''(0) = 2!c_2$$
$$f'''(0) = 3!c_3$$
$$\vdots$$
$$f^{(n)}(0) = n!c_n$$

so

$$c_0 = f(0), \ c_1 = f'(0), \ c_2 = \frac{f''(0)}{2!}, \ c_3 = \frac{f'''(0)}{3!}, \ldots, c_n = \frac{f^{(n)}(0)}{n!}$$

Substituting these values in (1) yields a polynomial, called the *n*th *Maclaurin* polynomial for f.*

11.9.1 DEFINITION   If *f* can be differentiated *n* times at 0, then we define the *n*th **Maclaurin polynomial** for *f* to be

$$p_n(x) = f(0) + f'(0)x + \frac{f''(0)}{2!}x^2 + \frac{f'''(0)}{3!}x^3 + \cdots + \frac{f^{(n)}(0)}{n!}x^n \quad (3)$$

This polynomial has the property that its value and the values of its first *n* derivatives match the value of $f(x)$ and its first *n* derivatives when $x = 0$.

▶ Example 1

Find the Maclaurin polynomials p_0, p_1, p_2, p_3, and p_n for e^x.

Solution.   Let $f(x) = e^x$. Thus

$$f'(x) = f''(x) = f'''(x) = \cdots = f^{(n)}(x) = e^x$$

and

$$f(0) = f'(0) = f''(0) = f'''(0) = \cdots = f^{(n)}(0) = e^0 = 1$$

Therefore

$$p_0(x) = f(0) = 1$$
$$p_1(x) = f(0) + f'(0)x = 1 + x$$

---

*COLIN MACLAURIN (1698–1746) Scottish mathematician. Maclaurin's father, a minister, died when the boy was only six months old, and his mother when he was nine years old. He was then raised by an uncle who was also a minister.

Maclaurin entered Glasgow University as a divinity student, but transferred to mathematics after one year. He received his Master's degree at age 17 and, in spite of his youth, began teaching at Marischal College in Aberdeen, Scotland.

Maclaurin met Isaac Newton during a visit to London in 1719 and from that time on he became Newton's disciple. During that era, some of Newton's analytic methods were bitterly attacked by major mathematicians and much of Maclaurin's important mathematical work resulted from his efforts to defend Newton's ideas geometrically. Maclaurin's work, *Treatise on Fluxions* (1742), was the first systematic formulation of Newton's methods. The treatise was so carefully done that it was a standard of mathematical rigor in calculus until the work of Cauchy in 1821.

Maclaurin was an outstanding experimentalist. He devised numerous ingenious mechanical devices, made important astronomical observations, performed actuarial computations for insurance societies, and helped to improve maps of the islands around Scotland.

$$p_2(x) = f(0) + f'(0)x + \frac{f''(0)}{2!}x^2 = 1 + x + \frac{x^2}{2!} = 1 + x + \frac{1}{2}x^2$$

$$p_3(x) = f(0) + f'(0)x + \frac{f''(0)}{2!}x^2 + \frac{f'''(0)}{3!}x^3$$

$$= 1 + x + \frac{x^2}{2!} + \frac{x^3}{3!} = 1 + x + \frac{1}{2}x^2 + \frac{1}{6}x^3$$

$$p_n(x) = f(0) + f'(0)x + \frac{f''(0)}{2!}x^2 + \cdots + \frac{f^{(n)}(0)}{n!}x^n$$

$$= 1 + x + \frac{x^2}{2!} + \cdots + \frac{x^n}{n!} \qquad \blacktriangleleft$$

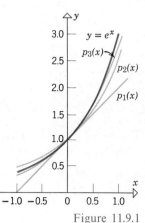

Figure 11.9.1

In Figure 11.9.1 we have sketched the graphs of e^x and its Maclaurin polynomials of degree 1, 2, and 3. Note that the graphs of e^x and $p_3(x)$ are virtually indistinguishable over the interval from -0.5 to $+0.5$. (In the next section we will investigate in detail the accuracy of Maclaurin polynomial approximations.)

▶ Example 2

Find the nth Maclaurin polynomial for $\ln(x + 1)$.

Solution. Let $f(x) = \ln(x + 1)$ and arrange the computations as follows:

$$f(x) = \ln(x + 1) \qquad\qquad f(0) = \ln 1 = 0$$

$$f'(x) = \frac{1}{x + 1} \qquad\qquad f'(0) = 1$$

$$f''(x) = -\frac{1}{(x + 1)^2} \qquad\qquad f''(0) = -1$$

$$f'''(x) = \frac{2}{(x + 1)^3} \qquad\qquad f'''(0) = 2$$

$$f^{(4)}(x) = -\frac{3 \cdot 2}{(x + 1)^4} \qquad\qquad f^{(4)}(0) = -3!$$

$$f^{(5)}(x) = \frac{4 \cdot 3 \cdot 2}{(x + 1)^5} \qquad\qquad f^{(5)}(0) = 4!$$

$$\vdots \qquad\qquad\qquad \vdots$$

$$f^{(n)}(x) = (-1)^{n+1}\frac{(n - 1)!}{(x + 1)^n} \qquad f^{(n)}(0) = (-1)^{n+1}(n - 1)!$$

Substituting these values in (3) yields

$$p_n(x) = x - \frac{x^2}{2} + \frac{x^3}{3} - \cdots + (-1)^{n+1}\frac{x^n}{n} \qquad\qquad \blacktriangleleft$$

▶ Example 3

In the Maclaurin polynomials for $\sin x$, only the odd powers of x appear explicitly. To see this, let $f(x) = \sin x$, thus

$$
\begin{aligned}
f(x) &= \sin x & f(0) &= 0 \\
f'(x) &= \cos x & f'(0) &= 1 \\
f''(x) &= -\sin x & f''(0) &= 0 \\
f'''(x) &= -\cos x & f'''(0) &= -1
\end{aligned}
$$

Since $f^{(4)}(x) = \sin x = f(x)$, the pattern $0, 1, 0, -1$ will repeat over and over as we evaluate successive derivatives at 0. Therefore, the successive Maclaurin polynomials for $\sin x$ are

$$p_1(x) = 0 + x = x$$

$$p_2(x) = 0 + x + 0 = x$$

$$p_3(x) = 0 + x + 0 - \frac{x^3}{3!} = x - \frac{x^3}{3!}$$

$$p_4(x) = 0 + x + 0 - \frac{x^3}{3!} + 0 = x - \frac{x^3}{3!}$$

$$p_5(x) = 0 + x + 0 - \frac{x^3}{3!} + 0 + \frac{x^5}{5!} = x - \frac{x^3}{3!} + \frac{x^5}{5!}$$

$$p_6(x) = 0 + x + 0 - \frac{x^3}{3!} + 0 + \frac{x^5}{5!} + 0 = x - \frac{x^3}{3!} + \frac{x^5}{5!}$$

$$p_7(x) = 0 + x + 0 - \frac{x^3}{3!} + 0 + \frac{x^5}{5!} + 0 - \frac{x^7}{7!} = x - \frac{x^3}{3!} + \frac{x^5}{5!} - \frac{x^7}{7!}$$

$$\vdots$$

In general, the Maclaurin polynomials for $\sin x$ are

$$p_{2n+1}(x) = p_{2n+2}(x) = x - \frac{x^3}{3!} + \frac{x^5}{5!} - \frac{x^7}{7!} + \cdots + (-1)^n \frac{x^{2n+1}}{(2n+1)!}$$

$(n = 0, 1, 2, \ldots)$. ◀

▶ Example 4

In the Maclaurin polynomials for $\cos x$, only the even powers of x appear explicitly. Using computations like those in Example 3, the reader should be able to show that the successive Maclaurin polynomials for $\cos x$ are

$$p_0(x) = p_1(x) = 1$$

$$p_2(x) = p_3(x) = 1 - \frac{x^2}{2!}$$

$$p_4(x) = p_5(x) = 1 - \frac{x^2}{2!} + \frac{x^4}{4!}$$

$$p_6(x) = p_7(x) = 1 - \frac{x^2}{2!} + \frac{x^4}{4!} - \frac{x^6}{6!}$$

$$p_8(x) = p_9(x) = 1 - \frac{x^2}{2!} + \frac{x^4}{4!} - \frac{x^6}{6!} + \frac{x^8}{8!}$$

$$\vdots$$

In general, the Maclaurin polynomials for $\cos x$ are

$$p_{2n}(x) = p_{2n+1}(x) = 1 - \frac{x^2}{2!} + \frac{x^4}{4!} - \cdots + (-1)^n \frac{x^{2n}}{(2n)!}$$

$(n = 0, 1, 2, \ldots)$. ◀

If we are interested in a polynomial approximation to $f(x)$ on an interval centered at $x = a$, then the idea is to choose the polynomial $p(x)$ so that the values of $p(x)$ and its first n derivatives match the values of $f(x)$ and its first n derivatives at $x = a$. The computations are simplest if the approximating polynomial is expressed in the form

$$p(x) = c_0 + c_1(x - a) + c_2(x - a)^2 + \cdots + c_n(x - a)^n \tag{4}$$

We leave it as an exercise for the reader to calculate the first n derivatives of $p(x)$ and show

$$p(a) = c_0, \; p'(a) = c_1, \; p''(a) = 2!c_2, \; p'''(a) = 3!c_3, \ldots, p^{(n)}(a) = n!c_n$$

Thus, if we want the values of $p(x)$ and its first n derivatives to match the values of $f(x)$ and its first n derivatives at $x = a$, we must have

$$c_0 = f(a), \; c_1 = f'(a), \; c_2 = \frac{f''(a)}{2!}, \; c_3 = \frac{f'''(a)}{3!}, \ldots, c_n = \frac{f^{(n)}(a)}{n!}$$

Substituting these values in (4) we obtain a polynomial called the *nth Taylor* polynomial about $x = a$ for f.*

---

*BROOK TAYLOR (1685–1731) English mathematician. Taylor was born of well-to-do parents. Musicians and artists were entertained frequently in the Taylor home, which undoubtedly had a lasting influence on young Brook. In later years, Taylor published a definitive work on the

11.9.2 DEFINITION   If f can be differentiated n times at a, then we define the ***nth Taylor polynomial for f about x = a*** to be

$$p_n(x) = f(a) + f'(a)(x - a) + \frac{f''(a)}{2!}(x - a)^2$$

$$+ \frac{f'''(a)}{3!}(x - a)^3 + \cdots + \frac{f^{(n)}(a)}{n!}(x - a)^n \quad (5)$$

Observe that the Taylor polynomials include the Maclaurin polynomials as a special case (let $a = 0$).

▶ Example 5

Find the Taylor polynomials $p_1(x)$, $p_2(x)$, and $p_3(x)$ for $\sin x$ about $x = \pi/3$.

Solution.   Let $f(x) = \sin x$. Thus

$$f(x) = \sin x \qquad f(\pi/3) = \sin \frac{\pi}{3} = \frac{\sqrt{3}}{2}$$

$$f'(x) = \cos x \qquad f'(\pi/3) = \cos \frac{\pi}{3} = \frac{1}{2}$$

$$f''(x) = -\sin x \qquad f''(\pi/3) = -\sin \frac{\pi}{3} = -\frac{\sqrt{3}}{2}$$

$$f'''(x) = -\cos x \qquad f'''(\pi/3) = -\cos \frac{\pi}{3} = -\frac{1}{2}$$

Substituting in (5) with $a = \pi/3$ yields

$$p_1(x) = f(\pi/3) + f'(\pi/3)\left(x - \frac{\pi}{3}\right) = \frac{\sqrt{3}}{2} + \frac{1}{2}\left(x - \frac{\pi}{3}\right)$$

---

mathematical theory of perspective and obtained major mathematical results about the vibrations of strings. There also exists an unpublished work, *On Musick,* that was intended to be part of a joint paper with Isaac Newton.

Taylor's life was scarred with unhappiness, illness, and tragedy. His first wife was not rich enough to suit his father; the two men argued bitterly and parted ways. Subsequently his wife died in childbirth. Then, after he remarried, his second wife also died in childbirth, though his daughter survived.

Taylor's most productive period was from 1714 to 1719, during which time he wrote on a wide range of subjects—magnetism, capillary action, thermometers, perspective, and calculus. In his final years, Taylor devoted his writing efforts to religion and philosophy. According to Taylor, the results that bear his name were motivated by coffeehouse conversations about works of Newton on planetary motion and works of Halley ("Halley's comet") on roots of polynomials.

Taylor's writing style was so terse and hard to understand that he never received credit for many of his innovations.

$$p_2(x) = f(\pi/3) + f'(\pi/3)\left(x - \frac{\pi}{3}\right) + \frac{f''(\pi/3)}{2!}\left(x - \frac{\pi}{3}\right)^2$$

$$= \frac{\sqrt{3}}{2} + \frac{1}{2}\left(x - \frac{\pi}{3}\right) - \frac{\sqrt{3}}{2\cdot 2!}\left(x - \frac{\pi}{3}\right)^2$$

$$p_3(x) = f(\pi/3) + f'(\pi/3)\left(x - \frac{\pi}{3}\right) + \frac{f''(\pi/3)}{2!}\left(x - \frac{\pi}{3}\right)^2 + \frac{f'''(\pi/3)}{3!}\left(x - \frac{\pi}{3}\right)^3$$

$$= \frac{\sqrt{3}}{2} + \frac{1}{2}\left(x - \frac{\pi}{3}\right) - \frac{\sqrt{3}}{2\cdot 2!}\left(x - \frac{\pi}{3}\right)^2 - \frac{1}{2\cdot 3!}\left(x - \frac{\pi}{3}\right)^3 \qquad \blacktriangleleft$$

Frequently, it is convenient to express the defining formula for the Taylor polynomial in sigma notation. To do this, we use the notation $f^{(k)}(a)$ to denote the kth derivative of f at $x = a$, and we make the added convention that $f^{(0)}(a)$ denotes $f(a)$. This enables us to write

$$\sum_{k=0}^{n} \frac{f^{(k)}(a)}{k!}(x - a)^k = f(a) + f'(a)(x - a)$$

$$+ \frac{f''(a)}{2!}(x - a)^2 + \cdots + \frac{f^{(n)}(a)}{n!}(x - a)^n$$

In particular, the nth Maclaurin polynomial for $f(x)$ may be written

$$\sum_{k=0}^{n} \frac{f^{(k)}(0)}{k!}x^k = f(0) + f'(0)x + \frac{f''(0)}{2!}x^2 + \cdots + \frac{f^{(n)}(0)}{n!}x^n$$

Because the value of f and its first n derivatives match the value of the Taylor polynomial and its first n derivatives at $x = a$, we might hope that as n increases the Taylor polynomials for f about $x = a$ will become better and better approximations to $f(x)$—at least in some interval centered at $x = a$. This raises the problem of finding those values of x for which the Taylor polynomials converge to $f(x)$ as $n \to +\infty$. In other words, for which values of x is it true that

$$f(x) = \lim_{n \to +\infty} \sum_{k=0}^{n} \frac{f^{(k)}(a)}{k!}(x - a)^k = \sum_{k=0}^{\infty} \frac{f^{(k)}(a)}{k!}(x - a)^k \qquad (6)$$

The power series in (6) is called the *Taylor series* about $x = a$ for the function f.

11.9.3 DEFINITION If f has derivatives of all orders at a, then we define the **Taylor Series for f about x = a** to be

$$\sum_{k=0}^{\infty} \frac{f^{(k)}(a)}{k!}(x-a)^k = f(a) + f'(a)(x-a)$$
$$+ \frac{f''(a)}{2!}(x-a)^2 + \cdots + \frac{f^{(k)}(a)}{k!}(x-a)^k + \cdots \quad (7)$$

11.9.4 DEFINITION If f has derivatives of all orders at 0, then we define the **Maclaurin series for f** to be

$$\sum_{k=0}^{\infty} \frac{f^{(k)}(0)}{k!}x^k = f(0) + f'(0)x + \frac{f''(0)}{2!}x^2 + \cdots + \frac{f^{(k)}(0)}{k!}x^k + \cdots \quad (8)$$

Observe that the Maclaurin series for f is just the Taylor series for f about $a = 0$.

▶ Example 6

In Example 1, we found the nth Maclaurin polynomial for e^x to be

$$\sum_{k=0}^{n} \frac{x^k}{k!} = 1 + x + \frac{x^2}{2!} + \cdots + \frac{x^n}{n!}$$

Thus the Maclaurin series for e^x is

$$\sum_{k=0}^{\infty} \frac{x^k}{k!} = 1 + x + \frac{x^2}{2!} + \frac{x^3}{3!} + \cdots + \frac{x^n}{n!} + \cdots$$

Similarly, from Example 3 the Maclaurin series for $\sin x$ is

$$\sum_{k=0}^{\infty} (-1)^k \frac{x^{2k+1}}{(2k+1)!} = x - \frac{x^3}{3!} + \frac{x^5}{5!} - \frac{x^7}{7!} + \cdots$$

and from Example 4 the Maclaurin series for $\cos x$ is

$$\sum_{k=0}^{\infty} (-1)^k \frac{x^{2k}}{(2k)!} = 1 - \frac{x^2}{2!} + \frac{x^4}{4!} - \frac{x^6}{6!} + \cdots \qquad ◀$$

▶ Example 7

Find the Taylor series about $x = 1$ for $1/x$.

Solution. Let $f(x) = 1/x$ so that

$$f(x) = \frac{1}{x} \qquad\qquad f(1) = 1$$

$$f'(x) = -\frac{1}{x^2} \qquad\qquad f'(1) = -1$$

$$f''(x) = \frac{2}{x^3} \qquad\qquad f''(1) = 2!$$

$$f'''(x) = -\frac{3 \cdot 2}{x^4} \qquad\qquad f'''(1) = -3!$$

$$f^{(4)}(x) = \frac{4 \cdot 3 \cdot 2}{x^5} \qquad\qquad f^{(4)}(1) = 4!$$

$$\vdots \qquad\qquad\qquad\qquad \vdots$$

$$f^{(k)}(x) = (-1)^k \frac{k!}{x^{k+1}} \qquad f^{(k)}(1) = (-1)^k k!$$

$$\vdots \qquad\qquad\qquad\qquad \vdots$$

Thus, substituting in (7) with $a = 1$ yields

$$\sum_{k=0}^{\infty} \frac{(-1)^k k!}{k!}(x - 1)^k = \sum_{k=0}^{\infty} (-1)^k (x - 1)^k$$

$$= 1 - (x - 1) + (x - 1)^2 - (x - 1)^3 + \cdots \quad ◀$$

Convergence of Taylor and Maclaurin series will be considered in the next section.

▶ Exercise Set 11.9

In Exercises 1–12, find the fourth Maclaurin polynomial $(n = 4)$ for the given function.

1. e^{-2x}.

2. $\dfrac{1}{1 + x}$.

3. $\sin 2x$.

4. $e^x \cos x$.

5. $\tan x$.

6. $x^3 - x^2 + 2x + 1$.

7. xe^x.

8. $\tan^{-1} x$.

9. $\sec x$.

10. $\sqrt{1 + x}$.

11. $\ln(3 + 2x)$.

12. $\sinh x$.

In Exercises 13–22, find the third Taylor polynomial $(n = 3)$ about $x = a$ for the given function.

13. e^x; $a = 1$.

14. $\ln x$; $a = 1$.

15. $\sqrt{x}$; $a = 4$.

16. $x^4 + x - 3$; $a = -2$.

17. $\cos x$; $a = \dfrac{\pi}{4}$.   18. $\tan x$; $a = \dfrac{\pi}{3}$.

19. $\sin \pi x$; $a = -\dfrac{1}{3}$.   20. $\csc x$; $a = \dfrac{\pi}{2}$.

21. $\tan^{-1} x$; $a = 1$.   22. $\cosh x$; $a = \ln 2$.

In Exercises 23–31, find the Maclaurin series for the given function. Express your answer in sigma notation.

23. e^{-x}.   24. e^{ax}.

25. $\dfrac{1}{1 + x}$.   26. xe^{x}.

27. $\ln(1 + x)$.   28. $\sin \pi x$.

29. $\cos\left(\dfrac{x}{2}\right)$.   30. $\sinh x$.

31. $\cosh x$.

In Exercises 32–39, find the Taylor series about $x = a$ for the given function. Express your answer in sigma notation.

32. $\dfrac{1}{x}$; $a = 3$.   33. $\dfrac{1}{x}$; $a = -1$.

34. e^{x}; $a = 2$.   35. $\ln x$; $a = 1$.

36. $\cos x$; $a = \dfrac{\pi}{2}$.   37. $\sin \pi x$; $a = \dfrac{1}{2}$.

38. $\dfrac{1}{x + 2}$; $a = 3$.   39. $\sinh x$; $a = \ln 4$.

40. Prove: The value of $p_n(x) = \displaystyle\sum_{k=0}^{n} \dfrac{f^{(k)}(a)}{k!}(x - a)^k$ and its first n derivatives match the value of $f(x)$ and its first n derivatives at $x = a$.

11.10 TAYLOR FORMULA WITH REMAINDER; CONVERGENCE OF TAYLOR SERIES

In this section we will analyze the error that results when a function f is approximated by a Taylor or Maclaurin polynomial. This work will enable us to investigate convergence properties of Taylor and Maclaurin series.

If we approximate a function f by its nth Taylor polynomial p_n, then the error at a point x is represented by the difference $f(x) - p_n(x)$. This difference is commonly called the ***nth remainder*** and is denoted by

$$R_n(x) = f(x) - p_n(x)$$

The following theorem gives an important explicit formula for this remainder.

11.10.1 THEOREM

Taylor's Theorem

Suppose that a function f can be differentiated $n + 1$ times at each point in an interval containing the point a, and let

$$p_n(x) = f(a) + f'(a)(x - a) + \frac{f''(a)}{2!}(x - a)^2 + \cdots + \frac{f^{(n)}(a)}{n!}(x - a)^n$$

be the nth Taylor polynomial about $x = a$ for f. Then for each x in the interval, there is at least one point c between a and x such that

$$R_n(x) = f(x) - p_n(x) = \frac{f^{(n+1)}(c)}{(n + 1)!}(x - a)^{n+1} \tag{1}$$

In this theorem, the statement that c is "between" a and x means that c is in the interval (a, x) if $a < x$, or in (x, a) if $x < a$, or $c = a = x$ if $a = x$.

Rewriting (1) as $f(x) = p_n(x) + R_n(x)$ yields the following result, called **Taylor's formula with remainder.**

$$f(x) = f(a) + f'(a)(x - a) + \frac{f''(a)}{2!}(x - a)^2 + \cdots$$
$$+ \frac{f^{(n)}(a)}{n!}(x - a)^n + \frac{f^{(n+1)}(c)}{(n + 1)!}(x - a)^{n+1} \quad (2)$$

This formula expresses $f(x)$ as the sum of its nth Taylor polynomial about $x = a$ plus a remainder or error term,

$$R_n(x) = \frac{f^{(n+1)}(c)}{(n + 1)!}(x - a)^{n+1} \tag{3}$$

in which c is some unspecified number between a and x. The value of c depends on a, x, and n. Historically, (3) was not due to Taylor, but rather to Joseph Louis Lagrange.* For this reason, (3) is frequently called **Lagrange's form of the remainder.** Other formulas for the remainder may be found in any advanced calculus texts. The proof of Taylor's theorem will be deferred until the end of the section.

---

*JOSEPH LOUIS LAGRANGE (1736–1813) French–Italian mathematician and astronomer. Lagrange, the son of a public official, was born in Turin, Italy. (Baptismal records list his name as Giuseppe Lodovico Lagrangia.) Although his father wanted him to be a lawyer, Lagrange was attracted to mathematics and astronomy after reading a memoir by the astronomer Halley. At age 16 he began to study mathematics on his own and by age 19 was appointed to a professorship at the Royal Artillery School in Turin. The following year Lagrange sent Euler solutions to some famous problems, using new methods that eventually blossomed into a branch of mathematics called calculus of variations. These methods and Lagrange's applications of them to problems in celestial mechanics were so monumental that by age 25 he was regarded by many of his contemporaries as the greatest living mathematician.

In 1776, on the recommendation of Euler, he was chosen to succeed Euler as the director of the Berlin Academy. During his stay in Berlin, Lagrange distinguished himself not only in celestial mechanics, but also in algebraic equations and the theory of numbers. After twenty years in Berlin, he moved to Paris at the invitation of Louis XVI. He was given apartments in the Louvre and treated with great honor, even during the revolution.

Napoleon was a great admirer of Lagrange and showered him with honors—count, senator, and Legion of Honor. The years Lagrange spent in Paris were devoted primarily to didactic treatises summarizing his mathematical conceptions. One of Lagrange's most famous works is a memoir, *Mecanique analytique* in which he reduced the theory of mechanics to a few general formulas from which all other necessary equations could be derived.

It is an interesting historical fact that Lagrange's father speculated unsuccessfully in several financial ventures, so his family was forced to live quite modestly. Lagrange himself stated that if his family had money, he would not have made mathematics his vocation.

In spite of his fame, Lagrange was always a shy and modest man. On his death, he was buried with honor in the Pantheon.

For Maclaurin polynomials ($a = 0$), (2) and (3) become:

$$f(x) = f(0) + f'(0)x + \frac{f''(0)}{2!}x^2 + \cdots + \frac{f^{(n)}(0)}{n!}x^n + \frac{f^{(n+1)}(c)}{(n+1)!}x^{n+1} \qquad \text{(2a)}$$

$$R_n(x) = \frac{f^{(n+1)}(c)}{(n+1)!}x^{n+1} \qquad \text{where } c \text{ is between } 0 \text{ and } x. \qquad \text{(3a)}$$

▶ Example 1

From Example 1 of the previous section, the nth Maclaurin polynomial for $f(x) = e^x$ is

$$1 + x + \frac{x^2}{2!} + \cdots + \frac{x^n}{n!}$$

Since $f^{(n+1)}(x) = e^x$ it follows that

$$f^{(n+1)}(c) = e^c$$

Thus from Taylor's formula with remainder (2a)

$$e^x = 1 + x + \frac{x^2}{2!} + \cdots + \frac{x^n}{n!} + \frac{e^c}{(n+1)!}x^{n+1} \qquad \text{(4)}$$

where c is between 0 and x. Note that (4) is valid for all real values of x since the hypotheses of Taylor's theorem are satisfied on the interval $(-\infty, +\infty)$ (verify). ◀

▶ Example 2

In Example 5 of the previous section, we found the Taylor polynomial of degree 3 about $x = \pi/3$ for $f(x) = \sin x$ to be

$$\frac{\sqrt{3}}{2} + \frac{1}{2}\left(x - \frac{\pi}{3}\right) - \frac{\sqrt{3}}{4}\left(x - \frac{\pi}{3}\right)^2 - \frac{1}{12}\left(x - \frac{\pi}{3}\right)^3$$

Since $f^{(4)}(x) = \sin x$ (verify) it follows that $f^{(4)}(c) = \sin c$. Thus from Taylor's formula with remainder (2) in the case $a = \pi/3$ and $n = 3$ we obtain

$$\sin x = \frac{\sqrt{3}}{2} + \frac{1}{2}\left(x - \frac{\pi}{3}\right) - \frac{\sqrt{3}}{4}\left(x - \frac{\pi}{3}\right)^2 - \frac{1}{12}\left(x - \frac{\pi}{3}\right)^3$$

$$+ \frac{\sin c}{4!}\left(x - \frac{\pi}{3}\right)^4$$

where c is some number between $\pi/3$ and x. ◀

In the previous section we raised the problem of finding those values of x for which the Taylor series for f about $x = a$ converges to $f(x)$, that is

$$f(x) = f(a) + f'(a)(x - a) + \frac{f''(a)}{2!}(x - a)^2$$

$$+ \cdots + \frac{f^{(k)}(a)}{k!}(x - a)^k + \cdots \quad (5)$$

To solve this problem, let us write (5) in sigma notation:

$$f(x) = \sum_{k=0}^{\infty} \frac{f^{(k)}(a)}{k!}(x - a)^k$$

This equality is equivalent to

$$f(x) = \lim_{n \to +\infty} \sum_{k=0}^{n} \frac{f^{(k)}(a)}{k!}(x - a)^k$$

or

$$\lim_{n \to +\infty}\left[f(x) - \sum_{k=0}^{n} \frac{f^{(k)}(a)}{k!}(x - a)^k\right] = 0 \quad (6)$$

But the bracketed expression in (6) is $R_n(x)$, since it is the difference between $f(x)$ and its nth Taylor polynomial about $x = a$. Thus (6) can be written

$$\lim_{n \to +\infty} R_n(x) = 0$$

which leads us to the following result.

11.10.2 THEOREM *The equality*

$$f(x) = \sum_{k=0}^{\infty} \frac{f^{(k)}(a)}{k!}(x - a)^k$$

holds if and only if $\lim_{n \to +\infty} R_n(x) = 0$.

To paraphrase this theorem, the *Taylor series for f converges to f(x) at precisely those points where the remainder approaches zero.*

▶ Example 3

Show that the Maclaurin series for e^x converges to e^x for all x.

Solution. We want to show

$$e^x = 1 + x + \frac{x^2}{2!} + \frac{x^3}{3!} + \cdots + \frac{x^n}{n!} + \cdots \tag{7}$$

holds for all x. From (4),

$$e^x = 1 + x + \frac{x^2}{2!} + \cdots + \frac{x^n}{n!} + R_n(x)$$

where

$$R_n(x) = \frac{e^c}{(n+1)!} x^{n+1}$$

and c is between 0 and x. Thus we must show that for all x

$$\lim_{n \to +\infty} R_n(x) = \lim_{n \to +\infty} \frac{e^c}{(n+1)!} x^{n+1} = 0 \tag{8}$$

Our proof of this result will hinge on the limit

$$\lim_{n \to +\infty} \frac{x^{n+1}}{(n+1)!} = 0 \tag{9}$$

which follows from Formula (1) of Section 11.8. We will consider three cases. If $x > 0$, then

$$0 < c < x$$

from which it follows that

$$0 < e^c < e^x$$

and consequently

$$0 < \frac{e^c}{(n+1)!} x^{n+1} < \frac{e^x}{(n+1)!} x^{n+1} \tag{10}$$

From (9) we obtain

$$\lim_{n \to +\infty} \frac{e^x}{(n+1)!} x^{n+1} = e^x \lim_{n \to +\infty} \frac{x^{n+1}}{(n+1)!} = e^x \cdot 0 = 0$$

Thus, from (10) and the pinching theorem (3.3.1), result (8) follows.

In the case $x < 0$, we have $c < 0$ since c is between 0 and x. Thus $0 < e^c < 1$ and consequently

$$0 < e^c \left| \frac{x^{n+1}}{(n+1)!} \right| < \left| \frac{x^{n+1}}{(n+1)!} \right|$$

or

$$0 < \left| \frac{e^c}{(n+1)!} x^{n+1} \right| < \left| \frac{x^{n+1}}{(n+1)!} \right|$$

or

$$0 < |R_n(x)| < \left| \frac{x^{n+1}}{(n+1)!} \right|$$

From (9) and the pinching theorem it follows that $\lim_{n \to +\infty} |R_n(x)| = 0$, and so $\lim_{n \to +\infty} R_n(x) = 0$.

Convergence in the case $x = 0$, is obvious since (7) reduces to

$$e^0 = 1 + 0 + 0 + 0 + \cdots$$
◀

▶ Example 4

Show that the Maclaurin series for $\sin x$ converges to $\sin x$ for all x.

Solution. We are trying to show that

$$\sin x = x - \frac{x^3}{3!} + \frac{x^5}{5!} - \frac{x^7}{7!} + \cdots$$

holds for all x. Let $f(x) = \sin x$, so that for all x either

$$f^{(n+1)}(x) = \pm \cos x \qquad \text{or} \qquad f^{(n+1)}(x) = \pm \sin x$$

In all of these cases $|f^{(n+1)}(x)| \leq 1$, so that for all possible values of c

$$|f^{(n+1)}(c)| \leq 1$$

Hence

$$0 \leq |R_n(x)| = \left| \frac{f^{(n+1)}(c)}{(n+1)!} x^{n+1} \right| \leq \frac{|x|^{n+1}}{(n+1)!} \qquad (11)$$

From (9) with $|x|$ replacing x,

$$\lim_{n \to +\infty} \frac{|x|^{n+1}}{(n+1)!} = 0$$

so that from (11) and the pinching theorem

$$\lim_{n \to +\infty} |R_n(x)| = 0$$

and consequently

$$\lim_{n \to +\infty} R_n(x) = 0$$

for all x. ◀

▶ Example 5

Find the Taylor series for $\sin x$ about $x = \pi/2$ and show that the series converges to $\sin x$ for all x.

Solution. Let $f(x) = \sin x$. Thus

$$f(x) = \sin x \qquad f\left(\frac{\pi}{2}\right) = \sin \frac{\pi}{2} = 1$$

$$f'(x) = \cos x \qquad f'\left(\frac{\pi}{2}\right) = \cos \frac{\pi}{2} = 0$$

$$f''(x) = -\sin x \qquad f''\left(\frac{\pi}{2}\right) = -\sin \frac{\pi}{2} = -1$$

$$f'''(x) = -\cos x \qquad f'''\left(\frac{\pi}{2}\right) = -\cos \frac{\pi}{2} = 0$$

Since $f^{(4)}(x) = \sin x$, the pattern $1, 0, -1, 0$ will repeat over and over as we evaluate successive derivatives at $\pi/2$. Thus the Taylor series representation about $x = \pi/2$ for $\sin x$ is

$$\sin x = 1 - \frac{1}{2!}\left(x - \frac{\pi}{2}\right)^2 + \frac{1}{4!}\left(x - \frac{\pi}{2}\right)^4 - \frac{1}{6!}\left(x - \frac{\pi}{2}\right)^6 + \cdots \qquad (12)$$

which we are trying to show is valid for all x. As in Example 4, $|f^{(n+1)}(c)| \leq 1$ so that

$$0 \leq |R_n(x)| = \left| \frac{f^{(n+1)}(c)}{(n+1)!} \left(x - \frac{\pi}{2} \right)^{n+1} \right| \leq \frac{\left| x - \frac{\pi}{2} \right|^{n+1}}{(n+1)!}$$

From (9) with $|x - \pi/2|$ replacing x it follows that

$$\lim_{n \to +\infty} \frac{\left| x - \frac{\pi}{2} \right|^{n+1}}{(n+1)!} = 0$$

so that by the same argument given in Example 4, $\lim_{n \to +\infty} R_n(x) = 0$ for all x.
This shows that (12) is valid for all x. ◀

REMARK. It can be shown that the Taylor series for e^x, $\sin x$, and $\cos x$ about any point $x = a$ converge to these functions for all x.

For reference, we have listed in Table 11.10.1 the Maclaurin series for a number of important functions, and we have indicated the interval over which the series converges to the function.

Table 11.10.1

MACLAURIN SERIES	INTERVAL OF VALIDITY
$e^x = \displaystyle\sum_{k=0}^{\infty} \frac{x^k}{k!} = 1 + x + \frac{x^2}{2!} + \frac{x^3}{3!} + \frac{x^4}{4!} + \cdots$	$-\infty < x < +\infty$
$\sin x = \displaystyle\sum_{k=0}^{\infty} (-1)^k \frac{x^{2k+1}}{(2k+1)!} = x - \frac{x^3}{3!} + \frac{x^5}{5!} - \frac{x^7}{7!} + \cdots$	$-\infty < x < +\infty$
$\cos x = \displaystyle\sum_{k=0}^{\infty} (-1)^k \frac{x^{2k}}{(2k)!} = 1 - \frac{x^2}{2!} + \frac{x^4}{4!} - \frac{x^6}{6!} + \cdots$	$-\infty < x < +\infty$
$\ln(1+x) = \displaystyle\sum_{k=0}^{\infty} (-1)^k \frac{x^{k+1}}{k+1} = x - \frac{x^2}{2} + \frac{x^3}{3} - \frac{x^4}{4} + \cdots$	$-1 < x \leq 1$
$\tan^{-1} x = \displaystyle\sum_{k=0}^{\infty} (-1)^k \frac{x^{2k+1}}{2k+1} = x - \frac{x^3}{3} + \frac{x^5}{5} - \frac{x^7}{7} + \cdots$	$-1 \leq x \leq 1$
$\dfrac{1}{1-x} = \displaystyle\sum_{k=0}^{\infty} x^k = 1 + x + x^2 + x^3 + \cdots$	$-1 < x < 1$
$\sinh x = \displaystyle\sum_{k=0}^{\infty} \frac{x^{2k+1}}{(2k+1)!} = x + \frac{x^3}{3!} + \frac{x^5}{5!} + \frac{x^7}{7!} + \cdots$	$-\infty < x < +\infty$
$\cosh x = \displaystyle\sum_{k=0}^{\infty} \frac{x^{2k}}{(2k)!} = 1 + \frac{x^2}{2!} + \frac{x^4}{4!} + \frac{x^6}{6!} + \cdots$	$-\infty < x < +\infty$

Sometimes Maclaurin series can be obtained by substituting in other Maclaurin series.

▶ Example 6

Using the Maclaurin series

$$e^x = 1 + x + \frac{x^2}{2!} + \frac{x^3}{3!} + \frac{x^4}{4!} + \cdots \qquad -\infty < x < +\infty$$

we can derive the Maclaurin series for e^{-x} by substituting $-x$ for x to obtain

$$e^{-x} = 1 + (-x) + \frac{(-x)^2}{2!} + \frac{(-x)^3}{3!} + \frac{(-x)^4}{4!} + \cdots$$
$$-\infty < -x < +\infty$$

or

$$e^{-x} = 1 - x + \frac{x^2}{2!} - \frac{x^3}{3!} + \frac{x^4}{4!} - \cdots \qquad -\infty < x < +\infty$$

From the Maclaurin series for e^x and e^{-x} we can obtain the Maclaurin series for $\cosh x$ by writing

$$\cosh x = \frac{1}{2}(e^x + e^{-x}) = \frac{1}{2}\left(\left[1 + x + \frac{x^2}{2!} + \frac{x^3}{3!} + \frac{x^4}{4!} + \cdots\right]\right.$$
$$\left. + \left[1 - x + \frac{x^2}{2!} - \frac{x^3}{3!} + \frac{x^4}{4!} + \cdots\right]\right)$$

or

$$\cosh x = 1 + \frac{x^2}{2!} + \frac{x^4}{4!} + \cdots \qquad -\infty < x < +\infty \qquad ◀$$

We could have also derived this Maclaurin series directly. However, sometimes indirect methods, like those in this example, are useful when it is messy to calculate the higher derivatives required for the direct calculation of a Maclaurin series. There is, however, a loose thread in the logic of this example. We have produced a power series in x that converges to $\cosh x$ for all x. But isn't it conceivable that we have produced a power series in x *different* from the Maclaurin series? In Section 11.12 we will show that if a power series in $x - a$ converges to $f(x)$ on some interval containing a, then the series must be the Taylor series for f about $x = a$. Thus, we are assured the series obtained for $\cosh x$ is in fact the Maclaurin series.

▶ Example 7

Using the Maclaurin series

$$\frac{1}{1-x} = 1 + x + x^2 + x^3 + \cdots \qquad -1 < x < 1$$

we can derive the Maclaurin series for $1/(1 - 2x^2)$ by substituting $2x^2$ for x to obtain

$$\frac{1}{1 - 2x^2} = 1 + (2x^2) + (2x^2)^2 + (2x^2)^3 + \cdots \qquad -1 < 2x^2 < 1$$

or

$$\frac{1}{1 - 2x^2} = 1 + 2x^2 + 4x^4 + 8x^6 + \cdots = \sum_{k=0}^{\infty} 2^k x^{2k}, \qquad -1 < 2x^2 < 1$$

Since $2x^2 \geq 0$ for all x, the convergence condition $-1 < 2x^2 < 1$ can be written in the equivalent form $0 \leq 2x^2 < 1$ or $0 \leq x^2 < 1/2$ or $-1/\sqrt{2} < x < 1/\sqrt{2}$. ◀

OPTIONAL

We conclude this section with a proof of Taylor's Theorem.

Proof of Theorem 11.10.1. By hypothesis, f can be differentiated $n + 1$ times at each point in an interval containing the point a. Choose any point b in this interval. To be specific, we will assume $b > a$. (The cases $b < a$ and $b = a$ are left to the reader.) Let $p_n(x)$ be the nth Taylor polynomial for $f(x)$ about $x = a$ and define

$$h(x) = f(x) - p_n(x) \qquad (13)$$
$$g(x) = (x - a)^{n+1} \qquad (14)$$

Because $f(x)$ and $p_n(x)$ have the same value and the same first n derivatives at $x = a$, it follows that

$$h(a) = h'(a) = h''(a) = \cdots = h^{(n)}(a) = 0 \qquad (15)$$

Also, we leave it for the reader to show that

$$g(a) = g'(a) = g''(a) = \cdots = g^{(n)}(a) = 0 \qquad (16)$$

and that $g(x)$ and its first n derivatives are nonzero when $x \neq a$.

It is straightforward to check that h and g satisfy the hypotheses of the extended mean value theorem (10.2.2) on the interval $[a, b]$, so that there is a point c_1 with $a < c_1 < b$ such that

$$\frac{h(b) - h(a)}{g(b) - g(a)} = \frac{h'(c_1)}{g'(c_1)} \tag{17}$$

or, from (15) and (16),

$$\frac{h(b)}{g(b)} = \frac{h'(c_1)}{g'(c_1)} \tag{18}$$

If we now apply the extended mean value theorem to h' and g' over the interval $[a, c_1]$, we may deduce that there is a point c_2 with $a < c_2 < c_1 < b$ such that

$$\frac{h'(c_1) - h'(a)}{g'(c_1) - g'(a)} = \frac{h''(c_2)}{g''(c_2)}$$

or, from (15) and (16),

$$\frac{h'(c_1)}{g'(c_1)} = \frac{h''(c_2)}{h''(c_2)}$$

which, when combined with (18), yields

$$\frac{h(b)}{g(b)} = \frac{h''(c_2)}{g''(c_2)}$$

It should now be clear that if we continue in this way, applying the extended mean value theorem to the successive derivatives of h and g, we will eventually obtain a relationship of the form

$$\frac{h(b)}{g(b)} = \frac{h^{(n+1)}(c_{n+1})}{g^{(n+1)}(c_{n+1})} \tag{19}$$

where $a < c_{n+1} < b$. However, $p_n(x)$ is a polynomial of degree n, so that its $(n + 1)$-st derivative is zero. Thus from (13)

$$h^{(n+1)}(c_{n+1}) = f^{(n+1)}(c_{n+1}) \tag{20}$$

Also, from (14), the $(n + 1)$-st derivative of $g(x)$ is the constant $(n + 1)!$, so that

$$g^{(n+1)}(c_{n+1}) = (n + 1)! \tag{21}$$

Substituting (20) and (21) in (19) yields

$$\frac{h(b)}{g(b)} = \frac{f^{(n+1)}(c_{n+1})}{(n+1)!}$$

Letting $c = c_{n+1}$, and using (13) and (14), it follows that

$$f(b) - p_n(b) = \frac{f^{(n+1)}(c)}{(n+1)!}(b-a)^{n+1}$$

But this is precisely (1) in Taylor's theorem, with the exception that the variable here is b rather than x. Thus, to finish we need only replace b by x. ∎

▶ Exercise Set 11.10

For the functions in Exercises 1–14, find:

(a) Lagrange's form of the remainder $R_n(x)$ for the given values of a and n;

(b) the largest open interval on which the remainder formula is valid;

(c) the interval of possible values for the number c in the remainder formula.

1. e^{2x}; $a = 0$; $n = 5$.

2. $\cos x$; $a = 0$; $n = 8$.

3. $\dfrac{1}{x+1}$; $a = 0$; $n = 4$.

4. $\tan x$; $a = 0$; $n = 2$.

5. xe^x; $a = 0$; $n = 3$.

6. $\ln(1+x)$; $a = 0$; $n = 5$.

7. $\tan^{-1} x$; $a = 0$; $n = 2$.

8. $\sinh x$; $a = 0$; $n = 6$.

9. $\sqrt{x}$; $a = 4$; $n = 3$.

10. $\dfrac{1}{x}$; $a = 1$; $n = 5$.

11. $\sin x$; $a = \dfrac{\pi}{6}$; $n = 4$.

12. $\cos \pi x$; $a = \dfrac{1}{2}$; $n = 2$.

13. $\dfrac{1}{(1+x)^2}$; $a = -2$; $n = 5$.

14. $\csc x$; $a = \dfrac{\pi}{2}$; $n = 1$.

In Exercises 15–18, find Lagrange's form of the remainder $R_n(x)$ when $a = 0$.

15. $f(x) = \dfrac{1}{1-x}$.

16. $f(x) = e^{-x}$.

17. $f(x) = e^{2x}$.

18. $f(x) = \ln(1+x)$.

19. Prove: The Maclaurin series for $\cos x$ converges to $\cos x$ for all x.

20. Prove: The Taylor series for $\sin x$ about $x = \pi/4$ converges to $\sin x$ for all x.

21. Prove: The Taylor series for e^x about $x = 1$ converges to e^x for all x.

22. Prove: The Maclaurin series for $\ln(1+x)$ converges to $\ln(1+x)$ if $0 \le x < 1$.

23. Prove: The Taylor series for e^x about any point $x = a$ converges to e^x for all x.

24. Prove: The Taylor series for $\sin x$ about any point $x = a$ converges to $\sin x$ for all x.

25. Prove: The Taylor series for $\cos x$ about any point $x = a$ converges to $\cos x$ for all x.

In Exercises 26–34, use the known Maclaurin series for e^x, $\sin x$, $\cos x$, and $1/(1-x)$ to find the Maclaurin series for the given function. In each case specify the interval on which your computations are valid.

26. xe^x.

27. e^{-2x}.

28. e^{x^2}.

29. $\dfrac{1}{1+x}$.

30. $\dfrac{1}{1-4x^2}$.

31. $\dfrac{x^2}{1+3x}$.

32. $\sin^2 x$ $\left[Hint:\ \sin^2 x = \dfrac{1}{2}(1 - \cos 2x). \right]$

33. $\cos^2 x$ $\left[Hint:\ \cos^2 x = \dfrac{1}{2}(1 + \cos 2x). \right]$

34. $\sinh x$.

35. Use the Maclaurin series for $1/(1-x)$ to express $1/x$ in powers of $x-1$.

36. The purpose of this exercise is to show that the Taylor series of a function f may possibly converge to a value different from $f(x)$ for certain x. Let

$$f(x) = \begin{cases} e^{-1/x^2}, & x \neq 0 \\ 0, & x = 0 \end{cases}$$

  (a) Use the definition of a derivative to show that $f'(0) = 0$.

  (b) With some difficulty it can be shown that $f^{(n)}(0) = 0$ for $n \geq 2$. Accepting this fact, show that the Maclaurin series of f converges for all x, but converges to $f(x)$ only at the point $x = 0$.

11.11 COMPUTATIONS USING TAYLOR SERIES

In this section we show how Taylor and Maclaurin series can be used to obtain approximate values for trigonometric functions and logarithms.

In the previous section we showed that for all x

$$e^x = 1 + x + \frac{x^2}{2!} + \frac{x^3}{3!} + \frac{x^4}{4!} + \cdots$$

In particular, if we let $x = 1$ we obtain the following expression for e as the sum of an infinite series

$$e = 1 + 1 + \frac{1}{2!} + \frac{1}{3!} + \frac{1}{4!} + \cdots$$

Thus we can approximate e to any degree of accuracy using an appropriate partial sum

$$e \approx 1 + 1 + \frac{1}{2!} + \frac{1}{3!} + \cdots + \frac{1}{n!} \tag{1}$$

The value of n in this formula is at our disposal, the larger we choose n the more accurate the approximation. For practical applications one would be interested in approximating e to a specified degree of accuracy, in which case it would be important to determine how large n should be chosen to attain the desired accuracy. For example, how large should n be taken in (1) to ensure an error of at most 0.00005? We will now show how Lagrange's remainder formula can be used to answer such questions.

According to Taylor's theorem, the absolute value of the error that results when $f(x)$ is approximated by its nth Taylor polynomial $p_n(x)$ about $x = a$ is

$$|R_n(x)| = |f(x) - p_n(x)| = \left| \frac{f^{(n+1)}(c)}{(n+1)!}(x-a)^{n+1} \right| \qquad (2)$$

In this formula, c is an unknown number between a and x, so that the value of $f^{(n+1)}(c)$ usually cannot be determined. However, it is frequently possible to determine an upper bound on the size of $|f^{(n+1)}(c)|$; that is, we can find a constant M such that $|f^{(n+1)}(c)| \leq M$. For such an M, it follows from (2) that

$$|R_n(x)| = |f(x) - p_n(x)| \leq \frac{M}{(n+1)!}|x-a|^{n+1} \qquad (3)$$

which gives an upper bound on the magnitude of the error $R_n(x)$. The examples to follow illustrate the usefulness of this result; but first let us introduce some terminology.

An approximation is said to be **accurate to n decimal places** if the magnitude of the error is less than 0.5×10^{-n}. For example,

DESCRIPTION	MAGNITUDE OF THE ERROR IS LESS THAN
1 decimal place accuracy	$0.05 = 0.5 \times 10^{-1}$
2 decimal place accuracy	$0.005 = 0.5 \times 10^{-2}$
3 decimal place accuracy	$0.0005 = 0.5 \times 10^{-3}$
$\vdots$	$\vdots$

APPROXIMATING e ▶ Example 1

Use (1) to approximate e to four decimal place accuracy.

Solution. From Taylor's formula with remainder (see Example 1, Section 11.10)

$$e^x = 1 + x + \frac{x^2}{2!} + \cdots + \frac{x^n}{n!} + \frac{e^c}{(n+1)!}x^{n+1}$$

where c is between 0 and x. Thus, in the case $x = 1$ we obtain

$$e = 1 + 1 + \frac{1}{2!} + \cdots + \frac{1}{n!} + \frac{e^c}{(n+1)!}$$

where c is between 0 and 1. This tells us that the magnitude of the error in approximation (1) is

$$|R_n| = \left|\frac{e^c}{(n+1)!}\right| = \frac{e^c}{(n+1)!} \tag{4}$$

where $0 < c < 1$. Since

$$c < 1$$

it follows that

$$e^c < e^1 = e$$

so that from (4)

$$|R_n| < \frac{e}{(n+1)!} \tag{5}$$

This inequality provides an upper bound on the magnitude of the error R_n. Unfortunately this inequality is not very useful since the right side involves the quantity e, which we are trying to estimate. However, if we use the fact that $e < 3$, then we can replace (5) with the more useful result

$$|R_n| < \frac{3}{(n+1)!}$$

It follows that if we choose n so that

$$|R_n| < \frac{3}{(n+1)!} < 0.5 \times 10^{-4} = 0.00005 \tag{6}$$

then approximation (1) will be accurate to four decimal places. An appropriate value for n may be found by trial and error. For example, using a handheld calculator or computer one can evaluate $3/(n+1)!$ for $n = 0, 1, 2, \ldots$ until a value of n satisfying (6) is obtained. We leave it for the reader to show that $n = 8$ is the first positive integer satisfying (6). Thus, to four decimal place accuracy

$$e \approx 1 + 1 + \frac{1}{2!} + \frac{1}{3!} + \frac{1}{4!} + \frac{1}{5!} + \frac{1}{6!} + \frac{1}{7!} + \frac{1}{8!} \approx 2.7183 \quad \blacktriangleleft$$

APPROXIMATING TRIGONOMETRIC FUNCTIONS

▶ Example 2

Use the Maclaurin series for $\sin x$ to approximate $\sin 3°$ to five decimal place accuracy.

Solution. In the Maclaurin series

$$\sin x = x - \frac{x^3}{3!} + \frac{x^5}{5!} - \frac{x^7}{7!} + \cdots \tag{7}$$

the angle x is assumed to be in radians (because the differentiation formulas for the trigonometric functions were derived with this assumption). Since $3° = \pi/60$ radians, it follows from (7) that

$$\sin 3° = \sin \frac{\pi}{60} = \left(\frac{\pi}{60}\right) - \frac{\left(\frac{\pi}{60}\right)^3}{3!} + \frac{\left(\frac{\pi}{60}\right)^5}{5!} - \frac{\left(\frac{\pi}{60}\right)^7}{7!} + \cdots \tag{8}$$

We must now decide how many terms in this series must be kept in order to obtain five decimal place accuracy. We will consider two possible approaches, one using Lagrange's remainder formula, and the other exploiting the fact that (8) satisfies the hypotheses of the alternating series test.

If we let $f(x) = \sin x$, then the magnitude of the error that results when $\sin x$ is approximated by its nth Maclaurin polynomial is

$$|R_n| = \left| \frac{f^{(n+1)}(c)}{(n+1)!} x^{n+1} \right|$$

where c is between 0 and x. Since $f^{(n+1)}(c)$ is either $\pm \sin c$ or $\pm \cos c$, it follows that $|f^{(n+1)}(c)| \leq 1$, so

$$|R_n| \leq \frac{|x|^{n+1}}{(n+1)!}$$

In particular, if $x = \pi/60$, then

$$|R_n| \leq \frac{\left(\frac{\pi}{60}\right)^{n+1}}{(n+1)!}$$

Thus, for five decimal place accuracy, we must choose n so that

$$\frac{\left(\frac{\pi}{60}\right)^{n+1}}{(n+1)!} < 0.5 \times 10^{-5} = 0.000005$$

By trial and error with the help of a handheld calculator, the reader can check that $n = 3$ is the smallest n that works. Thus in (8) we need only keep terms up to the third power for five decimal place accuracy, that is,

$$\sin 3° \approx \left(\frac{\pi}{60}\right) - \frac{\left(\frac{\pi}{60}\right)^3}{3!} \tag{9}$$
$$\approx 0.05234$$

An alternate approach to determining n uses the fact that (8) satisfies the hypotheses of the alternating series test, Theorem 11.7.1. (Verify.) Thus, by Theorem 11.7.2, if we use only those terms up to and including

$$\pm \frac{\left(\frac{\pi}{60}\right)^m}{m!} \qquad [m \text{ is an odd positive integer.}]$$

then the magnitude of the error will be at most

$$\frac{\left(\frac{\pi}{60}\right)^{m+2}}{(m+2)!}$$

Thus for five decimal place accuracy, we look for the first positive odd integer m such that

$$\frac{\left(\frac{\pi}{60}\right)^{m+2}}{(m+2)!} < 0.5 \times 10^{-5} = 0.000005$$

By trial and error, $m = 3$ is the first such integer. Thus to five decimal place accuracy

$$\sin 3° \approx \left(\frac{\pi}{60}\right) - \frac{\left(\frac{\pi}{60}\right)^3}{3!} \approx 0.05234 \tag{10}$$

which is the same as our earlier result. ◄

REMARK. It should be noted that there are two types of errors that result when computing with series. The first, called *truncation error,* is the error that results when the entire series is approximated by a partial sum. The second kind of error, called *round-off error,* results when decimal approximations are used. For example, (9) involves a truncation error of at most 0.5×10^{-5} (five decimal place accuracy). However, to obtain the numerical value (10) it was necessary to approximate π, thereby introducing a round-off error. If one seeks n decimal place accuracy in the final result, then it is common procedure to use $n + 1$ decimal places in each decimal approximation and then round off to n decimal places at the end of all computations. Although this procedure may occasionally not produce n decimal place accuracy, it works often enough that we will use it in this text. Better procedures are considered in a branch of mathematics called **numerical analysis.**

▶ Example 3

Approximate $\sin 92°$ to five decimal place accuracy.

Solution. We could proceed, as in the previous example, using the Maclaurin series

$$\sin x = x - \frac{x^3}{3!} + \frac{x^5}{5!} - \frac{x^7}{7!} + \cdots \qquad (11)$$

with $x = 92° = \frac{23}{45}\pi$ (radians). However, a better procedure is to work with the Taylor series about $x = \pi/2$ for $\sin x$ (Example 5, Section 11.10):

$$\sin x = 1 - \frac{1}{2!}\left(x - \frac{\pi}{2}\right)^2 + \frac{1}{4!}\left(x - \frac{\pi}{2}\right)^4 - \frac{1}{6!}\left(x - \frac{\pi}{2}\right)^6 + \cdots \qquad (12)$$

The advantage of this series over the Maclaurin series is its faster convergence at $x = 92° = \frac{23}{45}\pi$. In general, the rate of convergence of a Taylor series about $x = a$ decreases as one moves farther away from the point a. That is, the farther x becomes from a, the more terms that are required to achieve a given degree of accuracy. Therefore, for the purpose of approximating $\sin 92° = \sin \frac{23}{45}\pi$, series (12) is preferable to series (11) since $a = \pi/2$ is closer to $\frac{23}{45}\pi$ than $a = 0$.

Substituting $x = \frac{23}{45}\pi$ in (12) yields

$$\sin 92° = \sin \frac{23}{45}\pi = 1 - \frac{1}{2!}\left(\frac{\pi}{90}\right)^2 + \frac{1}{4!}\left(\frac{\pi}{90}\right)^4 - \frac{1}{6!}\left(\frac{\pi}{90}\right)^6 + \cdots \qquad (13)$$

If we let $f(x) = \sin x$, then the magnitude of the error in approximating $\sin x$ by its nth Taylor polynomial about $a = \pi/2$ is

$$|R_n| = \left| \frac{f^{(n+1)}(c)}{(n + 1)!}\left(x - \frac{\pi}{2}\right)^{n+1} \right|$$

where c is between $\pi/2$ and x. As in the previous example, $|f^{(n+1)}(c)| \leq 1$, so that

$$\cdot |R_n| \leq \frac{\left| x - \frac{\pi}{2} \right|^{n+1}}{(n+1)!}$$

In particular, if $x = \frac{23}{45}\pi$ then

$$|R_n| \leq \frac{\left(\frac{\pi}{90} \right)^{n+1}}{(n+1)!}$$

For five decimal place accuracy, we must choose n so that

$$\frac{\left(\frac{\pi}{90} \right)^{n+1}}{(n+1)!} < 0.5 \times 10^{-5} = 0.000005$$

By trial and error, the smallest such integer is $n = 3$. Thus in (13) we need only keep terms up to the third power for five decimal place accuracy, that is,

$$\sin 92° \approx 1 + 0 \cdot \left(\frac{\pi}{90} \right) - \frac{1}{2!} \left(\frac{\pi}{90} \right)^2 + 0 \cdot \left(\frac{\pi}{90} \right)^3$$

$$= 1 - \frac{1}{2!} \left(\frac{\pi}{90} \right)^2$$

$$\approx 0.99939$$

As in Example 2, we could also have obtained this result by exploiting the fact that (13) is an alternating series. ◀

APPROXIMATING LOGARITHMS   The Maclaurin series

$$\ln(1 + x) = x - \frac{x^2}{2} + \frac{x^3}{3} - \frac{x^4}{4} + \cdots \qquad -1 < x \leq 1 \qquad (14)$$

is the starting point for the approximation of natural logarithms. Unfortunately, this series is of limited usefulness because of its slow convergence and the restriction $-1 < x \leq 1$. However, if we replace x by $-x$ in this series we obtain

$$\ln(1 - x) = -x - \frac{x^2}{2} - \frac{x^3}{3} - \frac{x^4}{4} - \cdots \qquad -1 \leq x < 1 \qquad (15)$$

and on subtracting (15) from (14) we obtain

$$\ln \frac{1 + x}{1 - x} = 2\left(x + \frac{x^3}{3} + \frac{x^5}{5} + \frac{x^7}{7} + \cdots\right) \qquad -1 < x < 1 \qquad (16)$$

Series (16), first obtained by James Gregory* in 1668, can be used to compute the natural logarithm of any positive number y by letting

$$x = \frac{y - 1}{y + 1} \qquad\qquad (17)$$

so that

$$y = \frac{1 + x}{1 - x}$$

For example, to compute ln 2 we let $y = 2$ in (17), which yields $x = 1/3$. Substituting this value in (16) gives

$$\ln 2 = 2\left[(1/3) + \frac{(1/3)^3}{3} + \frac{(1/3)^5}{5} + \frac{(1/3)^7}{7} + \frac{(1/3)^9}{9} + \cdots\right]$$

$$= 2\left[\frac{1}{3} + \frac{1}{81} + \frac{1}{1,215} + \frac{1}{15,309} + \frac{1}{177,147} + \cdots\right]$$

$$\approx 2[0.33333 + 0.01235 + 0.00082 + 0.00007 + 0.00001 + \cdots]$$

Adding the five terms shown and then rounding to four decimal places at the end yields

$$\ln 2 \approx 2[0.34657] = 0.69314 \approx 0.6931$$

It is of interest to note that in the case $x = 1$, series (14) yields

$$\ln 2 = 1 - \frac{1}{2} + \frac{1}{3} - \frac{1}{4} + \frac{1}{5} - \cdots$$

---

*JAMES GREGORY (1638–1675) Scottish mathematician and astronomer. Gregory, the son of a minister, was famous in his time as the inventor of the Gregorian reflecting telescope, so named in his honor. Although he is not generally ranked with the great mathematicians, much of his work relating to calculus was studied by Leibniz and Newton and undoubtedly influenced some of their discoveries. There is a manuscript, discovered posthumously, which shows that Gregory had anticipated Taylor series well before Taylor.

This result is noteworthy because it is the first time we have been able to obtain the sum of the alternating harmonic series. However, this series converges too slowly to be of any computational value.

APPROXIMATING π If we let $x = 1$ in the Maclaurin series

$$\tan^{-1} x = x - \frac{x^3}{3} + \frac{x^5}{5} - \frac{x^7}{7} + \cdots \qquad -1 \le x \le 1 \qquad (18)$$

we obtain

$$\frac{\pi}{4} = \tan^{-1} 1 = 1 - \frac{1}{3} + \frac{1}{5} - \frac{1}{7} + \cdots$$

or

$$\pi = 4\left[1 - \frac{1}{3} + \frac{1}{5} - \frac{1}{7} + \cdots \right]$$

This famous series, obtained by Leibniz in 1674, converges too slowly to be of computational importance. A more practical procedure for approximating π uses the identity

$$\frac{\pi}{4} = \tan^{-1} \frac{1}{2} + \tan^{-1} \frac{1}{3} \qquad (19)$$

By using this identity and series (18) to approximate $\tan^{-1} \frac{1}{2}$ and $\tan^{-1} \frac{1}{3}$, the value of π can be effectively approximated to any degree of accuracy.

▶ Exercise Set 11.11
_____

In this exercise set, a handheld calculator will be useful.

1. How large should n be taken in (1) to approximate e to:
   (a) five decimal place accuracy?
   (b) ten decimal place accuracy?

In Exercises 2–8, apply Lagrange's form of the remainder.

2. Use $x = -1$ in the Maclaurin series for e^x to approximate $1/e$ to three decimal place accuracy.

3. Use $x = \frac{1}{2}$ in the Maclaurin series for e^x to approximate $\sqrt{e}$ to four decimal place accuracy.

4. Use the Maclaurin series for $\sin x$ to approximate $\sin 4°$ to five decimal place accuracy.

5. Use the Maclaurin series for $\cos x$ to approximate $\cos (\pi/20)$ to four decimal place accuracy.

6. Use an appropriate Taylor series for $\sin x$ to approximate $\sin 85°$ to four decimal place accuracy.

7. Use an appropriate Taylor series for $\cos x$ to approximate $\cos 58°$ to four decimal place accuracy.

8. Use an appropriate Taylor series for $\sin x$ to approximate $\sin 35°$ to four decimal place accuracy.

9. Use series (16) to approximate ln 1.25 to three decimal place accuracy.

10. Use series (16) to approximate ln 3 to four decimal place accuracy.

11. Use the Maclaurin series for $\tan^{-1} x$ to approximate $\tan^{-1} 0.1$ to three decimal place accuracy. [*Hint:* Use the fact that (18) is an alternating series.]

12. Use the Maclaurin series for $\sinh x$ to approximate $\sinh 0.5$ to three decimal place accuracy.

13. Use the Maclaurin series for $\cosh x$ to approximate $\cosh 0.1$ to four decimal place accuracy.

14. Use an appropriate Taylor series for $\sqrt[3]{x}$ to approximate $\sqrt[3]{28}$ to three decimal place accuracy.

15. For what range of x values can $\sin x$ be approximated by $x - x^3/3!$ with three decimal place accuracy?

16. Estimate the maximum error in the approximation $\cos x \approx 1 - x^2/2! + x^4/4!$ if $-0.2 \le x \le 0.2$.

17. For what range of x values can e^x be approximated by $1 + x + x^2/2!$ if the allowable error is at most 0.0005?

18. Estimate the maximum error in the approximation $\ln (1 + x) \approx x$ if $|x| < 0.01$.

19. (a) How many terms in the alternating harmonic series are needed to approximate ln 2 to six decimal place accuracy?
    (b) How many terms in series (16) with $x = \frac{1}{3}$ are needed to approximate ln 2 to six decimal place accuracy?

20. Prove identity (19).

21. Approximate $\tan^{-1} \frac{1}{2}$ and $\tan^{-1} \frac{1}{3}$ to three decimal place accuracy and then use identity (19) to approximate π.

22. The curve $y = \cos x$ is to be approximated by the parabola $x^2 = 2(1 - y)$ over the interval $[-0.2, 0.2]$. Estimate the maximum difference in the y-coordinates of the curves over this interval.

11.12 DIFFERENTIATION AND INTEGRATION OF POWER SERIES

If a function f is expressed as a power series

$$f(x) = \sum_{k=0}^{\infty} c_k(x - a)^k$$

for all x in some interval, then we say that the series *represents* $f(x)$ on that interval. Thus, for example, the geometric series $1 + x + x^2 + x^3 + \cdots$ represents the function $1/(1 - x)$ on the interval $(-1, 1)$ since

$$\frac{1}{1 - x} = 1 + x + x^2 + x^3 + \cdots \qquad -1 < x < 1$$

In this section we show how a power series representation of a function f can be used to obtain power series representations of $f'(x)$ and $\int f(x)\, dx$. Some applications of the results are also discussed.

The following theorem tells us that a power series representation of $f'(x)$ or $\int f(x)\, dx$ can be obtained by differentiating or integrating a power series representation of $f(x)$ term by term. We omit the proofs.

11.12.1 THEOREM *If the series $\sum_{k=0}^{\infty} c_k(x-a)^k$ has a nonzero radius of convergence R, and if for*

each x in the interval $(a-R, a+R)$ we have $f(x) = \sum_{k=0}^{\infty} c_k(x-a)^k$ then:

(a) *The series $\sum_{k=0}^{\infty} \dfrac{d}{dx}[c_k(x-a)^k] = \sum_{k=1}^{\infty} kc_k(x-a)^{k-1}$ has radius of conver-*

gence R, and for all x in the interval $(a-R, a+R)$

$$f'(x) = \sum_{k=0}^{\infty} \frac{d}{dx}[c_k(x-a)^k]$$

(b) *The series $\sum_{k=0}^{\infty} \left[\int c_k(x-a)^k \, dx\right] = \sum_{k=0}^{\infty} \dfrac{c_k}{k+1}(x-a)^{k+1}$ has radius of*

convergence R, and for all x in the interval $(a-R, a+R)$

$$\int f(x) \, dx = \sum_{k=0}^{\infty} \left[\int c_k(x-a)^k \, dx\right] + C$$

(c) *For all α and β in the interval $(a-R, a+R)$, the series*

$$\sum_{k=0}^{\infty} \left[\int_{\alpha}^{\beta} c_k(x-a)^k \, dx\right]$$

converges absolutely and

$$\int_{\alpha}^{\beta} f(x) \, dx = \sum_{k=0}^{\infty} \left[\int_{\alpha}^{\beta} c_k(x-a)^k \, dx\right]$$

REMARK. Note that in part (b) a separate constant of integration is not

introduced for each term in the series $\sum_{k=0}^{\infty} \left[\int c_k(x-a)^k \, dx\right]$; rather a single

constant C is added to the entire series.

▶ Example 1

To illustrate Theorem 11.12.1, we will obtain the familiar results

$$\frac{d}{dx}[\sin x] = \cos x \qquad \text{and} \qquad \int \cos x \, dx = \sin x + C$$

using power series. Since

$$\sin x = x - \frac{x^3}{3!} + \frac{x^5}{5!} - \frac{x^7}{7!} + \cdots \qquad -\infty < x < +\infty$$

$$\cos x = 1 - \frac{x^2}{2!} + \frac{x^4}{4!} - \frac{x^6}{6!} + \cdots \qquad -\infty < x < +\infty$$

it follows from part (a) of Theorem 11.12.1 that

$$\frac{d}{dx}[\sin x] = \frac{d}{dx}\left[x - \frac{x^3}{3!} + \frac{x^5}{5!} - \frac{x^7}{7!} + \cdots\right]$$

$$= 1 - 3\frac{x^2}{3!} + 5\frac{x^4}{5!} - 7\frac{x^6}{6!} + \cdots$$

$$= 1 - \frac{x^2}{2!} + \frac{x^4}{4!} - \frac{x^6}{6!} + \cdots$$

$$= \cos x$$

Also, from part (b) of Theorem 11.12.1

$$\int \cos x\, dx = \int\left[1 - \frac{x^2}{2!} + \frac{x^4}{4!} - \frac{x^6}{6!} + \cdots\right]dx$$

$$= \left[x - \frac{x^3}{3(2!)} + \frac{x^5}{5(4!)} - \frac{x^7}{7(6!)} + \cdots\right] + C$$

$$= \left[x - \frac{x^3}{3!} + \frac{x^5}{5!} - \frac{x^7}{7!} + \cdots\right] + C$$

$$= \sin x + C \qquad\qquad\qquad \blacktriangleleft$$

▶ Example 2

The integral

$$\int_0^1 e^{-x^2}\, dx$$

cannot be evaluated directly since there is no elementary antiderivative of e^{-x^2}. One possibility is to approximate the integral by some numerical technique like Simpson's Rule. However, another possibility is to represent e^{-x^2} by its Maclaurin series and then integrate term by term in accordance with part (c) of Theorem 11.12.1.

The simplest way to obtain the Maclaurin series for e^{-x^2} is to replace x by $-x^2$ in the Maclaurin series

$$e^x = 1 + x + \frac{x^2}{2!} + \frac{x^3}{3!} + \frac{x^4}{4!} + \cdots$$

to obtain

$$e^{-x^2} = 1 - x^2 + \frac{x^4}{2!} - \frac{x^6}{3!} + \frac{x^8}{4!} - \cdots$$

Therefore

$$\int_0^1 e^{-x^2}\, dx = \int_0^1 \left[1 - x^2 + \frac{x^4}{2!} - \frac{x^6}{3!} + \frac{x^8}{4!} - \cdots \right] dx$$

$$= \left[x - \frac{x^3}{3} + \frac{x^5}{5(2!)} - \frac{x^7}{7(3!)} + \frac{x^9}{9(4!)} - \cdots \right]_0^1$$

$$= 1 - \frac{1}{3} + \frac{1}{5 \cdot 2!} - \frac{1}{7 \cdot 3!} + \frac{1}{9 \cdot 4!} - \cdots \tag{1}$$

Thus we have found a series that converges to the value of the integral $\int_0^1 e^{-x^2}\, dx$. Using the first three terms in this series we obtain the approximation

$$\int_0^1 e^{-x^2}\, dx \approx 1 - \frac{1}{3} + \frac{1}{10} = \frac{23}{30} = 0.766 \ldots$$

Since series (1) satisfies the hypotheses of the alternating series test, it follows from Theorem 11.7.2 that the magnitude of the error in this approximation is at most $1/(7 \cdot 3!) = 1/42 \approx 0.024$. Greater accuracy can be obtained by using more terms in the series ◀

The following example illustrates how Theorem 11.12.1 can sometimes be used to obtain new Taylor series from known Taylor series.

► Example 3

Derive the Maclaurin series for $\tan^{-1} x$.

Solution. We could calculate this Maclaurin series directly. However we can also exploit Theorem 11.12.1 by first writing

$$\int \frac{1}{1 + x^2}\, dx = \tan^{-1} x + C$$

and then integrating the Maclaurin series for $1/(1 + x^2)$ term by term. Since

$$\frac{1}{1 - x} = 1 + x + x^2 + x^3 + x^4 + \cdots$$

it follows on replacing x by $-x^2$ that

$$\frac{1}{1 + x^2} = 1 - x^2 + x^4 - x^6 + x^8 - \cdots \tag{2}$$

Thus

$$\tan^{-1} x + C = \int \frac{1}{1 + x^2} \, dx = \int [1 - x^2 + x^4 - x^6 + x^8 - \cdots] \, dx$$

or

$$\tan^{-1} x = \left[x - \frac{x^3}{3} + \frac{x^5}{5} - \frac{x^7}{7} + \frac{x^9}{9} - \cdots \right] - C$$

The constant of integration may be evaluated by substituting $x = 0$ and using the condition $\tan^{-1} 0 = 0$. This gives $C = 0$, so that

$$\tan^{-1} x = x - \frac{x^3}{3} + \frac{x^5}{5} - \frac{x^7}{7} + \frac{x^9}{9} - \cdots \tag{3}$$

◄

Although it is conceivable at this point that series (3) is not the Maclaurin series for $\tan^{-1} x$, but rather some other power series converging to $\tan^{-1} x$, the following theorem tells us that this is not the case.

11.12.2 THEOREM *If*

$$f(x) = c_0 + c_1(x - a) + c_2(x - a)^2 + \cdots + c_n(x - a)^n + \cdots$$

for all x in some open interval containing a, then the series is the Taylor series for f about a.

Proof. By repeated application of Theorem 11.12.1a we obtain

$$f(x) = c_0 + c_1(x - a) + c_2(x - a)^2 + c_3(x - a)^3 + c_4(x - a)^4 + \cdots$$
$$f'(x) = c_1 + 2c_2(x - a) + 3c_3(x - a)^2 + 4c_4(x - a)^3 + \cdots$$
$$f''(x) = 2!c_2 + (3 \cdot 2)c_3(x - a) + (4 \cdot 3)c_4(x - a)^2 + \cdots$$
$$f'''(x) = 3!c_3 + (4 \cdot 3 \cdot 2)c_4(x - a) + \cdots$$
$$\vdots$$

On substituting $x = a$, all the powers of $x - a$ drop out leaving

$$f(a) = c_0 \qquad\qquad c_0 = f(a)$$
$$f'(a) = c_1 \qquad\qquad c_1 = f'(a)$$
$$f''(a) = 2!c_2 \quad \text{or} \quad c_2 = \frac{f''(a)}{2!}$$

$$f'''(a) = 3!c_3 \qquad\qquad c_3 = \frac{f'''(a)}{3!}$$
$$\vdots \qquad\qquad\qquad \vdots$$

which shows that the coefficients $c_0, c_1, c_2, c_3, \ldots$ are precisely the coefficients in the Taylor series about a for $f(x)$.

This theorem tells us that no matter how we arrive at a power series in $x - a$ converging to $f(x)$, be it by substitution, by integration, by differentiation, or by algebraic manipulation, the resulting series will be the Taylor series about a for $f(x)$.

REMARK. Parts (a) and (b) of Theorem 11.12.1 say nothing about the behavior of the differentiated and integrated series at the endpoints $a - R$ and $a + R$. Indeed, by differentiating termwise, convergence may be lost at one or both endpoints; and by integrating termwise, convergence may be gained at one or both endpoints. As an illustration, in Example 3 we derived the series (3) for $\tan^{-1} x$ by integrating series (2) for $1/(1 + x^2)$. We leave it as an exercise to show that the series for $\tan^{-1} x$ converges to $\tan^{-1} x$ on the interval $[-1, 1]$, while the series for $1/(1 + x^2)$ converges to $1/(1 + x^2)$ only on $(-1, 1)$. Thus convergence was gained at both endpoints by integrating.

We conclude this section with some tricks for obtaining Taylor series that would be messy to obtain directly. Our objective is to make the reader aware of these techniques; we will not attempt to justify the results.

▶ Example 4

Find the first three terms in the Maclaurin series for $e^{-x^2} \tan^{-1} x$.

Solution. Using the series for e^{-x^2} and $\tan^{-1} x$ obtained in Examples 2 and 3

$$e^{-x^2} \tan^{-1} x = \left(1 - x^2 + \frac{x^4}{2} - \cdots \right)\left(x - \frac{x^3}{3} + \frac{x^5}{5} + \cdots \right)$$

We now multiply out following a familiar format from elementary algebra

$$
\begin{array}{r}
1 - x^2 + \dfrac{x^4}{2} - \cdots \\[2mm]
\times \quad x - \dfrac{x^3}{3} + \dfrac{x^5}{5} + \cdots \\[1mm]
\hline
x - x^3 + \dfrac{x^5}{2} - \cdots \\[2mm]
-\dfrac{x^3}{3} + \dfrac{x^5}{3} - \dfrac{x^7}{6} + \cdots \\[2mm]
\dfrac{x^5}{5} - \dfrac{x^7}{5} + \cdots \\[1mm]
\hline
x - \tfrac{4}{3}x^3 + \tfrac{31}{30}x^5 - \cdots
\end{array}
$$

Thus,

$$e^{-x^2} \tan^{-1} x = x - \frac{4}{3}x^3 + \frac{31}{30}x^5 - \cdots$$

If desired, more terms in the series can be obtained by including more terms in the factors. ◄

► Example 5

Find the first three terms in the Maclaurin series for $\tan x$.

Solution. Instead of computing the series directly, we write

$$\tan x = \frac{\sin x}{\cos x} = \frac{x - \dfrac{x^3}{3!} + \dfrac{x^5}{5!} - \cdots}{1 - \dfrac{x^2}{2!} + \dfrac{x^4}{4!} - \cdots}$$

and follow the familiar "long division" process to obtain

$$
\begin{array}{r}
x + \dfrac{x^3}{3} + \dfrac{2x^5}{15} + \cdots \\[2ex]
1 - \dfrac{x^2}{2} + \dfrac{x^4}{24} - \cdots \overline{\smash{\big)}\; x - \dfrac{x^3}{6} + \dfrac{x^5}{120} - \cdots} \\[2ex]
x - \dfrac{x^3}{2} + \dfrac{x^5}{24} - \cdots \\[1ex]
\hline
\dfrac{x^3}{3} - \dfrac{x^5}{30} + \cdots \\[2ex]
\dfrac{x^3}{3} - \dfrac{x^5}{6} + \cdots \\[1ex]
\hline
\dfrac{2x^5}{15} + \cdots
\end{array}
$$

Thus

$$\tan x = x + \frac{x^3}{3} + \frac{2x^5}{15} + \cdots \qquad ◄$$

If m is a real number, then the Maclaurin series for $(1 + x)^m$ is called the **binomial series;** it is given by (verify)

$$1 + mx + \frac{m(m-1)}{2!}x^2 + \frac{m(m-1)(m-2)}{3!}x^3 + \cdots \qquad (4)$$

It is proved in advanced calculus that the binomial series converges to $(1 + x)^m$ if $|x| < 1$. Thus, for such values of x

$$(1 + x)^m = 1 + mx + \frac{m(m - 1)}{2!}x^2 + \cdots$$
$$+ \frac{m(m - 1)(m - 2) \cdots (m - k + 1)}{k!}x^k + \cdots \qquad (4)$$

or in sigma notation

$$(1 + x)^m = 1 + \sum_{k=1}^{\infty} \frac{m(m - 1) \cdots (m - k + 1)}{k!}x^k \qquad \text{if } |x| < 1$$

REMARK. If m is a positive integer, then $f(x) = (1 + x)^m$ is a polynomial of degree m so that

$$f^{(m+1)}(0) = f^{(m+2)}(0) = f^{(m+3)}(0) = \cdots = 0$$

and the binomial series reduces to the familiar binomial expansion

$$(1 + x)^m = 1 + mx + \frac{m(m - 1)}{2!}x^2 + \frac{m(m - 1)(m - 2)}{3!}x^3 + \cdots + x^m$$

▶ Example 6

Express $1/\sqrt{1 + x}$ as a binomial series.

Solution. Substituting $m = -\frac{1}{2}$ in (4) yields

$$\frac{1}{\sqrt{1 + x}} = 1 - \frac{1}{2}x + \frac{(-\frac{1}{2})(-\frac{1}{2} - 1)}{2!}x^2 + \frac{(-\frac{1}{2})(-\frac{1}{2} - 1)(-\frac{1}{2} - 2)}{3!}x^3$$
$$+ \cdots + \frac{(-\frac{1}{2})(-\frac{3}{2})(-\frac{5}{2}) \cdots (-\frac{1}{2} - k + 1)}{k!}x^k + \cdots$$
$$= 1 - \frac{1}{2}x + \frac{1 \cdot 3}{2^2 \cdot 2!}x^2 - \frac{1 \cdot 3 \cdot 5}{2^3 \cdot 3!}x^3 + \cdots$$
$$+ (-1)^k \frac{1 \cdot 3 \cdot 5 \cdots (2k - 1)}{2^k k!}x^k + \cdots \quad ◀$$

► Exercise Set 11.12

In Exercises 1–4, obtain the stated results by differentiating or integrating Maclaurin series term by term.

1. (a) $\dfrac{d}{dx}[e^x] = e^x;$ (b) $\displaystyle\int e^x\, dx = e^x + C.$

2. (a) $\dfrac{d}{dx}[\cos x] = -\sin x;$

   (b) $\displaystyle\int \sin x\, dx = -\cos x + C.$

3. (a) $\dfrac{d}{dx}[\sinh x] = \cosh x;$

   (b) $\displaystyle\int \sinh x\, dx = \cosh x + C.$

4. (a) $\dfrac{d}{dx}[\ln(1 + x)] = \dfrac{1}{1 + x};$

   (b) $\displaystyle\int \dfrac{1}{1 + x}\, dx = \ln(1 + x) + C.$

5. Derive the Maclaurin series for $1/(1 + x)^2$ by differentiating an appropriate Maclaurin series term by term.

6. By differentiating an appropriate series, show that

$$\sum_{k=1}^{\infty} kx^k = \frac{x}{(1 - x)^2}$$

7. By integrating an appropriate series, show that

$$\sum_{k=1}^{\infty} \frac{x^k}{k} = \ln\left(\frac{1}{1 - x}\right)$$

In Exercises 8–11, use a binomial series to find the first four nonzero terms in the Maclaurin series for the function and give the radius of convergence.

8. $\sqrt{1 + x}.$ 9. $(1 - x)^{1/3}.$

10. $(1 + x^2)^{-1/4}.$ 11. $\dfrac{x}{\sqrt[3]{1 - 2x}}.$

In Exercises 12–19, use series to approximate the value of the integral to three decimal place accuracy.

12. $\displaystyle\int_0^1 \sin x^2\, dx.$ 13. $\displaystyle\int_0^1 \cos \sqrt{x}\, dx.$

14. $\displaystyle\int_0^{0.1} \frac{\sin x}{x}\, dx.$ 15. $\displaystyle\int_0^{1/2} \frac{dx}{1 + x^4}.$

16. $\displaystyle\int_0^{1/2} \tan^{-1} 2x^2\, dx.$ 17. $\displaystyle\int_0^{0.1} e^{-x^3}\, dx.$

18. $\displaystyle\int_0^{0.2} \sqrt[3]{1 + x^4}\, dx.$ 19. $\displaystyle\int_0^{1/2} \frac{dx}{\sqrt[4]{x^2 + 1}}.$

In Exercises 20–27, use any method to find the first four nonzero terms in the Maclaurin series of the given function.

20. $x^4 e^x.$ 21. $e^{-x^2} \cos x.$

22. $\dfrac{x^2}{1 + x^4}.$ 23. $\dfrac{\sin x}{e^x}.$

24. $\tanh x.$ 25. $\sin^2 x.$

26. $\dfrac{\ln(1 + x)}{1 - x}.$ 27. $x^2 e^{4x} \sqrt{1 + x}.$

28. Obtain the familiar result, $\displaystyle\lim_{x \to 0} (\sin x)/x = 1$, by finding the Maclaurin series for $(\sin x)/x$, and taking the limit term by term.

29. Use the method of Exercise 28 to find the limits:

   (a) $\displaystyle\lim_{x \to 0} \frac{1 - \cos x}{\sin x}.$

   (b) $\displaystyle\lim_{x \to 0} \frac{\ln \sqrt{1 + x} - \sin 2x}{x}.$

30. (a) Use the relationship

$$\int \frac{1}{\sqrt{1 - x^2}}\, dx = \sin^{-1} x + C$$

   to find the first four nonzero terms in the Maclaurin series for $\sin^{-1} x.$

   (b) Express the series in sigma notation.

   (c) What is the radius of convergence?

31. (a) Use the relationship

$$\int \frac{1}{\sqrt{1 + x^2}}\, dx = \sinh^{-1} x + C$$

   to find the first four nonzero terms in the Maclaurin series for $\sinh^{-1} x.$

(b) Express the series in sigma notation.

(c) What is the radius of convergence?

32. Estimate the maximum error in the approximation $\sqrt{1 + x} \approx 1 + (x/2)$ if $|x| < 0.001$.

33. Derive (4). (Do not try to prove convergence.)

34. Prove: If the power series $\sum_{k=0}^{\infty} a_k x^k$ and $\sum_{k=0}^{\infty} b_k x^k$ have the same sum on an interval $(-r, r)$, then $a_k = b_k$ for all values of k.

▶ REVIEW EXERCISES

In Exercises 1–6, find $L = \lim_{n \to +\infty} a_n$ if it exists.

1. $a_n = (-1)^n/e^n$.

2. $a_n = e^{1/n}$.

3. $a_n = \dfrac{1}{\sqrt{n}} - \dfrac{1}{\sqrt{n+1}}$.

4. $a_n = \sin(\pi n)$.

5. $a_n = \sin\left(\dfrac{(2n-1)\pi}{2}\right)$.

6. $a_n = \dfrac{n+1}{n(n+2)}$.

7. Which of the sequences $\{a_n\}_{n=1}^{+\infty}$ in Exercises 1–6 are (a) decreasing? (b) nondecreasing? (c) alternating?

8. Suppose $f(x)$ satisfies $f'(x) > 0$ and $f(x) \leq 1 - e^{-x}$ for all $x \geq 1$. What can you conclude about the convergence of $\{a_n\}$ if $a_n = f(n)$, $n = 1, 2, \ldots$?

9. Use your knowledge of geometric series and p-series to determine all values of q for which the following series converge.

(a) $\sum_{k=0}^{\infty} \pi^k/q^{2k}$

(b) $\sum_{k=1}^{\infty} (1/k^q)^3$

(c) $\sum_{k=2}^{\infty} 1/(\ln q^k)$

(d) $\sum_{k=2}^{\infty} 1/(\ln q)^k$.

10. (a) Use a suitable test to find all values of q for which $\sum_{k=2}^{\infty} 1/[k (\ln k)^q]$ converges.

(b) Why can't you use the integral test for the series $\sum_{k=1}^{\infty} (2 + \cos k)/k^2$? Test for convergence using a test that does apply.

11. Express $1.3636\ldots$ as: (a) an infinite series in sigma-notation, (b) a ratio of integers.

12. In parts (a)–(d), use the comparison test to determine whether the series converges.

(a) $\sum_{k=1}^{\infty} \dfrac{2k - 1}{3k^2 - k}$

(b) $\sum_{k=1}^{\infty} \dfrac{2k + 1}{3k^2 + k}$

(c) $\sum_{k=1}^{\infty} \dfrac{2k - 1}{3k^3 - k^2}$

(d) $\sum_{k=1}^{\infty} \dfrac{2k + 1}{3k^3 + k^2}$.

13. Find the sum of the series (if it converges).

(a) $\sum_{k=1}^{\infty} \dfrac{2^k + 3^k}{6^{k+1}}$

(b) $\sum_{k=2}^{\infty} \ln\left(1 + \dfrac{1}{k}\right)$

(c) $\sum_{k=1}^{\infty} [k^{-1/2} - (k+1)^{-1/2}]$.

In Exercises 14–21, determine whether the series converges or diverges. You may use the following limits without proof:

$$\lim_{k \to +\infty} (1 + 1/k)^k = e, \quad \lim_{k \to +\infty} \sqrt[k]{k} = 1,$$

$$\lim_{k \to +\infty} \sqrt[k]{a} = 1$$

14. $\sum_{k=0}^{\infty} e^{-k}$.

15. $\sum_{k=1}^{\infty} ke^{-k^2}$.

16. $\sum_{k=1}^{\infty} \dfrac{k}{k^2 + 2k + 7}$.

17. $\sum_{k=1}^{\infty} \dfrac{\sqrt{k}}{k^2 + 7}$.

18. $\sum_{k=1}^{\infty} \left(\dfrac{k}{k+1}\right)^k$.

19. $\sum_{k=0}^{\infty} \dfrac{3^k k!}{(2k)!}$.

20. $\sum_{k=0}^{\infty} \dfrac{k^6 3^k}{(k+1)!}$.

21. $\sum_{k=1}^{\infty} \left(\dfrac{5k}{2k+1}\right)^{3k}$.

In Exercises 22-25, determine if the given series is absolutely convergent, conditionally convergent, or divergent.

22. $\sum_{k=1}^{\infty} (-1)^k/e^{1/k}$.

23. $\sum_{k=0}^{\infty} (-2)^k/(3^k + 1)$.

24. $\sum\limits_{k=0}^{\infty} (-1)^k/(2k+1)$.   **25.** $\sum\limits_{k=0}^{\infty} (-1)^k 3^k/2^{k+1}$.

26. Find the smallest value of n for which the nth partial sum approximates the sum of the series to the stated accuracy:

(a) $\sum\limits_{k=1}^{\infty} \dfrac{(-1)^k}{k^2+1}$; $|\text{error}| < 0.0001$.

(b) $\sum\limits_{k=1}^{\infty} \dfrac{(-1)^k}{5^k+1}$; $|\text{error}| < 0.00005$.

In Exercises 27–32, determine the radius of convergence and the interval of convergence of the given power series.

27. $\sum\limits_{k=1}^{\infty} \dfrac{(x-1)^k}{k\sqrt{k}}$.   **28.** $\sum\limits_{k=1}^{\infty} \dfrac{(2x)^k}{3k}$.

29. $\sum\limits_{k=1}^{\infty} \dfrac{(1-x)^{2k}}{4^k k}$.   **30.** $\sum\limits_{k=1}^{\infty} \dfrac{k^2(x+2)^k}{(k+1)!}$.

31. $\sum\limits_{k=1}^{\infty} \dfrac{k!(x-1)^k}{5^k}$.   **32.** $\sum\limits_{k=1}^{\infty} \dfrac{(2k)!x^k}{(2k+1)!}$.

In Exercises 33–35, find:
(a) the nth Taylor polynomial for f about $x = a$ (for the stated values of n and a);
(b) Lagrange's form of $R_n(x)$ (for the stated values of n and a);
(c) an upper bound on the absolute value of the error if $f(x)$ is approximated over the given interval by the Taylor polynomial obtained in part (a).

33. $f(x) = \ln(x-1)$; $a = 2$; $n = 3$; $[\tfrac{3}{2}, 2]$.

34. $f(x) = e^{x/2}$; $a = 0$; $n = 4$; $[-1, 0]$.

35. $f(x) = \sqrt{x}$; $a = 1$; $n = 2$; $[\tfrac{4}{9}, 1]$.

36. (a) Use the identity $a - x = a(1 - x/a)$ to find the Maclaurin series for $1/(a - x)$ from the geometric series. What is its radius of convergence?
(b) Find the Maclaurin series and radius of convergence of $1/(3 + x)$.
(c) Find the Maclaurin series and radius of convergence of $2x/(4 + x^2)$.

(d) Use partial fractions to find the Maclaurin series and radius of convergence of
$$\frac{1}{(1-x)(2-x)}.$$

37. Use the known Maclaurin series for $\ln(1 + x)$ to find the Maclaurin series and radius of convergence of $\ln(a + x)$ $(a > 0)$.

38. Use the identity $x = a + (x - a)$ and the known Maclaurin series for e^x, $\sin x$, $\cos x$, and $1/(1 - x)$ to find the Taylor series about $x = a$ for: (a) e^x (b) $\sin x$ and (c) $1/x$.

39. Use the series of Example 6 in Section 11.12 to find the Maclaurin series and radius of convergence of $1/\sqrt{9 + x}$.

In Exercises 40–44, use any method to find the first three nonzero terms of the Maclaurin series.

40. $e^{\tan x}$.   **41.** $\sec x$.

42. $(\sin x)/(e^x - 1)$.   **43.** $\sqrt{\cos x}$.

44. $e^x \ln(1 - x)$.

45. Find Maclaurin series for

(a) $\dfrac{1 - \cos x}{x^2}$ (b) $\dfrac{\ln(1 + x)}{x}$

(c) Use your answers to (a) and (b) to find
$$\lim_{x \to 0} \frac{1 - \cos x}{x^2} \text{ and } \lim_{x \to 0} \frac{\ln(1 + x)}{x}.$$

46. How many decimal places of accuracy can be guaranteed if we approximate $\cos x$ by $1 - x^2/2$ for $-0.1 < x < 0.1$?

47. For what values of x can $\sin x$ be replaced by $x - x^3/6 + x^5/120$ with an assured accuracy of 6×10^{-4}?

In Exercises 48–50, use a suitable Maclaurin polynomial to approximate the indicated quantity to three decimal place accuracy.

48. $\cos(10°)$.

49. $\displaystyle\int_0^1 \frac{(1 - e^{-t/2})}{t}\, dt$.   **50.** $\displaystyle\int_0^1 \frac{\sin x}{\sqrt{x}}\, dx$.

51. Show that $y = \sum\limits_{n=0}^{\infty} k^n x^n/n!$ satisfies $y' - ky = 0$ for any fixed k.

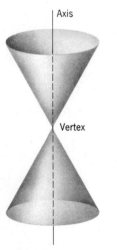

Axis

Vertex

Right circular cone

Figure 12.1.1

12

topics in
analytic geometry

12.1 INTRODUCTION TO THE CONIC SECTIONS

The surface shown in Figure 12.1.1 is called a ***double right circular cone*** or sometimes simply a ***cone.*** It is the three dimensional surface generated by a line that revolves about a fixed ***axis*** in such a way that the line always passes through a fixed point on the axis, called the ***vertex,*** and always makes the same angle with the axis. The cone consists of two parts, or ***nappes,*** that intersect at the vertex. The curves that can be obtained by intersecting a cone with a plane are called ***conics*** or ***conic sections.*** The most important conics are circles, ellipses, parabolas, and hyperbolas. A ***circle*** is obtained by intersecting a cone with a plane that is perpendicular to the axis and does not contain the vertex (Figure 12.1.2*a*). If the cutting plane is tilted slightly and intersects only one nappe, the resulting intersection is an ***ellipse*** (Figure 12.1.2*b*). If the cutting plane is tilted still further so that it is parallel to a line on the surface of the cone, but intersects only one nappe, the resulting intersection is a ***parabola*** (Figure 12.1.2*c*). Finally, if the plane intersects both nappes, but does not contain the vertex, the resulting intersection is a ***hyperbola*** (Figure 12.1.2*d*). By choosing the cutting plane to pass through the vertex, it is possible to obtain a point or a pair of lines for the intersection. These are called ***degenerate conic sections.***

According to the Alexandrian geographer and astronomer Eratosthenes, the conic sections were first discovered by Menaechmus, a geometer and astronomer in Plato's academy. While it is not known for certain what motivated the discovery of the conic sections, it is commonly believed that they resulted from the study of construction problems. Menaechmus, for example, used them to solve the problem of "doubling the cube," that is, constructing a cube whose volume is twice that of a given cube. Another theory suggests that the conic sections may have originated as a result of work on sundials. With the advent of analytic geometry and calculus, conic

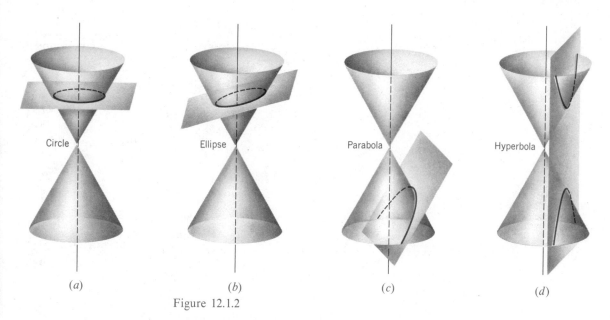

(a) Circle   (b) Ellipse   (c) Parabola   (d) Hyperbola

Figure 12.1.2

sections gained importance in the physical sciences. In 1609 Johannes Kepler published a book known as *Astronomia Nova* in which he presented his landmark discovery that the path of each planet about the sun is an ellipse. Galileo and Newton showed that objects subject to gravitational forces could also move along paths that are parabolas or hyperbolas.

Today, properties of conic sections are used in the construction of telescopes, radar antennas, and navigational systems, and in the determination of satellite orbits. In this chapter we will develop the basic geometric properties and equations of conic sections. For simplicity, our working definitions of the conic sections will be based on their geometric properties rather than their interpretation as intersections of a plane with a cone. This will enable us to keep our work in a two-dimensional setting.

12.2 THE PARABOLA; TRANSLATION OF COORDINATE AXES

12.2.1 DEFINITION   A *parabola* is the set of all points in the plane that are equidistant from a given line and a given point not on the line.

The given line is called the *directrix* of the parabola, and the given point the *focus* (Figure 12.2.1). A parabola is symmetric about the line that passes through the focus at right angles to the directrix. This line of symmetry, called the *axis* of the parabola, meets the parabola at a point called the *vertex.*

The equation of a parabola is simplest if the coordinate axes are positioned so that the vertex is at the origin and the axis of symmetry is along the x-axis or y-axis. The four possible such orientations are shown in Table 12.2.1. In

Figure 12.2.1

Parabola

Focus

Axis

Vertex

Directrix

All points on the parabola are equidistant from the focus and directrix.

Table 12.2.1

ORIENTATION	DESCRIPTION	EQUATION
	Vertex at the origin. Parabola opens in the positive x-direction. Symmetric about the x-axis.	$y^2 = 4px$ (p positive)
	Vertex at the origin. Parabola opens in the negative x-direction. Symmetric about the x-axis.	$y^2 = 4px$ (p negative)
	Vertex at the origin. Parabola opens in the positive y-direction. Symmetric about the y-axis.	$x^2 = 4py$ (p positive)
	Vertex at the origin. Parabola opens in the negative y-direction. Symmetric about the y-axis.	$x^2 = 4py$ (p negative)

the first two orientations, the focus is assumed to be at the point $(p, 0)$, where p is positive in one case and negative in the other. Since the vertex is equidistant from focus and directrix, it follows that the directrix has equation $x = -p$ in these cases. In the last two orientations, the focus is assumed to be at $(0, p)$; and the directrix has equation $y = -p$, where again p may be positive or negative.

To illustrate how the equations in Table 12.2.1 are obtained, let us begin with the first entry in the table—the parabola with focus at $(p, 0)$, directrix $x = -p$, and p positive.

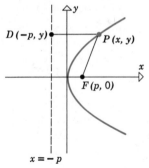

$D(-p, y)$

$P(x, y)$

$F(p, 0)$

$x = -p$

Figure 12.2.2

Let $P(x, y)$ be any point on the parabola. Since P is equidistant from focus and directrix, the distances PF and PD in Figure 12.2.2 are equal; that is,

$$PF = PD \tag{1}$$

where $D(-p, y)$ is the foot of the perpendicular from P to the directrix. From the distance formula, the distances PF and PD are

$$PF = \sqrt{(x - p)^2 + y^2} \quad \text{and} \quad PD = \sqrt{(x + p)^2} \tag{2}$$

Substituting in (1) and squaring yields

$$(x - p)^2 + y^2 = (x + p)^2 \tag{3}$$

and after simplifying

$$y^2 = 4px \tag{4}$$

Conversely, any point $P(x, y)$ satisfying (4) also satisfies (3) (reverse the steps in the simplification); thus from (2), $PF = PD$, which shows that P is equidistant from focus and directrix. Therefore, each point satisfying (4) lies on the parabola.

The remaining equations in Table 12.2.1 have similar derivations; they are left as exercises.

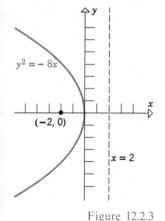

$y^2 = -8x$

$(-2, 0)$

$x = 2$

Figure 12.2.3

► Example 1

Find the focus and directrix of the parabola with equation $y^2 = -8x$.

Solution. The equation is of the form $y^2 = 4px$ with $4p = -8$ or $p = -2$. Since y occurs only to an even power, the parabola is symmetric about the x-axis. Thus the focus is on the x-axis at the point $(-2, 0)$ and the directrix has equation $x = 2$ (Figure 12.2.3). ◄

► Example 2

Find an equation for the parabola that is symmetric about the y-axis, has its vertex at the origin, and passes through the point $(5, 2)$.

Solution. Since the parabola is symmetric about the y-axis and has vertex at the origin, the equation is of the form

$$x^2 = 4py \tag{5}$$

Since the parabola passes through (5, 2) we must have $5^2 = 4p(2)$ or $4p = \frac{25}{2}$. Therefore (5) becomes

$$x^2 = \frac{25}{2} y$$ ◀

If a parabola has its axis of symmetry parallel to one of the coordinate axes, then the location of the focus and directrix may be found by introducing an auxiliary coordinate system with origin at the vertex and axes parallel to the original axes. Before we can discuss the details, we need some preliminary results.

TRANSLATION OF
AXES

In Figure 12.2.4a we have translated the axes of an xy-coordinate system to obtain a new $x'y'$-coordinate system whose origin O' is at the point $(x, y) = (h, k)$. As a result, a point P in the plane will have both (x, y)-coordinates and (x', y')-coordinates. As suggested by Figure 12.2.4b, these coordinates are related by

$$x' = x - h, \qquad y' = y - k \qquad\qquad (6)$$

or, equivalently,

$$x = x' + h, \qquad y = y' + k \qquad\qquad (7)$$

These are called the **translation equations.**

Figure 12.2.4   (a)   (b)

As an illustration, if the new origin is at $(h, k) = (4, -1)$ and the xy-coordinates of a point P are $(2, 5)$, then the $x'y'$-coordinates of P are

$$x' = x - h = 2 - 4 = -2 \quad \text{and} \quad y' = y - k = 5 - (-1) = 6$$

Let us now consider a parabola with vertex V at (h, k) and axis of symmetry parallel to the y-axis. If, as in Figure 12.2.5, we translate the axes so that

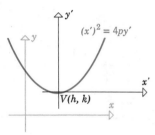

$(x')^2 = 4py'$

$V(h, k)$

Figure 12.2.5

the vertex V is at the origin of an $x'y'$-coordinate system, then in $x'y'$-coordinates the equation of the parabola will be

$$(x')^2 = 4py'$$

and from the translation equations (6), the corresponding equation in xy-coordinates will be:

Parabola with Vertex (h, k) and Axis Parallel to y-axis

$$(x - h)^2 = 4p(y - k) \tag{8}$$

Thus, in xy-coordinates, an equation of form (8) represents a parabola with vertex at (h, k) and axis of symmetry parallel to the y-axis. The parabola opens in the positive or negative y-direction according to whether p is positive or negative. Similarly

---

Parabola with Vertex (h, k) and Axis Parallel to x-axis

$$(y - k)^2 = 4p(x - h) \tag{9}$$

---

represents a parabola with vertex at (h, k) and axis of symmetry parallel to the x-axis. The parabola opens in the positive or negative x-direction according to whether p is positive or negative.

▶ Example 3

Sketch the parabola

$$(y - 3)^2 = 8(x + 4)$$

and locate the focus and directrix.

Vertex $(-4, 3)$

Focus $(-2, 3)$

Directrix $x = -6$

Figure 12.2.6

Solution. From (9), the equation represents a parabola with vertex at $(h, k) = (-4, 3)$ and axis of symmetry parallel to the x-axis. Moreover $4p = 8$ or $p = 2$. Since p is positive, the parabola opens in the positive x-direction. This places the focus 2 units to the right of the vertex or at the point $(-2, 3)$. The directrix is 2 units to the left of the vertex (and parallel to the y-axis), so its equation is $x = -6$. The parabola is sketched in Figure 12.2.6. ◀

Sometimes (8) and (9) appear in expanded form, in which case some algebraic manipulations are required to identify the parabola.

▶ Example 4

Show that the curve

$$y = 6x^2 - 12x + 8$$

is a parabola.

Solution. Because the equation involves x to the second power and y to the first power, and because the expanded form of (8) has the same property, we will try to rewrite the equation in form (8). To do this, we first divide by the coefficient of x^2 and collect all the x-terms on one side:

$$\frac{1}{6}y - \frac{8}{6} = x^2 - 2x$$

Next, we complete the square on the x-terms by adding 1 to both sides:

$$\frac{1}{6}y - \frac{2}{6} = (x - 1)^2$$

Finally we factor out the coefficient of the y-term to obtain

$$(x - 1)^2 = \frac{1}{6}(y - 2)$$

which has form (8) with

$$h = 1, \qquad k = 2, \qquad 4p = \frac{1}{6}, \qquad p = \frac{1}{24}$$

Thus the curve is a parabola with vertex at $(1, 2)$ and axis of symmetry parallel to the y-axis. Since $p = \frac{1}{24}$ is positive, the parabola opens in the positive y-direction, the focus is $\frac{1}{24}$ unit above the vertex, and the directrix $\frac{1}{24}$ unit below the vertex. Thus the focus is at $(1, \frac{49}{24})$ and the directrix has equation $y = \frac{47}{24}$ (Figure 12.2.7).   ◀

Figure 12.2.7

The procedure used in Example 4 can be used to prove the following general result.

12.2.2   THEOREM   *The graph of*

$$y = Ax^2 + Bx + C \qquad (A \neq 0)$$

is a parabola with axis of symmetry parallel to the y-axis; the parabola opens in the posiive y-direction if $A > 0$ and in the negative y-direction if $A < 0$. The graph of

$$x = Ay^2 + By + C \quad (A \neq 0)$$

is a parabola with axis of symmetry parallel to the x-axis; the parabola opens in the positive x-direction if $A > 0$ and in the negative x-direction if $A < 0$.

▶ Example 5

Find an equation for the parabola with vertex $(1, 2)$ and focus $(4, 2)$.

Solution. Since the focus and vertex are on a horizontal line, and since the focus is to the right of the vertex, the parabola opens to the right and its equation has the form

$$(y - k)^2 = 4p(x - h) \quad (p \text{ positive})$$

Since the vertex and focus are three units apart we have $p = 3$, and since the vertex is at $(h, k) = (1, 2)$ we obtain

$$(y - 2)^2 = 12(x - 1) \qquad\qquad ◀$$

REFLECTION PROPERTIES OF PARABOLAS

Parabolas have important applications in the design of telescopes, radar antennas, and lighting systems. This is due to the following property of parabolas (Exercise 42).

12.2.3 THEOREM
A Geometric Property of Parabolas

The tangent line at a point P on a parabola makes equal angles with the line through P parallel to the axis of symmetry and the line through P and the focus (Figure 12.2.8).

It is a principle of physics that when light is reflected from a point P on a surface, the angle of incidence equals the angle of reflection, that is, the angle between the incoming ray and the tangent line at P equals the angle between the outgoing ray and the tangent line at P. Therefore, if the reflecting surface has parabolic cross sections with a common focus, then it follows from Theorem 12.2.3 that all light rays entering parallel to the axis will be reflected through the focus (Figure 12.2.9a). In reflecting telescopes, this principle is used to reflect the (approximately) parallel rays of light from the stars or planets off a parabolic mirror to an eyepiece at the focus of the parabola. Conversely, if a light source is located at the focus of a parabolic reflector, it follows from Theorem 12.2.3 that the reflected rays will form a beam parallel to the axis (Figure 12.2.9b). The parabolic reflectors in flashlights and

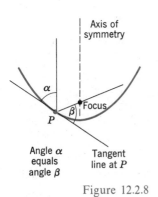

Angle α equals angle β

Figure 12.2.8

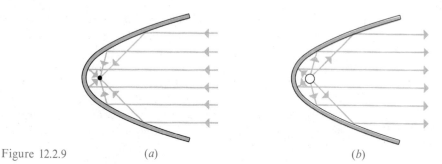

Figure 12.2.9   (a)   (b)

automobile headlights utilize this principle. The optical principles just discussed also apply to radar signals, which explains the parabolic shape of many radar antennas.

► Exercise Set 12.2

In Exercises 1–16, sketch the parabola. Show the focus, vertex, and directrix.

1. $y^2 = 6x$.

2. $y^2 = -10x$.

3. $x^2 = -9y$.

4. $x^2 = 4y$.

5. $5y^2 = 12x$.

6. $x^2 - 40y = 0$.

7. $(y - 3)^2 = 6(x - 2)$.

8. $(y + 1)^2 = -7(x - 4)$.

9. $(x + 2)^2 = -(y + 2)$.

10. $(x - \frac{1}{2})^2 = 2(y - 1)$.

11. $x^2 - 4x + 2y = 1$.

12. $y^2 - 6y - 2x + 1 = 0$.

13. $x = y^2 - 4y + 2$.

14. $y = 4x^2 + 8x + 5$.

15. $-y^2 + 2y + x = 0$.

16. $y = 1 - 4x - x^2$.

In Exercises 17–31, find an equation for the parabola satisfying the given conditions.

17. Vertex $(0, 0)$; focus $(3, 0)$.

18. Vertex $(0, 0)$; focus $(0, -4)$.

19. Vertex $(0, 0)$; directrix $x = .7$.

20. Vertex $(0, 0)$; directrix $y = \frac{1}{2}$.

21. Vertex $(0, 0)$; symmetric about the x-axis; passes through $(2, 2)$.

22. Vertex $(0, 0)$; symmetric about the y-axis; passes through $(-1, 3)$.

23. Focus $(0, -3)$; directrix $y = 3$.

24. Focus $(6, 0)$; directrix $x = -6$.

25. Axis $y = 0$; passes through $(3, 2)$ and $(2, -3)$.

26. Axis $x = 0$; passes through $(2, -1)$ and $(-4, 5)$.

27. Focus $(3, 0)$; directrix $x = 0$.

28. Vertex $(4, -5)$; focus $(1, -5)$.

29. Vertex $(1, 1)$; directrix $y = -2$.

30. Focus $(-1, 4)$; directrix $x = 5$.

31. Vertex $(5, -3)$; axis parallel to the y-axis; passes through $(9, 5)$.

32. Use the definition of a parabola (12.2.1) to find the equation of the parabola with focus $(2, 1)$ and directrix $x + y + 1 = 0$. [*Hint:* Use the result of Exercise 29, Section 1.5.]

33. (a) Find an equation for the parabola with axis parallel to the y-axis and passing through $(0, 3)$, $(2, 0)$, and $(3, 2)$.

   (b) Find an equation for the parabola with axis parallel to the x-axis and passing through the points in (a).

34. Prove: The line tangent to the parabola $x^2 = 4py$ at (x_0, y_0) is

$$y = \frac{x_0}{2p}x - y_0.$$

35. Find the vertex, focus, and directrix of the parabola $y = Ax^2 + Bx + C$ $(A \neq 0)$.

36. (a) Find an equation for the following parabolic arch with base b and height h.

(b) Find the area under the arch.

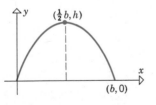

37. Prove: The vertex is the closest point on a parabola to the focus.

38. A comet moves in a parabolic orbit with the sun at its focus. When the comet is 40 million miles from the sun, the line from the sun to the comet makes an angle of $60°$ with the axis of the parabola. How close will the comet come to the sun? [*Hint:* See Exercise 37.]

39. How far from the vertex should a light source be placed on the axis of the following parabolic reflector to produce a beam of parallel rays?

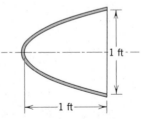

40. Derive the equation $x^2 = 4py$ (p positive) in Table 12.2.1.

41. Derive the equation $y^2 = 4px$ (p negative) in Table 12.2.1.

42. Prove Theorem 12.2.3. [*Hint:* Choose coordinate axes so the parabola has the equation $x^2 = 4py$. Show that the tangent line at $P(x_0, y_0)$ intersects the y-axis at $Q(0, -y_0)$ and that the triangle whose vertices are at P, Q, and the focus is isosceles.]

12.3 THE ELLIPSE

12.3.1 DEFINITION   An *ellipse* is the set of all points in the plane, the sum of whose distances from two fixed points is a given positive constant.

For all points on the ellipse, the sum of the distances to the foci is the same

Figure 12.3.1

The two fixed points are called the *foci* (plural of "focus"), and the midpoint of the line segment joining the foci is called the *center* of the ellipse (Figure 12.3.1).

To help visualize Definition 12.3.1, imagine that two ends of a string are tacked to the foci and a pencil traces a curve as it is held tight against the string (Figure 12.3.2). The resulting curve will be an ellipse since the sum of the distances to the foci is a constant, namely the total length of the string. Note that if the foci coincide, the ellipse reduces to a circle.

Figure 12.3.2

Figure 12.3.5

Figure 12.3.3

Figure 12.3.4

The line segment through the foci and across the ellipse is called the *major axis* (Figure 12.3.3), while the line segment across the ellipse, through the center, and perpendicular to the major axis is called the *minor axis*. It is traditional in the study of ellipses to denote the length of the major axis by $2a$, the length of the minor axis by $2b$, and the distance between the foci by $2c$ (Figure 12.3.4). The numbers a and b are called the *semiaxes* of the ellipse.

There is a basic relationship between the numbers a, b, and c that can be obtained by considering a point P at the end of the major axis and a point Q at the end of the minor axis. Because P and Q both lie on the ellipse, the sum of the distances from each of them to the foci will be the same. If we express this fact as an equation we obtain (see Figure 12.3.5):

$$2\sqrt{b^2 + c^2} = (a - c) + (a + c)$$

or

$$a = \sqrt{b^2 + c^2} \tag{1}$$

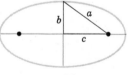

Figure 12.3.6

Relationship (1) shows that the distance from a focus to an end of the minor axis is a (Figure 12.3.6), which implies that for *all* points on the ellipse the sum of the distances to the foci is $2a$. (Why?).

It also follows from (1) that $a \geq b$ with the equality holding only when $c = 0$. Geometrically, this means that the major axis of an ellipse is at least as large as the minor axis, and that the two axes have equal length only when the foci coincide, in which case the ellipse is a circle.

The equation of an ellipse is simplest if the coordinate axes are positioned so the center of the ellipse is at the origin and the foci are on the x-axis or y-axis. The two possible such orientations are shown in Table 12.3.1.

To illustrate how the equations in Table 12.3.1 are derived, let us consider the first entry in the table. Let $P(x, y)$ be any point on the ellipse. Since the sum of the distances from P to the foci is $2a$ it follows (Figure 12.3.7) that

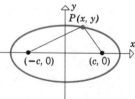

Figure 12.3.7

$$\sqrt{(x + c)^2 + y^2} + \sqrt{(x - c)^2 + y^2} = 2a$$

Transposing the second radical to the right side of the equation and squaring yields

Table 12.3.1

ORIENTATION	DESCRIPTION	EQUATION
	Foci and major axis on the x-axis. Minor axis on the y-axis. Center at the origin.	$\dfrac{x^2}{a^2} + \dfrac{y^2}{b^2} = 1$
	Foci and major axis on the y-axis. Minor axis on the x-axis. Center at the origin.	$\dfrac{x^2}{b^2} + \dfrac{y^2}{a^2} = 1$

$$(x + c)^2 + y^2 = 4a^2 - 4a \sqrt{(x - c)^2 + y^2} + (x - c)^2 + y^2$$

and, on simplifying,

$$\sqrt{(x - c)^2 + y^2} = a - \frac{c}{a}x$$

Squaring again and simplifying yields

$$\frac{x^2}{a^2} + \frac{y^2}{a^2 - c^2} = 1$$

which, by virtue of (1), may be written

$$\frac{x^2}{a^2} + \frac{y^2}{b^2} = 1 \tag{2}$$

Conversely, it can be shown that any point whose coordinates satisfy (2), has $2a$ as the sum of its distances from the foci, so that such a point is on the ellipse.

The second equation in Table 12.3.1 has a similar derivation.

▶ Example 1

Does the ellipse

$$\frac{x^2}{9} + \frac{y^2}{4} = 1$$

have its major axis on the x-axis or the y-axis?

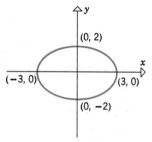

Figure 12.3.8

Solution. Let us locate the intersections of the ellipse with the coordinate axes. Setting $y = 0$ yields $x^2 = 9$, or $x = \pm 3$, so the ellipse intersects the x-axis at $(-3, 0)$ and $(3, 0)$. Setting $x = 0$ yields $y^2 = 4$, or $y = \pm 2$, so the ellipse intersects the y-axis at $(0, -2)$ and $(0, 2)$. Plotting the points of intersection and making a rough sketch shows that the major axis is along the x-axis (Figure 12.3.8).

Alternate Solution. Since it is always true that $a^2 \geq b^2$, it follows that $a^2 = 9$ and $b^2 = 4$, so that the given equation is of the form

$$\frac{x^2}{a^2} + \frac{y^2}{b^2} = 1$$

which is an ellipse with major axis along the x-axis (Table 12.3.1).   ◀

▶ Example 2

Show that the graph of the equation

$$2x^2 + y^2 = 4$$

is an ellipse, and locate the foci.

Solution. Dividing through by 4 yields

$$\frac{x^2}{2} + \frac{y^2}{4} = 1 \tag{3}$$

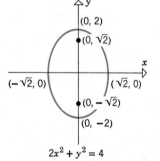

Figure 12.3.9

which is the equation of an ellipse having $a^2 = 4$ ($a = 2$) and $b^2 = 2$ ($b = \sqrt{2}$). Since the a^2 is under the y^2 in (3), the major axis is along the y-axis. From relationship (1)

$$c = \sqrt{a^2 - b^2} = \sqrt{4 - 2} = \sqrt{2}$$

so the foci are $(0, \sqrt{2})$ and $(0, -\sqrt{2})$. The ellipse is shown in Figure 12.3.9.   ◀

▶ Example 3

Find an equation for the ellipse with foci $(0, \pm 2)$ and major axis with endpoints $(0, \pm 4)$.

Solution. From Table 12.3.1, the equation has the form

$$\frac{x^2}{b^2} + \frac{y^2}{a^2} = 1$$

and from the given information, $a = 4$ and $c = 2$. It follows from (1) that

$$b^2 = a^2 - c^2 = 16 - 4 = 12$$

so the equation of the ellipse is

$$\frac{x^2}{12} + \frac{y^2}{16} = 1 \qquad \blacktriangleleft$$

If the axes of an ellipse are parallel to the coordinate axes but the center is not at the origin, then the equation of the ellipse may be determined by translation of axes. Specifically, if the center of the ellipse is at the point (h, k), and if we translate the xy-axes so the origin of the $x'y'$-system is at the center (h, k), then the equation of the ellipse in the $x'y'$-system will be

$$\frac{(x')^2}{a^2} + \frac{(y')^2}{b^2} = 1 \qquad \text{or} \qquad \frac{(x')^2}{b^2} + \frac{(y')^2}{a^2} = 1$$

depending on the location of the major and minor axes (Figure 12.3.10). Thus from the translation equations $x' = x - h$, $y' = y - k$, the equation of the ellipse in the xy-system will be one of the following:

Figure 12.3.10

Ellipse with Center (h, k) and Major Axis Parallel to x-axis

$$\frac{(x - h)^2}{a^2} + \frac{(y - k)^2}{b^2} = 1 \tag{4}$$

Ellipse with Center (h, k) and Major Axis Parallel to y-axis

$$\frac{(x - h)^2}{b^2} + \frac{(y - k)^2}{a^2} = 1 \tag{5}$$

Sometimes (4) and (5) appear in expanded form, in which case it is necessary to complete the squares before the ellipse can be analyzed.

▶ Example 4

Show that the curve

$$16x^2 + 9y^2 - 64x - 54y + 1 = 0$$

is an ellipse. Sketch the ellipse and show the location of the foci.

Solution. Our objective is to rewrite the equation in one of the forms, (4) or (5). To do this, group the x-terms and y-terms and take the constant to the right side:

$$(16x^2 - 64x) + (9y^2 - 54y) = -1$$

Next, factor out the coefficients of x^2 and y^2 and complete the squares:

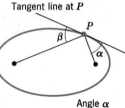

(2, 7)
(2, 3+√7)
(2, 3)
(−1, 3)   (5, 3)
(2, 3−√7)   x
(2, −1)

Figure 12.3.11

$$16(x^2 - 4x + 4) + 9(y^2 - 6y + 9) = -1 + 64 + 81$$

or

$$16(x - 2)^2 + 9(y - 3)^2 = 144$$

Finally, divide through by 144 to introduce a 1 on the right side:

$$\frac{(x - 2)^2}{9} + \frac{(y - 3)^2}{16} = 1$$

This is an equation of form (5), with $h = 2$, $k = 3$, $a^2 = 16$, and $b^2 = 9$. Thus the given equation is an ellipse with center $(2, 3)$ and major axis parallel to the y-axis. Since $a = 4$, the major axis extends 4 units above and 4 units below the center, so its endpoints are $(2, 7)$ and $(2, -1)$ (Figure 12.3.11). Since $b = 3$, the minor axis extends 3 units to the left and 3 units to the right of the center, so its endpoints are $(-1, 3)$ and $(5, 3)$. Since

$$c = \sqrt{a^2 - b^2} = \sqrt{16 - 9} = \sqrt{7}$$

the foci lie $\sqrt{7}$ units above and below the center, placing them at the points $(2, 3 + \sqrt{7})$ and $(2, 3 - \sqrt{7})$.   ◀

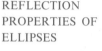

Tangent line at *P*

P

β

α

Angle α
equals angle
β

Figure 12.3.12

REFLECTION
PROPERTIES OF
ELLIPSES

In the exercises the reader is asked to prove the following result.

12.3.2   THEOREM

A Geometric Property
of Ellipses

A line tangent to an ellipse at a point P makes equal angles with the lines through P and the foci (Figure 12.3.12).

It follows from Theorem 12.3.2 that a ray of light emanating from one focus of an ellipse will be reflected through the other focus. This property of ellipses is used in "whispering galleries." Such rooms have ceilings whose cross sections are elliptical in shape with common foci. As a result, if a person standing at one focus whispers, the sound waves are reflected by the ceiling to the other focus, making it possible for a person at that focus to hear the whispered sound.

▶ Exercise set 12.3

In Exercises 1–14, sketch the ellipse. Label the foci and the ends of the major and minor axes.

1. $\dfrac{x^2}{16} + \dfrac{y^2}{9} = 1.$ 2. $\dfrac{x^2}{4} + \dfrac{y^2}{25} = 1.$

3. $9x^2 + y^2 = 9.$ 4. $4x^2 + 9y^2 = 36.$

5. $x^2 + 3y^2 = 2.$ 6. $16x^2 + 4y^2 = 1.$

7. $9(x - 1)^2 + 16(y - 3)^2 = 144.$

8. $(x + 3)^2 + 4(y - 5)^2 = 16.$

9. $3(x + 2)^2 + 4(y + 1)^2 = 12.$

10. $\frac{1}{4}x^2 + \frac{1}{8}(y + 2)^2 - 1 = 0.$

11. $x^2 + 9y^2 + 2x - 18y + 1 = 0.$

12. $9x^2 + 4y^2 + 18x - 24y + 9 = 0.$

13. $4x^2 + y^2 + 8x - 10y = -13.$

14. $5x^2 + 9y^2 - 20x + 54y = -56.$

In Exercises 15–26, find an equation for the ellipse satisfying the given conditions.

15. Ends of major axis $(\pm 3, 0)$; ends of minor axis $(0, \pm 2)$.

16. Ends of major axis $(0, \pm \sqrt{5})$; ends of minor axis $(\pm 1, 0)$.

17. Length of major axis 26; foci $(\pm 5, 0)$.

18. Length of minor axis 16; foci $(0, \pm 6)$.

19. Foci $(\pm 1, 0)$; $b = \sqrt{2}$.

20. Foci $(\pm 3, 0)$; $a = 4$.

21. $c = 2\sqrt{3}$; $a = 4$; center at the origin; foci on a coordinate axis (two answers).

22. $b = 3$; $c = 4$; center at the origin; foci on a coordinate axis (two answers).

23. Ends of major axis $(\pm 6, 0)$; passes through $(2, 3)$.

24. Center at $(0, 0)$; major and minor axes along the coordinate axes; passes through $(3, 2)$ and $(1, 6)$.

25. Foci $(1, 2)$ and $(1, 4)$; minor axis of length 2.

26. Foci $(2, 1)$ and $(2, -3)$; major axis of length 6.

27. Find the equation of the ellipse traced by a point that moves so the sum of its distances to $(4, 1)$ and $(4, 5)$ is 12.

28. Find the equation of the ellipse traced by a point that moves so the sum of its distances to $(0, 0)$ and $(1, 1)$ is 4.

In Exercises 29–32, find all intersections of the given curves and make a sketch of the curves that shows the points of intersection.

29. $x^2 + 4y^2 = 40$ and $x + 2y = 8.$

30. $y^2 = 2x$ and $x^2 + 2y^2 = 12.$

31. $x^2 + 9y^2 = 36$ and $x^2 + y^2 = 20.$

32. $16x^2 + 9y^2 = 36$ and $5x^2 + 18y^2 = 45.$

33. Find two values of k such that the line $x + 2y = k$ is tangent to the ellipse $x^2 + 4y^2 = 8$. Find the points of tangency.

34. Prove: The line tangent to the ellipse $x^2/a^2 + y^2/b^2 = 1$ at (x_0, y_0) has equation $xx_0/a^2 + yy_0/b^2 = 1.$

35. Find the area enclosed by the ellipse $b^2x^2 + a^2y^2 = a^2b^2.$

36. Find the volume of the solid generated when the region enclosed by the ellipse $b^2x^2 + a^2y^2 = a^2b^2$ is

(a) revolved about the x-axis

(b) revolved about the y-axis.

37. Derive the equation $x^2/b^2 + y^2/a^2 = 1$ in Table 12.3.1.

38. Prove Theorem 12.3.2. [*Hint:* Introduce coordinate axes so the ellipse has the equation $x^2/a^2 + y^2/b^2 = 1$, and use the result of Exercise 17 of Section 1.4.]

12.4 THE HYPERBOLA

12.4.1 DEFINITION   A *hyperbola* is the set of all points in the plane, the difference of whose distances from two fixed points is a given positive constant.

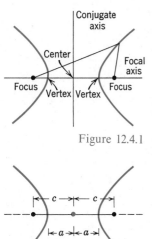

Figure 12.4.1

In this definition the "difference" of the distances is understood to mean the distance to the farther point minus the distance to the closer point.

The two fixed points are called the *foci*, and the midpoint of the line segment joining the foci is called the *center* of the hyperbola (Figure 12.4.1). The line through the foci is called the *focal axis* (or *transverse axis*) and the line through the center and perpendicular to the focal axis is called the *conjugate axis*. The hyperbola intersects the focal axis at two points, called *vertices*. The two separate parts of a hyperbola are called the *branches*.

It is traditional in the study of hyperbolas to denote the distance between the vertices by $2a$, the distance between the foci by $2c$ (Figure 12.4.2), and to define the quantity b as

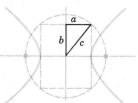

Figure 12.4.2

$$b = \sqrt{c^2 - a^2} \qquad (1)$$

Figure 12.4.3

(See Figure 12.4.3.)

If V is one vertex of a hyperbola, then, as illustrated in Figure 12.4.4, the distance from V to the farther focus minus the distance from V to the closer focus is

$$[(c - a) + 2a] - (c - a) = 2a$$

Thus for *all* points on a hyperbola, the distance to the farther focus minus the distance to the closer focus is $2a$. (Why?)

The equation of a hyperbola is simplest if the coordinate axes are positioned so the center of the hyperbola is at the origin and the foci are on the x-axis or y-axis. The two possible such orientations are shown in Table 12.4.1.

To illustrate how the equations in Table 12.4.1 are derived, let us consider the first entry in the table. Let $P(x, y)$ be any point on the hyperbola, so that the distance from P to the farther focus minus the distance from P to the closer focus is $2a$. Depending on which focus is farther from P, this condition leads either to the equation

Figure 12.4.4

$$\sqrt{(x + c)^2 + y^2} - \sqrt{(x - c)^2 + y^2} = 2a$$

or to

$$\sqrt{(x - c)^2 + y^2} - \sqrt{(x + c)^2 + y^2} = 2a$$

Table 12.4.1

ORIENTATION	DESCRIPTION	EQUATION
	Foci on the x-axis. Conjugate axis on the y-axis. Center at the origin.	$\dfrac{x^2}{a^2} - \dfrac{y^2}{b^2} = 1$
	Foci on the y-axis Conjugate axis on the x-axis. Center at the origin.	$\dfrac{y^2}{a^2} - \dfrac{x^2}{b^2} = 1$

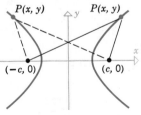

Figure 12.4.5

(See Figure 12.4.5.) In either case,

$$\left| \sqrt{(x + c)^2 + y^2} - \sqrt{(x - c)^2 + y^2} \right| = 2a \tag{2}$$

Squaring both sides of (2) yields

$$(x + c)^2 + y^2 - 2\sqrt{[(x + c)^2 + y^2][(x - c)^2 + y^2]} + (x - c)^2 + y^2 = 4a^2$$

Simplifying, then isolating the radical, then squaring again to remove the radical, ultimately yields (verify)

$$\frac{x^2}{a^2} - \frac{y^2}{c^2 - a^2} = 1$$

which, by virtue of (1), may be written

$$\frac{x^2}{a^2} - \frac{y^2}{b^2} = 1 \tag{3}$$

Conversely, it can be shown that for any point P whose coordinates satisfy (3), the distance from P to the farther focus minus the distance from P to the closer focus is $2a$, so that such a point must lie on the hyperbola.

The derivation of the second equation in Table 12.4.1 is similar.

$$\frac{x^2}{4} - \frac{y^2}{9} = 1$$

Figure 12.4.6

▶ Example 1

Does the hyperbola

$$\frac{x^2}{4} - \frac{y^2}{9} = 1$$

have its focal axis on the x-axis or the y-axis?

Solution. Unlike the case of the ellipse, the orientation of a hyperbola is not determined by examining the relative sizes of a^2 and b^2, but rather by noting where the minus sign occurs in the equation. From Table 12.4.1 we see that the focal axis is on the x-axis when the minus precedes the y^2-term, and it is on the y-axis when the minus precedes the x^2-term. In the given equation the minus precedes the y^2-term so that the focal axis is on the x-axis.

Alternately, the orientation of a hyperbola can be obtained by locating its vertices. If we let $x = 0$ in the given equation, we obtain

$$-\frac{y^2}{9} = 1 \quad \text{or} \quad y^2 = -9$$

This equation has no real solutions, so the hyperbola does not intersect the y-axis. This tells us that the focal axis is along the x-axis. The same conclusion can be reached by setting $y = 0$ in the given equation to obtain

$$\frac{x^2}{4} = 1 \quad \text{or} \quad x = \pm 2$$

This tells us that the vertices are at the points $(-2, 0)$ and $(2, 0)$, so again we conclude that the focal axis is on the x-axis (Figure 12.4.6). ◀

The following theorem states that hyperbolas have asymptotes (see Figure 12.4.7).

12.4.2 THEOREM
Asymptotes of Hyperbolas

(a) *The hyperbola*

$$\frac{x^2}{a^2} - \frac{y^2}{b^2} = 1$$

has asymptotes

$$y = \frac{b}{a}x \quad \text{and} \quad y = -\frac{b}{a}x$$

$$y = -\frac{b}{a}x \qquad y = \frac{b}{a}x$$

$(-a, b)$ (a, b)
$(-c, 0)$ $(c, 0)$

$$\frac{x^2}{a^2} - \frac{y^2}{b^2} = 1$$

$$y = -\frac{a}{b} \qquad y = \frac{a}{b}x$$

$(0, c)$
$(-b, a)$ (b, a)

$(0, -c)$

$$\frac{y^2}{a^2} - \frac{x^2}{b^2} = 1$$

Figure 12.4.7

$$y = \frac{b}{a}\sqrt{x^2 - a^2}$$
$$y = \frac{b}{a}x$$

Figure 12.4.8

(b) *The hyperbola*

$$\frac{y^2}{a^2} - \frac{x^2}{b^2} = 1$$

has asymptotes

$$y = \frac{a}{b}x \qquad and \qquad y = -\frac{a}{b}x$$

Proof. We will prove part (a); the proof of (b) is similar. If we write the equation $x^2/a^2 - y^2/b^2 = 1$ in the form

$$y^2 = \frac{b^2}{a^2}(x^2 - a^2)$$

then, in the first quadrant, the vertical distance between the line $y = (b/a)x$ and the hyperbola may be written (Figure 12.4.8).

$$\frac{b}{a}x - \frac{b}{a}\sqrt{x^2 - a^2}$$

But this distance tends to zero as $x \rightarrow +\infty$ since

$$\lim_{x \to +\infty} \left(\frac{b}{a}x - \frac{b}{a}\sqrt{x^2 - a^2} \right) = \lim_{x \to +\infty} \frac{b}{a}(x - \sqrt{x^2 - a^2})$$

$$= \lim_{x \to +\infty} \frac{b}{a}\frac{(x - \sqrt{x^2 - a^2})(x + \sqrt{x^2 - a^2})}{x + \sqrt{x^2 - a^2}}$$

$$= \lim_{x \to +\infty} \frac{ab}{x + \sqrt{x^2 - a^2}}$$

$$= 0$$

The analysis in the remaining quadrants is similar. ∎

REMARKS. There is a trick that can be used to avoid memorizing the equations of the asymptotes of a hyperbola. They can be obtained, when needed, by substituting 0 for the 1 on the right side of the hyperbola equation, and then solving for y in terms of x. For example, for the hyperbola

$$\frac{x^2}{a^2} - \frac{y^2}{b^2} = 1$$

we would write

$$\frac{x^2}{a^2} - \frac{y^2}{b^2} = 0 \qquad \text{or} \qquad y^2 = \frac{b^2}{a^2}x^2 \qquad \text{or} \qquad y = \pm\frac{b}{a}x$$

which are the equations for the asymptotes. As indicated in Figure 12.4.7, the asymptotes of a hyperbola are along the diagonals of a box extending a units on each side of the origin along the focal axis, and b units on each side of the origin along the conjugate axis.

▶ Example 2

Find the foci, vertices, and asymptotes of the hyperbola

$$\frac{y^2}{16} - \frac{x^2}{9} = 1$$

and sketch the graph.

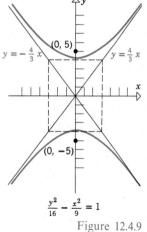

$$\frac{y^2}{16} - \frac{x^2}{9} = 1$$

Figure 12.4.9

Solution. Since the minus precedes the x^2-term, this hyperbola has its foci on the y-axis. Also, $a^2 = 16$ and $b^2 = 9$. Since $a = 4$, the vertices are at $(0, -4)$ and $(0, 4)$. Moreover, from (1), $c = \sqrt{a^2 + b^2} = \sqrt{25} = 5$, so the foci are at $(0, -5)$ and $(0, 5)$.

To obtain the asymptotes, we substitute 0 for 1 in the given equation, which yields

$$\frac{y^2}{16} - \frac{x^2}{9} = 0 \qquad \text{or} \qquad y = \pm\frac{4}{3}x$$

With the aid of a box extending $a = 4$ units above and below the origin in the y-direction and $b = 3$ units to the left and right of the origin in the x-direction, we obtain the sketch in Figure 12.4.9.   ◀

▶ Example 3

Find the equation of the hyperbola with vertices $(0, \pm 8)$ and asymptotes $y = \pm\frac{4}{3}x$.

Solution. Since the vertices are on the y-axis, the equation of the hyperbola has the form $y^2/a^2 - x^2/b^2 = 1$ and the asymptotes are

$$y = \pm\frac{a}{b}x$$

From the location of the vertices we have $a = 8$, so that the given equations of the asymptotes yield

$$y = \pm \frac{a}{b}x = \pm \frac{8}{b}x = \pm \frac{4}{3}x$$

from which it follows that $b = 6$. Thus the hyperbola has the equation

$$\frac{y^2}{64} - \frac{x^2}{36} = 1 \qquad \blacktriangleleft$$

If the center of a hyperbola is at the point (h, k) and its focal and conjugate axes are parallel to the coordinate axes then its equation has one of the forms:

Hyperbola with Center (h, k) and Focal Axis Parallel to x-axis.

$$\frac{(x - h)^2}{a^2} - \frac{(y - k)^2}{b^2} = 1 \qquad (4)$$

Hyperbola with Center (h, k) and Focal Axis Parallel to y-axis.

$$\frac{(y - k)^2}{a^2} - \frac{(x - h)^2}{b^2} = 1 \qquad (5)$$

The derivations are similar to those for the parabola and ellipse and will be omitted. As before, the equations of the asymptotes can be obtained by replacing the 1 by a 0 on the right side of (4) or (5), and then solving for y in terms of x.

▶ Example 4

Show that the curve

$$x^2 - y^2 - 4x + 8y - 21 = 0$$

is a hyperbola. Sketch the hyperbola and show the foci, vertices, and asymptotes.

Solution. Our objective is to rewrite the equation in form (4) or (5) by completing the squares. To do this, we first group the x-terms and y-terms and take the constant to the right side:

$$(x^2 - 4x) - (y^2 - 8y) = 21$$

Next, complete the squares:

$$(x^2 - 4x + 4) - (y^2 - 8y + 16) = 21 + 4 - 16$$

or

$$(x - 2)^2 - (y - 4)^2 = 9$$

Finally, divide by 9 to introduce a 1 on the right side:

$$\frac{(x - 2)^2}{9} - \frac{(y - 4)^2}{9} = 1 \qquad (6)$$

This is an equation of form (4) with $h = 2$, $k = 4$, $a^2 = 9$, and $b^2 = 9$. Thus the equation represents a hyperbola with center (2, 4) and focal axis parallel to the x-axis. Since $a = 3$, the vertices are located 3 units to the left and 3 units to the right of the center, or at the points $(-1, 4)$ and $(5, 4)$. From (1), $c = \sqrt{a^2 + b^2} = \sqrt{9 + 9} = 3\sqrt{2}$, so that the foci are located $3\sqrt{2}$ units to the left and right of the center or at the points $(2 - 3\sqrt{2}, 4)$ and $(2 + 3\sqrt{2}, 4)$.

Figure 12.4.10

The equations of the asymptotes may be found by substituting 0 for 1 in (6) to obtain

$$\frac{(x - 2)^2}{9} - \frac{(y - 4)^2}{9} = 0 \qquad \text{or} \qquad y - 4 = \pm(x - 2)$$

which yields the asymptotes

$$y = x + 2 \qquad \text{and} \qquad y = -x + 6$$

With the aid of a box extending $a = 3$ units left and right of the center and $b = 3$ units above and below the center we obtain the sketch in Figure 12.4.10. ◄

Figure 12.4.11

A hyperbola, like that in the last example, is called *equilateral* if $a = b$. For such hyperbolas, the asymptotes are perpendicular. (Verify.)

HYPERBOLIC NAVIGATION

By measuring the difference in reception times of synchronized radio signals from two widely spaced transmitters, a ship's electronic equipment can determine the difference $2a$ in its distances from the two transmitters. This information places the ship somewhere on the hyperbola whose foci are at the transmitters and whose points have $2a$ as their difference in distances from the foci. By using two pairs of transmitters, the position of a ship can be determined as the intersection of two hyperbolas (Figure 12.4.11).

REFLECTION
PROPERTIES OF
HYPERBOLAS

In the exercises, the reader is asked to prove the following result.

12.4.3 THEOREM
A Geometric Property of Hyperbolas

A line tangent to a hyperbola at a point P makes equal angles with the lines through P and the foci (Figure 12.4.12).

Tangent
line at P

Angle α
equals
angle β

Figure 12.4.12

Angle
of
incidence

Light
ray

Angle of
reflection

Figure 12.4.13

It follows from Theorem 12.4.3 that a ray of light emanating from one focus of a hyperbola will be reflected back along the line from the opposite focus (Figure 12.4.13). Light reflection properties of hyperbolas are used to advantage in the design of high-quality telescopes.

► Exercise Set 12.4

In Exercises 1–16, sketch the hyperbola. Find the coordinates of the vertices and foci, and find equations for the asymptotes.

1. $\dfrac{x^2}{16} - \dfrac{y^2}{4} = 1.$

2. $\dfrac{y^2}{9} - \dfrac{x^2}{25} = 1.$

3. $9y^2 - 4x^2 = 36.$

4. $16x^2 - 25y^2 = 400.$

5. $8x^2 - y^2 = 8.$

6. $3x^2 - y^2 = -9.$

7. $x^2 - y^2 = 1.$

8. $4y^2 - 4x^2 = 1.$

9. $\dfrac{(x-2)^2}{9} - \dfrac{(y-4)^2}{4} = 1.$

10. $\dfrac{(y+4)^2}{3} - \dfrac{(x-2)^2}{5} = 1.$

11. $(y+3)^2 - 9(x+2)^2 = 36.$

12. $16(x+1)^2 - 8(y-3)^2 = 16.$

13. $x^2 - 4y^2 + 2x + 8y - 7 = 0.$

14. $4x^2 - 9y^2 + 16x + 54y - 29 = 0.$

15. $16x^2 - y^2 - 32x - 6y = 57.$

16. $4y^2 - x^2 + 40y - 4x = -60.$

In Exercises 17–30, find an equation for the hyperbola satisfying the given conditions.

17. Vertices $(\pm 2, 0)$; foci $(\pm 3, 0)$.

18. Vertices $(0, \pm 3)$; foci $(0, \pm 5)$.

19. Vertices $(\pm 1, 0)$; asymptotes $y = \pm 2x$.

20. Vertices $(0, \pm 3)$; asymptotes $y = \pm x$.

21. Asymptotes $y = \pm \frac{3}{2}x$; $b = 4$.

22. Foci $(0, \pm 5)$; asymptotes $y = \pm 2x$.

23. Passes through $(5, 9)$; asymptotes $y = \pm x$.

24. Asymptotes $y = \pm \frac{3}{4}x$; $c = 5$.

25. Vertices $(\pm 2, 0)$; passes through $(4, 2)$.

26. Foci $(\pm 3, 0)$; asymptotes $y = \pm 2x$.

27. Vertices $(4, -3)$ and $(0, -3)$; foci 6 units apart.

28. Foci $(1, 8)$ and $(1, -12)$; vertices 4 units apart.

29. Vertices $(2, 4)$ and $(10, 4)$; foci 10 units apart.

30. Asymptotes $y = 2x + 1$ and $y = -2x + 3$; passes through the origin.

31. Find the equation of the hyperbola traced by a point that moves so the difference between its distances to $(0, 0)$ and $(1, 1)$ is 1.

32. Find the equation of the hyperbola traced by a point that moves so the difference between its distances to $(4, -3)$ and $(-2, 5)$ is 6.

In Exercises 33–36, find all intersections of the given curves, and make a sketch of the curves that shows the points of intersection.

33. $x^2 - 4y^2 = 36$ and $x - 2y - 20 = 0$.

34. $y^2 - 8x^2 = 5$ and $y - 2x^2 = 0$.

35. $3x^2 - 7y^2 = 5$ and $9y^2 - 2x^2 = 1$.

36. $x^2 - y^2 = 1$ and $x^2 + y^2 = 7$.

37. A point moves so that the product of its distances from the lines $y = mx$ and $y = -mx$ is a constant k^2. Show that the point moves along a hyperbola having these lines as asymptotes.

38. Prove: The line tangent to the hyperbola $x^2/a^2 - y^2/b^2 = 1$ at (x_0, y_0) has the equation $xx_0/a^2 - yy_0/b^2 = 1$.

39. A line tangent to the hyperbola $4x^2 - y^2 = 36$ intersects the y-axis at the point $(0, 4)$. Find the point(s) of tangency.

40. Let R be the region enclosed between the hyperbola $b^2x^2 - a^2y^2 = a^2b^2$ and the line $x = c$ through the focus $(c, 0)$. Find the volume of the solid generated by revolving R about
(a) the x-axis   (b) the y-axis.

41. Derive the equation $y^2/a^2 - x^2/b^2 = 1$ in Table 12.4.1.

42. Prove Theorem 12.4.3. [*Hint:* Introduce coordinate axes so the hyperbola has the equation $x^2/a^2 - y^2/b^2 = 1$, and use the result of Exercise 17 of Section 1.4.]

12.5 ROTATION OF AXES; SECOND DEGREE EQUATIONS

In previous sections we obtained equations of conic sections with axes parallel to the coordinate axes. In this section we study the equations of conics that are "tilted" relative to the coordinate axes. This will lead us to investigate rotations of coordinate axes.

Each of the equations

$$(y - k)^2 = 4p(x - h) \qquad (x - h)^2 = 4p(y - k)$$

$$\frac{(x - h)^2}{a^2} + \frac{(y - k)^2}{b^2} = 1 \qquad \frac{(x - h)^2}{b^2} + \frac{(y - k)^2}{a^2} = 1$$

$$\frac{(x - h)^2}{a^2} - \frac{(y - k)^2}{b^2} = 1 \qquad \frac{(y - k)^2}{a^2} - \frac{(x - h)^2}{b^2} = 1$$

represents a conic section with axis or axes parallel to the coordinate axes. By squaring out the quadratic terms and simplifying, each of these equations can be rewritten in the form

$$Ax^2 + Cy^2 + Dx + Ey + F = 0 \tag{1}$$

For example, $(y - k)^2 = 4p(x - h)$ can be written as

$$y^2 - 4px - 2ky + (k^2 + 4ph) = 0$$

which is of form (1) with

$$A = 0, \quad C = 1, \quad D = -4p, \quad E = -2k, \quad F = k^2 + 4ph$$

If a conic section is tilted so that its axes are not parallel to the x- and y-axis, then its equation will contain a "cross-product" term Bxy and the equation will have the form

$$Ax^2 + Bxy + Cy^2 + Dx + Ey + F = 0 \tag{2}$$

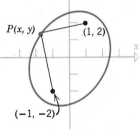

Figure 12.5.1

An equation of form (2) in which A, B, and C are not all zero is called a *second degree equation* or *quadratic equation* in x and y.

As an illustration, consider the ellipse with foci $F_1(1, 2)$ and $F_2(-1, -2)$ and such that the sum of the distances from each point $P(x, y)$ on the ellipse to the foci is 6 units. Expressing this condition as an equation, we obtain (Figure 12.5.1):

$$\sqrt{(x - 1)^2 + (y - 2)^2} + \sqrt{(x + 1)^2 + (y + 2)^2} = 6$$

Squaring both sides, then isolating the remaining radical, then squaring again, ultimately yields

$$8x^2 - 4xy + 5y^2 = 36$$

as the equation of the ellipse. This is an equation of form (2) with $A = 8$, $B = -4$, $C = 5$, $D = 0$, $E = 0$, $F = -36$.

To study conics that are tilted relative to the coordinate axes, it is frequently helpful to rotate the coordinate axes, so that the rotated coordinate axes are parallel to the axes of the conic. Before we can discuss the details, we need to develop some ideas about rotation of coordinate axes.

ROTATION OF AXES In Figure 12.5.2*a* the axes of an xy-coordinate system have been rotated about the origin through angle θ to produce a new $x'y'$-coordinate system. As shown in the figure, each point P in the plane has coordinates (x', y') as well as coordinates (x, y). To see how the two are related, let r be the distance

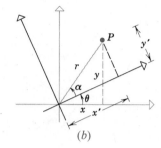

Figure 12.5.2   (a)   (b)

from the common origin to the point P, and let α be the angle shown in Figure 12.5.2b. It follows that

$$x = r \cos (\theta + \alpha), \qquad y = r \sin (\theta + \alpha) \tag{3}$$

and

$$x' = r \cos \alpha, \qquad y' = r \sin \alpha \tag{4}$$

Using familiar trigonometric identities, the relationships in (3) can be written

$$x = r \cos \theta \cos \alpha - r \sin \theta \sin \alpha$$
$$y = r \sin \theta \cos \alpha + r \cos \theta \sin \alpha$$

and on substituting (4) in these equations we obtain the following relationships:

> **The Rotation Equations**
>
> $$x = x' \cos \theta - y' \sin \theta$$
> $$y = x' \sin \theta + y' \cos \theta \tag{5}$$

▶ Example 1

Suppose the axes of an xy-coordinate system are rotated through an angle of $\theta = 45°$ to obtain an $x'y'$-coordinate system. Find the equation of the curve

$$x^2 - xy + y^2 - 6 = 0$$

in $x'y'$-coordinates.

Solution. Substituting the values $\sin \theta = \sin 45° = 1/\sqrt{2}$ and $\cos \theta = \cos 45° = 1/\sqrt{2}$ in (5) yields the rotation equations

$$x = \frac{x'}{\sqrt{2}} - \frac{y'}{\sqrt{2}}$$

$$y = \frac{x'}{\sqrt{2}} + \frac{y'}{\sqrt{2}}$$

Substituting these into the given equation yields

$$\left(\frac{x'}{\sqrt{2}} - \frac{y'}{\sqrt{2}}\right)^2 - \left(\frac{x'}{\sqrt{2}} - \frac{y'}{\sqrt{2}}\right)\left(\frac{x'}{\sqrt{2}} + \frac{y'}{\sqrt{2}}\right) + \left(\frac{x'}{\sqrt{2}} + \frac{y'}{\sqrt{2}}\right)^2 - 6 = 0$$

or

$$\frac{x'^2 - 2x'y' + y'^2 - x'^2 + y'^2 + x'^2 + 2x'y' + y'^2}{2} = 6$$

or

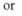

$$\frac{x'^2}{12} + \frac{y'^2}{4} = 1$$

which is the equation of an ellipse (Figure 12.5.3). ◀

Figure 12.5.3

If the rotation equations (5) are solved for x' and y' in terms of x and y, one obtains (Exercise 14):

$$x' = x \cos \theta + y \sin \theta$$

$$y' = -x \sin \theta + y \cos \theta \qquad \qquad (6)$$

▶ Example 2

Find the new coordinates of the point $(2, 4)$ if the coordinate axes are rotated through an angle of $\theta = 30°$.

Solution. From (6) with $x = 2$, $y = 4$, $\cos \theta = \cos 30° = \sqrt{3}/2$, and $\sin \theta = \sin 30° = 1/2$, we obtain

$$x' = 2(\sqrt{3}/2) + 4(1/2) = \sqrt{3} + 2$$
$$y' = -2(1/2) + 4(\sqrt{3}/2) = -1 + 2\sqrt{3}$$

Thus the new coordinates are $(\sqrt{3} + 2, -1 + 2\sqrt{3})$. ◀

In Example 1, we were able to identify the curve $x^2 - xy + y^2 - 6 = 0$ as an ellipse because the rotation of axes eliminated the xy-term, thereby reducing the equation to a familiar form. This occurred because the new

$x'y'$-axes were aligned with the axes of the ellipse. The following theorem tells how to determine an appropriate rotation of axes to eliminate the cross-product term of a second degree equation in x and y.

12.5.1 THEOREM *If the equation*

$$Ax^2 + Bxy + Cy^2 + Dx + Ey + F = 0 \qquad (7)$$

is such that $B \neq 0$, and if an $x'y'$-coordinate system is obtained by rotating the xy-axes through an angle θ satisfying

$$\cot 2\theta = \frac{A - C}{B} \qquad (8)$$

then, in $x'y'$-coordinates, equation (7) will have the form

$$A'x'^2 + C'y'^2 + D'x' + E'y' + F' = 0$$

Proof. Substituting (5) into (7) and simplifying yields,

$$A'x'^2 + B'x'y' + C'y'^2 + D'x' + E'y' + F' = 0$$

where

$$
\begin{aligned}
A' &= A \cos^2 \theta + B \cos \theta \sin \theta + C \sin^2 \theta \\
B' &= B (\cos^2 \theta - \sin^2 \theta) + 2(C - A) \sin \theta \cos \theta \\
C' &= A \sin^2 \theta - B \sin \theta \cos \theta + C \cos^2 \theta \\
D' &= D \cos \theta + E \sin \theta \\
E' &= -D \sin \theta + E \cos \theta \\
F' &= F
\end{aligned}
\qquad (9)
$$

(Verify.) To complete the proof we must show that $B' = 0$ if

$$\cot 2\theta = \frac{A - C}{B}$$

or equivalently

$$\frac{\cos 2\theta}{\sin 2\theta} = \frac{A - C}{B} \qquad (10)$$

However, by using the trigonometric double angle formulas, we can rewrite B' in the form

$$B' = B \cos 2\theta - (A - C) \sin 2\theta$$

Thus $B' = 0$ if θ satisfies (10). ∎

REMARK. It is always possible to satisfy (8) with an angle θ in the range $0 < \theta < \pi/2$. We will always use such a value of θ.

▶ Example 3

Identify and sketch the curve

$$xy = 1$$

Solution. As a first step, we will rotate the coordinate axes to eliminate the cross product term. Comparing the given equation to (7), we have

$$A = 0, \quad B = 1, \quad C = 0$$

Thus the desired angle of rotation must satisfy

$$\cot 2\theta = \frac{A - C}{B} = \frac{0 - 0}{1} = 0$$

This condition can be met by taking $2\theta = \pi/2$ or $\theta = \pi/4 = 45°$. Substituting $\cos \theta = \cos 45° = 1/\sqrt{2}$ and $\sin \theta = \sin 45° = 1/\sqrt{2}$ in the rotation equations (5) yields

$$
\begin{aligned}
x &= \frac{x'}{\sqrt{2}} - \frac{y'}{\sqrt{2}} \\
y &= \frac{x'}{\sqrt{2}} + \frac{y'}{\sqrt{2}}
\end{aligned}
\tag{11}
$$

and substituting these in the equation $xy = 1$ yields

$$\left(\frac{x'}{\sqrt{2}} - \frac{y'}{\sqrt{2}}\right)\left(\frac{x'}{\sqrt{2}} + \frac{y'}{\sqrt{2}}\right) = 1$$

or on simplifying

$$\frac{x'^2}{2} - \frac{y'^2}{2} = 1 \tag{12}$$

This is an equation of the form

$$\frac{x'^2}{a^2} - \frac{y'^2}{b^2} = 1$$

with $a = b = \sqrt{2}$. With the exception that the variables are x' and y' rather than x and y, this is the familiar equation of a hyperbola (see Table 12.4.1). This hyperbola has its focal axis on the x'-axis, its conjugate axis on the y'-axis. Moreover, $c = \sqrt{a^2 + b^2} = 2$, so that the vertices, foci, and asymptotes are:

vertices: $(x', y') = (\pm a, 0) = (\pm \sqrt{2}, 0)$

foci: $(x', y') = (\pm c, 0) = (\pm 2, 0)$

asymptotes: $y' = \pm \dfrac{b}{a} x'$ or $y' = \pm x'$

The vertices in xy-coordinates may be obtained by substituting $x' = \sqrt{2}$, $y' = 0$ and $x' = -\sqrt{2}$, $y' = 0$ in (11), and the foci may be obtained by substituting $x' = 2$, $y' = 0$ and $x' = -2$, $y' = 0$. The results are

vertices: $(x, y) = (1, 1)$ and $(x, y) = (-1, -1)$

foci: $(x, y) = (\sqrt{2}, \sqrt{2})$ and $(x, y) = (-\sqrt{2}, -\sqrt{2})$

To obtain the xy-equations of the asymptotes, it is desirable to use form (6) of the rotation equations. Substituting $\cos \theta = \cos 45° = 1/\sqrt{2}$ and $\sin \theta = \sin 45° = 1/\sqrt{2}$ in (6) yields

$$x' = \frac{1}{\sqrt{2}} x + \frac{1}{\sqrt{2}} y$$

$$y' = -\frac{1}{\sqrt{2}} x + \frac{1}{\sqrt{2}} y$$

(13)

Substituting (13) in the asymptote equations $y' = x'$ and $y' = -x'$ and simplifying shows that the asymptotes are

$$x = 0 \quad \text{and} \quad y = 0.$$

The curve $xy = 1$ is sketched in Figure 12.5.4. ◄

Figure 12.5.4

In problems where it is inconvenient to solve

$$\cot 2\theta = \frac{A - C}{B}$$

for θ, the values of $\sin\theta$ and $\cos\theta$ needed for the rotation equations may be obtained by first calculating $\cos 2\theta$ and then computing $\sin\theta$ and $\cos\theta$ from the identities

$$\sin\theta = \sqrt{\frac{1-\cos 2\theta}{2}} \qquad \cos\theta = \sqrt{\frac{1+\cos 2\theta}{2}}$$

▶ Example 4

Identify and sketch the curve

$$153x^2 - 192xy + 97y^2 - 30x - 40y - 200 = 0$$

Solution. We have $A = 153$, $B = -192$, and $C = 97$ so that

Figure 12.5.5

$$\cot 2\theta = \frac{A-C}{B} = -\frac{56}{192} = -\frac{7}{24}$$

Since θ is to be chosen in the range $0 < \theta < \pi/2$, this relationship is represented by the triangle in Figure 12.5.5. From that triangle we obtain

$$\cos 2\theta = -\frac{7}{25}$$

which implies that

$$\cos\theta = \sqrt{\frac{1+\cos 2\theta}{2}} = \sqrt{\frac{1-7/25}{2}} = \frac{3}{5}$$

$$\sin\theta = \sqrt{\frac{1-\cos 2\theta}{2}} = \sqrt{\frac{1+7/25}{2}} = \frac{4}{5}$$

Substituting these values in (5) yields the rotation equations

$$x = \frac{3}{5}x' - \frac{4}{5}y'$$

$$y = \frac{4}{5}x' + \frac{3}{5}y'$$

Substituting these expressions in the given equation yields

$$\frac{153}{25}(3x' - 4y')^2 - \frac{192}{25}(3x' - 4y')(4x' + 3y') + \frac{97}{25}(4x' + 3y')^2$$

$$- \frac{30}{5}(3x' - 4y') - \frac{40}{5}(4x' + 3y') - 200 = 0$$

which simplifies to

$$25x'^2 + 225y'^2 - 50x' - 200 = 0$$

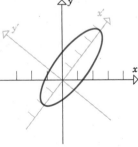

or

$$x'^2 + 9y'^2 - 2x' - 8 = 0$$

Completing the square yields

$$\frac{(x' - 1)^2}{9} + y'^2 = 1$$

Figure 12.5.6   This is an ellipse whose center is $(1, 0)$ in $x'y'$-coordinates (Figure 12.5.6).
◀

OPTIONAL (THE DISCRIMINANT)

It is possible to describe the graph of a second degree equation without rotating coordinate axes.

12.5.2   THEOREM   *Consider a second degree equation*

$$Ax^2 + Bxy + Cy^2 + Dx + Ey + F = 0 \tag{14}$$

(a)   *If $B^2 - 4AC < 0$ the equation represents an ellipse, a circle, a point, or else has no graph.*
(b)   *If $B^2 - 4AC > 0$ the equation represents a hyperbola or a pair of intersecting lines.*
(c)   *If $B^2 - 4AC = 0$ the equation represents a parabola, a line, a pair of parallel lines, or else has no graph.*

The quantity $B^2 - 4AC$ in this theorem is called the ***discriminant*** of the quadratic equation. To see why Theorem 12.5.2 is true, we need a fact about the discriminant. It can be shown (Exercise 17) that if the coordinate axes are rotated through any angle θ, and if

$$A'x'^2 + B'x'y' + C'y'^2 + D'x' + E'y' + F' = 0 \tag{15}$$

is the equation resulting from (14) after rotation, then

$$B^2 - 4AC = B'^2 - 4A'C' \tag{16}$$

In other words, the discriminant of a quadratic equation is not altered by rotating the coordinate axes; for this reason the discriminant is said to be *invariant* under a rotation of coordinate axes. In particular, if we choose the angle of rotation to eliminate the cross-product term, then (15) becomes

$$A'x'^2 + C'y'^2 + D'x' + E'y' + F' = 0 \tag{17}$$

and since $B' = 0$, (16) tells us that

$$B^2 - 4AC = -4A'C' \tag{18}$$

Part (*a*) of Theorem 12.5.2 can now be proved as follows. If $B^2 - 4AC < 0$, then from (18), $A'C' > 0$, so that (17) may be divided through by $A'C'$ and written in the form

$$\frac{1}{C'}\left(x'^2 + \frac{D'}{A'}x'\right) + \frac{1}{A'}\left(y'^2 + \frac{D'}{C'}y'\right) = -\frac{F'}{A'C'}$$

By completing the squares, this equation may be rewritten in the form

$$\frac{1}{C'}(x' - h)^2 + \frac{1}{A'}(y' - k)^2 = K \tag{19}$$

But A' and C' have the same sign (since $A'C' > 0$), so that by multiplying (19) through by -1, if necessary, we can assume A' and C' to be positive. Thus (19) can be written

$$\frac{(x' - h)^2}{(\sqrt{C'})^2} + \frac{(y' - k)^2}{(\sqrt{A'})^2} = K \tag{20}$$

If $K < 0$, this equation has no graph (the left side is nonnegative for all x' and y'); if $K = 0$ the equation is satisfied only by $x' = h$ and $y' = k$, so the graph is the single point (h, k); finally, if $K > 0$, we can rewrite (20) in the form

$$\frac{(x' - h)^2}{(\sqrt{C'}\sqrt{K})^2} + \frac{(y' - k)^2}{(\sqrt{A'}\sqrt{K})^2} = 1$$

which is an ellipse or a circle. The proofs of parts (*b*) and (*c*) require a similar kind of analysis.

▶ Example 5

Use the discriminant to identify the graph of

$$8x^2 - 3xy + 5y^2 - 7x + 6 = 0$$

Solution. We have

$$B^2 - 4AC = (-3)^2 - 4(8)(5) = -151$$

Since the discriminant is negative, the equation represents an ellipse, a point, or else has no graph. (Why can't the graph be a circle?) ◄

In cases where a quadratic equation represents a point, a pair of parallel lines, a pair of intersecting lines or has no graph, we say that equation represents a ***degenerate conic section.*** Thus if we allow for possible degeneracy, it follows from Theorem 12.5.2 that *every quadratic equation has a conic section as its graph.*

▶ Exercise Set 12.5

1. Let an $x'y'$-coordinate system be obtained by rotating an xy-coordinate system through an angle of $\theta = 60°$.
    (a) Find the $x'y'$-coordinates of the point whose xy-coordinates are $(-2, 6)$.
    (b) Find the xy-coordinates of the point whose $x'y'$-coordinates are $(5, 2)$.

2. Let an $x'y'$-coordinate system be obtained by rotating an xy-coordinate system through an angle of $\theta = 135°$.
    (a) Find the $x'y'$-coordinates of the point whose xy-coordinates are $(5\sqrt{2}, \sqrt{2})$.
    (b) Find the xy-coordinates of the point whose $x'y'$-coordinates are $(-3\sqrt{2}, \sqrt{2})$.

In Exercises 3–12, rotate the coordinate axes to remove the xy-term. Then name the conic and sketch its graph.

3. $xy = -9$.

4. $x^2 - xy + y^2 - 2 = 0$.

5. $x^2 + 4xy - 2y^2 - 6 = 0$.

6. $31x^2 + 10\sqrt{3}xy + 21y^2 - 144 = 0$.

7. $x^2 + 2\sqrt{3}xy + 3y^2 + 2\sqrt{3}x - 2y = 0$.

8. $34x^2 - 24xy + 41y^2 - 25 = 0$.

9. $9x^2 - 24xy + 16y^2 - 80x - 60y + 100 = 0$.

10. $5x^2 - 6xy + 5y^2 - 8\sqrt{2}x + 8\sqrt{2}y = 8$.

11. $52x^2 - 72xy + 73y^2 + 40x + 30y - 75 = 0$.

12. $6x^2 + 24xy - y^2 - 12x + 26y + 11 = 0$.

13. Let an $x'y'$-coordinate system be obtained by rotating an xy-coordinate system through an angle θ. Prove: For every choice of θ, the equation $x^2 + y^2 = r^2$ becomes $x'^2 + y'^2 = r^2$. Give a geometric explanation of this result.

14. Solve the rotation equations (5) for x' and y' in terms of x and y.

15. Let an $x'y'$-coordinate system be obtained by rotating an xy-coordinate system through an angle of $45°$. Use (6) to find the equation in xy-coordinates of the curve $3x'^2 + y'^2 = 6$.

16. Derive the expression for B' in (9).

17. Use (9) to prove that $B^2 - 4AC = B'^2 - 4A'C'$ for all values of θ.

18. Use (9) to prove that $A + C = A' + C'$ for all values of θ.

19. Prove: If $A = C$ in (7), then the cross-product term can be eliminated by rotating through $45°$.

20. Prove: If $B \neq 0$ then the graph of $x^2 + Bxy + F = 0$ is a hyperbola if $F \neq 0$ and two intersecting lines if $F = 0$.

In Exercises 21–25, use the discriminant to identify the graph of the given equation.

21. $x^2 - xy + y^2 - 2 = 0$.

22. $x^2 + 4xy - 2y^2 - 6 = 0$.

23. $x^2 + 2\sqrt{3}xy + 3y^2 + 2\sqrt{3}x - 2y = 0$.

24. $6x^2 + 24xy - y^2 - 12x + 26y + 11 = 0$.

25. $34x^2 - 24xy + 41y^2 - 25 = 0$.

26. Each of the following represents a degenerate conic section. Where possible, sketch the graph.
  (a) $x^2 - y^2 = 0$
  (b) $x^2 + 3y^2 + 7 = 0$

  (c) $8x^2 + 7y^2 = 0$
  (d) $x^2 - 2xy + y^2 = 0$
  (e) $9x^2 + 12xy + 4y^2 - 36 = 0$
  (f) $x^2 + y^2 - 2x - 4y = -5$.

27. Prove parts (b) and (c) of Theorem 12.5.2.

▶ REVIEW EXERCISES

In Exercises 1–8, identify the curve as a parabola, ellipse, or hyperbola, and give the following information:
Parabola: the coordinates of the vertex and focus, and the equation of the directrix.
Ellipse: the coordinates of the center and foci, and the lengths of the major and minor axes.
Hyperbola: the coordinates of the center, foci, and vertices, and equations of the asymptotes.

1. $y^2 + 12x - 6y + 33 = 0$.

2. $x^2 - 4y^2 = -1$.

3. $9x^2 + 4y^2 + 36x - 8y + 4 = 0$.

4. $6x + 8y - x^2 - 4y^2 = 12$.

5. $x^2 - 9y^2 - 4x + 18y - 14 = 0$.

6. $4y = x^2 + 2x - 7$.

7. $3x + 2y^2 - 4y - 7 = 0$.

8. $4x^2 = y^2 - 4y$.

In Exercises 9–16, find an equation for the curve described.

9. The parabola with vertex at $(1, 3)$ and directrix $x = -3$.

10. The ellipse with major axis of length 12 and foci at $(2, 7)$ and $(2, -1)$.

11. The hyperbola with foci $(0, \pm 5)$ and vertices 6 units apart.

12. The parabola with axis $y = 1$, vertex $(2, 1)$, and passing through $(3, -1)$.

13. The ellipse with foci $(\pm 3, 0)$ and such that the distances from the foci to $P(x, y)$ on the ellipse add up to 10 units.

14. The hyperbola with vertices $(0, \pm 2)$ and asymptotes $y = \pm 3x$.

15. The curve C with the property that the distance between the point $(3, 4)$ and any point $P(x, y)$ on C is equal to the distance between P and the line $y = 2$.

16. The hyperbola with vertices $(-3, 2)$ and $(1, 2)$ and perpendicular asymptotes.

In Exercises 17–20, sketch the curve whose equation is given in the stated exercise.

17. Exercise 1.   **18.** Exercise 2.

19. Exercise 3.   **20.** Exercise 6.

In Exercises 21–26, find the rotation angle θ needed to remove the xy-term; then name the conic and give its equation in $x'y'$-coordinates after the xy-term is removed.

21. $3x^2 - 2xy + 3y^2 = 4$.

22. $7x^2 - 8xy + y^2 = 9$.

23. $11x^2 + 10\sqrt{3}xy + y^2 = 4$.

24. $x^2 + 4xy + 4y^2 - 2\sqrt{5}x + \sqrt{5}y = 0$.

25. $16x^2 - 24xy + 9y^2 - 60x - 80y + 100 = 0$.

26. $73x^2 - 72xy + 52y^2 - 100 = 0$.

In Exercises 27–29, sketch the graph of the equation in the stated exercise, showing the xy- and $x'y'$-axes.

27. Exercise 21.   **28.** Exercise 23.

29. Exercise 24.

30. For the curve $xy + x - y = 1$,
  (a) find the rotation angle necessary to remove the xy-term;
  (b) find the equation in $x'y'$-coordinates obtained by rotating the xy-coordinate axes through an angle of $\theta = \tan^{-1}(3/4)$ radians and verify the identities $A' + C' = A + C$ and $B'^2 - 4A'C' = B^2 - 4AC$.

13 polar coordinates and parametric equations

13.1 POLAR COORDINATES

Figure 13.1.1

Up to now, we have specified the location of points in the plane by means of coordinates relative to a rectangular coordinate system consisting of two perpendicular coordinate axes. In this section, we introduce **polar coordinate systems.** This new kind of coordinate system will be easier to work with in many problems.

To form a polar coordinate system in a plane, we pick a fixed point O, called the **origin** or **pole;** and using the origin as an endpoint, we construct a ray, called the **polar axis.** After selecting a unit of measurement, we may associate with any point P in the plane a pair of **polar coordinates**

$$(r, \theta)$$

where r is the distance from P to the origin, and θ measures the angle from the polar axis to the line segment OP (Figure 13.1.1). The number r is called the **radial distance** of P and θ is called a **polar angle** of P. In Figure 13.1.2, the points $(6, 45°)$, $(3, 225°)$, $(5, 120°)$, and $(4, 330°)$ are plotted in polar coordinate systems.

Figure 13.1.2   (a)   (b)

The polar coordinates of a point are not unique. For example, the polar coordinates

$$(1, 315°), \quad (1, -45°), \quad \text{and} \quad (1, 675°)$$

all represent the same point (Figure 13.1.3). In general, if a point P has polar coordinates (r, θ), then for any integer $n = 0, 1, 2, \ldots$

$$(r, \theta + n \cdot 360°) \quad \text{and} \quad (r, \theta - n \cdot 360°)$$

are also polar coordinates of P.

(1, 315°)   (1, −45°)   (1, 675°)

All three pairs of polar coordinates
represent the same point.

Figure 13.1.3

In the case where P is the origin, the line segment OP in Figure 13.1.1b reduces to a point, since $r = 0$. Because there is no clearly defined polar angle in this case, we will agree that an arbitrary polar angle θ may be used. Thus, for arbitrary θ, the point $(0, \theta)$ is the origin.

Frequently, it is helpful to use both polar and rectangular coordinates in the same problem. To do this, we let the positive x-axis of the rectangular coordinate system serve as the polar axis for the polar coordinate system. When this is done, each point P has both polar coordinates (r, θ) and rectangular coordinates (x, y). As suggested by Figure 13.1.4, these coordinates are related by the equations

Figure 13.1.4

$$x = r \cos \theta \tag{1}$$
$$y = r \sin \theta$$

and

$$r^2 = x^2 + y^2 \tag{2a}$$

$$\tan \theta = \frac{y}{x} \tag{2b}$$

▶ Example 1

Find the rectangular coordinates of the point P whose polar coordinates are $(6, 135°)$.

The point P has polar
coordinates $(6, 135°)$ and
rectangular coordinates
$(-3\sqrt{2}, 3\sqrt{2})$

Figure 13.1.5

Solution. Substituting the polar coordinates $r = 6$ and $\theta = 135°$ in (1) yields

$$x = 6 \cos 135° = 6\left(-\frac{\sqrt{2}}{2}\right) = -3\sqrt{2}$$

$$y = 6 \sin 135° = 6\left(\frac{\sqrt{2}}{2}\right) = 3\sqrt{2}$$

Thus the rectangular coordinates of P are $(-3\sqrt{2}, 3\sqrt{2})$ (Figure 13.1.5).

◄

When we start graphing curves in polar coordinates, it will be desirable to allow negative values for r. This will require a special definition. For motivation, consider the point P with polar coordinates $(3, 225°)$. We can reach this point by rotating the polar axis $225°$ and then moving *forward* from the origin 3 units along the terminal side of the angle (Figure 13.1.6a). On the other hand, we can also reach the point P by rotating the polar axis $45°$ and then moving *backward* 3 units from the origin along the *extension* of the terminal side of the angle (Figure 13.1.6b). This suggests that the point $(3, 225°)$ might also be denoted by $(-3, 45°)$, with the minus sign serving to indicate that the point is on the extension of the angle's terminal side rather than on the terminal side itself.

Figure 13.1.6   (a)   (b)

Since the terminal side of the angle $\theta + 180°$ is the extension of the terminal side of the angle θ, we will define

$$(-r, \theta) \quad \text{and} \quad (r, \theta + 180°) \tag{3}$$

to be polar coordinates for the same point. With $r = 3$ and $\theta = 45°$ in (3) it follows that $(-3, 45°)$ and $(3, 225°)$ represent the same point.

In the exercises we ask the reader to show that relationships (1) and (2) remain valid when r is negative.

REMARK.   For many purposes it does not matter whether polar angles are represented in degrees or radians. However, later on, when we are concerned with differentiation problems, it will be essential to measure polar angles in radians, since the derivative formulas for the trigonometric functions were derived under this assumption. From here on, we will use radian measure.

▶ Example 2

Find polar coordinates of the point P whose rectangular coordinates are $(-2, 2\sqrt{3})$.

Solution. We will find polar coordinates (r, θ) of P such that $r > 0$ and $0 \leq \theta < 2\pi$. From (2a),

$$r^2 = x^2 + y^2 = (-2)^2 + (2\sqrt{3})^2 = 4 + 12 = 16$$

so $r = 4$. From (2b),

$$\tan \theta = \frac{y}{x} = \frac{2\sqrt{3}}{-2} = -\sqrt{3}$$

From this and the fact that $(-2, 2\sqrt{3})$ lies in the second quadrant, it follows that

$$\theta = \frac{2\pi}{3}$$

Thus $(4, 2\pi/3)$ are polar coordinates of P. All other polar coordinates of P have the form

$$\left(4, \frac{2\pi}{3} + 2n\pi\right) \quad \text{or} \quad \left(-4, \frac{5\pi}{3} + 2n\pi\right)$$

where n is an integer. ◀

Polar coordinates provide a new way of graphing certain equations. As an example, consider the equation

$$r = \sin \theta \tag{4}$$

By substituting values for θ at increments of $\pi/6$ ($=30°$), and calculating r we can construct the following table:

Table 13.1.1

θ (radians)	0	$\frac{\pi}{6}$	$\frac{\pi}{3}$	$\frac{\pi}{2}$	$\frac{2\pi}{3}$	$\frac{5\pi}{6}$	π	$\frac{7\pi}{6}$	$\frac{4\pi}{3}$	$\frac{3\pi}{2}$	$\frac{5\pi}{3}$	$\frac{11\pi}{6}$	2π
$r = \sin \theta$	0	$\frac{1}{2}$	$\frac{\sqrt{3}}{2}$	1	$\frac{\sqrt{3}}{2}$	$\frac{1}{2}$	0	$-\frac{1}{2}$	$-\frac{\sqrt{3}}{2}$	-1	$-\frac{\sqrt{3}}{2}$	$-\frac{1}{2}$	0

Figure 13.1.7

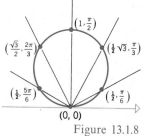

Figure 13.1.8

The pairs of values listed in this table may be plotted in one of two ways. One possibility is to view each pair (r, θ) as *rectangular* coordinates of a point and plot each point in the $r\theta$-plane. This yields points on the familiar sine curve Figure 13.1.7). Alternatively, we can view each pair (r, θ) as *polar* coordinates of a point and plot each point in a polar coordinate system. (Figure 13.1.8). Note that there are 13 pairs listed in Table 13.1.1, but only 6 points plotted in Figure 13.1.8. This is because the pairs from $\theta = \pi$ on yield duplicates of the preceding points. For example $(-1/2, 7\pi/6)$ and $(1/2, \pi/6)$ represent the same point.

The points in Figure 13.1.8 appear to lie on a circle. That this is indeed the case may be seen by expressing (4) in terms of x and y. To do this we first multiply (4) through by r to obtain

$$r^2 = r \sin \theta$$

which may be rewritten using (1) and (2) as

$$x^2 + y^2 = y$$

or

$$x^2 + y^2 - y = 0$$

or on completing the square

$$x^2 + \left(y - \frac{1}{2}\right)^2 = \frac{1}{4}$$

This is a circle of radius $\frac{1}{2}$ centered at the point $(0, \frac{1}{2})$ in the xy-plane.

▶ Example 3

Sketch the curve $r = \sin 2\theta$ in polar coordinates.

Solution. A rough sketch of the graph may be obtained without plotting points by observing how the value of r changes with θ.

If θ increases from 0 to $\pi/4$, then 2θ increases from 0 to $\pi/2$ and $r = \sin 2\theta$ increases from 0 to 1. (Figure 13.1.9a).

If θ increases from $\pi/4$ to $\pi/2$, then 2θ increases from $\pi/2$ to π and $r = \sin 2\theta$ decreases from 1 back to 0. Thus the curve completes a loop (Figure 13.1.9b).

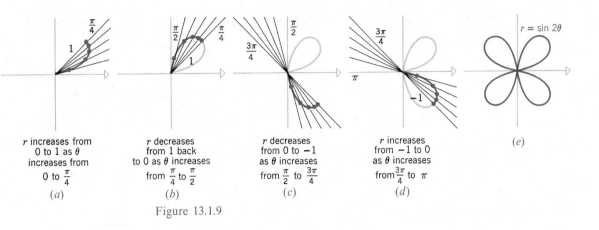

r increases from
0 to 1 as θ
increases from
0 to $\dfrac{\pi}{4}$

(a)

r decreases
from 1 back
to 0 as θ increases
from $\dfrac{\pi}{4}$ to $\dfrac{\pi}{2}$

(b)

r decreases
from 0 to -1
as θ increases
from $\dfrac{\pi}{2}$ to $\dfrac{3\pi}{4}$

(c)

r increases
from -1 to 0
as θ increases
from $\dfrac{3\pi}{4}$ to π

(d)

(e)

Figure 13.1.9

If θ increases from $\pi/2$ to $3\pi/4$ then 2θ increases from π to $3\pi/2$ and r decreases from 0 to -1. Because r is negative, the resulting portion of the graph is in the fourth quadrant although the values of θ are in the second quadrant (Figure 13.1.9c).

If θ increases from $3\pi/4$ to π, then 2θ increases from $3\pi/2$ to 2π and $r = \sin 2\theta$ increases from -1 to 0, and the curve completes a loop in the fourth quadrant (Figure 13.1.9d).

We leave it for the reader to continue the analysis for θ in the range π to 2π and show that the remaining portion of the curve consists of two loops in the second and third quadrants (Figure 13.1.9e). No new points are generated by allowing θ to vary past 2π or through negative values (verify) so that Figure 13.1.9e is the complete graph. ◄

▶ Exercise Set 13.1

1. Plot the following points in polar coordinates.
  (a) $(3, \pi/4)$ (b) $(5, 2\pi/3)$
  (c) $(1, \pi/2)$ (d) $(4, 7\pi/6)$
  (e) $(2, 4\pi/3)$ (f) $(0, \pi)$.

2. Plot the following points in polar coordinates.
  (a) $(2, -\pi/3)$ (b) $(3/2, -7\pi/4)$
  (c) $(-3, 3\pi/2)$ (d) $(-5, -\pi/6)$
  (e) $(-6, -\pi)$ (f) $(-1, 9\pi/4)$.

3. Find rectangular coordinates of the points whose polar coordinates are given.
  (a) $(6, \pi/6)$ (b) $(7, 2\pi/3)$
  (c) $(8, 9\pi/4)$ (d) $(5, 0)$
  (e) $(7, 17\pi/6)$ (f) $(0, \pi)$.

4. Find rectangular coordinates of the points whose polar coordinates are given.
  (a) $(-8, \pi/4)$ (b) $(7, -\pi/4)$
  (c) $(-6, -5\pi/6)$ (d) $(0, -\pi)$
  (e) $(-2, -3\pi/2)$ (f) $(-5, 0)$.

5. The following points are given in rectangular coordinates. Express the points in polar coordinates with $r \geq 0$ and $0 \leq \theta < 2\pi$.
  (a) $(-5, 0)$ (b) $(2\sqrt{3}, -2)$
  (c) $(0, -2)$ (d) $(-8, -8)$
  (e) $(-3, 3\sqrt{3})$ (f) $(1, 1)$.

6. Express the points in Exercise 5 in polar coordinates with $r \geq 0$ and $-\pi < \theta \leq \pi$.

7. Express the points in Exercise 5 in polar coordinates with $r \leq 0$ and $0 \leq \theta < 2\pi$.

In Exercises 8–13, identify the curve by transforming to rectangular coordinates.

8. $r = 3$.

9. $r \sin \theta = 4$.

10. $r = 2 \sin \theta$.

11. $r = \dfrac{6}{2 - \cos \theta}$.

12. $r = 5 \sec \theta$.

13. $r + 4 \cos \theta = 0$.

In Exercises 14–19, express the equations in polar coordinates.

14. $x^2 + y^2 = 9$.

15. $x = 7$.

16. $4xy = 9$.

17. $(x^2 + y^2)^2 = 16(x^2 - y^2)$.

18. $x^2 - y^2 = 4$.

19. $x^2 = 9y$.

In Exercises 20–23, use the method of Example 3 to sketch the curve in polar coordinates.

20. $r = \cos 2\theta$.

21. $r = 2(1 + \sin \theta)$.

22. $r = 4 \cos 3\theta$.

23. $r = 1 - \cos \theta$.

24. Prove that the distance between the points with polar coordinates (r_1, θ_1) and (r_2, θ_2) is

$$d = \sqrt{r_1{}^2 + r_2{}^2 - 2r_1 r_2 \cos(\theta_1 - \theta_2)}$$

25. Prove: In polar coordinates, the equation $r = a \sin \theta + b \cos \theta$ represents a circle.

26. Prove: The area of the triangle whose vertices have polar coordinates $(0, 0)$, (r_1, θ_1) and (r_2, θ_2) is $A = \frac{1}{2} r_1 r_2 \sin(\theta_2 - \theta_1)$. [Assume $0 \leq \theta_1 < \theta_2 \leq \pi$ and r_1 and r_2 are positive.]

27. Prove: Equations (1) and (2) hold if r is negative.

28. Prove that $rr_1 \sin(\theta - \theta_1) + rr_2 \sin(\theta_2 - \theta) + r_1 r_2 \sin(\theta_1 - \theta_2) = 0$ is the polar equation of the line through the points (r_1, θ_1) and (r_2, θ_2).

13.2   GRAPHS IN POLAR COORDINATES

In this section we will graph some curves defined in terms of polar coordinates. Since we will be working with both rectangular xy-coordinates and polar coordinates, we assume throughout that the positive x-axis coincides with the polar axis.

LINES IN POLAR
COORDINATES

A line perpendicular to the x-axis and passing through the point with xy-coordinates $(a, 0)$ has the equation

$$x = a$$

To express this equation in polar coordinates we substitute $x = r \cos \theta$ [see (1) in the previous section]; this yields

$$r \cos \theta = a \qquad (1)$$

Figure 13.2.1

This result makes sense geometrically since each point $P(r, \theta)$ on this line will yield the value a for $r \cos \theta$ (Figure 13.2.1).

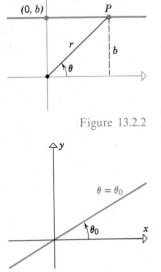

Figure 13.2.2

Figure 13.2.3

A line parallel to the *x*-axis that meets the *y*-axis at the point with *xy*-coordinates $(0, b)$ has the equation

$$y = b$$

Substituting $y = r \sin \theta$ yields

$$r \sin \theta = b \tag{2}$$

as the polar equation of this line. This makes sense geometrically since each point $P(r, \theta)$ on this line will yield the value b for $r \sin \theta$ (Figure 13.2.2). For any constant θ_0, the equation

$$\theta = \theta_0 \tag{3}$$

is satisfied by the coordinates of all points of the form $P(r, \theta_0)$, regardless of the value of r. Thus, the equation represents the line through the origin making an angle of θ_0 (radians) with the polar axis (Figure 13.2.3).

▶ Example 1

Sketch the graphs of the following equations in polar coordinates.

(a) $r \cos \theta = 3$ (b) $r \sin \theta = -2$ (c) $\theta = \dfrac{3\pi}{4}$

Solution. From (1), (2), and (3), each graph is a line; they are shown in Figure 13.2.4. ◀

Figure 13.2.4   (*a*)   (*b*)   (*c*)

By substituting $x = r \cos \theta$ and $y = r \sin \theta$ in the equation $Ax + By + C = 0$, we obtain the **general polar form of a line,**

$$r(A \cos \theta + B \sin \theta) + C = 0$$

CIRCLES IN POLAR
COORDINATES

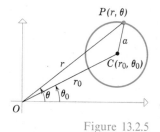

$P(r, \theta)$

Figure 13.2.5

Let us try to find the polar equation of a circle whose radius is a and whose center has polar coordinates (r_0, θ_0) (Figure 13.2.5). If we let $P(r, \theta)$ be an arbitrary point on the circle, and if we apply the law of cosines to the triangle OCP in Figure 13.2.5 we obtain

$$r^2 - 2rr_0 \cos(\theta - \theta_0) + r_0^2 = a^2 \tag{4}$$

This general equation is rarely used since the corresponding equation in rectangular coordinates is easier to work with. However, we will now consider some special cases of (4) that arise frequently.

A circle of radius a, centered at the origin, has an especially simple polar equation. If we let $r_0 = 0$ in (4) we obtain

$$r^2 = a^2$$

or since $a \geq 0$

$$r = a \tag{5}$$

This equation makes sense geometrically since the circle of radius a, centered at the origin, consists of all points $P(r, \theta)$ for which $r = a$, regardless of the value of θ (Figure 13.2.6a).

If a circle of radius a has its center on the x-axis and passes through the origin, then the polar coordinates of the center are either

$$(a, 0) \quad \text{or} \quad (a, \pi)$$

depending on whether the center is to the right or left of the origin (Figure 13.2.6b). For the circle to the right of the origin we have $r_0 = a$ and $\theta_0 = 0$, so (4) reduces to

$$r^2 - 2ra \cos \theta = 0$$

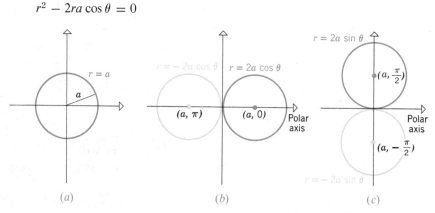

Figure 13.2.6

(a)

(b)

(c)

or

$$r = 2a \cos \theta \qquad (6a)$$

For the circle to the left of the origin, $r_0 = a$ and $\theta_0 = \pi$, so (4) reduces to

$$r = -2a \cos \theta \qquad (6b)$$

because $\cos(\theta - \pi) = -\cos \theta$.

If a circle of radius a passes through the origin and has its center on the y-axis, then the polar coordinates of the center are $(a, \pi/2)$ or $(a, -\pi/2)$, and (4) reduces to

$$r = 2a \sin \theta \qquad (7a)$$

or

$$r = -2a \sin \theta \qquad (7b)$$

depending on whether the center is above or below the origin (Figure 13.2.6c).

Equations (6a) and (7a) are interpreted geometrically in Figures 13.2.7a and 13.2.7b. In each figure, OPA is a right triangle (why?) so that for each point $P(r, \theta)$ on the circle, we have $r = 2a \cos \theta$ or $r = 2a \sin \theta$, depending on the orientation of the circle.

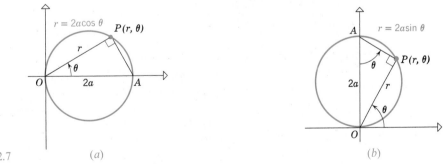

Figure 13.2.7

(a)

(b)

► Example 2

Sketch the graphs of the following equations in polar coordinates

(a) $r = 4 \cos \theta$ (b) $r = -5 \sin \theta$ (c) $r = 3$

Solution. These equations are of forms (6a), (7b), and (5), respectively, and therefore represent circles. The graphs are shown in Figure 13.2.8. ◄

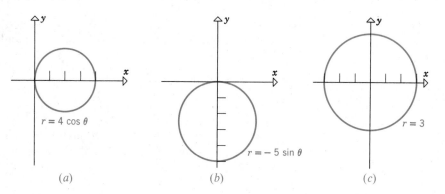

Figure 13.2.8   (a)   (b)   (c)

The points with polar coordinates (r, θ) and $(r, -\theta)$ are symmetric about the x-axis (Figure 13.2.9a). Thus, if a polar equation is unchanged on replacing θ by $-\theta$, then its graph is symmetric about the x-axis. The following tests for symmetry will be useful.

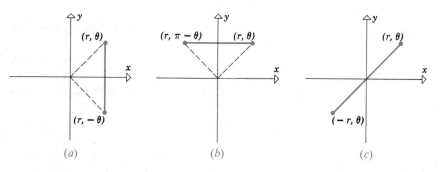

Figure 13.2.9   (a)   (b)   (c)

13.2.1 THEOREM (a) *A curve in polar coordinates is symmetric about the x-axis if replacing θ*
Symmetry Tests   *by $-\theta$ in its equation produces an equivalent equation (Figure 13.2.9a).*
                      (b) *A curve in polar coordinates is symmetric about the y-axis if replacing θ*
                           by $\pi - \theta$ in its equation produces an equivalent equation (Figure 13.2.9b).
                      (c) *A curve in polar coordinates is symmetric about the origin if replacing r*
                           by $-r$ in its equation produces an equivalent equation (Figure 13.2.9c).

CARDIOIDS AND   Equations of the form
LIMAÇONS

$$r = a + b \sin \theta \qquad r = a - b \sin \theta \qquad\qquad\qquad (8a\text{--}8b)$$
$$r = a + b \cos \theta \qquad r = a - b \cos \theta \qquad\qquad\qquad (8c\text{--}8d)$$

produce polar curves called *limaçons* (from the Latin word, "limax" for a slug, a snail-like creature.) If a and b are positive, then the shape of the limaçon is as shown in Figure 13.2.10.

Figure 13.2.10 $a > b$ $a = b$ $a < b$

The orientation of the limaçon relative to the polar axis depends on whether $\sin \theta$ or $\cos \theta$ appears in the equation and whether the $+$ or $-$ occurs. Because of the heart-shaped appearance of the curve in the case $a = b$, limaçons of this type are usually called *cardioids.*

▶ Example 3

Assuming a to be a positive constant, sketch the curve

$$r = a(1 - \cos \theta)$$

in polar coordinates.

Solution. This is an equation of form (8d) with $a = b$ and, therefore, represents a cardioid. Since $\cos(-\theta) = \cos \theta$ the equation is unaltered when θ is replaced by $-\theta$; thus the cardioid is symmetric about the x-axis. This being the case, we can obtain the entire curve by first sketching the portion of the cardioid above the x-axis and then reflecting this portion about the x-axis.

As θ varies from 0 to π, $\cos \theta$ decreases steadily from 1 to -1, and $1 - \cos \theta$ increases steadily from 0 to 2. Thus, as θ varies from 0 to π, the value of $r = a(1 - \cos \theta)$ will increase steadily from an initial value of $r = 0$ to a final value of $r = 2a$. Using this information, and by plotting the points in the following table, we obtain the graph in Figure 13.2.11a.

Table 13.2.1

θ	0	$\dfrac{\pi}{3}$	$\dfrac{\pi}{2}$	$\dfrac{2\pi}{3}$	π
$r = a(1 - \cos \theta)$	0	$\dfrac{a}{2}$	a	$\dfrac{3a}{2}$	$2a$

(a)

$r = a(1 - \cos \theta)$

(b)

Figure 13.2.11

On reflecting the curve in Figure 13.2.11*a* about the *x*-axis, we obtain the entire cardioid (Figure 13.2.11*b*). ◀

REMARK. We have sketched the cardioid in Figure 13.2.11 with the line $\theta = 0$ tangent to the curve at the origin. While it is not evident from our discussion that this is the case, we will prove in Section 13.5 that if a polar curve passes through the origin when $\theta = \theta_0$, then the line $\theta = \theta_0$ is tangent to the curve at the origin.

LEMNISCATES   Equations of the form

$$r^2 = a^2 \cos 2\theta \qquad r^2 = -a^2 \cos 2\theta \qquad \text{(9a–9b)}$$
$$r^2 = a^2 \sin 2\theta \qquad r^2 = -a^2 \sin 2\theta \qquad \text{(9c–9d)}$$

A lemniscate

Figure 13.2.12

represent propeller-shaped curves, called **lemniscates** (Figure 13.2.12). The lemniscates are centered at the origin, but the orientation relative to the polar axis depends on the sign preceding the a^2 and whether $\sin 2\theta$ or $\cos 2\theta$ appears in the equation.

▶ Example 4

Sketch the curve

$$r^2 = 4 \cos 2\theta$$

in polar coordinates.

Solution. The equation is of type (9a) with $a = 2$, and thus represents a lemniscate. We leave it for the reader to apply the symmetry tests in Theorem 13.2.1 and show that the curve is symmetric about the *x*-axis and the *y*-axis.

Therefore, we can obtain the entire curve by first sketching the portion of the lemniscate in the range $0 \leq \theta \leq \pi/2$ and then reflecting this portion about the x- and y-axes.

For θ in the range $0 \leq \theta \leq \pi/4$, the quantity $\cos 2\theta$ is nonnegative, so that for each such θ, there are two values of r satisfying $r^2 = 4 \cos 2\theta$, namely $r = 2\sqrt{\cos 2\theta}$ and $r = -2\sqrt{\cos 2\theta}$. As θ varies from 0 to $\pi/4$, the value of $\cos 2\theta$ decreases steadily from 1 to 0, so that $r = 2\sqrt{\cos 2\theta}$ decreases steadily from 2 to 0 and $r = -2\sqrt{\cos 2\theta}$ increases steadily from -2 to 0. With this information and the points plotted from the following table, we obtain the sketch in Figure 13.2.13a.

$r = -\sqrt{2}$

$r = \sqrt{2}$

(a)

Table 13.2.2

θ	0	$\dfrac{\pi}{6}$	$\dfrac{\pi}{4}$
$r = \pm 2\sqrt{\cos 2\theta}$	± 2	$\pm\sqrt{2}$	0

(Note that the curve passes through the origin when $\theta = \pi/4$, so that the line $\theta = \pi/4$ is tangent to the curve at the origin by the remark following Example 3.)

For θ in the range $\pi/4 < \theta < \pi/2$, the quantity $4 \cos 2\theta$ is negative, so there are no real values of r satisfying $r^2 = 4 \cos 2\theta$. Thus there are no points on the graph for such θ. The entire lemniscate is obtained by reflecting the curve in Figure 13.2.13a about the x-axis and then about the y-axis (Figure 13.2.13b). (The reflection about the y-axis adds no new points, however.) ◀

$r^2 = 4 \cos 2\theta$

(b)

Figure 13.2.13

SPIRALS   A curve that "winds around the origin" infinitely many times in such a way that r increases (or decreases) steadily as θ increases is called a **spiral**. The most common example is the **spiral of Archimedes**, which has an equation of the form

$$r = a\theta \qquad (\theta \geq 0) \tag{10a}$$

or

$$r = a\theta \qquad (\theta \leq 0) \tag{10b}$$

In these equations, θ is in radians and a is positive.

▶ Example 5

Sketch the curve
$$r = \theta \qquad (\theta \geq 0)$$
in polar coordinates.

Solution. This is an equation of form (10a) with $a = 1$ and, thus, represents an Archimedean spiral. Since $r = 0$ when $\theta = 0$, the origin is on the curve and the polar axis is tangent to the spiral.

A reasonably accurate sketch may be obtained by plotting the intersections of the spiral with the x- and y-axes and noting that r increases steadily as θ increases. The intersections with the x-axis occur when

$$\theta = 0, \pi, 2\pi, 3\pi, \ldots$$

at which points r has the values

$$r = 0, \pi, 2\pi, 3\pi, \ldots$$

and the intersections with the y-axis occur when

$$\theta = \frac{\pi}{2}, \frac{3\pi}{2}, \frac{5\pi}{2}, \frac{7\pi}{2}, \ldots$$

at which points, r has the values

$$r = \frac{\pi}{2}, \frac{3\pi}{2}, \frac{5\pi}{2}, \frac{7\pi}{2}, \ldots$$

The graph is shown in Figure 13.2.14. ◀

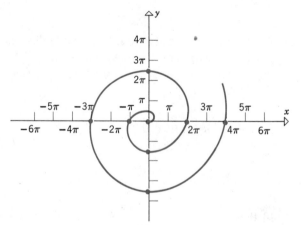

Figure 13.2.14

Starting from the origin, the spiral in Figure 13.2.14 loops counterclockwise around the origin. This is true of all Archimedean spirals of form (10a); Archimedian spirals of form (10b) loop clockwise around the origin.

ROSE CURVES   Equations of the form

$$r = a \sin n\theta \tag{11a}$$
$$r = a \cos n\theta \tag{11b}$$

represent flower-shaped curves called **roses.** The rose has n equally spaced petals or loops if n is odd and $2n$ equally spaced petals if n is even (Figure 13.2.15). The orientation of the rose relative to the polar axis depends on the sign of the constant a and whether $\sin \theta$ or $\cos \theta$ appears in the equation. (A rose curve occurred in Example 3 of the previous section.)

A three–petal rose
($n = 3$)

A four–petal rose
($n = 2$)

Figure 13.2.15

If $n = 1$ in (11a) or (11b) then we obtain the equation of a circle, which can be regarded as a one-petal rose.

▶ Exercise Set 13.2

In Exercises 1–32, sketch the curve in polar coordinates and give its name.

1. $\theta = \pi/6$.
2. $\theta = -3\pi/4$.
3. $r = 5$.
4. $r = 4 \sin \theta$.
5. $r = -6 \cos \theta$.
6. $r = -3 \sin \theta$.
7. $2r = \cos \theta$.
8. $r = 1 + \sin \theta$.
9. $r = 3(1 - \sin \theta)$.
10. $r - 2 = 2 \cos \theta$.
11. $r = 4 - 4 \cos \theta$.
12. $r = -5 + 5 \sin \theta$.
13. $r = -1 - \cos \theta$.
14. $r = 1 + 2 \sin \theta$.
15. $r = 2 + \sin \theta$.
16. $r = 4 + 3 \cos \theta$.
17. $r = 3 + 4 \cos \theta$.
18. $r = -3 - 4 \sin \theta$.
19. $r = 5 - 2 \cos \theta$.
20. $r^2 = 9 \cos 2\theta$.
21. $r^2 = -9 \cos 2\theta$.
22. $r^2 = \sin 2\theta$.
23. $r^2 = -16 \sin 2\theta$.
24. $r = 4\theta \ (\theta \geq 0)$.

25. $r = 4\theta \ (\theta \leq 0)$.
26. $r = 4\theta$.
27. $r = \cos 2\theta$.
28. $r = 3 \sin 2\theta$.
29. $r = \sin 3\theta$.
30. $r = 2 \cos 3\theta$.
31. $r = 9 \sin 4\theta$.
32. $r = \cos 5\theta$.

In Exercises 33–40, sketch the curve in polar coordinates.

33. $r = 4 \cos \theta + 4 \sin \theta$.
34. $r = e^\theta$ (logarithmic spiral).
35. $r = 4 \tan \theta$ (kappa curve).
36. $r = 2 + 2 \sec \theta$ (conchoid of Nichomedes).
37. $r = \sin (\theta/2)$.
38. $r^2 = \theta$ (parabolic spiral).
39. $r = 2 \sin \theta \tan \theta$ (cissoid).
40. $r\theta = 1 \ (\theta > 0)$ (hyperbolic spiral).

13.3   AREA IN POLAR COORDINATES

In this section we will be concerned with the following version of the area problem.

13.3.1   PROBLEM

The Area Problem in Polar Coordinates

Find the area of the region R between a polar curve $r = f(\theta)$ and two lines, $\theta = \alpha$ and $\theta = \beta$ (Figure 13.3.1).

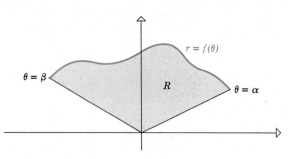

Figure 13.3.1

We will assume that $\alpha < \beta$ and $f(\theta)$ is continuous and nonnegative for $\alpha \leq \theta \leq \beta$. As a first step toward finding the area of the region R, let $\theta_1, \theta_2, \ldots, \theta_{n-1}$ be any numbers such that

$$\alpha < \theta_1 < \theta_2 < \cdots < \theta_{n-1} < \beta$$

and construct the lines

$$\theta = \theta_1, \ \theta = \theta_2, \ldots, \theta = \theta_{n-1}$$

These lines divide the angle from α to β into n subangles, and they divide the region R into n subregions (Figure 13.3.2). As indicated in this figure, we will denote the subangles by

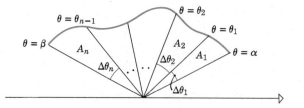

Figure 13.3.2

$$\Delta\theta_1, \Delta\theta_2, \ldots, \Delta\theta_n$$

and the areas of the subregions by

$$A_1, A_2, \ldots, A_n$$

Then the area A of the entire region may be written

$$A = A_1 + A_2 + \cdots + A_n = \sum_{k=1}^{n} A_k$$

Let us now concentrate on the areas $A_1, A_2, \ldots, A_n$ of the subregions. If $\Delta\theta_k$ is not too large, we can approximate the area A_k by the area of a *sector* having central angle $\Delta\theta_k$ and radius $f(\theta_k^*)$, where θ_k^* is an arbitrary angle terminating in the kth subregion (Figure 13.3.3).

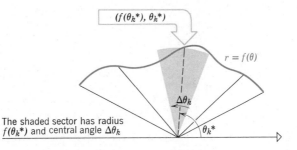

Figure 13.3.3

The shaded sector has radius $f(\theta_k^*)$ and central angle $\Delta\theta_k$

Thus from the area formula for a sector [Formula (13), Section 3.3],

$$A_k \approx \frac{1}{2} [f(\theta_k^*)]^2 \, \Delta\theta_k$$

and

$$A = \sum_{k=1}^{n} A_k \approx \sum_{k=1}^{n} \frac{1}{2} [f(\theta_k^*)]^2 \, \Delta\theta_k \qquad (1)$$

If we now increase n in such a way that max $\Delta\theta_k \to 0$, then the sectors will become better and better approximations to the subregions and (1) will approach the exact value of the area A; that is,

$$A = \lim_{\max \Delta\theta_k \to 0} \sum_{k=1}^{n} \frac{1}{2} [f(\theta_k^*)]^2 \, \Delta\theta_k \qquad (2)$$

Since the limit on the right side of (2) is the definite integral

$$\int_{\alpha}^{\beta} \frac{1}{2} [f(\theta)]^2 \, d\theta$$

we are led to the following definition:

13.3.2 DEFINITION If $f(\theta)$ is continuous and nonnegative for $\alpha \leq \theta \leq \beta$, then the area A enclosed by the polar curve $r = f(\theta)$ and the lines $\theta = \alpha$ and $\theta = \beta$ is

$$A = \int_\alpha^\beta \frac{1}{2} [f(\theta)]^2 \, d\theta$$

or equivalently

$$A = \int_\alpha^\beta \frac{1}{2} r^2 \, d\theta \tag{3}$$

The hardest part of applying (3) is determining the limits of integration. This can be done as follows:

> **Step 1.** Sketch the region R whose area is to be determined.
>
> **Step 2.** Draw an arbitrary "radial line" from the origin to the boundary curve $r = f(\theta)$.
>
> **Step 3.** Ask, "Over what interval of values must θ vary in order for the radial line to sweep out the region R?"
>
> **Step 4.** Your answer in Step 3 will determine the lower and upper limits of integration.

▶ Example 1

Find the area of the region in the first quadrant within the cardioid $r = 1 - \cos\theta$.

Solution. The region and a typical radial line are shown in Figure 13.3.4. For the radial line to sweep out the region, θ must vary from 0 to $\pi/2$. Thus from (3) with $\alpha = 0$ and $\beta = \pi/2$,

$$A = \int_0^{\pi/2} \frac{1}{2} r^2 \, d\theta = \frac{1}{2} \int_0^{\pi/2} (1 - \cos\theta)^2 \, d\theta$$

$$= \frac{1}{2} \int_0^{\pi/2} (1 - 2\cos\theta + \cos^2\theta) \, d\theta$$

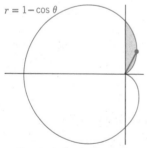

$r = 1 - \cos\theta$

The shaded region is swept out by the radial line as θ varies from 0 to $\frac{\pi}{2}$

Figure 13.3.4

With the help of the identity $\cos^2\theta = \frac{1}{2}(1 + \cos 2\theta)$, this may be rewritten

$$A = \frac{1}{2} \int_0^{\pi/2} \left(\frac{3}{2} - 2\cos\theta + \frac{1}{2}\cos 2\theta \right) d\theta$$

$$= \frac{1}{2}\left[\frac{3}{2}\theta - 2\sin\theta + \frac{1}{4}\sin 2\theta\right]_0^{\pi/2}$$

$$= \tfrac{3}{8}\pi - 1$$ ◀

▶ Example 2

Find the entire area within the cardioid of Example 1.

Solution. For the radial line to sweep out the entire cardioid, θ must vary from 0 to 2π. Thus from (3) with $\alpha = 0$ and $\beta = 2\pi$

$$A = \int_0^{2\pi} \frac{1}{2} r^2 \, d\theta = \frac{1}{2}\int_0^{2\pi} (1 - \cos\theta)^2 \, d\theta$$

If we proceed as in Example 1, this reduces to

$$A = \frac{1}{2}\int_0^{2\pi} \left(\frac{3}{2} - 2\cos\theta + \frac{1}{2}\cos 2\theta\right) d\theta = \frac{3\pi}{2}$$

Alternate Solution. Since the cardioid is symmetric about the *x*-axis, we can calculate the portion of area above the *x*-axis and double the result. In the portion of the cardioid above the *x*-axis, θ ranges from 0 to π, so that

$$A = 2\int_0^{\pi} \frac{1}{2} r^2 \, d\theta = \int_0^{\pi} (1 - \cos\theta)^2 \, d\theta = \frac{3\pi}{2}$$ ◀

▶ Example 3

Find the area of the region that is inside the cardioid $r = 4 + 4\cos\theta$ and outside the circle $r = 6$.

Solution. To sketch the region, we need to know where the circle and cardioid intersect. To find these points, we equate the given expressions for r. This yields

$$4 + 4\cos\theta = 6$$

from which we obtain $\cos\theta = \tfrac{1}{2}$ or

$$\theta = \frac{\pi}{3} \quad \text{and} \quad \theta = -\frac{\pi}{3}$$

The region is sketched in Figure 13.3.5.

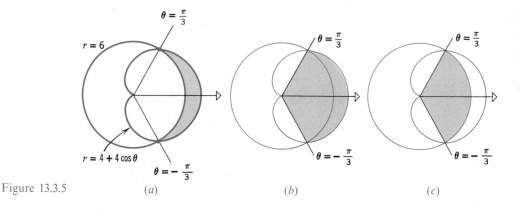

Figure 13.3.5   (a)   (b)   (c)

The desired area can be obtained by subtracting the shaded areas in parts (b) and (c) of Figure 13.3.5. Thus

$$A = \int_{-\pi/3}^{\pi/3} \frac{1}{2}(4 + 4\cos\theta)^2\,d\theta - \int_{-\pi/3}^{\pi/3} \frac{1}{2}(6)^2\,d\theta$$

$$= \int_{-\pi/3}^{\pi/3} \frac{1}{2}[(4 + 4\cos\theta)^2 - 36]\,d\theta$$

$$= \int_{-\pi/3}^{\pi/3} (16\cos\theta + 8\cos^2\theta - 10)\,d\theta \tag{4}$$

$$= \left[16\sin\theta + (4\theta + 2\sin 2\theta) - 10\theta\right]_{-\pi/3}^{\pi/3}$$

$$= 18\sqrt{3} - 4\pi$$

Alternate Solution.   Using symmetry, we may calculate the portion of area above the x-axis and double it to obtain the entire area A. This yields

$$A = 2\int_{0}^{\pi/3} \frac{1}{2}[(4 + 4\cos\theta)^2 - 36]\,d\theta$$

rather than (4). However, the end result will be the same.   ◀

WARNING.   In the last example we determined the intersections of the curves $r = 4 + 4\cos\theta$ and $r = 6$ by equating the right-hand sides and solving for θ. However, it may not be possible to find *all* the intersections of two polar curves, $r = f_1(\theta)$ and $r = f_2(\theta)$, by solving the equation $f_1(\theta) = f_2(\theta)$ for θ. This is because every point has infinitely many pairs of polar coordinates. It is possible, therefore, that a point of intersection may have no single pair of polar coordinates that satisfies both equations. For example, the cardioids

$$r = 1 - \cos\theta \quad\text{and}\quad r = 1 + \cos\theta \tag{5}$$

intersect at three points, the origin, the point $(1, \pi/2)$, and the point $(1, 3\pi/2)$ (Figure 13.3.6).

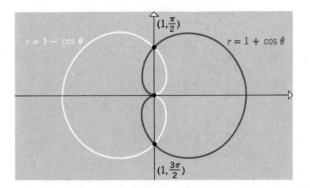

Figure 13.3.6

Equating the right-hand sides in (5) yields $1 - \cos\theta = 1 + \cos\theta$ or $\cos\theta = 0$, so

$$\theta = \frac{\pi}{2} + k\pi, \qquad k = 0, \pm 1, \pm 2, \ldots$$

Substituting any of these values in (5) yields $r = 1$, so that we have found only two distinct points of intersection, $(1, \pi/2)$ and $(1, 3\pi/2)$; the origin has been missed.

When calculating intersections in polar coordinates, it is a good idea to make a sketch of the curves to determine how many intersections there should be.

▶ Exercise Set 13.3

In Exercises 1–17, find the area of the region described.

1. The region in the first quadrant enclosed by the first loop of the spiral $r = \theta$ ($\theta \geq 0$), and the lines $\theta = \pi/6$ and $\theta = \pi/3$.

2. The region in the first quadrant within the cardioid $r = 1 + \sin\theta$.

3. The region enclosed by the cardioid $r = 2 + 2\cos\theta$.

4. The region outside the cardioid $r = 2 - 2\cos\theta$ and inside the circle $r = 4$.

5. The region inside the circle $r = 5\sin\theta$ and outside the limaçon $r = 2 + \sin\theta$.

6. The region enclosed by the inner loop of the limaçon $r = 1 + 2\cos\theta$.

7. The region inside the cardioid $r = 2 + 2\cos\theta$ and outside the circle $r = 3$.

8. The region inside the circle $r = 2a\sin\theta$.

9. The region enclosed by the curve $r^2 = \sin 2\theta$.

10. The region common to the circles $r = 4 \cos \theta$ and $r = 4 \sin \theta$.

11. The region enclosed by the rose $r = 4 \cos 3\theta$.

12. The region inside the rose $r = 2a \cos 2\theta$ and outside the circle $r = a\sqrt{2}$.

13. The region common to the circle $r = 3 \cos \theta$ and the cardioid $r = 1 + \cos \theta$.

14. The region between the loops of the limaçon $r = \frac{1}{2} + \cos \theta$.

15. The region inside the circle $r = 10$ and to the right of the line $r \cos \theta = 6$.

16. The region inside the cardioid $r = a(1 + \sin \theta)$ and outside the circle $r = a \sin \theta$.

17. The region enclosed by $r = a \sec^2 \frac{1}{2}\theta$ and the rays whose polar angles are $\theta = 0$ and $\theta = \frac{1}{2}\pi$.

18. (a) Find the error: The area inside the lemniscate $r^2 = a^2 \cos 2\theta$ is

$$A = \int_0^{2\pi} \frac{1}{2} r^2 \, d\theta = \int_0^{2\pi} \frac{1}{2} a^2 \cos 2\theta \, d\theta$$

$$= \frac{1}{4} a^2 \sin 2\theta \Big]_0^{2\pi} = 0$$

(b) Find the correct area.

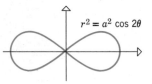

19. Find the area inside the lemniscate $r^2 = 4 \cos 2\theta$ and outside the circle $r = \sqrt{2}$.

20. A radial line is drawn from the origin to the spiral $r = a\theta$ ($a > 0$ and $\theta \geq 0$). Find the area swept out during the second revolution of the radial line that was not swept out during the first revolution.

13.4 PARAMETRIC EQUATIONS

In order to study a curve using calculus, we need some way to represent the curve mathematically. Often, the representation is an equation $y = f(x)$ whose graph is the given curve. However, graphs of functions are specialized curves in the sense that no vertical line can cut such a curve more than once. In order to represent curves that are not graphs of functions, different methods are required. In this section we discuss one of these methods.

Curves that are not graphs of functions can often be specified by using a pair of equations

$$x = x(t)$$
$$y = y(t)$$

to express the coordinates of a point (x, y) on the curve as functions of an auxiliary variable t. These are called ***parametric equations*** for the curve, and the variable t is called a ***parameter.***

Parametric equations arise naturally if one imagines a plane curve C to be traced by a moving point. If we use the parameter t to denote time, then the parametric equations $x = x(t), y = y(t)$ specify how the x- and y-coordinates of the moving point vary with time.

▶ Example 1

A particle moves in the plane so its x- and y-coordinates vary with time according to the equations

$$x = \tfrac{1}{2}t^3 - 6t$$
$$y = \tfrac{1}{2}t^2$$

Sketch the path of the particle over the time interval $0 \le t \le 4$.

Solution. The path of the particle, shown in Figure 13.4.1, was obtained by calculating the x- and y-coordinates of the particle for $t = 0, 1, 2, 3, 4$ (Table 13.4.1), plotting the points (x, y), and connecting successive points with a smooth curve. The arrows on the curve serve to indicate the direction of motion. ◀

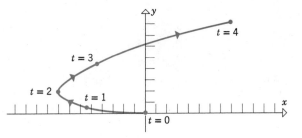

Figure 13.4.1

Table 13.4.1

t	0	1	2	3	4
$x = \tfrac{1}{2}t^3 - 6t$	0	$-\tfrac{11}{2}$	-8	$-\tfrac{9}{2}$	8
$y = \tfrac{1}{2}t^2$	0	$\tfrac{1}{2}$	2	$\tfrac{9}{2}$	8
(x, y)	$(0, 0)$	$(-\tfrac{11}{2}, \tfrac{1}{2})$	$(-8, 2)$	$(-\tfrac{9}{2}, \tfrac{9}{2})$	$(8, 8)$

Although time is a common parameter, it is not the only possibility. Other parameters are illustrated in the following examples.

▶ Example 2

Let C be a circle of radius a centered at the origin. If $P(x, y)$ is any point on the circle and θ is the angle measured counterclockwise from the positive x-axis to the line segment OP (Figure 13.4.2a), then

$$x = a \cos \theta \qquad \text{and} \qquad y = a \sin \theta$$

As θ varies from 0 to 2π, the point $P(x, y)$ makes one complete revolution. Thus

$$\begin{array}{l} x = a \cos \theta \\ y = a \sin \theta \end{array}, \; 0 \le \theta \le 2\pi$$

are parametric equations for the circle.

By changing the interval over which θ is allowed to vary, we can obtain different portions of the circle. For example

$$\begin{array}{l} x = a \cos \theta \\ y = a \sin \theta \end{array}, \; 0 \le \theta \le \pi$$

represents just the upper half of the circle (Figure 13.4.2b). ◀

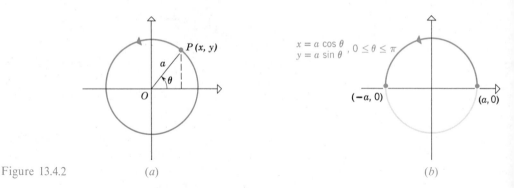

Figure 13.4.2 (a) (b)

A curve, which is represented parametrically, is traced in a certain direction as the parameter increases. This is called the **direction of increasing parameter** or sometimes the **orientation** of the curve. For example the circle

$$\begin{array}{l} x = a \cos \theta \\ y = a \sin \theta \end{array}, \; 0 \le \theta \le 2\pi$$

is traced counterclockwise as θ increases from 0 to 2π, so this is the direction of increasing parameter. Phrased another way, the circle is oriented counterclockwise. We have indicated this in Figure 13.4.2a with an arrow on the circle.

Sometimes a curve given parametrically may be recognized by eliminating the parameter.

► Example 3

Sketch the curve

$$x = 2t - 3$$
$$y = 6t - 7$$

Solution. Solving each equation for t, we obtain

$$t = \frac{x + 3}{2}$$

$$t = \frac{y + 7}{6}$$

Thus

$$\frac{x + 3}{2} = \frac{y + 7}{6}$$

or

$$y = 3x + 2$$

which is the straight line shown in Figure 13.4.3. The direction of increasing parameter, shown in the figure, was obtained from the original parametric equations by observing that x and y both increase as t increases. ◄

Figure 13.4.3

REMARK. In the last example, there were no conditions imposed on the parameter t. When no conditions are stated explicitly, it is understood that the parameter varies over the interval $(-\infty, +\infty)$.

► Example 4

Sketch the curve

$$\begin{aligned} x &= a \cos t \\ y &= b \sin t \end{aligned} \quad , \quad 0 \le t \le 2\pi$$

where $a > 0$ and $b > 0$.

Solution. We can eliminate t by writing

$$\frac{x}{a} = \cos t \qquad \text{and} \qquad \frac{y}{b} = \sin t$$

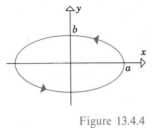

Figure 13.4.4

from which it follows that

$$\frac{x^2}{a^2} + \frac{y^2}{b^2} = 1$$

This is the equation of the ellipse shown in Figure 13.4.4. We leave it for the reader to show that the ellipse has counterclockwise orientation. ◀

▶ Example 5

Let C be the parabola

$$y = x^2$$

and let $m = dy/dx$ be the slope of the tangent to C at the point (x, y). Find parametric equations for C using m as a parameter.

Solution. Our objective is to express the coordinates (x, y) of an arbitrary point on the parabola in terms of the parameter m. But

$$\frac{dy}{dx} = 2x$$

so

$$x = \frac{1}{2}\frac{dy}{dx} = \frac{1}{2}m$$

and

$$y = x^2 = \left(\frac{1}{2}m\right)^2 = \frac{1}{4}m^2$$

which leads to the parametric equations

$$x = \frac{1}{2}m$$

$$y = \frac{1}{4}m^2$$

$x = \frac{1}{2}m$
$y = \frac{1}{4}m^2$

Figure 13.4.5

The direction of increasing parameter, indicated in Figure 13.4.5, is obtained by noting that x increases as the slope m increases, since $x = \frac{1}{2}m$. ◀

▶ Example 6

If $r = f(\theta)$ is a polar curve, we can obtain parametric equations for the curve in terms of θ by substituting $r = f(\theta)$ in the relationships

$$x = r \cos \theta$$
$$y = r \sin \theta$$

This yields

$$x = f(\theta) \cos \theta$$
$$y = f(\theta) \sin \theta$$

For example, the cardioid $r = 1 - \cos \theta$ may be represented by the parametric equations

$$x = (1 - \cos \theta) \cos \theta$$
$$y = (1 - \cos \theta) \sin \theta$$

$x = (1 - \cos \theta) \cos \theta$
$y = (1 - \cos \theta) \sin \theta$
$0 \leq \theta \leq 2\pi$

Figure 13.4.6

If we impose the added restriction, $0 \leq \theta \leq 2\pi$, then the point (x, y) will traverse the cardioid exactly once as θ varies over $[0, 2\pi]$ (Figure 13.4.6). ◀

If the parameter is eliminated from a pair of parametric equations, the graph of the resulting equation in x and y may possibly extend beyond the graph of the original parametric equations. For example, consider the curve with parametric equations

$$x = \cosh t$$
$$y = \sinh t \tag{1}$$

We may eliminate the parameter by using the identity

$$\cosh^2 t - \sinh^2 t = 1$$

(Formula (1) of Section 7.6) to obtain

$$x^2 - y^2 = 1 \tag{2}$$

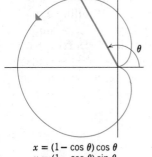

$x^2 - y^2 = 1$

Figure 13.4.7

which is the equation of the hyperbola shown in Figure 13.4.7. However, the original parametric equations represent only the right branch of this hyperbola since $x = \cosh t$ is never less than 1. To exclude the unwanted left branch we would have to replace (2) by

$$x^2 - y^2 = 1, \qquad x \geq 1$$

or more simply

$$x = \sqrt{1 + y^2}$$

To summarize, whenever the parameter is eliminated from parametric equations, it is important to check that the graph of the resulting equation does not include too much. If it does, added conditions must be imposed.

While parametric equations are of special importance for curves that are not graphs of functions, curves of the form $y = f(x)$ or $x = g(y)$ may also be represented parametrically. This can always be done by introducing a parameter t that is equal to the independent variable.

▶ Example 7

The curve $y = 4x^2 - 1$ may be represented parametrically as

$$x = t$$
$$y = 4t^2 - 1$$

and the curve $x = 5y^3 - y$ may be represented parametrically as

$$x = 5t^3 - t$$
$$y = t$$

◀

Sometimes it is difficult to eliminate the parameter from a pair of parametric equations and express y directly in terms of x. Nevertheless, by using the chain rule, it is possible to find the derivative of y with respect to x without eliminating the parameter. We start with the relationship

$$\frac{dy}{dt} = \frac{dy}{dx}\frac{dx}{dt}$$

Provided $dx/dt \neq 0$, this can be rewritten as

$$\frac{dy}{dx} = \frac{dy/dt}{dx/dt} \tag{3}$$

▶ Example 8

Given the curve

$$x = t^2$$
$$y = t^3$$

find dy/dx and d^2y/dx^2 at the point $(1, 1)$ without eliminating the parameter t.

Solution

$$\frac{dy}{dx} = \frac{dy/dt}{dx/dt} = \frac{3t^2}{2t} = \frac{3}{2}t \tag{4a}$$

To find the second derivative, denote dy/dx by y' and use (3) to write

$$\frac{d^2y}{dx^2} = \frac{dy'}{dx} = \frac{dy'/dt}{dx/dt} = \frac{3/2}{2t} = \frac{3}{4t} \tag{4b}$$

Note that these calculations express dy/dx and d^2y/dx^2 in terms of the parameter t. Since the point $(1, 1)$ on the curve corresponds to $t = 1$, it follows from (4a) and (4b) that

$$\frac{dy}{dx}\bigg|_{t=1} = \frac{3}{2}(1) = \frac{3}{2}$$

and

$$\frac{d^2y}{dx^2}\bigg|_{t=1} = \frac{3}{4}$$

The curve and the tangent at $(1, 1)$ are shown in Figure 13.4.8. Note that the tangent line has slope $\frac{3}{2}$ at $(1, 1)$ and the curve is concave up at $(1, 1)$, in keeping with our calculations. ◀

Figure 13.4.8

REMARK. Formula (3) does not apply if $dx/dt = 0$. However, at points where $dx/dt = 0$ and $dy/dt \neq 0$ there is generally a **vertical tangent** line. Points (x, y) where both $dx/dt = 0$ and $dy/dt = 0$ are called **singular points.** The behavior of a curve at a singular point must be determined on a case-by-case basis.

In Section 6.4 we obtained the following formula for the arc length L of a curve $y = f(x)$ from $x = a$ to $x = b$

$$L = \int_a^b \sqrt{1 + [f'(x)]^2}\, dx = \int_a^b \sqrt{1 + \left(\frac{dy}{dx}\right)^2}\, dx \tag{5}$$

This formula is a special case of the following general result about arc length of parametric curves

If $x'(t)$ and $y'(t)$ are continuous functions for $a \le t \le b$, then the curve

$$x = x(t) \atop y = y(t) \quad , \quad a \le t \le b$$

has arc length L given by

$$L = \int_a^b \sqrt{[x'(t)]^2 + [y'(t)]^2} \, dt \tag{6}$$

or equivalently

$$L = \int_a^b \sqrt{\left(\frac{dx}{dt}\right)^2 + \left(\frac{dy}{dt}\right)^2} \, dt \tag{7}$$

The derivation of these formulas is similar to the derivation of (5) and will be omitted. Observe that (5) follows from (7) by expressing the curve $y = f(x)$ in the parametric form

$$x = t \atop y = f(t)$$

in which case

$$\frac{dx}{dt} = 1 \quad \text{and} \quad \frac{dy}{dt} = f'(t) = f'(x) = \frac{dy}{dx}$$

▶ Example 9

Use (7) to find the arc length of the circle

$$x = a \cos t \atop y = a \sin t \quad , \quad 0 \le t \le 2\pi \quad (a > 0)$$

Solution.

$$L = \int_0^{2\pi} \sqrt{\left(\frac{dx}{dt}\right)^2 + \left(\frac{dy}{dt}\right)^2}$$

$$= \int_0^{2\pi} \sqrt{(-a \sin t)^2 + (a \cos t)^2} \, dt$$

$$= \int_0^{2\pi} a\, dt$$

$$= at \Big]_0^{2\pi}$$

$$= 2\pi a$$

◀

OPTIONAL

If a wheel rolls along a straight line without slipping, then a point on the rim of the wheel traces a curve called a *cycloid* (Figure 13.4.9).

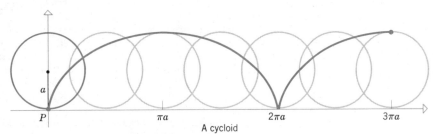

Figure 13.4.9

A cycloid

The cycloid is of special interest because it provides the solution to two famous mathematical problems: the ***brachistochrone problem*** (from Greek words meaning "shortest time") and the ***tautochrone problem*** (from Greek words meaning "equal time"). In June of 1696, Johann Bernoulli* posed the brachistochrone problem in the form of a challenge to other mathematicians. The problem was to determine the shape of a wire down which a bead might

---

*BERNOULLI. An amazing Swiss family that included several generations of outstanding mathematicians and scientists. The family fled from Antwerp to escape religious persecution and ultimately settled in Basel, Switzerland. There, Nikolaus Bernoulli (1623–1708), a druggist, had three sons Jakob I (also called Jacques or James), Nikolaus, and Johann I (also called Jean or John). (The roman numerals are used to distinguish family members with identical names.)

Nikolaus Bernoulli
(1623–1708)

Jakob I	Nikolaus	Johann I
(1654–1705)		(1667–1748)
(Jacques, James)		(Jean, John)

Nikolaus I	Nikolaus II	Daniel	Johann II
(1687–1759)	(1695–1726)	(1700–1782)	(1710–1790)

Johann III	Jakob II
(1744–1807)	(1759–1789)

Figure 13.4.10

slide from a point P to another point Q, not directly below, in the *shortest time*. At first, one might conjecture that the wire should form a straight line, since that shape yields the shortest distance from P to Q. However, the correct answer is half of one arch of an inverted cycloid (Figure 13.4.10). In essence, this shape allows the bead to fall rapidly at first, building up sufficient initial speed to reach Q in the shortest time, even though the path does not provide the shortest distance from P to Q. This problem was solved by Newton, Leibniz, L'Hôpital, Johann Bernoulli, and his older brother Jacob Bernoulli. It was formulated and incorrectly solved years earlier by Galileo who gave the arc of a circle as the answer.

Following Newton and Leibniz, the Bernoulli brothers, Jakob I and Johann I, are considered by some to be the two most important founders of calculus. Jakob I was self-taught in mathematics. His father wanted him to study for the ministry, but he turned to mathematics and in 1686 became a professor at the University of Basel. When he started working in mathematics, he knew nothing of Newton's and Leibniz' work. He eventually became familiar with Newton's results, but because so little of Leibniz' work was published, Jakob duplicated many of Leibniz' results.

Jakob's younger brother Johann I was urged to enter into business by his father. Instead, he turned to medicine and studied mathematics under the guidance of his older brother. He eventually became a mathematics professor at Gröningen in Holland and then, when Jakob died in 1705, Johann succeeded him as mathematics professor at Basel. Throughout their lives, Jakob I and Johann I had a mutual passion for criticizing each other's work, which frequently erupted into ugly confrontations. Leibniz tried to mediate the disputes, but Jakob, who resented Leibniz' superior intellect, accused him of siding with Johann and thus Leibniz became entangled in the arguments. The brothers often worked on common problems that they posed as challenges to one another. Johann, interested in gaining fame, often used unscrupulous means to make himself appear the originator of his brother's results; Jakob occasionally retaliated. Thus, it is often difficult to determine who deserves credit for many results. However, both men made major contributions to the development of calculus. In addition to his work on calculus, Jakob helped establish fundamental principles in probability including, the Law of Large Numbers, which is a cornerstone of modern probability theory. Johann was Euler's teacher at Basel and L'Hôpital's tutor in France.

Among the other members of the Bernoulli family, Daniel, son of Johann I, is the most famous. He was a professor of mathematics at St. Petersburg Academy in Russia and subsequently a professor of anatomy and then physics at Basel. He did work in calculus and probability, but is best known for his work in physics. A basic law of fluid flow, called Bernoulli's principle, is named in his honor. He won the annual prize of the French Academy 10 times for work on vibrating strings, tides of the sea, and kinetic theory of gases.

Johann II succeeded his father as professor of mathematics at Basel. His research was on the theory of heat and sound. Nikolaus I was a mathematician and law scholar who worked on probability and series. On the recommendation of Leibniz, he was appointed professor of mathematics at Padua and then went to Basel as a professor of logic and then law. Nikolaus II was professor of jurisprudence in Switzerland and then professor of mathematics at St. Petersburg Academy. Johann III was a professor of mathematics and astronomy in Berlin and Jakob II succeeded his uncle Daniel as professor of mathematics at St. Petersburg Academy in Russia. Truly an incredible family!

In the tautochrone problem the object is to find the shape of a wire from P to Q such that two beads started above Q at *any* points on the wire reach Q in the same amount of time. Again, the answer is half of one arch of a cycloid.

We conclude this section by deriving parametric equations for a cycloid.

▶ Example 10

Let P be a fixed point on the rim of a wheel of radius a. Suppose that the wheel rolls along the positive x-axis of a rectangular coordinate system, and that P is initially at the origin. When the wheel rolls a given distance, the

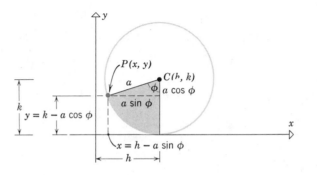

Figure 13.4.11

radial line from the center C to the point P rotates through an angle ϕ (Figure 13.4.11), which we will use as our parameter. (It is customary to regard ϕ as a positive angle, even though it is generated by a clockwise rotation.) Our objective is to express the coordinates of $P(x, y)$ in terms of the angle ϕ. Figure 13.4.11 suggests that the coordinates of $P(x, y)$ and the coordinates of the wheel's center $C(h, k)$ are related by

$$x = h - a \sin \phi$$
$$y = k - a \cos \phi \qquad (8)$$

Since the height of the wheel's center is the radius of the wheel, we have $k = a$; and since the distance h moved by the center is the same as the circular arc length subtended by ϕ (Why?), we have $h = a\phi$. Therefore, (8) yields

$$x = a\phi - a \sin \phi$$
$$y = a - a \cos \phi \qquad , \quad 0 \le \phi < +\infty$$

These are the parametric equations of the cycloid. ◀

► Exercise Set 13.4

1. (a) Sketch the curve $x = t$, $y = t^2$ by plotting the points corresponding to $t = -3, -2, -1, 0, 1, 2, 3$.

   (b) Show that the curve is a parabola by eliminating the parameter t.

2. Sketch the curve $x = t^3 - 4t$, $y = 2t$, $-3 \leq t \leq 3$ by plotting some appropriate points.

In Exercises 3–15, sketch the curve by eliminating the parameter t, and label the direction of increasing t.

3. $x = \cos t$, $y = \sin t$, $0 \leq t \leq 2\pi$.

4. $x = 1 + \cos t$, $y = 3 - \sin t$, $0 \leq t \leq 2\pi$.

5. $x = 3t - 4$, $y = 6t + 2$.

6. $x = t - 3$, $y = 3t - 7$, $0 \leq t \leq 3$.

7. $x = 2 \cos t$, $y = 5 \sin t$, $0 \leq t \leq 2\pi$.

8. $x = \sqrt{t}$, $y = 2t + 4$.

9. $x = 3 + 2 \cos t$, $y = 2 + 4 \sin t$, $0 \leq t \leq 2\pi$.

10. $x = 2 \cosh t$, $y = 4 \sinh t$.

11. $x = 4 \sin 2\pi t$, $y = 4 \cos 2\pi t$, $0 \leq t \leq 1$.

12. $x = \sec t$, $y = \tan t$, $\pi \leq t < \dfrac{3\pi}{2}$.

13. $x = \cos 2t$, $y = \sin t$, $0 \leq t \leq 2\pi$.

14. $x = 4t + 3$, $y = 16t^2 - 9$.

15. $x = t^2$, $y = 2 \ln t$, $t \geq 1$.

In Exercises 16–22, find dy/dx at the point corresponding to the given value of the parameter without eliminating the parameter.

16. $x = t^2 + 4$, $y = 8t$; $t = 2$.

17. $x = \cos t$, $y = \sin t$; $t = 3\pi/4$.

18. $x = t + 5$, $y = 5t - 7$; $t = 1$.

19. $x = \sqrt{t}$, $y = 2t + 4$; $t = 9$.

20. $x = \sec \theta$, $y = \tan \theta$; $\theta = \pi/3$.

21. $x = 4 \cos 2\pi s$, $y = 3 \sin 2\pi s$; $s = -1/4$.

22. $x = \sinh t$, $y = \cosh t$; $t = 0$.

In Exercises 23–26, find d^2y/dx^2 at the point corresponding to the given value of the parameter without eliminating the parameter.

23. $x = \frac{1}{2}t^2$, $y = \frac{1}{3}t^3$; $t = 2$.

24. $x = \cos \phi$, $y = \sin \phi$; $\phi = \pi/4$.

25. $x = \sqrt{t}$, $y = 2t + 4$; $t = 1$.

26. $x = \sec t$, $y = \tan t$; $t = \pi/3$.

In Exercises 27–35, find the arc length of the curve.

27. $x = 4t + 3$, $y = 3t - 2$, $0 \leq t \leq 2$.

28. $x = \cos^3 t$, $y = \sin^3 t$, $0 \leq t \leq \pi/2$.

29. $x = \frac{1}{3}t^3$, $y = \frac{1}{2}t^2$, $0 \leq t \leq 1$.

30. $x = \frac{1}{3}t^3$, $y = \frac{1}{2}t^2$, $-1 \leq t \leq 0$.

31. $x = \cos 2t$, $y = \sin 2t$, $0 \leq t \leq \pi/2$.

32. $x = e^t (\sin t + \cos t)$, $y = e^t (\cos t - \sin t)$ $1 \leq t \leq 4$.

33. $x = (1 + t)^2$, $y = (1 + t)^3$, $0 \leq t \leq 1$.

34. $x = e^t \cos t$, $y = e^t \sin t$, $0 \leq t \leq \pi/2$.

35. One arch of the cycloid
$x = a(t - \sin t)$, $y = a(1 - \cos t)$.

36. Find parametric equations for the rose $r = 2 \cos 2\theta$ using θ as the parameter.

37. Find parametric equations for the limaçon $r = 2 + 3 \sin \theta$ using θ as the parameter.

38. Find the equation of the tangent line to the curve $x = 2t + 4$, $y = 8t^2 - 2t + 4$ at the point where $t = 1$.

39. Find the equation of the tangent line to the curve $x = e^t$, $y = e^{-t}$ at the point where $t = 2$.

40. Find all values of t at which the curve $x = 2 \cos t$, $y = 4 \sin t$ has a tangent that is
    (a) horizontal   (b) vertical

41. Find all values of t at which the curve $x = 2t^3 - 15t^2 + 24t + 7$, $y = t^2 + t + 1$ has a tangent that is
    (a) horizontal   (b) vertical

42. Show that the curve $x = t^3 - 4t$, $y = t^2$ intersects itself at the point $(0, 4)$, and find equations for two tangent lines to the curve at the point of intersection.

43. Show that the curve $x = t^2 - 3t + 5$, $y = t^3 + t^2 - 10t + 9$ intersects itself at the point $(3, 1)$, and find equations for two tangent lines to the curve at the point of intersection.

44. A point traces the circle $x^2 + y^2 = 25$ so that $dx/dt = 8$ when the point reaches $(4, 3)$. Find dy/dt there.

45. Describe the curve whose parametric equations are $x = a \cos t + h$, $y = b \sin t + k$, $0 \le t \le 2\pi$.

46. A *hypocycloid* is a curve traced by a point P on the circumference of a circle that rolls inside a larger fixed circle. Suppose that the fixed circle has radius a, the rolling circle has radius b, and the fixed circle is centered at the origin. Let ϕ be the angle shown in the following figure, and assume that the point P is at $(a, 0)$ when $\phi = 0$. Show that the hypocycloid generated is given by the parametric equations

$$x = (a - b) \cos \phi + b \cos \left(\frac{a - b}{b} \phi \right)$$

$$y = (a - b) \sin \phi - b \sin \left(\frac{a - b}{b} \phi \right)$$

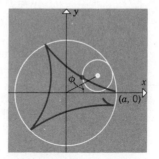

47. If $b = \frac{1}{4}a$ in Exercise 46, then the resulting curve is called a four-cusped hypocycloid.
 (a) Sketch this curve.
 (b) Show that the curve is given by the parametric equations $x = a \cos^3 \phi$, $y = a \sin^3 \phi$.
 (c) Show that curve is given by the equation $x^{2/3} + y^{2/3} = a^{2/3}$ in rectangular coordinates.

13.5 TANGENT LINES AND ARC LENGTH IN POLAR COORDINATES (OPTIONAL)

In this section we will use our results on parametric equations to derive formulas for arc length and slopes of tangent lines to polar curves.

Let

$$r = f(\theta)$$

be a curve in polar coordinates, for which f is a differentiable function of θ. We can obtain parametric equations for this curve in terms of θ by substituting $r = f(\theta)$ in the relationships

$$x = r \cos \theta$$
$$y = r \sin \theta$$

This yields the parametric equations

$$x = f(\theta) \cos \theta$$
$$y = f(\theta) \sin \theta \tag{1}$$

If we assume the existence of a tangent line to the curve $r = f(\theta)$ at the point $P(r, \theta)$, then the slope of this tangent line is

$$m = \tan \phi = \frac{dy}{dx}$$

where ϕ is the angle of inclination. From (1) we have

$$\frac{dx}{d\theta} = -f(\theta) \sin \theta + f'(\theta) \cos \theta$$
$$\frac{dy}{d\theta} = f(\theta) \cos \theta + f'(\theta) \sin \theta \tag{2}$$

Thus, if $dx/d\theta \neq 0$, it follows from Equation (3) of Section 13.4 that

$$m = \tan \phi = \frac{dy}{dx} = \frac{dy/d\theta}{dx/d\theta} \tag{3}$$

or

$$m = \tan \phi = \frac{f(\theta) \cos \theta + f'(\theta) \sin \theta}{-f(\theta) \sin \theta + f'(\theta) \cos \theta} = \frac{r \cos \theta + \sin \theta \dfrac{dr}{d\theta}}{-r \sin \theta + \cos \theta \dfrac{dr}{d\theta}} \tag{4}$$

which is a formula for the slope of the tangent line to $r = f(\theta)$ at $P(r, \theta)$.

Formula (4) was derived by assuming that $dx/d\theta \neq 0$. If $dx/d\theta = 0$ and $dy/d\theta \neq 0$, we will agree that the curve has a **vertical tangent.** This is reasonable since $\tan \phi$ becomes infinite in this case [see (3)], indicating that $\phi = \pi/2$. Points where both $dx/d\theta = 0$ and $dy/d\theta = 0$ are called **singular points.** No general statement can be made about the behavior of a polar curve at a singular point. Each case requires its own analysis.

If $\cos \theta \neq 0$, we may divide the numerator and denominator of (4) by $\cos \theta$ to obtain the following alternative formula for the slope of a tangent line.

$$m = \tan \phi = \frac{r + \tan \theta \dfrac{dr}{d\theta}}{-r \tan \theta + \dfrac{dr}{d\theta}} \tag{5}$$

▶ Example 1

Find the slope of the tangent line to the circle

$$r = 4 \cos \theta \tag{6}$$

at the point where $\theta = \pi/4$.

Solution.   Differentiating (6) yields

$$\frac{dr}{d\theta} = -4 \sin \theta \tag{7}$$

If $\theta = \pi/4$ it follows from (6) and (7) that

$$r = 4 \cos \frac{\pi}{4} = 2\sqrt{2} \quad \text{and} \quad \frac{dr}{d\theta} = -4 \sin \frac{\pi}{4} = -2\sqrt{2}$$

Substituting these values and $\tan \theta = \tan \frac{\pi}{4} = 1$ in (5) yields

$$\tan \phi = \frac{2\sqrt{2} + (1)(-2\sqrt{2})}{-(2\sqrt{2})(1) + (-2\sqrt{2})} = 0$$

Thus the circle has a horizontal tangent line when $\theta = \pi/4$ (Figure 13.5.1).
◀

Figure 13.5.1

▶ Example 2

At what points does the cardioid $r = 1 - \cos \theta$ have a vertical tangent line?

Solution.   A vertical tangent line will occur at points where $dx/d\theta = 0$ and $dy/d\theta \neq 0$. The cardioid is given parametrically by the equations

$$\begin{aligned} x &= r \cos \theta = (1 - \cos \theta) \cos \theta \\ y &= r \sin \theta = (1 - \cos \theta) \sin \theta \end{aligned} \quad , \quad 0 \leq \theta \leq 2\pi$$

Differentiating and then simplifying we obtain (verify)

$$\frac{dx}{d\theta} = \sin \theta (2 \cos \theta - 1) \tag{8}$$

$$\frac{dy}{d\theta} = (1 - \cos \theta)(1 + 2 \cos \theta) \tag{9}$$

$r = 1 - \cos \theta$

$\frac{5\pi}{3}$ $\frac{\pi}{3}$

Figure 13.5.2

From (8), $dx/d\theta = 0$ if $\sin \theta = 0$ or $\cos \theta = \frac{1}{2}$; this occurs where $\theta = 0$, π, $\pi/3$, $5\pi/3$, or 2π. But from (9), $dy/d\theta \neq 0$ if $\theta = \pi$, $\pi/3$, and $5\pi/3$, so vertical tangents occur at these points. Since $dy/d\theta = 0$ if $\theta = 0$ or $\theta = 2\pi$, these are singular points. Although we will not prove it, the cardioid has a horizontal tangent at the origin (Figure 13.5.2). ◀

Another way to find the tangent line to a polar curve at a point $P(r, \theta)$ is to find the angle ψ between the **radial line** OP and the tangent line (Figure 13.5.3). By definition, ψ is measured counterclockwise from the radial line OP to the tangent line, and is selected so

$$0 \leq \psi < \pi$$

If we choose the polar angle θ in the range $0 \leq \theta < 2\pi$, there is a simple algebraic relationship between θ, ϕ, and ψ that depends on the relative sizes of ϕ and θ. If $\phi \geq \theta$, then

$$\psi = \phi - \theta \tag{10}$$

(Figure 13.5.3a), and if $\phi < \theta$, then the relationship is

$$\psi = \phi + (\pi - \theta) = (\phi - \theta) + \pi \tag{11}$$

(Figure 13.5.3b). In either case we may write

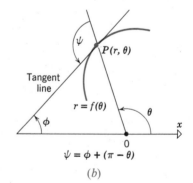

Figure 13.5.3

$r = f(\theta)$ ψ $P(r, \theta)$ Tangent line ϕ θ O x

$\psi = \phi - \theta$

(a)

ψ $P(r, \theta)$ Tangent line $r = f(\theta)$ ϕ θ 0 x

$\psi = \phi + (\pi - \theta)$

(b)

$$\tan \psi = \tan (\phi - \theta) \tag{12}$$

from which it follows that

$$\tan \psi = \frac{\tan \phi - \tan \theta}{1 + \tan \phi \tan \theta} \tag{13}$$

Substituting (5) into (13) yields

$$\tan \psi = \frac{\left[\left(r + \tan \theta \, \dfrac{dr}{d\theta}\right)\Big/\left(-r \tan \theta + \dfrac{dr}{d\theta}\right)\right] - \tan \theta}{1 + \left[\left(r + \tan \theta \, \dfrac{dr}{d\theta}\right)\Big/\left(-r \tan \theta + \dfrac{dr}{d\theta}\right)\right]\tan \theta}$$

$$= \frac{r + \tan \theta \, \dfrac{dr}{d\theta} + r \tan^2 \theta - \tan \theta \, \dfrac{dr}{d\theta}}{-r \tan \theta + \dfrac{dr}{d\theta} + r \tan \theta + \tan^2 \theta \, \dfrac{dr}{d\theta}}$$

$$= \frac{r(1 + \tan^2 \theta)}{(1 + \tan^2 \theta)\dfrac{dr}{d\theta}}$$

or

$$\tan \psi = \frac{r}{dr/d\theta} \tag{14}$$

▶ Example 3

For the cardioid $r = 1 - \cos \theta$, find the angle ψ between the tangent line and radial line at the points where the cardioid crosses the y-axis.

Solution: For the given cardioid we have,

$$\tan \psi = \frac{r}{dr/d\theta} = \frac{1 - \cos \theta}{\sin \theta}$$

so that from the trigonometric identity

$$\tan \frac{\theta}{2} = \frac{1 - \cos \theta}{\sin \theta}$$

(derive) it follows that

$$\tan \psi = \tan \frac{\theta}{2}$$

If $0 \le \theta < 2\pi$, this implies that

$$\psi = \frac{\theta}{2}$$

since $0 \leq \psi < \pi$. In particular, where the cardioid crosses the y-axis we have $\theta = \pi/2$ and $\theta = 3\pi/2$, so the angles from the radial lines to the tangent lines at these points are $\psi = \pi/4$ and $\psi = 3\pi/4$, respectively (Figure 13.5.4).

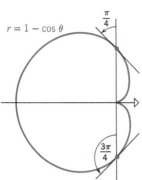

$r = 1 - \cos\theta$

$\dfrac{\pi}{4}$

$\dfrac{3\pi}{4}$

Figure 13.5.4

◀

In Section 13.2 we remarked informally that if a polar curve $r = f(\theta)$ passes through the origin where $\theta = \theta_0$, then the line $\theta = \theta_0$ is tangent to the curve at the origin. To see that this is so, we need only observe that in order for the curve $r = f(\theta)$ to pass through the origin when $\theta = \theta_0$, we must have $r = 0$ for this value of θ. Substituting $r = 0$ in (14) yields

$$\tan\psi = 0$$

(provided $dr/d\theta \neq 0$) which tells us that the angle ψ between the radial line $\theta = \theta_0$ and the tangent line is zero. Thus the tangent line coincides with the line $\theta = \theta_0$.

The arc length formula for a parametric curve can be used to obtain a formula for the arc length of a polar curve.

13.5.1 THEOREM

Arc Length of a Polar Curve

If a curve has the polar equation $r = f(\theta)$, where $f'(\theta)$ is continuous for $a \leq \theta \leq b$, then its arc length L from $\theta = a$ to $\theta = b$ is

$$L = \int_a^b \sqrt{[f(\theta)]^2 + [f'(\theta)]^2}\, d\theta \tag{15}$$

or equivalently

$$L = \int_a^b \sqrt{r^2 + \left(\frac{dr}{d\theta}\right)^2}\, d\theta \tag{16}$$

Proof. It follows from (2) that (verify):

$$\left(\frac{dx}{d\theta}\right)^2 + \left(\frac{dy}{d\theta}\right)^2 = [f(\theta)]^2 + [f'(\theta)]^2$$

Thus from (7) of Theorem 13.4.1 with θ in place of t, we obtain (15). ∎

▶ **Example 4**

Find the arc length of the spiral $r = e^\theta$ between $\theta = 0$ and $\theta = 1$.

Solution.

$$L = \int_a^b \sqrt{r^2 + \left(\frac{dr}{d\theta}\right)^2}\, d\theta = \int_0^1 \sqrt{(e^\theta)^2 + (e^\theta)^2}\, d\theta$$

$$= \int_0^1 \sqrt{2} e^{\theta} \, d\theta$$

$$= \sqrt{2} e^{\theta} \Big]_0^1$$

$$= \sqrt{2}(e - 1)$$ ◀

▶ Example 5

Find the total arc length of the cardioid $r = 1 + \cos \theta$.

Solution. The cardioid is traced out once as θ varies from $\theta = 0$ to $\theta = 2\pi$. Thus

$$L = \int_a^b \sqrt{r^2 + \left(\frac{dr}{d\theta}\right)^2} \, d\theta = \int_0^{2\pi} \sqrt{(1 + \cos \theta)^2 + (-\sin \theta)^2} \, d\theta$$

$$= \sqrt{2} \int_0^{2\pi} \sqrt{1 + \cos \theta} \, d\theta$$

Using the identity $1 + \cos \theta = 2 \cos^2 \tfrac{1}{2}\theta$ we obtain

$$L = 2 \int_0^{2\pi} \sqrt{\cos^2 \tfrac{1}{2}\theta} \, d\theta = 2 \int_0^{2\pi} |\cos \tfrac{1}{2}\theta| \, d\theta \qquad (17)$$

Since

$$\cos \tfrac{1}{2}\theta \geq 0 \qquad \text{when} \qquad 0 \leq \theta \leq \pi$$

and

$$\cos \tfrac{1}{2}\theta \leq 0 \qquad \text{when} \qquad \pi \leq \theta \leq 2\pi$$

we may rewrite (17) as

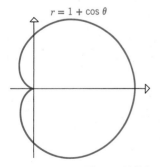

$r = 1 + \cos \theta$

$$L = 2 \left[\int_0^{\pi} \cos \tfrac{1}{2}\theta \, d\theta - \int_{\pi}^{2\pi} \cos \tfrac{1}{2}\theta \, d\theta \right]$$

$$= 4 \sin \tfrac{1}{2}\theta \Big]_0^{\pi} - 4 \sin \tfrac{1}{2}\theta \Big]_{\pi}^{2\pi}$$

$$= 8$$

Alternate Solution. Since the cardioid is symmetric about the x-axis (Figure 13.5.5), the entire arc length is twice the arc length from $\theta = 0$ to $\theta = \pi$. Thus, proceeding as above, we obtain

Figure 13.5.5

$$L = 2\sqrt{2} \int_0^{\pi} \sqrt{1 + \cos \theta} \, d\theta = 4 \int_0^{\pi} \cos \tfrac{1}{2}\theta \, d\theta = 8$$ ◀

▶ Exercise Set 13.5

In Exercises 1–6, find the slope of the tangent to the curve at the point with the given value of θ.

1. $r = 2 \cos \theta$; $\theta = \pi/3$.

2. $r = 1 + \sin \theta$; $\theta = \pi/4$.

3. $r = 1/\theta$; $\theta = 2$.

4. $r = a \sec 2\theta$; $\theta = \pi/6$.

5. $r = \cos 3\theta$; $\theta = 3\pi/4$.

6. $r = 4 - 3 \sin \theta$; $\theta = \pi$.

In Exercises 7–12, find $\tan \psi$ at the point on the curve with the given value of θ.

7. $r = 3 (1 - \cos \theta)$; $\theta = \pi/2$.

8. $r = 2\theta$; $\theta = 1$.

9. $r = 5 \sin \theta$; $\theta = \pi$.

10. $r = \tan \theta$; $\theta = 3\pi/4$.

11. $r = 4 \sin^2 \theta$; $\theta = 5\pi/6$.

12. $r = \sin \theta /(1 + \cos \theta)$; $\theta = \pi/3$.

In Exercises 13–19, find the arc length of the curve.

13. $r = e^{3\theta}$ from $\theta = 0$ to $\theta = 2$.

14. The entire circle $r = a$.

15. The entire circle $r = 2a \cos \theta$.

16. $r = \sin^2 (\theta/2)$ from $\theta = 0$ to $\theta = \pi$.

17. $r = a\theta^2$ from $\theta = 0$ to $\theta = \pi$.

18. $r = \sin^3 (\theta/3)$ from $\theta = 0$ to $\theta = \pi/2$.

19. The entire cardioid $r = a (1 - \cos \theta)$.
    [*Hint:* See Example 5.]

20. Find all points on the cardioid $r = a (1 + \cos \theta)$ where the tangent is
    (a) horizontal   (b) vertical.

21. Find all points on the limaçon $r = 1 - 2 \sin \theta$ where the tangent is horizontal.

22. Prove: If the polar curves $r = f_1(\theta)$ and $r = f_2(\theta)$ intersect at a point P, and if β is the smallest nonnegative angle between the tangent lines at P, then

$$\tan \beta = \left| \frac{\tan \psi_2 - \tan \psi_1}{1 + \tan \psi_1 \tan \psi_2} \right|$$

where ψ_1 and ψ_2 are evaluated at P.

23. Show that the curves $r = \sin 2\theta$ and $r = \cos \theta$ intersect at $(\sqrt{3}/2, \pi/6)$ and use the result of Exercise 22 to find the smallest nonnegative angle between their tangent lines at the point of intersection.

24. Find all points of intersection of the curves $r = \cos \theta$ and $r = 1 - \cos \theta$, and use the result of Exercise 22 to find the smallest nonnegative angle between the tangent lines at each point of intersection.

25. Prove that ψ is the same at each point of the curve $r = e^{a\theta}$.

26. Prove: At all points of intersection of the cardioids $r = a(1 + \cos \theta)$ and $r = b(1 - \cos \theta)$ (excluding the origin), the tangent lines are perpendicular.

▶ REVIEW EXERCISES

Find the rectangular coordinates of the points with the given polar coordinates.

1. (a) $(-2, 4\pi/3)$ (b) $(2, -\pi/2)$
   (c) $(0, -\pi)$ (d) $(-\sqrt{2}, -\pi/4)$
   (e) $(3, \pi)$ (f) $(1, \tan^{-1}(-\tfrac{4}{3}))$.

2. In parts (a)–(c), points are given in rectangular coordinates. Express them in polar coordinates in three ways:
   (i) with $r \geq 0$ and $0 \leq \theta < 2\pi$;
   (ii) with $r \geq 0$ and $-\pi < \theta \leq \pi$;
   (iii) with $r \leq 0$ and $0 \leq \theta < 2\pi$.
   (a) $(-\sqrt{3}, -1)$ (b) $(-3, 0)$
   (c) $(1, -1)$.

3. Sketch the region in polar coordinates determined by the given inequalities.
   (a) $1 \leq r \leq 2$, $\cos \theta \leq 0$
   (b) $-1 \leq r \leq 1$, $\pi/4 \leq \theta \leq \pi/2$.

In Exercises 4–11, identify the curve by transforming to rectangular coordinates.

4. $r = 2/(1 - \cos \theta)$.

5. $r^2 \sin (2\theta) = 1$.

6. $r = \pi/2$.

7. $r = -4 \csc \theta$.

8. $r = 6/(3 - \sin \theta)$.

9. $\theta = \pi/3$.

10. $r = 2 \sin \theta + 3 \cos \theta$.

11. $r = 0$.

In Exercises 12–15, express the given equation in polar coordinates.

12. $x^2 + y^2 = kx$.

13. $x = -3$.

14. $y^2 = 4x$.

15. $y = 3x$.

In Exercises 16–23, sketch the curve in polar coordinates.

16. $r = -4 \sin 3\theta$.

17. $r = -1 - 2 \cos \theta$.

18. $r = 5 \cos \theta$.

19. $r = 4 - \sin \theta$.

20. $r = 3(\cos \theta - 1)$.

21. $r = \theta/\pi$ $(\theta \geq 0)$.

22. $r = \sqrt{2} \cos (\theta/2)$.

23. $r = e^{-\theta/\pi}$ $(\theta \geq 0)$.

In Exercises 24–26, sketch the curves in the same polar coordinate system, and find all points of intersection.

24. $r = 3 \cos \theta$, $r = 1 + \cos \theta$.

25. $r = a \cos (2\theta)$, $r = a/2$ $(a > 0)$.

26. $r = 2 \sin \theta$, $r = 2 + 2 \cos \theta$.

In Exercises 27–29, set up, but *do not evaluate*, definite integrals for the stated area and arc length.

27. (a) The area inside both the circle and cardioid in Exercise 24.

(b) The arc length of that part of the cardioid outside the circle in Exercise 24.

28. (a) The area inside the rose and outside the circle in Exercise 25.

(b) The arc length of that part of the rose lying inside the circle in Exercise 25.

29. (a) The area inside the circle and outside the cardioid in Exercise 26.

(b) The arc length of that portion of the circle lying inside the cardioid in Exercise 26.

In Exercises 30–33, find the area of the region described.

30. One leaf of the rose $r = a \sin 3\theta$.

31. The region outside the circle $r = a$ and inside the lemniscate $r^2 = 2a^2 \cos 2\theta$.

32. The region in part (a) of Exercise 28.

33. The region in part (a) of Exercise 29.

In Exercises 34–37:

(a) Sketch the curve and label the direction of increasing parameter.

(b) Use the parametric equations to find dy/dx, d^2y/dx^2, and the equation of the tangent line at the point on the curve corresponding to the parameter value t_0 (or θ_0).

34. $x = 3 - t^2$, $y = 2 + t$, $0 \leq t \leq 3$; $t_0 = 1$.

35. $x = 1 + 3 \cos \theta$, $y = -1 + 2 \sin \theta$, $0 \leq \theta \leq \pi$; $\theta_0 = \pi/2$.

36. $x = 2 \tan \theta$, $y = \sec \theta$, $-\pi/2 < \theta < \pi/2$; $\theta_0 = \pi/3$.

37. $x = 1/t$, $y = \ln t$, $1 \leq t \leq e$; $t_0 = 2$.

In Exercises 38–43, find the arc length of the curve described.

38. $x = 2t^3$, $y = 3t^2$, $-4 \leq t \leq 4$.

39. $x = \ln \cos 2t$, $y = 2t$, $0 \leq t \leq \pi/6$.

40. $x = 3 \cos t - 1$, $y = 3 \sin t + 4$, $0 \leq t \leq \pi$.

41. $x = 3t^2$, $y = t^3 - 3t$, $0 \leq t \leq 1$.

42. $x = 1 - \cos t$, $y = t - \sin t$, $-\pi \leq t \leq \pi$.

43. $r = e^\theta$, $0 \leq \theta \leq 2\pi$.

In Exercises 44–46, $x(t)$ and $y(t)$ describe the motion of a particle. Find the coordinates of the particle when the instantaneous direction of motion is: (a) horizontal (b) vertical.

44. $x = -2t^2$, $y = t^3 - 3t + 5$.

45. $x = 1 - 2 \sin t$, $y = t + 2 \cos t$, $0 \leq t \leq \pi$.

46. $x = \ln t$, $y = t^2 - 4t$, $t > 0$.

47. At what instant does the trajectory described in Exercise 46 have a point of inflection?

48. Find a set of parametric equations for
(a) the line $y = 2x + 3$
(b) the ellipse $4(x - 2)^2 + y^2 = 4$.

Exercises 49–51 refer to the optional section 13.5.

49. For the cardioid $r = 2(1 + \cos \theta)$, find the slope of the tangent line and the angle ψ when $\theta = \pi/2$.

50. For the circle $r = 4 \sin \theta$, show that $\psi = \theta$ if $0 \leq \theta < \pi$ and find a formula for the inclination angle ϕ as a function of θ for $0 \leq \theta < \pi/2$.

appendix 1 trigonometry review

UNIT I

The purpose of this section and the next is to review the fundamentals of trigonometry.

We shall be concerned with angles in the plane that are generated by rotating a ray (or half-line) about its endpoint. The starting position of the ray is called the ***initial side*** of the angle and the final position of the ray is called the ***terminal side.*** The initial and terminal sides meet at a point called the ***vertex*** of the angle (Figure A.1).

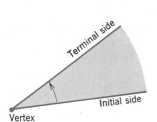

Figure A.1

An angle is considered ***positive*** if it is generated by a counterclockwise rotation and ***negative*** if it is generated by a clockwise rotation. In a rectangular coordinate system, an angle is said to be in ***standard position*** if its vertex is at the origin and its initial side is along the positive *x*-axis (Figure A.2).

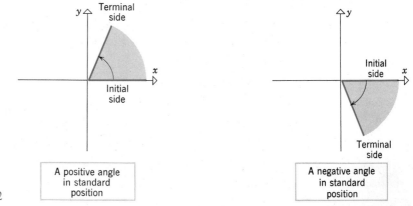

Figure A.2

A positive angle in standard position

A negative angle in standard position

Figure A.3

As shown in Figure A.3, an angle may be generated by making more than one complete revolution.

The *sine, cosine, tangent, cosecant, secant,* and *cotangent* of a positive acute angle θ can be defined as ratios of the sides of a right triangle. Using the notation from Figure A.4, these definitions take the form:

$$\sin \theta = \frac{\text{side opposite } \theta}{\text{hypotenuse}} = \frac{y}{r} \tag{1}$$

$$\cos \theta = \frac{\text{side adjacent to } \theta}{\text{hypotenuse}} = \frac{x}{r} \tag{2}$$

$$\tan \theta = \frac{\text{side opposite } \theta}{\text{side adjacent to } \theta} = \frac{y}{x} \tag{3}$$

$$\csc \theta = \frac{\text{hypotenuse}}{\text{side oppposite } \theta} = \frac{r}{y} \tag{4}$$

$$\sec \theta = \frac{\text{hypotenuse}}{\text{side adjacent to } \theta} = \frac{r}{x} \tag{5}$$

$$\cot \theta = \frac{\text{side adjacent to } \theta}{\text{side opposite to } \theta} = \frac{x}{y} \tag{6}$$

We will call sin, cos, tan, csc, sec, and cot the *trigonometric functions.* Because similar triangles have proportional sides, the values of the trigonometric functions depend only on the size of θ and not on the particular right triangle used to compute the ratios.

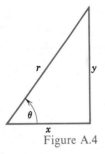

Figure A.4

▶ Example 1

Recall from high school geometry that the two legs of a $45°$–$45°$–$90°$ triangle are of equal size. This fact and the Theorem of Pythagoras yield Figure A.5a. Recall also that the hypotenuse of a $30°$–$60°$–$90°$ triangle is twice the size of the shorter leg and the shorter leg is opposite the $30°$ angle. These facts and the Theorem of Pythagoras yield Figure A.5b.

Figure A.5   (a)   (b)

From Figure A.5 we obtain

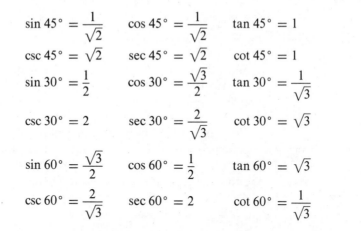

$$\sin 45° = \frac{1}{\sqrt{2}} \qquad \cos 45° = \frac{1}{\sqrt{2}} \qquad \tan 45° = 1$$

$$\csc 45° = \sqrt{2} \qquad \sec 45° = \sqrt{2} \qquad \cot 45° = 1$$

$$\sin 30° = \frac{1}{2} \qquad \cos 30° = \frac{\sqrt{3}}{2} \qquad \tan 30° = \frac{1}{\sqrt{3}}$$

$$\csc 30° = 2 \qquad \sec 30° = \frac{2}{\sqrt{3}} \qquad \cot 30° = \sqrt{3}$$

$$\sin 60° = \frac{\sqrt{3}}{2} \qquad \cos 60° = \frac{1}{2} \qquad \tan 60° = \sqrt{3}$$

$$\csc 60° = \frac{2}{\sqrt{3}} \qquad \sec 60° = 2 \qquad \cot 60° = \frac{1}{\sqrt{3}} \qquad \blacktriangleleft$$

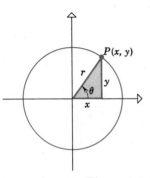

Figure A.6

Since a right triangle cannot have an angle larger than 90°, formulas (1)–(6) are not applicable if θ is obtuse. To obtain a definition that applies to all angles, we take the following approach: Given an angle θ, introduce a coordinate system so that the angle is in standard position. Then construct a circle of arbitrary radius r centered at the origin and let $P(x, y)$ be the point where the terminal side of the angle intersects the circle (Figure A.6). We make the following definition:

DEFINITION 1

$$\sin \theta = \frac{y}{r}, \qquad \cos \theta = \frac{x}{r}, \qquad \tan \theta = \frac{y}{x}$$

$$\csc \theta = \frac{r}{y}, \qquad \sec \theta = \frac{r}{x}, \qquad \cot \theta = \frac{x}{y}$$

The formulas in Definition 1 agree with formulas (1)–(6). However, Definition 1 applies to all angles—positive, negative, acute, or obtuse. Note that $\tan \theta$ and $\sec \theta$ are undefined if the terminal side of θ is on the y-axis (since $x = 0$), while $\csc \theta$ and $\cot \theta$ are undefined if the terminal side of θ is on the x-axis (since $y = 0$).

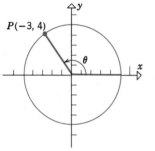

▶ Example 2

In Figure A.7,

$$x = -3, \qquad y = 4, \qquad \text{and} \qquad r = 5$$

Figure A.7   so that for the angle θ shown,

$$\sin\theta = \frac{4}{5} \qquad \cos\theta = -\frac{3}{5} \qquad \tan\theta = -\frac{4}{3}$$

$$\csc\theta = \frac{5}{4} \qquad \sec\theta = -\frac{5}{3} \qquad \cot\theta = -\frac{3}{4}$$ ◀

Since the value of a trigonometric function is determined by the point where the terminal side of the angle intersects a circle of radius r, it follows that the value of a trigonometric function will be the same for any two angles with the same terminal side (like those in Figure A.3.)

Using properties of similar triangles, it can be shown that the values of the trigonometric functions in Definition 1 are the same for every value r. In particular, for the *unit circle* (the circle of radius $r = 1$ centered at the origin) the formulas for $\sin\theta$ and $\cos\theta$ in Definition 1 become

$$x = \cos\theta, \qquad y = \sin\theta$$

Thus we have the following geometric interpretation of $\sin\theta$ and $\cos\theta$ (Figure A.8):

THEOREM 2
Geometric Interpretation of sin θ and cos θ

If an angle θ is placed in standard position, then its terminal side intersects the unit circle at the point $(\cos\theta, \sin\theta)$

The following useful theorem is suggested by Figure A.9. We omit the proof.

THEOREM 3 *If an angle θ is generated by a ray with endpoint (x_0, y_0) and initial side extending in the positive x-direction, then the intersection of the terminal side with the circle of radius r centered at (x_0, y_0) is*

$$(x_0 + r\cos\theta, y_0 + r\sin\theta)$$

Figure A.8

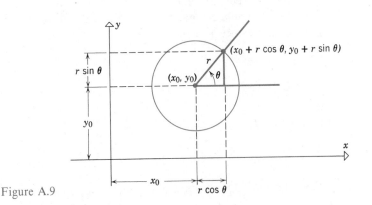

Figure A.9

RADIAN MEASURE In calculus, angles are measured in *radians* rather than degrees, minutes, and seconds because it simplifies many important formulas. Before we can define radian measure, we need some preliminary ideas.

Let Q be an arbitrary point on a circle of radius r and imagine that a particle, initially at Q, moves around the circle and stops at a point P (Figure A.10). During its motion the point will travel a distance d along the circle. To distinguish between clockwise and counterclockwise motions we introduce the **signed distance** or **signed arc length** s traveled by the point; it is defined by:

$$s = d \qquad \text{if the motion is counterclockwise,}$$
$$s = -d \qquad \text{if the motion is clockwise,}$$
$$s = 0 \qquad \text{if there is no motion}$$

Figure A.10

For example, if $s = 3$, the point has traveled 3 units in the counterclockwise direction from Q; and, if $s = -\pi$, the point has traveled π units in the clockwise direction from Q.

DEFINITION 4 To define the radian measure of an angle, we place the vertex of the angle at the center of a circle of arbitrary radius r. As a ray sweeps out the angle, the intersection of the ray with the circle moves some signed distance s along the circle (Figure A.11). We define

$$\theta = \frac{s}{r} \tag{7}$$

to be the **radian measure** of the angle.

Figure A.11

REMARK. We leave it as an exercise to show that the radian measure of an angle depends only on the size of the angle, and not on the radius r selected for the circle.

▶ Example 3

Express the angle 90° in radian measure.

Solution. An angle of 90° placed with its vertex at the center of a circle of radius r intercepts one fourth of the circumference (Figure A.12). Since the full circumference is $2\pi r$, we obtain

$$s = \tfrac{1}{4}(2\pi r) = \tfrac{1}{2}\pi r$$

Figure A.12 Thus the radian measure of the angle is

$$\theta = \frac{s}{r} = \frac{\frac{1}{2}\pi r}{r} = \frac{\pi}{2} \qquad \text{(radians)} \qquad \blacktriangleleft$$

▶ Example 4

From the fact that $90°$ corresponds to $\pi/2$ radians, the reader should be able to obtain the relationships in the following table

Degrees	30°	45°	60°	90°	120°	135°	150°	180°	270°	360°
Radians	$\frac{\pi}{6}$	$\frac{\pi}{4}$	$\frac{\pi}{3}$	$\frac{\pi}{2}$	$\frac{2\pi}{3}$	$\frac{3\pi}{4}$	$\frac{5\pi}{6}$	π	$\frac{3\pi}{2}$	2π

Some angles and their radian measure are shown in Figure A.13. ◀

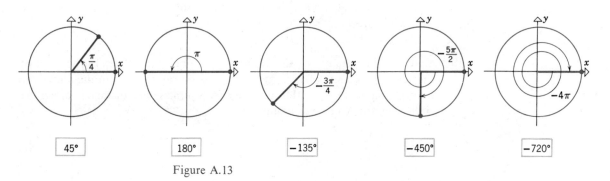

Figure A.13

The following table will be useful in the next example.

$$360° = 2\pi \text{ radians}$$
$$180° = \pi \text{ radians} \approx 3.14159 \ldots \text{ radians}$$
$$\left(\frac{180}{\pi}\right)° = 1 \text{ radian} \approx 57°\ 17'\ 44.8''$$
$$1° = \frac{\pi}{180} \text{ radian} \approx 0.01745 \text{ radian}$$

▶ Example 5

   (a) Express $146°$ in radians.

   (b) Express 3 radians in degrees.

Solution (a). Since

$$1° = \frac{\pi}{180} \text{ radians}$$

if follows that

$$146° = \frac{\pi}{180} \cdot 146 \text{ radians} = \frac{73\pi}{90} \text{ radians}$$

Solution (b). Since

$$1 \text{ radian} = \left(\frac{180}{\pi}\right)°$$

it follows that

$$3 \text{ radians} = \left(3 \cdot \frac{180}{\pi}\right)° = \left(\frac{540}{\pi}\right)° \approx 171.9° \qquad \blacktriangleleft$$

REMARK. By convention, an angle written without any units is understood to be measured in radians. Thus the statement $\theta = 3.2$ means $\theta = 3.2$ radians, *not* $\theta = 3.2°$.

▶ Example 6

Evaluate the trigonometric functions at $\theta = 5\pi/6$.

Solution. Construct a unit circle and place the angle $\theta = 5\pi/6 = 150°$ in standard position (Figure A.14).

Triangle OAP is a 30°–60°–90° triangle, so the leg opposite the 30° angle is half the length of the hypotenuse. Since the hypotenuse has length 1, the leg opposite the 30° angle has length 1/2; and by the Theorem of Pythagoras, the other leg has length $\sqrt{3}/2$. Thus, the terminal side of $\theta = 5\pi/6$ intersects the unit circle at $P(-\sqrt{3}/2, 1/2)$. From Definition 1 with

$$x = -\frac{\sqrt{3}}{2}, \qquad y = \frac{1}{2}, \qquad r = 1$$

we obtain

$$\sin\frac{5\pi}{6} = \frac{1}{2}, \qquad \cos\frac{5\pi}{6} = -\frac{\sqrt{3}}{2}, \qquad \tan\frac{5\pi}{6} = -\frac{1}{\sqrt{3}}$$

$$\csc\frac{5\pi}{6} = 2, \qquad \sec\frac{5\pi}{6} = -\frac{2}{\sqrt{3}}, \qquad \cot\frac{5\pi}{6} = -\sqrt{3} \qquad \blacktriangleleft$$

$P\left(-\frac{\sqrt{3}}{2}, \frac{1}{2}\right)$

Figure A.14

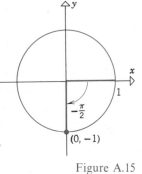

Figure A.15

▶ Example 7

Evaluate the trigonometric functions at $\theta = -\pi/2$.

Solution. Construct a unit circle and place the angle $\theta = -\pi/2 = -90°$ in standard position (Figure A.15). The terminal side of θ intersects the circle at $P(0, -1)$ so that from Definition 1 with

$$x = 0, \quad y = -1, \quad r = 1$$

we obtain

$$\sin\left(-\frac{\pi}{2}\right) = -1, \quad \cos\left(-\frac{\pi}{2}\right) = 0, \quad \tan\left(-\frac{\pi}{2}\right) = \text{undefined}$$

$$\csc\left(-\frac{\pi}{2}\right) = -1, \quad \sec\left(-\frac{\pi}{2}\right) = \text{undefined}, \quad \cot\left(-\frac{\pi}{2}\right) = 0 \quad ◀$$

Table 1 lists the values of the trigonometric functions for some frequently used angles. A more extensive table is given in Appendix 3. Dashes denote undefined quantities.

Table 1

	$\theta = 0$	$\frac{\pi}{6}$	$\frac{\pi}{4}$	$\frac{\pi}{3}$	$\frac{\pi}{2}$	$\frac{2\pi}{3}$	$\frac{3\pi}{4}$	$\frac{5\pi}{6}$	π	$\frac{3\pi}{2}$	2π
$\sin\theta$	0	$1/2$	$1/\sqrt{2}$	$\sqrt{3}/2$	1	$\sqrt{3}/2$	$1/\sqrt{2}$	$1/2$	0	-1	0
$\cos\theta$	1	$\sqrt{3}/2$	$1/\sqrt{2}$	$1/2$	0	$-1/2$	$-1/\sqrt{2}$	$-\sqrt{3}/2$	-1	0	1
$\tan\theta$	0	$1/\sqrt{3}$	1	$\sqrt{3}$	—	$-\sqrt{3}$	-1	$-1/\sqrt{3}$	0	—	0
$\csc\theta$	—	2	$\sqrt{2}$	$2/\sqrt{3}$	1	$2/\sqrt{3}$	$\sqrt{2}$	2	—	-1	—
$\sec\theta$	1	$2/\sqrt{3}$	$\sqrt{2}$	2	—	-2	$-\sqrt{2}$	$-2/\sqrt{3}$	-1	—	1
$\cot\theta$	—	$\sqrt{3}$	1	$1/\sqrt{3}$	0	$-1/\sqrt{3}$	-1	$-\sqrt{3}$	—	0	—

TRIGONOMETRIC IDENTITIES

It follows from Definition 1 that

$$\csc\theta = \frac{1}{\sin\theta}, \quad \sec\theta = \frac{1}{\cos\theta}, \quad \cot\theta = \frac{1}{\tan\theta} \tag{8a}$$

and

$$\tan\theta = \frac{\sin\theta}{\cos\theta}, \quad \cot\theta = \frac{\cos\theta}{\sin\theta} \tag{8b}$$

Moreover, from Figure A.6 and the Theorem of Pythagoras, we have

$$x^2 + y^2 = r^2$$

Dividing both sides by r^2 and using the definitions of $\sin\theta$ and $\cos\theta$, we obtain the following fundamental result

$$\sin^2\theta + \cos^2\theta = 1 \tag{9}$$

REMARK. The relationships in (8a), (8b) and (9) are *identities*, which means that the equalities hold for all values of θ where both sides are defined.

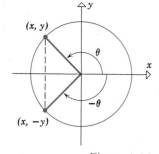

Figure A.16 shows two angles, θ and $-\theta$, having equal magnitude but opposite sign. By symmetry, the terminal sides of these angles intersect a circle centered at the origin at points whose x-coordinates are equal and whose y-coordinates differ only in sign. Thus

$$\sin(-\theta) = \frac{-y}{r} = -\frac{y}{r} = -\sin\theta$$

$$\cos(-\theta) = \frac{x}{r} = \cos\theta$$

Figure A.16 In summary, we have the identities

$$\sin(-\theta) = -\sin\theta \tag{10a}$$
$$\cos(-\theta) = \cos\theta \tag{10b}$$

As previously noted, a trigonometric function has the same value for two angles with the same terminal side. In radian measure, the angles θ, $\theta + 2\pi$, and $\theta - 2\pi$ all have the same terminal side, so

$$\sin\theta = \sin(\theta + 2\pi) = \sin(\theta - 2\pi)$$
$$\cos\theta = \cos(\theta + 2\pi) = \cos(\theta - 2\pi) \tag{11}$$

More generally, any multiple of 2π can be added to or subtracted from an angle θ in radians without changing the terminal side. Thus

$$\sin\theta = \sin(\theta \pm 2n\pi), \qquad n = 0, 1, 2, \ldots$$
$$\cos\theta = \cos(\theta \pm 2n\pi), \qquad n = 0, 1, 2, \ldots \tag{12}$$

The following result, called the **law of cosines,** generalizes the Theorem of Pythagoras:

THEOREM 5 *If the sides of a triangle have lengths a, b, and c, and if θ is the angle between*
Law of Cosines *the sides with lengths a and b, then*

$$c^2 = a^2 + b^2 - 2ab \cos \theta$$

Proof. Introduce a coordinate system so that θ is in standard position and the side of length a falls along the positive x-axis (Figure A.17).

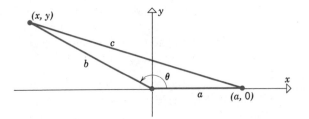

Figure A.17

As shown in Figure A.17, the side of length a extends from the origin to $(a, 0)$ and the side of length b extends from the origin to some point (x, y). From the definition of $\sin \theta$ and $\cos \theta$ we have

$$\sin \theta = \frac{y}{b}, \qquad \cos \theta = \frac{x}{b}$$

so

$$y = b \sin \theta, \qquad x = b \cos \theta \tag{13}$$

From the distance formula in Theorem 1.3.1 we obtain

$$c^2 = (x - a)^2 + (y - 0)^2$$

so that, from (13),

$$\begin{aligned} c^2 &= (b \cos \theta - a)^2 + b^2 \sin^2 \theta \\ &= a^2 + b^2 (\cos^2 \theta + \sin^2 \theta) - 2ab \cos \theta \\ &= a^2 + b^2 - 2ab \cos \theta \end{aligned}$$

which completes the proof. ▮

Using the law of cosines we will now be able to establish the following identities, called the **addition formulas** for sine and cosine.

$$\sin(\alpha - \beta) = \sin\alpha\cos\beta - \cos\alpha\sin\beta \qquad (14a)$$
$$\cos(\alpha - \beta) = \cos\alpha\cos\beta + \sin\alpha\sin\beta \qquad (14b)$$

$$\sin(\alpha + \beta) = \sin\alpha\cos\beta + \cos\alpha\sin\beta \qquad (15a)$$
$$\cos(\alpha + \beta) = \cos\alpha\cos\beta - \sin\alpha\sin\beta \qquad (15b)$$

Figure A.18

Identities (15a) and (15b) can be obtained by substituting $-\beta$ for β in (14a) and (14b) and using the identities

$$\sin(-\beta) = -\sin\beta, \quad \cos(-\beta) = \cos\beta$$

We shall derive (14b) and then deduce (14a) from (14b). In our derivation of (14b), we will assume that $0 \le \beta < \alpha < 2\pi$ (Figure A.18). The proofs for the remaining cases will be omitted.

As shown in Figure A.18, the terminal sides of α and β intersect the unit circle at the points $P_1(\cos\alpha, \sin\alpha)$ and $P_2(\cos\beta, \sin\beta)$. If we denote the lengths of the sides of triangle OP_1P_2 by OP_1, P_1P_2, and OP_2, then $OP_1 = OP_2 = 1$ and, from the distance formula in Theorem 1.3.1,

$$(P_1P_2)^2 = (\cos\beta - \cos\alpha)^2 + (\sin\beta - \sin\alpha)^2$$
$$= (\sin^2\alpha + \cos^2\alpha) + (\sin^2\beta + \cos^2\beta)$$
$$\qquad\qquad - 2(\cos\alpha\cos\beta + \sin\alpha\sin\beta)$$
$$= 2 - 2(\cos\alpha\cos\beta + \sin\alpha\sin\beta)$$

But angle $P_2OP_1 = \alpha - \beta$, so that the law of cosines yields

$$(P_1P_2)^2 = (OP_1)^2 + (OP_2)^2 - 2(OP_1)(OP_2)\cos(\alpha - \beta)$$
$$= 2 - 2\cos(\alpha - \beta).$$

Equating the two expressions for $(P_1P_2)^2$ and simplifying, we obtain

$$\cos(\alpha - \beta) = \cos\alpha\cos\beta + \sin\alpha\sin\beta$$

which completes the derivation of (14b).

To derive identity (14a), we will need the identities

$$\cos\left(\frac{\pi}{2} - \alpha\right) = \sin\alpha \qquad (16a)$$

$$\sin\left(\frac{\pi}{2} - \alpha\right) = \cos\alpha \qquad (16b)$$

which state that *the sine of an angle is the cosine of its complement* and *the cosine of an angle is the sine of its complement*. The derivations of (16a) and (16b) are left as exercises.

We can now derive (14a) as follows:

$$\begin{aligned}
\sin(\alpha - \beta) &= \cos\left[\frac{\pi}{2} - (\alpha - \beta)\right] \\
&= \cos\left[\left(\frac{\pi}{2} - \alpha\right) + \beta\right] \\
&= \cos\left[\left(\frac{\pi}{2} - \alpha\right) - (-\beta)\right] \\
&= \cos\left(\frac{\pi}{2} - \alpha\right)\cos(-\beta) + \sin\left(\frac{\pi}{2} - \alpha\right)\sin(-\beta) \\
&= \cos\left(\frac{\pi}{2} - \alpha\right)\cos\beta - \sin\left(\frac{\pi}{2} - \alpha\right)\sin\beta \\
&= \sin\alpha\cos\beta - \cos\alpha\sin\beta
\end{aligned}$$

In the special case where $\alpha = \beta$, (15a) and (15b) reduce to the identities

$$\sin 2\alpha = 2\sin\alpha\cos\alpha \qquad (17a)$$
$$\cos 2\alpha = \cos^2\alpha - \sin^2\alpha \qquad (17b)$$

These are called the **double-angle formulas.** By using the identity $\sin^2\alpha + \cos^2\alpha = 1$, (17b) can be rewritten in the alternate forms

$$\cos 2\alpha = 2\cos^2\alpha - 1 \qquad (17c)$$
$$\cos 2\alpha = 1 - 2\sin^2\alpha \qquad (17d)$$

If we replace α by $\alpha/2$ in (17c) and (17d) and use some algebra, we obtain the **half-angle formulas.**

$$\cos^2\frac{\alpha}{2} = \frac{1 + \cos\alpha}{2} \qquad (18a)$$

$$\sin^2\frac{\alpha}{2} = \frac{1 - \cos\alpha}{2} \qquad (18b)$$

Identities (14a) and (15a) and some algebra yield the first of the following *product formulas.*

$$\sin \alpha \cos \beta = \frac{1}{2} [\sin (\alpha - \beta) + \sin (\alpha + \beta)] \qquad (19\text{a})$$

$$\sin \alpha \sin \beta = \frac{1}{2} [\cos (\alpha - \beta) - \cos (\alpha + \beta)] \qquad (19\text{b})$$

$$\cos \alpha \cos \beta = \frac{1}{2} [\cos (\alpha - \beta) + \cos (\alpha + \beta)] \qquad (19\text{c})$$

The derivations of (19b) and (19c) are left as exercises.

We can deduce some important identities involving tangent, cotangent, secant, and cosecant from (8) and (9). The first of the identities

$$\tan^2 \theta + 1 = \sec^2 \theta \qquad (20\text{a})$$
$$1 + \cot^2 \theta = \csc^2 \theta \qquad (20\text{b})$$

can be obtained by dividing

$$\sin^2 \theta + \cos^2 \theta = 1$$

by $\cos^2 \theta$ and the second by dividing by $\sin^2 \theta$.

In the exercises we have asked the reader to deduce the first of the identities

$$\tan (\alpha + \beta) = \frac{\tan \alpha + \tan \beta}{1 - \tan \alpha \tan \beta} \qquad (21\text{a})$$

$$\tan (\alpha - \beta) = \frac{\tan \alpha - \tan \beta}{1 + \tan \alpha \tan \beta} \qquad (21\text{b})$$

from the relationship $\tan (\alpha + \beta) = \sin (\alpha + \beta)/\cos (\alpha + \beta)$. Identity (21b) is obtained from (21a) by substituting $-\beta$ for β and using the relationship

$$\tan (-\beta) = \frac{\sin (-\beta)}{\cos (-\beta)} = \frac{-\sin \beta}{\cos \beta} = -\tan \beta \qquad (22)$$

If we let $\alpha = \beta$ in (21a) we obtain the double-angle formula:

$$\tan 2\alpha = \frac{2 \tan \alpha}{1 - \tan^2 \alpha} \qquad (23)$$

If we let $\beta = \pi$ in (21a) and (21b) and use the fact that $\tan \pi = 0$ we obtain the identities

$$\tan (\alpha + \pi) = \tan \alpha \qquad (24a)$$

$$\tan (\alpha - \pi) = \tan \alpha \qquad (24b)$$

The trigonometric identities obtained in this section are the ones most commonly used. In later sections other identities will be derived as they are needed.

▶ Trigonometry Exercises I

1. Express the following angles in radians.

   (a) $270°$ (b) $390°$ (c) $20°$

   (d) $138°$ (e) $\left(\dfrac{\pi}{10}\right)°$.

2. Express the following angles in degrees.

   (a) $\dfrac{\pi}{10}$ (b) 3.2 (c) $\dfrac{\pi}{6}$ (d) $\dfrac{2\pi}{5}$

   (e) 2.

3. Sketch the following angles in standard position.

   (a) $\dfrac{\pi}{3}$ (b) $-\dfrac{5\pi}{6}$ (c) $\dfrac{9\pi}{4}$ (d) $-\pi$

   (e) 4π.

4. Evaluate $\sin \theta$, $\cos \theta$, $\tan \theta$, $\csc \theta$, $\sec \theta$, $\cot \theta$ for the following values of θ.

   (a) $\theta = \dfrac{\pi}{3}$ (b) $\theta = -\dfrac{5\pi}{6}$ (c) $\theta = \dfrac{9\pi}{4}$

   (d) $\theta = -\pi$ (e) $\theta = 4\pi$

5. Find all values of θ (in radians) such that
   (a) $\sin \theta = 0$ (b) $\cos \theta = 0$ (c) $\tan \theta = 0$

6. Find all values of θ (in radians) such that
   (a) $\tan \theta$ is undefined (b) $\csc \theta$ is undefined.

7. Find all values of θ (in radians) such that
   (a) $\cot \theta$ is undefined (b) $\sec \theta$ is undefined.

8. Find all values of θ (in radians) such that
   (a) $\sin \theta = 1$ (b) $\cos \theta = -\dfrac{1}{\sqrt{2}}$.

9. Find all values of θ (in radians) such that
   (a) $\sin \theta = -1$ (b) $\cos \theta = \dfrac{1}{2}$.

10. Find $\sin \theta$, $\cos \theta$, $\tan \theta$, $\csc \theta$, $\sec \theta$, $\cot \theta$ for the angles sketched in Figures A.19a and A.19b.

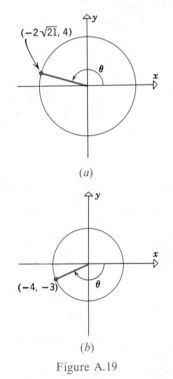

(a)

(b)

Figure A.19

11. Use identities and Table 1 to find the values of
    (a) $\cos 15°$ (b) $\sin 22.5°$ (c) $\sin 37.5°$.

12. Use identities and Table 1 to find the values of
    (a) $\sin 75°$ (b) $\tan 75°$.

13. (a) Let θ be an acute angle such that $\sin \theta = a/3$. Express the remaining trigonometric functions in terms of a by using the triangle in Figure A.20.

Figure A.20

   (b) Let θ be an acute angle such that $\tan \theta = a/5$. Construct an appropriate triangle, and use it to express the remaining trigonometric functions in terms of a.

   (c) Let θ be an acute angle such that $\sec \theta = a$. Construct an appropriate triangle, and use it to express the remaining trigonometric functions in terms of a.

14. How could you use a ruler and protractor to estimate $\sin 17°$ and $\cos 17°$?

15. Let ABC be a triangle, whose angles at A and B are $30°$ and $45°$. If the side opposite the angle B has length 9, find the lengths of the remaining sides and the size of the angle C.

Exercises 16–18 refer to an arbitrary triangle ABC in which side a is opposite angle A, side b is opposite angle B, and side c is opposite angle C.

16. Prove: The area of a triangle ABC can be written

$$\text{area} = \frac{1}{2}bc \sin A$$

Find two other similar formulas for the area.

17. Prove the **law of sines:** In any triangle, the sides are proportional to the sines of the opposite angles. That is,

$$\frac{a}{\sin A} = \frac{b}{\sin B} = \frac{c}{\sin C}$$

18. (a) Use Exercises 16 and 17 to prove that the area of a triangle ABC can be written

$$\text{area} = \frac{b^2 \sin A \sin C}{2 \sin B}$$

   (b) Find two other similar formulas for the area.

19. Prove: If θ is positive and measured in radians, then the area of a sector with angle θ and radius r is

$$\text{area} = \frac{1}{2}r^2\theta$$

20. Derive the identities

$$\cos\left(\frac{\pi}{2} - \alpha\right) = \sin \alpha$$

$$\sin\left(\frac{\pi}{2} - \alpha\right) = \cos \alpha$$

21. From the identities in Exercise 20, obtain the identities

$$\cos\left(\frac{\pi}{2} + \alpha\right) = -\sin \alpha$$

$$\sin\left(\frac{\pi}{2} + \alpha\right) = \cos \alpha$$

22. Derive identities (18a) and (18b).

23. Derive identities (19b) and (19c).

24. Derive identity (21a).

25. Express

$$\sin\left(\frac{3\pi}{2} - \theta\right) \quad \text{and} \quad \cos\left(\frac{3\pi}{2} + \theta\right)$$

in terms of $\sin \theta$ and $\cos \theta$.

In Exercises 26–37, verify the given identities.

26. $\tan \dfrac{\theta}{2} = \dfrac{1 - \cos \theta}{\sin \theta}$.

27. $\tan \dfrac{\theta}{2} = \dfrac{\sin \theta}{1 + \cos \theta}$.

28. $2 \csc 2\theta = \sec \theta \csc \theta$.

29. $\tan \theta + \cot \theta = 2 \csc 2\theta$.

30. $\sin 3\theta + \sin \theta = 2 \sin 2\theta \cos \theta$.

31. $\sin 3\theta - \sin \theta = 2 \cos 2\theta \sin \theta$.

32. $\dfrac{\cos \theta \sec \theta}{1 + \tan^2 \theta} = \cos^2 \theta$.

33. $\dfrac{\cos \theta \tan \theta + \sin \theta}{\tan \theta} = 2 \cos \theta$.

34. $\dfrac{\sin 2\theta}{\sin \theta} - \dfrac{\cos 2\theta}{\cos \theta} = \sec \theta.$

35. $\dfrac{\sin \theta + \cos 2\theta - 1}{\cos \theta - \sin 2\theta} = \tan \theta.$

36. $\cos \left(\dfrac{\pi}{3} + \theta\right) + \cos \left(\dfrac{\pi}{3} - \theta\right) = \cos \theta.$

37. $\sin \left(\dfrac{3\pi}{2} - \theta\right) + \sin \left(\dfrac{3\pi}{2} + \theta\right) = -2 \cos \theta.$

38. Beginning with identity $\sin^2 \theta + \cos^2 \theta = 1$, obtain the identity

$$1 + \cot^2 \theta = \csc^2 \theta$$

39. Express $\sin 3\theta$ and $\cos 3\theta$ in terms of $\sin \theta$ and $\cos \theta$.

40. (a) Express $3 \sin \alpha + 5 \cos \alpha$ in the form $C \sin (\alpha + \phi)$.
    (b) Show that a sum of the form $A \sin \alpha + B \cos \alpha$ can be rewritten in the form $C \sin (\alpha + \phi)$.

41. Show that the length of the diagonal of the parallelogram in Figure A.21 is

$$d = \sqrt{a^2 + b^2 + 2ab \cos \theta}$$

Figure A.21

42. Prove that the values of the trigonometric functions, as defined in Definition 1, do not depend on the value of r.

43. Prove that the ratio s/r in formula (7) depends only on the angle and not on the radius r selected for the circle. [*Hint:* Use the result from plane geometry that states that the arc subtended on a circle by a central angle is proportional to the angle.]

UNIT II

Recall that an angle written without any units is understood to be measured in radians. Thus the expression

$$\sin 3$$

means, the sine of 3 radians. More generally, for any real number x, the expression

$$\sin x \tag{1}$$

means, the sine of x radians (that is, the sine of an angle whose radian measure is x).

REMARK. The use of the letter x in expression (1) rather than θ, α, β, or some other symbol is dangerous but virtually unavoidable in many problems. The danger is that x had a different meaning in the definition of the trigonometric functions (Definition 1 of Unit I). While it is obviously not correct to give a symbol two different meanings in the *same* problem, there is nothing wrong with using a symbol differently in *different* problems as long as we clearly understand how it is being used.

The expression sin x is defined for every real number x so that

$$f(x) = \sin x$$

defines a function f with domain $(-\infty, +\infty)$. By definition, the graph of sin x is the graph of the equation

$$y = \sin x$$

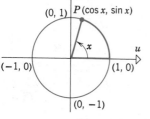

Figure A.22

The general shape of this graph can be obtained geometrically by constructing an angle of x radians in standard position and recalling from the previous section that the terminal side of this angle intersects the unit circle at the point

$$(\cos x, \sin x)$$

This is illustrated in Figure A.22, where we have used u and v to label the coordinate axes so we can use x for the angle. Thus, by observing the ordinate of the point P in Figure A.22, we can see how the value of sin x varies with x. When $x = 0$, P is the point $(1, 0)$, so that sin $x = 0$. As x increases from 0 to $\pi/2$, P moves counterclockwise along the circle from $(1, 0)$ to the point $(0, 1)$; thus, sin x increases from 0 to 1 (Figure A.23a). As x increases from $\pi/2$ to π, P moves counterclockwise along the circle from $(0, 1)$ to $(-1, 0)$; thus, sin x decreases from 1 to 0 (Figure A.23b). As x increases from π to $3\pi/2$, P moves counterclockwise along the circle from $(-1, 0)$ to $(0, -1)$; thus, sin x decreases from 0 to -1 (Figure A.23c). As x increases from $3\pi/2$ to 2π, P completes one full revolution by moving counterclockwise from $(0, -1)$ to $(1, 0)$; thus, sin x increases from -1 to 0 (Figure A.23d). As x increases past 2π, the point P begins a second revolution around the circle so that the values of sin x begin to repeat. In fact, the graph of

$$y = \sin x$$

will repeat itself every 2π units, since P makes a complete revolution whenever x changes by 2π. For negative values of x, the graph of sin x can be obtained by observing the ordinate of P during a *clockwise* rotation around the unit circle. Alternatively we can use the identity

$$\sin(-x) = -\sin x$$

(a)

(b)

Figure A.23   (c)

(d)

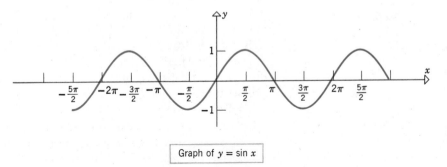

Graph of $y = \sin x$

Figure A.24

to observe that the values of the sine function at x and $-x$ differ only in sign. Either way we are led to the graph in Figure A.24.

    The fact that the graph of $y = \sin x$ repeats itself every 2π units could have been anticipated from the identities

$$\sin(x + 2\pi) = \sin x, \qquad \sin(x - 2\pi) = \sin x$$

It is this *periodic* or repetitive nature of trigonometric functions that makes them useful for studying repetitive physical phenomena such as vibrations and wave motion.

    By observing the abscissa of the point P in Figure A.22, we can see how the value of $\cos x$ varies with x. An analysis similar to the one we used to graph $\sin x$ yields the graph of $y = \cos x$ shown in Figure A.25.

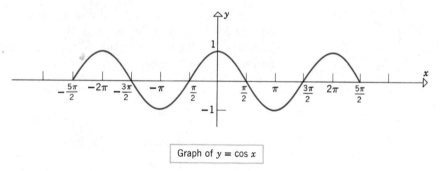

Figure A.25

Graph of $y = \cos x$

Observe that the graph of $\cos x$ is symmetric about the y-axis. This could have been anticipated from the identity

$$\cos(-x) = \cos x$$

which shows that for all x the cosine function assigns the same value to $-x$ and x.

The graph of

$$y = \tan x$$

can also be obtained geometrically. To begin, let us assume that

$$0 \leq x < \frac{\pi}{2}$$

If we construct the angle x in standard position and let Q be the intersection of its terminal side with the vertical line L through $(1, 0)$ (Figure A.26), then the coordinates of Q are

$$(1, \tan x)$$

Figure A.26

To see this, observe that triangles OAP and OBQ are similar, so that

$$\frac{BQ}{OB} = \frac{AP}{OA} \tag{2}$$

But $OB = 1$, $OA = \cos x$, and $AP = \sin x$, so that from (2)

$$BQ = \frac{\sin x}{\cos x} = \tan x$$

Thus, as shown in Figure A.26, the ordinate of Q is $\tan x$.

As x increases from 0 toward $\pi/2$, the point Q travels from $(1, 0)$ up the vertical line L getting higher and higher without any bound. Thus, as x increases from 0 toward $\pi/2$, the value of tan x increases without bound from an initial value of zero (Figure A.27a).

From the identity

$$\tan(-x) = -\tan x$$

and Figure A.27a, we can obtain the portion of the graph shown in Figure A.27b.

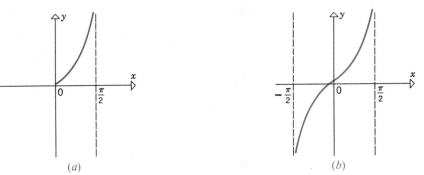

Figure A.27   (a)   (b)

Finally, from the identities

$$\tan(x + \pi) = \tan x \quad \text{and} \quad \tan(x - \pi) = \tan x$$

we see that the graph of $y = \tan x$ repeats itself every π units. This fact together with Figure A.27b leads us to the complete graph of $y = \tan x$ shown in Figure A.28.

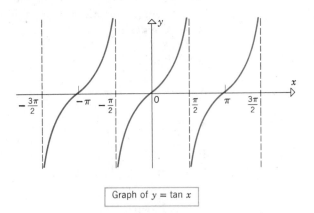

Figure A.28

Graph of $y = \tan x$

REMARK. Unlike the sine and cosine functions that have domain $(-\infty, +\infty)$, the tangent function is not defined for all x. Since $\tan x = \sin x/\cos x$, the tangent function is undefined whenever $\cos x = 0$; this occurs when x is an odd integer multiple of $\pi/2$, that is, when

$$x = \pm\frac{\pi}{2}, \pm\frac{3\pi}{2}, \pm\frac{5\pi}{2}, \ldots$$

DEFINITION 1   A function f is called **periodic** if there is a positive number p such that

$$f(x + p) = f(x) \tag{3}$$

whenever x and $x + p$ lie in the domain of f. We call p a **period** of the function. The smallest positive period is called the **fundamental period** of f or sometimes the **period** of f.

In words, (3) states that the values of f repeat themselves every p units.

▶ Example 1

For the sine and cosine functions, 2π is a period since

$$\sin(x + 2\pi) = \sin x, \qquad \cos(x + 2\pi) = \cos x$$

Also, 4π, 6π, 8π, etc. are periods for sine and cosine since

$$\sin(x + 4\pi) = \sin x, \ \sin(x + 6\pi) = \sin x, \ \sin(x + 8\pi) = \sin x$$
$$\cos(x + 4\pi) = \cos x, \ \cos(x + 6\pi) = \cos x, \ \cos(x + 8\pi) = \cos x$$

and so forth. The fundamental period of sine and cosine is 2π.
For the tangent function, π is a period since

$$\tan(x + \pi) = \tan x$$

Also, 2π, 3π, 4π, etc. are periods; but π is the fundamental period.   ◀

Functions of the form

$$a \sin bx \qquad \text{and} \qquad a \cos bx \tag{4}$$

occur frequently in applications. The graphs of such functions are considered in the following examples.

▶ Example 2

Sketch the graphs of

(a) $y = 3 \sin x$,   (b) $y = \sin 4x$,   (c) $y = 3 \sin 4x$.

Solution (a). The factor of 3 in

$$y = 3 \sin x$$

has the effect of multiplying each y-coordinate in the graph of $y = \sin x$ by 3, thereby stretching the graph of $y = \sin x$ vertically by a factor of three (Figure A.29a).

(a)

(b)

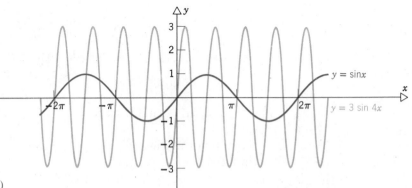

Figure A.29 (c)

Solution (b). The graph of $y = \sin x$ repeats itself when x changes by 2π. Thus, the graph of $y = \sin 4x$ repeats itself whenever $4x$ changes by 2π, or equivalently when x changes by $\frac{1}{4}(2\pi) = \pi/2$. Thus, the factor of 4 in

$$y = \sin 4x$$

has the effect of reducing the fundamental period by a factor of $\frac{1}{4}$. This leads to the graph in Figure A.29b.

Solution (*c*).   The graph of

$$y = 3 \sin 4x$$

combines both the stretching effect discussed in (a) and the period effect discussed in (b). The graph is shown in Figure A.29c.   ◄

The last example illustrates the following general result.

THEOREM 2   *If* $a \neq 0$ *and* $b \neq 0$, *then the functions* $a \sin bx$ *and* $a \cos bx$ *have fundamental period*

$$\frac{2\pi}{|b|}$$

and their graphs oscillate between $-a$ *and* a.

We leave the proof as an exercise.

► Example 3

The graph of the function $6 \cos 7x$ oscillates between -6 and 6 and has fundamental period $2\pi/7$; the function $\frac{1}{3} \sin \frac{1}{4}x$ oscillates between $-\frac{1}{3}$ and $\frac{1}{3}$ and has fundamental period 8π.   ◄

We leave it as an exercise to prove the following result.

THEOREM 3   *If* $b \neq 0$, *then the function* $\tan bx$ *has fundamental period*

$$\frac{\pi}{|b|}$$

► Trigonometry Exercises II.

1. Find the fundamental period of

   (a) $\sin 5x$ (b) $\cos \frac{1}{3}x$ (c) $\tan \frac{1}{8}x$

   (d) $\tan 7x$ (e) $\sin \pi x$ (f) $\sec \frac{\pi}{5}x$

   (g) $\cot \pi x$ (h) $\csc \frac{x}{k}$.

2. As stated in Theorem 2 of this section, the graphs of $a \sin bx$ and $a \cos bx$ oscillate between $-a$ and a. The number $|a|$ is called the **amplitude** of the function. Find the amplitudes of

   (a) $4 \sin 6x$ (b) $-8 \cos \pi x$

   (c) $\frac{1}{2} \sin x$ (d) $-\frac{1}{8} \cos 2x$.

3. Sketch the graphs of
   (a) $2 \sin x$ (b) $\sin 3x$ (c) $2 \sin 3x$.

4. Sketch the graphs of
   (a) $\frac{1}{2} \cos \pi x$ (b) $-4 \cos x$ (c) $\cos(-4x)$.

5. Use the graphs of $\sin x$, $\cos x$, $\tan x$, and the relationships

   $$\cot x = \frac{1}{\tan x}, \quad \sec x = \frac{1}{\cos x}, \quad \csc x = \frac{1}{\sin x}$$

   to help sketch the graphs of
   (a) $\cot x$ (b) $\sec x$ (c) $\csc x$.

6. Sketch the graphs of
   (a) $3 \cot 2x$ (b) $\frac{1}{4} \sec \pi x$ (c) $2 \csc \frac{x}{3}$.

7. Sketch the graphs of
   (a) $3 \sin \left(2x + \frac{\pi}{4}\right)$ (b) $1 + \sin x$

   (c) $|\cos x|$ (d) $\sin x + \cos x$.

8. Sketch the graphs of
   (a) $\tan \left(\pi x - \frac{1}{2}\right)$ (b) $2 \sin x + 3 \cos 2x$

   (c) $|\tan x|$ (d) $2 - \cos x$.

9. Classify the following functions as even, odd or neither. (See Exercise 22 in Section 2.2.)
   (a) $\sin x$ (b) $\cos x$ (c) $\tan x$
   (d) $|\sec x|$ (e) $\cot(x^2)$ (f) $3 \sec(-2x)$
   (g) $-4 \csc x$ (h) $\frac{\sin x}{x}$.

10. Prove: If $0 < x < \pi/2$, then

    $$\sin x < x < \tan x$$

    [*Hint:* In Figure A.26, compare the areas of triangle OAP, sector OBP, and triangle OBQ. The area of a sector is discussed in Exercise 19 of Unit I of the trigonometry review.]

11. Prove: For all real numbers x,

    $$|\sin x| \leq |x|$$

    [*Hint:* First consider the cases $|x| \geq \pi/2$ and $x = 0$; these are easy. Then use the left-hand inequality in Exercise 10 to complete the proof.]

12. Prove Theorem 2 of this section.

13. Prove Theorem 3 of this section.

appendix 2 supplementary material

I. ONE-SIDED AND INFINITE LIMITS: A RIGOROUS APPROACH

In Section 2.4 we discussed the one-sided limits

$$\lim_{x \to a^-} f(x) = L \qquad \text{and} \qquad \lim_{x \to a^+} f(x) = L$$

from an intuitive viewpoint. We interpreted the first statement to mean that the value of $f(x)$ approaches L as x approaches a from the left side, and we interpreted the second statement to mean that the value of $f(x)$ approaches L as x approaches a from the right side. We can formalize these ideas mathematically as follows:

DEFINITION 1 Let f be defined on some open interval extending to the left from a. We will write

$$\lim_{x \to a^-} f(x) = L$$

if given any number $\epsilon > 0$, we can find a number $\delta > 0$ such that $f(x)$ satisfies

$$|f(x) - L| < \epsilon$$

whenever x satisfies

$$a - \delta < x < a$$

DEFINITION 2   Let f be defined on some open interval extending to the right from a. We will write

$$\lim_{x \to a^+} f(x) = L$$

if given any number $\epsilon > 0$, we can find a number $\delta > 0$ such that $f(x)$ satisfies

$$|f(x) - L| < \epsilon$$

whenever x satisfies

$$a < x < a + \delta$$

Before discussing some examples, let us compare Definitions 1 and 2 to the definition of $\lim_{x \to a} f(x) = L$ as stated in Definition 2.6.1. Given $\epsilon > 0$, the definition of $\lim_{x \to a} f(x) = L$ requires that we be able to find a $\delta > 0$ such that

$$|f(x) - L| < \epsilon \tag{1}$$

whenever x satisfies

$$0 < |x - a| < \delta$$

or equivalently, whenever x lies in the set

$$(a - \delta, a) \cup (a, a + \delta)$$

(See Figure A.30a.)

Figure A.30

$\|f(x) - L\| < \epsilon$ if $0 < \|x - a\| < \delta$	$\|f(x) - L\| < \epsilon$ if $a - \delta < x < a$	$\|f(x) - L\| < \epsilon$ if $a < x < a + \delta$
(a)	(b)	(c)

On the other hand, the definition of $\lim\limits_{x \to a^-} f(x) = L$ requires only that (1) hold for x in the set

$(a - \delta, a)$

(Figure A.30*b*), while the definition of $\lim\limits_{x \to a^+} f(x) = L$ requires only that (1) be satisfied on the set

$(a, a + \delta)$

(See Figure A.30*c*.)

▶ Example 1

Because the function

$$f(x) = \sqrt{x}$$

is undefined for $x < 0$, we cannot consider the limits

$$\lim_{x \to 0} \sqrt{x} \qquad \text{or} \qquad \lim_{x \to 0^-} \sqrt{x}$$

The expression

$$\lim_{x \to 0^+} \sqrt{x}$$

is the only one that makes sense. Prove that

$$\lim_{x \to 0^+} \sqrt{x} = 0$$

Solution. Let $\epsilon > 0$ be given. We must find a $\delta > 0$ such that $f(x) = \sqrt{x}$ satisfies

$$|\sqrt{x} - 0| < \epsilon \tag{2}$$

whenever x satisfies

$$0 < x < 0 + \delta$$

or equivalently

$$0 < x < \delta \tag{3}$$

To find δ, we rewrite (2) as

$$\sqrt{x} < \epsilon \qquad (4)$$

(since $\sqrt{x} = |\sqrt{x}|$). We must choose δ so that (4) holds whenever (3) does. We can do this by taking

$$\delta = \epsilon^2$$

To see that this choice of δ works, assume that x satisfies (3). Since we are letting $\delta = \epsilon^2$, it follows from (3) that

$$0 < x < \epsilon^2$$

or, equivalently,

$$0 < \sqrt{x} < \epsilon \qquad (5)$$

Thus (4) holds since it is just the right-hand inequality in (5). This proves that

$$\lim_{x \to 0^+} \sqrt{x} = 0 \qquad \blacktriangleleft$$

▶ Example 2

Let

$$f(x) = \begin{cases} 2x, & x < 1 \\ x^3, & x \geq 1 \end{cases}$$

Prove that

$$\lim_{x \to 1^-} f(x) = 2$$

Solution. Let $\epsilon > 0$ be given. We must find a $\delta > 0$ such that $f(x)$ satisfies

$$|f(x) - 2| < \epsilon \qquad (6)$$

whenever x satisfies

$$1 - \delta < x < 1 \qquad (7)$$

Because we are considering only x satisfying (7), it follows from the formula for $f(x)$ that

$$f(x) = 2x$$

Thus we can rewrite (6) as

$$|2x - 2| < \epsilon$$

or

$$|x - 1| < \frac{\epsilon}{2}$$

or

$$1 - \frac{\epsilon}{2} < x < 1 + \frac{\epsilon}{2} \tag{8}$$

We must choose δ so that (8) is satisfied whenever (7) is satisfied. We can do this by taking

$$\delta = \frac{\epsilon}{2}$$

To see that this choice of δ works, assume that x satisfies (7). With $\delta = \epsilon/2$, it follows from (7) that

$$1 - \frac{\epsilon}{2} < x < 1 \tag{9}$$

and this implies that

$$1 - \frac{\epsilon}{2} < x < 1 + \frac{\epsilon}{2}$$

since ϵ is positive.

Thus we have proved

$$\lim_{x \to 1^-} f(x) = 2 \qquad\qquad\blacktriangleleft$$

In Section 2.4 we discussed the limits

$$\lim_{x \to -\infty} f(x) = L \quad \text{and} \quad \lim_{x \to +\infty} f(x) = L$$

from an intuitive viewpoint. We interpreted the first statement to mean that the value of $f(x)$ approaches L as x moves indefinitely far from the origin along the negative x-axis, and we interpreted the second statement to mean that the value of $f(x)$ approaches L as x moves indefinitely far from the origin along the positive x-axis.

The phrase, "x moves indefinitely far from the origin along the positive x-axis" is intended to convey the idea that given *any* positive number N (no matter how large) eventually x is larger than N (Figure A.31a). Similarly, the phrase, "x moves indefinitely far from the origin along the negative x-axis" is intended to convey the idea that given any negative number N (no matter how small) eventually x is smaller than N (Figure A.31b).

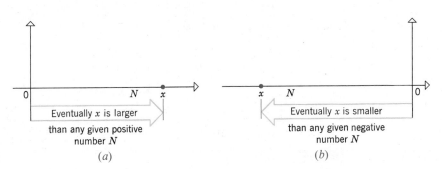

Figure A.31   (a)   (b)

Thus, when we say that $\lim\limits_{x \to +\infty} f(x) = L$, we mean that for any $\epsilon > 0$, we can find a positive number N (sufficiently large) so that $f(x)$ satisfies

$$|f(x) - L| < \epsilon$$

when x is greater than N (Figure A.32). Similarly, when we say that $\lim\limits_{x \to -\infty} f(x) = L$ we mean that for any $\epsilon > 0$, we can find a negative number N so that $f(x)$ satisfies

$$|f(x) - L| < \epsilon$$

when x is smaller than N (Figure A.33).

f(x) satisfies $|f(x) - L| < \epsilon$
when $x > N$

f(x) satisfies $|f(x) - L| < \epsilon$
when $x < N$

Figure A.32   Figure A.33

These ideas suggest the following formal definitions.

DEFINITION 3 Let f be a function that is defined on some infinite open interval $(x_0, +\infty)$. We will write

$$\lim_{x \to +\infty} f(x) = L$$

if given any number $\epsilon > 0$, there corresponds a positive number N such that

$$|f(x) - L| < \epsilon$$

whenever x satisfies

$$x > N$$

DEFINITION 4 Let f be a function that is defined on some infinite open interval $(-\infty, x_0)$. We will write

$$\lim_{x \to -\infty} f(x) = L$$

if given any number $\epsilon > 0$, there corresponds a negative number N such that

$$|f(x) - L| < \epsilon$$

whenever x satisfies

$$x < N$$

▶ Example 3

Prove that

$$\lim_{x \to +\infty} \frac{1}{x} = 0$$

Solution. We must show that for any $\epsilon > 0$ we can find a positive number N such that $f(x) = 1/x$ satisfies

$$\left| \frac{1}{x} - 0 \right| < \epsilon \tag{10}$$

whenever x satisfies

$$x > N \tag{11}$$

To find N, rewrite (10) as

$$\frac{1}{|x|} < \epsilon$$

or

$$|x| > \frac{1}{\epsilon} \tag{12}$$

Thus we must find a positive number N such that x satisfies (12) whenever x satisfies (11). We can do this by taking

$$N = \frac{1}{\epsilon}$$

To prove that this choice of N works, assume that x satisfies (11). Since we are letting $N = 1/\epsilon$, (11) becomes

$$x > \frac{1}{\epsilon} \tag{13}$$

But $1/\epsilon$ is positive (since $\epsilon > 0$) so that the x in (13) is positive. Thus (13) can be written

$$|x| > \frac{1}{\epsilon}$$

which is precisely inequality (12). Thus we have proved

$$\lim_{x \to +\infty} \frac{1}{x} = 0. \qquad \blacktriangleleft$$

In Section 2.4 we discussed the limits

$$\lim_{x \to a} f(x) = +\infty \qquad \lim_{x \to a} f(x) = -\infty \tag{14}$$

$$\lim_{x \to a^+} f(x) = +\infty \qquad \lim_{x \to a^+} f(x) = -\infty \tag{15}$$

$$\lim_{x \to a^-} f(x) = +\infty \qquad \lim_{x \to a^-} f(x) = -\infty \tag{16}$$

$$\lim_{x \to +\infty} f(x) = +\infty \qquad \lim_{x \to +\infty} f(x) = -\infty \tag{17}$$

$$\lim_{x \to -\infty} f(x) = +\infty \qquad \lim_{x \to -\infty} f(x) = -\infty \tag{18}$$

from an intuitive viewpoint. Recall that each of these expressions describes a particular way in which the limit fails to exist. The $+\infty$ to the right of the $=$

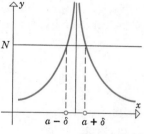

Figure A.34

sign indicates that the limit fails to exist because $f(x)$ is increasing without bound, and the $-\infty$ indicates that the limit fails to exist because $f(x)$ is decreasing without bound.

To make these ideas mathematically precise, we must clarify the meanings of the phrases, "$f(x)$ is increasing without bound" and "$f(x)$ is decreasing without bound." We will limit our discussion to the limits in (14); formal definitions of the remaining limits are discussed in the exercises.

When we say that $f(x)$ increases without bound as x approaches a from either side, we mean that given any positive number N (no matter how large) the value of $f(x)$ eventually exceeds N as x approaches a from either side. Figure A.34 illustrates this idea. For the curve in that figure we have

$$\lim_{x \to a} f(x) = +\infty$$

We picked an arbitrary positive number N and marked it on the y-axis. As x approaches a from either side, x will eventually lie between the points

$$a - \delta \quad \text{and} \quad a + \delta$$

shown in the figure. When this occurs the value of $f(x)$ will satisfy

$$f(x) > N \tag{19}$$

It is evident from the figure that no matter how large we make N, we will always be able to find a sufficiently small positive number δ such that $f(x)$ satisfies (19) when x is between $a - \delta$ and $a + \delta$ (but different from a.) This suggests the following definition.

DEFINITION 5   Let f be defined in some open interval containing the number a, except that f need not be defined at a. We will write

$$\lim_{x \to a} f(x) = +\infty$$

if given any positive number N, we can find a number $\delta > 0$ such that $f(x)$ satisfies

$$f(x) > N$$

whenever x satisfies

$$0 < |x - a| < \delta$$

When we say that $f(x)$ decreases without bound as x approaches a from either side, we mean that given any negative number N (no matter how

Figure A.35

small) the value of $f(x)$ will eventually be smaller than N as x approaches a from either side. Figure A.35 illustrates this idea. For the curve in that figure we have

$$\lim_{x \to a} f(x) = -\infty$$

We picked an arbitrary negative number N and marked it on the y-axis. As x approaches a from either side, x will eventually lie between the points

$$a - \delta \quad \text{and} \quad a + \delta$$

shown in the figure. When this occurs the value of $f(x)$ will satisfy

$$f(x) < N$$

This suggests the following definition.

DEFINITION 6   Let f be defined in some open interval containing the number a, except that f need not be defined at a. We will write

$$\lim_{x \to a} f(x) = -\infty$$

if given any negative number N we can find a number $\delta > 0$ such that $f(x)$ satisfies

$$f(x) < N$$

whenever x satisfies

$$0 < |x - a| < \delta$$

▶ Example 4

Prove that

$$\lim_{x \to 0} \frac{1}{x^2} = +\infty$$

Solution. We must show that given any positive number N, we can find a positive number δ such that $f(x) = 1/x^2$ satisfies

$$\frac{1}{x^2} > N \tag{20}$$

whenever x satisfies

$$0 < |x - 0| < \delta$$

or equivalently

$$0 < |x| < \delta \tag{21}$$

To find δ we will rewrite (20) as

$$x^2 < \frac{1}{N} \tag{22}$$

Thus we must find a positive number δ such that x satisfies (22) whenever x satisfies (21). We can do this by taking

$$\delta = \frac{1}{\sqrt{N}}$$

To prove that this choice of δ works, assume that x satisfies (21). Since we are letting $\delta = 1/\sqrt{N}$, (21) becomes

$$0 < |x| < \frac{1}{\sqrt{N}}$$

or equivalently

$$0 < |x|^2 < \frac{1}{N}$$

or since $|x|^2 = x^2$,

$$0 < x^2 < \frac{1}{N} \tag{23}$$

Thus (22) is satisfied, since it is just the right-hand inequality in (23). This proves that

$$\lim_{x \to +\infty} \frac{1}{x^2} = 0 \qquad \blacktriangleleft$$

▶ Exercises

In Exercises 1–6, use Definitions 1 and 2 to prove that the stated limit is correct.

1. $\lim\limits_{x \to 2^+} (x + 1) = 3.$

2. $\lim\limits_{x \to 1^-} (3x + 2) = 5.$

3. $\lim\limits_{x \to 4^+} \sqrt{x - 4} = 0.$

4. $\lim\limits_{x \to 0^-} \sqrt{-x} = 0.$

5. $\lim\limits_{x \to 2^+} f(x) = 2$, where $f(x) = \begin{cases} x, & x > 2 \\ 3x, & x \le 2 \end{cases}$.

6. $\lim\limits_{x \to 2^-} f(x) = 6$, where $f(x) = \begin{cases} x, & x > 2 \\ 3x, & x \le 2 \end{cases}$.

In Exercises 7–14, use Definitions 3 and 4 to prove that the stated limit is correct.

7. $\lim\limits_{x \to +\infty} \dfrac{1}{x^2} = 0.$

8. $\lim\limits_{x \to -\infty} \dfrac{1}{x} = 0.$

9. $\lim\limits_{x \to -\infty} \dfrac{1}{x + 2} = 0.$

10. $\lim\limits_{x \to +\infty} \dfrac{1}{x + 2} = 0.$

11. $\lim\limits_{x \to +\infty} \dfrac{x}{x + 1} = 1.$

12. $\lim\limits_{x \to -\infty} \dfrac{x}{x + 1} = 1.$

13. $\lim\limits_{x \to -\infty} \dfrac{4x - 1}{2x + 5} = 2.$

14. $\lim\limits_{x \to +\infty} \dfrac{4x - 1}{2x + 5} = 2.$

In Exercises 15–20, use Definitions 5 and 6 to prove that the stated limit is correct.

15. $\lim\limits_{x \to 3} \dfrac{1}{(x - 3)^2} = +\infty.$

16. $\lim\limits_{x \to 3} \dfrac{-1}{(x - 3)^2} = -\infty.$

17. $\lim\limits_{x \to 0} \dfrac{1}{|x|} = +\infty.$

18. $\lim\limits_{x \to 1} \dfrac{1}{|x - 1|} = +\infty.$

19. $\lim\limits_{x \to 0} \left(-\dfrac{1}{x^4} \right) = -\infty.$

20. $\lim\limits_{x \to 0} \dfrac{1}{x^4} = +\infty.$

21. Define limit statements (15) and (16).

22. Use the definitions in Exercise 21 to prove:

    (a) $\lim\limits_{x \to 0^+} \dfrac{1}{x} = +\infty$

    (b) $\lim\limits_{x \to 0^-} \dfrac{1}{x} = -\infty.$

23. Use the definitions in Exercise 21 to prove:

    (a) $\lim\limits_{x \to 1^+} \dfrac{1}{1 - x} = -\infty$

    (b) $\lim\limits_{x \to 1^-} \dfrac{1}{1 - x} = +\infty.$

24. Define limit statements (17) and (18).

25. Use the definitions in Exercise 24 to prove:

    (a) $\lim\limits_{x \to +\infty} (x + 1) = +\infty$

    (b) $\lim\limits_{x \to -\infty} (x + 1) = -\infty.$

26. Use the definitions in Exercise 24 to prove:

    (a) $\lim\limits_{x \to +\infty} (x^2 - 3) = +\infty$

    (b) $\lim\limits_{x \to -\infty} (x^3 + 5) = -\infty.$

II. SOME PROOFS OF LIMIT THEOREMS

In this section we will prove some fundamental theorems about limits. Our first result states that a function cannot have two different limits at the same point.

THEOREM 1   If $\lim\limits_{x \to a} f(x) = L_1$ and $\lim\limits_{x \to a} f(x) = L_2$, then $L_1 = L_2$.

REMARK.   In this section we will only consider the cases where a and the limits are finite (that is, not equal to $+\infty$ or $-\infty$.) Other cases are considered in the exercises.

Proof. We shall assume that $L_1 \neq L_2$ and show that this assumption leads to a contradiction. It will then follow that $L_1 = L_2$. Let

$$\epsilon = \frac{|L_1 - L_2|}{2} \qquad (1)$$

Since we are assuming $L_1 \neq L_2$, it follows that $\epsilon > 0$.
  From the assumption

$$\lim_{x \to a} f(x) = L_1$$

we can find $\delta_1 > 0$ such that

$$|f(x) - L_1| < \epsilon \qquad (2)$$

whenever

$$0 < |x - a| < \delta_1 \qquad (3)$$

and since

$$\lim_{x \to a} f(x) = L_2 \qquad (4)$$

we can find $\delta_2 > 0$ such that

$$|f(x) - L_2| < \epsilon \qquad (5)$$

whenever

$$0 < |x - a| < \delta_2 \qquad (6)$$

Let δ be the smaller of the numbers δ_1 and δ_2, and choose any x satisfying

$$0 < |x - a| < \delta$$

Since $\delta \leq \delta_1$ and $\delta \leq \delta_2$, this x satisfies both (3) and (6) and, therefore, $f(x)$ satisfies both (2) and (5). But

$$
\begin{aligned}
2\epsilon &= |L_1 - L_2| && \text{[from (1)]} \\
&= |L_1 - f(x) + f(x) - L_2| \\
&= |(L_1 - f(x)) + (f(x) - L_2)| \\
&\leq |L_1 - f(x)| + |f(x) - L_2| && \text{[the triangle inequality]}
\end{aligned}
$$

$$< \epsilon + \epsilon \qquad\qquad\qquad\qquad \text{[(2) and (5)]}$$
$$= 2\epsilon$$

Thus we have shown that

$$2\epsilon < 2\epsilon$$

which is a contradiction.

Next, we prove the result discussed in Example 1 of Section 2.5.

THEOREM 2 *For any constant k,*

$$\lim_{x \to a} k = k$$

Proof. Let $\epsilon > 0$ be given. We must find $\delta > 0$ such that

$$|k - k| < \epsilon \qquad\qquad\qquad\qquad (7)$$

whenever

$$0 < |x - a| < \delta \qquad\qquad\qquad\qquad (8)$$

But (7) can be written

$$0 < \epsilon$$

which holds for all values of x. Thus for *any* positive δ whatever, (7) will be satisfied when (8) is satisfied.

Next we will prove part of Theorem 2.5.1a.

THEOREM 3 If $\lim_{x \to a} f(x) = L_1$ *and* $\lim_{x \to a} g(x) = L_2$ *then*

$$\lim_{x \to a} [f(x) + g(x)] = \lim_{x \to a} f(x) + \lim_{x \to a} g(x) = L_1 + L_2$$

Proof. Let $\epsilon > 0$ be given. To prove that

$$\lim_{x \to a} [f(x) + g(x)] = L_1 + L_2$$

we must show that given $\epsilon > 0$ we can find $\delta > 0$ such that

$$|[f(x) + g(x)] - [L_1 + L_2]| < \epsilon \tag{9}$$

whenever

$$0 < |x - a| < \delta \tag{10}$$

Since

$$\lim_{x \to a} f(x) = L_1$$

we can find $\delta_1 > 0$ such that

$$|f(x) - L_1| < \frac{\epsilon}{2} \tag{11}$$

whenever

$$0 < |x - a| < \delta_1 \tag{12}$$

and since

$$\lim_{x \to a} g(x) = L_2$$

we can find $\delta_2 > 0$ such that

$$|g(x) - L_2| < \frac{\epsilon}{2} \tag{13}$$

whenever

$$0 < |x - a| < \delta_2 \tag{14}$$

Let δ be the smaller of the numbers δ_1 and δ_2. If x satisfies (10), then x will also satisfy both (12) and (14) since $\delta \leq \delta_1$ and $\delta \leq \delta_2$. Consequently, $f(x)$ will satisfy both (11) and (13). Therefore

$$
\begin{aligned}
|[f(x) + g(x)] - [L_1 + L_2]| &= |[f(x) - L_1] + [g(x) - L_2]| \\
&\leq |f(x) - L_1| + |g(x) - L_2| \\
&< \frac{\epsilon}{2} + \frac{\epsilon}{2} \\
&= \epsilon
\end{aligned}
$$

Thus for the δ we have selected, (9) is satisfied whenever (10) is satisfied. This completes the proof. ∎

Next we will prove part of Theorem 2.5.1c. This proof is a little more complicated than the previous proofs.

THEOREM 4   *If* $\lim_{x \to a} f(x) = L_1$ *and* $\lim_{x \to a} g(x) = L_2$ *then*

$$\lim_{x \to a} [f(x)g(x)] = \lim_{x \to a} f(x) \lim_{x \to a} g(x) = L_1 L_2$$

Proof.   Let $\epsilon > 0$ be given. We must find $\delta > 0$ such that

$$|f(x)g(x) - L_1 L_2| < \epsilon \tag{15}$$

whenever

$$0 < |x - a| < \delta \tag{16}$$

To find δ we will express (15) in a different form. We can write

$$f(x) = L_1 + [f(x) - L_1] \quad \text{and} \quad g(x) = L_2 + [g(x) - L_2]$$

When we multiply these expressions and subtract $L_1 L_2$, we obtain

$$f(x)g(x) - L_1 L_2 = L_1[g(x) - L_2] + L_2[f(x) - L_1] \\ + [f(x) - L_1][g(x) - L_2]$$

so that (15) can be rewritten

$$|L_1[g(x) - L_2] + L_2[f(x) - L_1] + [f(x) - L_1][g(x) - L_2]| < \epsilon \tag{17}$$

Thus we must find $\delta > 0$ such that (17) holds whenever (16) does. Since ϵ is positive, the numbers

$$\sqrt{\epsilon/3}, \quad \frac{\epsilon}{3(1 + |L_1|)}, \quad \frac{\epsilon}{3(1 + |L_2|)}$$

are also positive. Therefore, since

$$\lim_{x \to a} f(x) = L_1 \quad \text{and} \quad \lim_{x \to a} g(x) = L_2$$

we can find positive numbers δ_1, δ_2, δ_3, and δ_4 such that

$$
\left.
\begin{aligned}
|f(x) - L_1| &< \sqrt{\epsilon/3} & \text{when} & \quad 0 < |x - a| < \delta_1 \\
|f(x) - L_1| &< \frac{\epsilon}{3(1 + |L_2|)} & \text{when} & \quad 0 < |x - a| < \delta_2 \\
|g(x) - L_2| &< \sqrt{\epsilon/3} & \text{when} & \quad 0 < |x - a| < \delta_3 \\
|g(x) - L_2| &< \frac{\epsilon}{3(1 + |L_1|)} & \text{when} & \quad 0 < |x - a| < \delta_4
\end{aligned}
\right\} \quad (18)
$$

Let δ be the smallest of the numbers δ_1, δ_2, δ_3, and δ_4. Then $\delta \leq \delta_1$, $\delta \leq \delta_2$, $\delta \leq \delta_3$, and $\delta \leq \delta_4$. Thus, if x satisfies (16), then x will also satisfy the four conditions on the right side of (18). Consequently, $f(x)$ and $g(x)$ will satisfy the four conditions on the left side of (18). Therefore

$$
\begin{aligned}
|L_1[g(x) - L_2] &+ L_2[f(x) - L_1] + [f(x) - L_1][g(x) - L_2]| \\
&\leq |L_1[g(x) - L_2]| + |L_2[f(x) - L_1]| + |[f(x) - L_1][g(x) - L_2]| \\
&= |L_1| \, |g(x) - L_2| + |L_2| \, |f(x) - L_1| + |f(x) - L_1| \, |g(x) - L_2| \\
&< |L_1| \frac{\epsilon}{3(1 + |L_1|)} + |L_2| \frac{\epsilon}{3(1 + |L_2|)} + \sqrt{\epsilon/3} \, \sqrt{\epsilon/3} \quad \text{[From (18)]} \\
&= \frac{\epsilon}{3} \frac{|L_1|}{1 + |L_1|} + \frac{\epsilon}{3} \frac{|L_2|}{1 + |L_2|} + \frac{\epsilon}{3} \\
&< \frac{\epsilon}{3} + \frac{\epsilon}{3} + \frac{\epsilon}{3}. \quad \left[\text{Since } \frac{|L_1|}{1 + |L_1|} < 1 \text{ and } \frac{|L_2|}{1 + |L_2|} < 1. \right] \\
&= \epsilon
\end{aligned}
$$

Thus, for the δ we have selected, (17) holds whenever (16) does. This completes the proof. ∎

In Section 2.4 we interpreted the statement

$$\lim_{x \to a} f(x) = L$$

to mean

$$\lim_{x \to a^+} f(x) = L \quad \text{and} \quad \lim_{x \to a^-} f(x) = L$$

The following theorem shows that this interpretation is consistent with our formal limit definitions.

THEOREM 5 *If* $\lim\limits_{x \to a^+} f(x) = \lim\limits_{x \to a^-} f(x) = L$, *then* $\lim\limits_{x \to a} f(x) = L$ *and conversely.*

Proof. Assume

$$\lim_{x \to a^+} f(x) = \lim_{x \to a^-} f(x) = L \tag{19}$$

To prove

$$\lim_{x \to a} f(x) = L$$

we must show that given $\epsilon > 0$ there exists a $\delta > 0$ such that

$$|f(x) - L| < \epsilon \tag{20}$$

whenever

$$0 < |x - a| < \delta \tag{21}$$

Because of (19), there exists a number $\delta_1 > 0$ such that (20) is satisfied whenever

$$a < x < a + \delta_1 \tag{22}$$

and there exists a number $\delta_2 > 0$ such that (20) is satisfied whenever

$$a - \delta_2 < x < a \tag{23}$$

Let δ be the smaller of δ_1 and δ_2, so that $\delta \leq \delta_1$ and $\delta \leq \delta_2$. Thus, if x satisfies (21), x will satisfy both (22) and (23) (Figure A.36) and consequently $f(x)$ will satisfy (20). Thus for the δ we have selected, (20) holds whenever (21) does. This completes this part of the proof.

The proof of the converse is left as an exercise. ▋

Figure A.36

▶ Exercises

1. Prove: For any constant k, $\lim\limits_{x \to +\infty} k = k$.

2. Prove: For any constant k, $\lim\limits_{x \to -\infty} k = k$.

3. Prove: If $\lim\limits_{x \to -\infty} f(x) = L_1$ and $\lim\limits_{x \to -\infty} g(x) = L_2$, then

$$\lim_{x \to -\infty} [f(x) + g(x)] = L_1 + L_2.$$

4. Prove: If $\lim\limits_{x \to +\infty} f(x) = L_1$ and $\lim\limits_{x \to +\infty} g(x) = L_2$, then

$$\lim_{x \to +\infty} [f(x) + g(x)] = L_1 + L_2.$$

5. Use Theorems 2, 3, and 4 of this section to prove:
    If $\lim\limits_{x \to a} f(x) = L_1$ and $\lim\limits_{x \to a} g(x) = L_2$, then

$$\lim_{x \to a} [f(x) - g(x)] = \lim_{x \to a} f(x) - \lim_{x \to a} g(x)$$
$$= L_1 - L_2.$$

6. Suppose $\lim_{x \to a} f(x) = +\infty$ and $\lim_{x \to a} g(x) = +\infty$.

(a) Prove: $\lim_{x \to a} [f(x) + g(x)] = +\infty$.

(b) Is it true that $\lim_{x \to a} [f(x) - g(x)] = 0$?

7. Suppose $\lim_{x \to a} f(x) = -\infty$ and $\lim_{x \to a} g(x) = +\infty$.

(a) Prove: $\lim_{x \to a} [f(x) - g(x)] = -\infty$

(b) Is it true that $\lim_{x \to a} [f(x) + g(x)] = 0$?

8. Use Theorems 2 and 4 of this section to prove: If $\lim_{x \to a} f(x) = L$ and k is a constant, then $\lim_{x \to a} [kf(x)] = kL$.

9. Prove: $\lim_{x \to a} f(x) = L$ if and only if $\lim_{x \to a} [f(x) - L] = 0$.

10. Prove: If $\lim_{x \to a} f(x) = L$, then $\lim_{x \to a} |f(x)| = |L|$.

11. Finish the proof of Theorem 5 of this section.

appendix 3 tables

Table 1
Table of Trigonometric Functions

DEGREES	RADIANS	SIN	COS	TAN	DEGREES	RADIANS	SIN	COS	TAN
0°	0.000	0.000	1.000	0.000					
1°	0.017	0.017	1.000	0.017	21°	0.367	0.358	0.934	0.384
2°	0.035	0.035	0.999	0.035	22°	0.384	0.375	0.927	0.404
3°	0.052	0.052	0.999	0.052	23°	0.401	0.391	0.921	0.424
4°	0.070	0.070	0.998	0.070	24°	0.419	0.407	0.914	0.445
5°	0.087	0.087	0.996	0.087	25°	0.436	0.423	0.906	0.466
6°	0.105	0.105	0.995	0.105	26°	0.454	0.438	0.899	0.488
7°	0.122	0.122	0.993	0.123	27°	0.471	0.454	0.891	0.510
8°	0.140	0.139	0.990	0.141	28°	0.489	0.469	0.883	0.532
9°	0.157	0.156	0.988	0.158	29°	0.506	0.485	0.875	0.554
10°	0.175	0.174	0.985	0.176	30°	0.524	0.500	0.866	0.577
11°	0.192	0.191	0.982	0.194	31°	0.541	0.515	0.857	0.601
12°	0.209	0.208	0.978	0.213	32°	0.559	0.530	0.848	0.625
13°	0.227	0.225	0.974	0.231	33°	0.576	0.545	0.839	0.649
14°	0.244	0.242	0.970	0.249	34°	0.593	0.559	0.829	0.675
15°	0.262	0.259	0.966	0.268	35°	0.611	0.574	0.819	0.700
16°	0.279	0.276	0.961	0.287	36°	0.628	0.588	0.809	0.727
17°	0.297	0.292	0.956	0.306	37°	0.646	0.602	0.799	0.754
18°	0.314	0.309	0.951	0.325	38°	0.663	0.616	0.788	0.781
19°	0.332	0.326	0.946	0.344	39°	0.681	0.629	0.777	0.810
20°	0.349	0.342	0.940	0.364	40°	0.698	0.643	0.766	0.839

Table 1
Table of Trigonometric Functions (Continued)

DEGREES	RADIANS	SIN	COS	TAN	DEGREES	RADIANS	SIN	COS	TAN
41°	0.716	0.656	0.755	0.869	66°	1.152	0.914	0.407	2.246
42°	0.733	0.669	0.743	0.900	67°	1.169	0.921	0.391	2.356
43°	0.750	0.682	0.731	0.933	68°	1.187	0.927	0.375	2.475
44°	0.768	0.695	0.719	0.966	69°	1.204	0.934	0.358	2.605
45°	0.785	0.707	0.707	1.000	70°	1.222	0.940	0.342	2.748
46°	0.803	0.719	0.695	1.036	71°	1.239	0.946	0.326	2.904
47°	0.820	0.731	0.682	1.072	72°	1.257	0.951	0.309	3.078
48°	0.838	0.743	0.669	1.111	73°	1.274	0.956	0.292	3.271
49°	0.855	0.755	0.656	1.150	74°	1.292	0.961	0.276	3.487
50°	0.873	0.766	0.643	1.192	75°	1.309	0.966	0.259	3.732
51°	0.890	0.777	0.629	1.235	76°	1.326	0.970	0.242	4.011
52°	0.908	0.788	0.616	1.280	77°	1.344	0.974	0.225	4.332
53°	0.925	0.799	0.602	1.327	78°	1.361	0.978	0.208	4.705
54°	0.942	0.809	0.588	1.376	79°	1.379	0.982	0.191	5.145
55°	0.960	0.819	0.574	1.428	80°	1.396	0.985	0.174	5.671
56°	0.977	0.829	0.559	1.483	81°	1.414	0.988	0.156	6.314
57°	0.995	0.839	0.545	1.540	82°	1.431	0.990	0.139	7.115
58°	1.012	0.848	0.530	1.600	83°	1.449	0.993	0.122	8.144
59°	1.030	0.857	0.515	1.664	84°	1.466	0.995	0.105	9.514
60°	1.047	0.866	0.500	1.732	85°	1.484	0.996	0.087	11.43
61°	1.065	0.875	0.485	1.804	86°	1.501	0.998	0.070	14.30
62°	1.082	0.883	0.469	1.881	87°	1.518	0.999	0.052	19.08
63°	1.100	0.891	0.454	1.963	88°	1.536	0.999	0.035	28.64
64°	1.117	0.899	0.438	2.050	89°	1.553	1.000	0.017	57.29
65°	1.134	0.906	0.423	2.145	90°	1.571	1.000	0.000	

Table 2
Table of Exponential and Hyperbolic Functions

x	e^x	e^{-x}	SINH x	COSH x	TANH x	x	e^x	e^{-x}	SINH x	COSH x	TANH x
0.00	1.0000	1.00000	0.0000	1.0000	.00000	0.40	1.4918	.67032	0.4108	1.0811	.37995
0.01	1.0101	0.99005	0.0100	1.0001	.01000	0.41	1.5068	.66365	0.4216	1.0852	.38847
0.02	1.0202	.98020	0.0200	1.0002	.02000	0.42	1.5220	.65705	0.4325	1.0895	.39693
0.03	1.0305	.97045	0.0300	1.0005	.02999	0.43	1.5373	.65051	0.4434	1.0939	.40532
0.04	1.0408	.96079	0.0400	1.0008	.03998	0.44	1.5527	.64404	0.4543	1.0984	.41364
0.05	1.0513	.95123	0.0500	1.0013	.04996	0.45	1.5683	.63763	0.4653	1.1030	.42190
0.06	1.0618	.94176	0.0600	1.0018	.05993	0.46	1.5841	.63128	0.4764	1.1077	.43008
0.07	1.0725	.93239	0.0701	1.0025	.06989	0.47	1.6000	.62500	0.4875	1.1125	.43820
0.08	1.0833	.92312	0.0801	1.0032	.07983	0.48	1.6161	.61878	0.4986	1.1174	.44624
0.09	1.0942	.91393	0.0901	1.0041	.08976	0.49	1.6323	.61263	0.5098	1.1225	.45422
0.10	1.1052	.90484	0.1002	1.0050	.09967	0.50	1.6487	.60653	0.5211	1.1276	.46212
0.11	1.1163	.89583	0.1102	1.0061	.10956	0.51	1.6653	.60050	0.5324	1.1329	.46995
0.12	1.1275	.88692	0.1203	1.0072	.11943	0.52	1.6820	.59452	0.5438	1.1383	.47770
0.13	1.1388	.87809	0.1304	1.0085	.12927	0.53	1.6989	.58860	0.5552	1.1438	.48538
0.14	1.1503	.86936	0.1405	1.0098	.13909	0.54	1.7160	.58275	0.5666	1.1494	.49299
0.15	1.1618	.86071	0.1506	1.0113	.14889	0.55	1.7333	.57695	0.5782	1.1551	.50052
0.16	1.1735	.85214	0.1607	1.0128	.15865	0.56	1.7507	.57121	0.5897	1.1609	.50798
0.17	1.1853	.84366	0.1708	1.0145	.16838	0.57	1.7683	.56553	0.6014	1.1669	.51536
0.18	1.1972	.83527	0.1810	1.0162	.17808	0.58	1.7860	.55990	0.6131	1.1730	.52267
0.19	1.2092	.82696	0.1911	1.0181	.18775	0.59	1.8040	.55433	0.6248	1.1792	.52990
0.20	1.2214	.81873	0.2013	1.0201	.19738	0.60	1.8221	.54881	0.6367	1.1855	.53705
0.21	1.2337	.81058	0.2115	1.0221	.20697	0.61	1.8404	.54335	0.6485	1.1919	.54413
0.22	1.2461	.80252	0.2218	1.0243	.21652	0.62	1.8589	.53794	0.6605	1.1984	.55113
0.23	1.2586	.79453	0.2320	1.0266	.22603	0.63	1.8776	.53259	0.6725	1.2051	.55805
0.24	1.2712	.78663	0.2423	1.0289	.23550	0.64	1.8965	.52729	0.6846	1.2119	.56490
0.25	1.2840	.77880	0.2526	1.0314	.24492	0.65	1.9155	.52205	0.6967	1.2188	.57167
0.26	1.2969	.77105	0.2629	1.0340	.25430	0.66	1.9348	.51685	0.7090	1.2258	.57836
0.27	1.3100	.76338	0.2733	1.0367	.26362	0.67	1.9542	.51171	0.7213	1.2330	.58498
0.28	1.3231	.75578	0.2837	1.0395	.27291	0.68	1.9739	.50662	0.7336	1.2402	.59152
0.29	1.3364	.74826	0.2941	1.0423	.28213	0.69	1.9937	.50158	0.7461	1.2476	.59798
0.30	1.3499	.74082	0.3045	1.0453	.29131	0.70	2.0138	.49659	0.7586	1.2552	.60437
0.31	1.3634	.73345	0.3150	1.0484	.30044	0.71	2.0340	.49164	0.7712	1.2628	.61068
0.32	1.3771	.72615	0.3255	1.0516	.30951	0.72	2.0544	.48675	0.7838	1.2706	.61691
0.33	1.3910	.71892	0.3360	1.0549	.31852	0.73	2.0751	.48191	0.7966	1.2785	.62307
0.34	1.4049	.71177	0.3466	1.0584	.32748	0.74	2.0959	.47711	0.8094	1.2865	.62915
0.35	1.4191	.70469	0.3572	1.0619	.33638	0.75	2.1170	.47237	0.8223	1.2947	.63515
0.36	1.4333	.69768	0.3678	1.0655	.34521	0.76	2.1383	.46767	0.8353	1.3030	.64108
0.37	1.4477	.69073	0.3785	1.0692	.35399	0.77	2.1598	.46301	0.8484	1.3114	.64693
0.38	1.4623	.68386	0.3892	1.0731	.36271	0.78	2.1815	.45841	0.8615	1.3199	.65271
0.39	1.4770	.67706	0.4000	1.0770	.37136	0.79	2.2034	.45384	0.8748	1.3286	.65841

Table 2
A47
Table of Exponential and Hyperbolic Functions (Continued)

x	e^x	e^{-x}	SINH x	COSH x	TANH x	x	e^x	e^{-x}	SINH x	COSH x	TANH x
0.80	2.2255	.44933	0.8881	1.3374	.66404	3.00	20.086	.04979	10.018	10.068	.99505
0.81	2.2479	.44486	0.9015	1.3464	.66959	3.10	22.198	.04505	11.076	11.122	.99595
0.82	2.2705	.44043	0.9150	1.3555	.67507	3.20	24.533	.04076	12.246	12.287	.99668
0.83	2.2933	.43605	0.9286	1.3647	.68048	3.30	27.113	.03688	13.538	13.575	.99728
0.84	2.3164	.43171	0.9423	1.3740	.68581	3.40	29.964	.03337	14.965	14.999	.99777
0.85	2.3396	.42741	0.9561	1.3835	.69107	3.50	33.115	.03020	16.543	16.573	.99818
0.86	2.3632	.42316	0.9700	1.3932	.69626	3.60	36.598	.02732	18.286	18.313	.99851
0.87	2.3869	.41895	0.9840	1.4029	.70137	3.70	40.447	.02472	20.211	20.236	.99878
0.88	2.4109	.41478	0.9981	1.4128	.70642	3.80	44.701	.02237	22.339	22.362	.99900
0.89	2.4351	.41066	1.0122	1.4229	.71139	3.90	49.402	.02024	24.691	24.711	.99918
0.90	2.4596	.40657	1.0265	1.4331	.71630	4.00	54.598	.01832	27.290	27.308	.99933
0.91	2.4843	.40252	1.0409	1.4434	.72113	4.10	60.340	.01657	30.162	30.178	.99945
0.92	2.5093	.39852	1.0554	1.4539	.72590	4.20	66.686	.01500	33.336	33.351	.99955
0.93	2.5345	.39455	1.0700	1.4645	.73059	4.30	73.700	.01357	36.843	36.857	.99963
0.94	2.5600	.39063	1.0847	1.4753	.73522	4.40	81.451	.01227	40.719	40.732	.99970
0.95	2.5857	.38674	1.0995	1.4862	.73978	4.50	90.017	.01111	45.003	45.014	.99975
0.96	2.6117	.38289	1.1144	1.4973	.74428	4.60	99.484	.01005	49.737	49.747	.99980
0.97	2.6379	.37908	1.1294	1.5085	.74870	4.70	109.95	.00910	54.969	54.978	.99983
0.98	2.6645	.37531	1.1446	1.5199	.75307	4.80	121.51	.00823	60.751	60.759	.99986
0.99	2.6912	.37158	1.1598	1.5314	.75736	4.90	134.29	.00745	67.141	67.149	.99989
1.00	2.7183	.36788	1.1752	1.5431	.76159	5.00	148.41	.00674	74.203	74.210	.99991
1.10	3.0042	.33287	1.3356	1.6685	.80050	5.10	164.02	.00610	82.008	82.014	.99993
1.20	3.3201	.30119	1.5095	1.8107	.83365	5.20	181.27	.00552	90.633	90.639	.99994
1.30	3.6693	.27253	1.6984	1.9709	.86172	5.30	200.34	.00499	100.17	100.17	.99995
1.40	4.0552	.24660	1.9043	2.1509	.88535	5.40	221.41	.00452	110.70	110.71	.99996
1.50	4.4817	.22313	2.1293	2.3524	.90515	5.50	244.69	.00409	122.34	122.35	.99997
1.60	4.9530	.20190	2.3756	2.5775	.92167	5.60	270.43	.00370	135.21	135.22	.99997
1.70	5.4739	.18268	2.6456	2.8283	.93541	5.70	298.87	.00335	149.43	149.44	.99998
1.80	6.0496	.16530	2.9422	3.1075	.94681	5.80	330.30	.00303	165.15	165.15	.99998
1.90	6.6859	.14957	3.2682	3.4177	.95624	5.90	365.04	.00274	182.52	182.52	.99998
2.00	7.3891	.13534	3.6269	3.7622	.96403	6.00	403.43	.00248	201.71	201.72	.99999
2.10	8.1662	.12246	4.0219	4.1443	.97045	6.25	518.01	.00193	259.01	259.01	.99999
2.20	9.0250	.11080	4.4571	4.5679	.97574	6.50	665.14	.00150	332.57	332.57	1.0000
2.30	9.9742	.10026	4.9370	5.0372	.98010	6.75	854.06	.00117	427.03	427.03	1.0000
2.40	11.023	.09072	5.4662	5.5569	.98367	7.00	1096.6	.00091	548.32	548.32	1.0000
2.50	12.182	.08208	6.0502	6.1323	.98661	7.50	1808.0	.00055	904.02	904.02	1.0000
2.60	13.464	.07427	6.6947	6.7690	.98903	8.00	2981.0	.00034	1490.5	1490.5	1.0000
2.70	14.880	.06721	7.4063	7.4735	.99101	8.50	4914.8	.00020	2457.4	2457.4	1.0000
2.80	16.445	.06081	8.1919	8.2527	.99263	9.00	8103.1	.00012	4051.5	4051.5	1.0000
2.90	18.174	.05502	9.0596	9.1146	.99396	9.50	13360.	.00007	6679.9	6679.9	1.0000
						10.00	22026.	.00005	11013.	11013.	1.0000

Table 3
Natural Logarithms

n	$\ln n$	n	$\ln n$	n	$\ln n$
0.0	—	4.5	1.5041	9.0	2.1972
0.1	-2.3026	4.6	1.5261	9.1	2.2083
0.2	-1.6094	4.7	1.5476	9.2	2.2192
0.3	-1.2040	4.8	1.5686	9.3	2.2300
0.4	-0.9163	4.9	1.5892	9.4	2.2407
0.5	-0.6931	5.0	1.6094	9.5	2.2513
0.6	-0.5108	5.1	1.6292	9.6	2.2618
0.7	-0.3567	5.2	1.6487	9.7	2.2721
0.8	-0.2231	5.3	1.6677	9.8	2.2824
0.9	-0.1054	5.4	1.6864	9.9	2.2925
1.0	0.0000	5.5	1.7047	10	2.3026
1.1	0.0953	5.6	1.7228	11	2.3979
1.2	0.1823	5.7	1.7405	12	2.4849
1.3	0.2624	5.8	1.7579	13	2.5649
1.4	0.3365	5.9	1.7750	14	2.6391
1.5	0.4055	6.0	1.7918	15	2.7081
1.6	0.4700	6.1	1.8083	16	2.7726
1.7	0.5306	6.2	1.8245	17	2.8332
1.8	0.5878	6.3	1.8405	18	2.8904
1.9	0.6419	6.4	1.8563	19	2.9444
2.0	0.6931	6.5	1.8718	20	2.9957
2.1	0.7419	6.6	1.8871	25	3.2189
2.2	0.7885	6.7	1.9021	30	3.4012
2.3	0.8329	6.8	1.9169	35	3.5553
2.4	0.8755	6.9	1.9315	40	3.6889
2.5	0.9163	7.0	1.9459	45	3.8067
2.6	0.9555	7.1	1.9601	50	3.9120
2.7	0.9933	7.2	1.9741	55	4.0073
2.8	1.0296	7.3	1.9879	60	4.0943
2.9	1.0647	7.4	2.0015	65	4.1744
3.0	1.0986	7.5	2.0149	70	4.2485
3.1	1.1314	7.6	2.0281	75	4.3175
3.2	1.1632	7.7	2.0142	80	4.3820
3.3	1.1939	7.8	2.0541	85	4.4427
3.4	1.2238	7.9	2.0669	90	4.4998
3.5	1.2528	8.0	2.0794	95	4.5539
3.6	1.2809	8.1	2.0919	100	4.6052
3.7	1.3083	8.2	2.1041	200	5.2983
3.8	1.3350	8.3	2.1163	300	5.7038
3.9	1.3610	8.4	2.1282	400	5.9915
4.0	1.3863	8.5	2.1401	500	6.2146
4.1	1.4110	8.6	2.1518	600	6.3069
4.2	1.4351	8.7	2.1633	700	6.5511
4.3	1.4586	8.8	2.1748	800	6.6846
4.4	1.4816	8.9	2.1861	900	6.8024

answers
to odd-
numbered
problems

▶ Exercise Set 1.1 (Page 15)

1. (a)

(b)

15. a, c

17. (a) false   (b) true   (c) true   (d) false
(e) true   (f) true   (g) true

19. $(-\infty, \frac{10}{3})$

(c)

(d)

21. $(-\infty, -\frac{11}{2}]$

(e)

(f)

23. $(-\frac{3}{2}, \frac{1}{2}]$

3. (a) rational   (b) integer, rational
(c) integer, rational   (d) rational
(e) integer, rational   (f) irrational
(g) rational   (h) integer, rational

25. $(-\infty, 3) \cup (4, +\infty)$

27. $(-\frac{3}{2}, 2)$

5. a, d, f

7. (a) $\frac{41}{333}$ (b) $\frac{115}{9}$ (c) $\frac{20943}{550}$ (d) $\frac{537}{1250}$

29. $(-\infty, -2] \cup (2, +\infty)$

9. (a) all values   (b) none

11. (a) yes   (b) no

31. $(-\infty, -3) \cup (3, +\infty)$

13. (a) $\{x: x$ is a positive odd integer$\}$
(b) $\{x: x$ is an even integer$\}$
(c) $\{x: x$ is irrational$\}$
(d) $\{x: x$ is an integer and $7 \leq x \leq 10\}$

33. $(-\infty, -2) \cup (4, +\infty)$

A49

35. $[4, 5]$

39. $(2, +\infty)$

37. $(-8, 0) \cup (4, +\infty)$

47. $(-\infty, -\frac{1}{2})$

► Exercise Set 1.2 (Page 23)

1. (a) 7   (b) $\sqrt{2}$ (c) k^2 (d) k^2

5. (a) 2   (b) 1   (c) 14
     (d) $3 + \sqrt{2}$ (e) 7   (f) 5

7. (a) -9 (b) 7   (c) 12

9. $-\frac{5}{6}, \frac{3}{2}$ **11.** $\frac{1}{2}, \frac{5}{2}$ **13.** $-\frac{11}{10}, \frac{11}{8}$

15. $1, \frac{17}{5}$ **17.** $(-9, -3)$

19. $(-\infty, \frac{1}{2}] \cup [\frac{9}{2}, +\infty)$ **21.** $(-\infty, \frac{5}{2})$ **23.** $[\frac{7}{10}, \frac{7}{2}]$

25. $(0, +\infty)$ **27.** $\alpha \in (-\infty, 2] \cup [3, +\infty)$

29. $ab \geq 0$

35. $\frac{1}{3}$

37. $\frac{13}{11}$ (other answers are possible)

► Exercise Set 1.3 (Page 32)

1.

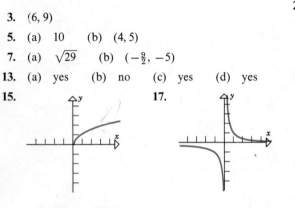

3. $(6, 9)$

5. (a) 10   (b) $(4, 5)$

7. (a) $\sqrt{29}$ (b) $(-\frac{9}{2}, -5)$

13. (a) yes   (b) no   (c) yes   (d) yes

15.   **17.**

19.   **21.**

23. The union of the graphs of $x - y = 0$ and $x + y = 0$.

25.

27. $(x - 3)^2 + (y + 2)^2 = 16$

29. $(x + 3)^2 + (y + 4)^2 = 25$

31. circle; $C(1, 2)$, $r = 4$

33. point

35. circle; $C(0, 0)$, $r = \frac{1}{3}$

37. $3x - 2y - 5 = 0$

39. 0

▶Exercise Set 1.4 (Page 40)

1. (a) $\frac{1}{2}$ (b) -1 (c) 0 (d) not defined

3. (a) $27°$ (b) $45°$ (c) $63°$ (d) $89°$

  (e) $153°$ (f) $135°$ (g) $120°$ (h) $91°$

5.

7. (a) 14 (b) $-\frac{1}{3}$

9. 29

11. (a) yes (b) no

19. (a) $\frac{7}{19}$ (b) $\frac{57}{25}$ (c) $\frac{4}{13}$

21. (a) $20°$ (b) $66°$ (c) $17°$

23. 11.09

▶Exercise Set 1.5 (Page 47)

1. (a) (b) (c) (d)

3. (a) (b) (c)

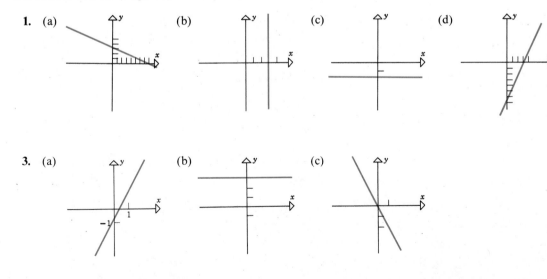

5.

	(a)	(b)	(c)	(d)	(e)
Slope	3	$-1/4$	$-3/5$	0	$-b/a$
y-intercept	2	3	$8/5$	1	b

7. (a) $60°$ (b) $117°$ **9.** $y = -2x + 4$

11. $y = 4x + 7$ **13.** $y = 11x - 18$

15. $y = \dfrac{1}{\sqrt{3}}x - 3$ **17.** $y = \frac{1}{2}x + 2$

19. $y = 1$ **21.** $x = 5$

23. (a) parallel   (b) perpendicular
(c) parallel   (d) perpendicular
(e) neither

25. (a) $(4, -1)$ (b) $(1, -2)$

27. 3   **31.** 4

35. (a) $(\frac{9}{2}, \frac{7}{2})$ (b) $x - 9y + 27 = 0$ **37.** $(-\frac{29}{8}, -\frac{2}{4})$

39. (a) $F = \frac{9}{5}C + 32$ (b) $\frac{5}{9}$

▶ Chapter 1 Review Exercises (Page 48)

1. (a) $(-3, 5]$ (b) $[-3, 3]$
(c) $(-\infty, -\frac{1}{2}] \cup [\frac{1}{2}, +\infty)$

3. (a) $(-\infty, -\frac{1}{2}) \cup (3, +\infty)$ (b) $[1, 4]$

5. (a) $(-2, -1] \cup [1, 2)$
(b) $(-\infty, -5] \cup [-1, +\infty)$

7. (a) Take $a = -2, b = 1$.
(b) $a + b > 0$

9. Both legs of a right triangle are no longer than the hypotenuse.

11. (a) all points on the y-axis or the line $y = x$
(b) all points on the line $x = 1$ or the line $y = x + 1$

13. (a) all points on or outside the circle of radius 4 about $(1, 3)$
(b)

15. at $(2, 4)$ and $(-1, 1)$

17.

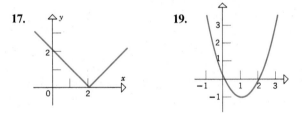

19.

21. $(x - 1)^2 + (y - 2)^2 = 25$

23. four circles: $(x - h)^2 + (y - k)^2 = 25$, where $h = 1$ or 11 and $k = 2$ or 12

25. point $(-2, -1)$ **27.** no graph

29. (a) $y = 4x/3$; 10; $(0, 0)$
(b) $x = 3$; 8; $(3, 0)$
(c) $y = 4$; 6; $(0, 4)$
(d) $y = 7 - x$; $\sqrt{2}$; $(\frac{7}{2}, \frac{7}{2})$

31. $(4, -3)$ and $(-4, 3)$

33. (a) $90°$ (b) $0°$ (c) $135°$ (d) $60°$

35. $y = -3$ **37.** $y = -\frac{1}{2}x$

39. $L: y - 0 = (-2)(x - 1)$
$L': y = \frac{1}{2}x - 3$; $(2, -2)$

41. $L: y - 1 = (\frac{2}{5})(x - 3)$
$L': 3x + 2y = -8$; $(-2, -1)$ **43.** no

▶ Exercise Set 2.1 (Page 62)

1. (a) 14   (b) 50   (c) 2
    (d) 11   (e) $3t^2 + 2$

3. (a) 4   (b) -8 (c) $\frac{1}{3}$
    (d) $\frac{1}{3.1}$ (e) 5.8

5. (a) 1   (b) 1   (c) -1
    (d) -1 (e) not defined

7. $(-\infty, -\frac{7}{5}) \cup (-\frac{7}{5}, +\infty)$

9. $(-\infty, -2] \cup [1, +\infty)$

11. $(-\infty, +\infty)$ **13.** $[5, +\infty)$

15. $(-\infty, -3] \cup [2, +\infty)$ **17.** $(-\infty, 3) \cup (3, +\infty)$

19.

21.

23.

25.

27.

29.

31.

33.

35. $\dfrac{1 - (1/x)}{1 + (1/x)} = \dfrac{x - 1}{x + 1}, x \neq 0.$

37. (a) $y = \dfrac{1}{x^2}$ (b) $y = \dfrac{1 - x}{1 + x}$ (c) $y = -x$

39. a, d, f

▶ Exercise Set 2.2 (Page 73)

1. (a) $t^2 + 1$ (b) $t^2 + 4t + 5$
    (c) $x^2 + 4x + 5$ (d) $\dfrac{1}{x^2} + 1$
    (e) $x^2 + 2hx + h^2 + 1$ (f) $x^2 + 1$
    (g) $x + 1, x \geq 0$ (h) $9x^2 + 1$

3. (a) $x^2 + 2x + 1$ (b) $-x^2 + 2x - 1$
    (c) $2x(x^2 + 1)$ (d) $\dfrac{2x}{x^2 + 1}$
    (e) $2(x^2 + 1)$ (f) $4x^2 + 1$
    (g) $2kx$

5. (a) $\sqrt{x + 1} + x - 2$ (b) $\sqrt{x + 1} - x + 2$
    (c) $(x - 2)\sqrt{x + 1}$ (d) $\dfrac{\sqrt{x + 1}}{x - 2}$
    (e) $\sqrt{x - 1}$ (f) $\sqrt{x + 1} - 2$
    (g) $k\sqrt{x + 1}$

7. (a) $\sqrt{x - 2} + \sqrt{x - 3}$
    (b) $\sqrt{x - 2} - \sqrt{x - 3}$
    (c) $\sqrt{x - 2}\sqrt{x - 3}$ (d) $\dfrac{\sqrt{x - 2}}{\sqrt{x - 3}}$
    (e) $\sqrt{\sqrt{x - 3} - 2}$ (f) $\sqrt{\sqrt{x - 2} - 3}$

(g) $k\sqrt{x-2}$

9. (a) $4x - 15$ (b) $4x^2 - 20x + 25$

11. $g(x) = \sqrt{x},\ h(x) = x + 2$

13. $g(x) = x^7,\ h(x) = x - 5$

15. $g(x) = |x|,\ h(x) = x^2 - 3x + 5$

19. (a) explicit algebraic
 (b) rational, explicit algebraic
 (c) explicit algebraic
 (d) polynomial, rational, explicit algebraic

21. (a) explicit algebraic
 (b) rational, explicit algebraic
 (c) monomial, polynomial, rational, explicit algebraic
 (d) explicit algebraic

23. (a)   (b)

25. yes

27. $\sqrt{\tfrac{1}{2}x - 3}$

► Exercise Set 2.3 (Page 83)

1. (a) $\tfrac{7}{2}$ (b) 3, $y = 3x - \tfrac{9}{2}$

(c)

3. (a) $-\tfrac{1}{6}$ (b) $-\tfrac{1}{4},\ y = -\tfrac{1}{4}x + 1$

(c)

5. (b) $y = 75x - 250$ (c) $y = 3x_0^2 x - 2x_0^3$

7. (a) t_0 (b) 0   (c) speeding up
 (d) slowing down

9. It is a straight line.

11. (a) 320,000 ft   (b) 8000 ft/sec
 (c) 45 ft/sec   (d) 24,000 ft/sec

13. (a) 720 ft/min   (b) 192 ft/min

► Exercise Set 2.4 (Page 93)

1. (a) -1 (b) 3
 (c) does not exist   (d) 1
 (e) -1 (f) 3

3. (a) 1   (b) 1
 (c) 1   (d) 1
 (e) $-\infty$ (f) $+\infty$

5. (a) 0   (b) 0

     (c) 0   (d) 3

     (e) $+\infty$ (f) $+\infty$

7. (a) $-\infty$ (b) $+\infty$

     (c) does not exist   (d) not defined

     (e) 2   (f) 0

9. (a) $-\infty$ (b) $-\infty$

     (c) $-\infty$ (d) 1

     (e) 2   (f) 2

11. (a) 0   (b) 0

     (c) 0   (d) 0

     (e) does not exist   (f) does not exist

13. all values except -4

▶ Exercise Set 2.5 (Page 108)

1. 7   **3.** π **5.** 36

7. $\sqrt{109}$ **9.** 14   **11.** 0

13. 8   **15.** 4   **17.** $-\frac{4}{5}$

19. $\frac{3}{2}$ **21.** 0   **23.** 0

25. $-\sqrt{5}$ **27.** $\dfrac{1}{\sqrt{6}}$ **29.** $\sqrt{3}$

31. $+\infty$ **33.** does not exist   **35.** $-\infty$

37. $+\infty$ **39.** does not exist   **41.** $+\infty$

43. $-\infty$ **45.** $-\frac{1}{7}$ **47.** -1

49. (a) 2   (b) 2   (c) 2

51. 4

55. if $r(a)$ is defined

▶ Exercise Set 2.6 (Page 119)

Note: there are other possible answers for Exercises 1–21, 27.

1. 0.05   **3.** 1/700   **5.** 0.05

7. 1/9000   **9.** 1   **11.** $\delta = \dfrac{\epsilon}{3}$

13. $\delta = \epsilon$ **15.** $\delta = \min\left(\dfrac{\epsilon}{6}, 1\right)$

17. $\delta = \min\left(\dfrac{\epsilon}{36}, \dfrac{1}{4}\right)$ **19.** $\delta = \min(2\epsilon, 4)$

21. $\delta = \epsilon$ **27.** $\delta = \min\left(\dfrac{\epsilon}{8}, 2\right)$

▶ Chapter 2 Review Exercises (Page 120)

1. $|x| \le 2$; $\sqrt{2}, 2, 1$

3. $x \ne -2, 1$; $\frac{1}{2}$, undefined, $\frac{1}{4}$

5. all x; $-1, 3, \sqrt{3}$

7. (a) $(-6 + 6x - 2x^2)/x^2$

     (b) $-9/[x(x + 3)]$

     (c) $(3x^2 - 9x + 3)/(x - 3)$

     (d) $(4x - 3)/(3 - x)$

9. the horizontal line $y = -\pi$; domain: all x; range: $\{-\pi\}$

11. domain: $x \ne -2$; range $y \ne -2$

13. domain: $x \geq -\frac{1}{3}$; range: $y \leq 0$

15. domain: $x \neq \pm 2$; range: $y \neq 0, \frac{1}{2}$

17. Some possible answers are:
   (a) $h(x) = x^3$, $g(x) = x^2 + 3$;
        $h(x) = x^6$, $g(x) = x + 3$
   (b) $h(x) = x^2 + 1$, $g(x) = \sqrt{x}$;
        $h(x) = x^2$, $g(x) = \sqrt{x + 1}$
   (c) $h(x) = 3x + 2$, $g(x) = \sin x$;
        $h(x) = 3x$, $g(x) = \sin (x + 2)$

19. $y = -1$

21. 5

23. (a) -1 (b) does not exist   (c) 1
   (d) 0   (e) $-\infty$ (does not exist)   (f) 0
   (g) 0   (h) $-\infty$ (does not exist)

25. 2, 1, 0, does not exist, $+\infty$ (does not exist), $-\infty$ does not exist

27. 5, 10, 0, 10, 0, $-\infty$ (does not exist), $+\infty$ (does not exist)

► Exercise Set 3.1 (Page 128)

1. $6x$

3. 5

5. $\dfrac{1}{2\sqrt{x}}$

7. $-\dfrac{2}{x^3}$

9. $2ax$

11. $18, y = 18x - 27$

13. $5, y = 5x - 2$

15. $\frac{1}{6}, y = \frac{1}{6}x + \frac{3}{2}$

17. (a) $8x$ (b) 8

19.

21.

23.

25.

27.

29. $-\dfrac{1}{2x^{3/2}}$

▶ Exercise Set 3.2 (Page 139)

1. $28x^6$

3. $24x^7 + 2$

5. 0

7. $-\frac{1}{3}(7x^6 + 2)$

9. $3ax^2 + 2bx + c$

11. $24x^{-9}$

13. $-3x^{-4} - 7x^{-8}$

15. $18x^2 - \frac{3}{2}x + 12$

17. $-15x^{-2} - 14x^{-3} + 48x^{-4} + 32x^{-5}$

19. $10x^9 + 24x^5 + 8x$

21. $\dfrac{3}{(2x + 1)^2}$

23. $\dfrac{(2x^7 + 4x + 2)(24x^7 - 12x) - (3x^8 - 6x^2 + 1)(14x^6 + 4)}{(2x^7 + 4x + 2)^2}$

25. $\left(\dfrac{3x + 2}{x}\right)(-5x^{-6}) + (x^{-5} + 1)\left(-\dfrac{2}{x^2}\right)$

27. $\dfrac{1}{4}\left(\dfrac{x}{x + 1}\right)(-2x^{-3}) + \dfrac{1}{4}(x^{-2} + 5) \cdot \dfrac{1}{(x + 1)^2}$

29. $32t$

31. $3\pi r^2$

33. $\dfrac{7 - 2t^3}{(t^3 + 7)^2}$

35. $(1, \frac{5}{6}), (2, \frac{2}{3})$

37. $a = 3, b = 2$

41. (a) $2\left(1 + \dfrac{1}{x}\right)(x^{-3} + 7)$

$\qquad + (2x + 1)\left(-\dfrac{1}{x^2}\right)(x^{-3} + 7)$

$\qquad + (2x + 1)\left(1 + \dfrac{1}{x}\right)(-3x^{-4})$

(b) $(-5x^{-6})(x^2 + 2x)(4 - 3x)(2x^9 + 1)$

$\qquad + x^{-5}(2x + 2)(4 - 3x)(2x^9 + 1)$

$\qquad + x^{-5}(x^2 + 2x)(-3)(2x^9 + 1)$

$\qquad + x^{-5}(x^2 + 2x)(4 - 3x)(18x^8)$

(c) $3(7x^6 + 2)(x^7 + 2x - 3)^2$

(d) $100x(x^2 + 1)^{49}$

▶ Exercise Set 3.3 (Page 149)

1. $-2 \sin x - 3 \cos x$

3. $\dfrac{x \cos x - \sin x}{x^2}$

5. $\sec^3 x + \sec x \tan^2 x$

7. $-\dfrac{\csc x}{1 + \csc x}$

9. 0

13. 3

15. $\frac{7}{3}$

17. 1

19. 2

21. 0

23. $-\frac{25}{49}$

25. (a) $y = x$ (b) $y = -x + \pi$

(c) $y = \dfrac{1}{\sqrt{2}}x - \dfrac{\pi}{4\sqrt{2}} + \dfrac{1}{\sqrt{2}}$

31. $\dfrac{\pi}{4}$

39. 2

▶ Exercise Set 3.4 (Page 158)

1. (a) 5 (b) 4

(c)

3. (a) $-\frac{1}{3}$ (b) -0.5

(c)

5. $dy = 3x^2\, dx$

$\Delta y = 3x^2\, \Delta x + 3x(\Delta x)^2 + (\Delta x)^3$

7. $dy = (2x - 2)\,dx$
$\Delta y = 2x\,\Delta x + (\Delta x)^2 - 2\,\Delta x$

9. $dy = (12x^2 - 14x + 2)\,dx$

11. $dy = (\cos x - x\sin x)\,dx$

13. (a) 10 (b) 4

15. (a) 3π (b) 4π

17. (a) ± 2 square feet
(b) $\pm 1\%$ (side), $\pm 2\%$ (area)

19. $\pm 10\%$

▶ Exercise Set 3.5 (Page 166)

1. $37(x^3 + 2x)^{36}(3x^2 + 2)$

3. $8x\sec^2(4x^2)$

5. $-x^3\sec\left(\dfrac{1}{x}\right)\tan\left(\dfrac{1}{x}\right) + 5x^4\sec\left(\dfrac{1}{x}\right)$

7. $\sin x\sin(\cos x)$

9. $12(5x + 8)^{13}(x^3 + 7x)^{11}(3x^2 + 7)$
$\qquad\qquad + 65(x^3 + 7x)^{12}(5x + 8)^{12}$

11. $\dfrac{-64x(2x + 1)^{-3}(4x^2 - 1)^{-9} + 6(4x^2 - 1)^{-8}(2x + 1)^{-4}}{(2x + 1)^{-6}}$

13. $y = -x$ **15.** $y = -1$

17. $\frac{2}{25}(x - 2)$

19. $-\dfrac{9\sin^2(1/x)\cos(1/x)}{x^2}$

21. $3\cot^2\theta\csc^2\theta$

23. $\pi(b - a)\sin 2\pi\omega$

25. (b) $\begin{cases} \cos x, & 0 < x < \pi \\ -\cos x, & -\pi < x < 0 \end{cases}$ for both

27. $\dfrac{d}{dx}[f(g(x))] = f'(g(x))g'(x)$

29. 6

▶ Exercise Set 3.6 (Page 174)

1. $\frac{3}{2}x^{-1/4}$

3. $\dfrac{3x^2}{2\sqrt{1 + x^3}}$

5. $\dfrac{\sec^2\sqrt{x}}{2\sqrt{x}}$

7. $-\dfrac{1}{7}\left(1 - \dfrac{1}{x^2}\right)\left(x + \dfrac{1}{x}\right)^{-8/7}$

9. $\dfrac{1}{2\sqrt{x}} + (3x)^{-2/3} + \dfrac{5}{4}(5x)^{-3/4}$

11. $-\dfrac{x}{y}$

13. $-\dfrac{y^2}{x^2}$

15. $-\dfrac{\sqrt{y}}{\sqrt{x}}$

17. $\dfrac{1 - 70x(x^2 + 3y^2)^{34}}{210y(x^2 + 3y^2)^{34}}$

19. $\sec y$

21. $\dfrac{1 - 3y^2\tan^2(xy^2 + y)\sec^2(xy^2 + y)}{3(2xy + 1)\tan^2(xy^2 + y)\sec^2(xy^2 + y)}$

23. $\dfrac{3y^2\sin^2(xy^2)\cos(xy^2)}{2\sqrt{1 + \sin^3(xy^2)} - 6xy\sin^2(xy^2)\cos(xy^2)}$

25. $-\dfrac{14}{13}$

27. 2

29. -2

31. 1

33. $\dfrac{6}{5}$

35. $\dfrac{2t^3 + 3a^2}{2a^3 - 6at}$

37. $-\dfrac{b^2\lambda}{a^2\omega}$

39. $-\dfrac{2y^3 + 3t^2y}{(6ty^2 + t^3)\cos t}$

41. $y = \dfrac{\sqrt{3}}{3}x,\; y = -\dfrac{\sqrt{3}}{3}x$

43. (a) $(0, 0), \left(-\dfrac{5}{\sqrt{2}}, 0\right), \left(\dfrac{5}{\sqrt{2}}, 0\right)$
(b) $9x + 13y = 40$

▶ Exercise Set 3.7 (Page 187)

1. continuous on $(1, 2)$, $[2, 3]$, $(2, 3)$;
   discontinuous on $[1, 3]$, $(1, 3)$, and $[1, 2]$ at $x = 2$

3. continuous on $(1, 3)$, $(1, 2)$, $(2, 3)$;
   discontinuous on $[1,3]$ at $x = 1$ and $x = 3$; on $[1,2]$ at $x = 1$, and on $[2,3]$ at $x = 3$

5. none

7. none

9. $x = \pm 4$

11. none

13. $x = n\pi$, $n = 0, \pm 1, \pm 2, \ldots$

15. $x = n\pi$, $n = 0, \pm 1, \pm 2, \ldots$

17. none

19. $x = 0, 1$

21. (a) 5   (b) 1

33. not differentiable at $x = 1$

35. (a) $x = \frac{2}{3}$ (b) $x = -2, 2$

▶ Chapter 3 Review Exercises (Page 188)

1. k

3. $-2/\sqrt{9 - 4x}$ for $x < 9/4$

5. 0

7. $y + 1 = 5(x - 3)$

9. (a) 12   (b) -7 (c) 9
   (d) $-9/4$ (e) 5   (f) 21
   (g) -35 (h) -7 (i) -126
   (j) -12 (k) $3/2$ (l) $-3/2$

11. $2(x - 3)^3(x^2 + 6x + 3)/(x^2 + 2x)^2$; $3, -3 \pm \sqrt{6}$

13. $-3(3x + 1)^2(3x + 2)/x^7$; $-1/3, -2/3$

15. $(7x^2 + 5x + 3)x^{-1/2}(x^2 + x + 1)^{-2/3}/6$; none

17. $-2\sqrt{2}/x^3 + 2x^{-2}/5$

19. $\sqrt{3}$ $(z = \sin^2 2r)$

21. $2(x - 1)/x^3$ $\left(u = \left(1 - \frac{1}{x}\right)^2\right)$

23. 0

25. 2 $(F(x) = 2x)$

27. ± 1

29. ± 2

31. odd integer multiples of $\pi/4$

33. $\Delta x = \pi/4, \Delta y = 1, dy = \pi/2$

35. (a) $-97/48$ (b) $1 - (\pi/90)$

37. $dh = -\pi/30$ ft

39. (a) 2000 gal/min
    (b) 2500 gal/min

41. $\dfrac{dy}{dx} = -\dfrac{y^2 - \cos (x + 2y)}{2 \cos (x + 2y) - 2xy}$;
    $y = -\frac{1}{2}x, y = -\frac{1}{2}(x - \pi)$

45. a/b

47. does not exist

49. The sine function is continuous.

▶ Exercise Set 4.1 (Page 198)

1. (a) $\dfrac{dA}{dt} = 2x \dfrac{dx}{dt}$ (b) 12 ft²/min

3. (a) $\dfrac{dV}{dt} = \pi \left(r^2 \dfrac{dh}{dt} + 2rh \dfrac{dr}{dt}\right)$
   (b) -20π in.³/sec; decreasing

5. (a) $\dfrac{d\theta}{dt} = \dfrac{\cos^2 \theta}{x^2} \left(x \dfrac{dy}{dt} - y \dfrac{dx}{dt}\right)$
   (b) $-\dfrac{5}{16}$ radians/sec; decreasing

7. $\dfrac{2}{9\pi}$ cm/sec

9. $\dfrac{3}{2\pi}$ ft/min

11. 250 mph

13. $600\sqrt{7}$ mph

15. $\frac{1}{12}$ radians/sec

17. (a) $-\dfrac{60}{7}$ units/sec   (b) falling

19. $\dfrac{11\pi}{150}$ in./min

▶ Exercise Set 4.2 (Page 207)

1. (a) $x = \frac{5}{2}$ (stationary)
    (b) $x = 0$ (not differentiable)

3. (a) $x = \pm\sqrt{2}$ (stationary)
    (b) $x = n\frac{\pi}{3}$, $n = 0, \pm1, \pm2, \ldots$ (stationary)

5. (a) $x = n\frac{\pi}{4}$, $n = 1, 2, 3, 4, 5, 6, 7$ (stationary)
    (b) $x = 0$ (stationary)

7. maximum value 1 when $x = 0, 1$;
    minimum value 0 when $x = \frac{1}{2}$

9. maximum value 27 when $x = 4$;
    minimum value -1 when $x = 0$

11. maximum value $3/\sqrt{5}$ when $x = 1$;
    minimum value $-3/\sqrt{5}$ when $x = -1$

13. maximum value 48 when $x = 8$;
    minimum value 0 when $x = 0, 20$

15. maximum value $1 - (\pi/4)$ when $x = -\pi/4$;
    minimum value $(\pi/4) - 1$ when $x = \pi/4$

17. maximum value 2 when $x = 0$;
    minimum value $\sqrt{3}$ when $x = \pi/6$

19. maximum value 17 when $x = -5$;
    minimum value 1 when $x = -3$

21. maximum value $3\sqrt{3}/2$ when $x = (\pi/6) + n\pi$
    $(n = 0, \pm1, \pm2, \ldots)$
    minimum value $-3\sqrt{3}/2$ when $x = (5\pi/6) + n\pi$
    $(n = 0, \pm1, \pm2, \ldots)$

23. maximum value 2; minimum value $-\frac{1}{4}$

▶ Exercise Set 4.3 (Page 219)

1. $5 + 5$

3. 500 ft by 750 ft

5. 5 in. by $\frac{12}{5}$ in.

7. (a) Use all of the wire for the circle.
    (b) $\dfrac{12\pi}{\pi + 4}$ in. for the circle and $\dfrac{48}{\pi + 4}$ in. for the square.

9. $L/12$ by $L/12$ by $L/12$

11. (a) 7000   (b) yes

13. height $L/\sqrt{3}$, radius $\sqrt{2/3}\,L$

15. height $2\sqrt{(5 - \sqrt{5})/10}\,R$, radius $\sqrt{(5 + \sqrt{5})/10}\,R$

17. $\dfrac{\pi}{3}$

19. $2\pi R^3/(9\sqrt{3})$

21. (a) 24   (b) $24   (c) $24.10
    (d) $R'(x) = 10$, $P'(x) = 6 - 0.2x$

25. (a) $\frac{3}{4}$ (b) $\frac{3}{16}$

29. (c) $\frac{1}{4}$ mi downstream from the house

▶ Exercise Set 4.4 (Page 232)

1. (a) $42x - 10$ (b) 24
    (c) $-\frac{1}{4}(x + 3)^{-3/2}$ (d) $2\sec^2 x \tan x$

3. (a) $4\sin^2 x - 4\cos^2 x$
    (b) $-\frac{10}{27}x^{-4/3} - \frac{32}{27}x^{-7/3}$

5. (a) 0   (b) $-\frac{1}{2}$

7. (a) $\dfrac{1}{2x^{3/2}}$
    (b) $-\dfrac{2\sin y \cos y(1 + x\sin y) + x\cos^3 y}{(1 + x\sin y)^3}$

9. (a) $(-\infty, -2), (-2, +\infty)$ (b) none
    (c) $(-2, +\infty)$ (d) $(-\infty, -2)$ (e) -2

11. (a) $(1, +\infty)$ (b) $(-\infty, 0), (0, 1)$
    (c) $(-\infty, 0), (\frac{2}{3}, +\infty)$ (d) $(0, \frac{2}{3})$
    (e) $x = 0, \frac{2}{3}$

13. (a) $(\pi, 2\pi)$ (b) $(0, \pi)$
    (c) $(\pi/2, 3\pi/2)$ (d) $(0, \pi/2), (3\pi/2, 2\pi)$
    (e) $\pi/2, 3\pi/2$

15. (a) $(-\pi/2, \pi/2)$ (b) none
    (c) $(0, \pi/2)$ (d) $(-\pi/2, 0)$ (e) 0

17. (a) $(-\infty, -2), (-2, +\infty)$ (b) none
     (c) $(-\infty, -2)$ (d) $(-2, +\infty)$
     (e) -2

19. (a)   (b)

     (c)

21. (a)   (b)

23. (a) $x = a$ (b) none   (c)

27. (c) $\dfrac{3n - 2}{3}$

29. (a) $(-1)^n \dfrac{n(n - 1)(n - 2) \cdots 1}{x^{n+1}}$

     (b) $\dfrac{2^n n(n - 1)(n - 2) \cdots 1}{(1 - 2x)^{n+1}}$

▶ Exercise Set 4.5 (Page 242)

1.

3. $(-2, 9)$ 10 $(-1, 7)$ 5 $(0, 5)$

9. $y = 2$ $x = 3$

5. $(1, 0)$

7. $(0, -1)$ $(-1, -2)$ $\left(-\frac{3}{2}, -\frac{43}{16}\right)$

11. $x = -1$ $x = 1$ $y = 1$ $(0, 0)$

13.

15.

17.

19.

21.

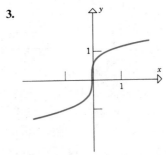

▶ Exercise Set 4.6 (Page 247)

1.

3.

5.

7.

9.

11.

13.

15. **17.** **19.**

▶ Exercise Set 4.7 (Page 253)

1. relative max of 5 at $x = -2$

3. relative min of 0 at $x = \pi$;
   relative max of 1 at $x = \pi/2, 3\pi/2$

5. none

7. relative min of 0 at $x = 1$;
   relative max of $\frac{4}{27}$ at $x = \frac{1}{3}$

9. relative min of 0 at $x = 0$;
   relative max of 1 at $x = -1, 1$

11. relative min of 0 at $x = 0$

13. relative min of 0 at $x = 0$

15. relative min of 0 at $x = -2, 2$;
   relative max of 4 at $x = 0$

17. relative min of 0 at $x = \pm\pi/2, \pm 3\pi/2,$
   $\pm 5\pi/2, \ldots$;
   relative max of 1 at $x = 0, \pm\pi, \pm 2\pi, \ldots$

19. relative min of $\tan(1)$ at $x = 0$

21. relative min of 0 at $x = \pi/2, \pi, 3\pi/2$;
   relative max of 1 at $x = \pi/4, 3\pi/4, 5\pi/4, 7\pi/4$

23. minimum value $-\frac{13}{4}$

25. maximum value -4

27. (a) relative min of 0 at $x = a$
   (b) none

29. (b) 125

31. $a_0 = 9, a_1 = -8, a_2 = 2$

▶ Exercise Set 4.8 (Page 258)

1. (a) 10, 10   (b) no minimum

3. $40\sqrt{2}$ ft ($1 fencing), $20\sqrt{2}$ ft ($2 fencing)

5. base: $5\sqrt[3]{4}$ cm square
   height: $10\sqrt[3]{4}$ cm
   7. radius: $\sqrt[6]{\dfrac{450}{\pi^2}}$ cm
   height: $\dfrac{30}{\pi}\sqrt[3]{\dfrac{\pi^2}{450}}$ cm

9. $(-\sqrt{2}, 1), (\sqrt{2}, 1)$

11. (a) no maximum   (b) -3

13. closest: (4, 8)
   farthest: $(-4, -8)$

15. $\left(-\dfrac{1}{\sqrt{3}}, \dfrac{3}{4}\right)$

17. height: $4R$
   radius: $\sqrt{2}R$

19. $4(1 + 2^{2/3})^{3/2}$ ft

▶ Exercise Set 4.9 (Page 265)

1. 3

3. π

5. 1

7. $\frac{5}{4}$

▶ Chapter 4 Review Exercises (Page 271)

1. decreasing 39π m^3/sec

3. 60 ft/sec (toward pole)

5. $m = -2 = f(-1)$; $M = 27/256 = f(3/4)$

7. no m; $M = 1/\sqrt{3} = f(\sqrt{3})$

9. $m = -3 = f(3)$; $M = 0 = f(2)$

13. $(-\frac{1}{\sqrt{3}}, \frac{3}{4})$ $(\frac{1}{\sqrt{3}}, \frac{3}{4})$ **15.**

17. **19.**

21. at $(1, 2)$, $y = 2x$; at $(2, 3)$, $y = 3$

23. rel. max at $x = \pi/2$ and $3\pi/2$;
rel. min at $x = 7\pi/6$ and $11\pi/6$

25. rel. min at $x = 9$

27. rel. max at $x = 2\pi/3$ and $4\pi/3$;
rel. min at $x = \pi$

29. $2\sqrt{8}$ by $3\sqrt{2}$

31. $r = 2P/(3\pi + 8)$ ft. **33.** $r/h = \pi/8$

35. satisfied; $c = 0$ **37.** satisfied; $c = \sqrt{\pi/2}$

39. satisfied; $c = 1$ **41.** satisfied; $c = \frac{1}{2}$, $\sqrt{2}$

▶ Exercise Set 5.1 (Page 278)

1. 3

3. 2

5. 14/3

7. 1/2

9. yes

▶ Exercise Set 5.2 (Page 286)

1. $\frac{1}{9}x^9 + C$ **3.** $\frac{7}{12}x^{12/7} + C$

5. $8\sqrt{t} + C$ **7.** $\frac{2}{9}x^{9/2} + C$

9. $\frac{1}{2}x^2 + \frac{1}{5}x^5 + C$ **11.** $-4\cos x + 2\sin x + C$

13. $4\tan x - \csc x + C$ **15.** $\sec x + x + C$

17. $\sec x + C$ **19.** $\theta - \cos\theta + C$

21. $\frac{1}{2}x^2 - \frac{2}{x} + \frac{1}{3x^3} + C$ **23.** $F(x) = \frac{3}{4}x^{4/3} + \frac{5}{4}$

25. $f(x) = \frac{4}{15}x^{5/2} + C_1 x + C_2$

▶ Exercise Set 5.3 (Page 293)

1. (a) $\frac{1}{24}(x^2 + 1)^{24} + C$ (b) $-\frac{1}{4}\cos^4 x + C$
(c) $-2\cos\sqrt{x} + C$
(d) $\frac{3}{4}\sqrt{4x^2 + 5} + C$

3. (a) $-\frac{1}{2}\csc^2 x + C$ (b) $\frac{1}{10}(1 + \sin t)^{10} + C$
(c) $\frac{2}{7}(1 + x)^{7/2} - \frac{4}{5}(1 + x)^{5/2} +$
$\frac{2}{3}(1 + x)^{3/2} + C$
(d) $-\cot(\sin x) + C$

5. $x + \sin x + C$ **7.** $\frac{1}{18}(x - 2)^{18} + C$

9. $\frac{5}{\pi}\sin(\pi x - 3) + C$ **11.** $\frac{2}{3}\sqrt{3s + 1} + C$

13. $\dfrac{n}{b(n + 1)}(a + bx)^{(n+1)/n} + C$

15. $\frac{1}{7}\sin^7 x + C$ **17.** $\frac{2}{3}\sqrt{x^3 + 1} + C$ **29.** $\frac{2}{3}(y + 1)^{3/2} - 2(y + 1)^{1/2} + C$

19. $\frac{2}{3}(\tan x)^{3/2} + C$ **21.** $\frac{1}{3}\tan^3 x + C$ **31.** $-\cos x + \frac{1}{3}\cos^3 x + C$

23. $-\cos(\sin\theta) + C$ **33.** $\tan x - x + C$

25. $\dfrac{1}{b(n + 1)}\sin^{n+1}(a + bx) + C$ **36.** (a) $\frac{1}{2}\sin^2 x + C$; $-\frac{1}{2}\cos^2 x + C$

27. $\frac{3}{10}(x^2 + 5)^{5/3} + C$ (b) The two answers differ by a constant. (Subtract the answers and use a trigonometric identity.)

▶ Exercise Set 5.4 (Page 302)

1.

t	s	v	$|v|$	a	direction; motion
1	-5	-9	9	-6	left; speeding up
2	-16	-12	12	0	left; neither
3	-27	-9	9	6	left; slowing down
4	-32	0	0	12	stopped
5	-25	15	15	18	right; speeding up

3. (a) 12   (b) $t = 2.2$, $s = -24.2$

5.

7.

9.

11. (a) $v(3) = 16$ ft/sec, $v(5) = -48$ ft/sec
     (b) 196 ft   (c) 112 ft/sec

13. $80\sqrt{10}$ ft/sec

15. (a) t_1 (b) speeding up   (c) slowing down

▶ Exercise Set 5.5 (Page 311)

1. (a) 36   (b) 55   (c) 40   (d) 6

3. (a) $\displaystyle\sum_{k=1}^{10} k$ (b) $\displaystyle\sum_{k=1}^{20} 3k$

     (c) $\displaystyle\sum_{k=1}^{49} k(k + 1)$ (d) $\displaystyle\sum_{k=0}^{4} 2^k$

     (e) $\displaystyle\sum_{k=1}^{10} 2k$ (f) $\displaystyle\sum_{k=1}^{8} (2k - 1)$

5. (a) $\displaystyle\sum_{k=1}^{5} (-1)^{k+1}a_k$ (b) $\displaystyle\sum_{k=0}^{5} (-1)^{k+1}b_k$

     (c) $\displaystyle\sum_{k=0}^{n} a_k x^k$ (d) $\displaystyle\sum_{k=0}^{5} a^{5-k}b^k$

7. 214,365

9. (a) n^2 (b) -3 (c) $\frac{1}{2}n(n + 1)x$
     (d) $(n - m + 1)c$

11. (a) $\displaystyle\sum_{k=0}^{14} (k+4)(k+1)$ (b) $\displaystyle\sum_{k=5}^{19} (k-1)(k-4)$

13. $\displaystyle\sum_{k=1}^{18} k \sin\frac{\pi}{k}$

15. both are valid

17. (a) $\frac{3}{2}(3^{20}-1)$ (b) $2^{31}-2^5$ (c) $-\frac{2}{3}\left(1+\frac{1}{2^{101}}\right)$

19. 110

▶ Exercise Set 5.6 (Page 321)

1. (a) 46 (b) 58

3. (a) $\dfrac{\sqrt{2}\pi}{4} \approx 1.111$ (b) $\dfrac{(\sqrt{2}+2)\pi}{4} \approx 2.682$

5. $\frac{15}{4}$

7. $\frac{1}{3}$

9. 320

13. $\frac{1}{3}$

17. $\frac{1}{2}(b^2-a^2)$

11. $\frac{15}{4}$

15. (b) $\frac{1}{4}(b^4-a^4)$

▶ Exercise Set 5.7 (Page 331)

1. (a) $\frac{71}{6}$ (b) 2

3. (a) $-\frac{117}{16}$ (b) 3

7. (a) $\displaystyle\lim_{\max \Delta x_k \to 0} \sum_{k=1}^{n} 2x_k^* \Delta x_k$; $a=1, b=2$

(b) $\displaystyle\lim_{\max \Delta x_k \to 0} \sum_{k=1}^{n} (1+\cos x_k^*) \Delta x_k$;

$\quad a=-\pi/2, b=\pi/2$

(c) $\displaystyle\lim_{\max \Delta x_k \to 0} \sum_{k=1}^{n} \frac{x_k^*}{x_k^*+1} \Delta x_k$; $a=0, b=1$

9. (a) -1 (b) $\frac{5}{2}$ (c) $\frac{21}{2}$ (d) $\frac{1}{2}k^2-k$

11. $-\frac{15}{4}$

▶ Exercise Set 5.8 (Page 339)

1. $\frac{65}{4}$

5. $\frac{844}{5}$

9. 0

13. $\frac{2}{3}$

3. $\frac{81}{10}$

7. 0

11. $\dfrac{\pi^2}{9}+2\sqrt{3}$

15. 0

17. $\frac{2}{3}(\sqrt{10}-2\sqrt{2})$

21. 12

25. $\frac{203}{2}$

19. $-\frac{5}{3}a^{3/2}$

23. $\frac{3}{4}(2-\sqrt{2})$

29. b, c

▶ Exercise Set 5.9 (Page 347)

1. (a) $\displaystyle\int_1^3 u^7\,du$ (b) $-\dfrac{1}{2}\displaystyle\int_7^4 u^{1/2}\,du$ **3.** $2 - \dfrac{\sqrt{2}}{2}$ **5.** $\frac{929}{12}$

                                       7. $\frac{1192}{15}$ **9.** $-\frac{1}{48}$

(c) $\dfrac{1}{\pi}\displaystyle\int_{-\pi}^{\pi} \sin u\,du$ (d) $\displaystyle\int_0^1 u^2\,du$ **11.** $\frac{106}{405}$ **13.** 0

                                       15. $2(\sqrt{7} - \sqrt{3})$ **17.** -4

(e) $\dfrac{1}{2}\displaystyle\int_3^4 (u - 3)u^{1/2}\,du$ **19.** $\frac{2}{3}$, 3   **21.** 0, 2

                                       23. $\frac{9}{4}$, $\frac{11}{4}$ **25.** $-\frac{10}{3}$, $\frac{17}{3}$

(f) $\displaystyle\int_2^0 (u + 5)u^{20}\,du$ **27.** $\frac{204}{25}$, $\frac{204}{25}$

▶ Exercise Set 5.10 (Page 351)

1. (a) $\frac{4}{3}$ (b) $\dfrac{2}{\sqrt{3}}$

3. $f_{\text{ave}} = \frac{4}{3}$; $x^* = \frac{16}{9}$

5. $f_{\text{ave}} = \frac{1}{2}\alpha(x_0 + x_1) + \beta$; $x^* = \frac{1}{2}(x_0 + x_1)$ if $\alpha \neq 0$;
     x^* is any point in $[x_0, x_1]$ if $\alpha = 0$.

9. 1404π lb

▶ Chapter 5 Review Exercises (Page 352)

1. $-\frac{1}{2}x^{-2} + 2\sqrt{x} + 5\cos x + C$

3. $\frac{2}{9}(\sqrt{x} + 2)^9 + C$

5. $-\frac{1}{2}\cos\sqrt{2x^2 - 5} + C$

7. $2x^{3/2} + \frac{6}{17}x^{17/6} + C$

9. $\frac{1}{5}\tan(\sin 5t) + C$

11. (a) $\frac{1}{6}y^6 + y^4 + 2y^2 + C$
     (b) $\frac{1}{6}(y^2 + 2)^3 + C$

13. $\displaystyle\int_0^1 u^4\,du = 1/5$

15. $\displaystyle\int_1^4 \dfrac{u - 1}{\sqrt{u}}\,du = \dfrac{8}{3}$

17. $\displaystyle\int_0^{\pi/2} \dfrac{4}{\pi}\cos u\,du = 4/\pi$

19. $17/2$

21. 128 ft/sec; 256 ft

23. (a) 20   (b) 8   (c) $4n$
     (d) $8/15$ (e) $61/24$ (f) 9
     (g) $1 + \sqrt{2}$ (h) $3(\sqrt{2} + 1)/4$

25. (a) $\displaystyle\sum_{k=1}^{9} (-1)^{k-1}\left(\dfrac{k}{k + 1}\right)^2 = \sum_{k=2}^{10} (-1)^k\left(\dfrac{k - 1}{k}\right)^2$

     (b) $\displaystyle\sum_{k=1}^{11} \dfrac{(-\pi)^{k+1}}{k} = \sum_{k=2}^{12} (-1)^k \dfrac{\pi^k}{k - 1}$

27. (a) $64\left[1 - \dfrac{(n + 1)(2n + 1)}{6n^2}\right]$

     (b) $64\left[1 - \dfrac{(n - 1)(2n - 1)}{6n^2}\right]$

     (c) $128/3$

29. (a) 12   (b) 12   (c) 12

31. (a) -4 (b) $-9\pi/4$
     (c) -2 (d) 0
     (e) 0   (f) 30
     (g) 2

33. (a) 10   (b) 1
     (c) -4 (let $u = -x$)
     (d) 4

35. $f_{\text{ave}} = 7$; $x^* = -\sqrt{7/3}$

37. $f_{\text{ave}} = 13/4$; $x^* = -5/4$

▶ Exercise Set 6.1 (Page 358)

1. $\frac{32}{3}$ 3. $\frac{49}{192}$

5. $\pi - 2$ 7. $\frac{9}{2}$

9. $\frac{355}{6}$ 11. $\frac{1}{2}$

13. $y = \dfrac{9}{\sqrt[3]{4}}$

▶ Exercise Set 6.2 (Page 365)

1. (a) $\dfrac{32\pi}{5}$ (b) 8π

3. (a) $\dfrac{373\pi}{14}$ (b) $\dfrac{77\pi}{5}$

5. $\dfrac{648\pi}{5}$ 7. $\frac{4}{3}\pi ab^2$

9. $\frac{1}{3}\pi r^2 h$ 11. $40{,}000\pi$ ft^3

13. $36\sqrt{3}$ 15. $\frac{1}{4}(\pi + 2)$

17. $\frac{2}{3}r^3 \tan\theta$

▶ Exercise Set 6.3 (Page 373)

1. $2\pi/5$ 3. $3\pi\sqrt[3]{4}\,(1 + 3\sqrt[3]{3})$

5. $20\pi/3$ 7. $\pi/2$

9. $\pi/5$ 11. (a) $7\pi/30$

13. $9\pi/14$ 15. $\frac{1}{3}\pi r^2 h$

17. $\dfrac{4\pi}{3}[r^3 - (r^2 - a^2)^{3/2}]$

▶ Exercise Set 6.4 (Page 377)

1. $\sqrt{5}$

3. $\frac{1}{243}(85\sqrt{85} - 8)$

5. $\frac{17}{6}$

7. (c) $\frac{1}{27}(13\sqrt{13} + 80\sqrt{10} - 16)$

▶ Exercise Set 6.5 (Page 382)

1. $35\sqrt{2}\,\pi$ 3. 8π

5. $\dfrac{16\pi}{9}$ 7. $40\pi\sqrt{82}$

9. 24π 11. $\dfrac{16{,}911\pi}{1024}$

▶ Exercise Set 6.6 (Page 387)

1. (a) 210 in.-lb   (b) $\frac{5}{6}$ in.-lb

3. $\frac{9}{5}$ ft-ton

5. 20 lb/ft

7. $900\pi\rho$ ft-lb

9. (a) $926{,}640$ ft-lb   (b) 0.468 horsepower

11. $75{,}000$ ft-lb

13. (a) 96×10^9 (b) $4{,}800{,}000$ mile-lb

▶ Exercise Set 6.7 (Page 393)

1. (a) 2808 lb   (b) 3600 lb

3. 998.4 lb   **5.** 6988.8 lb

7. 5200 lb

9. $\dfrac{\sqrt{2}}{2}\rho a^3$

11. $14{,}976\sqrt{17}$ lb

13. (b) $80\rho_0$ lb/min

▶ Chapter 6 Review Exercises (Page 394)

1. (a) $\displaystyle\int_0^2 (x + 2 - x^2)\,dx$

   (b) $\displaystyle\int_0^2 \sqrt{y}\,dy + \int_2^4 [\sqrt{y} - (y - 2)]\,dy$

3. (a) $\displaystyle\int_0^9 2\sqrt{x}\,dx$ (b) $\displaystyle\int_{-3}^3 (9 - y^2)\,dy$

5. (a) $\displaystyle\int_0^2 2\pi x(x + 2 - x^2)\,dx$

   (b) $\displaystyle\int_0^2 \pi y\,dy + \int_2^4 \pi[y - (y - 2)^2]\,dy$

7. (a) $\displaystyle\int_0^4 2\pi x[\tfrac{1}{2}x - (2 - \sqrt{4 - x})]\,dx$

   (b) $\displaystyle\int_0^2 \pi[(4y - y^2)^2 - 4y^2]\,dy$

9. (a) $\displaystyle\int_0^9 4\pi x^{3/2}\,dx$

   (b) $\displaystyle\int_{-3}^3 \pi(81 - y^4)\,dy$

11. (a) $16/3$ (b) 8π

13. $11/4$ **15.** $\pi a^2 b/24$

17. (a) $256\pi/15$ (b) $40\pi/3$

19. $61/27$ **21.** $779/240$

23. $\tfrac{1}{27}\pi(145^{3/2} - 103^{3/2})$ sq. units

25. $1017\pi/5$ sq. units   **27.** $28\pi\sqrt{3}/5$ sq. units

29. 3 in.-lb   **31.** 10,600 ft-lb

33. 6656π ft-lb   **35.** $28\rho/3$ lb

▶ Exercise Set 7.2 (Page 403)

1. $\dfrac{1}{x}$

3. $\dfrac{2\ln x}{x}$

5. $\dfrac{\sec^2 x}{\tan x}$

7. $\dfrac{1 - x^2}{x(1 + x^2)}$

9. $\dfrac{3x^2 - 14x}{x^3 - 7x^2 - 3}$

11. $\dfrac{1}{2x\sqrt{\ln x}}$

13. $\dfrac{x(1 + 2\ln x)}{(1 + \ln x)^2}$

15. $-\dfrac{y}{x(y + 1)}$

17. $\tfrac{1}{2}\ln|x| + C$

19. $\tfrac{1}{3}\ln|x^3 - 4| + C$

21. $\ln|\tan x| + C$

23. $-\tfrac{1}{3}\ln(1 + \cos 3\theta) + C$

25. $\tfrac{1}{2}x^2 - \tfrac{1}{2}\ln(x^2 + 1) + C$ **27.** $\tfrac{1}{4}(\ln y)^4 + C$

29. $x^3 + 1$

31. (a) $\displaystyle\int_1^x \dfrac{1}{1 + t^2}\,dt$ (b) $\displaystyle\int_2^x \dfrac{1}{t - 1}\,dt$

   (c) $\displaystyle\int_{-3}^x \dfrac{1}{t - 1}\,dt$

35. (a) $\dfrac{3}{x}$

   (b) $\dfrac{\cos x}{1 + \sin^2 x}$

37. (a) $3x^2 \sin^2(x^3) - 2x \sin^2(x^2)$

   (b) $\dfrac{1}{1 + x} - \dfrac{1}{x(1 + \ln x)}$

▶ Exercise Set 7.3 (Page 410)

1. (a) $r + s$ (b) $s - r$ (c) $-r$
(d) $2s$ (e) $\frac{1}{5}s$ (f) $-2(r + s)$

5.

7.

9. $\dfrac{1}{2x} + \dfrac{1}{3(x + 3)} + \dfrac{3}{5(3x - 2)}$

11. $\frac{3}{2}(\ln x)^2 + C$

13. $-\frac{1}{2}(\ln x)^2 + C$

15. $x\sqrt[3]{1 + x^2}\left[\dfrac{1}{x} + \dfrac{2x}{3(1 + x^2)}\right]$

17. $\dfrac{(x^2 - 8)^{1/3}\sqrt{x^3 + 1}}{x^6 - 7x + 5}\left[\dfrac{2x}{3(x^2 - 8)} + \dfrac{3x^2}{2(x^3 + 1)}\right.$

$\left. - \dfrac{6x^5 - 7}{x^6 - 7x + 5}\right]$

▶ Exercise Set 7.4 (Page 419)

1. (a) $\dfrac{1}{x}$, $x > 0$

(b) x^2, $x \neq 0$
(c) $-x^2$, $-\infty < x < +\infty$
(d) $-x$, $-\infty < x < +\infty$
(e) x^3, $x > 0$
(f) $x + \ln x$, $x > 0$
(g) $x - \sqrt[3]{x}$, $-\infty < x < +\infty$
(h) e^x/x, $x > 0$

3. (a) 3.3 (b) 15 (c) 30

5. $-10xe^{-5x^2}$

7. $x^2e^x(x + 3)$

9. $\dfrac{4}{(e^x + e^{-x})^2}$

11. $(x \sec^2 x + \tan x)e^{x\tan x}$

13. $(1 - 3e^{3x})e^{(x - e^{3x})}$

15. (a) Not of the form a^x, where a is a constant.
(b) $x^x(1 + \ln x)$

17. $\dfrac{2x}{(x^2 - 3y)e^y + 3}$

19. $\frac{1}{7}e^{7x} + C$

21. $e^{\sin x} + C$

23. $\ln(1 + e^x) + C$

25. $\frac{1}{3}(1 + e^{2t})^{3/2} + C$

27. $2e^{\sqrt{y}} + C$

29. $\dfrac{\pi^{\sin x}}{\ln \pi} + C$

31. -36

33. $3 + e - e^2$

37. $ck^n e^{kx}$

39. (a) $(x^3 - 2x)^{\ln x}\left[\dfrac{(3x^2 - 2)\ln x}{x^3 - 2x} + \dfrac{\ln(x^3 - 2x)}{x}\right]$

(b) exe^{-1}

(c) $x^{(e^x)}e^x\left(\dfrac{1}{x} + \ln x\right)$

41. $\dfrac{-qk_0}{2T^2}\exp\left[-(q/2)\left(\dfrac{T - T_0}{T_0 T}\right)\right]$

49. $e^x - 3\ln(e^x + 3) + C$

▶ Exercise Set 7.5 (Page 423)

1.

3.

5.

7.

9. (a) $+\infty$ (b) 0

11. (a) 1 (b) -1

13. (a) yes (b) no

(c)

17. (d)

19. (a)

(b)

(c)

21. (a)

(b)

(c)

23. $\dfrac{3-e}{2e}$

▶ Exercise Set 7.6 (Page 430)

1.

	$\sinh x_0$	$\cosh x_0$	$\tanh x_0$	$\coth x_0$	$\operatorname{sech} x_0$	$\operatorname{csch} x_0$
(a)	-2	$\sqrt{5}$	$-2/\sqrt{5}$	$-\sqrt{5}/2$	$1/\sqrt{5}$	$-1/2$
(b)	$-3/4$	$5/4$	$-3/5$	$-5/3$	$4/5$	$-4/3$
(c)	$-4/3$	$5/3$	$-4/5$	$-5/4$	$3/5$	$-3/4$
(d)	$1/\sqrt{3}$	$2/\sqrt{3}$	$1/2$	2	$\sqrt{3}/2$	$\sqrt{3}$
(e)	$8/15$	$17/15$	$8/17$	$17/8$	$15/17$	$15/8$
(f)	-1	$\sqrt{2}$	$-1/\sqrt{2}$	$-\sqrt{2}$	$1/\sqrt{2}$	-1

17. $4\cosh(4x - 8)$

19. $-\dfrac{\operatorname{csch}^2(\ln x)}{x}$

21. $\dfrac{\operatorname{csch}\left(\frac{1}{x}\right)\coth\left(\frac{1}{x}\right)}{x^2}$

23. $\frac{1}{7}\sinh^7 x + C$

25. $\frac{2}{3}(\tanh x)^{3/2} + C$

27. $\ln\cosh x + C$

29. $-\frac{1}{3}\operatorname{sech}^3 x + C$

37. $a\sinh\left(\dfrac{x_1}{a}\right)$

▶ Exercise Set 7.7 (Page 445)

1. $y = Cx$

3. $y = Ce^{-\sqrt{1+x^2}} - 1$

5. $y = \ln(\sec x + C)$

7. $y = e^{-2x} + Ce^{-3x}$

9. $y = e^{-x}\sin(e^x) + Ce^{-x}$

11. $y = -\frac{2}{7}x^4 + Cx^{-3}$

13. $y = -1 + 4e^{x^2/2}$

15. $y = 2 - e^{-t}$

17. $y = \sqrt[3]{3t - 3\ln t + 24}$

19. $y = x + \dfrac{1}{x}$

21. $y^2 = -x^2 + 6x - 5$

23. (a) $y = 200 - 175e^{-t/25}$
   (b) 136 lb

25. 25 lb

27. (a) $y = 10e^{-0.005t}$
   (b) 7 mg

29. 196 days

31. 6.8 years

33. (a) 14,400
   (b) 38 years

▶ Chapter 7 Review Exercises (Page 447)

1. (a) $-(2r + s)$ (b) $2s - \frac{3}{2}r$
   (c) $\frac{1}{4}(3r - s)$

3. (a) $\ln 3/(2\ln 5 + \ln 3)$
   (b) $\frac{1}{2}(\ln 5 - \ln 3)$

5. (a) $\sqrt{34}/5$ (b) $-3/\sqrt{34}$
   (c) $-6\sqrt{34}/25$

7. $-1/(2\sqrt{e^x})$

9. $(\ln x - 1)/(\ln x)^2$

11. 0

13. $\ln 10 - \cot x$

15. $x^4 e^{\tan x}\left(\sec^2 x + \dfrac{4}{x}\right)$

17. $(x^2 + a^2)^{-1/2}$

19. $6x\exp(3x^2)$

21. $1/(4x\sqrt{\frac{1}{2}\ln x})$

23. $x^{\pi-1}\pi^x(\pi + x\ln\pi)$

25. $5\cosh(\tanh(5x))\operatorname{sech}^2(5x)$

27. $3e^{3x} + 4e^{2x} + e^x$

31. (a) $dy = -e^{-x}\,dx$ (b) $dy = dx/(1 + x)$
   (c) $dy = 2x(\ln 2)2^{x^2}\,dx$

33. (a) $\dfrac{1}{x\sqrt{4 + x}}$ (b) $5e^{5x}\sqrt{5x + e^{5x}}$

35. $Y = kt + b,\ b = \ln C$ **37.** $\ln(1 + e^x) + C$

39. $\dfrac{x^{e+1}}{e + 1} + C$ **41.** $4\ln|x| + 3/x + C$

43. $\frac{1}{2}\ln|2\sec x - 1| + C$ **45.** $\frac{1}{2}e^{\sin 2x} + C$

47. $\frac{1}{2}\tanh^2 x + C_1 = -\frac{1}{2}\operatorname{sech}^2 x + C_2$

49. ln 2   **51.** 1/ln 2

55. inflection points at 0, 3 − $\sqrt{3}$, and 3 + $\sqrt{3}$; relative extremum at 3

59. $A = \frac{1}{2}(1 - e^{-2b}) = \frac{1}{4}$ when $b = \ln\sqrt{2}$; $A \to \frac{1}{2}$ as $b \to +\infty$

61. $\frac{1}{2}\pi(2 + \sinh 2)$

63. (a) about 352 million
    (b) the year 2058

▶ Exercise Set 8.1 (Page 457)

1. (a) yes   (b) no   (c) yes   (d) no

3. $\frac{1}{7}(x + 6)$

5. $\left(\dfrac{x + 5}{3}\right)^{1/3}$

7. $\ln(x^2 - 1)$, $x > 1$ **9.** 8

11. $-\dfrac{1}{(x - 1)^2}$

15. $1 - \dfrac{\sqrt{3}}{3}$

17. (b) $\dfrac{1}{\sqrt[3]{2}}$

19.

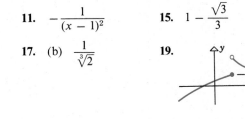

▶ Exercise Set 8.2 (Page 461)

1. (a) $-\dfrac{\pi}{2}$ (b) π (c) $-\dfrac{\pi}{4}$

(d) $\dfrac{\pi}{4}$ (e) 0   (f) $\dfrac{\pi}{2}$

3. $\dfrac{1}{2}, -\sqrt{3}, -\dfrac{1}{\sqrt{3}}, 2, -\dfrac{2}{\sqrt{3}}$ **5.** $\frac{4}{5}, \frac{3}{5}, \frac{3}{4}, \frac{5}{3}, \frac{5}{4}$

7. (a) $\dfrac{\pi}{7}$ (b) 0   (c) $\dfrac{2\pi}{7}$

9. (a) $0 \le x \le \pi$ (b) $-1 \le x \le 1$

(c) $-\dfrac{\pi}{2} < x < \dfrac{\pi}{2}$ (d) $-\infty < x < +\infty$

(e) $0 < x \le \dfrac{\pi}{2}$, $\pi < x \le \dfrac{3\pi}{2}$ (f) $|x| \ge 1$

11. $\frac{24}{25}$

13. $\dfrac{\pi}{2}$

15. $-4\sqrt{5}$

17. (a)   (b)

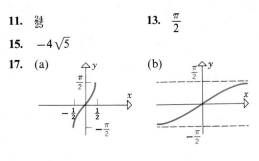

▶ Exercise Set 8.3 (Page 468)

1. (a) $\dfrac{1}{\sqrt{9 - x^2}}$ (b) $-\dfrac{2}{\sqrt{1 - (2x + 1)^2}}$

3. (a) $\dfrac{7}{x\sqrt{x^{14} - 1}}$ (b) $-\dfrac{1}{\sqrt{e^{2x} - 1}}$

5. (a) $-\dfrac{1}{|x|\sqrt{x^2 - 1}}$ (b) $\begin{cases} 1, & \sin x > 0 \\ -1, & \sin x < 0 \end{cases}$

7. (a) $\dfrac{e^x}{x\sqrt{x^2 - 1}} + e^x \sec^{-1} x$

(b) $\dfrac{3x^2(\sin^{-1} x)^2}{\sqrt{1 - x^2}} + 2x(\sin^{-1} x)^3$

9. (a) $-\dfrac{1}{x^2 + 1}$

(b) $10(1 + x\csc^{-1} x)^9 \left(-\dfrac{1}{\sqrt{x^2 - 1}} + \csc^{-1} x\right)$

11. $\dfrac{y\sqrt{1-(x-y)^2}+\sqrt{1-x^2y^2}}{\sqrt{1-x^2y^2}-x\sqrt{1-(x-y)^2}}$

13. $\dfrac{\pi}{2}$

15. $-\dfrac{\pi}{12}$

17. $\frac{1}{4}\tan^{-1}4x+C$

19. $\tan^{-1}(e^x)+C$

21. $\sin^{-1}(\tan x)+C$

23. $\sin^{-1}(\ln x)+C$

25. (a) $\sin^{-1}\left(\dfrac{x}{3}\right)+C$

(b) $\dfrac{1}{\sqrt{5}}\tan^{-1}\left(\dfrac{x}{\sqrt{5}}\right)+C$

(c) $\dfrac{1}{\sqrt{\pi}}\sec^{-1}\left(\dfrac{x}{\sqrt{\pi}}\right)+C$

29. $\dfrac{\pi^2}{4}$

31. $1+2\sqrt{2}$

33. $\dfrac{52\pi}{3}$ mi/min

35. $2\sqrt{6}$ ft

▶ Exercise Set 8.4 (Page 474)

5. (a) $\dfrac{1}{\sqrt{9+x^2}}$

(b) $\dfrac{2}{\sqrt{(2x+1)^2-1}}$

7. (a) $-\dfrac{7}{x\sqrt{1-x^{14}}}$

(b) $-\dfrac{1}{\sqrt{1+e^{2x}}}$

9. (a) $-\dfrac{1}{|x|\sqrt{x^2+1}}$

(b) $\begin{cases} 1, & x>0 \\ -1, & x<0 \end{cases}$

11. (a) $-\dfrac{e^x}{x\sqrt{1-x^2}}+e^x\,\mathrm{sech}^{-1}x$

(b) $\dfrac{3x^2(\sinh^{-1}x)^2}{\sqrt{1+x^2}}+2x(\sinh^{-1}x)^3$

13. (a) $-\dfrac{1}{2x}$

(b) $10(1+x\,\mathrm{csch}^{-1}x)^9\left(-\dfrac{x}{|x|\sqrt{1+x^2}}+\mathrm{csch}^{-1}x\right)$

15. $\cosh^{-1}\left(\dfrac{x}{\sqrt{2}}\right)+C$

17. $-\mathrm{sech}^{-1}(e^x)+C$

19. $-\frac{1}{3}\,\mathrm{csch}^{-1}|x^3|+C$

21. -0.2028

23. (a) (b)

▶ Chapter 8 Review Exercises (Page 475)

1. b, d, e

3. does not exist

5. $\frac{1}{2}\ln(x-1)$

7. If $ad-bc\neq 0$, then $f^{-1}(x)$
$\quad= (-dx+b)/(cx-a)$

9. (a) $(-\infty,5/2)$ (b) $(-2,+\infty)$
(c) $(-\pi/3, 2\pi/3)$

11. $-3/(x+1)^2$

13. $\frac{1}{2}\sqrt{e^x}$

15. (a) $2\pi/3$ (b) $3/4$
(c) $3/5$ (d) $3/5$

17. (a) $\pi/4$ (b) $-\pi/4$
(c) $2\sqrt{6}$ (d) $\pi/3$

19. (a) $33/65$
(b) $56/65$
(c) 7

21. $5/4,\ -3/4,\ 34/16$

23. (a)

(b) $y=f(x)=\pi/2$ (constant), $|x|\leq 1$

25. $-2/[x(\sec^{-1}x^2)^2\sqrt{x^4-1}]$

27. $(\sec x\tan x)/|\tan x|$

29. $\dfrac{1}{1-(\ln x)^2}+\tanh^{-1}(\ln x)$

31. $3/(2\sqrt{(1-9x^2)}\sin^{-1}3x)$

33. $\exp(\sec^{-1}x)/(x\sqrt{x^2-1})$

35. $(\pi^{\sin^{-1}x})\ln\pi/\sqrt{1-x^2}$

37. 1 (since $y=x$)

39. Show $y'=\cos^2 y$.

41. $\frac{1}{2}\ln|(1 + e^x)/(1 - e^x)| + C$

43. $\cosh^{-1}(\ln x) + C$ **45.** $\pi/18$

47. $2\tan^{-1}\sqrt{x} + C$

49. $\pi/3\ (u = \sqrt{x})$ **51.** $1/\sqrt{2}$

53. $\sin^{-1} x = \tan^{-1}(x/\sqrt{1 - x^2})$

▶ Exercise Set 9.2 (Page 487)

1. $-xe^{-x} - e^{-x} + C$

3. $x^2 e^x - 2xe^x + 2e^x + C$

5. $\dfrac{e^x}{2}(\sin x - \cos x) + C$

7. $x^2 \sin x + 2x \cos x - 2\sin x + C$

9. $x\tan x + \ln|\cos x| + C$

11. $-\frac{1}{3}x\cos(3x + 1) + \frac{1}{9}\sin(3x + 1) + C$

13. $\dfrac{e^{2x}}{13}(2\cos 3x + 3\sin 3x) + C$

15. $\frac{1}{2}x[\sin(\ln x) - \cos(\ln x)] + C$

17. $\frac{1}{3}(2 - \sqrt{2})$

19. $-\dfrac{\pi}{8}$

21. (a) $-\frac{1}{3}\sin^2 x\cos x - \frac{2}{3}\cos x + C$

(b) $\frac{1}{32}(3\pi - 8)$

25. (b) $x^3 e^x - 3x^2 e^x + 6xe^x - 6e^x + C$

▶ Exercise Set 9.3 (Page 494)

1. $-\frac{1}{6}\cos^6 x + C$ **3.** $\dfrac{1}{2a}\sin^2 ax + C$

5. $\frac{1}{2}\theta - \frac{1}{20}\sin 10\theta + C$

7. $\dfrac{3}{8}x + \sin\left(\dfrac{x}{2}\right) + \dfrac{1}{8}\sin x + C$

9. $\sin\theta - \frac{2}{3}\sin^3\theta + \frac{1}{5}\sin^5\theta + C$

11. $\frac{1}{6}\sin^3 2t - \frac{1}{10}\sin^5 2t + C$

13. $-\frac{1}{5}\cos^5 x + \frac{1}{7}\cos^7 x + C$

15. $-\frac{1}{5}\cos^5\theta + \frac{2}{7}\cos^7\theta - \frac{1}{9}\cos^9\theta + C$

17. $\frac{1}{8}x - \frac{1}{32}\sin 4x + C$

19. $-\frac{1}{6}\cos 3x + \frac{1}{2}\cos x + C$

21. $-\dfrac{1}{3}\cos\left(\dfrac{3x}{2}\right) - \cos\left(\dfrac{x}{2}\right) + C$

23. $\dfrac{1}{7\cos^7 x} + C$ **25.** $\dfrac{5\sqrt{2}}{12}$

27. 0 **29.** $\frac{1}{24}$

▶ Exercise Set 9.4 (Page 500)

1. $\frac{1}{3}\tan(3x + 1) + C$

3. $\frac{1}{3}\tan^3 x + C$

5. $\frac{1}{16}\tan^4(4x) + \frac{1}{24}\tan^6(4x) + C$

7. $\frac{1}{7}\sec^7 x - \frac{1}{5}\sec^5 x + C$

9. $\frac{1}{4}\sec^3 x\tan x - \frac{5}{8}\sec x\tan x$
$+ \frac{3}{8}\ln|\sec x + \tan x| + C$

11. $\frac{1}{6}\sec^3(2t) + C$

13. $\frac{1}{3}\sec^2 x\tan x + \frac{2}{3}\tan x + C$

15. $\dfrac{1}{5\pi}\sec^4(\pi x)\tan(\pi x) + \dfrac{4}{15\pi}\sec^2(\pi x)\tan(\pi x)$

$+ \dfrac{8}{15\pi}\tan(\pi x) + C$

17. $\frac{1}{3}\tan^3 x - \tan x + x + C$

19. $\frac{1}{6}\tan^3(x^2) + C$

21. $-\frac{1}{5}\csc^5 x + \frac{1}{3}\csc^3 x + C$

23. $-\frac{1}{2}\csc^2 x - \ln|\sin x| + C$

25. $\frac{2}{3}\tan^{3/2}x + \frac{2}{7}\tan^{7/2}x + C$

27. $\frac{\sqrt{3}}{2} - \frac{\pi}{6}$

33. $-\dfrac{1}{\sqrt{a^2 + b^2}}\ln|\csc(x+\theta) + \cot(x+\theta)| + C$,

where θ satisfies

$$\cos\theta = \frac{a}{\sqrt{a^2+b^2}} \text{ and } \sin\theta = \frac{b}{\sqrt{a^2+b^2}}$$

29. $-\frac{1}{2} + \ln 2$

▶ Exercise Set 9.5 (Page 506)

1. $\dfrac{9}{2}\sin^{-1}\left(\dfrac{x}{3}\right) - \dfrac{1}{2}x\sqrt{9-x^2} + C$

3. $-\sqrt{4-x^2} + C$

5. $\dfrac{\sqrt{4x^2-9}}{9x} + C$

7. $\dfrac{x}{\sqrt{1-x^2}} + C$

9. $x - \tan^{-1}x + C$

11. $-\dfrac{\sqrt{9-4x^2}}{9x} + C$

13. $-\dfrac{x}{\sqrt{9x^2-1}} + C$

15. $\frac{1}{2}\sin^{-1}(e^x) + \frac{1}{2}e^x\sqrt{1-e^{2x}} + C$

17. $\frac{2048}{15}$

19. $\dfrac{\sqrt{3}}{2} - \dfrac{\sqrt{2}}{2}$

21. $\dfrac{\pi}{8}$

23. (a) $\sinh^{-1}\left(\dfrac{x}{3}\right) + C$

(b) $\ln\left(\dfrac{\sqrt{x^2+9}}{3} + \dfrac{x}{3}\right) + C$

▶ Exercise Set 9.6 (Page 510)

1. $\dfrac{1}{3}\tan^{-1}\left(\dfrac{x-2}{3}\right) + C$

3. $\sin^{-1}\left(\dfrac{x-1}{3}\right) + C$

5. $\sinh^{-1}(x-3) + C$

7. $2\sin^{-1}\left(\dfrac{x+1}{2}\right) + \dfrac{1}{2}(x+1)\sqrt{3-2x-x^2} + C$

9. $\dfrac{1}{\sqrt{10}}\tan^{-1}\dfrac{\sqrt{2}(x+1)}{\sqrt{5}} + C$

11. $\ln(x^2+2x+5) + \dfrac{3}{2}\tan^{-1}\left(\dfrac{x+1}{2}\right) + C$

13. $\sqrt{x^2+2x+2} + 2\sinh^{-1}(x+1) + C$

▶ Exercise Set 9.7 (Page 519)

1. $-2\ln|x-2| + 3\ln|x-3| + C$

3. $\frac{5}{2}\ln|2x-1| + 3\ln|x+4| + C$

5. $3x + 12\ln|x-2| - \dfrac{2}{x-2} + C$

7. $-\dfrac{1}{2(x-2)^2} + C$

9. $3\ln|x| - \ln|x-1| - \dfrac{5}{x-1} + C$

11. $\dfrac{1}{6}\ln\left|\dfrac{\sin\theta-1}{\sin\theta+5}\right| + C$

13. $\dfrac{1}{32}\ln\left|\dfrac{x-2}{x+2}\right| - \dfrac{1}{16}\tan^{-1}\dfrac{x}{2} + C$

15. $\dfrac{1}{\sqrt{2}}\tan^{-1}\left(\dfrac{x+1}{\sqrt{2}}\right) + \dfrac{1}{x^2+2x+3} + C$

17. $-\frac{7}{84}\ln|4x-1| + \frac{6}{17}\ln(x^2+1) + \frac{3}{17}\tan^{-1}x + C$

19. (a) $(x-1)(x-2)(x-3)$
(b) $(x-4)(x^2+x+5)$
(c) $(x-2)(x-3)(x^2+1)$

21. $\frac{1}{8}\ln|x-1| - \frac{1}{5}\ln|x-2|$
$\qquad + \frac{1}{12}\ln|x-3| - \frac{1}{120}\ln|x+3| + C$

▶ Exercise Set 9.8 (Page 525)

1. $2x^{1/2} - 3x^{1/3} + 6x^{1/6} - 6\ln(x^{1/6} + 1) + C$

3. $4\ln\dfrac{v^{1/4}}{|1 - v^{1/4}|} + C$

5. $2t^{1/2} + 3t^{1/3} + 6t^{1/6} + 6\ln|t^{1/6} - 1| + C$

7. $\frac{1}{3}(1 + x^2)^{3/2} - (1 + x^2)^{1/2} + C$

9. $\ln\left|\tan\left(\dfrac{x}{2}\right) + 1\right| + C$

11. 1

13. $\dfrac{4}{\sqrt{3}}\tan^{-1}\left(\sqrt{3}\tan\dfrac{x}{2}\right) - x + C$

17. $\dfrac{2}{\sqrt{3}}\tan^{-1}\left(\dfrac{2\tanh(x/2) + 1}{\sqrt{3}}\right) + C$

▶ Exercise Set 9.9 (Page 534)

1. (a) 0.7366 (b) 0.6532
  (c) 0.6949 (d) 0.6932

3. (a) 1.8961 (b) 1.8961
  (c) 1.8961 (d) 2.0045

5. (a) 1.1261 (b) 0.6352
  (c) 0.8806 (d) 0.8818

7. 3.1416

9. 0.6928

▶ Chapter 9 Review Exercises (Page 535)

1. $\frac{1}{2}x\sin 2x + \frac{1}{4}\cos 2x + C$

3. $\frac{1}{3}\sec^3 x - \sec x + C$

5. $\frac{1}{9}\tan^3 3t + C$

7. $x - \sin x + C$

9. $\frac{1}{6}x^3 + \frac{1}{4}(x^2 - \frac{1}{2})\sin 2x + \frac{1}{4}x\cos 2x + C$

11. $\frac{1}{4}\sec^4 x + C$

13. $-(\sin^3 2x)/48 - (\sin 4x)/64 + x/16 + C$

15. $\frac{1}{4}$

17. $\frac{1}{2}$

19. $\frac{1}{2}$

21. $x/\sqrt{1 + x^2} + C$

23. $\ln|\sec(e^x) + \tan(e^x)| + C$

25. $\sqrt{e^{2x} + 1} + C$

27. $\frac{1}{13}e^{3x}(3\sin 2x - 2\cos 2x) + C$

29. $\frac{5}{6}\pi - \sqrt{3}$

31. $-(\ln x + 1)/x + C$

33. $\sqrt{x^2 - 9} + C$

35. $3\ln(3 + \sqrt{8}) - \sqrt{8}$

37. $\frac{1}{5}\sqrt{2x + 3}(x^2 - 2x + 6) + C$

39. $\frac{1}{2}\sin^{-1}\left(\dfrac{2t + 1}{2}\right) + C$

41. $-\sqrt{a^2 - x^2}/(a^2 x) + C$

43. $\frac{1}{2}[x\sqrt{a^2 - x^2} + a^2\sin^{-1}(x/a)] + C$

45. $-\sqrt{4x - x^2} + C$

47. $\ln|(2x + 1)/(x + 1)| + C$

49. $-\dfrac{\ln|x|}{6} - \dfrac{2}{15}\ln|x + 3| + \dfrac{3}{10}\ln|x - 2| + C$

51. $\frac{1}{3}\ln|x^3 - 3x| + C$

53. $\dfrac{7}{4}\ln\left|\dfrac{x - 1}{x + 1}\right| - \dfrac{3}{2}\tan^{-1}x + C$

55. $\dfrac{1}{26}\left(\ln\left(\dfrac{(x - 3)^2}{x^2 + 4}\right) - 3\tan^{-1}\left(\dfrac{x}{2}\right)\right) + C$

57. $\frac{7}{8}\tan^{-1}\left(\dfrac{x}{2}\right) - \dfrac{24 + 5x}{4(4 + x^2)} + C$

59. $\frac{1}{2}(\ln(x^2 + 2x + 5) - \tan^{-1}(\frac{1}{2}(x + 1))) + C$

61. $\dfrac{1}{\sqrt{2}}\sin^{-1}\sqrt{\frac{2}{3}}x + C$

63. $2(\sqrt{t} - \tan^{-1}\sqrt{t}) + C$

65. $-\frac{7}{4} + 3\ln 2$

67. $\frac{1}{2}(x - \ln|\sin x - \cos x|) + C$

69. $\frac{1}{4}\cot^2(\frac{1}{2}x) + \frac{1}{2}\ln|\tan(\frac{1}{2}x)| + C$

73. (a) $\frac{1}{2}\tan^{-1}(1) = \pi/8$

(b) $\pi(\pi + 2)/64$ (c) $\pi \ln 2$

75. (a) $e^{2x}(4x^3 - 6x^2 + 6x - 3)/8 + C$

(b) $(\pi - 2)/125$

(c) $\frac{1}{6}(\sin x \cos x)^3 + \frac{1}{16}x - \frac{1}{64}\sin(4x) + C$

77. (a) $\frac{1}{4}(\sec^4 \theta - 2\sec^2 \theta) + C_1$

(b) $\frac{1}{4}\tan^4 \theta + C_2, C_2 = C_1 - \frac{1}{4}$

79. (a) 58.9276   (b) 54.7328

81. (a) 1.8277   (b) 1.8279

83. 0.36972

▶ Exercise Set 10.1 (Page 543)

1. 1

5. $\ln \frac{5}{3}$

9. divergent

13. divergent

3. divergent

7. $-\frac{1}{4}$

11. 0

15. $\dfrac{\pi}{2}$

17. $\frac{9}{2}$

21. (b) $\frac{1}{2}$

29. $\dfrac{2\pi NI}{kr}\left(1 - \dfrac{a}{\sqrt{r^2 + a^2}}\right)$

19. 2

25. $+\infty$

31. (b) 24,000,000

▶ Exercise Set 10.2 (Page 552)

1. 1

5. 1

9. -1

3. 1

7. $\dfrac{1}{\pi}$

11. 0

13. $\frac{1}{2}$

17. $+\infty$

21. (b) 2

25. 2

15. 2

19. $\frac{1}{9}$

23. $k = -1, l = \pm 2\sqrt{2}$

▶ Exercise Set 10.3 (Page 560)

1. 0

5. $+\infty$

9. 0

13. e^2

17. 1

21. 0

27. (a) 0   (b) $+\infty$

(c) 0   (d) $-\infty$

(e) $+\infty$ (f) $+\infty$

(g) $-\infty$ (h) $-\infty$

3. $-\infty$

7. 0

11. π

15. 1

19. $-\frac{1}{2}$

23. 2

29.

31. -1

▶ Chapter 10 Review Exercises (Page 561)

1. $\pi/2$

3. 1

5. 6

7. $\frac{1}{2}$

9. diverges

11. 1

13. $\pi/4$

15. $n > -1;\ -1/(n+1)^2$

17. 3/2

19. $+\infty$

21. 0

23. 0

25. $-\dfrac{1}{4\pi^2}$

27. 0

29. 0

31. 1

33. (a) $+\infty$ (b) $3\pi/2$

35. 2

▶ Exercise Set 11.1 (Page 570)

1. $\frac{1}{3}, \frac{2}{4}, \frac{3}{5}, \frac{4}{6}, \frac{5}{7}$; converges to 1

3. 2, 2, 2, 2, 2; converges to 2

5. $\dfrac{\ln 1}{1}, \dfrac{\ln 2}{2}, \dfrac{\ln 3}{3}, \dfrac{\ln 4}{4}, \dfrac{\ln 5}{5}$; converges to 0

7. 0, 2, 0, 2, 0; diverges

9. $-1, \frac{16}{9}, -\frac{54}{28}, \frac{128}{65}, -\frac{250}{126}$; diverges

11. $\frac{6}{2}, \frac{12}{8}, \frac{20}{18}, \frac{30}{32}, \frac{42}{50}$; converges to $\frac{1}{2}$

13. $1, 2^{1/2}, 3^{1/3}, 4^{1/4}, 5^{1/5}$; converges to 1

15. $\left\{\dfrac{2n-1}{2n}\right\}_{n=1}^{+\infty}$; converges to 1

17. $\left\{\dfrac{1}{3^n}\right\}_{n=1}^{+\infty}$; converges to 0

19. $\left\{\dfrac{1}{n} - \dfrac{1}{n+1}\right\}_{n=1}^{+\infty}$; converges to 0

21. $\{\sqrt{n+1} - \sqrt{n+2}\}_{n=1}^{+\infty}$; converges to 0

23. (a) 3, 6, 12, 24, 48, 96, 192, 384
(b) 1, 1, 2, 3, 5, 8, 13, 21

25. (a) $1, \frac{3}{4}, \frac{2}{3}, \frac{5}{8}$ (b) $\frac{1}{2}$

27. (a) 3 (b) 11 (c) 1001

▶ Exercise Set 11.2 (Page 578)

1. increasing

3. decreasing

5. decreasing

7. increasing

9. nonincreasing

11. not monotone

13. converges

15. diverges

17. converges

19. (a) 0 (b) does not exist

▶ Exercise Set 11.3 (Page 587)

1. (a) converges to $\frac{5}{2}$ (b) converges to $\frac{1}{2}$
(c) diverges

3. $\frac{4}{7}$

5. 6

7. diverges

9. $\frac{1}{3}$

11. $\frac{1}{6}$

15. $-\frac{1}{3}$

19. $\frac{532}{99}$

23. diverges

13. $\frac{448}{3}$

17. $\frac{4}{9}$

21. $\frac{869}{1111}$

► Exercise Set 11.4 (Page 596)

1. $\frac{4}{3}$
3. $-\frac{1}{36}$
5. diverges
7. diverges
9. converges
11. diverges
13. diverges
15. diverges
17. converges
19. diverges

21. converges
23. converges
25. diverges
31. (a) diverges (b) diverges
(c) diverges (d) converges

► Exercise Set 11.5 (Page 604)

1. converges
3. inconclusive
5. diverges
7. diverges
9. converges
11. diverges
13. converges
15. converges
17. converges
19. converges
21. diverges
23. converges
25. converges
27. diverges
29. converges
31. converges
33. converges

► Exercise Set 11.6 (Page 612)

13. converges
15. converges
17. diverges
19. converges
21. diverges
23. converges
25. diverges
27. converges
29. diverges
31. converges
33. converges
35. converges
37. $p > 1$
39. converges

► Exercise Set 11.7 (Page 624)

1. converges
3. diverges
5. converges
7. converges absolutely
9. diverges
11. converges absolutely
13. conditionally
15. divergent
17. conditionally
19. absolutely
21. conditionally
23. divergent
25. conditionally
27. absolutely
29. conditionally
31. 0.125
33. 0.1
35. 9999
37. 39,999

► Exercise Set 11.8 (Page 631)

1. $1, [-1, 1)$
3. $+\infty, (-\infty, +\infty)$
5. $\frac{1}{5}, [-\frac{1}{5}, \frac{1}{5}]$
7. $1, [-1, 1]$
9. $1, (-1, 1]$
11. $+\infty, (-\infty, +\infty)$
13. $+\infty, (-\infty, +\infty)$
15. $1, [-1, 1]$
17. $1, (-2, 0]$
19. $\frac{4}{3}, (-\frac{19}{3}, -\frac{11}{3})$
21. $1, [-2, 0]$
23. $+\infty, (-\infty, +\infty)$
25. $(-\infty, +\infty)$
27. $(a - b, a + b)$
29. $+\infty$

▶ Exercise Set 11.9 (Page 641)

1. $1 - 2x + 2x^2 - \frac{4}{3}x^3 + \frac{2}{3}x^4$

3. $2x - \frac{4}{3}x^3$　　　**5.** $x + \frac{1}{3}x^3$

7. $x + x^2 + \frac{x^3}{2!} + \frac{x^4}{3!}$　**9.** $1 + \frac{1}{2}x^2 + \frac{5}{24}x^4$

11. $\ln 3 + \frac{2}{3}x - \frac{2}{9}x^2 + \frac{8}{81}x^3 - \frac{4}{81}x^4$

13. $e + e(x - 1) + \frac{e}{2!}(x - 1)^2 + \frac{e}{3!}(x - 1)^3$

15. $2 + \frac{1}{4}(x - 4) - \frac{1}{64}(x - 4)^2 + \frac{1}{512}(x - 4)^3$

17. $\dfrac{\sqrt{2}}{2} - \dfrac{\sqrt{2}}{2}\left(x - \dfrac{\pi}{4}\right) - \dfrac{\sqrt{2}}{4}\left(x - \dfrac{\pi}{4}\right)^2$

$\qquad + \dfrac{\sqrt{2}}{12}\left(x - \dfrac{\pi}{4}\right)^3$

19. $-\dfrac{\sqrt{3}}{2} + \dfrac{\pi}{2}\left(x + \dfrac{1}{3}\right) + \dfrac{\sqrt{3}\,\pi^2}{4}\left(x + \dfrac{1}{3}\right)^2$

$\qquad - \dfrac{\pi^3}{12}\left(x + \dfrac{1}{3}\right)^3$

21. $\dfrac{\pi}{4} + \dfrac{1}{2}(x - 1) - \dfrac{1}{4}(x - 1)^2 + \dfrac{1}{12}(x - 1)^3$

23. $\displaystyle\sum_{k=0}^{\infty} (-1)^k \frac{x^k}{k!}$　　**25.** $\displaystyle\sum_{k=0}^{\infty} (-1)^k x^k$

27. $\displaystyle\sum_{k=1}^{\infty} (-1)^{k+1} \frac{x^k}{k}$　**29.** $\displaystyle\sum_{k=0}^{\infty} (-1)^k \frac{x^{2k}}{4^k(2k)!}$

31. $\displaystyle\sum_{k=0}^{\infty} \frac{x^{2k}}{(2k)!}$　　**33.** $\displaystyle\sum_{k=0}^{\infty} (-1)(x + 1)^k$

35. $\displaystyle\sum_{k=1}^{\infty} (-1)^{k+1} \frac{(x - 1)^k}{k}$

37. $\displaystyle\sum_{k=0}^{\infty} (-1)^k \frac{\pi^{2k}}{(2k)!}\left(x - \frac{1}{2}\right)^{2k}$

39. $\displaystyle\sum_{k=0}^{\infty} \left(\frac{16 + (-1)^{k+1}}{8}\right)\frac{(x - \ln 4)^k}{k!}$

▶ Exercise Set 11.10 (Page 653)

1. (a) $\dfrac{2^6 e^{2c}}{6!}x^6$　(b) $(-\infty, +\infty)$

　　(c) between 0 and x

3. (a) $-\dfrac{x^5}{(c + 1)^6}$　(b) $(-1, +\infty)$

　　(c) between 0 and x

5. (a) $\dfrac{(4 + c)e^c}{4!}x^4$　(b) $(-\infty, +\infty)$

　　(c) between 0 and x

7. (a) $-\dfrac{(1 - 3c^2)}{3(1 + c^2)^3}x^3$　(b) $(-\infty, +\infty)$

　　(c) between 0 and x

9. (a) $-\dfrac{5}{128c^{7/2}}(x - 4)^4$　(b) $(0, +\infty)$

　　(c) between 4 and x

11. (a) $\dfrac{\cos c}{5!}\left(x - \dfrac{\pi}{6}\right)^5$　(b) $(-\infty, +\infty)$

　　(c) between $\dfrac{\pi}{6}$ and x

13. (a) $\dfrac{7}{(1 + c)^8}(x + 2)^6$　(b) $(-\infty, -1)$

　　(c) between -2 and x

15. $\dfrac{x^{n+1}}{(1 - c)^{n+2}}$

17. $\dfrac{2^{n+1} e^{2c}}{(n + 1)!}x^{n+1}$

27. $\displaystyle\sum_{k=0}^{\infty} (-1)^k \frac{2^k x^k}{k!},\ (-\infty, +\infty)$

29. $\displaystyle\sum_{k=0}^{\infty} (-1)^k x^k,\ (-1, 1)$

31. $\displaystyle\sum_{k=0}^{\infty} (-1)^k 3^k x^{k+2},\ \left(-\frac{1}{3}, \frac{1}{3}\right)$

33. $1 + \displaystyle\sum_{k=1}^{\infty} (-1)^k \frac{2^{2k-1}}{(2k)!}x^{2k},\ (-\infty, +\infty)$

35. $\displaystyle\sum_{k=0}^{\infty} (-1)^k (x - 1)^k$

▶ Exercise Set 11.11 (Page 662)

1. (a) 9   (b) 13

3. 1.6487

5. 0.9877

7. 0.5299

9. 0.223

11. 0.100

13. 1.0050

15. $|x| < 0.57$

17. $-0.144 < x < 0.137$

19. (a) 1,999,999   (b) 8

▶ Exercise Set 11.12 (Page 671)

5. $\displaystyle\sum_{k=1}^{\infty} (-1)^{k+1} k x^{k-1}$

9. $1 - \frac{1}{3}x - \frac{1}{9}x^2 - \frac{5}{81}x^3 + \cdots;$ 1

11. $x + \frac{2}{3}x^2 + \frac{8}{9}x^3 + \frac{112}{81}x^4 + \cdots;$ $\frac{1}{2}$

13. 0.764

15. 0.494

17. 0.100

19. 0.491

21. $1 - \frac{3}{2}x^2 + \frac{25}{24}x^4 - \frac{211}{720}x^6 + \cdots$

23. $x - x^2 + \frac{1}{3}x^3 - \frac{1}{30}x^5 + \cdots$

25. $x^2 - \frac{1}{3}x^4 + \frac{2}{45}x^6 - \frac{1}{315}x^8 + \cdots$

27. $x^2 + \frac{9}{2}x^3 + \frac{79}{8}x^4 + \frac{683}{48}x^5 + \cdots$

29. (a) 0   (b) $-\frac{3}{2}$

31. (a) $x - \frac{1}{6}x^3 + \frac{3}{40}x^5 - \frac{5}{112}x^7$

(b) $x + \displaystyle\sum_{k=1}^{\infty} (-1)^k \frac{1\cdot 3\cdot 5 \cdots (2k-1)}{2^k k!(2k+1)} x^{2k+1}$

(c) 1

▶ Chapter 11 Review Exercises (Page 672)

1. $L = 0$

3. $L = 0$

5. does not exist

7. (a) 2, 3, 6   (b) 4   (c) 1, 5

9. (a) $|q| > \sqrt{\pi}$ (b) $q > 1/3$
    (c) none $(p = 1)$ (d) $q > e$ or $0 < q < 1/e$

11. (a) $1 + 36 \displaystyle\sum_{k=1}^{\infty} (.01)^k$

     (b) 15/11

13. (a) 1/4   (b) diverges   (c) 1

15. converges

17. converges

19. converges

21. diverges (general term

23. converges absolutely   does not approach zero)

25. diverges

27. $R = 1; \; 0 \le x \le 2$

29. $R = 2; \; -1 < x < 3$ **31.** $R = 0; \; x = 1$

33. (a) $(x-2) - (x-2)^2/2 + (x-2)^3/3$

(b) $\dfrac{-1}{4(c-1)^4}(x-2)^4$, c between 2 and x

(c) $\dfrac{(\frac{1}{2})^4}{4(\frac{1}{2})^4} = \frac{1}{4}$

35. (a) $1 + (x-1)/2 - (x-1)^2/8$

(b) $\dfrac{(x-1)^3}{16c^{5/2}}$, c between 1 and x

(c) $\dfrac{(5/9)^3}{16(2/3)^5} \approx 0.08$

37. $\ln a + \displaystyle\sum_{k=0}^{\infty} (-1)^k \frac{(x/a)^{k+1}}{k+1}; \; R = a$

39. $\dfrac{1}{3}\left\{1 + \displaystyle\sum_{k=1}^{\infty} (-1)^k \frac{1\cdot 3 \cdots (2k-1)}{2\cdot 4 \cdots (2k)}(x/9)^k\right\};$
$R = 9$

41. $1 + x^2/2 + 5x^4/24$

43. $1 - x^2/6 - x^4/72$

47. $|x| \le 1.17$

49. 0.444

45. (a) $\displaystyle\sum_{k=0}^{\infty} \frac{(-1)^k x^{2k}}{(2k+2)!}$ (b) $\displaystyle\sum_{k=0}^{\infty} \frac{(-1)^k x^k}{k+1}$

(c) $\frac{1}{2}$, 1

▶ Exercise Set 12.2 (Page 682)

1.

3.

5.

7.

9.

11.

13.

15.

17. $y^2 = 12x$

19. $y^2 = -28x$

21. $y^2 = 2x$

23. $x^2 = -12y$

25. $y^2 = -5(x - \frac{19}{5})$

27. $y^2 = 6(x - \frac{3}{2})$

29. $(x - 1)^2 = 12(y - 1)$

31. $(x - 5)^2 = 2(y + 3)$

33. (a) $y = \frac{7}{6}x^2 - \frac{23}{6}x + 3$

   (b) $x = -\frac{7}{6}y^2 + \frac{17}{6}y + 2$

35. vertex: $\left(-\dfrac{B}{2A}, \dfrac{4AC - B^2}{4A}\right)$

   focus: $\left(-\dfrac{B}{2A}, \dfrac{4AC - B^2 + 1}{4A}\right)$

   directrix: $y = \dfrac{4AC - B^2 - 1}{4A}$

39. $\frac{1}{16}$ ft

► Exercise Set 12.3 (Page 689)

1.

3.

5.

7.

9.

11.

13.

15. $\dfrac{x^2}{9} + \dfrac{y^2}{4} = 1$

17. $\dfrac{x^2}{169} + \dfrac{y^2}{144} = 1$

19. $\dfrac{x^2}{3} + \dfrac{y^2}{2} = 1$

21. $\dfrac{x^2}{16} + \dfrac{y^2}{4} = 1, \dfrac{x^2}{4} + \dfrac{y^2}{16} = 1$

23. $\dfrac{x^2}{36} + \dfrac{y^2}{81/8} = 1$

25. $(x - 1)^2 + \dfrac{(y - 3)^2}{2} = 1$

27. $\dfrac{(x - 4)^2}{32} + \dfrac{(y - 3)^2}{36} = 1$

29.

31.

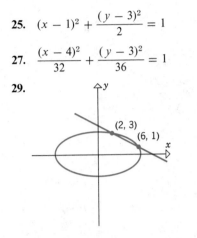

33. $k = -4$ at $(-2, -1)$
$k = 4$ at $(2, 1)$

35. πab

▶ Exercise Set 12.4 (Page 697)

1.

3.

5.

7.

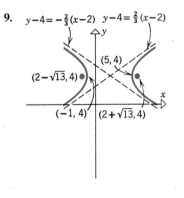

9. $y - 4 = -\frac{2}{3}(x - 2)$ $y - 4 = \frac{2}{3}(x - 2)$

11. $y + 3 = -3(x + 2)$ $y + 3 = 3(x + 2)$

13.

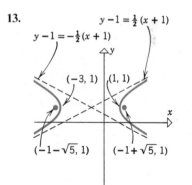

$y - 1 = \frac{1}{2}(x + 1)$

$y - 1 = -\frac{1}{2}(x + 1)$

$(-3, 1)$ $(1, 1)$

$(-1 - \sqrt{5}, 1)$ $(-1 + \sqrt{5}, 1)$

15. $y + 3 = -4(x - 1)$

$y + 3 = 4(x - 1)$

$(-1 - 3)$ $(3 - 3)$

$(-1 - 2\sqrt{17}, -3)$ $(1 + 2\sqrt{17}, -3)$

17. $\dfrac{x^2}{4} - \dfrac{y^2}{5} = 1$ **19.** $x^2 - \dfrac{y^2}{4} = 1$

21. $\dfrac{x^2}{64/9} - \dfrac{y^2}{16} = 1, \dfrac{y^2}{36} - \dfrac{x^2}{16} = 1$

23. $\dfrac{y^2}{56} - \dfrac{x^2}{56} = 1$ **25.** $\dfrac{x^2}{4} - \dfrac{y^2}{4/3} = 1$

27. $\dfrac{(x - 2)^2}{4} - \dfrac{(y + 3)^2}{5} = 1$

29. $\dfrac{(x - 6)^2}{16} - \dfrac{(y - 4)^2}{9} = 1$

31. $8xy - 4x - 4y + 1 = 0$

33.

$\left(\dfrac{109}{10}, -\dfrac{91}{20}\right)$

35.

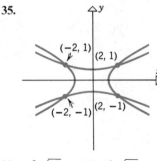

$(-2, 1)$ $(2, 1)$

$(-2, -1)$ $(2, -1)$

39. $\left(\frac{3}{2}\sqrt{13}, -9\right), \left(-\frac{3}{2}\sqrt{13}, -9\right)$

▶ Exercise Set 12.5 (Page 708)

1. (a) $(-1 + 3\sqrt{3}, \sqrt{3} + 3)$

(b) $\left(\frac{5}{2} - \sqrt{3}, \frac{5}{2}\sqrt{3} + 1\right)$

3.

5.

7.

9.

11.

15. $x^2 + xy + y^2 = 3$

21. ellipse, point, or no graph

23. parabola, line, pair of parallel lines, or no graph

25. ellipse, point, or no graph

▶ Chapter 12 Review Exercises (Page 709)

1. parabola: $V(-2, 3)$, $F(-5, 3)$, directrix $x = 1$

3. ellipse: $C(-2, 1)$, $F(-2, 1 \pm \sqrt{5})$, axis lengths 6 and 4

5. hyperbola: $C(2, 1)$, $F(2 \pm \sqrt{10}, 1)$, $V(2 \pm 3, 1)$, asymptotes $y - 1 = \pm(x - 2)/3$

7. parabola: $V(3, 1)$, $F(21/8, 1)$, directrix $x = 27/8$

9. $(y - 3)^2 = 16(x - 1)$ **11.** $y^2/9 - x^2/16 = 1$

13. $x^2/25 + y^2/16 = 1$ **15.** $(x - 3)^2 = 4(y - 3)$

17.

19.

21. $\pi/4$; ellipse; $(x'/\sqrt{2})^2 + (y')^2 = 1$

23. $\pi/6$; hyperbola; $4x'^2 - y'^2 = 1$;

25. $\tan^{-1}(4/3)$; parabola; $y'^2 = 4(x' - 1)$

27.

29.

$\tan^{-1}(2) \approx 63.4°$

▶ Exercise Set 13.1 (Page 715)

1.

$\left(5, \frac{2\pi}{3}\right)$
$\left(3, \frac{\pi}{4}\right)$
$\left(1, \frac{\pi}{2}\right)$
$(0, \pi)$
$\left(4, \frac{7\pi}{6}\right)$
$\left(2, \frac{4\pi}{3}\right)$

3. (a) $(3\sqrt{3}, 3)$ (b) $\left(-\frac{7}{2}, \frac{7\sqrt{3}}{2}\right)$

(c) $(4\sqrt{2}, 4\sqrt{2})$ (d) $(5, 0)$

(e) $\left(-\frac{7\sqrt{3}}{2}, \frac{7}{2}\right)$ (f) $(0, 0)$

5. (a) $(5, \pi)$ (b) $\left(4, \frac{11\pi}{6}\right)$

(c) $\left(2, \frac{3\pi}{2}\right)$ (d) $\left(8\sqrt{2}, \frac{5\pi}{4}\right)$

(e) $\left(6, \frac{2\pi}{3}\right)$ (f) $\left(\sqrt{2}, \frac{\pi}{4}\right)$

7. (a) $(-5, 0)$ (b) $\left(-4, \frac{5\pi}{6}\right)$

(c) $\left(-2, \frac{\pi}{2}\right)$ (d) $\left(-8\sqrt{2}, \frac{\pi}{4}\right)$

(e) $\left(-6, \frac{5\pi}{3}\right)$ (f) $\left(-\sqrt{2}, \frac{5\pi}{4}\right)$

9. line **11.** ellipse

13. circle **15.** $r \cos\theta = 7$

17. $r^2 = 16 \cos 2\theta$ **19.** $r = 9 \sec\theta \tan\theta$

21.

23.

▶ Exercise Set 13.2 (Page 725)

1.

Line

3.

Circle

9.

Cardioid

11.

Cardioid

5.

Circle

7.

Circle

13.

Cardioid

15.

Limaçon

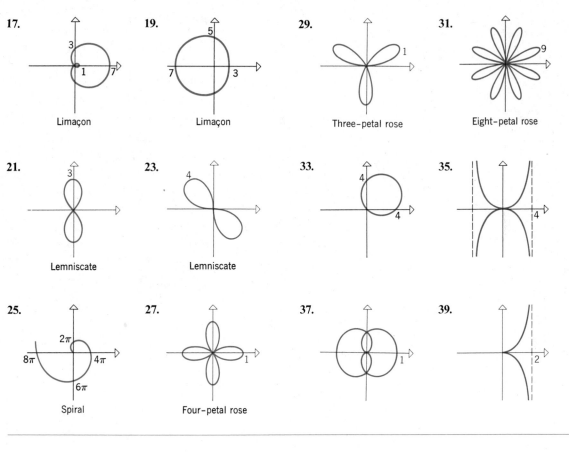

17. Limaçon **19.** Limaçon **29.** Three-petal rose **31.** Eight-petal rose

21. Lemniscate **23.** Lemniscate **33.** **35.**

25. Spiral **27.** Four-petal rose **37.** **39.**

▶ Exercise Set 13.3 (Page 731)

1. $\dfrac{7\pi^3}{1296}$

3. 6π

13. $\dfrac{5\pi}{4}$

15. $100 \cos^{-1}\left(\tfrac{3}{5}\right) - 48$

5. $\dfrac{8\pi}{3} + \sqrt{3}$

7. $\dfrac{9\sqrt{3}}{2} - \pi$

17. $\tfrac{4}{3}a^2$

19. $2\sqrt{3} - \dfrac{2\pi}{3}$

9. 1

11. 4π

▶ Exercise Set 13.4 (Page 744)

1. **3.** **5.** **7.**

9.

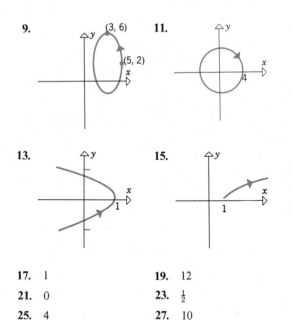

11.

13.

15.

17. 1

19. 12

21. 0

23. $\frac{1}{2}$

25. 4

27. 10

29. $\frac{1}{3}(2\sqrt{2} - 1)$

31. π

33. $\frac{1}{27}(80\sqrt{10} - 13\sqrt{13})$

35. $8a$

37. $x = (2 + 3\sin\theta)\cos\theta$
$y = (2 + 3\sin\theta)\sin\theta$

39. $y = -e^{-4}x + 2e^{-2}$

41. (a) $-\frac{1}{2}$ (b) 1, 4

43. $y = 5x - 14$, $y = 6x - 17$

45. If $|a| = |b|$, a circle with center at (h, k) and radius $|a|$; if $|a| \neq |b|$, an ellipse with center at (h, k) and vertices $(h \pm |a|, k)$ if $|a| > |b|$ and vertices $(h, k \pm |b|)$ if $|a| < |b|$

47. (a)

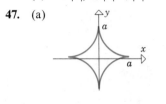

▶ Exercise Set 13.5 (Page 752)

1. $\dfrac{1}{\sqrt{3}}$

3. $\dfrac{\tan 2 - 2}{2\tan 2 + 1}$

5. -2

7. 1

9. 0

11. $-\dfrac{1}{2\sqrt{3}}$

13. $\dfrac{\sqrt{10}}{3}(e^6 - 1)$

15. $2\pi a$

17. $\dfrac{a}{3}[(\pi^2 + 4)^{3/2} - 8]$ **19.** $8a$

21. $\theta = \dfrac{\pi}{2}, \dfrac{3\pi}{2}, \sin^{-1}\frac{1}{4}, \pi - \sin^{-1}\frac{1}{4}$

23. $\tan^{-1} 3\sqrt{3}$

▶ Chapter 13 Review Exercises (Page 752)

1. (a) $(1, \sqrt{3})$ (b) $(0, -2)$
(c) $(0, 0)$ (d) $(-1, 1)$
(e) $(-3, 0)$ (f) $(3/5, -4/5)$

3. (a) (b)

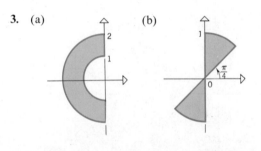

5. $2xy = 1$ (hyperbola)   **7.** $y = -4$ (line)

9. $y = \sqrt{3}x$ (line)   **11.** $x = y = 0$ (point)

13. $r\cos\theta = -3$ **15.** $\tan\theta = 3$

17.   **19.**

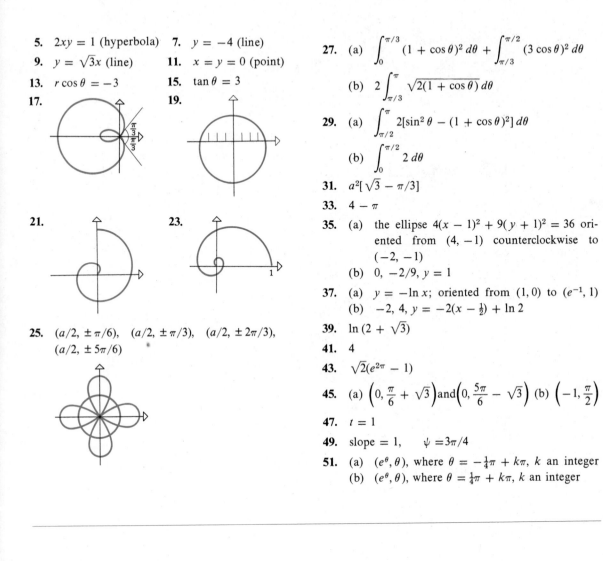

21.   **23.**

25. $(a/2, \pm\pi/6)$, $(a/2, \pm\pi/3)$, $(a/2, \pm 2\pi/3)$, $(a/2, \pm 5\pi/6)$

27. (a) $\displaystyle\int_0^{\pi/3} (1 + \cos\theta)^2\, d\theta + \int_{\pi/3}^{\pi/2} (3\cos\theta)^2\, d\theta$

(b) $\displaystyle 2\int_{\pi/3}^{\pi} \sqrt{2(1 + \cos\theta)}\, d\theta$

29. (a) $\displaystyle\int_{\pi/2}^{\pi} 2[\sin^2\theta - (1 + \cos\theta)^2]\, d\theta$

(b) $\displaystyle\int_0^{\pi/2} 2\, d\theta$

31. $a^2[\sqrt{3} - \pi/3]$

33. $4 - \pi$

35. (a) the ellipse $4(x - 1)^2 + 9(y + 1)^2 = 36$ oriented from $(4, -1)$ counterclockwise to $(-2, -1)$

(b) $0, -2/9, y = 1$

37. (a) $y = -\ln x$; oriented from $(1, 0)$ to $(e^{-1}, 1)$

(b) $-2, 4, y = -2(x - \frac{1}{2}) + \ln 2$

39. $\ln(2 + \sqrt{3})$

41. 4

43. $\sqrt{2}(e^{2\pi} - 1)$

45. (a) $\left(0, \frac{\pi}{6} + \sqrt{3}\right)$ and $\left(0, \frac{5\pi}{6} - \sqrt{3}\right)$ (b) $\left(-1, \frac{\pi}{2}\right)$

47. $t = 1$

49. slope $= 1$, $\psi = 3\pi/4$

51. (a) (e^θ, θ), where $\theta = -\frac{1}{4}\pi + k\pi$, k an integer

(b) (e^θ, θ), where $\theta = \frac{1}{4}\pi + k\pi$, k an integer

---

▶ Exercise Set 14.1 (Page 761)

1.

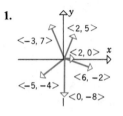

3. (a) $\langle -1, 3\rangle$ (b) $\langle -7, 2\rangle$
(c) $\langle 2, 1\rangle$ (d) $\langle -8, 7\rangle$

5. $(4, -4)$

7. (a) 5   (b) $5\sqrt{2}$ (c) 3
(d) $\sqrt{2}$ (e) 2   (f) 3

9. $\langle -\frac{2}{3}, 1\rangle$

15. circle of radius 1, center at (x_0, y_0)

17. (a) $\left\langle \dfrac{1}{2}, \dfrac{\sqrt{3}}{2}\right\rangle$

(b) $\langle -2\sqrt{2}, 2\sqrt{2}\rangle$

19. $(1, 1), (-1, -1)$

▶ Trigonometry Exercises I (Page A14)

1. (a) $\dfrac{3\pi}{2}$ (b) $\dfrac{13\pi}{6}$ (c) $\dfrac{\pi}{9}$

(d) $\dfrac{23\pi}{30}$ (e) $\dfrac{\pi^2}{1800}$

3. (a)   (b)

(c)   (d)

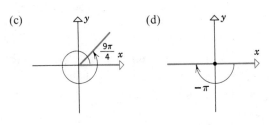

(e)

5. (a) $n\pi,\, n = 0, \pm 1, \pm 2, \ldots$

(b) $\dfrac{\pi}{2} + n\pi,\, n = 0, \pm 1, \pm 2, \ldots$

(c) $n\pi,\, n = 0, \pm 1, \pm 2, \ldots$

7. (a) $n\pi,\, n = 0, \pm 1, \pm 2, \ldots$

(b) $\dfrac{\pi}{2} + n\pi,\, n = 0, \pm 1, \pm 2, \ldots$

9. (a) $\dfrac{3\pi}{2} + 2n\pi,\, n = 0, \pm 1, \pm 2, \ldots$

(b) $\dfrac{\pi}{3} + 2n\pi$ and $\dfrac{5\pi}{3} + 2n\pi,\, n = 0, \pm 1, \pm 2, \ldots$

11. (a) $\frac{1}{4}(\sqrt{2} + \sqrt{6})$

(b) $\frac{1}{2}\sqrt{2 - \sqrt{2}}$

(c) $\sqrt{2 - \sqrt{2 - \sqrt{3}}}$

13.

	$\sin\theta$	$\cos\theta$	$\tan\theta$	$\csc\theta$	$\sec\theta$	$\cot\theta$
(a)	$\dfrac{a}{3}$	$\dfrac{\sqrt{9 - a^2}}{3}$	$\dfrac{a}{\sqrt{9 - a^2}}$	$\dfrac{3}{a}$	$\dfrac{3}{\sqrt{9 - a^2}}$	$\dfrac{\sqrt{9 - a^2}}{a}$
(b)	$\dfrac{a}{\sqrt{a^2 + 25}}$	$\dfrac{5}{\sqrt{a^2 + 25}}$	$\dfrac{a}{5}$	$\dfrac{\sqrt{a^2 + 25}}{a}$	$\dfrac{\sqrt{a^2 + 25}}{5}$	$\dfrac{5}{a}$
(c)	$\dfrac{\sqrt{a^2 - 1}}{a}$	$\dfrac{1}{a}$	$\sqrt{a^2 - 1}$	$\dfrac{a}{\sqrt{a^2 - 1}}$	a	$\dfrac{1}{\sqrt{a^2 - 1}}$

15. $\frac{9}{2}\sqrt{2}, \frac{9}{2}(1 + \sqrt{3}), 105°$

25. $\sin\left(\dfrac{3\pi}{2} - \theta\right) = -\cos\theta,$

$\cos\left(\dfrac{3\pi}{2} + \theta\right) = \sin\theta$

39. $\sin 3\theta = 3\sin\theta \cos^2\theta - \sin^3\theta,$

$\cos 3\theta = \cos^3\theta - 3\sin^2\theta \cos\theta$

► Trigonometry Exercises II (Page A23)

1. (a) $\dfrac{2\pi}{5}$ (b) 6π (c) 8π (d) $\dfrac{\pi}{7}$

(e) 2   (f) 10   (g) 1   (h) $2\pi k$

3. (a) (b) (c)

5. (a) (b) (c)

7. (a) (b)

(c) (d)

9. (a) odd   (b) even   (c) odd   (d) even

(e) even   (f) even   (g) odd   (h) even

index

44. $\displaystyle\int \frac{u^2\,du}{\sqrt{a^2-u^2}} = -\frac{u}{2}\sqrt{a^2-u^2} + \frac{a^2}{2}\sin^{-1}\frac{u}{a} + C$

45. $\displaystyle\int \frac{du}{u\sqrt{a^2-u^2}} = -\frac{1}{a}\ln\left|\frac{a+\sqrt{a^2-u^2}}{u}\right| + C$

$\qquad = -\frac{1}{a}\cosh^{-1}\frac{a}{u} + C$

46. $\displaystyle\int \frac{du}{u^2\sqrt{a^2-u^2}} = -\frac{\sqrt{a^2-u^2}}{a^2 u} + C$

47. $\displaystyle\int (a^2-u^2)^{3/2}\,du = -\frac{u}{8}(2u^2-5a^2)\sqrt{a^2-u^2} + \frac{3a^4}{8}\sin^{-1}\frac{u}{a} + C$

48. $\displaystyle\int \frac{du}{(a^2-u^2)^{3/2}} = \frac{u}{a^2\sqrt{a^2-u^2}} + C$

Integrals Containing $2au - u^2$

49. $\displaystyle\int \sqrt{2au-u^2}\,du = \frac{u-a}{2}\sqrt{2au-u^2} + \frac{a^2}{2}\cos^{-1}\left(1-\frac{u}{a}\right) + C$

50. $\displaystyle\int u\sqrt{2au-u^2}\,du = \frac{2u^2-au-3a^2}{6}\sqrt{2au-u^2}$

$\qquad\qquad + \frac{a^3}{2}\cos^{-1}\left(1-\frac{u}{a}\right) + C$

51. $\displaystyle\int \frac{\sqrt{2au-u^2}\,du}{u} = \sqrt{2au-u^2} + a\cos^{-1}\left(1-\frac{u}{a}\right) + C$

52. $\displaystyle\int \frac{\sqrt{2au-u^2}\,du}{u^2} = -\frac{2\sqrt{2au-u^2}}{u} - \cos^{-1}\left(1-\frac{u}{a}\right) + C$

53. $\displaystyle\int \frac{du}{\sqrt{2au-u^2}} = \cos^{-1}\left(1-\frac{u}{a}\right) + C$

54. $\displaystyle\int \frac{u\,du}{\sqrt{2au-u^2}} = -\sqrt{2au-u^2} + a\cos^{-1}\left(1-\frac{u}{a}\right) + C$

55. $\displaystyle\int \frac{u^2\,du}{\sqrt{2au-u^2}} = -\frac{(u+3a)}{2}\sqrt{2au-u^2} + \frac{3a^2}{2}\cos^{-1}\left(1-\frac{u}{a}\right) + C$

56. $\displaystyle\int \frac{du}{u\sqrt{2au-u^2}} = -\frac{\sqrt{2au-u^2}}{au} + C$

57. $\displaystyle\int \frac{du}{(2au-u^2)^{3/2}} = \frac{u-a}{a^2\sqrt{2au-u^2}} + C$

58. $\displaystyle\int \frac{u\,du}{(2au-u^2)^{3/2}} = \frac{u}{a\sqrt{2au-u^2}} + C$

Integrals Containing Trigonometric Functions

59. $\displaystyle\int \sin u\,du = -\cos u + C$

60. $\displaystyle\int \cos u\,du = \sin u + C$

61. $\displaystyle\int \tan u\,du = \ln|\sec u| + C$

62. $\displaystyle\int \cot u\,du = \ln|\sin u| + C$

63. $\displaystyle\int \sec u\,du = \ln|\sec u + \tan u| + C$

$\qquad = \ln|\tan(\tfrac14\pi + \tfrac12 u)| + C$

64. $\displaystyle\int \csc u\,du = \ln|\csc u - \cot u| + C$

$\qquad = \ln|\tan \tfrac12 u| + C$

65. $\displaystyle\int \sec^2 u\,du = \tan u + C$

66. $\displaystyle\int \csc^2 u\,du = -\cot u + C$

67. $\displaystyle\int \sec u \tan u\,du = \sec u + C$

68. $\displaystyle\int \csc u \cot u\,du = -\csc u + C$

69. $\displaystyle\int \sin^2 u\,du = \tfrac12 u - \tfrac14\sin 2u + C$

70. $\displaystyle\int \cos^2 u\,du = \tfrac12 u + \tfrac14\sin 2u + C$

71. $\displaystyle\int \tan^2 u\,du = \tan u - u + C$

72. $\displaystyle\int \cot^2 u\,du = -\cot u - u + C$

73. $\displaystyle\int \sin^n u\,du = -\frac{1}{n}\sin^{n-1} u \cos u + \frac{n-1}{n}\int \sin^{n-2} u\,du$

74. $\displaystyle\int \cos^n u\,du = \frac{1}{n}\cos^{n-1} u \sin u + \frac{n-1}{n}\int \cos^{n-2} u\,du$

75. $\displaystyle\int \tan^n u\,du = \frac{1}{n-1}\tan^{n-1} u - \int \tan^{n-2} u\,du$

76. $\displaystyle\int \cot^n u\,du = -\frac{1}{n-1}\cot^{n-1} u - \int \cot^{n-2} u\,du$

77. $\displaystyle\int \sec^n u\,du = \frac{1}{n-1}\sec^{n-2} u \tan u + \frac{n-2}{n-1}\int \sec^{n-2} u\,du$

78. $\displaystyle\int \csc^n u\,du = -\frac{1}{n-1}\csc^{n-2} u \cot u + \frac{n-2}{n-1}\int \csc^{n-2} u\,du$

79. $\displaystyle\int \sin mu \sin nu\,du = -\frac{\sin(m+n)u}{2(m+n)} + \frac{\sin(m-n)u}{2(m-n)} + C$

80. $\displaystyle\int \cos mu \cos nu\,du = \frac{\sin(m+n)u}{2(m+n)} + \frac{\sin(m-n)u}{2(m-n)} + C$

81. $\displaystyle\int \sin mu \cos nu\,du = -\frac{\cos(m+n)u}{2(m+n)} - \frac{\cos(m-n)u}{2(m-n)} + C$

82. $\displaystyle\int u \sin u\,du = \sin u - u \cos u + C$

83. $\displaystyle\int u \cos u\,du = \cos u + u \sin u + C$

84. $\displaystyle\int u^2 \sin u\,du = 2u \sin u + (2-u^2)\cos u + C$